Passive Microwave Remote Sensing of Land-Atmosphere Interactions

Passive Microwave Remote Sensing of Land-Atmosphere Interactions

Editors:
B.J. Choudhury, Y.H. Kerr,
E.G. Njoku and P. Pampaloni

///VSP///
Utrecht, The Netherlands, 1995

VSP BV
P.O. Box 346
3700 AH Zeist
The Netherlands

First published in 1995

ISBN 90-6764-186-3

CIP-DATA KONINKLIJKE BIBLIOTHEEK, DEN HAAG

Passive

Passive microwave remote sensing of land-atmosphere
interactions / ed. by B.J. Choudhury, Y.H. Kerr
E.G. Njoku, P. Pampaloni. - Utrecht : VSP
ESA/NASA international workshop held at Saint Lary
(France) from January 11-15, 1993. - With index, ref.
ISBN 90-6764-186-3 bound
Subject headings: remote sensing / climatology.

Printed in The Netherlands by Koninklijke Wöhrmann BV, Zutphen.

Contents

Soil moisture

Surface temperature

Atmospheric correction and precipitation

Instrumentation

Working group reports

ESA/NASA International Workshop, pp. ix–x
B. J. Choudhury, Y. H. Kerr, E. G. Njoku and P. Pampaloni (Eds)

Preface

Current research in monitoring, understanding and predicting land–atmosphere phenomena related to global change has emphasized the need for a better description of the land surface parameters governing energy, mass and momentum exchanges between the land surface and the atmosphere, and for long-term measurements of these parameters over a range of spatial and temporal scales. Recent advances in theory and observations using passive microwave remote sensing have highlighted the potential of spaceborne sensors for contributing to the required land surface measurements of soils, vegetation, snow cover and precipitation. Furthermore, the spatial resolution of passive microwave observations matches the spatial scales of large-scale models of land–atmosphere interactions both for data assimilation and validation.

Three approved future missions, namely the Tropical Rainfall Measuring Mission (TRMM), the Earth Observing System (EOS) and ESA/EU metsat (METOP1), in addition to the ongoing US Defense Mapping Satellite Program (DMSP) mission, will carry passive microwave sensors collectively operating in the frequency range 6.8 to 90 GHz, and will provide land surface measurements. Also, studies are underway for a spaceborne sensor in the 1–3 GHz range to provide improved soil moisture measurements. New research is needed to develop algorithms for extracting useful geophysical parameters from these microwave data for input into large-scale models of land–atmosphere interactions. This research should include development of improved models for microwave emission from the land–atmosphere system, improved coupling between remote sensing observables and large-scale land surface models, and development of retrieval/assimilation schemes for incorporating spaceborne observations into the models.

In order to stimulate and focus this research, the European Space Agency (ESA) and the National Aeronautics and Space Administration (NASA) sponsored a workshop, under the framework of the International Satellite Land Surface Climatology Project (ISLSCP), to bring together researchers in the passive microwave remote sensing and land–atmosphere interaction modeling communities. The goals of the workshop were to review the state of the art in microwave radiometry related to land applications, to exchange ideas leading to new research directions, and to promote coordination of future research activities. The workshop took place at Saint Lary (France) from January 11–15, 1993, and was attended by 43 scientists from 12 countries. The workshop was held as a follow-on to a previous ESA workshop at San Miniato (Italy), whose objective was to review the science requirements and design of the Multifrequency Imaging Microwave Radiometer (MIMR), under development by ESA for operation as part of the METOP1 and EOS missions.

The specific objectives of the Saint Lary workshop are outlined as follows.

1. To review the existing state of knowledge, both theory and observations, pertaining to passive microwave remote sensing of land surface and atmospheric parameters related to exchanges of water and energy at the land–atmosphere boundary.
2. To identify the potential for developing coordinated plans, where appropriate, for (a) future experiments, (b) model and algorithm intercomparisons and validation for both forward modeling and inversion methods, and (c) approaches for processing and assimilating passive microwave data from space into useful scientific products for environmental monitoring and for land–atmosphere interaction models.

These objectives were addressed at the workshop by a combination of review and contributed papers and by discussions in working groups. Plenary sessions were held at intervals to discuss the reports of each working group and to prepare an overall summary for the workshop.

This book is the outcome of the Saint Lary workshop. All papers in this book have been reviewed by two anonymous reviewers and one of the editors. Each working group report has been edited by the participants of that group, the group chairman and rapporteur, with overall editing by Dr. Barry E. Goodison.

Financial support to the participants was provided by the ESA, the NASA and the Data and Information System (DIS) office of the International Geosphere Biosphere Program (IGBP). Funding for the publication of this book was provided by the EOS Program Office of NASA.

Special thanks go to the Conference Secretary, Mrs Jeanine Batlle, LERTS, Toulouse, France, who contributed substantially to the organization of the meeting logistics, to Mrs Patricia Brasure, NASA/Goddard, USA, for her expert organizational assistance, and to Ms Susan Gagliano, Center for Microwave Remote Sensing, Florence, Italy, for her skillful collaboration in the editing of this volume.

The editors are in alphabetical order:

Bhaskar J. Choudhury (Greenbelt, USA)
Yann H. Kerr (Toulouse, France)
Eni G. Njoku (Pasadena, USA)
Paolo Pampaloni (Firenze, Italia)

ESA/NASA International Workshop, pp. xi–xii
B. J. Choudhury, Y. H. Kerr, E. G. Njoku and P. Pampaloni (Eds)

List of Authors

Armstrong R. L.
Barrett E. C.
Basist A.
Calvet J.-C.
Chang A. T. C.
Chanzy A.
Chen Z.
Choudhury B. J.
Dahl P. A.
Davis D.
England A. W.
Entekhabi D.
Essery R.
Fedor L. S.
Ferrazzoli P.
Foster J. L.
Galantowicz J. F.
Golovachev S. P.
Goodison B. E.
Grody N. C.
Guerriero L.
Gurney R. J.
Hewer F.
Hsu C.
Hwang J.-N.
Ijjas G.
Jackson T. J.
Jin Y.-Q.
Kerr Y. H.
Kim E. J.
Kniveton D.
Kong J. A.
Kustas W. P.
Lin D.-S.
Liou Y. A.
Mancini M.
Mätzler Ch.
Menard Y.
Mugnai A.
Nagler Th.
Nakamura H.
Njoku E.
Njoku E. G.
Noilhan J.
Pak K.
Paloscia S.
Pampaloni P.
Rao K. S.
Rao Y. S.
Reutov E. A.
Robinson D. A.
Rott H.
Rowntree P. R.
Saatchi S. S.
Schmugge T.
Shutko A. M.
Smith E. A.
Snider J. B.
Spies T. E.
Tripoli G.
Troch P. A.
Tsang L.

Walker A. E.
Wang J. R.
Wegmüller U.
West R.
Westwater E. R.
Wigneron J.-P.
Winebrenner D. P.
Wood E. F.

ESA/NASA International Workshop, p. xiii
B. J. Choudhury, Y. H. Kerr, E. G. Njoku and P. Pampaloni (Eds)

List of Reviewers

We wish to thank the following reviewers who have contributed their time and talent to the success of this book:

R. Armstrong
G. Asrar
E. Barrett
B. Burns
A. Chang
A. Chanzy
B. Choudhury
P. Ciotti
A. England
E. Engman
D. Entekhabi
P. Ferrazzoli
J. Foster
B. Goodison
N. Grody
L. Guerriero
R. Gurney
G. Ijjas
T. Jackson
Y. Q. Jin
Y. Kerr
J. Kong
C. Maetzler
J. Noilhan
E. Njoku
S. Paloscia
P. Pampaloni
S. Rasool
K. S. Rao
D. Robinson
H. Rott
S. Saatchi
T. Schmugge
A. Shutko
E. Smith
J. Snider
H. Suess
L. Tsang
J. Wang
U. Wegmueller
J. Weinman
J.-P. Wigneron
E. Wood

ESA/NASA International Workshop, pp. xv–xvi
B. J. Choudhury, Y. H. Kerr, E. G. Njoku and P. Pampaloni (Eds)

Subject index

Global and mesoscale modeling and data needs

ESA/NASA International Workshop, pp. 3–15
B. J. Choudhury, Y. H. Kerr, E. G. Njoku and P. Pampaloni (Eds)

Large-scale land–atmosphere models and data needs

P. R. ROWNTREE

Hadley Centre, Meteorological Office, Bracknell, RG12 2SY, England

Abstract—The structure of large-scale models is outlined with particular emphasis on the energy balance at the land–atmosphere interface. The data needed for large-scale land–atmosphere models are discussed under three main headings. Firstly, extensive global data on land surface characteristics are needed to run the models. The relative importance of different characteristics is considered in terms of the sensitivity of the modelled processes to their specification. Two approaches are considered — direct observation of characteristics from space and the indirect approach of identifying vegetation and soil types from space and using ground-based data to provide the quantitative characteristics.

A second major group of data requirements are those for the validation of climate models; many of these also contribute to the initialisation of the models used for forecasting over periods of a few days. Data on soil moisture, snow cover and precipitation are particularly needed. Finally, the role of remote sensing in the monitoring of environmental change is considered.

1. INTRODUCTION

The aim of this paper is to outline the needs of weather forecasting and climate simulation and prediction models for data. The increasing complexity of these models and their global character mean that many of the data needs can only be met comprehensively using remote sensing from satellites. This paper is not concerned with how the needs may be met; that is the role of other papers in this volume. I shall focus on what the needs are. To appreciate these, an outline of the model structure is helpful so the first section of the paper describes this.

2. MODEL STRUCTURE

The physical processes in all these models are represented on a 3-dimensional grid; for climate models the horizontal mesh is about 300 km or more, while there are about 10 to 20 layers in the vertical; for example, the current version of the Meteorological Office (MO) model has a horizontal mesh of 2.5° latitude by 3.75° longitude with 19 layers in the vertical. For global forecast models, the gridmesh is about 100 km, with around 20 layers or more; regional or mesoscale models are covered in the paper by Noilhan. Apart from the spatial extent and resolution, most or all of the following will also apply to the mesoscale models.

The variables represented in the atmospheric model are the wind components, temperature, humidity and surface pressure, while cloud water and ice are increasingly being included. The land surface state is described by three main groups of variables, those describing the temperature, the hydrology and the snowcover. Except when snow is present, these variables mainly represent the soil; the role of the vegetation is principally to modify the amounts of energy available from radiative inputs and to determine the partitioning of that energy between the fluxes of energy into the soil and, in latent and sensible form, into the atmosphere. Because vegetation and climate interact, a need is perceived to model vegetation in climate models. This will introduce additional needs — for data on land biomass and its seasonal variation and on carbon content.

Physically-based equations for the rates of change of these atmosphere and land surface variables are used to advance their values from their initial state to that at the next timestep. For a short-range forecast, this process is repeated until the forecast of the atmospheric state a few days ahead is complete; for a climate simulation, the model is run for many years to provide an estimate of the unperturbed climate which can be validated against that observed. To predict climate change in response to a change in forcing such as increased CO_2 or deforestation, the model simulation is repeated with appropriately modified model parameters to represent the changes in atmospheric composition or surface conditions.

For some variables, including most of those representing the surface state, the exact physical equations cannot be used directly because important processes occur on spatial scales much too small to be represented explicitly. An atmospheric example of this is that of convection, in which the vertical transfer of energy, of great importance on large scales, is organised on scales of a kilometer or less. An example at the surface is runoff; for this the heterogeneity of the distribution of soil water due to variations of elevation, slope and soil and vegetation characteristics, as well as the variations in the input of water by rainfall and snowmelt, all play a role; runoff occurs without the grid-averaged soil becoming saturated, yet the runoff depletes the grid-average soil moisture. To cope with this, the process must be parametrized; that is, the change in the large scale mean due to small scale processes is estimated in terms of the large scale variables.

3. NEEDS FOR MODEL DATA

The requirements for data for the models can be considered under three main headings:

A. Data to define those characteristics of the land surface which are not predicted.

B. Data to
   (a) validate climate simulations;
   (b) initialise forecasts;
   (c) develop parametrizations;
   mostly the data needs for (a), (b) and (c) are similar.

C. Data to define environmental changes, in particular changes in land use; these are needed as input data for model experiments to estimate the response of climate to such changes, e.g. tropical deforestation.

Group B of variables combines validation of climate models and initialisation of forecast models. Winds, temperatures and humidities in the atmosphere, and snow cover and soil moisture are simulated by climate models but a short-period forecast model should use such data as are available in the initialisation stage. The distinction between groups A and B is flexible: A includes those features of the land surface which are not predicted and so must be prescribed, but the range of variables which is predicted is slowly expanding. For example, some early climate models specified soil moisture and how it affects evaporation whereas today climate models usually simulate it. The surface albedo is specified in current models for each vegetation type; models are now being developed to predict the vegetation type; for these albedo will be a predicted variable.

3.1. Land surface characteristics

3.1.1. Data requirements. The land surface characteristics fall into three main groups — those dependent on topography, vegetation and soils. The topographic ones basically concern elevation, but we need not only the mean elevation over the model gridbox, but also the variability on all scales as a measure of the roughness. This is important for its frictional effects on the flow, both for generating small scale turbulence and for generating gravity waves which can carry the frictional effects of the surface out of the turbulent boundary layer to the middle and upper troposphere and even into the stratosphere.

Those characteristics which are dependent on vegetation and soil are listed in Table 1 for the Hadley Centre model. At this point we should consider why these data are needed. The energy balance of the land surface can be expressed by:

$$R_N - G = H + LE \tag{1}$$

where G is the heat flux into the soil, formulated using a diffusive formulation for which the soil thermal properties are needed. The other terms are defined by equations (2) to (4) for the surface net downward radiation

$$R_N = (1 - \alpha_s)R_{SD} + \varepsilon_D R_{LD} - \varepsilon_U \sigma T_*^4 \tag{2}$$

where α_s is the albedo for solar radiation, R_{SD}, R_{LD} are downward solar and longwave radiation, ε_D, ε_U are the emissivities for upward and downward longwave radiation, σ is the Stefan–Boltzmann constant and T_* is the surface temperature — and the sensible and latent turbulent heat fluxes

$$H = \rho c_p C_H V \delta\theta = \rho c_p \,\delta\theta / r_A \tag{3}$$

$$LE = \rho L C_E V \delta q = \rho L \,\delta q / (r_A + r_s) \tag{4}$$

Here, ρ is the density of air, c_p the specific heat of air, C_H and C_E are turbulence transfer coefficients, V the near-surface wind speed and L is the latent heat coefficient. The other terms are discussed below.

The albedos help to determine the solar radiative input, while the surface and sub-surface energy budget determines the surface temperature which modifies the longwave contribution. The moisture-related variables control the partitioning of the radiative input between the sensible and latent heat fluxes. Note that both fluxes can be expressed as differences $\delta\theta$ and δq (of potential temperature and specific humidity respectively) between the surface and the near-surface atmosphere, divided by a resistance. Part of this resistance for the latent flux, and all of it for the sensible flux, is an atmospheric resistance r_A determined by the surface roughness, wind speed and atmospheric stability. However, in the absence of surface water on the surface of the leaf or canopy, the transfer of water vapour must also overcome a stomatal resistance r_s which is often expressed, following [1], in terms of a minimum stomatal resistance determined by the plant physiology, the water stress at the roots and atmospheric factors — radiation, humidity and temperature. Hence the need for data on canopy stomatal resistance, on root depth and soil hydrological characteristics — to estimate the soil's available water capacity — and on the canopy capacity — to estimate how much water can be held on, and evaporated from, the canopy.

Most of the above, especially the soil characteristics, will be difficult to estimate from space. Some, such as albedo, may be directly measurable after allowance for atmospheric absorption and the geometry involving the solar zenith angle, viewing angle and the bidirectional reflectance of the surface. It may only be possible to estimate others, such as the stomatal resistance r_s, by indirect methods such as deducing the evaporation from the diurnal range of temperature and inferring r_s. In such cases, the alternative of estimating the characteristics on the basis of identification of the vegetation type has considerable appeal. The status of remote sensing of the vegetation type was reviewed by Townshend *et al.* [2]. For current models, the vegetation data sets have been defined on the basis of atlas data by Wilson and Henderson-Sellers [3] and Matthews [4]. The former data set is used in the MO model; it specifies a primary and secondary vegetation type for each 1° land gridbox; the secondary type is considered to cover 25% of the box. A similar data set of soil texture, colour and drainage characteristics derived from the FAO (Food and Agriculture Organisation) soils map by Wilson and Henderson-Sellers [3] is used in the MO model.

The problem of defining the characteristics for each vegetation type and soil type including their seasonal variation, still remains. Most of these are measured directly, even with ground-based or airborne equipment, only on a very selective basis. The values currently used in models rely on the assumption that if the vegetation type is known, characteristics such as root depth, albedo and stomatal resistance follow on the basis of relations established using the limited data available. As examples, values for three vegetation types and two soil types as used in the MO model are shown in Table 1.

The variation with type can be large: albedo (especially in the presence of snow), roughness length, root depth, infiltration factor (the amount by which vegetation is assumed to enhance the maximum infiltration rate of a soil), and saturated hydraulic conductivity show particularly wide ranges.

Even if the atlas-based data sets are adequate for current models — a view disputed by Townshend *et al.* [2] — the surface of the Earth is continually changing, partly as a result of man's activities. To maintain an up-to-date global land surface data set in a rapidly changing world clearly invites the use of remote sensing. Probably the surface characteristic most accessible to remote sensing is the surface albedo. Becker *et al.* [5] reviewed the status of the science in this area. They found considerable differences between remotely-sensed and ground-based data though there was no general way of saying which was correct. A surprising feature of the remotely-sensed data was the large contrast between albedos over tropical Africa between September and January, particularly the higher values over tropical savanna and rain forest in the wet season, with differences of about 0.07. The seasonal variation of albedo is undoubtedly complicated [6] partly due to the effects of cloud on the spectrum of the solar radiation received at the ground (note that these effects of cloud on the albedo are not directly observable from space though they should be accessible through a combination of clear

Table 1.
Vegetation and soil-dependent characteristics in the Meteorological Office climate model, with values for some specific types for the column version

Vegetation type	Conifer forest	Pasture	Desert
Albedo: snowfree	0.144	0.194	0.35
deep snow	0.255	0.65	0.80
Roughness length (m)	0.76	0.02	0.003
Canopy resistance (s/m)	84.5	79	100
Canopy store (mm)	1.13	0.63	0.50
Root depth (m)	0.84	0.57	0.10
Infiltration factor	5.5	1.68	0.50
Soil type	Clay	Medium	Sand
x_s† (saturation)	0.49	0.46	0.42
x_c† (critical)	0.25	0.17	0.087
x_w† (wilting)	0.18	0.105	0.055
Saturated conductivity (K_s) (10^{-3} kg/m^2/s)	0.361	1.81	8.06
c‡	11	8	5
Heat capacity (10^6 J/m^3/K)	2.87	2.22	1.92
Heat conductivity (W/m/K)	0.73	0.47	0.49

† the x values are soil moisture concentrations by volume; x_c is the value below which transpiration is limited by soil moisture
‡ c is an exponent controlling the dependence of conductivity K on soil moisture.

sky observations and radiative calculations). However, a review by Arino *et al.* [7] of the remote sensing techniques used indicates that a substantial part of the observed variation could be due to observational error particularly directional effects.

The soils data needs in Table 1 pose a major problem for remote sensing. In the foreseeable future, the soil texture and drainage data will probably have to be estimated using ground-based data, though some inferences, particularly concerning soil colour, may be possible through remote sensing.

3.1.2. Evidence on sensitivities. In this section I consider the question: how sensitive are the modelled land surface processes to the surface characteristics? This question needs to be answered to assess relative priorities for obtaining the data. There is some evidence on which to base an answer. I have used a number of methods. The best estimate may come from sensitivity tests with a General Circulation Model (GCM), though even then the sensitivity will depend on the area over which the surface parameter is changed. This approach has been used for the most obvious sensitivities such as those due to albedo and has also been used to demonstrate the sensitivity to surface moisture availability (e.g. [8]). With a GCM an estimate can be made not only of the response of evaporation which I shall mostly use to measure sensitivity, but also rainfall whose response can be modified by the dynamics for which a 3-dimensional model is needed. Responses remote from the perturbation can also be estimated. For albedo, the sensitivity in the tropics where the rainfall depends strongly on surface forcing — in contrast to middle latitudes in winter where land precipitation is mainly forced dynamically — is such that a 1% albedo decrease increases rainfall and evaporation by from 2% to 4% [9]. Part of this sensitivity arises from positive feedbacks, such as:

(a) decreased albedo increasing low level moisture convergence into heated land areas and so enhancing rainfall, and
(b) increased moisture convergence moistening the soil and so delaying the limitation of evaporation by deficient soil moisture in the dry season.

In the absence of GCM experiments, it is possible to make a comparative assessment of the role of changes in different parameters by using a simpler one-dimensional or column model in which the dynamics of horizontal interactions are either ignored or are represented statistically by randomly varying advective forcing [10]; generally a realistic climate can only be maintained in a column model by use of such advective forcing. This approach can also give an estimate of the effects on rainfall but the magnitude of the estimate is dependent on the relative strength of the advective forcing; this effectively defines the area over which the land surface change is assumed to be applied.

An analytical estimate of the effect on evaporation for some parameters can also be made [11] using the Penman–Monteith equation, derived from equations (1) to (4); this approach has the advantage of improving understanding of the mechanisms, but assumptions may be needed for parameters which would be modelled with the other methods. Each of these approaches has contributed to the assessment which follows, though the column experiments are the major contributor, following the earlier estimates by [12].

The results in Table 2 are expressed as the change needed to increase evaporation by 5%. They were mostly obtained from experiments for conditions typical of the south of England.

They are compared with the range of values allotted to the different land surface types. Except for albedo, the range is at most 8 times the change needed to modify evaporation by 5%; for albedo the ratio is up to 30 if variations due to snow are included, 15 in the absence of snow. Longwave emissivity appears to have quite low importance; indeed, variations in longwave emissivity are usually ignored in models. Some of the sensitivities are discussed further below.

The sensitivity to roughness requires some comment. Analysis of the Penman–Monteith equation [11] shows that the sign of the dependence changes between low and high values of the stomatal resistance r_s: if r_s is zero due to the presence of water on the canopy, evaporation increases with increasing roughness. On the other hand if r_s equals a critical value, which for typical conditions lies in the middle of the range of

Table 2.
Change (dX) in land surface characteristics needed to increase evaporation by 5%

	Range	dX
Stomatal resistance	160 s/m	−40 s/m
Root depth	1.5 m	+0.2 m (only if limited by evaporation)
Roughness length	*4000	? see text (can be either sign)
Canopy capacity	0.6 mm	+0.6 mm (no evidence on linearity)
Snowy albedo	0.6	−0.02
Snowfree albedo	0.3	ditto
Hydraulic conductivity	*25	*5 (opposite signs possible)
Longwave emissivity	0.02	0.04

Table 3.
Runoff (mm/yr) for three values of saturated hydraulic conductivity (K_S) (as Table 1) in 3-year experiments with the MO column model

Year	Surface runoff			Drainage			Total runoff		
	Fine	Med	Crse	Fine	Med	Crse	Fine	Med	Crse
1	88	11	0.0	53	113	150	140	125	150
2	100	13	0.2	1	21	44	101	34	44
3	106	15	0.5	6	114	139	111	128	140
Ave.	98	13	0.2	20	83	111	117	96	111

observed r_s for a dry canopy, variations in z_0 have no effect on evaporation. For larger values of r_s, an increase in roughness reduces evaporation.

The sensitivity to the soil hydraulic conductivity is complicated because this property affects both the fast runoff occurring when rainfall exceeds the soil's surface infiltration capacity, and the drainage which occurs with high soil moisture concentrations. The three soil textures used in the MO model [13] chosen as appropriate for coarse, medium and fine soils are such that for typical rainfall rates, the main contrast in infiltration rates occurs between the medium and fine soils with both medium and coarse soils having relatively low surface runoff (Table 1). The drainage term tends to compensate for this both through the consequent differences in soil wetness, and also because coarser-textured soils drain faster; the total effect can easily be that medium soil with the intermediate value of K_s has the least total runoff. This occurred in two years of a three year run with the MO column model driven with forcing appropriate for southern England (Table 3).

The above discussion is in terms of the response of evaporation. Other variables may be much more sensitive; for example, the difference between precipitation and evaporation, which over a long period can be equated to runoff, varies by a much larger fraction because it is the difference between two large quantities. For southern England, a typical rainfall of about 600 mm/yr contributes partly to evaporation (450 mm/yr) and partly to runoff (150 mm/yr). Thus, if rainfall is the same, the runoff will be 3 times more sensitive than the evaporation.

3.2. Validation of climate simulations

Climate models simulate a wide range of quantities, which need to be validated. As noted earlier, many of these are needed as initial data for assimilation into weather forecasting models; obvious examples of possible uses for observed data include soil moisture and snowcover; less obviously, precipitation rates and top-of-the-atmosphere (TOA) radiation could be useful in the assimilation process for forcing atmospheric humidity and cloud and surface albedo and temperature towards reality. By providing information on model errors and atmospheric behaviour, such data also contribute to the development of model parametrizations.

A list of some of the climate model products needing validation is shown in Table 4. The knowledge of even the longterm mean values of these quantities, from ground-based data or already available satellite data, varies widely. It is generally good for near-surface temperatures and snow extents over land, and atmospheric winds and temperatures, fair

Table 4.
Data needed for climate model validation

Atmospheric temperature, moisture, clouds, winds, radiation
Precipitation rates
Land near-surface temperatures (above and below surface)
Soil moisture
Runoff
Snow cover (mass, frequency, spectral albedo)
Ice sheet and glacier elevation and extent
Future:
Vegetation and soil –
type, characteristics including carbon content

for snowmass and precipitation rates over land and poor for soil moisture. In the rest of this paper I will discuss the present state of validation for these last 3 quantities.

3.2.1. Land precipitation. Several estimates exist for precipitation over land. Most are constructed using ground-based data, but they can differ because of differences in the selections of observations or the spatial averaging scales or the need to estimate the precipitation where data are lacking. Analyses have also been attempted using OLR (top-of-the-atmosphere outgoing longwave radiance) data from satellites (e.g. [14] for the tropics for 1986–89). Comparison of different estimates of the precipitation for June–August along 90°E over Asia made for this Workshop (not shown) revealed major differences. For example, the analysis by Schutz and Gates [15] gives maximum values over northern India of less than 10 mm/day and the rainfall decreases southward along the east coast of the Bay of Bengal. The analysis by Jaeger [16] shows a peak of 15 mm/day on the east coast of the Bay near 15°N, while Shea [17] has a similar peak but at about 20°N. The satellite-based estimates for 1986–88 resemble Jaeger's but with intensities about 5 mm/day greater. Further north, the sparsity of data is a major problem; for example near 35°N, the ground-based estimates range between 1 and 6 mm/day with the satellite analysis indicating 6 mm/day. All the ground-based estimates show a minimum near 40°N; the values over Siberia at 50–60°N vary between 1.5 and 3 mm/day. The simulation by a recent version of the MO model lies generally within the range of the observed estimates.

3.2.2. Snowmass. The main sources of data for snow cover are ground-based data on snow frequency and depth, and satellite data on snow extent from visible data and snowmass and snowdepth from microwave data. Note that models generally simulate and predict snowmass, though conversion of snowdepths to snowmass is possible using estimates of the snow density (not necessarily with great confidence as snow density ranges from about 70 kg/m^2 for some new snow to 500 kg/m^2 for hard-packed late season snow). Foster (personal communication) has compared the seasonal cycle of the MO model's simulated snowmass with two observational sources, the NOAA snow depth climatology (SDC) and SMMR microwave data for 1987 (Figure 1). The model winter snowmass agrees well with the estimates from the SDC data, which over North America are higher than the SMMR data by a factor of 2 to 3. Over Eurasia, differences have the same sign but are much smaller — part of these differences may arise from

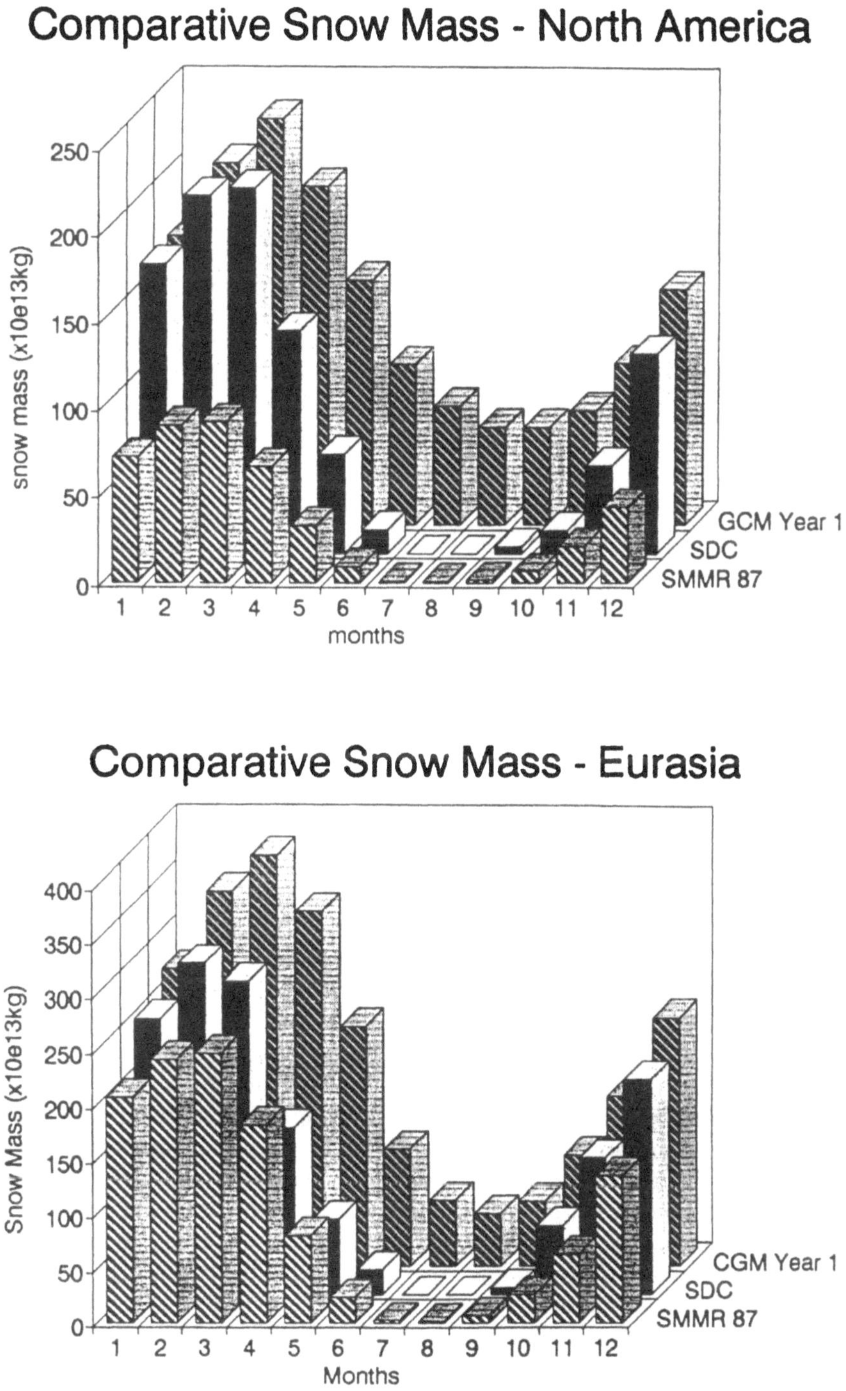

Figure 1. Annual cycle of snowmass from the MO model (CGM Year 1) and as estimated from the NOAA snowdepth climatology (SDC) and from microwave observations from satellite for 1987 (SMMR 87) for (a) North America and (b) Eurasia (D. J. Foster, personal communication).

Table 5.
Analysis of available soil moisture as percentage of capacity averaged for 30–60°E as: (a) measured [18] and (b) estimated using precipitation and temperature data [20] (from [19])

Data	(a)		(b)	
Season	DJF	JJA	DJF	JJA
62°N	100	100	100	55
58°N	100	98	97	41
54°N	85	57	83	27
50°N	63	24	57	12

sampling error; there is considerable interannual variability in snow cover, especially for individual continents. Both observational data sets showed that the modelled snowmass is clearly excessive in summer; this fault is much reduced in the latest version of the model.

3.2.3. Soil moisture. One of the most important indicators of the effects of increased greenhouse gases on the climate is the soil moisture; findings that summer aridity (as represented by soil moisture) in the Great Plains area of N. America was exacerbated by increased CO_2 attracted considerable attention in the late 1980s. Any model variable should be validated but it is especially important for one whose changes are of such concern and which has such a central role in determining evaporation in the models. Unfortunately the data available to validate soil moisture are very limited. However, there are published ground-based data for Asia between about 45 and 60°N and 25 and 100°E, analysed by Vinnikov and Yeserkepova [18]. In the 1990 IPCC Report [19], these data were compared with the estimates made by Mintz and Serafini [20] (see also [21]) on the basis of observations of precipitation and temperature. Whilst there are many realistic features in these estimates, Table 5 shows there is a substantial difference between them and the Russian data in summer. It appears likely that the estimation of evaporation from the readily available air temperature data is not sufficiently accurate at least in high latitudes.

3.3. Environmental change

As discussed earlier, it is desirable to maintain an up-to-date record of the state of the land surface for use in estimating land surface characteristics for weather forecast and climate simulation models. Another reason for doing this is to monitor land surface changes; it appears from climate change experiments already completed that tropical and mid-latitude deforestation can both affect the climate (e.g. [22]–[25]). An example of the effects of tropical deforestation is shown here from Lean and Rowntree [26] (Fig. 2). The effects of such changes in vegetation are not likely to be only local. The decreases in rainfall shown in Fig. 2 are associated with decreases in ascent which must be compensated by increases elsewhere. Experiments so far have not been of long enough duration to allow a confident analysis as to where they will occur, but they do suggest that both adjacent regions (e.g. parts of eastern South America) and also remote areas such as the western Pacific are likely to be affected.

Experiments published so far on South American tropical deforestation assume total deforestation of the Amazon basin rain forests. To make more realistic experiments in

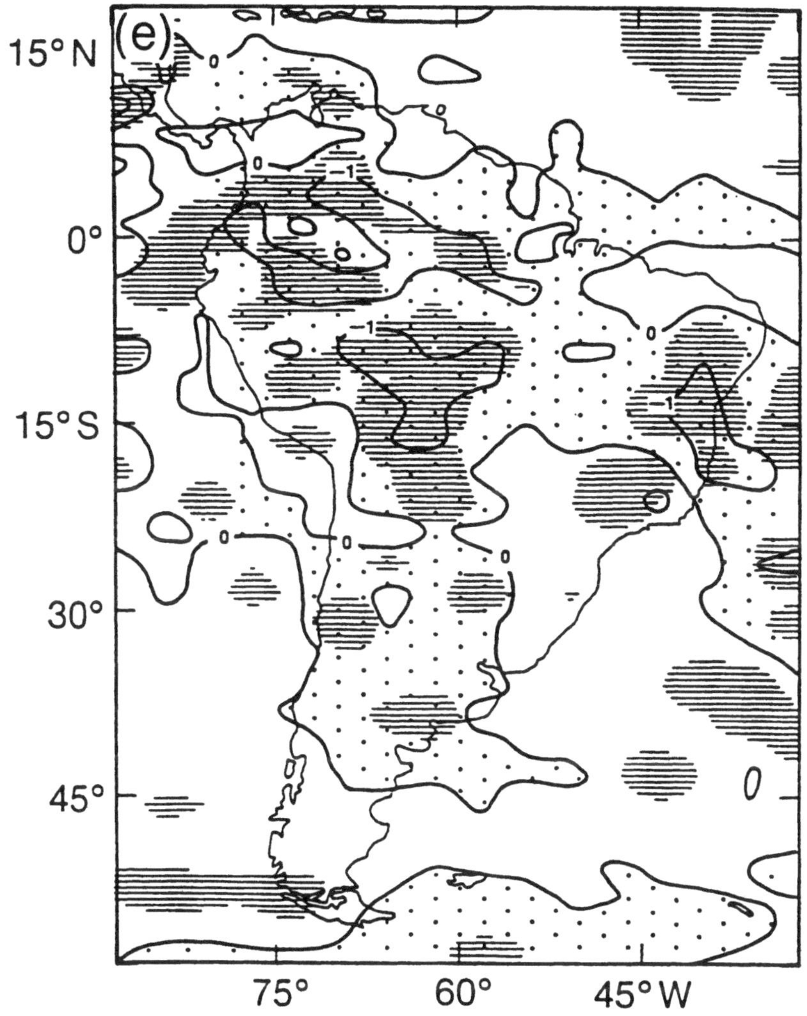

Figure 2. Simulated annual mean change in precipitation (mm/day) over South America due to deforestation (from [26]). Areas of decrease are stippled and changes significant at the 90% level or above are shaded.

which partial deforestation is assumed, we need to know the actual deforestation from remote sensing. Otherwise, it will not be possible to make assessments of the effects of vegetation changes either due to man's modifications of the land surface (intentional or otherwise) or due to natural vegetation adaptations to climate change. For example, if a record existed now of the vegetation changes over the last 2000 years or more, we could use them in climate models to assess the contribution of man's land surface modifications to climate change.

Where future modifications are large, the use of such records started from as early a date as possible will allow validation of the models' ability to predict climate change. A current use of such a record would be to allow a best estimate to be made of the extent of recent changes, including but not confined to tropical deforestation. Such data could be used in climate change experiments and so allow estimation of the extent of the change in 'current climate' relative to say the mean of the last 30 years. This period is commonly used in climate normals for guidance in the building, engineering, insurance etc industries; estimates of the change in present expectations relative to those normals should allow more reliable guidance.

Valuable additional data for environmental monitoring would be data on biomass to allow analysis of changes in the land carbon content. This is a possible major contributor to changes in the global carbon budget acting as a sink to compensate for the input of CO_2 to the atmosphere from fossil fuel burning, but little is known of the magnitude of the contribution. Reliable data would allow assessment of the present status and of temporal changes, and would also provide data for validation of models of vegetation change.

4. SUMMARY

The data needed for large-scale land–atmosphere models may be summarized under a number of headings. Extensive global data on land surface characteristics are needed. The models appear to be particularly sensitive to the specifications of surface albedo, roughness and characteristics such as root depth and soil hydraulic conductivity which affect the moisture availability. Because of the difficulty of observing these directly from space, the indirect approach of identifying vegetation and soil types from space and using ground-based data to provide the quantitative characteristics should also be attempted.

A second major group of data requirements are those for the validation of climate models; many of these also contribute to the initialisation of the models used for forecasting over periods of a few days. Data on soil moisture and snow depth and mass are particularly needed. Precipitation data are also needed over the oceans, and over land areas where data are sparse relative to the scale of spatial variability.

Finally, remote sensing can play an important role in the monitoring of environmental change. The importance of this for assessments of the development of climate change due to land surface modifications was noted, as well as the potential value, for studies of the carbon cycle, of observations on the carbon in the land biomass.

REFERENCES

1. P. G. Jarvis. The interpretation of the variations in leaf water potential and stomatal conductance found in canopies in the field. *Phil. Trans. Roy. Soc. London, Ser. B*, **273**, 593–610 (1976).
2. J. Townshend, C. Justice, Wei Li, C. Gurney and J. McManus. Global land cover classification by remote sensing: present capabilities and future possibilities. *Remote Sens. Environ.*, **35**, 243–255 (1991).
3. M. F. Wilson and A. Henderson-Sellers. A global archive of land cover and soils data for use in general circulation climate models. *J. Climatology*, **5**, 119–143 (1985).
4. E. Matthews. Global vegetation and land use: new high resolution data base for climate studies. *J. Climate Appl. Meteor.*, **22**, 474–487 (1983).
5. F. Becker, H.-J. Bolle and P. R. Rowntree. *The International Satellite Land–Surface Climatology Project*. ISLSCP Report No. 10, UNEP (1988).
6. J. S. Oguntoyinbo. Reflection coefficient of natural vegetation, crops and urban surfaces in Nigeria. *Q. J. R. Meteorol. Soc.*, **96**, 430–441 (1970).
7. O. Arino, G. Dedieu and P. Y. Deschamps. Accuracy of satellite land surface reflectance determination. *J. Appl. Meteor.*, **30**, 960–972 (1991).

8. Y. Mintz. The sensitivity of numerically simulated climates to land surface boundary conditions. In: *The Global Climate*, J. T. Houghton (Ed.), Cambridge University Press, pp. 79–105 (1984).
9. M. F. Mylne and P. R. Rowntree. Modelling the effects of albedo change associated with tropical deforestation. *Climatic Change*, **21**, 317–343 (1992).
10. D. A. Warrilow. Indications of the sensitivity of European climate to land use variations using a one dimensional model. In: *Proceedings of ISLSCP Conference, Rome, 2–6 December 1985*, European Space Agency ESA SP-248, pp. 159–166 (1986).
11. P. R. Rowntree. *Atmospheric parametrization schemes for evaporation over land: basic concepts and climate modelling aspects*. In: *Section 2 of Land Surface Evaporation: Measurement and Parametrization*, T. J. Schmugge and J.-C. Andre (Eds), Springer-Verlag, New York, pp. 5–29 (1991).
12. D. A. Warrilow, A. B. Sangster and A. Slingo. *Modelling of Land Surface Processes and Their Influence on European Climate*. Meteorological Office, Met O 20 Tech Note DCTN 38 (1986).
13. D. A. Warrilow. *The sensitivity of the UK Meteorological Office atmospheric general circulation model to recent changes in the parametrization of hydrology*. In: *Proceedings of ISLSCP Conference, Rome, 2–6 December 1985*, European Space Agency ESA SP-248, pp. 143–149 (1986).
14. J. E. Janowiak and P. A. Arkin. Rainfall variations in the tropics during 1986–1989, as estimated from observations of cloud-top temperature. *J. Geophys Res.*, **96**, 3359–3373 (1991).
15. C. Schutz and W. L. Gates. *Global Climatic Data for Surface, 800 mb, 400 mb: July*. Report R-1029-ARPA, Rand, Santa Monica (1972).
16. L. Jaeger. *Monatskarten des Niederschlags für die Ganze Erde*. Bericht Deutscher Wetterdienst, 18, No. 139, pp. 38 (1976).
17. D. J. Shea. *Climatological atlas: 1950–1979*. NCAR Technical Note/TN-269+STR, NCAR, Boulder, Col. (1986).
18. K. Ya. Vinnikov and I. B. Yeserkepova. Soil moisture: empirical data and model results. *J. Climate*, **4**, 66–79 (1991).
19. W. L. Gates, P. R. Rowntree and Q.-C. Zeng. *Validation of climate models*. In: *Climate Change – the IPCC Scientific Assessment*, J. T. Houghton, G. J. Jenkins and J. J. Ephraums (Eds), Cambridge University Press, pp. 93–130 (1990).
20. Y. Mintz and Y. V. Serafini. *Global monthly climatology of soil moisture and water balance*. LMD Internal Report No. 148, Laboratoire de Météorologie Dynamique, Paris (1989).
21. Y. Mintz and Y. V. Serafini. A global monthly climatology of soil moisture and water balance. *Climate Dynamics*, **8**, 13–27 (1992).
22. J. Lean and D. A. Warrilow. Simulation of the regional climatic impact of Amazon deforestation. *Nature*, **342**, 411–413 (1989).
23. C. A. Nobre, P. J. Sellers and J. Shukla. Amazonian deforestation and regional climate change. *J. Climate*, **10**, 957–988 (1991).
24. G. Thomas and P. R. Rowntree. The boreal forests and climate. *Q. J. R. Meteorol. Soc.*, **118**, 469–497 (1992).
25. G. B. Bonan, D. Pollard and S. L. Thompson. Effects of boreal forest vegetation on global climate. *Nature*, **359**, 716–718 (1992).
26. J. Lean and P. R. Rowntree. A GCM simulation of the impact of Amazonian deforestation on climate using an improved canopy representation. *Q. J. R. Meteorol. Soc.*, **119**, 509–530 (1993).

ESA/NASA International Workshop, pp. 17–54
B. J. Choudhury, Y. H. Kerr, E. G. Njoku and P. Pampaloni (Eds)

Mesoscale land–atmosphere models and usefulness of microwave remote-sensing data

J. NOILHAN and J.-C. CALVET

Météo-France, CNRM, 31057 Toulouse, France

1. INTRODUCTION

During the last decade, a lot of evidence has been accumulated from mesoscale integrations that the atmosphere at mesoscale is sensitive to horizontal variations in land surface processes. This sensitivity has been found to occur on scales of several tens of kilometers and time periods of several hours. The short term influence of the surface energy and water budgets on the structure of the planetary boundary layer (PBL) is of great importance for mesoscale models, at least for meso-β-scale models using horizontal grid size about 10 km and time steps of a few minutes. Short term forecasts are highly dependent on initial surface conditions, particularly on the soil moisture status which determines to a large extent the prediction of near-surface atmospheric fields.

This paper is divided into three parts. The first section summarizes observational and numerical studies reporting on evidence of atmospheric sensitivity to land surface inhomogeneities. A brief review of essential representations in land surface transfer schemes to be used in mesoscale models is then provided. The evaluation of the accuracy of surface flux predictions is an imperative requirement for limited-area atmospheric models. Examples of comparison between observed and predicted surface quantities are briefly presented. Finally, some specific needs for implementing surface parameterizations in mesoscale models are indicated.

The second section concerns the retrieval of geophysical variables from satellite microwave radiometry. Methods for extracting surface variables (surface temperature and surface moisture) from spaceborne microwave radiometers are discussed.

An application of temperature and humidity retrieval of the Amazon forest is presented in the last section. A first attempt of comparison between remotely sensed and predicted surface variables concludes the paper.

2. NEEDS OF MESOSCALE ATMOSPHERIC MODELS

2.1. *Mesoscale circulations induced by surface forcing*

Pielke [1] and more recently Bougeault [2] gave a detailed review of mesoscale flows generated by spatial contrast conditions, with emphasis on differences in surface temperature and heat fluxes. The most widely studied case related to large spatial temperature contrast was the sea-breeze mechanism. The numerical integrations over Florida described by the previous authors showed how the enhanced humidity convergence by thermal contrast at the coastline can induce convective rainfalls [3].

Ookouchi *et al.* [4] were the first to concentrate on the generation of mesoscale circulation produced by spatial differences in soil moisture. In the study of Ookouchi *et al.* [4], as well as in the works of Mahfouf *et al.* [5] and Pinty *et al.* [6] presented hereafter, the analysis was restricted to 2D flows, often over flat terrain, and in the absence of atmospheric forcing which could mask the atmospheric circulations driven by surface properties. Using a 2D atmospheric model with a crude representation of surface hydrology (the surface moisture availability being kept constant during the integration), Ookouchi *et al.* [4] investigated the characteristics of the perturbation induced in the PBL as a result of soil moisture contrasts between two adjacent regions. The patterns of the induced flow were found to be very similar to the sea-breeze case, but with some attenuation because the intensity of the thermal contrast between the two zones was weaker. As an example, Fig. 1(a) shows the computed maximum wind speed at the ground level for various sensitivity experiments corresponding to differences in soil moisture, and subsequently in surface temperature (Fig. 1b). For large differences of soil moisture between the two adjacent areas (case F5 in Fig. 1), the intensity of the resulting thermal circulation in the atmosphere can be comparable to that of a sea-breeze (case SB in Fig. 1). Ookouchi *et al.* also examined the flow induced by a symmetric mountain, having a wet and a dry slope. They showed that the combination between orography and differential heating between the two mountain sides acted efficiently to transport moisture upward and favoured the formation of convective clouds. Mahfouf *et al.* [5] conducted a similar analysis in a more satisfactory way, using a more realistic surface parameterization including an explicit description of transfers in the soil [7] and within a layer of vegetation [8]. Using again a 2D mesoscale model (grid size of 10 km, 7 vertical levels below 2 km), the approach consisted in computing the full interactions of the land–surface–atmosphere system by only specifying the soil and vegetation types as surface boundary conditions. The authors showed the large influence of the soil texture on surface moisture availability. Mahfouf *et al.* [5] also noted the difference of evaporation regimes between a vegetated and a bare soil area. In the vegetated area, the transpiration was mainly determined by the deep soil moisture and did not vary very much on a day time scale. On the other hand, in the bare soil case, evaporation was strongly linked to the diurnal variation of the near-surface moisture content (see also [9]), depending on the soil texture and on the soil moisture itself. These differences were responsible for a mesoscale flow generated by contiguous bare and vegetated areas. Mahfouf *et al.* [5] indicated that the transition zone was a preferred location for the initiation of moist convection. A similar conclusion was made by Segal *et al.* [10] who found that the intensity of the mesoscale circulation was not too much dependent upon the spatial extent of the transition zone between two distinct land-use categories. Mahfouf *et al.* [5] discussed also the vegetation shielding effect on the ground surface. They showed that the presence of an unstressed vegetation canopy reduces the spatial contrast between turbulent fluxes and consequently the intensity of the induced mesoscale flow. As an example, Fig. 2 presents the numerical results from 2D experiments combining the influence of a soil texture discontinuity with the presence of an overlying vegetation canopy. The vegetation cover was varied from 0 (bare ground) to 0.5 and 1 (dense canopy layer). Because of large near-surface soil

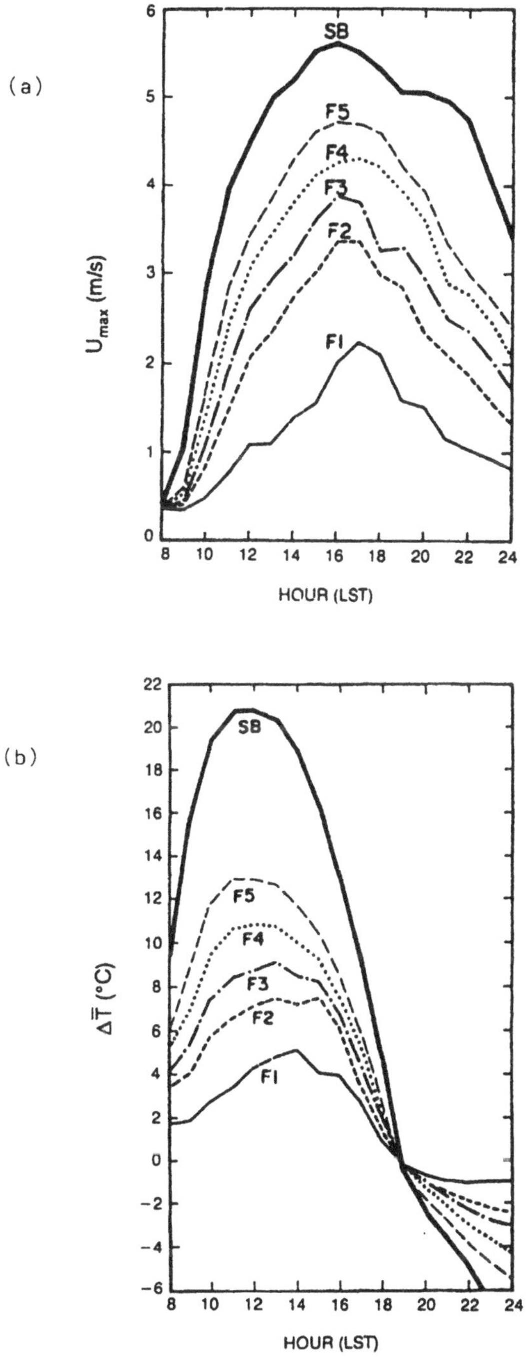

Figure 1. Maximum wind speed (m/s) at 5 m height along the domain cross section (a) and surface temperature difference ΔT between dry and wet domains (b) for various moisture availability contrasts as a function of time. (Reproduced from [4]).

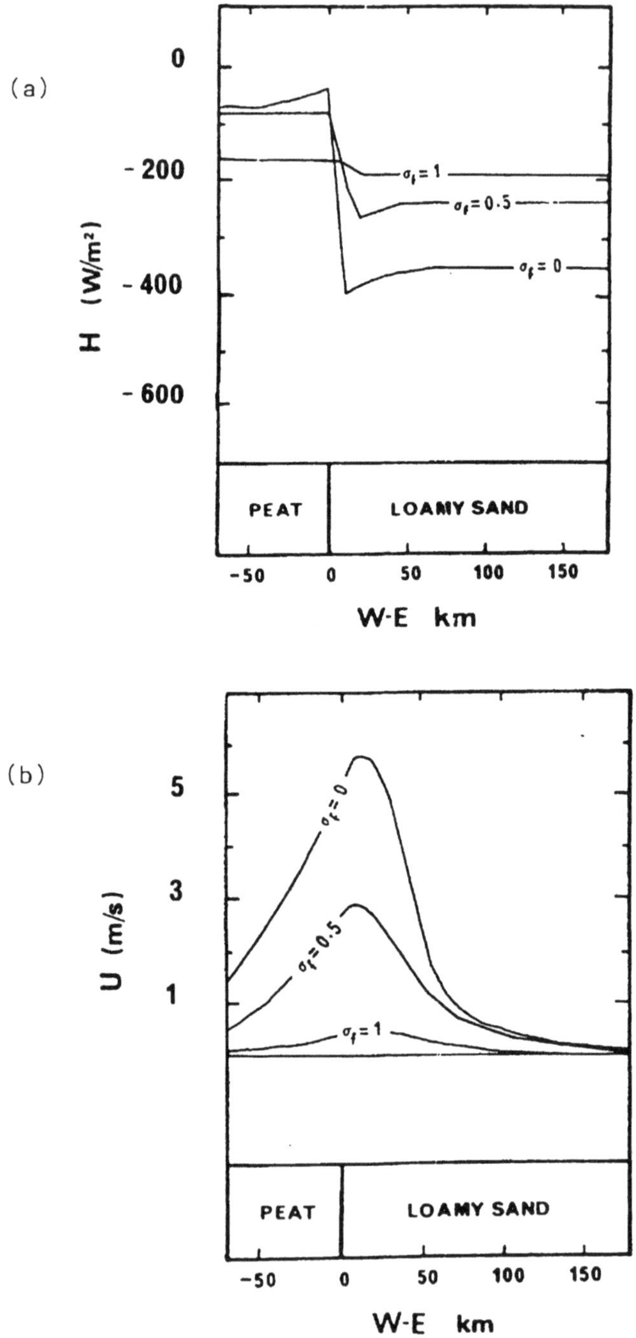

Figure 2. Horizontal variations of surface sensible heat flux (predicted at 1400 LST) (a) and of horizontal wind speed at 150 m height (predicted at 1800 LST) (b) over a soil texture discontinuity (peaty and loamy soils) and for different vegetation shielding factors σ_f varying from 1 (dense vegetative cover) to 0.5 and 0 (bare soil). (Reproduced from [5]).

moisture contrast in the bare ground case, a sensible heat flux difference of 200 W/m^2 was computed by the 2D model at 1400 LST. The maximum predicted wind at 150 m was 6 m/s, a value comparable with the Ookouchi *et al.* results. On the other hand, when the two types of soil were covered with a dense well-watered canopy, the spatial contrast in surface exchanges became very weak (Fig. 2a) and only a weak current circulated from the wet peat soil zone to the somewhat dryer loamy sandy one.

The next step in the analysis of inland circulations was to introduce more realism in the 2D numerical context by using *in situ* observations. Such a study was performed by Pinty *et al.* [6] who considered the idealized circulation (no background wind speed) that might be produced by a land–surface discontinuity present in the HAPEX–MOBILHY (1986) experimental area [11]: an extensive pine forest over a sandy soil surrounded by an agricultural area over loamy soil (this analysis was also described in [12]). Pinty *et al.* [6] used the same 2D model as Mahfouf *et al.* [5]. In the 2D model, the western side was covered by a typical cereal crop and the eastern side by a forest. The soil and vegetation properties in the model were assigned according to observations and to calibrations of the surface scheme in a stand-alone mode against fluxes measurements available for both types of vegetation canopies. The initial soil water content was also taken from the observations as well as the initial profiles of temperature and specific humidity in the PBL. An inland breeze was predicted (Fig. 3) because of a difference of sensible heat of about 200 W/m^2 between the two zones. This was mainly due to differences in plant stomatal responses: the surface resistance of the pine forest was higher than those of the growing cereal crop because of its higher dependence on the atmospheric specific humidity deficit (see section 2.2). The atmospheric response to the sensible heat flux horizontal distribution is depicted by the cross section of predicted atmospheric quantities at the end of the afternoon (after 12 hours of integration) (Fig. 3). Horizontal wind of about 4 m/s blew from the cereal field to a depth of approximately 1 km with a return flow aloft of the order of 2 m/s. The horizontal flow was associated with an ascending motion above the forest and a weaker subsiding flow over the cereal zone, forming a close circulation analogous to a classic sea breeze flow. Because of the surface differential heating, the depth of the boundary layer was about 500 m higher over the forest than over the crop area. A more turbulent PBL was predicted over the forest as a result of a higher roughness length as well as a larger transfer of heat into the atmosphere. Finally, the atmosphere over the forest was drier because of stronger vertical mixing. From these atmospheric fields, the authors concluded that the turbulent region over the transition zone should contribute to preferred cloud formation.

More recently, Bechtold *et al.* [13] considered the same case but investigated the influence of a large-scale background wind on the pattern of the inland breeze. Bechtold *et al.* [13] concluded that the intensity and spatial extent of the breeze is not too much influenced by the value of the background wind and indicated that such a mechanism might actually be observed. Unlike the sea breeze circulation, the presence of turbulence over the whole domain considered was found to lead to a more horizontally uniform boundary layer.

The HAPEX–MOBILHY programme provided a remarkable source of *in situ* observations to study the ability of three dimensional mesoscale models to reproduce evidence of the impact of a forest canopy on the regional meteorology [12]. Taking full advantage of improved meteorological analysis from supplementary ground and upper-air observations, three-dimensional runs were conducted by two groups using similar

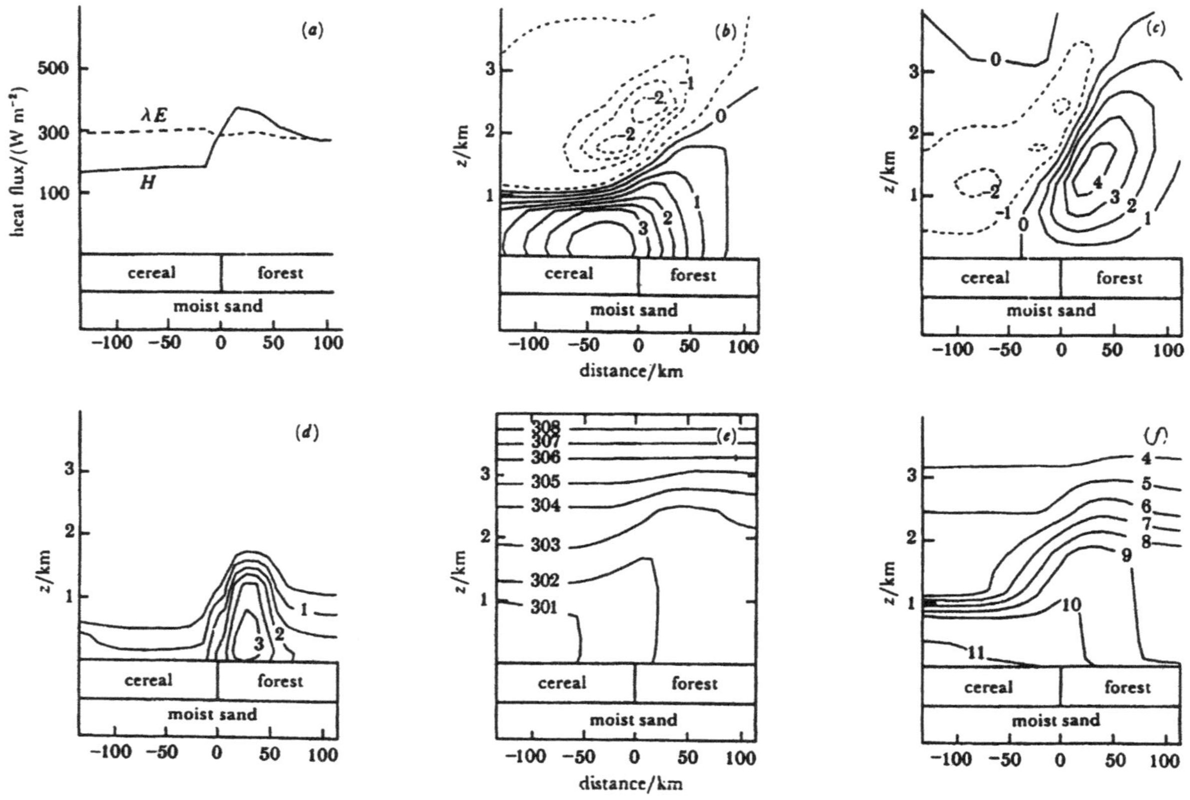

Figure 3. Two-dimensional transition between a cereal crop (left part of the domain) and a forest (right part of the domain) over sandy soil close to saturation: (a) sensible (H) and latent (LE) heat fluxes, (b) cross section of horizontal wind velocity (m/s), (c) of vertical velocity (cm/s), (d) of eddy kinetic energy (m^2/s^2), (e) of potential temperature (K), and (f) of water–vapor mixing ratio (g/kg). (Reproduced from [12]).

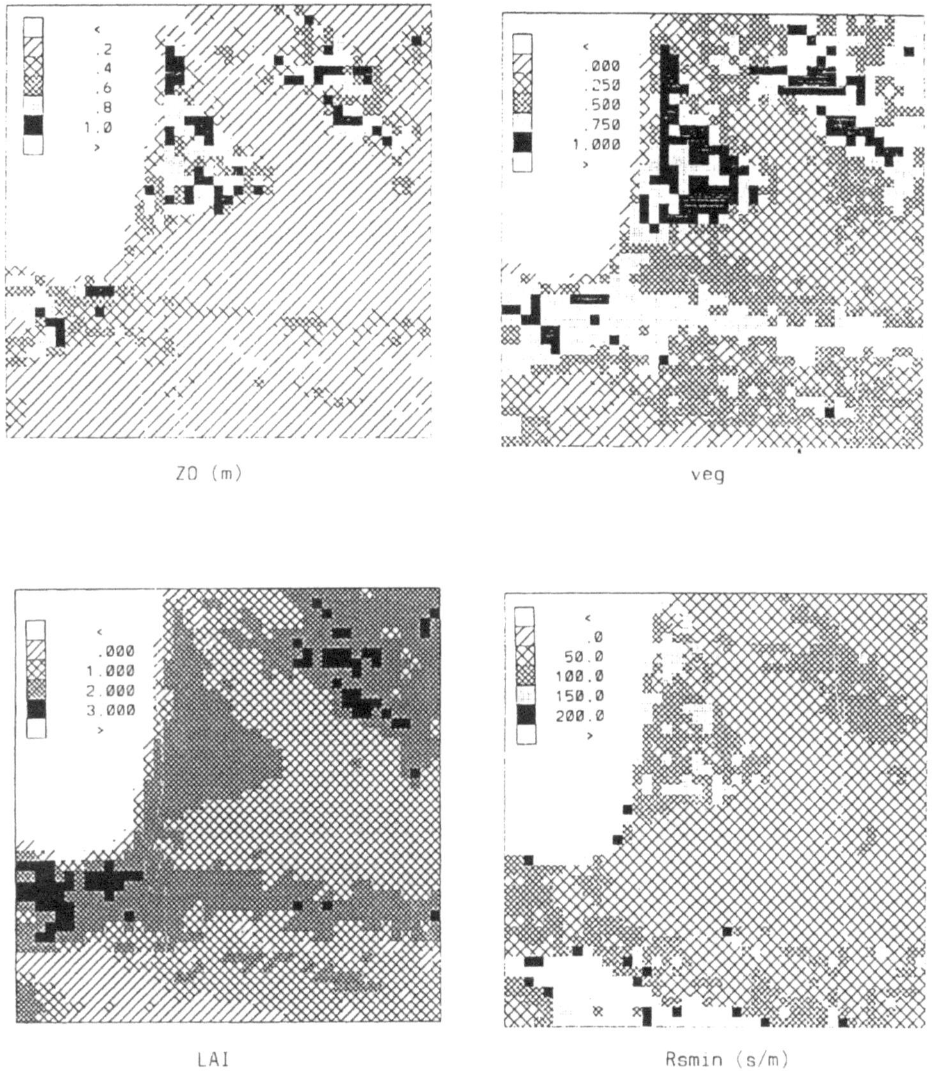

Figure 4. Maps of the model grid values of the surface parameters veg, Z_0, $R_{s\,min}$ and LAI depending upon the vegetation. (Reproduced from [21]).

mesoscale models, differing only by distinct land–surface parameterizations. Results reported by Pinty *et al.* [14] and by Mascart *et al.* [15] were obtained using a transfer scheme solving heat and moisture diffusion at 13 vertical levels within the first meter of soil, and separate energy budgets for the soil surface and the canopy layer. On the other hand, the studies of Bougeault *et al.* [16, 17] and Noilhan *et al.* [18, 19] made use of a simpler land surface scheme, the ISBA parameterization briefly described in section 2.2. The results obtained by the two groups behave rather similarly since the same large-scale weather analysis were used as well as identical maps for the spatial distribution of soil and vegetation types within a domain of 400 km × 400 km. Moreover, both surface schemes were calibrated against daily variation of surface fluxes observed

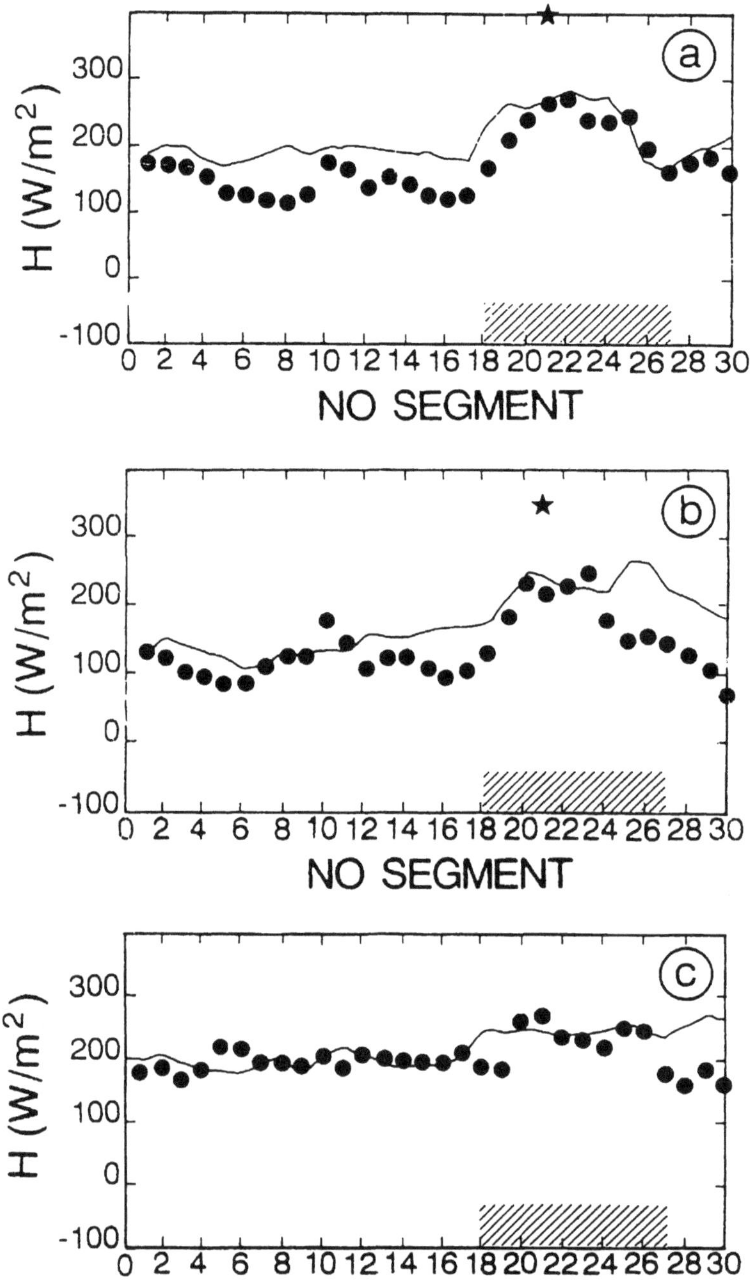

Figure 5. Comparison between the King Air measurements (full circle) and the prediction (full line) of the horizontal variation of the sensible heat flux at 100 m above the surface on 16 June 1986 (a), 19 June 1986 (b), and 10 July 1986 (c). The observations at the HYDRA (full star) sites below the aircraft trajectory are reported in (a) and (b). The numbers in abscissa refer to flight sub-segment numbers. (Reproduced from [23]).

(a)

(b)

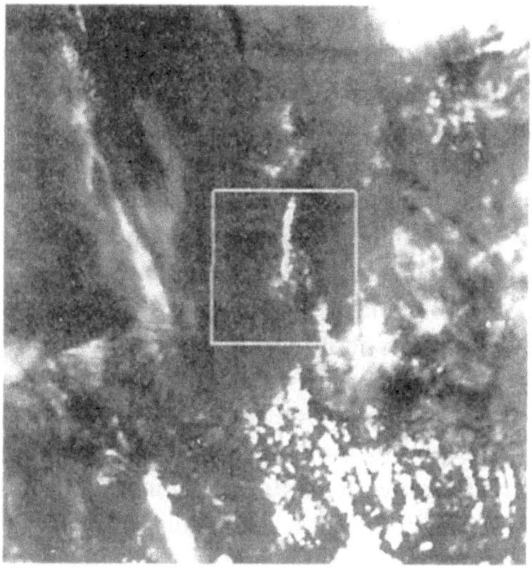

Figure 6. Prediction (a) and observation (b) (NOAA6/AVHRR) of the boundary-layer cloud cover at 1800 UTC on 16 June 1986 during HAPEX–MOBILHY. The forecast was realized with the PERIDOT mesoscale research model over a 400 km × 400 km domain encompassing the HAPEX–MOBILHY square indicated in (a) and (b). (Reproduced from [17]).

at an ensemble of twelve flux stations [20] prior to their implementation within the mesoscale models. However, the atmospheric response to surface inhomogeneities was found somewhat smaller in the Bougeault *et al.* [16, 17] studies because they tried to incorporate some representation of the surface sub-grid variability in the model, taking advantage of a vegetation classification available at the 1 km^2 AVHRR grid scale [21]. In the following, we shall restrict ourselves to the description of some results obtained by Bougeault *et al.* [17] and Noilhan *et al.* [18]. As far as surface processes are concerned, the mesoscale model was run with a realistic description of the land-use categories present in the domain. As an example, Fig. 4 shows the spatial distribution of surface parameters related to the vegetation, i.e. the vegetation cover veg, the minimum surface resistance $R_{s\,min}$, the leaf area index LAI and the roughness length Z_0. The prominent, triangular-shaped, coniferous Les Landes forest is clearly visible with significantly different values of surface quantities than the nearby agricultural zone. One-day integrations for four clear days have been carried out showing clearly the impact of surface processes on the prediction of the lower atmospheric fields at the regional scale. Evidence of atmospheric perturbations generated by the contrast between the forested and the agricultural areas was provided by repetitive flights with an instrumented aircraft which was flown close to the surface. Detailed comparisons of turbulent fluxes predicted by the mesoscale model with in situ measurements along the aircraft trajectory (Fig. 5) were used as an original way to validate the forecast within the area of interest [23]. Concerning the impact of surface inhomogeneities on the atmosphere, the main striking pattern at the beginning of the observations was the development of a mesoscale circulation between the pine forest and the surrounding crop area. Because of high values of $R_{s\,min}$ [24] and of the vegetation fraction veg over the forest, the sensible heat flux was higher than over the crop area both in the observations and in the simulations on 16 and 19 June 1986. On the other hand, at the end of the Special Observing Period (after 1 July), the contrast between the crops and the forest was reduced as a result of crop development, resulting in a decrease in the soil heat storage [25]. Thanks to a physically-based surface transfer scheme, particularly for the canopy resistance formulation, the mesoscale model was able to give a coherent picture of what happened and what was sensed during selected sunny days within the HAPEX square. A clear manifestation of the inland circulation between the forest and the crops was the cumulus convection over the forest, observed and predicted on 16 June (Fig. 6). The circulation in the HAPEX area could be illustrated further by the 3D prediction of the vertical velocity at the middle of the PBL showing an ascending motion over the forest and subsiding air in the surroundings (Fig. 7). The maximum velocities were +5 and −2 cm/s, with subsidence in the central part of the forest, where the clearings are frequent. This results in a characteristic, open-triangular-shaped pattern, which was in good qualitative agreement with the distribution of cumuli observed at that time. This result was considered by the authors one of the major successes of the simulation, since it proved that the mesoscale model correctly predicted the mesoscale circulation induced by the surface contrast. To the authors' knowledge, this analysis was the first study based on real observations. From a sensitivity study in which the surface resistance of the forest was reduced, André *et al.* [12] showed that the triggering of cloud convection was reduced, emphasizing the importance of the stomatal resistance parameterization.

Another important aspect of surface processes which may have an impact on the mesoscale meteorology is the interception mechanism. Water intercepted by the leaves

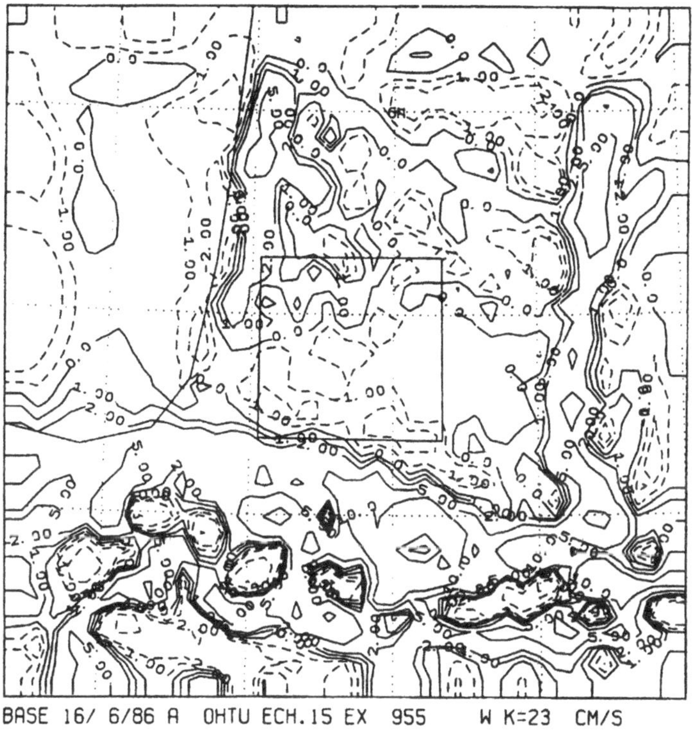

Figure 7. Mesoscale prediction of the vertical velocity at 500 m above ground for 1500 UTC on 16 June 1986. Units are cm/s and dashed lines correspond to subsiding (negative) velocity. A well-defined ascendance is predicted over Les Landes forest, surrounded by subsidence. (Reproduced from [17]).

evaporates back into the atmosphere at the potential rate, without reaching the ground. Interception loss can be significantly higher than transpiration rate, especially over forest canopies because of their high value of roughness length [26]. Forests have also a significant water storage capacity (several millimetres of water can be retained on the foliage leaf surfaces) as compared with short vegetative canopies. Interception also affects the vertical structure of the lower atmosphere. Stewart [27] proved experimentally that forests, when their surface is wet, can draw energy from the overlying air to sustain their evaporation of intercepted rainfall involving negative values of the Bowen ratio (the ratio between sensible and latent heat fluxes). From a mesoscale integration concerning a frontal intrusion in the HAPEX–MOBILHY area, Blyth *et al.* [28] analyzed the mechanism for maintaining negative sensible heat flux over forest, which is capable of sustaining wet canopy evaporation when net radiation is low. The main condition needed for this mechanism to occur was a strong wind shear within a stably stratified boundary layer providing downward transport of heat and upward evaporation. The link between interception processes and mesoscale meteorology was investigated by André *et al.* [12]. This study showed how the availability of additional water through interception can increase the regional rainfall through a positive feedback of atmospheric humidity. Blyth *et al.* [28] performed complementary sensitivity tests on the same situation. They showed that the predicted daily rainfall was 30% higher over forested area than when the surface was all bare soil. Blyth *et al.* [28] demonstrated that the higher evaporation rates from a wet forest during a rainfall event accounted for about half of that increase but did not contribute to an increase in soil moisture. The convergence of water vapour in the PBL due to the high roughness of the forest accounted for the other 50% of the increase in rainfall which increased the soil moisture. They concluded that the increased rainfall recycling from a positive humidity feedback influences the regional rainfall dynamics and patterns only. On the other hand, they noted that the roughness effect can influence the long term water balance of a region by providing a net gain of water by the land from the atmosphere.

Finally, another interesting feature of rainfall interception is the rapid transition between wet and dry canopies (less than 10 hours). Thus, a correct estimation in a Numerical Weather Prediction (NWP) model of the Bowen ratio and of the feedbacks between the boundary layer and the wetted surface will be only possible if the interception phenomenon is taken into account.

2.2. Essential representations in current land–surface parameterizations (LSP)

Given the large number of LSP developed during the last decade, we shall now restrict the discussion to the essential processes described by any surface parameterization to be used in mesoscale models. Unlike Global Circulation Models (GCM) integrations, short-term limited area NWP models are mainly determined by atmospheric and surface initial conditions. Therefore, a specific requirement of mesoscale models with regard to surface processes is to reduce as much as possible the number of input parameters while attempting to preserve the representation of the physics controlling the surface and subsurface processes. The challenge is then to capture the behaviour of soil and vegetation with a broad enough parameterization. During the course of the discussion, examples taken from integrations with the Interaction-Soil-Biosphere-Atmosphere scheme (ISBA: [29]) will be presented for demonstrating successes, failures and possible improvements of LSP.

The role of LSP is to compute the surface momentum, heat and water fluxes for atmospheric conditions provided at the first level of the atmospheric model. Roughly speaking, three subschemes are considered, respectively the surface layer, the soil and the vegetation layers.

Any mesoscale model pays a large attention to computations of surface boundary layer processes. The height of the lowest model level is usually in the first 50 m in order to allow for the computation of turbulent transfers of momentum, sensible and latent heat fluxes from the surface similarity theory (constant flux layer assumption). The transfer coefficients are determined from roughness lengths for momentum, heat and moisture exchanges and the thermal stability of the near-surface stability gradient. There are now a lot of experimental studies showing that it is necessary to distinguish between roughness lengths for momentum and heat, particularly in the case of bare ground and sparse vegetation canopies. Moreover, in some LSP including a description of the exchanges within the canopy, the classical surface layer theory is accomodated to solve separately the transfers within and above the vegetation layer (SiB: [30]).

Concerning the representation of soil and vegetation, most LSP use at least 2 soil layers and one canopy layer. Physically, the two soil layers are necessary since the top thin layer (some mm) responds faster with the diurnal cycle while the deeper layer controls the monthly changes [8]. Since the extraction of water in the deep soil by plant transpiration and rainfall interception are essential components of the surface water budget, it is also important to include a representation of the vegetation into LSP.

Most of LSP use the force-restore method for temperature and soil moisture (SiB, BATS, ISBA). The method is based on analytical solution of the diffusion of heat and moisture into the soil with the assumption of periodic surface forcing. The force restore method computes two prognostic variables: the surface one interacts rapidly with the surface forcing (radiation, precipitation, turbulent fluxes), the other one responds with the storage term. The method was improved in ISBA through the calibration of the time constant coefficients between the two reservoirs as a function of the soil moisture and the soil texture. A set of coefficients were obtained using a reference model with a large number of soil layers solving explicitly the heat and water diffusion equations [31]. When referring to thermal exchanges, the force-restore method considers the ground surface temperature and a daily mean soil temperature. On the other hand, when considering moisture exchanges, the prognostic variables are the water content of a thin surface layer and the mean soil moisture content over the total depth of the active soil considered. The depth of the active soil layer is assumed to include the whole root system and the water flux at the bottom of the layer is neglected (no contribution of the seasonal cycle). The main advantage of the method is its simplicity as well as the low number of parameters required. For instance, if the textural components of the soil are known (i.e. the soil type according the USDA classification or the percentage of sand and clay), ISBA computes the heat and mass transfers using a set of 7 calibrated coefficients. More complex methods (for instance [5, 7]) solve explicitly the heat and water transfer diffusion equations from a multi-layer soil model, using finite difference techniques. Despite the increasing accuracy, these soil schemes are less suitable for mesoscale modelling than the two-layers soil models because of the increased computational cost. Since the soil transfers are highly non-linear (especially for water diffusion), multi-layer soil models need a high vertical resolution, particularly close to the surface. Moreover, they require a lot of details on the initial vertical profiles of temperature and water content which are never available at the mesoscale. The large number of vertical levels prohibits also the development of an assimilation scheme for the soil moisture, a specific requirement of NWP models. However, it must be indicated that most of the pioneering works reported in the previous section were conducted using such detailed soil schemes.

The vegetation in LSP is usually treated following the 'big-leaf' concept designed by Deardorff [8]. Many improvements of this idea have been proposed but the general philosophy remains. The vegetation is considered as a single big-leaf, shielding partly the ground surface. Separate energy budgets are computed for the canopy layer and for the ground surface. The vegetation layer may intercept and evaporate rainfall or dew retained on the leaves. Canopy transpiration is the other physiological process which is represented. Transpiration is commonly expressed through the parameterization of a canopy surface resistance describing the physiological control of plants to extract water from the deep soil layer. There is now a large agreement to use the general expression of Jarvis [32] for the dependence of the stomatal resistance on environmental conditions and soil moisture. Sensible and latent heat fluxes are computed for the vegetation and for the ground surface and then combined into total resultant fluxes according to the respective fractions of vegetation and bare soil present in the grid box. Total evaporation is then split into three components: soil evaporation, canopy evaporation and transpiration. The big-leaf approach has at least four prognostic variables for the canopy-surface medium: the temperatures of the vegetation and of the ground surface, the near-surface water content or the surface specific humidity and the intercepted water.

Due to specific requirements for operational weather forecast, ISBA is a simplified version of the Deardorff's parameterization since it considers a single energy balance for the whole surface medium. Therefore, the number of prognostic variables is reduced because a single surface temperature is computed for the ground surface and the canopy layer. The original force-restore method was modified in order to incorporate the effect of the canopy heat capacity. Extensive calibrations of ISBA against a broad range of observations seem to indicate that there is no basis to complicate the scheme any more in the case of significant vegetation cover or bare ground. However, the method needs improvement to simulate correctly conditions of sparse canopy under dry conditions where significant differences of temperature are generally observed between the vegetation and the ground surface. On the other hand, when using the big-leaf concept with separate energy budgets, other difficulties arise concerning both the partitioning of available radiative energy and the parameterization of turbulent exchanges between the vegetation and the ground surface and the atmosphere within the canopy. A correct representation of these processes increases considerably the number of parameters leading to a complexity and uncertainties often incompatible with mesoscale model requirements.

We shall now present some examples of LSP calibration in a stand-alone mode (i.e. the LSP is constrained to follow the observed atmospheric forcing at the reference level), with special emphasis on integrations performed with ISBA using data sets collected over a broad range of land-uses. These examples illustrate some problems involved in the formulation of surface fluxes, respectively in the cases of bare ground and vegetated area (rainfall interception and transpiration). Obviously, calibrations of LSP at the local scale are necessary prior to their implementation in mesoscale models in order to acquire some confidence in those surface quantities or parameters which are not well known and to evaluate to what extent the predictions of surface fluxes are accurate, as a necessary surface boundary condition of mesoscale models.

2.2.1. Exchanges over bare ground. Given the sensitivity of mesoscale meteorology to surface thermal contrasts, special attention has been paid to ISBA predictions of exchanges over bare ground using the force restore method. Two data sets collected over bare ground under dry conditions were examined (AVIGNON84, [33] and EFEDA91, [34]). Of particular interest in the observations was the magnitude of daytime soil heat flux G which represented a significant fraction of net radiation ($\approx 30\%$). Figure 8(a) shows a time evolution (10 consecutive days) of observed and predicted soil heat flux G at a site of EFEDA91. Because the local measurements of thermal properties were directly introduced in the surface scheme, the predictions of G were in fair agreement with the observations. Note the daytime values of G which culminate at 200 $\mathrm{W/m^2}$. Previous runs using the parameterization of thermal conductivity proposed in the literature overestimated values of G by about 20%. A similar behaviour was observed by Mahfouf and Noilhan [9] using the AVIGNON84 data set. These two studies illustrated the possible large errors involved when using prescribed soil properties from existing soil classification.

Another important feature concerning heat exchanges over bare ground is the significant departure between the roughness length for momentum (Z_{0m}) and heat (Z_{0h})

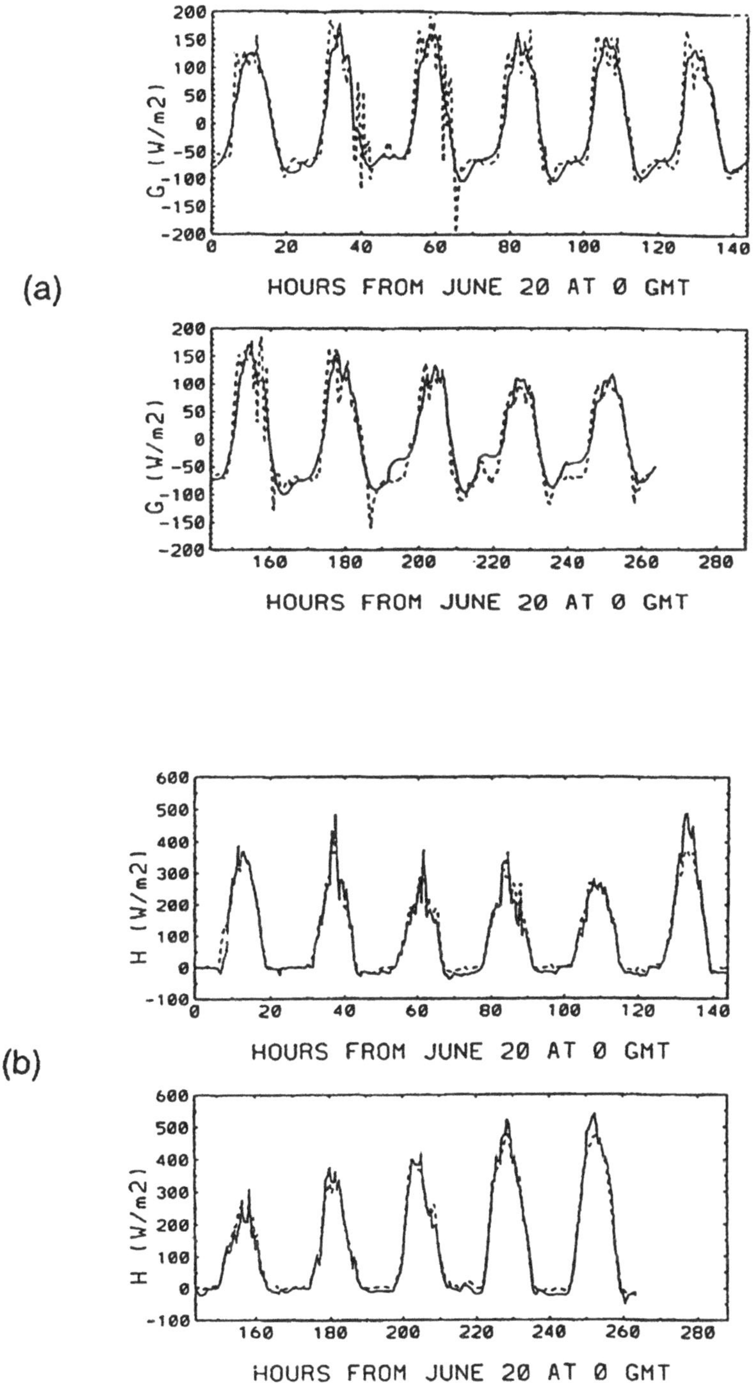

Figure 8. Comparison of the predicted soil heat flux *G* (a) and sensible heat *H* (b) with observations taken continuously during 11 days at the Barrax site during EFEDA. (Reproduced from [35]).

transfers. Although confirmed from observations many years ago, few land–surface schemes distinguish between these two roughness lengths. Observations show that over natural surfaces the ratio Z_{0m}/Z_{0h} is close to 10. Such an assumption was tested with the EFEDA91 data set [35] leading to a satisfactory prediction of the sensible heat flux H as shown in Fig. 8(b). Using the classical assumption that $Z_{0h} = Z_{0m}$, the sensible heat flux was clearly overestimated.

Formulation of surface evaporation in simple parameterization schemes is also an important issue. Indeed, evaporation over bare ground is a very complex mechanism because liquid water is not generally available at the surface but evaporation occurs at a certain soil depth depending on the near-surface soil moisture. Accurate prediction of surface evaporation requires the use of detailed multi-layer soil models with very high spatial and temporal resolutions [5, 31, 33]. In the parameterization context, the challenge is to reproduce the surface hydrology with a limited number of both soil variables (e.g. the surface temperature and the water content of a thin superficial layer) and surface parameters. Most of evaporation formulations in surface schemes are based on bulk aerodynamical expressions by using a parameterization of the surface specific humidity [8, 36]. Some parameterizations make use of threshold methods based on the concept of water supply and demand close to the surface [37, 75]. In this latter approach, considerable information about hydraulic properties in a thin layer is needed, making them very sensitive to the averaging depth to which the near-surface water content is known. Using the AVIGNON84 data set, Mahfouf and Noilhan [9] examined various formulations of surface evaporation (Fig. 9). Tests in a stand-alone mode were run, starting with observed soil water contents and changing only the formulation for surface evaporation. In the context of these observations, the diffusion-limiting moisture flux was strongly underestimated. On the other hand the most widely used bulk aerodynamic formulations (so-called α- and β-methods) provided comparable results at daytime. On the other hand, the β-methods [30] overestimated the nocturnal evaporation which was better predicted using the α-method [29].

In the case of very dry soils, effects of vapour phase diffusion have to be taken into account in the time variation of superficial soil moisture. A specific investigation

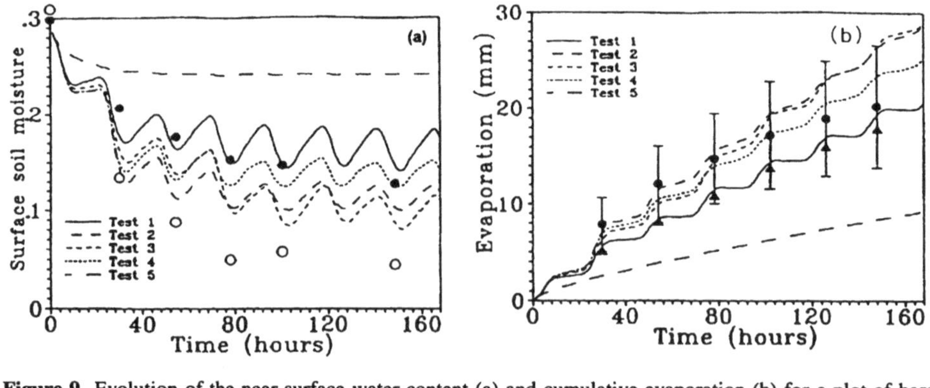

Figure 9. Evolution of the near-surface water content (a) and cumulative evaporation (b) for a plot of bare ground. (Reproduced from [9]). Daily observations of soil moisture are taken at 1.5 cm (full circles) and 0.5 cm (open circles). Estimations of evaporation are deduced from observed water budget (full circles with standard deviations) and aerodynamic measurements (full triangles) (AVIGNON84 data set). The different curves in (a) and (b) refer to ISBA predictions with various formulation of surface evaporation. Test 1 corresponds to the α-method, while Test 4 refers to a β-scheme for surface evaporation.

of heat and water exchanges for very dry soils was performed by Braud *et al.* [35] who improved ISBA by taking into account soil moisture transfers in vapour phase. This approach was tested favourably against observations of surface evaporation and near-surface water content at the EFEDA91 site.

2.2.2. Rainfall interception. Rutter *et al.* [38] derived the first predictive model of rainfall interception by computing the amount of water retained on a one-layer vegetation canopy taking into account interception of rain, drainage rate and potential evaporation from the wetted part of the canopy. Most of recent interception schemes in LSP are conceptually similar to the approach of Rutter *et al.* [38]. This is the case for ISBA, but the drainage rate from the canopy is simplified. Rainfall or dew are intercepted until the water storage reaches a critical saturation level $W_{r\,\max}$, at which stage excess water is assumed to drain immediately. The maximum reservoir capacity is set equal to $\alpha \times LAI \times veg$, where α is an empirical coefficient, often set to 0.2 mm [37] which seems valid for coniferous canopies. In many calibrations of interception schemes the α coefficient was tuned to agree with observations.

Validations of interception schemes have been carried out mainly over forest canopies for which long enough time series of precipitation and throughfall were recorded [39, 40]. The interception data obtained continuously during 25 months at Ducke Reserve Forest (2S57′, 59W57′) during ARME84 have been widely used for calibrating interception schemes. As an example, Fig. 10 shows different predictions of cumulative interception loss compared with observations [40]. Four models (Rutter, Gash, SiB and ISBA models) were applied to the same automatic weather station forcing during two years. The Rutter model and Gash analytical model were used with the observed state parameters, in particular for the interception capacity estimated to 0.74 mm. The four simulations described in Sellers *et al.* [41] are combinations of two leaf area indices and two values of $W_{r\,\max}$. Two simulations were run with ISBA: first of all, with $W_{r\,\max} = 0.1 \times LAI$ to be compared with one run of SiB; the second one with $W_{r\,\max} = 0.15 \times LAI = 0.78$ mm, for comparison with Rutter model results. From Fig. 10, all the schemes overestimate consistently the interception loss during the second year of measurements. It is difficult to conclude if these discrepancies are due to model imperfections or to measurement errors.

The SiB calculations show that $\alpha = 0.1$ yields a calculated interception loss in better agreement with observations than with $\alpha = 0.25$ (Fig. 10.1b). The integrations of Sellers *et al.* [41] illustrate also the large sensitivity of the predictions to the value of LAI.

In the Rutter model, LAI is implicitly included since the observed interception capacity is used. Moreover, the aerodynamic resistance to the transport of water vapor is calculated from a simplified expression ($R_a = f/U_a$, where U_a is the wind speed at the reference level and f is a constant).

The ISBA integration with $\alpha = 0.15$ compares with the Rutter results. On the other hand, ISBA predictions with $\alpha = 0.1$ are larger than the SiB results under comparable conditions. This difference may be due to different formulation of R_a in the two models. ISBA computes R_a from a single expression [42] applied between the top of the canopy and the reference level (assuming $Z_{0m} = 3$ m and a displacement height $d = 30$ m), whereas in the SiB model the aerodynamic resistance between canopy air space and reference height, and the canopy bulk resistance are obtained from momentum transfer calculations within and above the forest canopy.

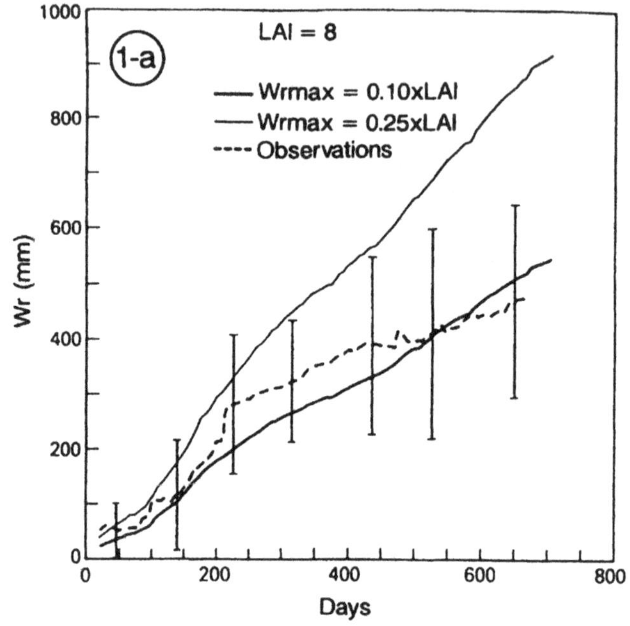

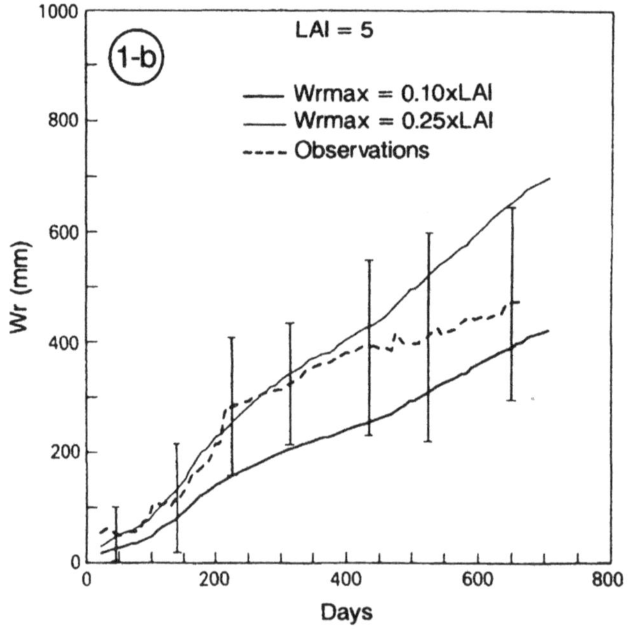

Figure 10. Prediction of the cumulative interception loss of the Amazonian rain forest over a 25-month period (ARME84) using four interception models: SiB [41] (10-1ab); Rutter model and Gash's analytical model (10-2ab) (integrations described in [40]) and ISBA (10-3). The observations are reported in 10-1 and 10-2. The values of LAI and of the reservoir capacity $W_{r\,max}$ (mm) considered in the SiB integrations are given in 10-1a and 10-1b. ISBA was run with $LAI = 5.0$, while in Rutter and Gash models, $W_{r\,max}$ (mm) was set equal to 0.78 mm.

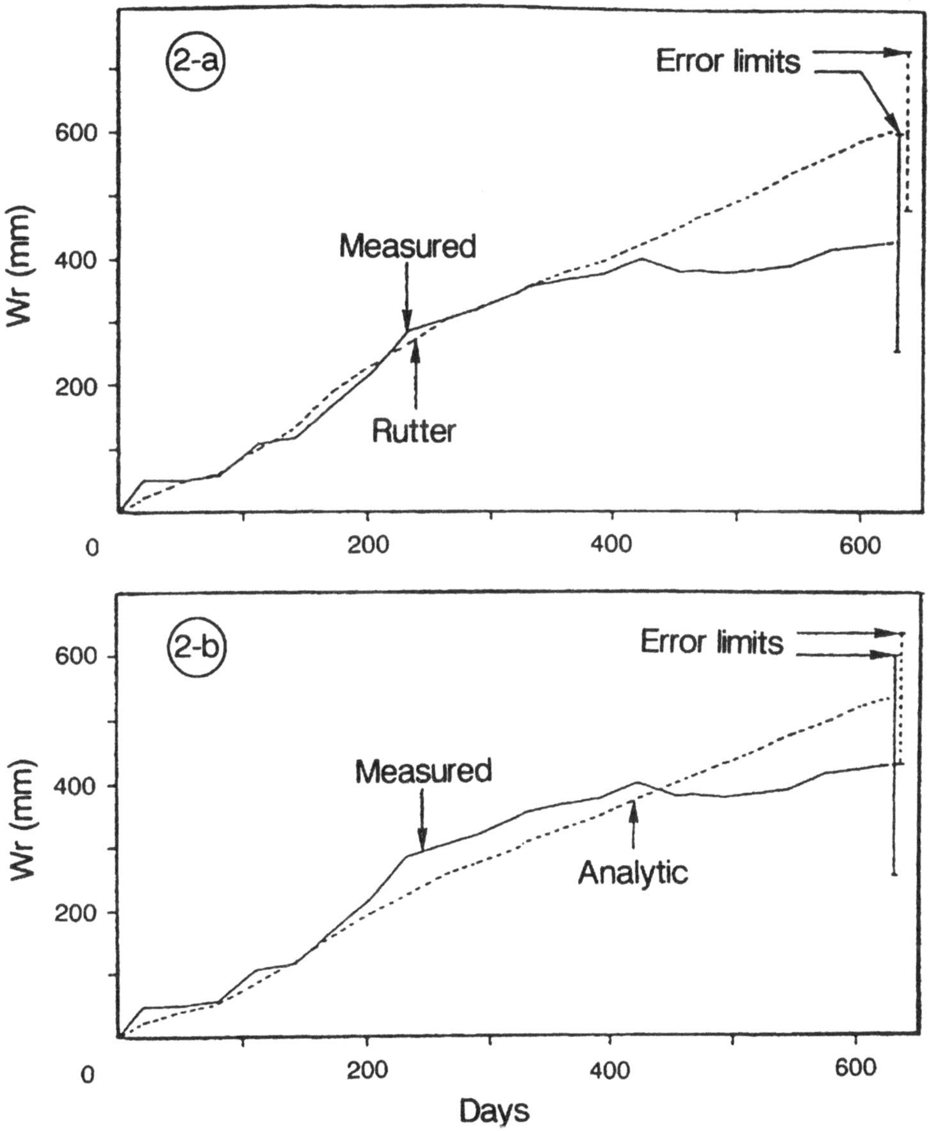

Figure 10. (Continued).

2.2.3. Plant transpiration and soil water budget. Plant transpiration is probably the main issue of LSP because of its link with the deep soil moisture. Water extraction from the rooting system is largely driven by environmental conditions and soil moisture availability: extraction of water is enhanced in conditions of full radiation, moderate atmospheric specific humidity deficit, optimum air temperature and freely available soil water in the rooting zone. Plants regulate the amount of water exchange by means of the saturated tissues inside the leaf stomates. Most of LSP express plant transpiration

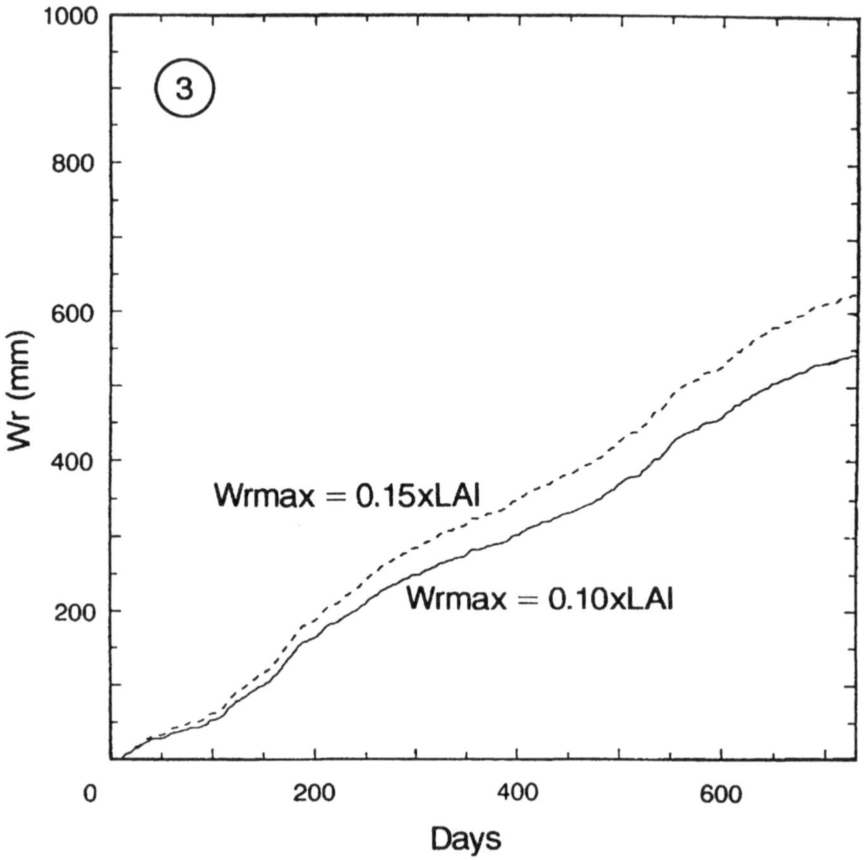

Figure 10. (Continued).

as:

$$E_{tr} = veg\,\rho_a \frac{(q_{sat}(T_s) - q_a)}{R_a + R_s}$$

where $q_{sat}(T_s)$ is the saturation specific humidity at the surface temperature T_s, q_a the air specific humidity and R_s the bulk surface resistance.

In this expression, water saturation is assumed to be maintained within the plant stomates through biological processes. Plants adaptation to their environment is tentatively described in parameterizing the surface resistances R_s. The general approach of Jarvis [32] is adopted by many LSP:

$$R_s = g(\text{plant})\, f_1(R_g)\, f_2(\text{soil moisture})\, f_3(\delta e)\, f_4(T_a).$$

The previous expression was initially developed for the resistance to water diffusion through a single leaf. The resistance for the whole canopy should be obtained by relating the profile of surface resistance through the canopy to the attenuation of solar radiation and atmospheric conditions within the canopy. This seems particularly important in the case of forests because of significant vertical gradients of temperature, humidity and wind speed. The study of Sellers *et al.* [41] illustrates how this can be achieved with SiB for a given vegetation architecture. The stress factors in the expression of R_s are not readily available and must be determined either from complex physiological experiments or calibration of LSP versus observations.

Because ISBA does not take into account processes inside the canopy, the parameterization of R_s attempts to describe the whole canopy. The function g(plant) is set to $R_{s\,min}/LAI$ where $R_{s\,min}$ is a minimum surface resistance depending on the plant type. Scaling up R_s to the whole canopy is simply attempted by attenuation of R_s with increasing leaf area index. Typically, ISBA calibration studies gave $R_{s\,min}$ of the order of 40 s/m for a developed crop canopy [43, 44], reaching higher values ($\approx$ 500 s/m) when plants experience maturation or senescence. A value of $R_{s\,min} = 100$ s/m was calibrated during HAPEX–MOBILHY over Les Landes pine forest for $LAI = 2.3$ m^2/m^2 [24], while $R_{s\,min} = 250$ s/m was found for the Amazon forest ($LAI = 5$ m^2/m^2) by Manzi and Planton [45].

The function $f_1(R_g)$ takes into account the effect of photosynthetically active radiation (PAR), assumed equal to 55% of the global radiation R_g in ISBA.

The function $f_3(\delta e)$ accounts for stomates closure in conditions of dry environment, a process particularly important for forest canopies:

$$f_3(\delta e) = \left[1 - h_e(e_{sat}(T_a) - e_a)\right]^{-1}$$

where $\delta e = e_{sat}(T_a) - e_a$ is the atmospheric vapour pressure deficit and h_e a parameter to be calibrated. The R_s dependence on $f_3(\delta e)$ is the most important environmental constraint for the transpiration of the Amazon forest and SiB calibration for the Amazon tropical forest provided $h_e = 0.0273$ hPa^{-1}. This strong dependence was also observed for Les Landes forest during HAPEX–MOBILHY, involving a lower latent heat flux over the pine forest than over the nearby crop area. Jacquemin and Noilhan [43] proposed $h_e = 0.037$ hPa^{-1} for Les Landes forest, in agreement with observations [24]. Such calibration of $f_3(\delta e)$ over Les Landes forest [43] leads to a fair prediction of the canopy surface conductance when compared with the observed conductance estimated

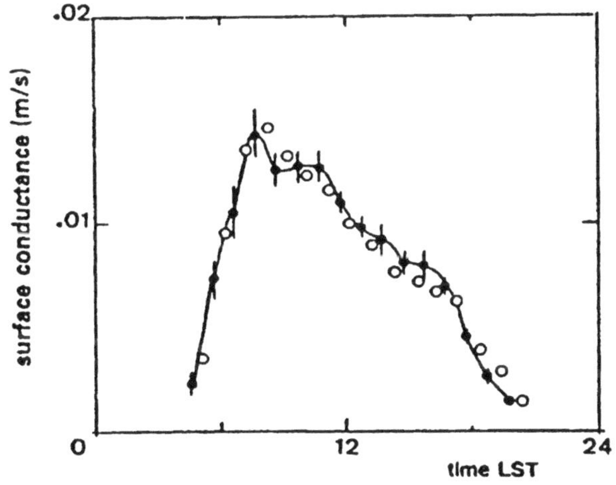

Figure 11. Daily variations of the mean observed (full circle) and predicted (open circle) surface conductance of Les Landes forest. (After [43]).

for the entire data set (Fig. 11). The maximum value of the canopy conductance is observed a few hours after sunrise, followed by a decrease all day long as a result of increasing atmospheric specific humidity deficit. The vapour deficit term $f_3(\delta e)$ seems less important for short canopies. However, it was found particularly significant for the prairie grass canopy in FIFE87 [46].

The function $f_4(T_a)$ used in ISBA follows Dickinson's [37] proposal. The parabolic shape of $f_4(T_a)$ is centred around an optimum air temperature of 25 °C where the stress is negligible. Transpiration stops when $T_a = 0$ °C or $T_a = 50$ °C, an assumption which sounds valid for temperate latitudes. At the moment, no clear validation of $f_4(T_a)$ is obtained with ISBA.

The function f_2(soil moisture) expresses the plant stress versus soil moisture in the root zone. In many LSP, this stress factor is computed from the leaf water potential which depends on plant physiology and soil moisture. In ISBA, f_2 is computed as:

$$f_2 = \left(\frac{w_2 - w_{\mathrm{wilt}}}{w_{\mathrm{fc}} - w_{\mathrm{wilt}}} \right)^{-1}$$

where the volumetric water contents w_2, w_{wilt} and w_{fc} refer respectively to soil moisture in the root zone, to the permanent wilting point and to field capacity. The values of w_{wilt} and w_{fc} are fixed for a given soil texture, w_{wilt} being computed for a soil moisture potential of −15 bar, below which plants cannot extract water any more from the soil matrix (see Table 1 in [43]). Choudhury [47] was one of the first to simulate the effects of soil water potential on a corn canopy. The main difficulty in calibrating the soil water stress function is to find a suitable data set, i.e. long enough time series of micrometeorological quantities together with soil moisture and soil hydraulic properties. These observations were collected during the 3 months HAPEX–MOBILHY Special Observing Period [20]. Unfortunately, only a moderate soil water stress was observed. However, this data set was used by Mahfouf [44] to evaluate the prediction of soil moisture by ISBA for periods of one or two months and for crop canopies. Figure 12 shows examples of prediction of cumulative evaporation and soil moisture for three agricultural sites. In all cases, the predictions agree reasonably well with observations, although some scatter is observed between observations and predictions of soil moisture. It is important to mention that this data set seems fully consistent, e.g. the variation of the amount of soil moisture matches reasonably well the soil-atmosphere interface exchanges (precipitation and evaporation) [20]. In the course of ISBA development, the result of Mahfouf [44] was particularly important since it gave substantial confidence in the ISBA functioning when used over long time periods.

2.3. *Specific needs for implementing and validating land–surface parameterizations*

We shall now briefly describe some specific requirements for implementing and validating LSP within mesoscale models. At this stage, it is convenient to identify input and output requirements. The input needs are:

(i) information on the land–surface at the mesoscale grid scale ($\approx$ 10 km),

(ii) initialization of the temperature and soil moisture prognostic variables.

As far as surface processes are concerned, model output of energy and water budgets (predicted surface temperature and humidity) need to be validated .

Any LSP requires information on the land-use. First of all, the knowledge of the soil and vegetation types for each grid cell is essential. This information is not generally

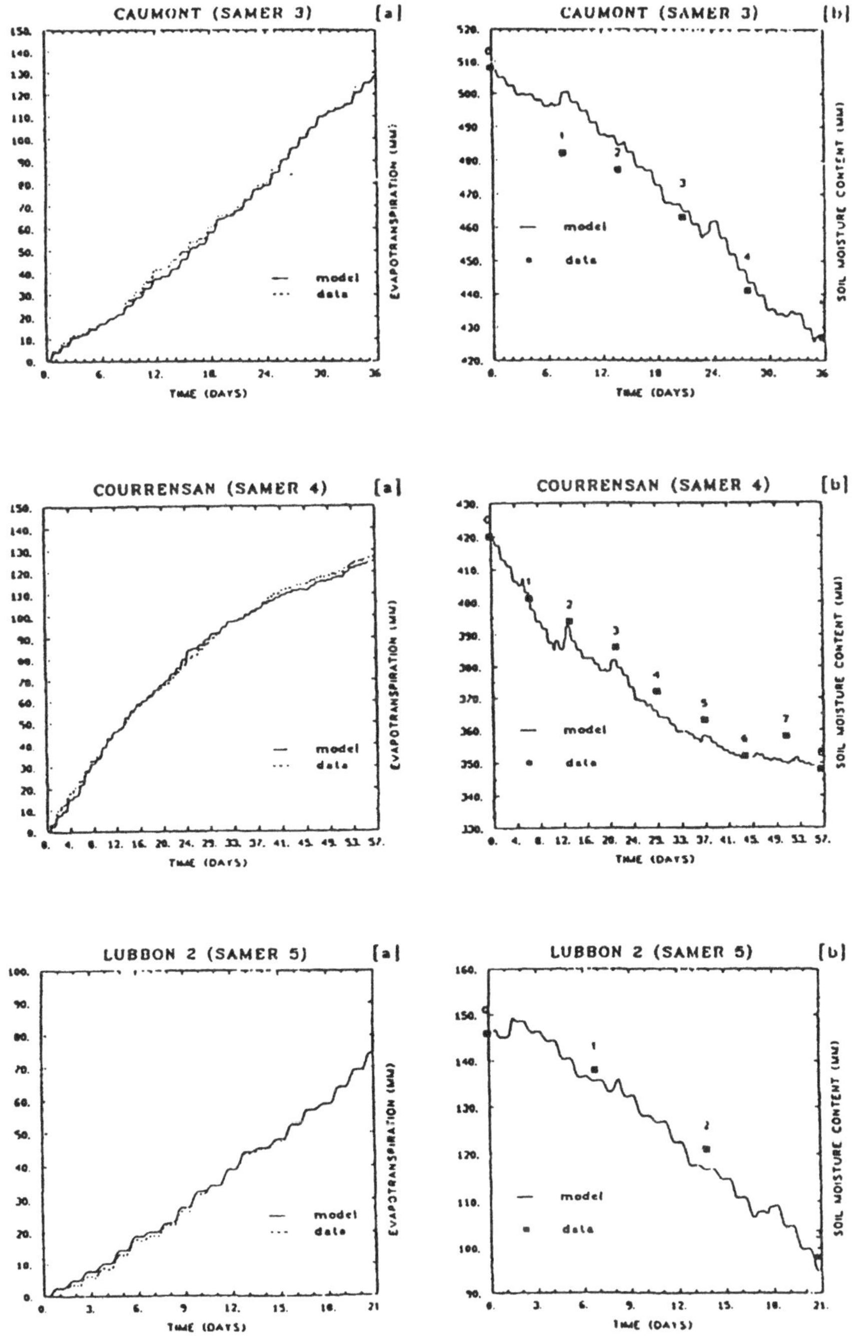

Figure 12. Time evolution of accumulated evaporation (left panels) and soil moisture (right panels) observed (dashed line) and simulated (solid line) at three sites of HAPEX–MOBILHY 1986. The neutron sounding measurements of soil moisture are represented by full squares. (Reproduced from [44]).

available at a resolution of 10 km and mesoscale modellers have to derive specific methods to provide them within the area of interest. Remote sensing represents a useful source of information for classifying the vegetation. An example of vegetation classification in the HAPEX–MOBILHY area from AVHRR data is described in [21]. From these dominant types of soil and vegetation in the grid box, a set of secondary parameters must be prescribed. Most of them can be estimated by numerical and field experiments while others can be inferred from satellite observations. Noilhan *et al.* [22] proposed a simple relationship for estimating the vegetation cover from NDVI measurements. Encouraging results for the retrieval of vegetation properties (surface resistance, leaf area index) from remote sensing were obtained during the last five years but much work remains to be done in that area.

As shown in the beginning of this section, the initial value of the deep soil moisture determines to a large extent the mesoscale prediction of surface fluxes. This point is probably the main difficulty in using LSP in NWP models. Because of the slow time scale evolution of water content, any erroneous initialization of soil moisture in NWP models will cause a poor evaluation of surface exchanges. This justifies the need for an initialization scheme of soil moisture for short-to-medium range atmospheric forecast. A sequential assimilation based on optimum interpolation was tested by Mahfouf [48] and

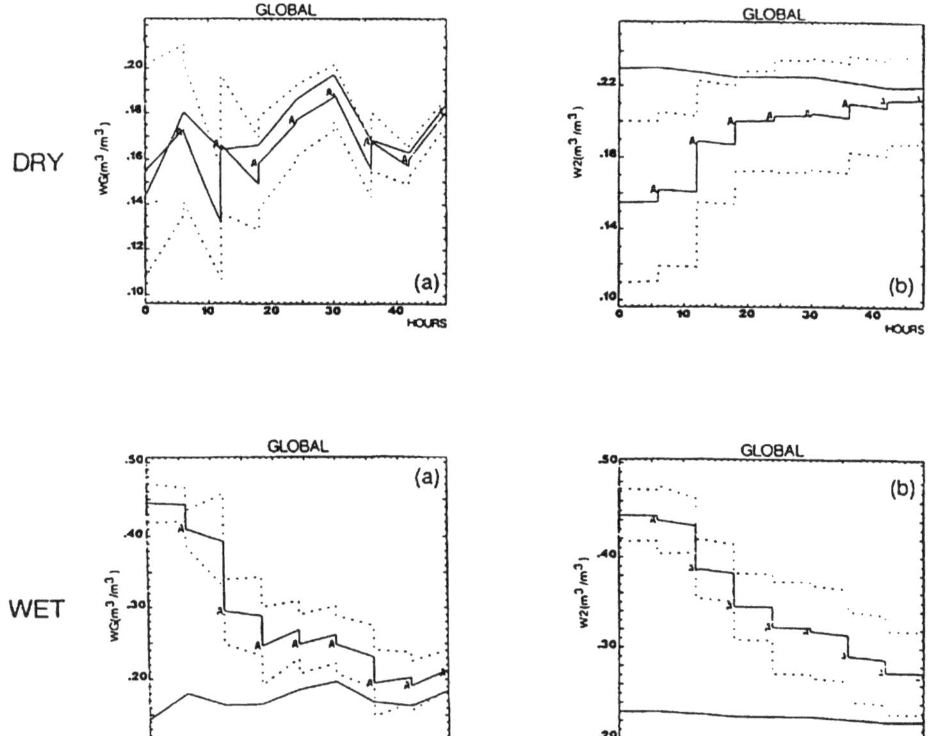

Figure 13. Evolution of the surface (a) and deep-soil (b) moisture contents during a 3D assimilation of 48 hours, starting from a dry guess and a wet guess. Dotted curves indicate the dispersion of the errors in the mesoscale domain. The evolution in the reference simulation is represented by the solid curves. (Reproduced from [49]).

Bouttier *et al.* [49, 50] with ISBA. The method relates linearly near-surface atmospheric forecast errors on air temperature and relative humidity to soil moisture increments for both the surface and deep reservoirs. One-dimensional sensitivity and validation studies [29, 48] showed that atmospheric quantities at 2 m above the ground surface contain a lot of information about the physics of soil–vegetation–atmosphere interactions as described by ISBA. The most critical aspect of the algorithm is the formulation of the optimum coefficients relating atmospheric forecast errors to soil moisture corrections, depending mainly upon local time and surface characteristics (vegetation cover and soil texture).

In a first attempt to test the method within a 3D model, a 48-hour clear-sky period from the HAPEX–MOBILHY experiment was selected. The results of the assimilation demonstrate that the convergence of the method is rapid either with a dry or a wet soil moisture guess (Fig. 13). However, some limitations of the method were observed for places where the atmosphere above the point of interest does not contain information about soil moisture. Such situations occured when mesoscale circulations generated large horizontal advections and at grid points with isolated surface properties influenced by their surrounding environment. This study opens the way for an operational use of this technique, even if more efficient variational techniques should also be developed in the future.

Passive microwaves can certainly provide useful information on the near-surface soil moisture which can be used to retrieve the deep soil water content following, for instance, the assimilation method described previously. Tentatively, the inversed surface temperature could be used for a similar goal.

3. GEOPHYSICAL VARIABLES RETRIEVAL FROM SATELLITE MICROWAVE RADIOMETRY

The variable driven by the energy budget is surface temperature, which can be estimated through the use of infrared or microwave radiometry, if the emissivities are known [51]. The variable driven by the hydrological budget is surface moisture. Passive and active microwave remote sensing are likely to gain access to this latter variable [52]. In recent years, several studies have focused on the use of spaceborne microwave radiometers data over land surfaces. They concern mainly the Nimbus-7 Scanning Multi-channel Microwave Radiometer (SMMR) and the DMSP Special Sensor Microwave/Imager (SSM/I). Generally, the methods developed consist in finding appropriate linear combinations of the available channels to allow the description, at least qualitatively, of some surface parameters. We propose a procedure for a refined interpretation of passive microwave data for use in land surface-atmosphere interaction modelling.

3.1. SMMR and SSM/I data as indices of the surface state

A first step in land parameters description is the identification of the pixel type: land/free water, rain/no rain, snow/soil/vegetation. Neale *et al.* [53], after an extensive study of SSM/I channels behavior, propose thresholds on channels combinations to classify land surface types. More specific studies focus on snowcover [54] and rainfall [55]. Whereas land and flooded areas can be easily discriminated, other pixel types are determined with difficulty [56].

Once the pixel type is determined, specific surface parameters are likely to be estimated through the analysis of the passive microwave signal. Land surface temperature as well as vegetation and soil moisture indices were proposed [76]. The early results for vegetation and hydrological variables monitoring are overviewed by Townshend *et al.* [57] and Choudhury [56]. In fact, up to now, few signal interpretation techniques based on rigorous modelling have been proposed. This is due to the fact that numerous factors presenting a rapid temporal evolution are liable to influence the signal, namely atmospheric conditions, surface structure (at various scales), temperature and moisture. This problem is particularly acute in the case of vegetated surfaces and is complicated by the coarse resolution of current spaceborne sensors. A way to cope with the complexity of the surface and atmospheric emission is to consider monthly averaged data or to limit the study to a few channels (see for instance the screening procedure proposed by Choudhury, [58]). This approach is probably the best for the interpretation of long time-series for biosphere monitoring. But as far as meteorological modelling is concerned the interpretation of snapshot images giving access to relevant physical parameters is necessary.

3.2. An improved method for geophysical variables retrieval

An approach compatible with meteorological modelling requirements consists in gathering and using as much information as possible so as to separate thoroughly all the factors influencing the signal through accurate modelling. Consequently, full advantage should be taken of the microwave signal itself (i.e. all the available channels must be studied) and external information must possibly be considered. In particular, a good description of the atmosphere through radiosoundings, geostationary satellites, GCMs analysis, etc., can be used to perform atmospheric corrections, necessary to interpret correctly the signal [59]. Modelled or satellite derived (through infrared radiometry) surface temperatures can enter the calibration procedure of microwave emission models. Thoroughly analysed, the microwave emission of land surfaces is a quite complex process depending on the temperature and moisture of the surface but also on the size and angular distribution of each element of the surface. For instance soil roughness can influence the signal to a large extent [60] and as far as vegetation is concerned, leaf size and leaf inclination effects can be significant [61]. Keeping in mind that the full complexity of the surface is unlikely to be retrieved through satellite measurements it is necessary to implement a simple (but with sound physics) model describing the surface structure by few parameters.

Once a proper modelling tool is available, the accuracy of the geophysical parameters retrieval depends on the quality of the treatment of the satellite data. Provided a sufficient number of channels (sensitive enough to physical parameters of the surface) properly calibrated and navigated are available, a six-steps method to extract most of the information of the signal can be proposed:

1. Obtain a good description of the atmosphere. The water density profile, the rain rate, the cloud coverage and the clouds altitude and thickness are particularly important parameters. They can be derived through radiosoundings, geostationary meteorological satellites and/or GCMs' analyses.
2. Perform atmospheric corrections. The detailed description of atmospheric radiation components can be found in [62, 63].

3. Get an independent source for surface temperature corresponding to one passive microwave measurement during a stable (in terms of surface structure) period.
4. If the heterogeneity of the surface requires different modelling approaches between one pixel and another the identification of the pixel type is necessary. Thresholds on relevant combinations of the multi-channel brightness temperatures can be applied.
5. Assess the structure parameters through available surface measurements and possibly calibrate the model for the unknown parameters. Use the external source of surface temperature to carry out the calibration with a number of parameters compatible with the number of available channels.
6. Assuming the structure parameters remain constant over the studied period, solve the inverse problem and retrieve surface temperature (possibly temperature profile) and humidity.

In order to overcome the inherent limitations of this type of study (i.e. 1 — the high number of vegetation parameters liable to influence the signal, and their rapid temporal evolution, 2 — the mixing of bare soil, covered soil and vegetation signals, 3 — the coarse spatial resolution of current spaceborne radiometers) one can focus, as a first step, on homogeneous or particularly well described areas. In the following section an example of application is given over the Amazon forest, a case study described in [64–66].

4. TEMPERATURE AND HUMIDITY OF THE AMAZON FOREST FROM SMMR DATA

The Amazon forest is (*a priori*) an easier study area than other types of canopy since structure parameters and soil moisture are expected to vary little. Furthermore it is a wide, homogeneously vegetated area whose expanse is compatible with spaceborne radiometers resolution. The method described in section 3.2 was applied to available SMMR temperature corrected tapes data (6.6, 10.7, 18, and 37 GHz frequencies), at daytime, 1200 Local Standard Time (LST), and nighttime (0000 LST), over a zone near Manaus (3S, 60W), Brazil. Each map corresponds to a unique measurement within a six days period. The exact day of measurement was not identified. Two periods are investigated: the wet (April–May) and dry (July–August) seasons of 1985.

4.1. Data processing

For all frequencies, the pixels are selected within a 300 × 300 km area covering the Ducke Reserve Forest (DRF) site (2S57′, 59W57′), where micrometeorological measurements from the Anglo-Brazilian Amazon Region Micrometeorology Experiment (ARME) [67] and the US–Brazilian Global Tropospheric Experiment (GTE)/Amazon Boundary Layer Experiment (ABLE) 2A mission [68] are available. Figure 14 presents the Amazon basin as well as the pixels selection area. The atmospheric corrections require information about the integrated water content (IWC) of the atmosphere, the altitude of the lower and upper boundaries and the liquid water density of the cloud layer, the rain rate. The rain rate is given by top-of-the-canopy measurements. The IWC, base and top of the cloud layer are estimated using the GTE/ABLE 2A radiosoundings. When radiosoundings are not available the values of IWC are evaluated

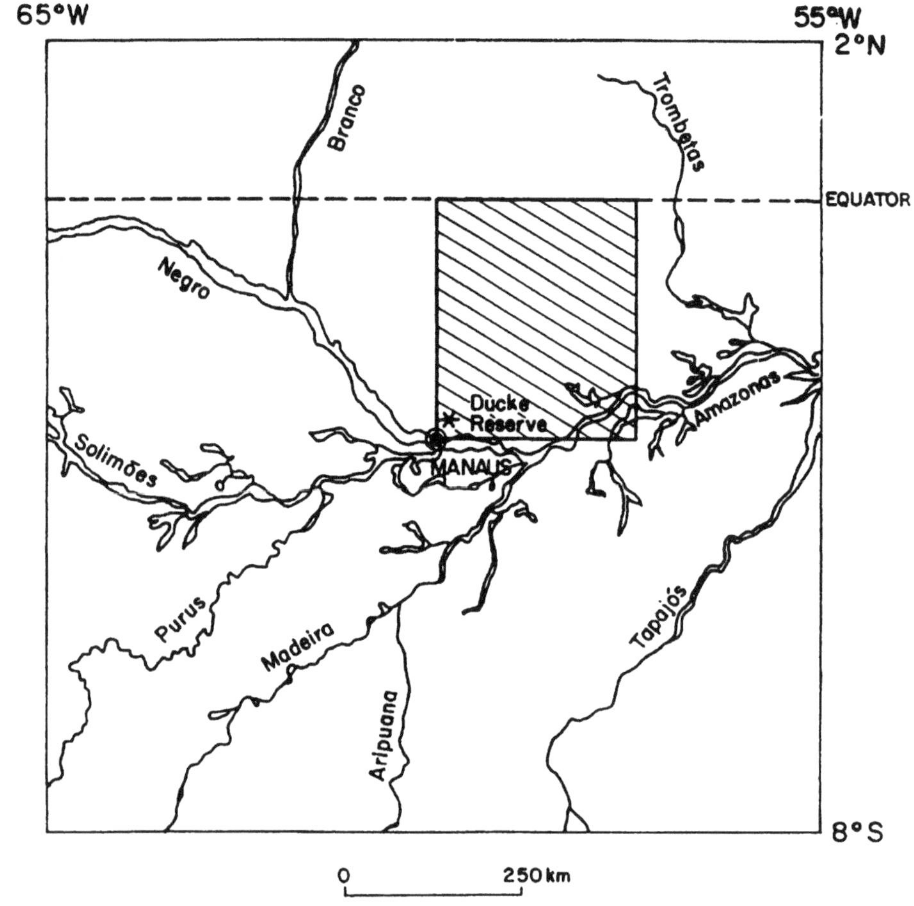

Figure 14. Ducke Reserve site and pixels selection zone in the Amazon basin (55W–65W × 2N–8S). (Reproduced from [66]).

through the monthly climatology given by Kousky and Kagano [69] and the cloud layer characteristics are determined from surface measurements of the solar incident radiation and from GOES infrared images [66]. Once the atmosphere description is complete, the effective brightness temperature of the canopy can be approximated as the radiometric brightness temperature corrected for the atmospheric upwelling emission, divided by the atmospheric extinction. These atmospheric radiation components are computed for each frequency following Waters [62] and Hofer and Njoku [63].

It is observed that the pixels which include a portion of a river, as well as neighbouring ones, show a sharp contrast with forest pixels. A pixel classification can be performed using the polarization ratio (PR) defined by Kerr and Njoku [59]. PR, for a given frequency, is the difference between the vertical and horizontal polarized raw brightness temperatures normalized by their sum. Rivers and flooded land surfaces present high PR values. For the Amazon basin, it is possible to establish a PR threshold of 0.015 for all the frequencies, above which the pixel can be considered as contaminated by a river signal.

4.2. The continuous model and the inverse problem

For a given set of entry variables a canopy microwave emission model must enable us to compute the effective brightness temperatures corresponding to the SMMR channels, provided the canopy structure parameters are known. Here, the entry variables are bioclimatic parameters with possible diurnal and seasonal variations, namely the gravimetric moisture (GM) and the vertical temperature profile of vegetation. The temperature profile, supposed to be linear, is characterised by the temperature at the bottom and at the top of the canopy (T_1 and T_2, respectively). Physically, effective brightness temperature depends on GM through the dielectric constant (permittivity) of the medium. In the case of a vegetation bioelement variations of permittivity depend on constraints on water molecules due to their interaction with solutes and structural elements of the cells [70]. Permittivity is therefore the intermediate variable between the water potential of vegetation (GM being a component of the water potential) and the microwave emission. As far as temperature is concerned, the link is quite straightforward: according to the Rayleigh–Jeans approximation of the Planck law, valid in the microwave region, the microwave radiation emitted by a body is proportional to its temperature. In order to restrict the number of structure parameters for an easier inversion, the continuous modelling approach seems to be the most appropriate. It was already successfully used to simulate field [71] or remote sensing [77] data. In the continuous approach, vegetation is considered as a whole. The canopy is reduced to a continuous medium, which is described in a statistical way through the spatial fluctuation of the permittivity [78, 79]. The model used in this study is described by Wigneron *et al.* [71]. The soil-vegetation interface is modeled through modified Fresnel reflectivities [60]. The continuous approach is applied to one vegetation layer in which the radiative transfer equations are solved by a Gauss quadrature method. Vertical temperature gradients can be simulated but GM is supposed homogeneous. However, the value of GM can differ from one frequency to another if a GM gradient exists. The whole canopy layer can affect the signal at low frequencies (6.6, 10.7 GHz) whereas only superficial vegetation affects high frequencies (37 GHz).

Compared with a discrete approach [61, 66], the continuous model has the advantage of describing the medium structure through relatively few parameters: height, equivalent vegetation volume fraction (Vf) and, for each frequency, two correlation lengths L_z and L_r describing, respectively, the vertical and horizontal fluctuations of the permittivity. Height was set at 35 m, as in [41]. The other structure parameters are not measurable and possible values must be assessed through a calibration procedure. In this study, calibration and inversion (see next section) are both performed by minimizing the rms difference between the modeled and measured effective brightness temperatures using the Powell's iterative optimization procedure described by Press *et al.* [72]. The calibration is carried out for the second April–May SMMR (six days) period, 1200 LST, for which T_1 and T_2, modeled from ARME data [80] are available. The optimized equivalent Vf is 173 ppm. Given biomass and structure data of the Amazon forest obtained from literature review [66], an actual Vf value can be estimated. An approximated value of total Vf is about 1400 ppm: 60 for leaves, 120 for small branches (diameter below 15 cm), and 1220 ppm for stems and big branches. The discrepancy between the optimized (173 ppm) and the actual (1400 ppm) values may have two origins:

1. Given the high SMMR incidence observation angle at the surface (50.3 degrees), the lowest part of the canopy, including most of the stem and big branches

Table 1.
Calibrated gravimetric moisture and correlation lengths (continuous model) for the 'terra firme' Amazon forest

Frequency	Continuous model		
(GHz)	GM	L_z (mm)	L_r (mm)
37.0	0.44	0.964	1.000
18.0	0.56	1.551	1.035
10.7	0.63	3.620	1.017
6.6	0.63	9.246	1.668

biomass, may have little influence on the signal, even for low frequencies. Incidentally, it must be noted that the sum of leaves and small branches actual Vf (180 ppm) is not far from the optimized value.

2. The leaves, though of less importance in terms of biomass, may have higher influence on the signal and thus impose a low Vf value.

Table 1 summarizes the values of L_z and L_r along with GM for each frequency. The high value of L_z obtained at low frequencies may denote a higher influence of vertical structures (stems and big branches).

4.3. Temperature and humidity retrieval — Comparison with ISBA simulations

A sensitivity study of the continuous model (calibrated as explained in the previous section) is described in [64–66]. It is shown that, in the case of the Amazon forest, both temperature and humidity of the surface have a strong influence on the signal. Assuming the structure parameters remain constant over the studied periods, the optimization procedure was applied to the three entry variables T_1, T_2 and GM.

In a first attempt to investigate how microwave radiometry can be used to validate/control the ISBA surface scheme, a comparison is provided between satellite derived surface properties and the corresponding quantities computed from ISBA at the same location.

In the simulation, ISBA was constrained by the atmospheric forcing (air temperature, humidity, wind speed, precipitation and incoming radiation) observed 15 m above the forest canopy at DRF [67]. Prior running ISBA for the two years of the observing period, a calibration was performed for selected periods where flux measurements were available. From this calibration, the most important surface parameters $R_{s\,min}$ and veg were set to 250 s/m and 0.99, respectively. The value of LAI was set to 6 m^2/m^2 following Robert's observations. A loamy soil texture was set according to USDA classification. The initial soil water content was taken to its value at field capacity (0.25 m^3/m^3), the saturation volumetric soil moisture being equal to 0.42 m^3/m^3 for this particular soil texture. ISBA was run for a 760 days period, starting on the first of September of 1983.

Figure 15 shows a comparison between the prediction of near-surface soil moisture and weekly measurements taken by neutron probe at 10 cm depth [73]. Remember that this comparison is made for the last part of the ISBA integration and, consequently, the results are no longer dependent on the initial soil moisture. The predictions (given every day at noon) are in fair agreement with the observations during the first 170 days of the considered period. Both estimates remain close to the saturation, given the prevailing heavy rainfalls at that time of the year. This agreement seems to indicate that the

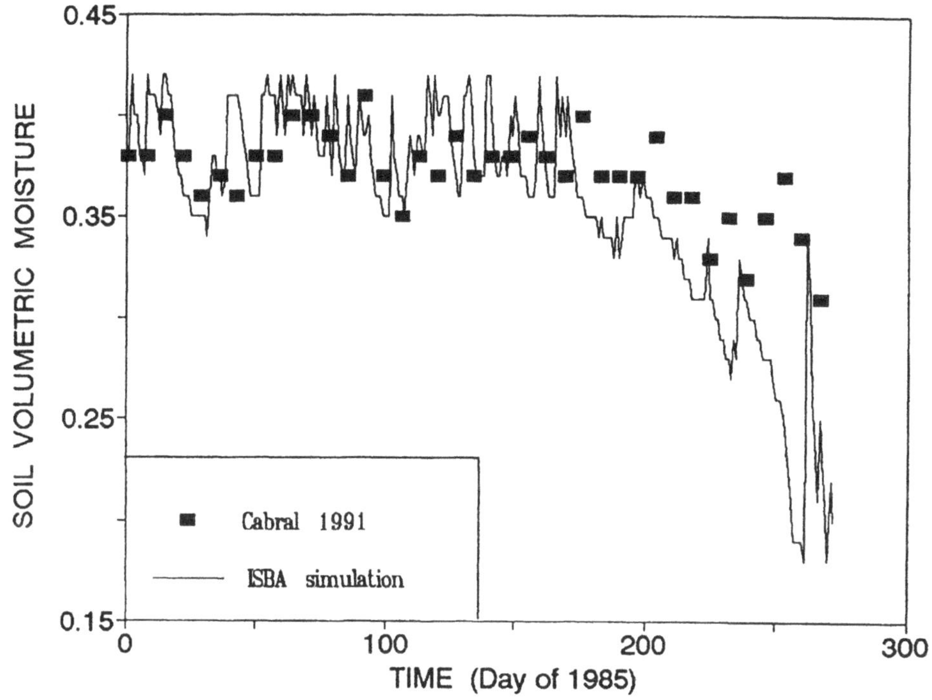

Figure 15. Surface soil moisture at DRF for 1985, 1200 LST: measurements [73] and ISBA simulations.

hydraulic soil properties associated with the chosen soil texture are correct, at least for the surface layer. Deeper measurements at 1.70 m depth suggest a higher value of the soil porosity. During the depletion period, the predicted surface moisture is significantly lower than the observations. This discrepancy may be due to the differences in the soil depth considered in the two approaches: a few mm in ISBA, 10 cm in the observations, as suggested by the time lag between the two time series.

4.3.1. Surface temperature of the forest canopy. Figure 16 shows the remotely sensed top and bottom surface temperatures of the forest canopy for the six days SMMR periods during the wet and dry seasons in addition to the daily ISBA predictions for the same period, at noon and midnight. Despite the uncertainties in the SMMR data timing, the ISBA predictions lay remarkably between the two microwave estimates. The agreement is particularly satisfactory during the dry season. At nightime and during the wet season, ISBA predictions are close to the temperature at the bottom of the canopy, which is systematically lower ($\approx 5\,^{\circ}$C) than the top canopy temperature (Fig. 16c). This latter feature is consistent with micrometeorological measurements described in [67]. Indeed, the vertical structure of the Amazonian forest favours a significant storage of heat in the upper part of the canopy as showed by [81]. Incidentally, this last study was used to estimate the heat capacity of the forest used in ISBA. Thanks to this calibration, ISBA predicts a reasonable variation of nocturnal thermal processes in the canopy-soil medium.

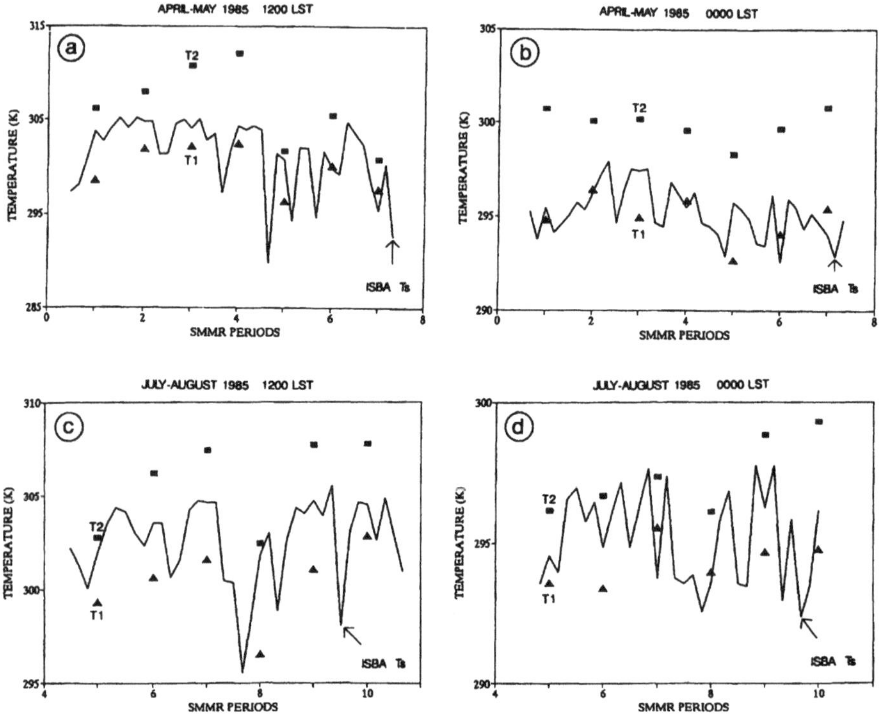

Figure 16. ISBA surface temperature and SMMR derived vegetation temperature at the top (T_2) and at the bottom (T_1) of the canopy: (a) and (b) wet season, (c) and (d) dry season; (a) and (c) 1200 LST, (b) and (d) 0000 LST.

4.3.2. Surface resistance. Another very important aspect of microwave capacity to derive geophysical variables is the retrieval of the amount of water in the vegetation layer. This information can be tentatively related to the canopy surface resistance. The moisture ratio MR is defined as the normalized difference between the retrieved water contents at 37 and 6.6 GHz [66]. The quantity MR is representative of the vertical gradient of humidity within the vegetation and therefore, should give some indication on the water status of the plants.

Figure 17 shows the correlation between MR and the surface resistances calculated from ISBA and from [74]. The Robert's estimate refers to the leaf stomatal resistance, while ISBA is supposed to represent the whole canopy behavior. This can explain the lower values of resistances computed from ISBA during unstressed conditions.

As explained in [66], the highest values of MR ($\geqslant 0.2$) may be due to the presence of intercepted water on the leaves, affecting the microwave signal. It is probably for the same reason that positive values of MR are found during the wet season where the source of humidity is located in the upper part of the canopy.

The surface resistances in Fig. 17 correspond to averages of noon values over the six days SMMR period. A negative correlation is found between MR and the surface resistances, for both wet and dry seasons. Despite this encouraging correlation, its seems difficult to derive a single relationship for the whole data set because of the MR scatter during unstressed conditions.

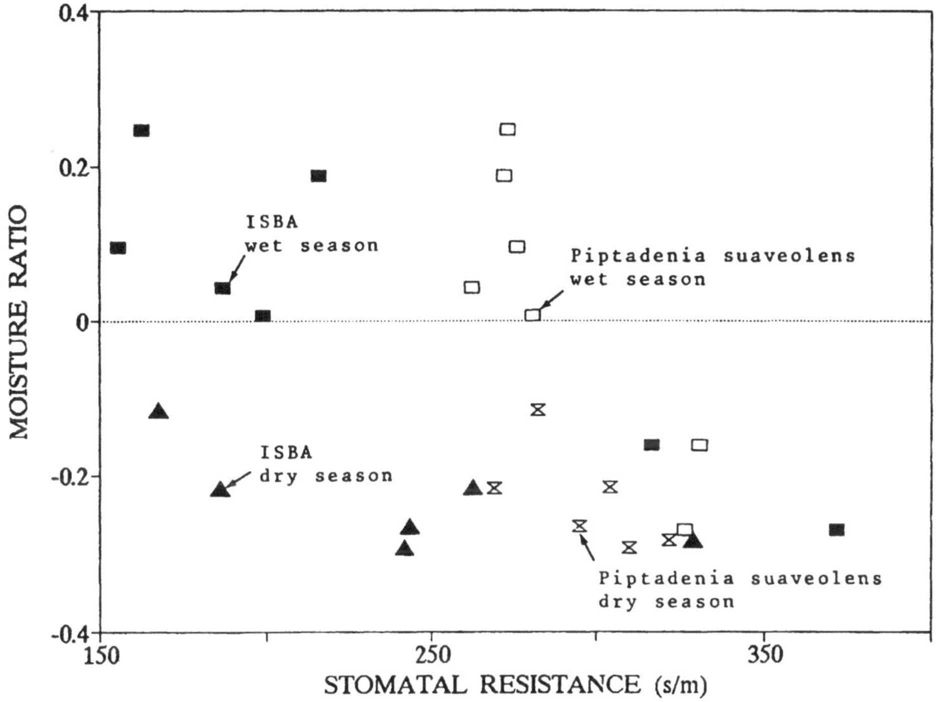

Figure 17. Continuous model moisture ratio versus the *Piptadenia suaveolens* stomatal resistance [74] and the ISBA surface resistance.

5. CONCLUSION

Land–surface patterns strongly influence the atmospheric boundary layer structures and mesoscale circulations. It has been shown from mesoscale modelling and observational studies that the partitioning of sensible and latent heat fluxes into different Bowen ratios as a result of spatially varying landscape can significantly influence the lower boundary layer and result in mesoscale flows as strong as a sea breeze. Inland circulations are due to spatial thermal contrasts between adjacent regions with well marked differences in land-use: transitions between dry and wet bare soils, between bare soil and vegetation and between low vegetation and forest canopies. Three dimensional mesoscale integrations using the HAPEX–MOBILHY data base provided evidence of mesoscale flow created by the contrast between an extensive pine forest and an agricultural area, leading to cloud formation over the forest. Consequently, there is a need for accurate description of the surface processes for use as boundary conditions in atmospheric mesoscale models. The main physical processes to be parameterized are the heat and water soil transfers and the exchanges within a vegetation layer (rainfall interception, transpiration). Implementation of LSP in mesoscale models needs important biomass and biophysical data sets for various vegetation types and hydrological data for correct initialization. Initial soil moisture estimation is a crucial issue because of its slow time scale evolution. This brings about a specific need for an initialization scheme for short-to-medium range atmospheric forecast. Microwave imagery may provide useful information on the near-surface water content which could be used to estimate the deep soil moisture. Whether surface temperature could also be used in the assimilation process of forecasting requires further investigation. A first step to evaluate the microwave potential should be to check the consistency between remotely sensed and predicted surface quantities (surface moisture, surface temperature and surface resistance).

Field campaigns and remote sensing data enable us to test and to validate mesoscale models and their surface schemes. From SMMR data it was possible to derive Amazon forest surface temperature and water status for the wet and dry seasons. They compare favourably with surface energy exchange simulations. This result is particularly significant. It shows that satellite microwave radiometry can give access to surface variables which are relevant for meteorological modelling. In the particular case of the Amazon forest, microwave information provided a valuable source of validation of surface transfer schemes.

REFERENCES

1. R. A. Pielke. *Mesoscale Meteorological Modelling*. Academic Press (1984).
2. P. Bougeault. Parameterization schemes of land–surface processes for mesoscale atmospheric models. In: *Land Surface Evaporation — Measurements and Parameterization*, pp. 55–92. Springer-Verlag, T. J. Schmugge and J. C. André (Ed.) (1991).
3. P. Bougeault. *Le Modèle Péridot: une Étude de Qualification à Méso Échelle*. Technical Report 168, EERM (1986).
4. Y. Ookouchi, M. Segal, R. C. Kessler and R. A. Pielke. Evaluation of soil moisture effects on generation and modification of mesoscale circulations. *Mon. Wea. Rev.*, **11**, 2281–2292 (1984).
5. J. F. Mahfouf, E. Richard and P. Mascart. The influence of soil and vegetation on the development of mesoscale circulations. *J. Climate Appl. Meteor.*, **26**, 1483–1495 (1987).
6. J. P. Pinty, P. Mascart, E. Richard and R. Rosset. An investigation of mesoscale flows induced by vegetation inhomogeneities using an evapotranspiration model calibrated against HAPEX–MOBILHY data. *J. Appl. Meteor.*, **28**, 976–992 (1989).
7. M. C. McCumber and R. A. Pielke. Simulation of the effects of surface fluxes of heat and moisture in a mesoscale numerical model. Part I: Soil layer. *J. Geophys. Res.*, **86**, 9929–9938 (1981).
8. J. W. Deardorff. Efficient prediction of ground temperature and moisture with inclusion of a layer of vegetation. *J. Geophys. Res.*, **83**, 1889–1903 (1978).
9. J. F. Mahfouf and J. Noilhan. Comparative study of various formulations of evaporation from bare soil using *in situ* data. *J. Appl. Meteor.*, **30** (9), 1354–1365 (1991).
10. M. Segal, R. Avissar, M. C. McCumber and R. A. Pielke. Evaluation of vegetation effects on the generation and modification of mesoscale circulations. *J. Atmos. Sci.*, **45**, 2268–2292 (1988).
11. J. C. André, J. P. Goutorbe, A. Perrier, F. Becker, P. Bessemoulin, P. Bougeault, Y. Brunet, W. Brutsaert, T. Carlson, R. Cuenca, J. Gash, J. Gelpe, P. Hildebrand, J. P. Lagouarde, C. Lloyd, L. Mahrt, P. Mascart, C. Mazaudier, J. Noilhan, C. Ottlé, M. Payen, T. Phulpin, R. Stull, J. Shuttleworth, T. Schmugge, O. Taconet, C. Tarrieu, R. M. Thépenier, C. Valencogne, D. Vidal-Madjar and A. Weill. Evaporation over land surfaces: first results from HAPEX–MOBILHY special observing period. *Annales Geophysicae*, **6** (5), 477–492 (1988).
12. J. C. André, P. Bougeault, J. F. Mahfof, P. Mascart, J. Noilhan and J. P. Pinty. Impact of forest on mesoscale meteorology. *Phil. Trans. Roy. Soc. Lond. B*, **324**, 407–422 (1989).
13. P. Bechtold, J. P. Pinty and P. Mascart. A numerical investigation of the influence of large-scale winds on sea-breeze and inland-type circulations. *J. Appl. Meteor.*, **30**, 1268–1270 (1991).
14. J. P. Pinty, P. Mascart, P. Bechtold and R. Rosset. An application of the vegetation–atmosphere coupling concept to the HAPEX–MOBILHY experiment. *Agricul. Forest Meteorol.*, **61**, 253–279 (1992).

15. P. Mascart, O. Taconet, J. P. Pinty and M. Ben Mehrez. Canopy resistance formulation and its effect in mesoscale model: a HAPEX perspective. *Agricul. Forest. Meteor.*, **54**, 319–351 (1991).

16. P. Bougeault, J. Noilhan, P. Lacarrère and P. Mascart. An experiment with an advanced surface parameterization in a meso-beta-model. Part I: Implementation. *Mon. Wea. Rev.*, **119** (10), 2358–2373 (1991).

17. P. Bougeault, B. Bret, P. Lacarrère and J. Noilhan. An experiment with an advanced surface parameterization in a meso-beta-model. Part II: The 16 June 1986 simulation. *Mon. Wea. Rev.*, **119** (10), 2374–2392 (1991).

18. J. Noilhan, P. Lacarrère and Ph. Bougeault. An experiment with an advanced surface parameterization in a meso-beta-model. Part III: Comparison with the HAPEX–MOBILHY data set. *Mon. Wea. Rev.*, **119** (10), 2393–2413 (1991).

19. J. Noilhan, J. C. André, Ph. Bougeault, J. P. Goutorbe and P. Laccarère. Some aspects of the HAPEX–MOBILHY programme: The data base and the modelling strategy. In: *Land Surface–Atmospheric Interactions for Climate Modelling: Observations, Models and Analysis*, Eric F. Wood (Ed.), pp. 31–61, Surveys in Geophysics, Vol. 12, Kluwer (1991).

20. J. P. Goutorbe, J. Noilhan, C. Valancogne and R. H. Cuenca. Soil moisture variations during HAPEX–MOBILHY. *Annales Geophysicae*, **7** (4), 415–426 (1989).

21. T. Phulpin and J. Noilhan. Use of AVHRR and METEOSAT data for the determination of land surface parameters. *Adv. Space Res.*, **9**, 269–274 (1989).

22. J. Noilhan, J. P. Jullien, P. Lacarrère, T. Phulpin, and M. Stoll. Quelques exemples d'utilisation des données satellitaires pour la modélisation atmosphériques à moyenne échelle. (Use of satellite data in mesoscale atmospheric modelling). In: *Physical Measurements and Signatures in Remote Sensing*, pp. 541–546, ESA SP-319 (1991).

23. J. Noilhan and P. Lacarrère. GCM gridscale evaporation from mesoscale modelling. *ECMWF Proceedings, 16–18 September 1991*, 245–274 (1991).

24. J. H. C. Gash, W. J. Shuttleworth, C. R. Lloyd, J. C. André, J. P. Goutorbe and J. Gelpe. Micrometeorological measurements in Les LANDES forest during HAPEX–MOBILHY. *Agri. Forest Meteorol.*, **46**, 131–147 (1989).

25. H. Giordani. *Représentation des Échanges de Surface dans les Modèles Météorologiques*. Technical Report, Météo-France CNRM, 31057, Toulouse, France (1991).

26. J. F. Mahfouf and B. Jacquemin. A study of rainfall interception using a land surface parameterization for mesoscale meteorological models. *J. Appl. Meteor.*, **28**, 1282–1302 (1989).

27. J. B. Stewart. Evaporation from the wet canopy of a pine forest. *Water Res. Research*, **13**, 915–921 (1977).

28. E. M. Blyth, J. Noilhan and A. J. Dolman. Sensitivity of mesoscale circulations to surface wetness: an example from HAPEX–MOBILHY. *J. Appl. Meteorol.*, (1994). (In press).

29. J. Noilhan and S. Planton. A simple parameterization of land surface processes for meteorological models. *Mon. Wea. Rev.*, **117**, 536–549 (1989).

30. P. J. Sellers, Y. Mintz, Y. C. Sud and A. Dalcher. A Simple Biosphere model (SiB) for use within general circulation models. *J. Atmos. Sci.*, **43**, 505–531 (1986).

31. J. Noilhan. *Une Modélisation Unidimensionnelle des Échanges Énergétiques dans le Sol, la Végétation et la Couche Limite Atmosphérique*. Technical Report 171, EERM (1987).

32. P. G. Jarvis. The interpretation of the variations in leaf water potential and stomatal conductance found in canopies in the field. *Phil. Trans. Roy. Soc. London*, **273**, 593–610 (1976).

33. A. Passerat, L. Bruckler, J. L. Thony and M. Vauclin. Numerical modelling of coupled heat and water flows during drying in a stratified bare soil. Comparison with field observations. *J. Hydrol.*, **105**, 109–138 (1989).

34. H. J. Bolle *et al.* EFEDA: European field experiment in a desertification-threatened area. *Annales Geophysicae*, **11**, 173–189 (1993).

35. I. Braud, J. Noilhan, P. Bessemoulin, P. Mascart, R. Haverkamp and M. Vauclin. Bare-ground surface heat and water exchanges under dry conditions: observations and parameterization. *Bound.-Layer Meteor.*, **66**, 173–200 (1993).

36. J. Kondo, N. Saigusa and T. Sato. A parameterization of evaporation from bare soil surfaces. *J. Appl. Meteor.*, **29**, 385–389 (1990).

37. R. E. Dickinson. Modelling evapotranspiration for three-dimensional global climate models. In: *Climate Processes and Climate Sensitivity*, Vol. 29, pp. 58–72. American Geophysical Union, Geophys. Monog. (1984).

38. A. J. Rutter, K. A. Kershaw, P. C. Robins and A. J. Morton. A predictive model of rainfall interception in forests. Part I: Derivation of the model from observations in a plantation of Corsican pine. *Agric. Meteor.*, **9**, 367–384 (1971).

39. J. H. C. Gash, I. R. Wright and C. R. Lloyd. Comparative estimates of interception loss from three coniferous forests in Great Britain. *J. Hydrol.*, **48**, 89–105 (1980).

40. C. R. Lloyd, J. H. C. Gash, W. J. Shuttleworth and A. de O. Marques. The measurement and modelling of rainfall interception by Amazonian rain forest. *Agri. Forest Meteorol.*, **43**, 277–294 (1988).

41. P. J. Sellers, W. J. Shuttleworth, J. L. Dorman, A. Dalcher and J. M. Roberts. Calibrating the Simple Biosphere model for Amazonian tropical forest using field and remote sensing data. Part I: Average calibration with field data. *J. Appl. Meteor.*, **28**, 727–759 (1989).

42. J. F. Louis. A parametric model of vertical eddy fluxes in the atmosphere. *Bound.-Layer Meteor.*, **17**, 187–202 (1979).

43. B. Jacquemin and J. Noilhan. Sensitivity study and validation of a land suface parameterization using the HAPEX–MOBILHY data set. *Bound.-Layer Meteor.*, **52**, 93–134 (1990).

44. J. F. Mahfouf. A numerical simulation of the surface moisture budget during HAPEX–MOBILHY. *Bound.-Layer Meteor.*, **53**, 201–222 (1990).

45. A. O. Manzi. *Introduction d'un schéma des transferts sol–végétation–atmosphère dans un modèle de circulation générale et application à la simulation de la déforestation amazonienne*. PhD Thesis, Université Paul Sabatier, Physique de l'Atmosphére, Toulouse, France (1993).

46. P. J. Sellers, F. G. Hall, D. E. Strebel, R. D. Kelly, S. B. Verma, B. L. Markham, B. L. Blad, D. S. Schimel, J. R. Wang and E. Kanemasu. *FIFE Interim Report. Experiment Execution — Results — Analyses*. Technical Report, ISLSCP (1990).

47. B. J. Choudhury. Simulating the effects of weather variables and soil water potential on a corn canopy temperature. *Agric. Meteor.*, **29**, 169–182 (1983).

48. J. F. Mahfouf. Analysis of soil moisture from near-surface parameters: a feasibility study. *J. Appl. Meteor.*, **30**, 1534–1547 (1991).

49. F. Bouttier, J. F. Mahfouf and J. Noilhan. Sequential assimilation of soil moisture from atmospheric low-level parameters. Part I: Sensitivity and calibration studies. *J. Appl. Meteor.*, **32**, 1335–1351 (1993).

50. F. Bouttier, J. F. Mahfouf and J. Noihan. Sequential assimilation of soil moisture from atmospheric low-level parameters. Part II: Implementation in a mesoscale model. *J. Appl. Meteor.*, **32**, 1353–1365 (1993).

51. F. T. Ulaby, R. K. Moore and A. K. Fung. *Microwave Remote Sensing*. Vol. 3, Addison-Wesley, Massachusetts (1986).

52. T. J. Schmugge. Remote sensing of soil moisture: recent advances. *IEEE Trans. Geosci. Remote Sensing*, **3**, 336–344 (1983).

53. C. M. U. Neale, M. J. McFarland and K. Chang. Land-surface-type classification using microwave brightness temperatures from the Special Sensor Microwave Imager. *IEEE Trans. Geosci. Remote Sensing*, **28**, 829–838 (1990).

54. A. T. C. Chang, J. L. Foster and D. K. Hall. Satellite sensor estimate of northern hemisphere snow volume. *Int. J. Rem. Sensing*, **10**, 167–171 (1990).

55. A. J. Negri, R. F. Adler and C. D. Kummerow. False-color display of SSM/I data. *Bull. Amer. Meteor. Soc.*, **70**, 146–151 (1989).

56. B. J. Choudhury. Passive microwave remote sensing contribution to hydrological variables. *Surveys Geophysics*, 63–84 (1991).

57. J. R. G. Townshend, B. J. Choudhury, L. Giddings, C. O. Justice, S. D. Prince and C. J. Tucker. Comparison of data from the scanning multifrequency microwave radiometer (SMMR) with data from the advanced very high resolution radiometer (AVHRR) for terrestrial environmental monitoring: an overview. *Int. J. Rem. Sensing*, **10**, 1687–1690 (1989).

58. B. J. Choudhury. Monitoring global land surface using Nimbus-7 37 GHz data: theory and examples. *Int. J. Rem. Sensing*, **10**, 1579–1605 (1989).

59. Y. H. Kerr and E. G. Njoku. On the use of passive microwaves at 37 GHz. In: *Proceedings of the Fifth International Symposium on Physical Measurements and Signatures in Remote Sensing, ESA-SP 319*, pp. 333–337, Courchevel (1991).

60. J. R. Wang and B. J. Choudhury. Remote sensing of soil moisture content over bare field at 1.4 GHz frequency. *J. Geophys. Res.*, **86**, 5277–5282 (1981).

61. J. P. Wigneron, J. C. Calvet, Y. K. Kerr, A. Chanzy and A. Lopes. Microwave emission of vegetation sensitivity to leaf characteristics. *IEEE Trans. Geosci. Remote Sensing*, **31** (3), 716–726 (1993).

62. J. W. Waters. Absorption and emission by atmospheric gases. In: *Methods of Experimental Physics*, M. L. Meeks (Ed.), New York Academic (1976).

63. R. Hofer and E. G. Njoku. Regression techniques for oceanographic parameter retrieval using space borne mocrowave radiometry. *IEEE Trans. Geosci. Remote Sensing*, **19**, 178–189 (1981).

64. J.-C. Calvet, J.-P. Wigneron, E. Mougin, Y. H. Kerr and J. L. S. Brito. Use of passive microwaves over the Amazon forest. *IGARSS'92, Houston*, **2**, 1673–1675 (1992).

65. J. L. S. Brito, J.-C. Calvet and J.-P. Wigneron. Obtencao de parametros micrometerologicos da floresta Amazonica a partir de dados Nimbus7/SMMR. In: *Congresso Brasileiro de Meteorologia*, pp. 168–172, Sao Paulo, SBM-C74a, 1 (1992).

66. J.-C. Calvet, J.-P. Wigneron, E. Mougin, Y. H. Kerr and J. L. S. Brito. Plant water content and temperature of the Amazon forest from satellite microwave radiometry. *IEEE Trans. Geosci. Remote Sensing*, (1994). (In press).

67. J. W. Shuttleworth *et al.* Eddy correlation of energy partition for Amazonian forest. *Quart. J. Roy. Met. Soc.*, **110**, 1143–1162 (1984).

68. Y. Viswanadham, L. C. B. Molion, A. O. Manzi, L. D. A. Sa and V. P. Silva Filho. Micrometeorological measurements in Amazon forest during GTE/ABLE 2a mission. *J. Geophys. Res.*, **95**, 13669–13682 (1990).

69. V. E. Kousky and M. T. Kagano. A climatological study of the tropospheric circulation over the Amazon region. *Acta Amazonica*, **10**, 743–758 (1981).

70. K. C. McDonald, M. C. Dobson and F. T. Ulaby. Using MIMICS to model L-band multiangle and multitemporal backscatter from a walnut orchard. *IEEE Trans. Geosci. Remote Sensing*, **28**, 477–491 (1990).

71. J.-P. Wigneron, Y. K. Kerr, A. Chanzy and Y. Q. Jin. Inversion of surface parameters from microwave measurements over a Soybean field. *Remote Sens. Environ.*, **46**, 61–72 (1993).

72. W. H. Press, B. P. Flannery, S. A. Teukolsky and W. T. Vetterling. Minimization or maximization of functions — Powell's method. In: *Numerical Recipes — Fortran Version*, pp. 294–301. Cambridge University Press (1989).

73. O. M. R. Cabral. *Water Storage in the Soil of a Terra Firme Forest and a Rubber Tree Plantation.* PhD Thesis, INPE, Sao Jose dos Campos, Brazil (1991).

74. J. Roberts, O. M. R. Cabral and L. F. de Aguiar. Stomatal and boundary-layer conductances measured in an Amazonian terra firme rain forest. *J. Appl. Ecol.* (1989).

75. P. J. Wetzel and J. T. Chang. Evapotranspiration from non-uniform surfaces: a first approach for short term numerical weather predictions. *Mon. Wea. Rev.*, **116**, 600–621 (1988).

76. M. J. McFarland, R. L. Miller and C. M. U. Neale. Land–surface temperature derived from the SSM/I passive microwave brightness temperatures. *IEEE Trans. Geosci. Remote Sensing*, **28** (5), 839–845 (1990).

77. R. G. Isaacs, Y. Q. Jin, R. D. Worsham, G. Deblonde and V. J. Falcone. The RADTRAN microwave surface emission models. *IEEE Trans. Geosci. Remote Sensing*, **27**, 433–440 (1989).

78. L. Tsang and J. A. Kong. Thermal microwave emission from half space random media. *Radio Sci.*, **11**, 599–609 (1976).

79. Y. Q. Jin. The radiative transfer equation for strongly-fluctuating continuous random media. *J. Quant. Spetrosc. Radiat. Transfer*, **42**, 529–537 (1989).

80. A. O. Marques Filho. (Instituto Nacional de Pesquisas da Amazônia, Brazil). Personal communication (1991).

81. C. J. Moore and G. Fisch. Estimating heat storage in Amazonian tropical forest. *Agricul. Forest Meteor.*, **38**, 147–169 (1986).

ESA/NASA International Workshop, pp. 55–76
B. J. Choudhury, Y. H. Kerr, E. G. Njoku and P. Pampaloni (Eds)

Snow cover and snow mass estimates from remote sensing, climatology and the United Kingdom Meteorological Office General Circulation Model

J. L. FOSTER,[1] A. T. C. CHANG,[1] R. J. GURNEY,[2] F. HEWER[3] and R. ESSERY[3]

[1]*NASA/Goddard Space Flight Center, Laboratory for Hydrospheric Processes, Hydrological Sciences Branch, Greenbelt, MD, USA*
[2]*Department of Geography, University of Reading, Reading, England*
[3]*United Kingdom Meteorological Office, Hadley Centre, Bracknell, England*

Abstract—Snow responds rapidly to changing meteorological and climatic conditions, has a different albedo than snow-free areas and a given climate forcing may influence the snow cover differently in different geographic regions. It is therefore important that snow extent and thickness are mapped and measured accurately from spacecraft and represented accurately in General Circulation Models (GCMs). A complete validation of the GCMs must include an assessment of their ability to reproduce both the climatological mean distribution and the interannual snow cover variability. Also, there must be a high degree of assurance that what is being observed remotely is reliable. Comparisons between long-term NOAA satellite visible data of snow cover extent, passive microwave satellite observations and snow extent data from the United Kingdom Meteorological Office's GCM show good correspondence, especially for Eurasia. From November through March, the difference in snow extent between the three data sets in Eurasia is less than 10%, and there is just a 3% difference between the model data and the microwave data for the December through February period. The greatest difference occurs during the month of October when there is a 48% difference in snow extent between the model data and the microwave data. For North America, from November through March the microwave values are about 18% less than the NOAA values, but the model produces about 26% more snow than that estimated from NOAA for the same period. In regard to snow mass, differences are considerable between the microwave and modeled data. For example, in Eurasia there is a 52% difference between the model and the microwave results for the month of March. For North America the difference is 156%. The microwave data tends to underestimate snow mass while on the other hand the model generates too much snow, particularly in the spring and summer. The snow mass is greater in April than January based on results from the model. Using a global snow depth climatology prepared by the US Air Force the snow mass for Eurasia in January is estimated to be approximately 252×10^{13} kg. According to the microwave and model data the snow mass is 206×10^{13} kg and 271×10^{13} kg, respectively. Underestimation of snow extent and snow mass with the microwave data seems to be related to the inability of microwave radiometric techniques to provide reliable information on snow conditions in areas of shallow snow and in densely forested regions. With the model data, overestimation of snow extent and snow mass results from the model continuing to accumulate snow at a few high latitude and high altitude grid points even during the period when the snow is expected to melt.

1. INTRODUCTION

Snow cover can influence atmospheric general circulation by interacting with and modifying overlying air masses. The effects of snow cover can be manifested in alterations of normal circulation patterns which result in anomalously cold winters, delayed spring warming or weaker than usual monsoons. The knowledge of snow cover distribution, extent, depth, and water equivalent is vital for improving medium and long-range weather forecasts and for validating global climate models.

Even though the present generations of General Circulation Models (GCMs) have different parameterizations, they generally agree on key results. For example, most models predict temperatures to increase by at least 1.5 °C with a doubling of CO_2 and other greenhouse gases. Also, most of the GCMs expect that the greenhouse warming will be more pronounced at higher latitudes.

A complete validation of the GCMs must include an assessment of their ability to reproduce both the climatological mean distribution and the interannual variability of snow cover. A dozen years ago data on global distributions of snow and ice-cover were unavailable, incomplete, or hard to work with. Visible and passive microwave snow data sets from satellite observations have now been compiled which have greatly helped to span the breach between observations and the users of GCM.

The observed water equivalent of snow is required to verify the surface snow mass simulated by GCMs. Such observations have been made locally for Europe, North America and elsewhere from climatological records and are archived in various reports. Passive microwave data from sensors such as the Scanning Multichannel Microwave Radiometer (SMMR) and the Special Sensor Microwave Imager (SSM/I) provide a more realistic synoptic representation of the snow water equivalent. In addition, the snow data derived from passive microwave radiometers on board satellites may be useful for input to GCMs because the scale of the microwave data is such that these data can easily be made compatible with the typical grid scales used in GCMs. Data can be also acquired through cloud cover and darkness [1].

The overall purpose of this study is to compare quantitatively the time series of snow cover and snow depth using data sets derived from SMMR passive microwave satellite data, visible satellite data, model data from the United Kingdom Meteorological Office (Hadley Centre for Climate Prediction and Research), and climatological data from the US Air Force. Data for both North America and Eurasia will be examined interannually and intra-annually in an effort to assess the magnitude of the temporal and spatial variations between these data sets. When, where and why discrepancies exist between the different data sets will be discussed and remedies will be proposed that may help resolve the observed discrepancies. Ultimately, the final goal is to improve the model's ability to simulate large-scale snow conditions. For this paper the NOAA snow data will be the standard that the Hadley GCM and the SMMR data sets are compared to in terms of snow cover extent, and the climatological data will serve as the standard for the snow depth comparisons with the model and microwave data.

Preliminary analysis using outputs from the Hadley Centre E5AC GCM and SMMR estimates of snow cover showed that there was large scale agreement of the seasonal variation in the position of snow boundaries. However, both the modeled and microwave snow depth conditions were found to be too shallow in many sub-arctic areas, such as in boreal forests, when compared to available climatological records of snow depth.

2. DATA AND APPROACH

The means by which snow information has been extracted from the NOAA visible data, the passive microwave data, the climatological data and the model data are described below.

2.1. NOAA snow cover data

Since 1966, NOAA has prepared a weekly snow and ice boundary chart for the Northern Hemisphere (Fig. 1). This chart is prepared on a 1 : 62 500 000 polar-stereographic base map centered on the North Pole. Satellite images from the visible scanning radiometers on board the NOAA polar-orbiting satellites are the main source of information used in making the charts. Secondary input comes from the system of Geostationary Satellites which, even though they have sub-satellite points on the equator, are capable of observing areas to about 60° north and south of the equator.

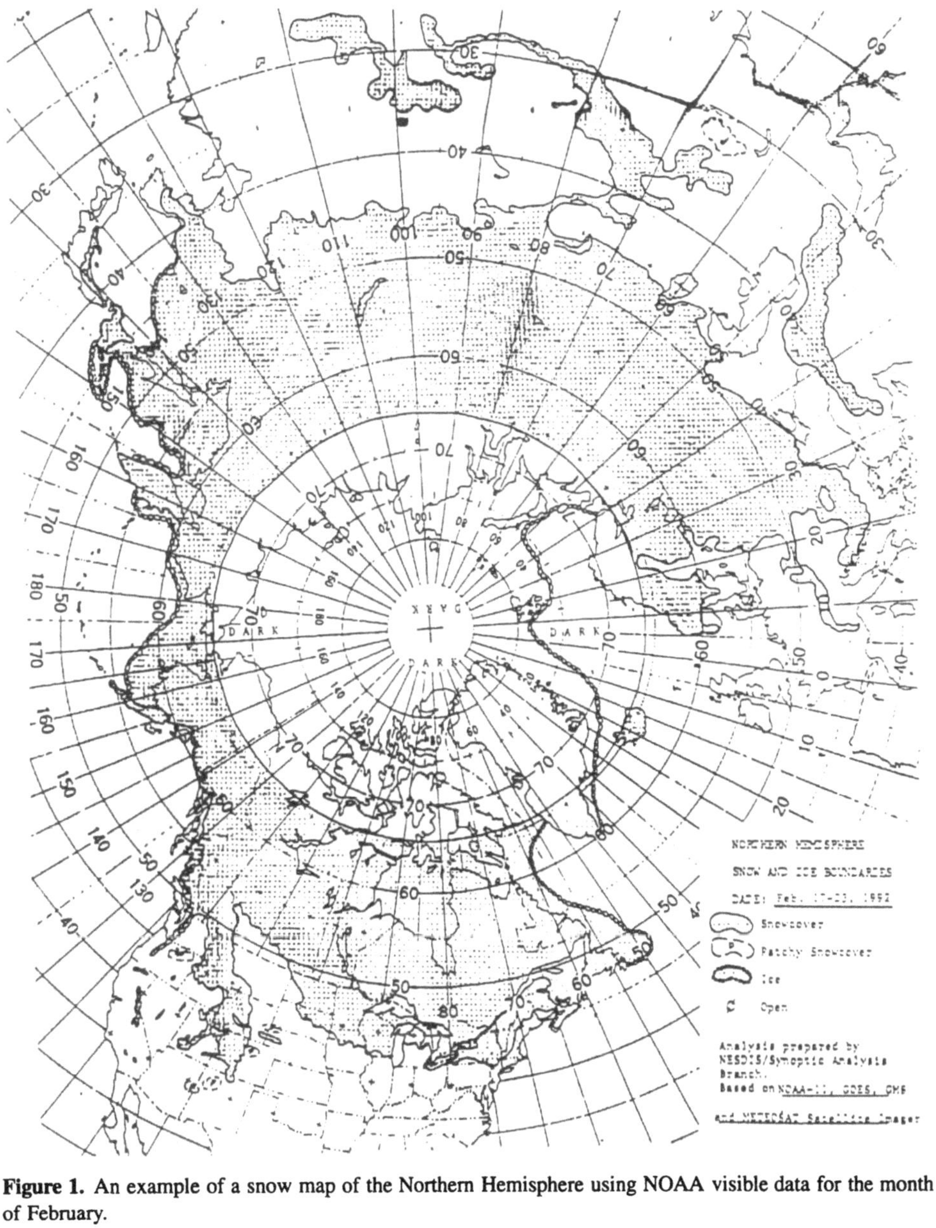

Figure 1. An example of a snow map of the Northern Hemisphere using NOAA visible data for the month of February.

Monthly mean snow cover charts have been constructed from the weekly charts. They are constructed by deriving a subjective average of the weekly chart boundaries of each month. The areal extent of continental snow cover within this average monthly snow cover boundary is then measured and recorded. Each chart is the latest cloud-free snow observation of the particular area of the world. If an area is cloud-covered for several days, the analysis of snow cover for that area will necessarily be several days old. The charts are compared with surface station reports to check for errors or to confirm drastic changes in snow cover as can result from a new snowfall or rapid snowmelt. In winter much of the far northern latitudes lies in total darkness. Consequently, the visual band sensors on the satellites are inoperative and the surface is assumed to be snow covered [2].

The snow and ice charts show three levels of increasing improvement in consistency and detail (1966–70, 1971–73, 1974–present) with the most recent charts being more reliable [3]. Himalayan snow cover was not mapped consistently during the 1966–1974 period.

2.2. *Passive microwave snow cover and snow mass data*

Fresh snow is easily recognized on satellite imagery and its continental and hemispheric extent have been mapped and measured routinely for more than 25 years. Snow cover extent can also be mapped using passive microwave data, although the method used to measure snow extent from microwave emission is not nearly as straightforward. Snow depth derivations are also amenable to microwave radiometry (Fig. 2). Between November 1978 and July 1987, the Scanning Multichannel Microwave Radiometer (SMMR) on the Nimbus-7 satellite and since August 1987 the Special Sensor Microwave Imager (SSM/I) on DMSP satellites, have been acquiring passive microwave data to estimate snow extent and depth [1].

Presently, a number of algorithms are available to evaluate and retrieve snow cover and snow depth parameters for specific regions and specific seasonal conditions. These algorithms have been derived from research using a combination of microwave sensors aboard satellites, aircraft, and trucks, as well as *in situ* field studies [4].

The algorithm developed by Chang *et al.* [5] to derive snow cover and snow depth prescribes a snow density 300 kg/m^3 and a snow grain size of 0.3 mm for the entire snowpack regardless of the condition or location of the pack. The difference between the SMMR 37 GHz and 18 GHz channels is used to derive a snow depth–brightness temperature relationship for a uniform snow field. This is approximated by a linear function that is valid up to about a snow depth of 75 cm:

$$\mathrm{SD} = 1.59\,(T_{\mathrm{B}}18\mathrm{H} - T_{\mathrm{B}}37\mathrm{H})$$

where SD is snow depth in cm, H is horizontal polarization, and 1.59 is a constant derived by using the linear portion of the 37 and 18 GHz simulated responses to obtain a linear fit of the difference between the 18 GHz and 37 GHz frequencies. If the 18 GHz T_{B} is less than the 37 GHz T_{B}, the snow depth is zero, and no snow cover is assumed.

Evaluation of the various algorithms that have been derived shows that only algorithms including the 37 GHz channel provide adequate agreement with the manually measured snow depth and snow water equivalent values. It may also be noted

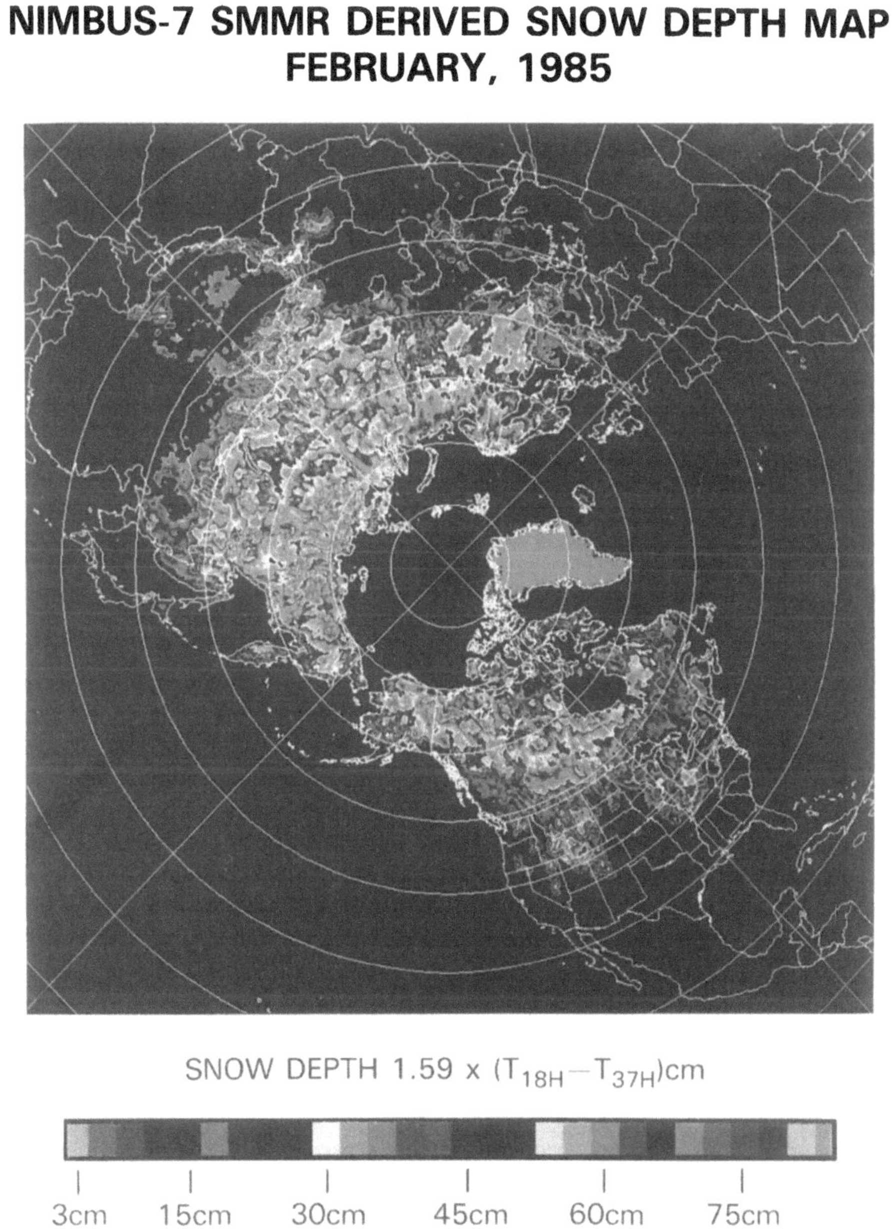

Figure 2. An example of a snow map produced using passive microwave data of the Northern Hemisphere for the month of February.

that algorithms using the T_B18H – T_B37H difference often give better results than the 37 GHz channel alone. Use of the 18 GHz channel helps to reduce the effects of the snow and ground temperatures and the atmospheric water vapor on changes in T_B [4]. Global SMMR snow maps have a scale of 0.5° latitude by 0.5° longitude.

2.3. Hadley Centre General Circulation Model snow cover and snow mass computations

The GCM used in this study is the United Kingdom Meteorological Office (Hadley Centre for Climate Prediction and Research) global climate model. The Hadley GCM is presently referred to as the unified model version 2.5 and has been run for 5 years. The resolution of this model is 2.5° latitude by 3.75° longitude. There are 20 atmospheric layers and 4 soil layers.

There are three snow water processes represented in the model: 1) sublimation/deposition due to turbulent transport in the near surface atmosphere, 2) accumulation of snowfall and 3) melting of snow when the surface temperature is above the melting point [6]. In this model precipitation falls to the surface as snow when the temperature of the lowest model layer is below the freezing point. Snow will accumulate as it falls as long as the surface layer of the model is below the freezing point (Fig. 3). The accumulated snow melts when the surface temperature is predicted to rise above the melting point. Snow cover modifies the surface albedo depending on the snow free albedo of the ground, the depth of the snow and its surface temperature [7]. The albedo is reduced as the snow ages. A simple linear scheme between the albedo of deep snow and the albedo of snow free land is used to represent the aging of the snow surface.

2.4. Global snow depth climatology

The US Air Force Environmental Technical Applications Center (USAF/ETAC) at Scott Air Force Base in Illinois has assembled a global snow depth climatology that incorporates current data sources, is fully documented and is capable of being updated easily (Fig. 4). This data base is now maintained at the National Climatic Data Center in Asheville, North Carolina. This global snow depth climatology uses a mesh reference grid that divides each hemisphere into 64 equal boxes. Each box is divided into 4096 grid points that are about 46 km apart. For each month, every box and every grid point, a snow depth value (taken to be representative of the middle of the month) is assigned based on results primarily from climatological records, literature searches, surface weather synoptic reports, and/or data obtained at snow course sites. Snow depth values were read at each grid point and digitized. Each grid value was compared from one month to the next to insure temporal consistency of the data [8].

3. RESULTS

A quantitative comparison between the remotely-sensed microwave observations and the GCM output is needed to show the differences between these two data sets and where there are areas of agreement and disagreement in a geographic context. For each 0.5° × 0.5° (longitude, latitude) grid block snow depth values are calculated using the aforementioned Chang *et al.* [5] SMMR snow algorithm. Continental snow mass is found by summing the values of the individual grid blocks. This is how the determination of the snow boundary is made. Snow cover area is assessed by summing the number of grid blocks which are snow covered irrespective of the snow depth. For the purpose of this study North America encompasses all land areas between 10°W and 170°W longitude (in a westward or clockwise direction around the pole). Eurasia is considered to include those land areas between 170°W and 10°W longitude. However,

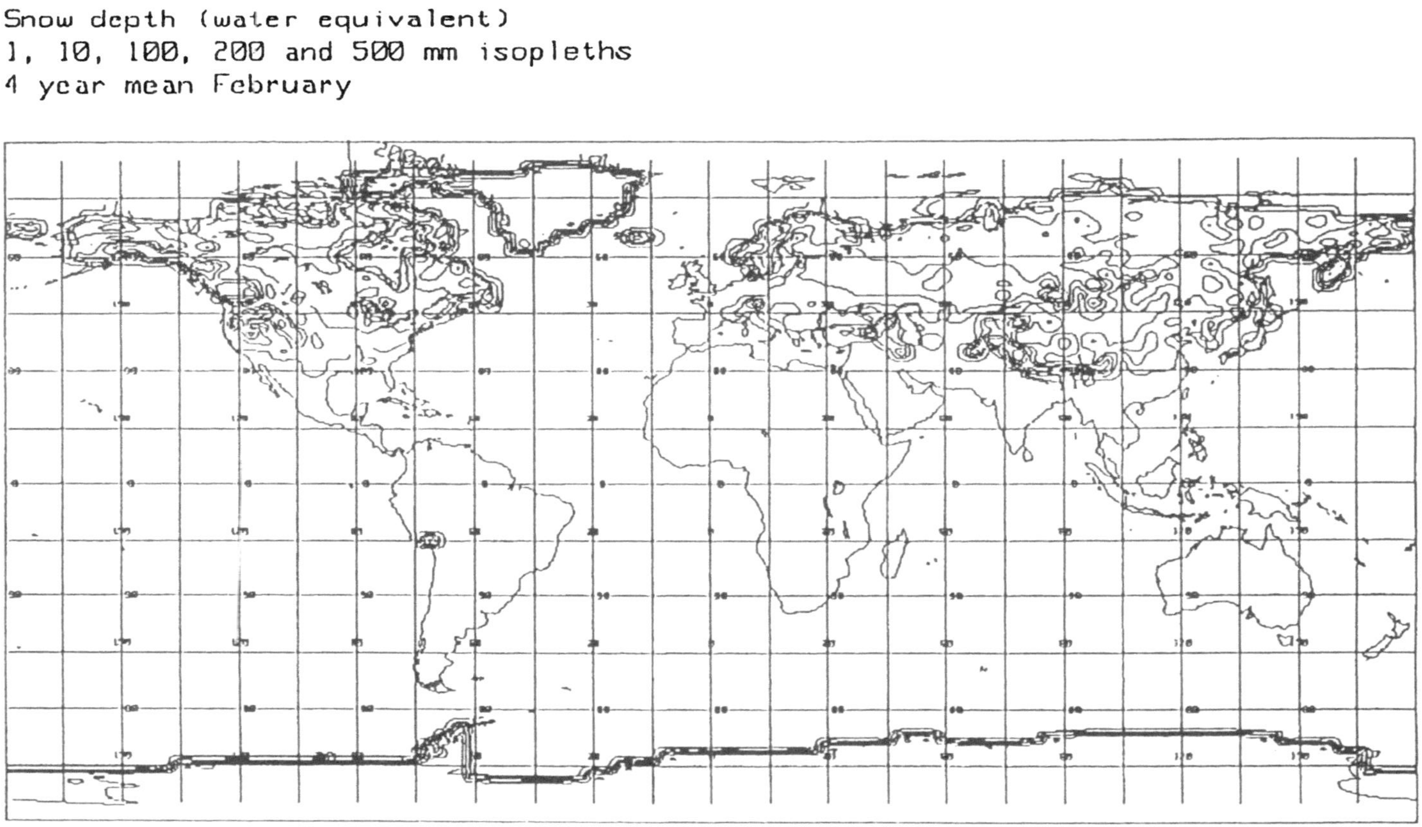

Figure 3. Snow cover in the Northern Hemisphere for the month of February from the Hadley Centre GCM.

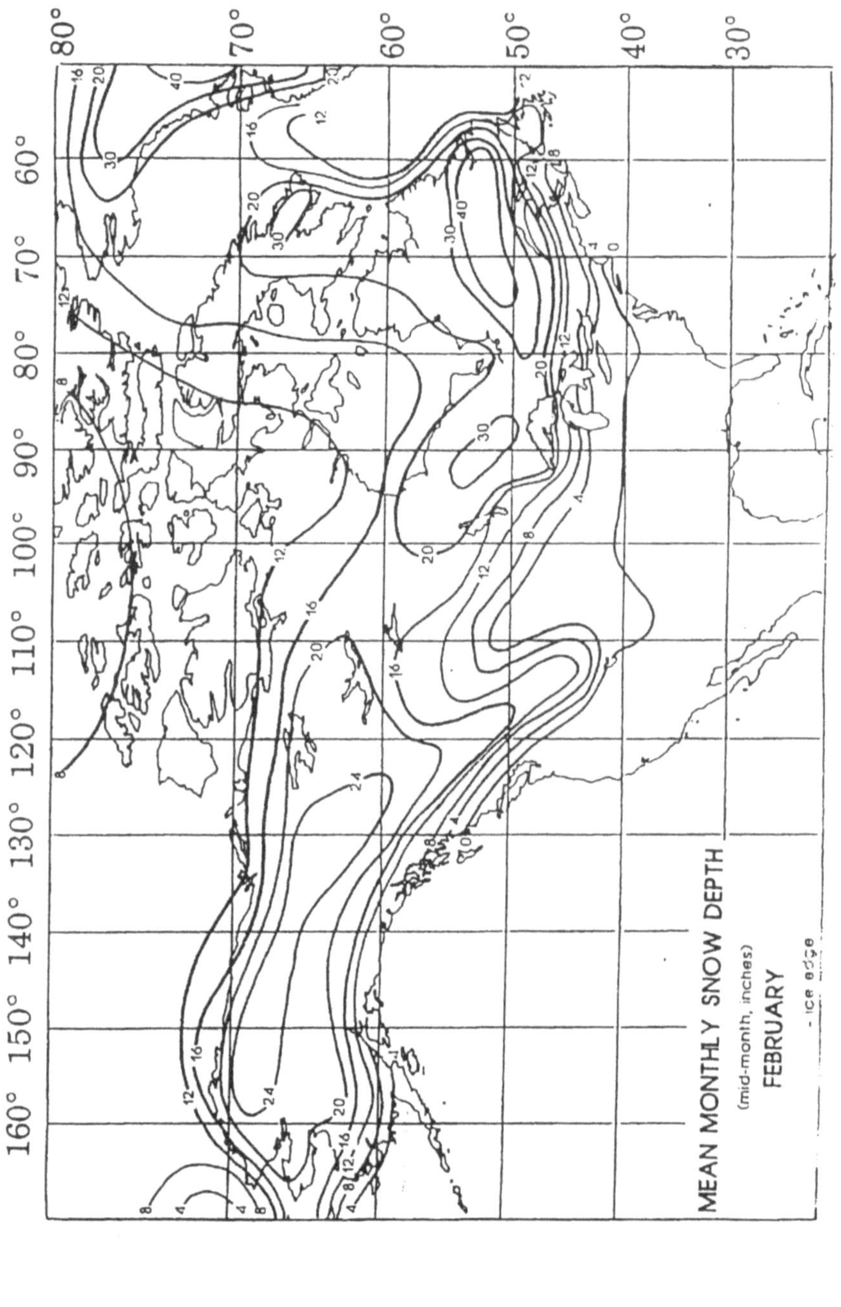

Figure 4. Snow cover map of the North America from snow depth climatology for the month of February.

ice sheets are not counted in the snow cover calculations since the emphasis here is seasonal snow only. Thus Greenland and parts of Ellsmere Island have been excluded.

In the Hadley GCM (version 2.5 of the unified model) the mass of lying snow, as assessed by accumulating millimeters (mm) of snow water equivalent (SWE), is predicted for each 2.5° × 3.75° latitude by longitude grid block. SWE is converted to snow depth by assuming a constant density of 250 kg/m^3. Snow contour depths of 1, 10, 100, 200 and 500 mm of SWE are used to compute the snow mass for each grid block: continental snow mass is, as with the microwave data, simply a summation of the snow covered grid blocks. Snow extent is crudely estimated by adding the areas of the grid boxes with more than 1 mm of SWE. This results in an area somewhat less than that bounded by the 1 mm isopleth produced by the contouring routine. In the model output, North America encompasses the area between 50°W and 170°W longitude and 15°N and 75°N latitude. For Eurasia coverage is between 180°E and 15°W longitude and 0°N and 90°N latitude. The model and SMMR defined boundaries of North America and Eurasia are somewhat different, but the total area is approximately the same. Greenland is also excluded in the model computations. Due to the coarser resolution of the modeled data, Iceland and most of the Canadian Archipelago are not included in the snow cover tabulations.

Snow mass is presented in units of 10^{13} kilograms, and snow extent is in units of 10^6 square kilometers. As a first test, year 1 of the GCM is compared with the year 1987 of the microwave data which, based on 9 years of SMMR data, is a fairly representative snow year in terms of snow extent, thickness and duration.

Looking first at the Eurasian continent, snow extent measurements from the microwave data and the model agree very well during the months of greatest snow cover (December–March) (Table 1). There is less than a 3% difference in snow extent between the two data sets for December, January and February. During the spring (April, May and June) differences increase to 17%, 12% and 15%, respectively. For the summer months of July, August and September the actual amount differences are small, but because of the greatly reduced snow extent at this time of year the percentage differences are considerable. In October and November the differences between the microwave and modeled data are 48% and 16%, respectively. Summing the monthly snow cover for all twelve months results in snow extent totals of 185.1 × 10^6 km^2 for the microwave data and 185.6 × 10^6 km^2 for the model data. The monthly average percentage difference during the cold season months (excluding June–September) between the two data sets is 14%. It is worth noting that for the first six months of the year the microwave values exceed the model values, however, for the remaining six months the model values surpass the microwave estimates.

In regards to snow mass, differences between the two data sets are high even during the winter season. The smallest differences occur in January (32%), and at the time when snowpacks normally have their greatest mass (March) the difference between the microwave and modeled values is 52% (Table 1). The modeled values are greater than the microwave values for each month except May, June and September. During May, October and November the modeled snow mass is 100% greater than the microwave snow mass. The average monthly percentage difference between the two data sets is 98.5% for the colder months (October–May).

For North America the differences between the microwave and model values of snow extent are considerably higher than they are for Eurasia. Here the springtime difference

Table 1.
Table comparing snow cover extent and snow mass for both Eurasia and North America using passive microwave satellite data from SMMR and output data from the UK Meteorological Organization (Hadley Centre) general circulation model

	Area ($\times 10^6$ km^2)				Mass ($\times 10^6$ km^2)			
	SMMR (1987)	GCM (Year 1)	Absolute difference	Percentage difference	SMMR (1987)	GCM (Year 1)	Absolute difference	Percentage difference
				Eurasia				
Jan.	29.5	28.5	−1.0	3	206.0	271.0	+65.0	32
Feb.	29.0	28.4	−0.6	2	241.0	344.0	+103.0	43
Mar.	28.3	25.2	−3.1	11	247.0	376.0	+129.0	52
Apr.	23.2	19.2	−4.0	17	182.0	325.0	+143.0	79
May	12.5	10.9	−1.6	12	80.1	219.0	+138.9	173
Jun.	4.8	4.1	−0.7	15	23.2	107.0	+83.8	361
Jul.	0.6	1.6	+1.0	171	2.0	60.0	+58.0	2841
Aug.	0.4	1.3	+0.9	197	1.2	48.0	+46.8	3933
Sep.	2.9	4.4	+1.5	48	6.4	59.0	+52.6	825
Oct.	10.2	15.1	+4.9	48	26.0	102.0	+76.0	294
Nov.	18.2	21.2	+2.9	16	63.2	157.0	+93.8	148
Dec.	25.5	25.9	+0.4	2	136.0	227.0	+91.0	67
Jan.–Dec.	185.1	185.6	22.8	14 excluding June–Sept.	1214.1	2295.0	1082.9	99 excluding June–Sept.
				North America				
Jan.	13.0	18.8	+5.8	45	71.1	166.0	+94.9	133
Feb.	13.1	19.1	+6.0	46	89.3	208.0	+118.7	133
Mar.	12.7	16.3	+3.6	28	91.4	234.0	+142.6	156
Apr.	9.4	11.2	+1.8	20	65.9	195.0	+129.1	196
May	5.7	6.5	+0.8	14	31.8	141.0	+109.2	343
Jun.	2.9	2.8	−0.1	2	8.8	92.0	+83.2	948
Jul.	0.2	1.1	+0.9	443	0.5	68.0	+67.5	12735
Aug.	0.1	1.3	+1.2	885	0.4	56.0	+55.6	14259
Sep.	0.6	3.0	+2.4	367	1.5	56.0	+54.5	3536
Oct.	2.4	7.9	+5.5	236	6.0	65.0	+59.0	976
Nov.	6.6	14.3	+7.9	118	20.5	93.0	+72.5	354
Dec.	10.4	16.6	+6.2	59	42.6	136.0	+93.4	219
Jan.–Dec.	77.1	119.0	42.2	71 excluding June–Sept.	429.8	1510.0	1082.1	278 excluding June–Sept.

is less than the difference during the winter season. For example, snow cover estimates in February are 46% greater with the model than with the microwave data. By May the differences have decreased to 14%. June is the only month when the microwave measurements exceed the modeled snow extent values. As with Eurasia, the actual amount differences in summer are not large, but the percentage differences are very great; almost 900% difference between the two data sets for the month of August. Even in October and November the model snow extent values are more than double the microwave values. The summed monthly snow extent values for the microwave data are 77×10^6 km^2 and for the model 119×10^6 km^2. The averaged (October–May) monthly percentage difference is 5 times higher than for Eurasia.

Looking at snow mass for North America, the percentage differences between the microwave and model values are greater than 100% for all months except May and June. In contrast to the snow extent data there are smaller differences in snow mass between the two data sets in January and February than in March or April. Here, as well, the differences are greater in the fall than in the winter or spring. The average monthly cold season percentage differences between the microwave and model data is 278% which is about three times more than the difference for Eurasia (Table 1).

Ratioing snow mass versus snow area is a simple way to gauge the reliability of the microwave and the model snow amounts (Table 2). For Eurasia, with the microwave data March and February have the highest ratios of 8.7 ($\times 10^7$ kg/km^2) and 8.3, respectively. September and October have the lowest ratio of 2.2 and 2.5, respectively. When the snow cover first expands southward in the fall a thin mantle of snow is accreted so that the extent is great in comparison to the thickness. Maximum thickness and area is attained in late winter, after which the snow extent and thickness both decrease through the spring and summer. With the model data the autumn is also the season having the lowest ratio (6.8 for October), however, the greatest ratios occur during summer rather than winter.

The microwave results for North America are similar to the results for Eurasia. March is again the month with the greatest ratio. September, October and July have the lowest ratios. With the exception of the months of August, September and October, the ratio is higher for each month in Eurasia than it is for North America. The model results for North America are also similar to the results for Eurasia. The highest ratio is in July; the lowest is in November. The model ratios for North America exceed those of Eurasia from April to October, but the model ratios for Eurasia are higher from November through March.

As mentioned earlier the NOAA snow cover extent data is the standard against which the microwave and model data are validated. Comparisons using NOAA visible 1987 data, SMMR microwave data also for 1987 and the GCM data for year 1 are shown in Table 3. For Eurasia there is a very good correspondence between all three data sets. During the period from November through March the difference in snow

Table 2.
Ratios of snow mass to snow extent using passive microwave data and GCM data

$$\frac{\text{Snow mass}}{\text{Snow extent}} \times 10^7 \text{kg/km}^2$$

	Eurasia		North America	
	SMMR (1987)	GCM (Year 1)	SMMR (1987)	GCM (Year 1)
Jan.	7.0	9.5	5.5	8.8
Feb.	8.3	11.9	6.8	10.9
Mar.	8.7	14.9	7.2	14.3
Apr.	7.8	16.9	7.0	17.4
May	6.4	20.0	5.6	21.8
Jun.	4.8	26.0	3.0	32.4
Jul.	3.4	37.5	2.5	59.6
Aug.	2.7	36.6	3.0	43.8
Sep.	2.2	13.5	2.4	18.7
Oct.	2.5	6.8	2.6	8.2
Nov.	3.5	7.4	3.1	6.5
Dec.	5.3	8.8	4.1	8.2

Table 3.
Table comparing NOAA visible snow cover extent estimates with passive microwave and GCM snow cover extent estimates

Snow cover extent ($\times 10^6$ km^2)

	NOAA (1987)	SMMR (1987)	Absolute difference	Percentage difference	GCM (Year 1)	Absolute difference	Percentage difference
			Eurasia				
Jan.	29.3	29.5	+0.2	1	28.5	−0.8	3
Feb.	28.7	29.0	+0.3	1	28.4	−0.3	1
Mar.	27.0	28.3	+1.3	5	25.2	−1.8	7
Apr.	19.2	23.2	+4.0	21	19.2	0.0	0
May	9.5	12.5	+3.0	32	10.9	+1.4	15
Jun.	5.7	4.8	−0.9	16	4.1	−1.6	28
Jul.	1.3	0.6	−0.7	54	1.6	+0.3	23
Aug.	0.4	0.4	0.0	0	1.3	+0.9	64
Sep.	2.0	2.9	+0.9	45	4.4	+2.4	60
Oct.	8.7	10.2	+1.5	17	15.1	+6.4	70
Nov.	19.9	18.2	−1.7	9	21.2	+1.3	7
Dec.	28.3	25.5	−2.8	10	25.9	−2.4	8
Jan.−Dec.	180.0	185.1	17.3	12 excluding June−Sept.	185.6	19.6	14 excluding June−Sept.
			North America				
Jan.	14.6	13.0	−1.6	11	18.8	+4.2	29
Feb.	13.8	13.1	−0.7	5	19.1	+5.3	38
Mar.	13.5	12.7	−0.8	6	16.3	+2.8	21
Apr.	9.3	9.4	+0.1	1	11.2	+1.9	20
May	6.0	5.7	−0.3	5	6.5	+0.5	8
Jun.	4.2	2.9	−1.3	31	2.8	−1.4	33
Jul.	1.9	0.2	−1.7	89	1.1	−0.8	42
Aug.	0.5	0.1	−0.4	80	1.3	+0.8	160
Sep.	2.1	0.6	−1.5	71	3.0	+0.9	43
Oct.	6.5	2.4	−4.1	63	7.9	+1.4	22
Nov.	12.3	6.6	−5.8	47	14.3	+2.0	16
Dec.	13.1	10.4	−2.7	21	16.6	+3.5	27
Jan.−Dec.	97.8	77.1	21.0	20 excluding June−Sept.	118.9	25.5	22.6 excluding June−Sept.

extent is less than or equal to 10%. In the model runs there appears to be considerably more snow in August, September and October than is derived from the remotely-sensed data sets. Adding each month's snow cover values for the visible, microwave and model data results in total annual amounts of 180.0, 185.1 and 185.6 $\times 10^6$ km^2, respectively. The summed percentage differences for the microwave and model data when compared against the NOAA data are also reasonably close. Excluding the months from June through September the percentage difference is 12% for the microwave data and 14% for the model data (Table 3).

For North America, the percentage differences are greater than for Eurasia in each month with the exception of April and May for the microwave data and May for the model data. During the period December through May snow extent differences are smaller with the microwave data than with the model data, though the model gives results that are in better agreement with the NOAA data from September through November. Summing the monthly snow cover values for the NOAA, microwave and model data sets gives totals of 97.8, 77.1 and 118.6 × 10^6 km^2, respectively. The microwave snow extent values are about 21% less than the NOAA snow extent observations, but the model produces about 23% more snow than is observed from the NOAA data during the period from October through May. The largest disparity between the microwave and model data, when compared to the NOAA data, occurs in October, November, April and May for Eurasia and in October, November and December for North America.

With the NOAA data record, looking at the years 1979–1990, October is the month having the greatest standard deviation in both Eurasia and North America (3.8×10^6 km^2 and 1.5×10^6 km^2, respectively). October and November are the months with the largest snow cover increase from one month to the next for both Eurasia and North America. April and May show the largest decrease in snow cover for Eurasia, but in North America May and June are the months which have the largest snow cover decrease from one month to the next.

With the microwave snow data record, for the years 1979–1987, the month with the largest standard deviation is October for both Eurasia and North America. October, November and December all have a significant increase in snow cover from one month to the next. For example, there was an increase of 8×10^6 km^2 from October to November in 1987 for Eurasia. May and June are the months having the largest snow cover decrease; again this is the case for both continents. In terms of snow mass, for both Eurasia and North America the month with the largest standard deviation is February. December and January are the months which show the greatest month to month increase in snow mass, and March and April are the months which show the greatest decrease in snow mass for Eurasia and North America.

A data record *per se* does not exist for the GCM snow data. The model is run for five years and three months, but the first three months data are not used so that averages are not influenced by the initial data. For the model data the year having the greatest mean snow cover extent is year one for Eurasia and year three for North America. In regards to snow mass, year four has the largest mass in Eurasia and in North America year four also has the largest mass. For both continents year two has the smallest snow extent and year one has the smallest mass.

The microwave and model data on snow mass were compared to the global snow depth climatology compiled by the US Air Force. Snow depth data for both North America and Eurasia have been acquired. Snow volume has been converted to mass by using 0.3 as a constant snow density. From Table 4, it can be seen that the climatological snow mass for Eurasia is in between the microwave and model estimates during the winter months (January–March). From April through September the microwave snow mass values are somewhat higher than the climatological values, but the model values are grossly higher than either the microwave or climatological values. During the fall (October–December) the climatological derived snow mass is again between the microwave and model results with the model data comparing more favorably to the climatological data. However, for the year, excluding June–September, the percentage difference from the climatological snow mass is 28% using the microwave data and 60% using the model data.

Table 4.
Table comparing snow mass estimated from snow depth climatology, passive microwave data and GCM data

Snow mass ($\times 10^{13}$ kg)

	Snow depth climatology	Passive microwave (1987)	Absolute difference	Percentage difference	GCM (Year 1)	Absolute difference	Percentage difference
			Eurasia				
Jan.	252.0	206.0	−46.0	18	271.0	+19.0	8
Feb.	303.0	241.0	−62.0	20	344.0	+41.0	14
Mar.	286.0	247.0	−39.0	14	376.0	+90.0	31
Apr.	153.0	182.0	+29.0	19	325.0	+172.0	112
May	68.8	80.1	+11.3	16	219.0	+150.2	218
Jun.	22.3	23.2	+0.9	4	107.0	+84.7	380
Jul.	0.0	2.0	+2.0		60.0	+60.0	
Aug.	0.0	1.2	+1.2		48.0	+48.0	
Sep.	6.1	6.4	+0.3	5	59.0	+52.9	867
Oct.	63.5	26.0	−37.5	59	102.0	+38.5	61
Nov.	126.0	63.2	−62.8	50	157.0	+31.0	25
Dec.	198.0	136.0	−62.0	31	227.0	+29.0	15
Jan.–Dec.	1478	1214	354.0	28 excluding June–Sept.	2295.0	816.3	60 excluding June–Sept.
			North America				
Jan.	166.0	71.1	−94.9	57	166.0	0.0	0
Feb.	206.0	89.3	−116.7	57	208.0	+2.0	1
Mar.	210.0	91.4	−118.6	56	234.0	+24.0	11
Apr.	128.0	65.9	−62.1	49	195.0	+67.0	52
May	56.5	31.8	+24.7	44	141.0	+84.5	150
Jun.	13.6	8.8	+4.8	35	92.0	+78.4	576
Jul.	0.0	0.5	+0.5		68.0	+68.0	
Aug.	0.0	0.4	+0.4		56.0	+56.0	
Sep.	3.9	1.5	−2.4	62	56.0	+52.1	1336
Oct.	13.1	6.0	−7.1	54	65.0	+51.9	396
Nov.	50.1	20.5	−29.6	59	93.0	+42.9	86
Dec.	115.0	42.6	−72.4	63	136.0	+21.0	18
Jan.–Dec.	962.2	429.8	534.2	55 excluding June–Sept.	1510.0	547.8	89 excluding June–Sept.

Table 4 also shows that, as with the snow cover extent comparisons, the passive microwave and modeled snow mass differences are greater for North America than for Eurasia. With the microwave data the direction of the monthly departures from the snow depth climatology for North America is similar to that for Eurasia, although April has much less snow mass and not more as is the case for Eurasia. The model indicates more snow mass than does the snow depth climatology for each month from March through December. Note the remarkable agreement between the modeled snow mass and the snow depth climatology for the months of January and February. Additionally, note that with the model data for both North America and Eurasia, the departures from the snow depth climatology are greater for each succeeding month from the winter through the summer.

4. DISCUSSION

Measurements from the NOAA series of satellites are for the most part based on observations using visible data so that direct rather than indirect snow extent values are obtainable. With microwave data the presence of snow is derived using a simple algorithm, and with the modeled data a more sophisticated mathematical formulation is used to determine where a snow cover exists. However, snow amounts as assessed from the NOAA data may be biased by a number of problems. Among them, subjective interpretation of what is snow and what is clouds, for instance, can effect the accuracy of the snow measurements. Other problems stem from variability in orbit time and slight sensor differences between the NOAA satellites. The scale of the charts also limits the precision of the snowline positioning to about 25 km. Moreover, there are small errors owing to the analog averaging of the weekly satellite data to derive the monthly snow extent values [9]. For example, snow cover data for the 29th through the 31st of October would be included in November's snow totals, which can lead to a bias towards smaller snow cover values for October [14]. While the NOAA monthly averages are based on weekly observations, microwave monthly averages result from six days of data collected near the center of the month which are assumed to be representative of a 30 day period.

There is generally a somewhat better agreement between the NOAA and microwave snow extent data than between the NOAA and modeled data, particularly during the coldest months of the year. This is to be expected since the NOAA and microwave data presented here are for a specific year (1987), and the model data is for the first year of a five year simulation.

For both the microwave data and the modeled data, in regards to percentage differences, comparisons with the NOAA snow extent data are in closer accord for Eurasia than for North America. The microwave and model data for Eurasia slightly overestimate the snow extent as observed from NOAA. However, for North America the microwave data underestimate and the model data overestimate snow cover area. A partial explanation for this is the difference in size between these two land masses. Eurasia is more than double the size of North America, and its snow cover at the time of maximum extent in January and February is about twice as large. Looking at Table 3, during the month of December in Eurasia there is a 2.8×10^6 km^2 difference in snow extent between the NOAA and microwave estimates, and for North America there is a 2.7×10^6 km^2 difference. Although, the percentage difference between the NOAA and microwave data for Eurasia is just 10%, while it is 21% for North America. Even when the size difference is accounted for though, the difference in snow cover extent between the NOAA and microwave data and between the NOAA and model data is still greater for North America than it is for Eurasia.

With the GCM data, excessive snow cover for North America as compared to the NOAA and microwave data could result from using an isopleth of 1 mm to compute the snow cover area. With a density of 250 kg/m^3, which is used in the model formulation, 1 mm of SWE translates to a snow depth of about 0.4 cm. Such a shallow covering is considered to be just a trace of snow by snow hydrologists and is usually discontinuous over large areas. On the model maps the 1 mm isopleth is continuous across the southern tier of the US during the coldest winter months and is positioned further to the south (Fig. 3) than is the snowline as mapped from the NOAA and microwave data. With the NOAA visible data 0.4 cm of snow is not easily discerned, and with

SMMR and SSMI data such a thin veneer of snow is readily penetrated by the upwelling microwave radiation. For Eurasia, the 1 mm contour seems to correspond more closely to the snowline as mapped from the visible and microwave data.

Difference in snow cover areal extent during the late fall and early spring between the NOAA, microwave and model data sets may be due to the positioning of the snowline in the boreal forests. The visible sensors on board the NOAA satellites are unable to penetrate dense forest covers and monitor the underlying snow. With the microwave data the emissivity of trees, especially conifers (Fig. 5) can overwhelm the scattering signal which results when upwelling microwave energy is redistributed by snow crystals. This will be discussed later. Thus, remotely-sensed snow observations may under-represent actual snow extent and snow mass values in forested regions. Also with the NOAA data, during the autumn months as the snow cover expands, decreasing amounts of daylight reduce the normally high contrast between snow covered and snow free surfaces. This may adversely effect how and where the snowline in drawn.

The microwave snow cover area for North America is less than the estimates from the NOAA data for every month except April. This may be attributable to the above mentioned ineffectiveness of microwave radiation in providing information about shallow snow cover (less than about 3 cm). Because North America has a larger portion of its surface area located at lower latitudes than does Eurasia it is reasonable that a larger portion of its snow cover is transitional and ephemeral in nature. This may help account for the differences in snow extent between NOAA and the microwave data for Eurasia and North America.

When compared to the climatological data (Table 4) snow depth as derived from the microwave data underestimates the actual thickness and mass of snow on the ground. This is consistent with earlier investigations using the Rand snow depth data [10]. The modeled data on the other hand tends to overestimate snow mass. As with snow cover area, there is closer agreement between the microwave data and the model data, in terms of percent difference, for Eurasia than for North America. However, as can be seen from the difference column in Table 1 the tabulated difference is virtually the same for each continent. Hence, when looking at the entire year the size difference between Eurasia and North America seems to explain a good deal of the relative percentage differences.

The most likely reason why the microwave data underestimates snow mass has to do with the effects of vegetation above snow fields. Forests not only absorb some of the radiation scattered by snow crystals, but trees are also emitters of microwave radiation. So in forested areas the signal received by a radiometer on board a satellite is produced by a combination of media. Generally, the denser the forest, the higher the microwave brightness temperature despite the type and condition of the media underlying the forest canopy. Furthermore, because the canopy shields the snow from direct solar radiation the deepest snows accumulate in the densest forests. The relationship typically observed for most snow fields; that thicker snowpacks having lower brightness temperatures than shallower snowpacks is therefore counterbalanced by the higher brightness temperatures observed in areas of higher biomass.

The boreal forests which stretch across the northern tier of both Eurasia and North America is perhaps the physiographic region where most of the difference occurs between the snow depth measurements based on climatological data and those based on microwave observations (Fig. 6). In addition to the boreal forest, strips of forest

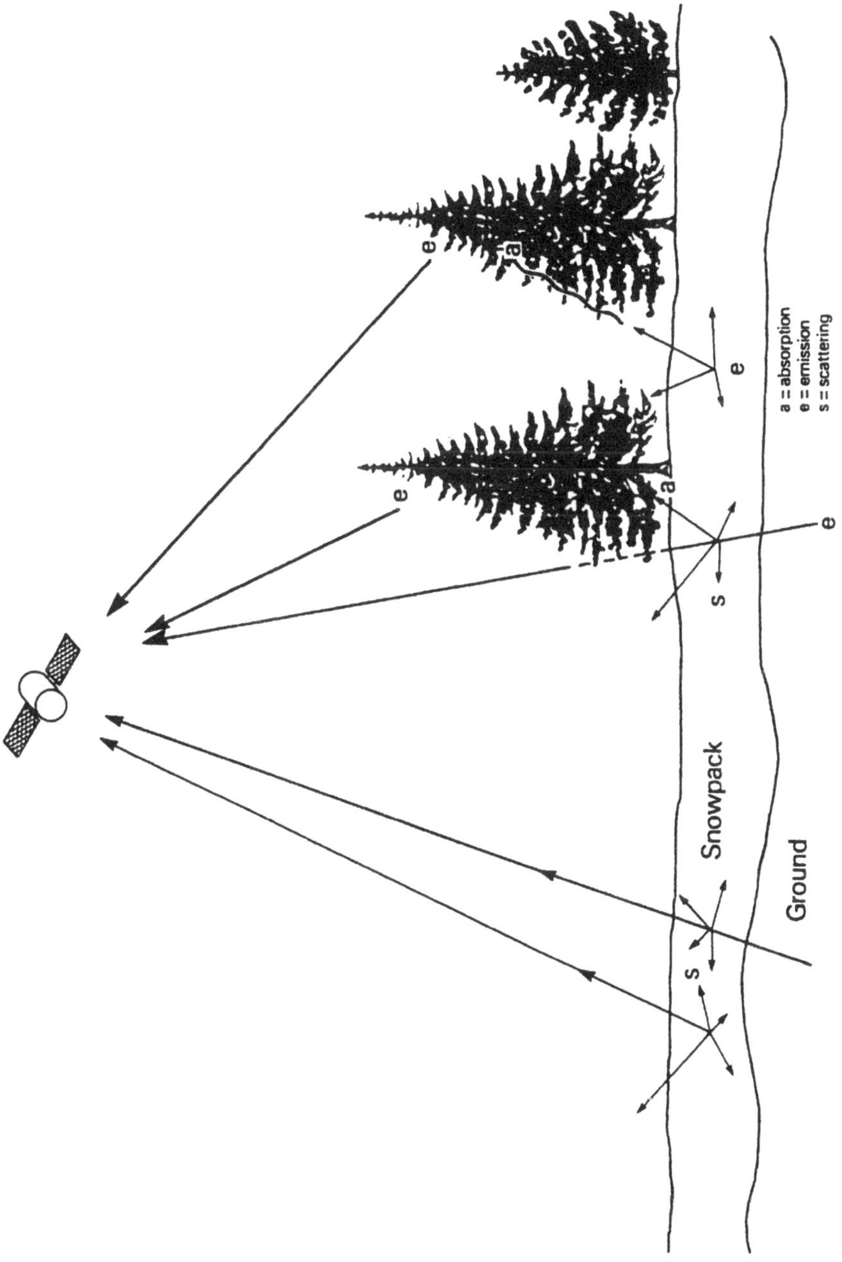

Figure 5. Schematic showing interaction of microwaves, snow and vegetation.

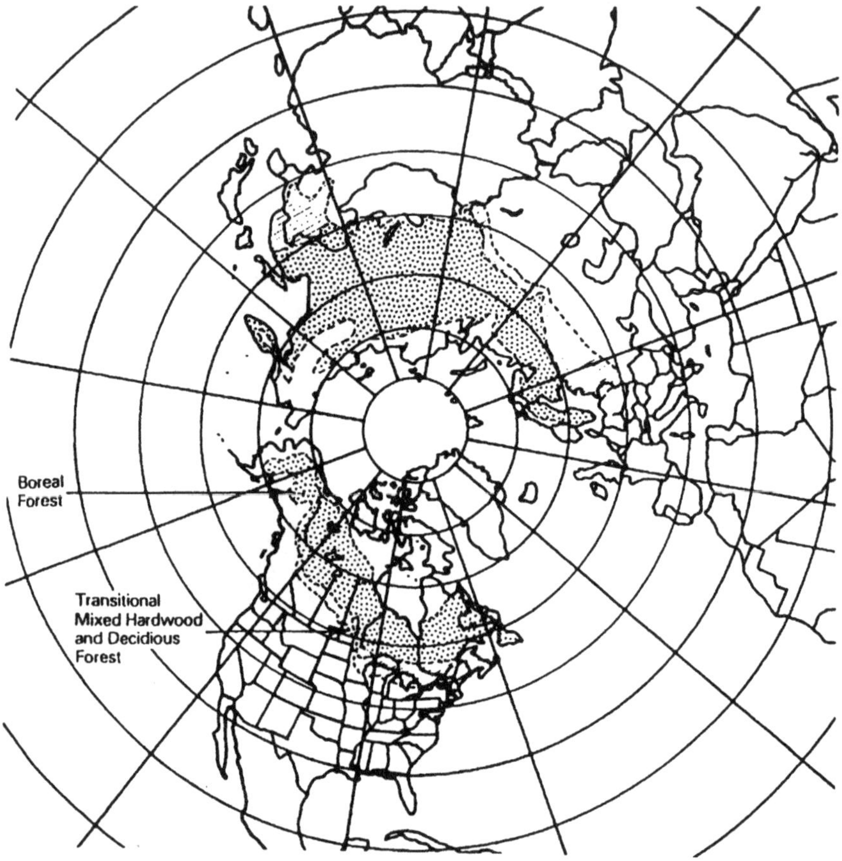

Figure 6. Map showing the location and extent of the boreal forests (from Foster *et al.*, [11]).

cover adjacent to the southern border of the boreal zone, such as transitional mixed hardwood/conifer forests in Germany, Poland, New England and the Great Lakes area, can alter the established snow/microwave relationships. Cordillera such as the Rocky Mountains and Alps Ranges, having thick forests covering their flanks, may limit the efficacy of microwave algorithms in these areas as well.

Another source of possible error is the transparency of microwave radiation to thin snow covers as noted earlier. If the band of snow near the southern limit of the continental snow line is sufficiently shallow (less than about 3 cm) then the radiation upwelling from the ground may pass through the snowpack virtually unimpeded. Although these shallow snowpacks make-up a rather small portion of the total snow mass, not including these areas would result in snow mass values that are too small.

The above discussed error sources are a function of how microwave energy interacts with matter. An additional source of error that should not be overlooked results from constraints imposed upon the algorithm. For example, the density of the snow is assumed to be 300 kg/m^3. New fallen snow may have a density that is closer to 100 kg/m^3, whereas aging snow in a ripe snowpack may have a density of 600 kg/m^3. Firn or snow that does not melt from one year to the next has a density of 800 kg/m^3.

The 300 kg/m^3 value used for the Chang *et al.* [5] algorithm is what might be measured in snowpacks during the months of January, February and March in much of the northern hemisphere. Higher densities would normally occur in April and May while lower densities would be the rule during the autumn season. Because snow depths are generally deeper in the spring than in the fall, using a constant density of 300 kg/m^3 may contribute to the lower values observed when using microwave radiometric techniques to infer snow mass.

In contrast to the microwave data the modeled data overestimate snow mass when compared against the climatological mean snow depth. The larger snow cover extent produced by the model during most months will of course result in larger values of snow mass. But this is not the sole reason why the modeled snow mass is larger than that estimated using the climatological or microwave data. For example, in Eurasia during the month of March the snow mass produced from the model is 31% larger than the estimated snow mass using climatological data (Table 4), even though the snow cover extent is in agreement with both the NOAA and microwave areal values. Moreover, for both Eurasia and North America the modeled snow mass during April is greater than in January by about 18%.

The model produces more snow than either the climatological or microwave data for every month and for each continent (Tables 1 and 4). The differences are relatively small during the December–February period, but become glaring during the summer months. This is obvious when looking at the snow mass to snow area ratios (Table 2). For the microwave data the monthly ratios are fairly constant, but for the model the ratios increase dramatically from May through August. Referring to Table 1, for Eurasia the month of June shows slightly less snow extent from the model than from the microwave data (4.12×10^6 km^2 vs. 4.84×10^6 km^2), however, the model produces more than four times as much snow mass (107×10^{13} kg vs. 23×10^{13} kg).

The reason for this has more to do with where snow is permitted to accumulate and melt than it does with how snow accumulates and melts. A 'runaway' effect seems to occur within the model. The model packages, i.e. radiation, precipitation and boundary layer processes, are ostensibly building-up and removing snow properly for much of the year. However, during the summer in certain preferred high altitude and high latitude locations, where there exists a perennial snow cover that does not melt during the summer, the snow is continuing to accumulate, and hence the snow mass is an order of magnitude higher than expected. Also, as can be seen from Fig. 7, there is an upward drift in snow mass from one year to the next which is caused by the aggregation of the perennial snow as stipulated by the model. It should be remembered that the microwave data are specified for 1987, whereas the data from the model and from the climatology are not necessarily representative of a given year. The remarkable agreement between the model and the climatological estimates for January in North America (Table 4) points out just how accurate the GCMs can be in simulating large scale conditions. However, the model is initialized using climatological data and as mentioned earlier there is a three month 'spin-up' before the model is run for a five year period. The initial snow condition will not be forgotten in a three month period, so it is expected that year one is remembering the initial conditions better than the subsequent years. Whether or not the models are sensitive enough to respond to actual interannual changes remains to be seen. It is evident though, that even with the assumptions, parameterizations, and gross resolution intrinsic to GCMs, they are nevertheless capable of representing snow cover and snow mass in a reasonable way.

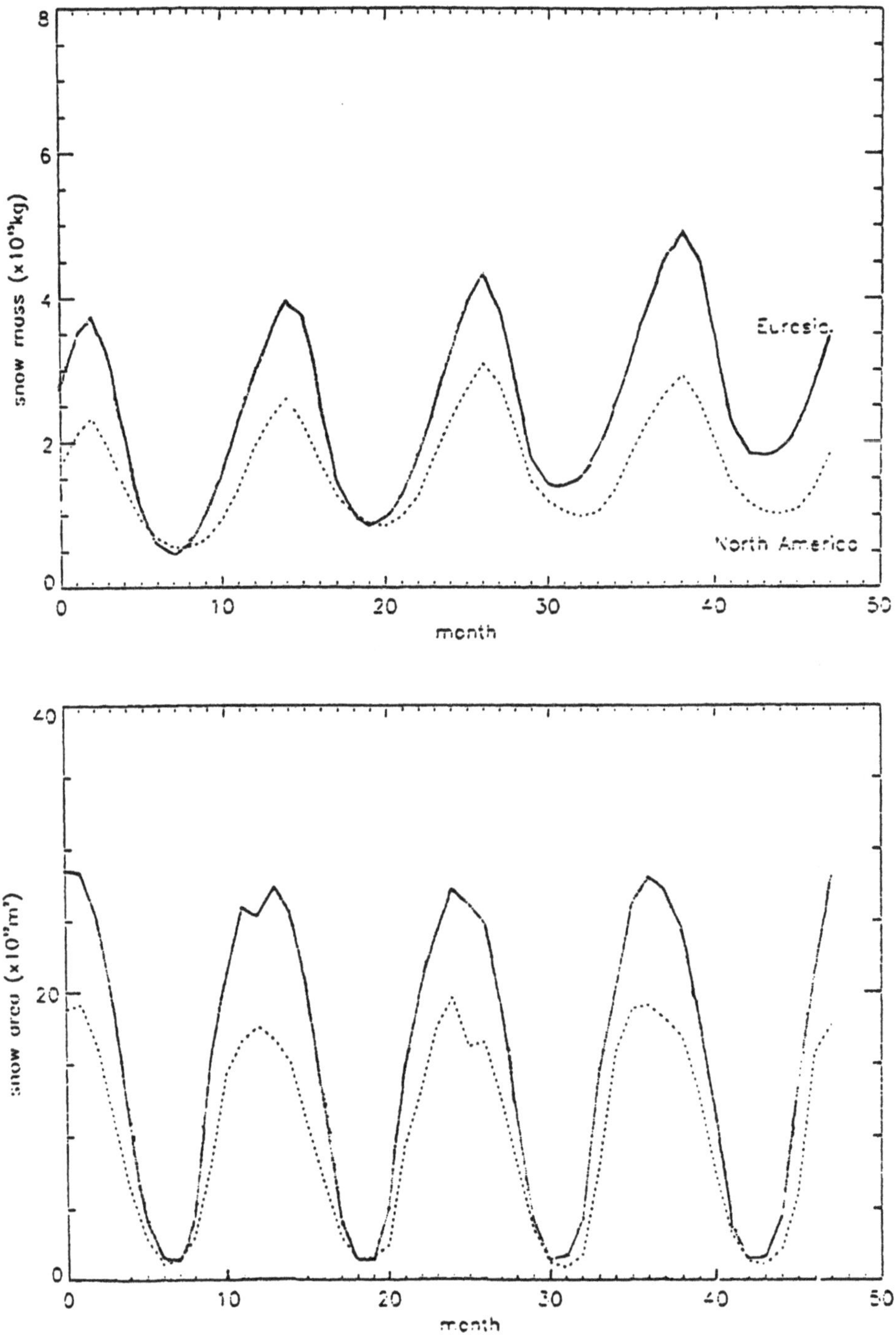

Figure 7. Graphs showing 4 year time series of snow extent and snow mass measurements from the Hadley GCM.

5. FUTURE DIRECTIONS

Based on the findings of this research effort there remain several areas that need to be investigated. The work here is in essence the first attempt to look at continental differences in snow conditions using several data sets. Although new information has been acquired concerning the temporal and spatial variations that exist amongst the data sets, more work needs to be done to learn if snow build-up and snowmelt processes are being adequately handled by the models and the microwave data sets in specific physiographic or climatic provinces. For instance, by looking at microwave snow data in the boreal forests and defining what the error bands are, it will be possible to increase the sensitivity of the algorithm and hence its accuracy in this region.

Forests present perhaps the biggest challenge to microwave snow retrieval algorithms, but progress is being made in revealing what the forests conceal. A recent study found that in the boreal forest of Saskatchewan the bias between the measured and remotely sensed snow water equivalent could be reduced by 22% by using an algorithm including some information about the percentage of forest cover [11]. Also, a study by Hallikainen and Jolma [12] in Finland found that using vertically polarized data and data from the 10 GHz channel together with information from the 37 GHz channel may provide better measurements of SWE in forested areas. These findings will be investigated to see whether they can be used to refine the Chang *et al.* [5], microwave algorithm, so that snow conditions can be assessed more accurately.

Work is also going on to determine the areal extent and thickness of the snow cover better in the vicinity of the snowline. On the SSM/I radiometer the 85 GHz data show a strong decrease of brightness temperature with snow depth up to about 7 cm of snow. With deeper snow depths, saturation is observed. The 37 GHz–85 GHz difference seems especially sensitive to thin snow covers, and for areas of central and eastern Europe this algorithm increased the area classified as snow by about 10% compared to using the 18 GHz–37 GHz algorithm [13].

Additionally, the effects of snow grain size information and topography within a microwave footprint need to be included, eventually, in the retrieval algorithm. Airborne microwave overflights (e.g. Advanced Microwave Precipitation Radiometer (AMPR), mm wave imaging radiometer (MIR), or Airborne Multichannel Microwave Radiometer (AMMR)), with a spatial resolution of the order of 2 km, are being planned in conjunction with the upcoming BOREAS experiment. This data will be used to test and evaluate and refine snow retrieval algorithms.

Pertaining to the model, how the accumulation and depletion of snow is being treated in those areas where snow fields can persist through the summer months is currently being addressed. This may be a matter of adjusting model parameters at a few high elevation or high latitude points or grid boxes. One way to assess where the model is running astray is to check the modeled air and surface temperatures at non-snow-covered-grid points, adjacent to snow-covered grid points, during the spring and summer months to see if the snow should be melting. Also, more research is required to learn if the model is performing properly in different physiographic provinces. Furthermore, the model snow cover is presently initialized by climatological snow data, and it is unlikely that this initial condition is forgotten in the three month 'spin-up' period used in this model. Running the model for 10 years rather than just five years should mollify this problem.

In regards to the snow depth climatology, this data would be improved with the inclusion of new snow data for the former Soviet Union, which are now available for the first time from 284 meteorological stations. If these new data can be incorporated into the present snow depth climatology data set then greater certitude will result in the actual location of the continental snowline and the drawing of the isonivals (lines of equal snow depth) across Eurasia.

Additionally, in order for accurate comparisons to be made and for meaningful statistical analysis, it is imperative that each of the four snow data sets use the same map projection and land mask. Currently, with the climatological data for instance, the gridding method used results in some high latitude peninsulas and islands being excluded from the snow cover totals. This needs to be rectified in future algorithm renditions.

REFERENCES

1. A. T. C. Chang, J. L. Foster and D. K. Hall. Snow volume measurements derived from SMMR. *Remote Sensing Letters*, **11**, (1), 167—171 (1990).
2. M. Matson and D. Wiesnet. New data base for climate studies. *Nature*, **289**, 451–456 (1981).
3. G. Kukla and D. Robinson. Climatic value of operational snow and ice charts. In: *World Data Center A for Glaciology [Snow and Ice]. Glaciological Data. Report GD-11*, pp. 103–120, Snow Watch 1980 (1981).
4. J. L. Foster, D. K. Hall and A. T. C. Chang. Remote sensing of snow. *Eos Trans.*, **AGU 68**, 681–684 (1987).
5. A. T. C. Chang, J. L. Foster and D. K. Hall. Nimbus-7 SMMR derived global snow cover parameters. *Annals Glaciology*, **9**, 39–44 (1987).
6. D. Gregory and R. N. B. Smith. *The Hadley Centre for Climate Prediction and Research. Canopy, Surface and Soil Hydrology*. Unified Model Documentation. Paper 25, version 1 (1990).
7. H. Cattle. Global climate models and Antarctic climate change. In: *Antarctic and Global Climate Change*, G. M. Harris and B. Stonehouse (Eds), pp. 21–34, Belhaven Press, London, England (1991).
8. US Air Force. Global Snow Depth Climatology, Foster, D. J. Jr and R. D. Davy, USAF publication USAFETAC/TN-88/006, Scott Air Force Base, Illinois, USA (1988)
9. D. Wiesnet and M. Matson. A possible forecasting technique for winter snow cover in the Northern Hemisphere and Eurasia. *Monthly Weather Review*, **104**, 828–835 (1976).
10. C. Schutz and L. Bregman. *Global Snow Depth Data, A Monthly Summary*. The Rand Corporation, Santa Monica, CA, USA (1975).
11. J. L. Foster, D. K. Hall, A. T. C. Chang and A. Rango. Derivation of snow water equivalent in boreal forests using microwave radiometry. *Arctic*, **44**, Suppl. 1, 147–152 (1991).
12. M. Hallikainen and P. Jolma. Comparisons of algorithms for retrieval of snow water equivalent from Nimbus-7 SMMR data in Finland. *IEEE Trans. Geosci. Remote Sensing*, **30** (1), 124–131 (1992).
13. T. Nagler. *Comparison of Spaceborne Snow Retrieval Algorithms*. PhD Dissertation (1991).
14. D. Robinson. Hemispheric snow cover from satellites. *Annals of Glaciology*, **17**, 367–371 (1993).

ESA/NASA International Workshop, pp. 77–92
B. J. Choudhury, Y. H. Kerr, E. G. Njoku and P. Pampaloni (Eds)

Soil moisture estimation: comparisons between hydrologic model estimates and remotely sensed estimates

E. F. WOOD,[1] DAH-SYANG LIN,[1] P. A. TROCH,[2] M. MANCINI[3] and T. J. JACKSON[4]

[1]*Department of Civil Engineering and Operations Research, Princeton University, Princeton, NJ 08544, USA*
[2]*Laboratory of Hydrology and Water Management, Ghent University, Coupure Links 653, B9000 Gent, Belgium*
[3]*DIIAR Politecnico di Milano, Piazza Leonardo da Vinci, Milano, Italy*
[4]*USDA/ARS Hydrology Laboratory, Beltsville, MD 20705, USA*

Abstract—Regionally averaged soil moisture estimates from the PBMR and a hydrologic model are compared with ground measurements over three agricultural areas characterized by different levels of spatial variability. The results successfully reflect the temporal variation of the soil moisture. Performance of the PBMR depends on the level of the spatial variability, and the accuracy of the estimated vegetation correction factor. A preliminary attempt to calibrate the model using the PBMR-derived soil moisture has yielded promising results.

1. INTRODUCTION

Soil moisture can be defined as the storage of precipitation within a shallow layer of the earth that is generally limited to the aeration zone. Despite the insignificant amount of water compared to the global water budget, soil moisture can govern the runoff generation processes in a variety of environments and provides the linkage between the hydrologic cycle and the energy cycle through evapotranspiration. It is, therefore, an important variable in many hydrologic and agricultural investigations. Recent studies with general circulation models (GCMs) using active land surface parameterization have shown that strong feedbacks exist between the soil moisture anomalies and climate [1].

As a result of the inhomogeneity of soil properties, vegetation and precipitation, soil moisture is highly variable both spatially and temporally. The measurement of soil moisture is traditionally conducted on a point basis. Spatial soil moisture distribution is often obtained through interpolation of point measurements. Acquiring a spatial soil moisture map over a heterogeneous area through conventional techniques is expensive and time-consuming.

Recent advances in microwave remote sensing have demonstrated the ability to measure soil moisture in the surface layer of approximately 5 cm under a variety of field conditions [2]. Despite the promising perspective of the remote sensing technique, its application to agricultural and hydrologic sciences has been hampered by several problems. First, existing hydrologic models are conventionally formulated on point processes. These models are not capable of using the remotely sensed data as direct input. Secondly, the algorithms that are currently used to extract soil moisture from the microwave measurements have limited ranges of validity and are subject to further verification. Finally, the error structures of the remote sensing techniques are not clear. A better understanding of these problems is needed before one can efficiently utilize the remotely sensed soil moisture data.

This paper compares model simulated and remotely sensed soil moisture with ground observations using the data collected in a multi-sensor aircraft campaign (MAC-HYDRO'90). MAC-HYDRO'90 was designed to study the sensors' performances and their application to hydrologic studies. Both passive and active microwave sensors were flown during the campaign which was carried out from July 9, 1990 to July 20, 1990. The remotely sensed data were complemented by concurrent ground measurements. The purpose of this paper is to examine the capabilities of microwave remote sensing techniques in soil moisture mapping, and to investigate how the remotely sensed soil moisture data can be applied to hydrologic modeling.

2. DATA AND METHODS OF ANALYSIS

2.1. Site description and weather conditions

MAC-HYDRO'90 was conducted over a portion of the Mahantango Creek located in central Pennsylvania. This catchment is operated by the Northeast Watershed Research Center of the Agricultural Research Service (ARS) and has a drainage area of 7.4 km^2. The average annual precipitation and evapotranspiration for the catchment are 1128 mm and 479 mm per year, respectively [3]. The soils within this area consist primarily of silt loams and loams, and contain $0.5 \sim 2.0$ % organic carbon. Rock fragments are present in the surface layer in some soils. Within the catchment, there are three intensive study areas, namely Paul's farm, Brown's farm and James' site (see Fig. 1). Paul's farm (0.6 km^2) is located on the eastern portion of the catchment and contains a mixture of land uses. Brown's farm and James' site are mainly covered by pastures.

The weather conditions during the experiment were dry initially. No rain was recorded during the preceding 5 days, resulting in nearly uniform dry soil conditions. After the first flight (July 10, 1990), there were approximately 52 mm of rainfall over a four-day period, followed by a strong dry down. These conditions generated a wide range of soil moisture conditions which provide an excellent test for the hydrologic model and remote sensing techniques. Table 1 summarizes the rainfall record and the dates of data collections during MAC-HYDRO'90.

Table 1.
MAC-HYDRO'90 data collection

Date	Rainfall accumulation (mm)	PBMR	SAR	Ground data
July 10	0	Yes	Yes	Yes
July 13	39	No	Yes	Yes
July 15	52	Yes	Yes	Yes
July 17	52	Yes	Yes	Yes
July 18	52	Yes	No	Yes
July 19	52	Yes	No	Yes
July 20	52	No	No	Yes

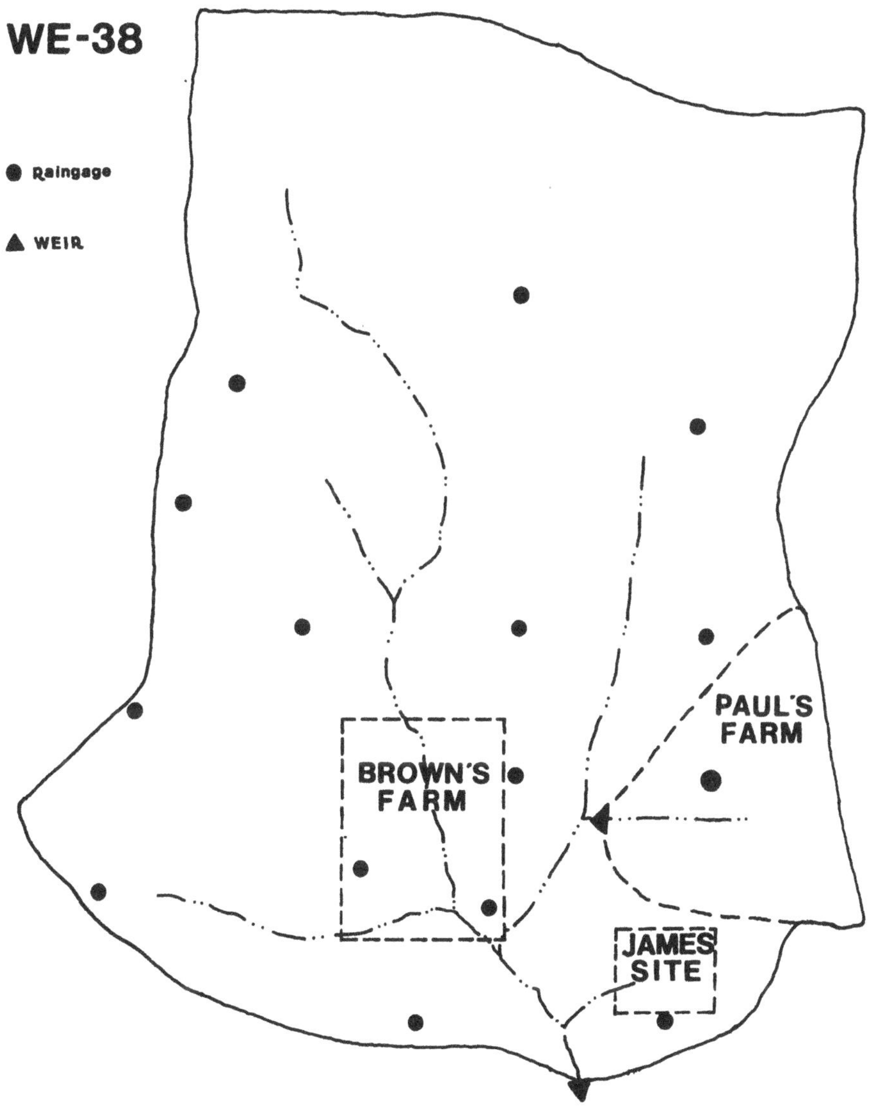

Figure 1. Topographic scale for the studied catchment. The three intensive study areas are outlined by dash lines. Circled letters represent the raingages.

2.2. Ground data

Soil samples were collected along 9 transects (B_1, B_2, B_3, B_4, B_5, B_6, P_1, P_2 and P_3) within the three intensive study areas at two depths, $0 \sim 5$ cm and $5 \sim 10$ cm, during the time of the overflights. Each sample consists of approximately 125 cm^3 of soils. Soil bulk densities were measured at the sites to infer the volumetric soil moisture contents from the gravimetric measurements. The transects were aligned at right angles to the streams. Rainfall records were collected from a network of 14 tipping-bucket raingages deployed over the catchment. The location of the raingage network is displayed in Fig. 1.

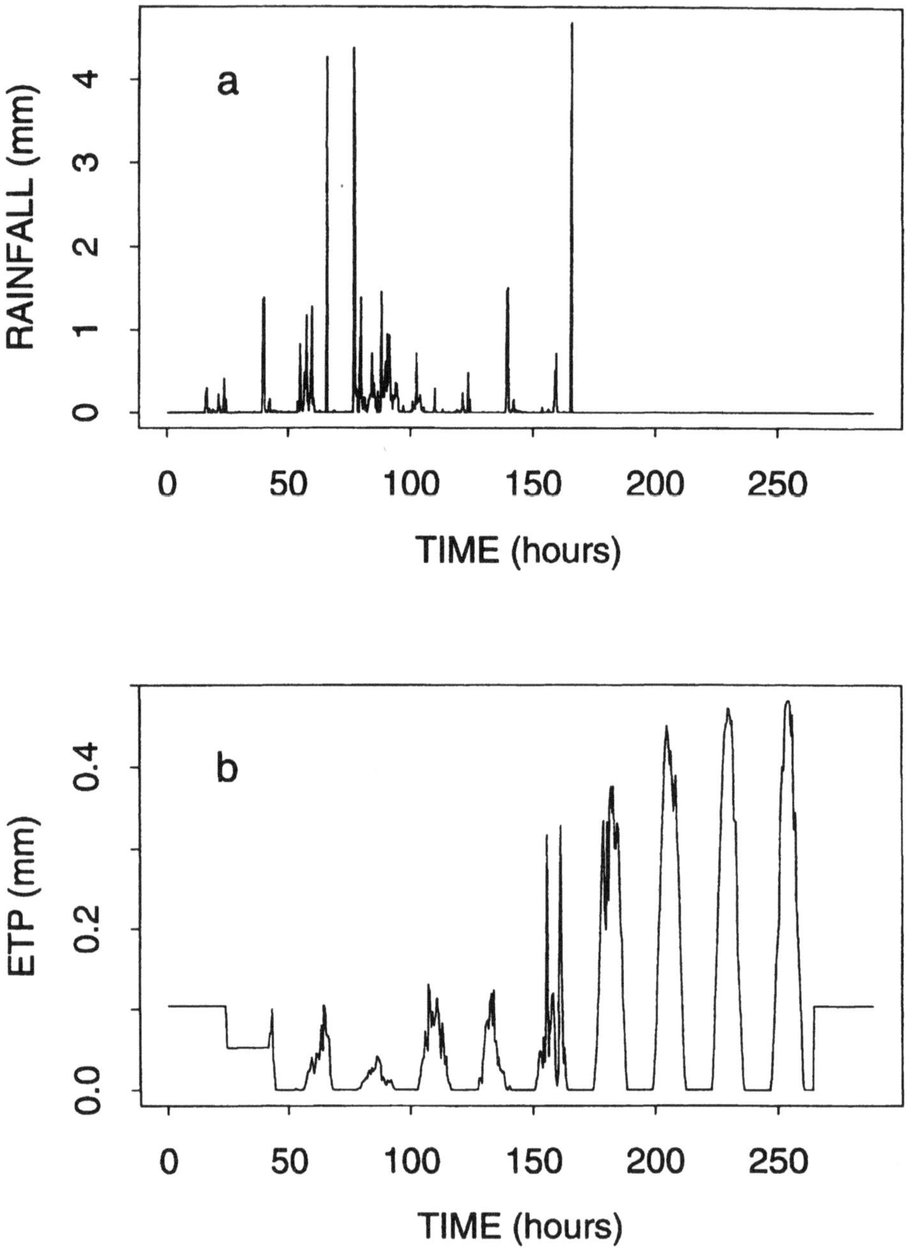

Figure 2. Time series of (a) areal precipitation, and (b) estimated potential evaporation during MAC-HYDRO'90.

Meteorological data were measured by using a micrometeorological station (located beside the raingage *H*). Time series of potential evaporation is calculated using the Priestley–Taylor method from observed net radiation at a sampling interval of 30 minutes. For July 9, 10 and 20, when the station was not operating, daily potential evaporation values of 5 mm, 2.5 mm and 5 mm, respectively, are estimated using the weather reports. Figure 2 shows the temporal variations of the areal average precipitation and estimated potential evaporation for the 11-day experimental period.

Land cover information was compiled for the studied area and was classified into 9 categories (see Fig. 3). Vegetation configurations (e.g. height, density, water contents, etc.) were gathered from a number of representative agricultural fields.

A detailed soil map was assembled from the literature [3, 4]. Fifteen soil types can be identified within the studied catchment. Topography of the area is represented by the USGS 7.5-min digital elevation model (DEM) data, which has a 30 m $\times$ 30 m resolution. A GIS system is used to integrate data from various sources.

2.3. Passive microwave radiometer

The passive microwave sensor used in MAC-HYDRO'90 was the push broom microwave radiometer (PBMR). The PBMR operates at L-band (1.42 GHz) and has four horizontally polarized beams pointing at $\pm 8°$ and $\pm 24°$ from nadir. The nominal ground resolution of a beam position is roughly 0.3 times the altitude of the aircraft, which was 300 m in MAC-HYDRO'90. As a result, the ground resolution of the MAC-HYDRO'90 PBMR measurements is approximately 90 m. For a detailed description of the PBMR, see [5].

Data collected from the PBMR were processed following the procedures that have been successfully employed in previous experiments [6]. A time recording is made on both the PBMR data and video tape of the flight coverage. Following the flight the video tape is reviewed and times are noted on a 1 : 24 000 topographic map of key geographic features. These points are used with the recorded microwave data times to interpolate an *x*, *y* location for each observation. Beam position and angular variations are corrected for using a deterministic technique described in [7]. Errors caused by aircraft (pitch, roll and yaw) are not accounted for in the above procedure.

After the mapping, a nearest neighbor weighting scheme is conducted to fill gaps between the flightlines and/or to take into account multiple coverage of an area. This process results in a uniform map product with a resolution of 20 m $\times$ 20 m. However, for an area that is characterized by large spatial variability such as Mahantango Creek, one has to be very careful in interpreting the resulting PBMR data.

Using the procedures described above, a brightness temperature map was created for each date. The resulting images are shown in Fig. 4. It was found that the July 10 image consisted of uniformly high brightness temperature values with very little correlation with geographic features. Following the rainfall event, the July 15 image appears much 'cooler' than the July 10 image. In addition, the areas that are forested retain their high brightness temperatures throughout the duration of the experiment. In general, the overall patterns of the PBMR measurements reflect the observed meteorological conditions, and display a variability not as high as that suggested by the field structure of the area because of the instrument's resolution.

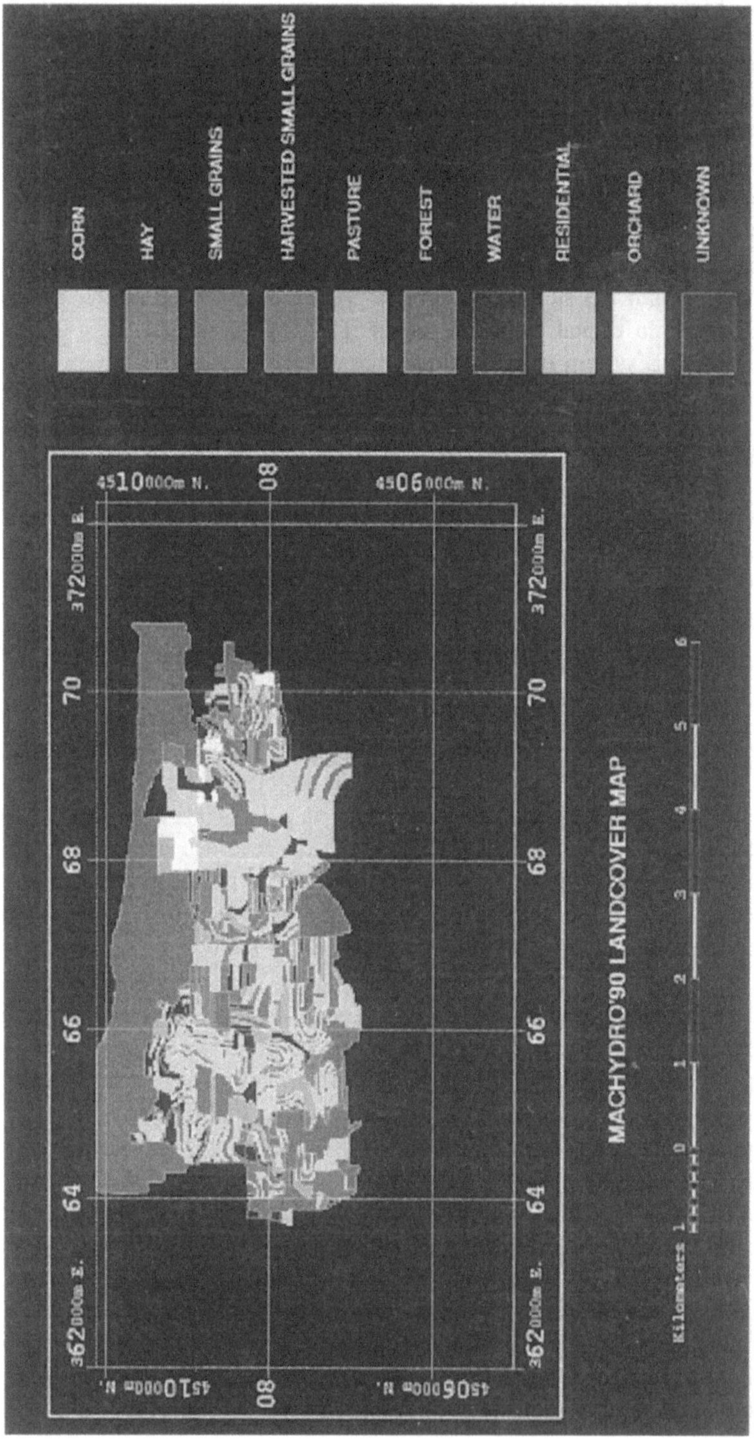

Figure 3. MAC-HYDRO'90 landcover map.

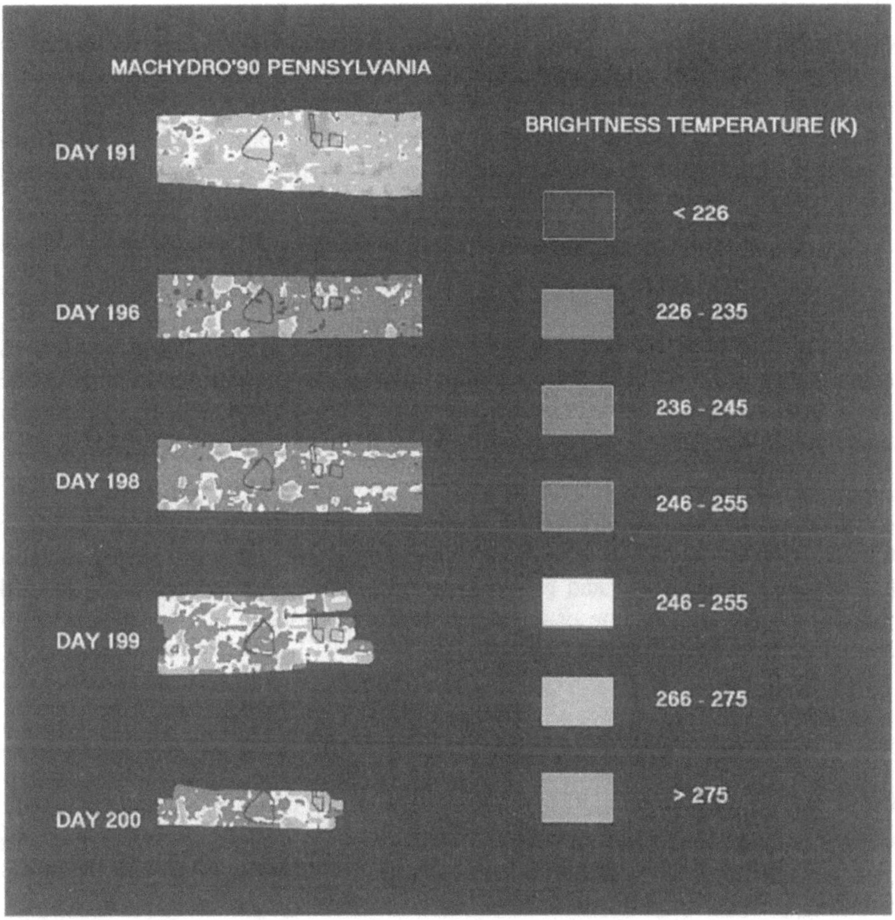

Figure 4. The PBMR brightness temperature maps for the MAC-HYDRO'90 experiment.

2.4. Hydrologic model

The hydrologic model employed in this study was developed by Famiglietti and Wood [8]. The distinguishing feature of this model is that the effect of topographic controls on the land surface fluxes is reflected by applying the local water and energy balance to each interval of the probability distribution of a topographic index, a concept advanced by Beven and Kirby [9] and used extensively in various versions of TOPMODEL. The water and energy balance equations for each interval of the topographic index distribution with drainage area A can be expressed as

$$\frac{\mathrm{d}S_{\mathrm{u}}}{\mathrm{d}t} + \frac{\mathrm{d}S_{\mathrm{g}}}{\mathrm{d}t} = Ap - Ae - Q \tag{1}$$

$$R_{\mathrm{n}} = LE + H + G \tag{2}$$

where t is time, e is the catchment average evaporation rate, p is the averaged rainfall intensity, Q is the streamflow at the basin outlet, S_{u} is the volume of water stored in the unsaturated zone, and S_g is the groundwater storage. R_n is the net radiation, LE and H are the latent heat and sensible heat, respectively. G represents the ground heat flux. Other contributions to the water and energy balance, such as deep percolation and the storage in surface water bodies, are assumed to be of minor importance and are neglected.

A three-parameter gamma distribution of the following form is used to approximate the probability distribution of the topographic index, which is determined from the DEM data:

$$f_T(x) = \frac{1}{\chi\Gamma(\phi)}\left(\frac{x-\mu}{\chi}\right)^{\phi-1} \exp\left\{-\frac{x-\mu}{\chi}\right\} \tag{3}$$

where $x = \ln(a/\tan\beta)$ is the topographic index, a is the local drainage area per unit contour length, and β is the land surface slope angle. $\Gamma(\cdot)$ represents the gamma function, ϕ, χ, μ are the three parameters of the gamma distribution, and are estimated by using the method of moments.

2.4.1. Streamflow generation. The hydrologic model predicts soil saturation and its relationship to both saturation excess and infiltration excess surface runoff generation, using spatially variable soil and topographic data. The initial storage capacity in the unsaturated zone is a function of the depth to the water table. The local depth to the water table, z_i, can be computed by [10]

$$z_i = \overline{z} - \frac{1}{f}\left\{\ln\left(\frac{aT_e}{T_i\tan\beta}\right) - \lambda\right\} \tag{4}$$

where $\overline{z}$ is areal average of z_i, T_i is the local value of the transmissivity coefficient $T = K_{s0}/f$, K_{s0} is the saturated hydraulic conductivity at the surface, f is the exponential decay of hydraulic conductivity with depth, $\ln(T_e)$ is the expected value of $\ln(T_i)$, λ is the expected value of the topographic index $\ln(a/\tan\beta)$.

The derivation of Equation (4) is based on the assumption that the water table is nearly parallel to the soil surface and that the saturated hydraulic conductivity, K_s, declines exponentially with depth z_d according to

$$K_s(z) = K_{s0}\exp(-fz_d). \tag{5}$$

Assuming a state of hydraulic equilibrium, the local storage capacity S_i can be expressed in terms of z_i and pertinent soil parameters as follows

$$S_i = (\theta_s - \theta_r)\left\{z_i - \psi_c - \frac{1}{1-B}\left[(\psi_c z_i)^B - \psi_c\right]\right\} \tag{6}$$

where θ_s is the saturation moisture content and θ_r is the residual moisture content. The air entry pressure head value ψ_c and the pore size distribution index B are the parameters in the soil moisture characteristic relationship formulated by Brooks and Corey [17]:

$$\theta(\psi) = \begin{cases} \theta_r + (\theta_s - \theta_r)\left(\frac{\psi_c}{\psi}\right)^B, & \psi \leqslant \psi_c \\ \theta_s, & \psi \geqslant \psi_c \end{cases} \tag{7}$$

$$K(\psi) = \begin{cases} K_s \left(\frac{\psi_c}{\psi}\right)^{2+3B}, & \psi \leqslant \psi_c \\ K_s, & \psi \geqslant \psi_c \end{cases} \tag{8}$$

where θ is volumetric moisture content, ψ is pressure head and K is hydraulic conductivity. Whenever the cumulative infiltration volume exceeds the local initial soil moisture storage capacity saturation excess runoff is generated.

The infiltration excess component of the model applies the time compression approximation to Philip's equation to compute the local infiltration rate, g_i, under local erratic rainfall, p_i [8]:

$$g_i = \min\left[g_i^*(F), p_i\right] \tag{9}$$

where F is cumulative infiltration. The local infiltration capacity g_i^* is calculated as

$$g_i^*(F) = cK_{s0}\left\{1 + \left[\left(1 + \frac{4cK_{s0}F}{S_r^2}\right)^{1/2} - 1\right]^{-1}\right\} \tag{10}$$

where S_r is sorptivity and cK_{s0} accounts for the effect of gravity on the infiltration rate. The parameter c ranges from 0.5 to 1.0 and depends on the saturation of the soil profile. Analytical expressions for c and S_r are given in Sivapalan *et al.* [10]. Infiltration excess runoff is generated on those parts of the catchment where $p_i > g_i^*$.

The model assumes an exponential relationship between baseflow and groundwater storage such that the baseflow contribution Q_b to total streamflow is given by:

$$Q_b = Q_o \exp(-f\overline{z}) \tag{11}$$

where Q_o represents subsurface flow at complete saturation. During a computational time interval, the average depth to the water table, $\overline{z}$, is assumed to be constant. The value of $\overline{z}$ is updated at the beginning of the next time interval, taking into account the total infiltrated and drained volume.

The total streamflow for the catchment is obtained by summing the contributions to surface runoff from saturation excess, infiltration excess and subsurface flow. In this version of the model overland flow or channel flow routing is not considered.

2.4.2. Evaporation. Saturated areas are allowed to evaporate at the potential rate e_p (atmosphere controlled stage). As the soil near the surface dries out the moisture delivery rate is limited by the properties of the soil profile (soil controlled stage). The model uses an analytical solution to the desorption problem and again applies the time compression approximations which results in the following expression:

$$e_a = \min\left[e_a^*(E_a), e_p\right] \tag{12}$$

in which e_a is the actual evaporation rate and e_a^* represents the evaporation capacity as a function of cumulative evaporation E_a:

$$e_a^*(E_a) = \frac{S_e^2}{E_a} \tag{13}$$

where S_e is the desorptivity, which varies with soil moisture content and other soil dependent parameters. Vegetation regulated transpiration is not accounted for in the model.

During the interstorm periods, evaporated water is extracted from the groundwater storage so that, for each time step, the water table is updated taking into account evaporation and drainage losses.

2.4.3. Initial and boundary conditions. Two sets of initial conditions are used in simulation. One is determined from the discharge records using the procedure proposed by Troch *et al.* [11]. Another is determined from the PBMR-derived soil moisture conditions of July 10, 1990. The lower and lateral boundaries are assumed to be impervious. Given the initial and boundary conditions, the model is run for the 11-day duration of the experiment. Solution includes the sensible and latent heats and the soil moisture distributions for each interval of the topographic index distribution.

2.5. Soil moisture estimation

Since the field sizes that produce the variation in the land cover are smaller than the footprint of the PBMR in the studied area, it is unlikely that, using the currently available technology for geolocation, each data cell can be accurately located to apply a vegetation correction. Hence, soil moistures are estimated on a regionally average basis. Using the observed vegetation water contents and the table presented in [12], optical thickness τ is estimated for each type of land cover. Then, the regionally average vegetation correction parameter is calculated by weighting τ's according to their areal percentages to the area. Forest and residential areas are excluded from the computation because, under these conditions, it is unlikely useful information concerning the soil moisture could be extracted from the PBMR measurements. Table 2 lists the weights of each category and the resulting averaged optical thickness $\bar{\tau}$ for the three intensive study areas.

Assuming the single scattering albedo of the vegetation is zero and the temperatures of the soil and the vegetation are equivalent, the PBMR brightness temperature T_B can be expressed as [14]:

$$T_B = \left(1 - R_H^2 \Upsilon^2\right) T_v \tag{14}$$

Table 2.
Weight for each land cover type for the studied areas

Land cover	τ[1]	Paul's farm	Brown's farm	James' site
Corn	0.35	41%	25%	50%
Hay	0.09	14%	20%	9%
Small grains	0.20	30%	21%	35%
Pasture	0.15	15%	34%	6%
$\bar{\tau} =$		0.24	0.20	0.26

[1] Estimated optical thickness at 1.42 GHz.

where T_v is the physical temperature of the vegetation, R_H^2 is the H-polarized reflectivity of the air-soil interface, and Υ is the transmissivity of the vegetation layer. Υ depends upon the vegetation dielectric properties, the plant shape or structure and the sensor's configuration, and can be approximated by

$$\Upsilon = \exp\left(-\overline{\tau}\sec\Omega\right) \tag{15}$$

where Ω is the look angle of the instrument. This angle is approximately equal to 10° for the PBMR.

The reflectivity of the air-soil interface at Ω is related to the relative dielectric constant of soil ε_r by

$$R_H^2 = \left|\frac{\cos\Omega - \sqrt{\varepsilon_r - \sin^2\Omega}}{\cos\Omega + \sqrt{\varepsilon_r - \sin^2\Omega}}\right|^2 . \tag{16}$$

Given the observed soil surface temperatures, the PBMR brightness temperatures, and $\overline{\tau}$, one can infer the relative soil dielectric constant by solving the above equations. In general, ε_r is a complex number consisting of the real and the imaginary parts. Since the error incurred by neglecting the imaginary part is usually very small, we will only use the real part of ε_r in the following computation.

To infer the volumetric soil moisture content θ from the relative soil dielectric constant, two dielectric mixing models are employed. The first one was developed by Dobson *et al.* [13] and can be expressed as

$$\varepsilon_r^{0.65} = 1 + \frac{\rho_b}{\rho_{ss}}\left(\varepsilon_{ss}^{0.65} - 1\right) + \theta^{\beta}\left(\gamma^{0.65} - 1\right) \tag{17}$$

where γ is the dielectric constant of pure water at 23°C, which is approximately equal to 78.5 for the PBMR's frequency. ρ_b is the soil bulk density, ρ_{ss} is the density of the solid soil material, and ε_{ss} is the solid soil dielectric constant. The exponent β is related to the sand (S) and clay (C) fractions (by weights) of the soil [15],

$$\beta = 1.09 - 0.11S + 0.18C. \tag{18}$$

The value of β for each studied area is estimated using a weighting procedure, similar to the one used in deriving the vegetation correction parameter, from the soil map.

The second model was empirically developed using the soil samples collected in MAC-HYDRO'90 and laboratory analysis results [16]. This relationship can be written as

$$\varepsilon_r = 2.583\exp\left(0.081\,\theta\right). \tag{19}$$

It should be pointed out that the model of Dobson *et al.* [13] requires additional input information (i.e. soil composition). For the range of soil moisture conditions encountered in MAC-HYDRO'90, the maximum difference between the estimated soil moistures from the above two mixing models is approximately 10 %. This discrepancy decreases with decreasing soil moisture content and the level of spatial heterogeneity of the studied area.

3. RESULTS AND DISCUSSION

Table 3 lists the soil moisture estimates from various sources for the Paul's farm. The hydrologic model was initialized using the streamflow records. Ground soil moisture measurements (0–5 cm) taken within the area are averaged and included for comparison. Similar comparisons are made for the Brown's farm and the James' site. Results are summarized in Tables 4 and 5, respectively. The number of ground samples used in averaging ranges between 27 and 71, depending on the date and site. Values enclosed within the parentheses are the sample standard deviations of the estimates, which can be viewed as a measure of the level of the spatial heterogeneity. It can be seen from the tables that the James' site has the smallest spatial soil moisture heterogeneity among the three intensive study areas. The high standard deviation values for the Paul's farm and the Brown's farm suggest that the mean of the ground measurements may not be an appropriate index for the regionally averaged soil moisture condition. Nonetheless, despite the large uncertainties, the temporal variations displayed by the ground measurements are found to be consistent with the weather conditions for all three sites. Thus, we will focus on the temporal soil moisture pattern in the following analysis.

Table 3.
Regional soil moisture estimates for the Paul's farm

Date	T_B (K)	T_v (K)	PBMR[1] (%)	PBMR[2] (%)	Model (%)	Ground (%)
July 10	263	298.2	12.8	11.2	28	10.3 (7.3)
July 13	–	–	–	–	38	20.4 (6.9)
July 15	228	294.4	31.2	22.6	36	16.8 (6.9)
July 17	231	295.8	29.9	21.9	33	19.2 (7.3)
July 18	246	297.2	21.1	17.1	32	19.6 (6.9)
July 19	245	298.3	22.2	17.7	30	17.9 (7.4)
July 20	–	–	–	–	26	15.9 (6.1)

[1] Using the model by Dobson *et al.* [13].
[2] Using the model by Jackson *et al.* [16].

Table 4.
Regional soil moisture estimates for the Brown's farm

Date	T_B (K)	T_v (K)	PBMR[1] (%)	PBMR[2] (%)	Model (%)	Ground (%)
July 10	264.2	299.2	11.4	10.0	30	15.3 (9.7)
July 13	–	293.2	–	–	40	24.9 (8.9)
July 15	248.1	295.8	17.3	14.6	39	26.6 (10.1)
July 17	242.3	296.4	20.5	16.6	37	25.7 (12.6)
July 18	247.9	297.4*	18.1	15.1	35	24.3 (10.9)
July 19	250.5	298.4	17.2	14.5	32	21.8 (10.7)
July 20	–	–	–	–	24	23.0 (12.3)

[1] Using the model by Dobson *et al.* [13].
[2] Using the model by Jackson *et al.* [16].
* Interpolated using data from July 17 and 20, 1990.

Table 5.
Regional soil moisture estimates for the James' site

Date	T_B (K)	T_v (K)	PBMR[1] (%)	PBMR[2] (%)	Model (%)	Ground (%)
July 10	270.5	296.4	9.1	8.0	24	9.7 (4.1)
July 13	–	–	–	–	36	20.1 (5.2)
July 15	246.0	294.7	21.1	17.0	34	20.4 (5.5)
July 17	251.1	295.3	18.5	15.3	33	18.4 (4.9)
July 18	254.8	296.5	17.0	14.3	30	18.2 (5.6)
July 19	251.5*	299.0	20.0	16.3	26	16.4 (3.9)
July 20	–	–	–	–	22	15.2 (4.0)

[1] Using the model by Dobson *et al.* [13]
[2] Using the model by Jackson *et al.* [16]
* Only 30% of the area was covered by the PBMR footprint.

It is found that, in general, soil moistures derived from the PBMR correctly reflect the temporal patterns of soil moisture by showing the high soil moisture contents on the wet days. In addition, the PBMR derived soil moisture contents are in good agreement with the ground measurements over the James' site. Notice that, on July 19, 1990, only 30 % of this area was covered in the PBMR footprint. Comparisons between the PBMR derived soil moisture and ground measurements indicate an overestimation over the Paul's farm and an underestimation for the Brown's farm. The above results seem to suggest that the performance of the vegetation correction procedure used in this study is related to the spatial variabilities of the area over which it is applied. It is, however, difficult to draw any conclusion because of the uncertainties associated with the ground measurements. Comprehensive ground sampling will be needed for this purpose.

The performance of the PBMR vegetation correction procedure is also affected by the accuracy in the estimated optical thickness for each type of land cover. A sensitivity analysis is performed to examine the impact of this variable. The optical thickness values listed in Table 2, which are derived from field measurements and/or previous experiments, are used as a reference state. The results of the sensitivity analysis indicate that a 15 % increase/decrease to each type of land cover can result in an approximately 10 % increase/decrease in terms of the PBMR regionally averaged soil moisture estimates for the three study areas. This implies that the effectiveness of the PBMR soil moisture retrieval algorithm is largely dependent upon the accuracy of the optical thickness for each type of land cover. Extra cautions should be exercised when estimating these values.

Estimates from the hydrologic model using the discharge-derived initial conditions appear to be wetter than other measurements. The temporal variation, however, reflects the weather conditions. The discrepancy between the model predictions and the ground measurements may be caused by errors in the initial soil moisture conditions. In applying the model to an experiment such as MAC-HYDRO'90, two fundamental difficulties arise that can affect the model's performance. The major difficulty occurs during the summer season when the upper soil layer effectively becomes disconnected from the lower portion of the soil column (the saturated portion of the column) which is draining and providing the observed base discharge in the

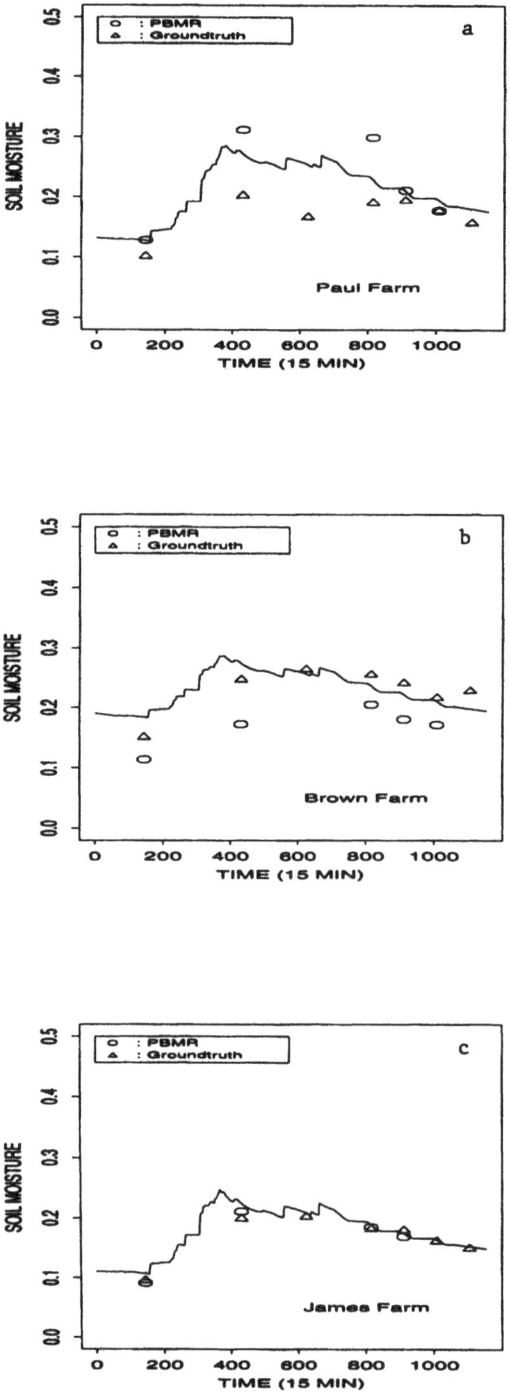

Figure 5. Regionally averaged soil moistures simulated by the hydrologic model using the PBMR derived initial conditions for (a) the Paul's farm, (b) the Brown's farm, and (c) the James' site.

stream. Essentially no vertical percolation is occurring. Thus the discharge initialization procedure provides little information about the state of the surface soil moisture. The second difficulty is related to the first and is concerned with the short duration of the experiment. The initial conditions persist until a sequence of storm and interstorm periods perturb the catchment sufficiently so that the water fluxes (infiltration and evaporation) and soil moisture represent the modeled processes. For MAC-HYDRO'90, errors in the model initialization still persisted throughout the two week simulation. Therefore, errors in this variable, due to the above reasons, can be expected.

Having recognized the initialization problems, we next attempt to apply the remotely sensed data to correct the initial state of the hydrologic model. A new set of initial conditions for each site are derived from the July 10, 1990 PBMR soil moisture estimates by integrating the hydrologic model backward in time. Fig. 5 shows the temporal variation of regionally averaged soil moisture predicted by the hydrologic model using the PBMR derived initial conditions. The forcing (i.e. precipitation and potential evaporation) and other parameters are kept unchanged. It is clear from the figure that the usage of the remotely sensed data, even in a rather simple fashion, has significantly improved the performance of the hydrologic model. This is especially the case for the James' site which has the smallest spatial variability among the three studied areas.

4. CONCLUSIONS

Regionally averaged soil moisture estimates from the PBMR and a hydrologic model are compared with ground measurements over three agricultural areas characterized by different level of spatial variability. Results can be summarized as follows:

(1) Both the PBMR and the hydrologic model successfully reflect the temporal variation of the soil moisture during MAC-HYDRO'90. Performance of the sensor is a function of the level of the spatial variability, and the accuracy of the estimated vegetation correction factor $\overline{\tau}$. For a homogeneous area such as the James' site, the soil moisture estimation procedure works satisfactorily. Comprehensive ground measurements are needed to evaluate the PBMR derived soil moisture patterns over the areas characterized by large spatial heterogeneity.

(2) Soil moisture estimates from the hydrologic model using the discharge-derived initial conditions are too wet. It is suspected that the discrepancy is caused by the disconnected status of the soil column during the period of the experiment. A preliminary attempt to calibrate the model using the PBMR derived watershed averaged soil moisture has significantly improved the model predictions.

Acknowledgements

This study was supported by NASA Grant NAG5-1628 and USDA Cooperative Agreement 58-3K47-0-039 for application of SIR-C Synthetic Aperture Radar to Hydrology. Support from NASA and USDA is highly appreciated.

REFERENCES

1. E. F. Wood. Global scale hydrology: advances in land surface modeling. *Rev. Geophysics*, Supplement, 193–201 (1991).
2. E. T. Engman. Progress in microwave remote sensing of soil moisture. *Cana. J. Remote Sensing*, **16**, 6–14 (1990).
3. A. S. Rogowski, E. T. Engman and E. L. Jacoby. *Transient Response of a Layered, Sloping Soil to Natural Rainfall in the Presence of a Shallow Water Table*. USDA, ARS Report ARS-NE-30, Hyatsville, MD (1974).
4. K. M. Loague and R. A. Freeze. A comparison of Rainfall–Runoff modeling techniques on small upland catchments. *Water Resources Res.*, **21**, 229–248 (1985).
5. T. J. Schmugge, J. R. Wang and G. Asrar. Results from the push broom microwave radiometer flights over the Konza Prairie in 1985. *IEEE Trans. Geosci. Rem. Sens.*, **26**, 590–596 (1988).
6. T. J. Schmugge, T. J. Jackson, W. P. Kustas and J. R. Wang. Passive microwave remote sensing of soil moisture: Results from HAPEX, FIFE and MONSOON'90. *J. Photogramm. Remote Sensing*, in press (1992).
7. T. J. Jackson, M. E. Hawley and P. E. O'Neill. Preplanting soil moisture using passive microwave sensors. *Water Res. Bull.*, **23**, 11–19 (1987).
8. J. Famiglietti and E. F. Wood. A catchment scale water balance model for FIFE. *J. Geophys. Res.*, **97**, 18997–19007 (1992).
9. K. J. Beven and M. J. Kirkby. A physically based variable contributing area model of basin hydrology. *Hydrol. Sci. Bull.*, **24** (1), 43–69 (1979).
10. M. Sivapalan, K. J. Beven and E. F. Wood. On hydrologic similarity, 2, a scaled model of storm runoff production. *Water Resources Res.*, **23**, 2266–2278 (1987).
11. P. A. Troch, F. P. De Troch and W. Brutsaert. Effective water table depth to describe initial conditions prior to storm rainfall in humid regions. *Water Resources Res.*, **29**, 427–434 (1992).
12. T. J. Jackson and T. J. Schmugge. Vegetation effects on the microwave emission of soils. *Remote Sens. Environ.*, **36**, 203–212 (1991).
13. M. C. Dobson, F. T. Ulaby, M. Hallikainen and M. El-Rayes. Microwave dielectric behavior of wet soil. Part II: Dielectric mixing models. *IEEE Trans. Geosci. Rem. Sens.*, **GE-23**, 35–46 (1985).
14. T. J. Jackson and P. E. O'Neill. Attenuation of soil microwave emissivity by corn and soybeans at 1.4 and 5 GHz. *IEEE Trans. Geosci. Rem. Sens.*, **GE-28**, 978–980 (1990).
15. F. T. Ulaby, R. K. Moore and A. K. Fung. *Microwave Remote Sensing, Active and Passive*. Vol. III, Artech House, Inc., Norwood, MA, (1986).
16. T. J. Jackson, E. T. Engman, D. Le Vine, T. J. Schmugge, S. Saatchi, R. Lang, E. Wood, R. Parry and W. Teng. MAC-HYDRO'90 multitemporal passive microwave mapping. submitted to *Water Resources Res.* (1992).
17. R. H. Brook and A. T. Corey. Hydraulic properties of porous media. In: *Hydrology Paper No. 3*, Colorado State University, Fort Collins, CO (1964).

Synergism

ESA/NASA International Workshop, pp. 95–115
B. J. Choudhury, Y. H. Kerr, E. G. Njoku and P. Pampaloni (Eds)

Sensitivity of active and passive microwave sensors to land parameters

PAOLO PAMPALONI

CNR/IROE, via Panciatichi 64, 50127 Firenze, Italy

Abstract—A comparative evaluation of the potential of active and passive microwave sensors in estimating some important geophysical parameters over land on the basis of experimental results obtained in the past years is presented. The sensitivity of backscattering coefficient and emissivity to some significant land parameters is investigated and evaluated with reference to sensor resolution at the present state of the art. The results show that both microwave backscattering and emission are sensitive to soil moisture, snow cover and vegetation biomass over a wide frequency range. However in some cases the measurement is better accomplished by a radiometer and in other cases by radar. Active systems, because of their high spatial resolution, appear to be the most suited sensors for all-weather land mapping, crop classification, surface roughness and slope detection. The major problem in using passive microwave sensors, which have a high sensitivity to water content of soil, snow and vegetation seems to be the coarse ground resolution. Most frequently the simultaneous use of passive and active sensors can be synergistic and allows improvement in the estimation accuracy.

Combinations of active and passive sensors can be synergistic in practical applications and in the understanding of the interaction mechanism between microwave radiation and land surfaces.

1. INTRODUCTION

Microwave emission and scattering from the Earth's surface are the two fundamental processes exploited by passive and active microwave sensors to detect land and ocean signatures. These two processes can be related to one another, according to Kirchhoff's law and the reciprocity theorem, as in Equation (1) [1]:

$$e(\theta,\phi,p) = 1 - \frac{1}{4\pi\cos\theta}\int_0^{2\pi}\int_0^{\pi/2}\big[\sigma^\circ(\theta,\phi;\ \theta_s,\phi_s;\ p,p) + \sigma^\circ(\theta,\phi;\ \theta_s,\phi_s;\ p,p_s)\big]\sin\theta_s\,d\theta_s\,d\phi_s \tag{1}$$

where $e(\theta,\phi,p)$ is the emissivity (i.e. the ratio between the radiation power emitted by a homogeneous 'grey' body, and the corresponding power from a blackbody at the same temperature) measured along the θ, ϕ direction at p polarization, and $\sigma^\circ(\theta,\phi;\ \theta_s,\phi_s;\ p,p_s)$ is the bistatic scattering coefficient (i.e. the ratio between the scattered to the incident power upon the surface) for a pair of incident (θ,ϕ) and scattering (θ_s,ϕ_s) directions and a pair of p and p_s polarizations (Fig. 1).

Equation (1) relates emissivity to the fraction of power which is globally scattered in the upper half space (*reflectivity*), while the two parameters used in microwave remote sensing are the emissivity $e(\theta,\phi)$ and backscattering coefficient $\sigma^\circ(\theta,\phi;\ \theta_s=\theta,$

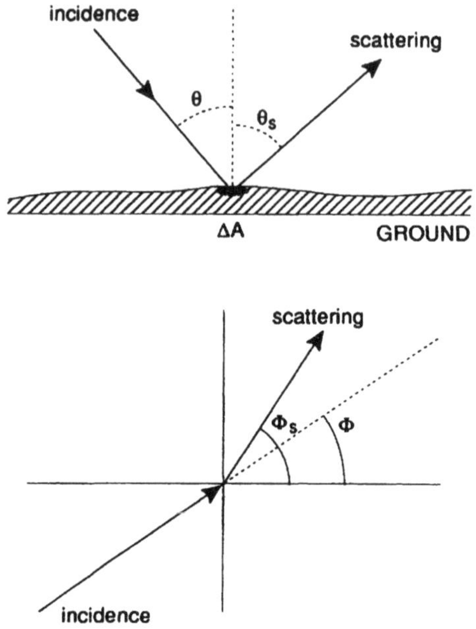

Figure 1. Geometry of the scattering. (From Ferrazzoli *et al.* [2]).

$\phi_s = \phi) = \sigma^\circ(\theta, \phi)$. As a consequence a full correlation between the measured emission and backscatter can be established only in very particular cases, whereas in general only a partial correlation occurs [2]. Indeed $\sigma^\circ(\theta, \phi)$ depends on the angular pattern of the scattered power which is generally unknown. For example, it should be recalled that scattering from a flat surface consists of only the specularly reflected component, so $\sigma^\circ(\theta, \phi) = 0$, $(\theta \neq 0)$ whatever the emissivity is, whereas for a 'perfectly rough' surface, where Lambert's law holds, emissivity and backscattering are biunivocally related and independent of polarization [3]. Another example of full correlation is the case of a half space of isotropic dissipative scatterers where the bistatic scattering coefficient is uniformly distributed in azimuth. An approximate solution has also been derived for a two layer medium assuming that volume scattering is relatively isotropic with respect to surface scattering and the transmissivity of the top interface is about 1 [4]. This case is sometimes approximated by natural surfaces such as soil covered by deep vegetation or dry snow.

The theoretical aspects of the relationship between emissivity and backscattering are treated in depth elsewhere [5]; this paper intends to contribute a comparative evaluation of the potential of active and passive microwave sensors in estimating some important geophysical parameters over land on the basis of experimental results. Indeed, in spite of the difficulty in explaining the emissivity and backscattering measurements with the same theoretical model, the combination of active and passive sensors can be synergistic in practical applications and in the understanding of interaction mechanisms between microwave radiation and land surfaces. In order to investigate the performance of active and passive microwave sensors in measuring land parameters, we have taken into consideration data collected over the past decade by airborne and ground-based active and passive sensors. According to the results of these investigations, radar systems, because of their high spatial resolution, appear to be the sensors best suited for all-weather land mapping, crop classification, surface roughness and slope detection. On the other hand, radiometers have a high sensitivity to water content of soil, vegetation and atmosphere, and their use can be of great interest in hydrology, meteorology, climatology and, although to a lesser extent, in agriculture as well. The information on land surface useful in these disciplines includes: soil moisture, vegetation cover, plant water content, snow extent, snow water equivalent and onset of snow melting.

The capability of active and passive microwave sensors to measure soil moisture content SMC and snow cover has been the subject of extensive research, carried out in order to understand the underlying physical principles and to determine the optimum observation parameters as well as the achievable accuracy of these measurements [6, 7]. Also, significant investigations on the vegetation biomass detection and plant water content measurement have been carried out since the early 1980s [6, 7]. It should be noted that most experiments were generally performed using truck-mounted sensors, and that only a relatively few aircraft measurements, allowing synchronic views of many different fields, are available.

2. EXPERIMENTAL RESULTS

The basic sensed parameters are the backscattering coefficient σ°, (or normalized radar cross section in m^2/m^2) and the emissivity. The latter can be approximated by the ratio T_n between microwave and infrared brightness temperature. Both σ° and T_n depend on frequency, polarization, and observation angle, and are functions of the characteristics of the observed medium.

In the standard case of a rough soil covered by snow or vegetation, the measured microwave emissivity or backscattering consists of more contributions [7]: the term due to snow or vegetation–air interface, the one from soil attenuated by the overlying layer, the direct contribution from snow or vegetation volume and the indirect one due to the interaction with air and soil boundaries. Since the average permittivity of vegetation canopies and dry snow is small, the reflection at the air interface can be neglected. However, the permittivity of snow increases with the presence of liquid water, thus the surface reflectivity of wet snow is much greater than that of dry snow and the upper wet snow boundary must be treated as a rough surface.

In most cases, scattering and emission from natural and agricultural vegetation are mainly influenced by soil parameters at small off nadir observation angles ($\theta \leqslant 20°$), and at the lower frequencies ($f \leqslant 3$ GHz). On the contrary, vegetation effects are dominant at the higher observation angles and frequencies. Dry snow is more transparent than vegetation and, in general, for snow depth of less than one meter and frequencies lower than 18 GHz, emission or scattering contribution from the soil below snowpack dominates even at incidence angles around 50 degrees, whereas for wet snow the importance of soil contribution is greatly reduced due to the high value of the dielectric loss factor of the wet snow layer.

2.1. Bare and vegetated soil

Investigations carried out since the 1970s have shown that both emission and backscattering from land surfaces are affected by soil characteristics (surface roughness and soil moisture) [8–12] and by the amount and type of biomass overlying soil [13–17]. The

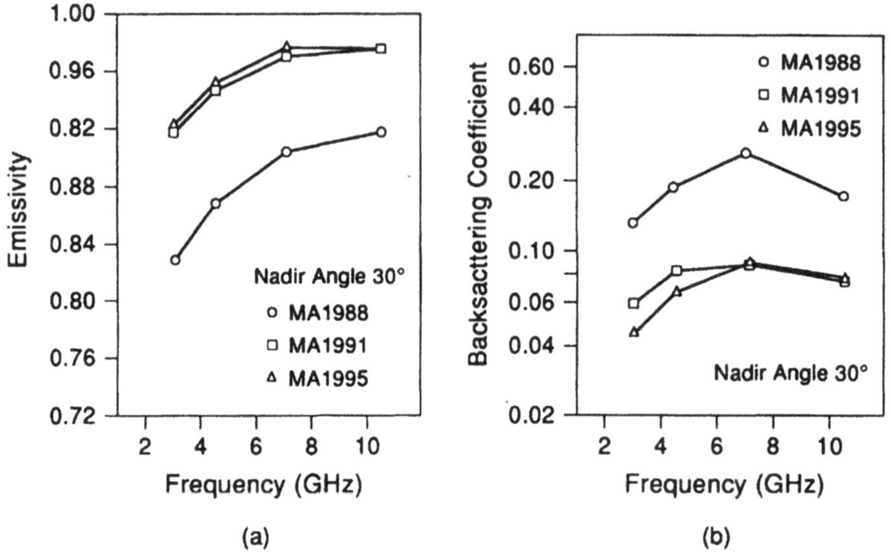

Figure 2. Spectra of emissivities and backscattering coefficients γ of bare soil with 0.82 cm rms and 2 cm correlation length for different volumetric soil moisture content (○ = 40%, □ = 36%, △ = 31%). (After Hüppi *et al.* [18]).

spectral behavior of emissivity and backscattering coefficient γ ($= \sigma^\circ / \cos\theta$) of rough bare soil in different moisture conditions, measured with radiometer–scatterometer RASAM, by Hüppi *et al.* [18], is represented in Fig. 2. The diagram shows a significant decrease of emissivities and an increase of γ, at all frequencies, as soil moisture increases.

A direct comparison of X-band backscatter and emissivity of agricultural fields, carried out using data collected by means of a Synthetic Aperture Radar (VARAN-S) and the X-band channel of the IROE-CNR Airborne Multi-frequency Passive Sensor, is represented in Fig. 3 [2]. Figure 3(a) shows the normalized temperature T_n of each field obtained by averaging all the samples collected in the same field during one or more overpasses, while Fig. 3(b) represents the backscattering coefficient σ° calculated as the average of all the pixels belonging to the field. The crops observed were alfalfa, corn, sugarbeet and wheat stubble (indicated as bare soil). A comparison of Figs 3(a) and 3(b) shows a general consistency of the normalized temperatures with the corresponding backscatter coefficient. Indeed, fields with higher backscattering generally have lower brightness, and vice versa. The same measured data are directly compared in Fig. 4 where the backscattering coefficient $\sigma^\circ_{\mathrm{is}}$, computed from the measured emissivity for a half space of isotropic dissipative scatterers, is also shown as a reference. As expected the $\sigma^\circ_{\mathrm{is}}$ curve is an upper limit which is not exceeded by experimental σ° values. Besides, data representing bare soil are far from the isotropic curve while points for vegetation are closer to $\sigma^\circ_{\mathrm{is}}$; in particular, sugarbeet plants which have large leaves experience the highest backscattering and the lowest emissivities, while alfalfa (small leaves) behaves in the opposite fashion and corn exibits an intermediate behavior. These results indicate that the simultaneous use of active and passive sensors simplifies the differentiation between vegetation and bare soil.

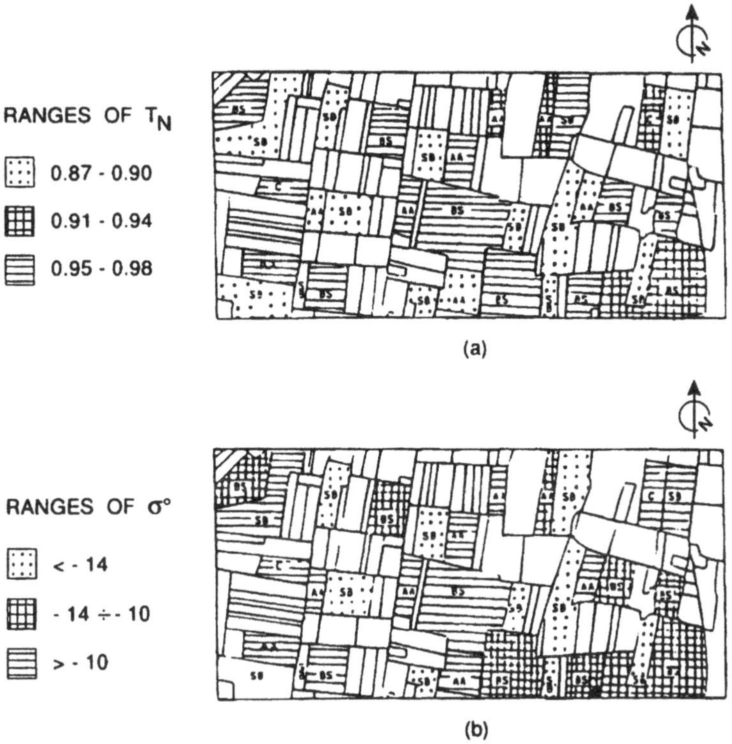

Figure 3. Maps of normalized temperature (a) and backscattering (b) at 10 GHz. AA = alfalfa, SB = sugarbeet, C = corn, BS = bare soil (wheat stubble). (From Ferrazzoli *et al.* [2]).

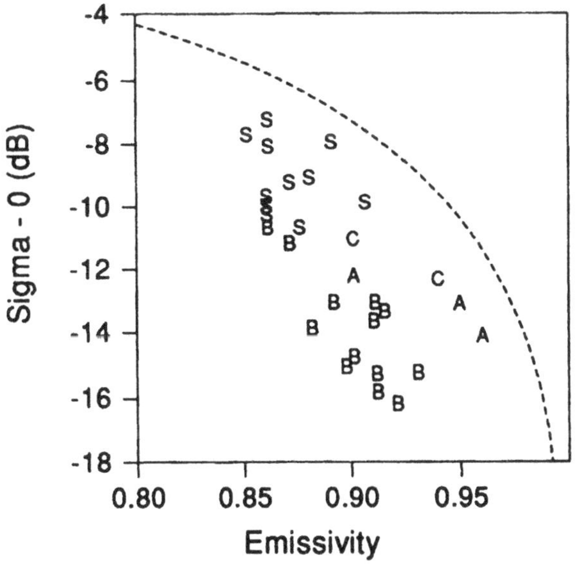

Figure 4. Experimental results of σ° vs. emissivity: B = bare soil, A = alfalfa, C = corn, and S = sugarbeet. The dashed line is the theoretical σ°_{is} versus emissivity. (After Ferrazzoli *et al.* [2]).

2.1.1. Soil moisture. The use of microwave sensors in soil moisture content measurements SMC has been the subject of much research over the last few years. Significant progress has been achieved in uncovering the relationship between microwave emission and backscattering and soil water content, and the best observation parameters have been determined as well as potentials and limitations for practical use. The experiments, performed at the University of Kansas by means of a multifrequency ground based scatterometer, allowed the definition of the best observation parameters for the measurement of SMC [8, 9]. According to these investigations, the highest sensitivity to SMC of a top 5 cm layer is obtained at a frequency $f \approx 4.5$ GHz, H polarization and incidence angle $\theta \approx 10°$. The sensitivity is about 0.15 dB (% of field capacity) on bare soils, and 0.13 dB (% of field capacity) on vegetation covered soils. For these observation parameters the sensitivity to surface roughness, which also influences radar response, is minimized. These results were confirmed by the measurements taken in France with a scatterometer developed by Centre National d'Etudes Spatiales (CNES) [19, 20]. Other measurements performed at the University of Bern from L- to X-band indicated that the correlation between $\sigma°$ and SMC of bare soil is similar to the 4.6 GHz regression line for frequencies from 2.5 to 11 GHz [18].

On the other hand, the potential of microwave radiometers in measuring SMC has been tested in much research performed mostly in the former Soviet Union [10, 11] and in the United States [12]. This research demonstrated that, at frequencies around 1.5 GHz or lower, microwave emission from soil is very well correlated to its moisture content even for vegetated fields [13, 14]. At 1.4 GHz and H polarization, the sensitivity of microwave emission to near surface SMC is about 3.5 K/0.01 $\mathrm{g\,cm^{-3}}$ for smooth bare surfaces and can decrease to about 1 K/0.01 $\mathrm{g\,cm^{-3}}$ due to the effect of surface roughness.

The characteristics of active and passive sensors in measuring soil moisture have been compared by Ulaby *et al.* [21] who pointed out that 'the L-band radiometer yields a significantly lower moisture estimate error at low soil moisture, while the C-band radar is found to be significantly more accurate at soil moisture greater than approximately 70% of field capacity'; the paper concludes that the combination of active and passive microwave sensors is expected to estimate the moisture of vegetation covered soil with errors less than ±30% of true value even under lossy crop canopy (Fig. 5). In more recent work by Oh *et al.* [22], SMC and surface roughness have been retrieved from polarimetric radar observation at L-, C- and X-band, using an empirical model. The result appears quite promising, although, as the authors recognize, additional experiments should be conducted over a wide range of soil conditions before this technique be widely applied.

Another comparison between active and passive systems has been made in a paper by Ferrazzoli *et al.* [17] on the basis of data collected on vegetated and bare soils in four aircraft surveys during the 1988 — AGRISCATT/AGRIRAD airborne Campaign. According to this investigation, scatterometric data at L-band shows a sensitivity of (0.25 db/0.01 $\mathrm{g/cm^3}$) and correlation coefficient $r = 0.84$ in the order of those obtained by Ulaby *et al.* [8] on bare soils. At C-band, the backscattering coefficient, measured at H polarization and 10 degree incidence angle, was found to be correlated to soil moisture with a sensitivity of 0.22 dB/0.01 $\mathrm{g/cm^3}$ and a correlation coefficient $r = 0.67$. The sensitivity is in fair agreement with Ulaby's data [8, 9], while the low value of r could be due to the wide range of surface roughness and to the variety of crop covers. The results obtained, on the same fields, with the L-band channel of Airborne

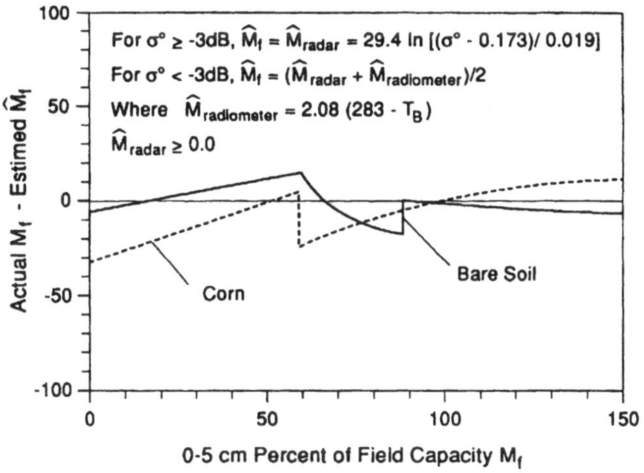

Figure 5. Soil moisture prediction error from a combined C-band radar and L-band radiometer observing bare soil and a mature crop canopy. (After Ulaby *et al.* [21]).

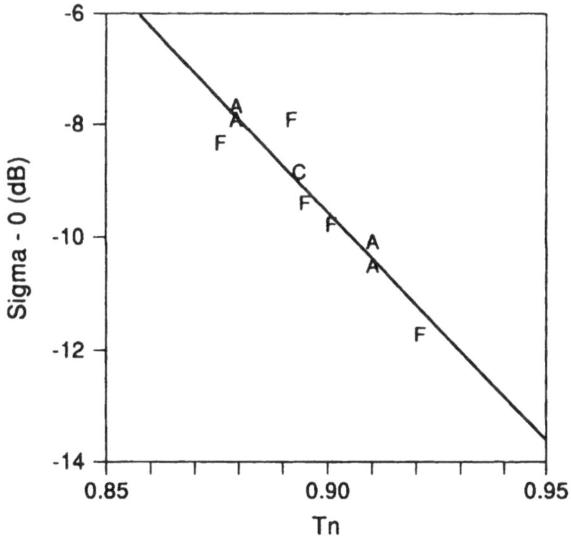

Figure 6. Backscattering coefficient σ° as a function of normalized temperature T_n both measured at L-band and $\theta = 10^\circ$ incidence angle. Figures and letters are the same as in Fig. 4. (After Ferrazzoli *et al.* [2]).

Multi-frequency Passive Sensor, show that the relation between the normalized temperature T_n and the SMC has a correlation coefficient (0.88) slightly higher than that obtained from L-band backscattering data and a sensitivity equal to 0.0023 (emissivity unit/0.01 g/cm^3), which corresponds to about 0.7 K/0.01 g/cm^3 and is slightly lower than the values generally found in literature for vegetated rough terrain. The direct comparison of active and passive systems, carried out by representing σ° as a function of T_n (Fig. 6) shows that, as expected, σ° (in dB) decreases almost linearly as T_n increases. From the slope of the regression line (−82.3 dB/emissivity unity) we find that the radar resolution necessary to obtain the same sensitivity to SMC of a radiometer with a resolution of 1 K, is 0.27 dB. In many practical applications the SMC should be estimated with a precision in the order of ±10% of the measured value (in the range 0.10 and 0.60 g/cm^3) and a resolution of about 0.01 g/cm^3. To achieve this objective an L-band radiometer should have a resolution and an absolute calibration in the order of about 0.7 K and 2 K respectively. At the present state of the art, these specifications can be more easily met than the corresponding figures (0.25 and 0.5 dB, respectively) of an active system at L- or C-band.

In summary, the simultaneous observations of emissivity and backscattering showed consistent values of σ° and T_n and pointed out that, leaving out the problem of ground resolution, the user's specifications in terms of sensitivity and precision can be more easily met by passive sensors. However the combined use of passive and active sensors can be synergistic and allows an improvement in estimation accuracy [21]. The major problem in using passive microwave sensors for SMC measurements seems to be the ground resolution which is in the order of a few tens of meters from airborne sensors and of many kilometers from space. The use of synthetic aperture antennas could reduce this drawback. Another serious restriction in populated areas of Western Europe is the strong man-made interference.

2.1.2. Vegetation biomass. All the plant constituents (stalks, leaves and fruits), influence emission and scattering in different manners depending on their shape, dimension, orientation and water content. An analytic description of the electromagnetic properties of a vegetation layer becomes very difficult, or impossible, to obtain, and even the more sophisticated models, which take into consideration plant structural parameters, must introduce strong simplifications. In the simplest approaches only the total biomass overlaying soil is taken into consideration. This quantity is usually represented by the plant water content PWC (kg/m^2) or by the leaf area index LAI which, during the growth phase of agricultural crops, are correlated to each other.

Investigations of crop canopies with radars, carried out since the 1970s, pointed out that the backscattering coefficient is correlated to plant moisture, structure and dry biomass, and indicate the possibility of using radar for crop monitoring and classification [16, 23–27]. At the same time, emission from vegetation was mainly studied with the intention of evaluating the sensitivity reduction to SMC measurement due to different types of vegetation (classified as small grains, broadleaf and forest) [13, 28]. Other observations of agricultural crops between 1.4 GHz and 36 GHz were able to detect quantitative relations between microwave emission and vegetation features [15, 17, 29]. A significant parameter for monitoring the growth of vegetation [13, 14] was found to be the polarization index (PI),

$$\mathrm{PI} = \frac{2(T_{\mathrm{bv}} - T_{\mathrm{bh}})}{T_{\mathrm{bv}} + T_{\mathrm{bh}}}$$

measured at off nadir observation angle [15, 30, 31].

Experimental data have shown a recurrent different behavior between 'wide leaf' crops such as corn, sugarbeet and sunflower, and 'small leaf' plants such as wheat, barley and alfalfa. Spectra of normalized temperature and backscattering collected during the AGRISCATT/AGRIRAD Campaigns in Central Italy, for two well developed crops representative of two categories are represented in Figs 7(a) and 7(b).

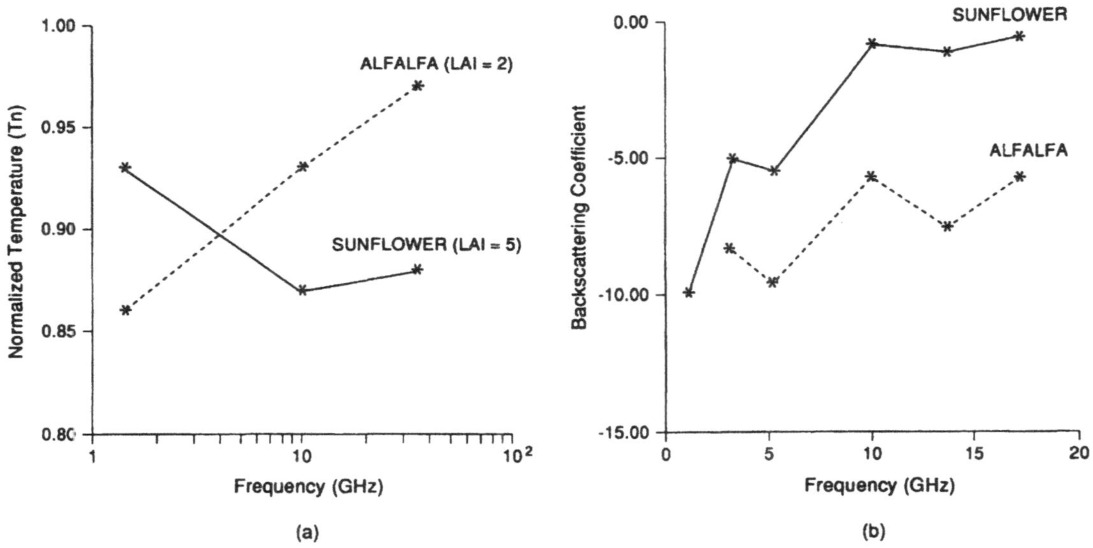

Figure 7. Spectra of normalized temperature (a) and backscattering coefficient (b) of small leaf (alfalfa) and wide leaf (sunflower) crops.

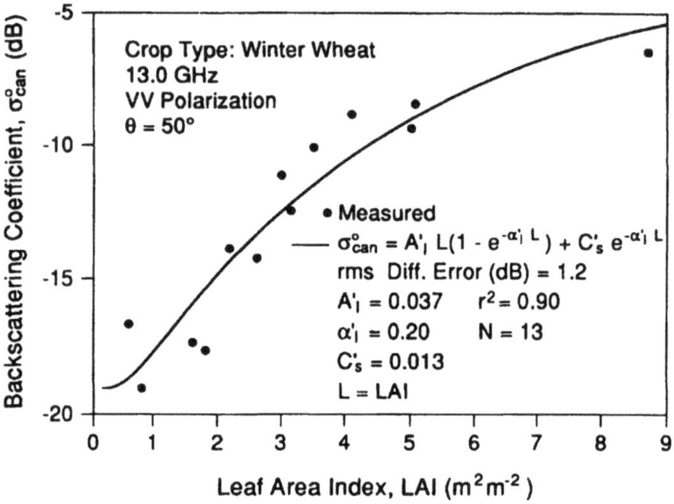

Figure 8. Backscattering coefficient at 13 GHz as a function of leaf area index of wheat. (From Ulaby *et al.* [7]).

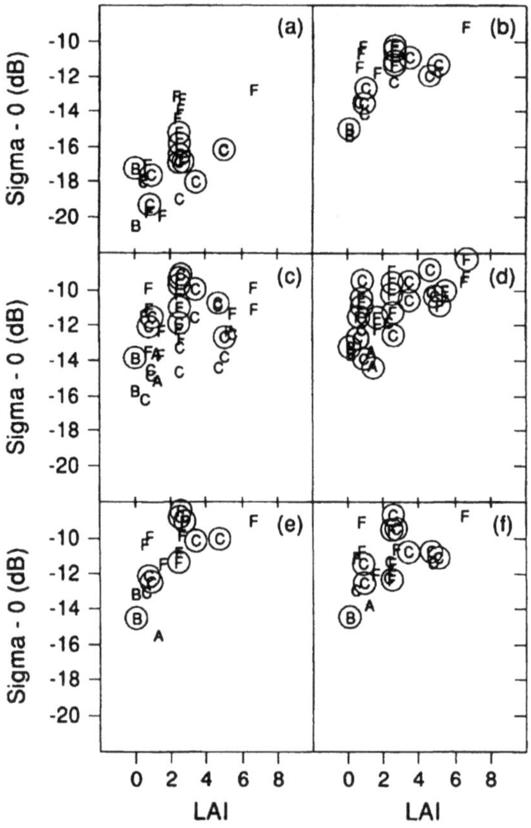

Figure 9. Backscattering coefficient σ° vs. leaf area index LAI. Frequency (GHz): 1.2 (a), 3.2 (b), 5.3 (c), 9.65 (d), 13.7 (e), 17.2 (f). Observation angle $\theta = 40^\circ$, H and V polarization. Crops: corn (C), sunflower (S), alfalfa (A), bare soil (averaged values for LAI $<$ 0.5) (B). (From Ferrazzoli *et al.* [17]).

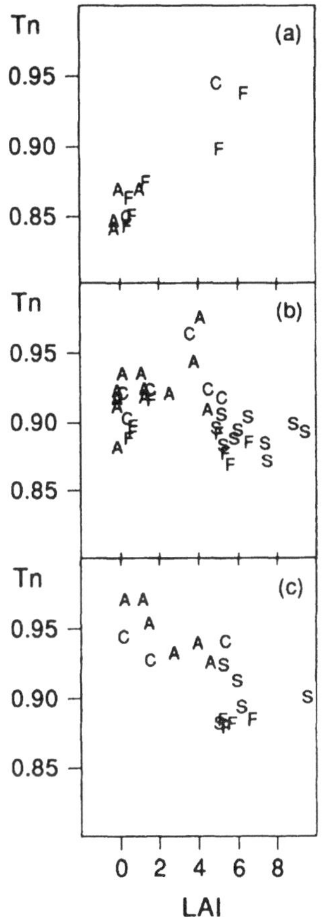

Figure 10. Normalized temperature as a function of leaf area index LAI at 1.4 GHz (L-band) (a), 10 GHz (X-band) (b) and 36 GHz (Ka-band) (c). Crops are the same as in Fig. 8, S = sugarbeet. (From Ferrazzoli *et al.* [17]).

Measurements carried out during the life cycle of crops have shown that as vegetation grows on moist soil, the backscattering at 13 GHz of wheat gradually increases as the crop grows (Fig. 8) [25], whereas 'wide leaf' crops exhibit a rapid saturation of the backscattering coefficient even at low frequency (S- and C-band) [7, 17] (Fig. 9). On the other hand, the normalized temperature T_n of various crops changes with leaf area index LAI in a different fashion according to the observation frequency (Fig. 10) [17, 32]. At 1.4 GHz (H polarization, $\theta = 40°$), T_n of corn and sunflower linearly increases as the vegetation grows, while an opposite behavior (decrease) has been observed at 36 GHz, where leaf area and thickness, in terms of wavelength, are generally large enough to show a relatively high reflectivity at H polarization. At 10 GHz the emissivity of 'small leaf' crops exhibits a trend similar to that observed at 1.4 GHz for wide leaf crops and reaches very high values on mature crops, whereas T_n of well developed 'wide leaf' plants shows a similar behavior to that observed at 36 GHz. The high emissivity of fully grown wheat at X-band is confirmed by data collected by Hüppi *et al.* [18].

A direct comparison of the dependence of active and passive microwave data at 10 GHz on plant water content PWC, is shown in Figs 11(a) and 11(b) which represent

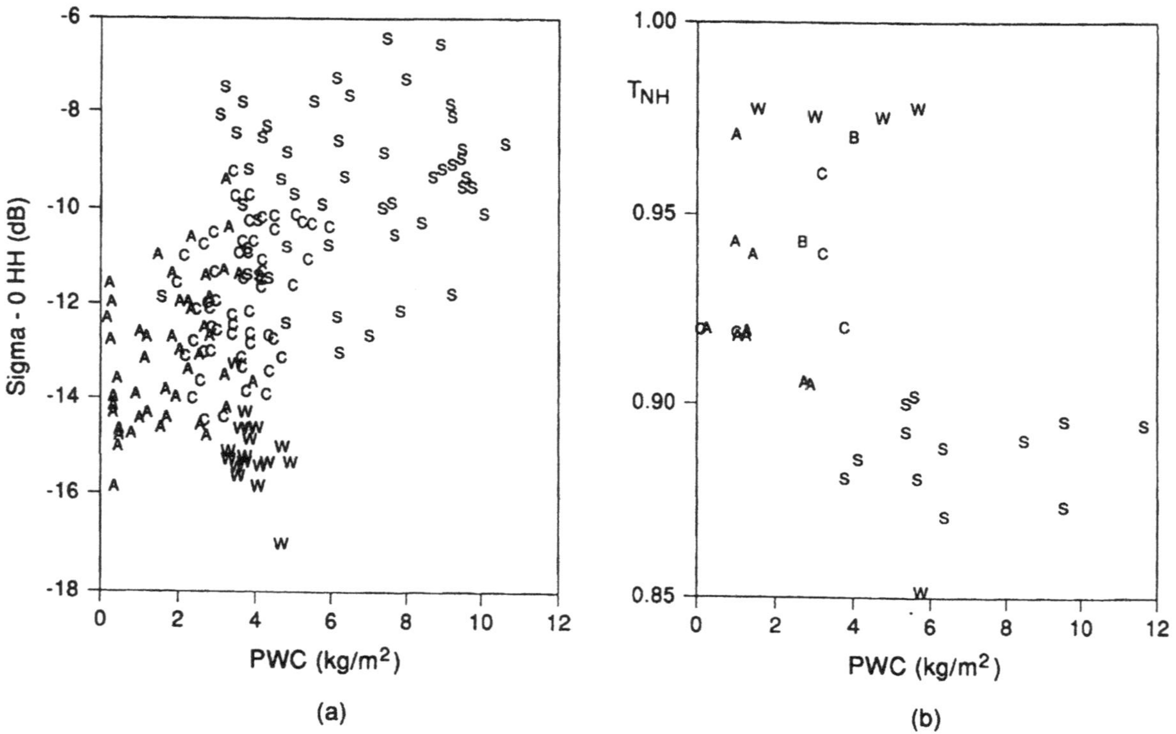

Figure 11. X-band backscattering coefficient σ° ($\theta = 45^\circ$, H polarization) (a), and normalized temperature T_n ($\theta = 40^\circ$, H polarization) (b) as a function of plant water content (PWC). (A = alfalfa, B = barley, C = corn, S = sugarbeet, W = wheat). (After Coppo *et al.* [16]).

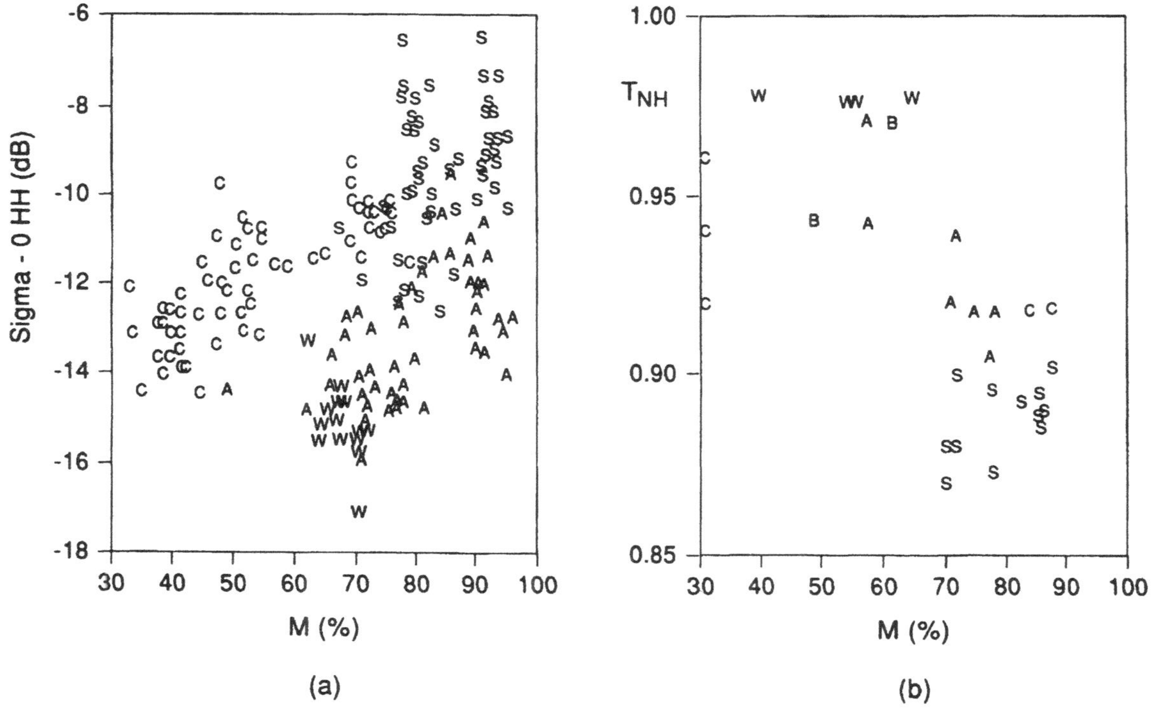

Figure 12. X-band backscattering coefficient σ° ($\theta = 45^\circ$, H polarization) (a), and normalized temperature T_n ($\theta = 40^\circ$, H polarization) (b) as a function of % (by weight) of water in plant. Crop code is the same of Fig. 11. (After Coppo *et al.* [16]).

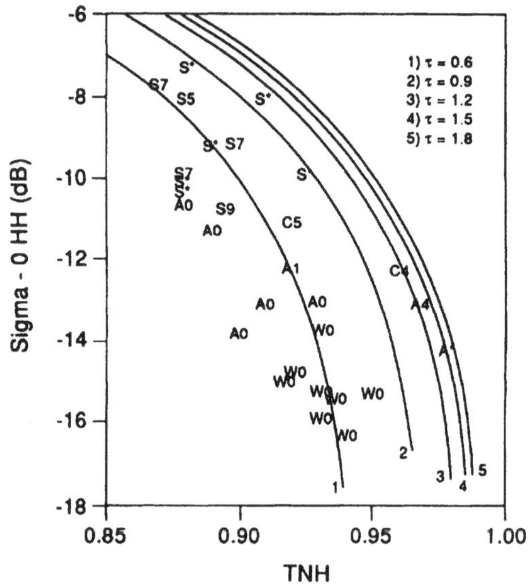

Figure 13. Backscattering coefficient σ° as a function of normalized temperature T_n at X-band ($\theta = 45°$, H polarization). The lines refer to the model of a uniform medium covered by a layer of isotropic scatterers by Tsang *et al.* [4] for a soil reflectivity equal to 0.25 and optical depth τ ranging from 0.6 to 1.8. Crop code is the same as in Fig. 11. Numbers represent rounded values of leaf area index LAI and * denotes well developed crop with unknown LAI. (After Coppo *et al.* [16]).

the backscattering coefficient and the normalized temperature of several crop types simultaneously measured with airborne sensors [16]. Although data are rather widely spread, we see that, in general, σ° increases as PWC increases, but ripe wheat and alfalfa seem to have a contrary behavior. In turn, the general trend of the normalized temperature is to decrease as PWC increases. Also in this case we can detect an opposite trend for 'small leaf' crops; however, the most evident feature is the high difference between T_n of wheat and sugarbeet, while corn and alfalfa have intermediate values depending on the growth stage. Since for fully grown crops and during the vegetation senescence the dry biomass increases at a low rate and vegetation begins to lose water, it appears significant to correlate microwave parameters to the percentage of water in the plants M. Diagrams of Figs 12(a) and 12(b) [16] show that σ° increases and T_n decreases as a function of M. Also in these diagrams we can recognize, for a given M, two ranges of σ° and T_n corresponding respectively to 'wide leaf' and 'small leaf' crops.

On the whole, experimental data show that, in general, the backscattering increases as the normalized temperature decreases; however, a general direct relation between these two parameters has not yet been found. A comparison with a model developed by Tsang *et al.* [4] for a uniform continuous medium (soil) covered by a layer of isotropic scatterers (leaves) show that the model reproduces fairly well the growth of alfalfa, while it is less consistent with data for corn (Fig. 13) [16].

The dependence of emissivity and backscattering on plant water content is fairly well established for the ensemble of crops during their life cycle; however, the sensitivity of both parameters appears quite low for detecting plant water conditions of a single crop in a small time interval.

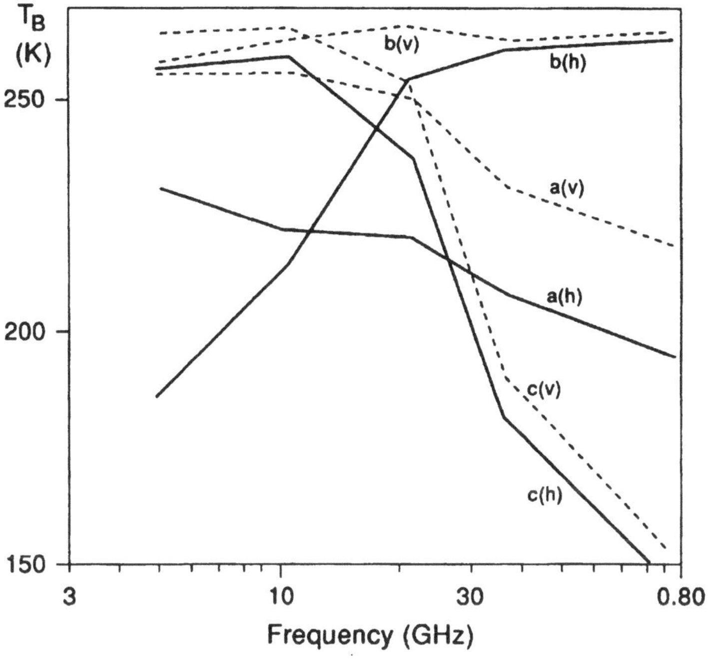

Figure 14. Average spectra of brightness temperatures observed at H and V polarizations at $\theta = 50°$. (a) During high winter condition, snow depths of all measurements normalized to 48 cm water content, (b) wet firn layer (several cm) on the surface of the snow cover, (c) thick crust of refrozen firn. (From Schanda *et al.* [39]).

2.2. *Snow cover*

The capability of active and passive microwave sensors to monitor seasonal variation of snow cover has been the subject of several experimental activities carried out with ground based and satellite systems [33–37]. Moreover, the dielectric characteristics of several snow types have been investigated by means of experiments and theoretical models and data are available at frequencies up to 90 GHz [38].

Measurements carried out between 3 GHz and 90 GHz with passive systems and between 1 GHz and 36 GHz with radars have pointed out the sensitivity of microwave emission and backscattering to snow type and water equivalent. The average spectra of brightness temperature T_b obtained by Schanda *et al.* [39] show that T_b of dry and refrozen snow decreases with frequency whereas T_b of wet spring snow increases (Fig. 14). Correspondingly σ°_{is} of dry snow increases as the frequency increases from 1 to 36 GHz (Fig. 15) [34]. In most cases the spectral behavior of wet snow shows a slight increase in frequency due to the increasing effect of surface roughness [39, 40].

Experiments conducted with ground based sensors in the Alps have shown that, up to X-band, emission and backscatter from a layer of dry snow are mostly influenced by the soil conditions below the snowpack and by snow layering, while at the higher frequencies of the microwave band both parameters appear sensitive to snow type and water equivalent [33, 37]. As an example the Ka-band emissivity of a homogeneous layer of dry snow decreases with increasing depth, and then with the water equivalent (Fig. 16a), while the backscattering at 17 GHz increases (Fig. 16b) [7]. This trend

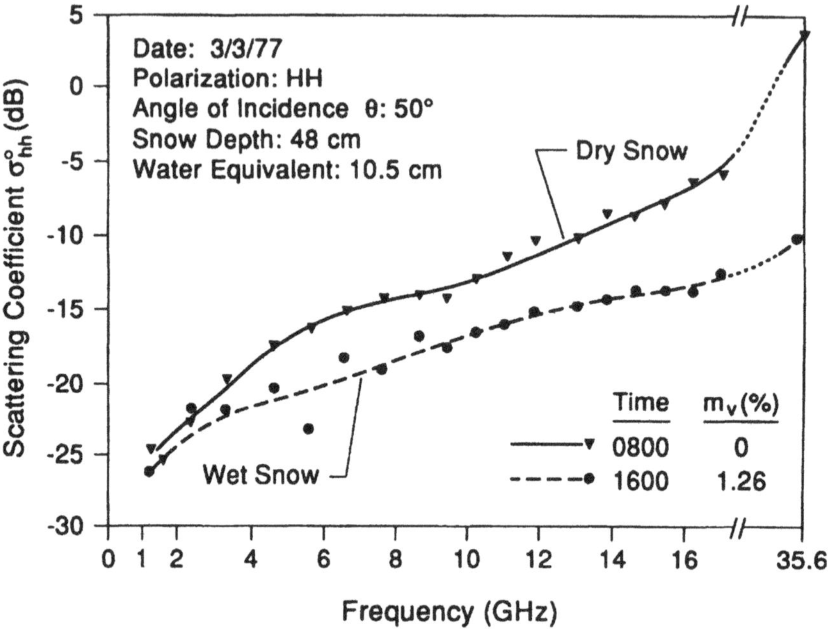

Figure 15. Spectral response of σ^0 for dry and wet snow. (From Stiles and Ulaby [34]).

can be explained considering the role played by volume scattering which increases as the frequency increases. If snow melts, the presence of liquid water in the surface layer determines an increase of emissivity [6] due to transition from volume to surface scattering. In this condition the contrast between the high frequency emissivity of wet snow and snow-free regions tends to be undetectable, in that it is in the range of variation due to soil moisture and vegetation cover. For a homogeneous layer of wet snow, the brightness temperature decreases as a function of liquid water content W. At 6.8 GHz the decrease is almost linear with W, while at higher frequencies a non linear behavior is found with a decreasing sensitiviy to W [41]. On the other hand, scatterometric measurements at 10 GHz of natural snow covers have pointed out the very low attenuation of a dry snow layer, and a contrast of about 10 dB between the backscattering of snow free and wet snow surface (Fig. 17) [42]. Other scatterometric data as well as the analysis of airborne (AIRSAR and E-SAR) and spaceborne (ERS-1) data pointed out that wet snow can be well imaged also at C-band frequencies [40].

A strong effect of snowpack stratification on microwave emission and backscatter from dry Antarctic firn has been detected by means of simultaneous measurements with active and passive systems at X- and C-band and confirmed by satellite SMMR and SMM/I data [37]. According to Rott and Sturm [37]: 'Due to long term stability of the signatures, the interior part of the Antarctica is an excellent target for external calibration and for intercomparison of microwave sensors.'

In summary, simultaneous use of active and passive microwave sensors at appropriate frequencies has a notable performance in snow cover monitoring. According to present knowledge, an optimum snow monitoring system should include a dual polarized radiometer operating at three frequencies (close to 18, 35 and 90 GHz) for

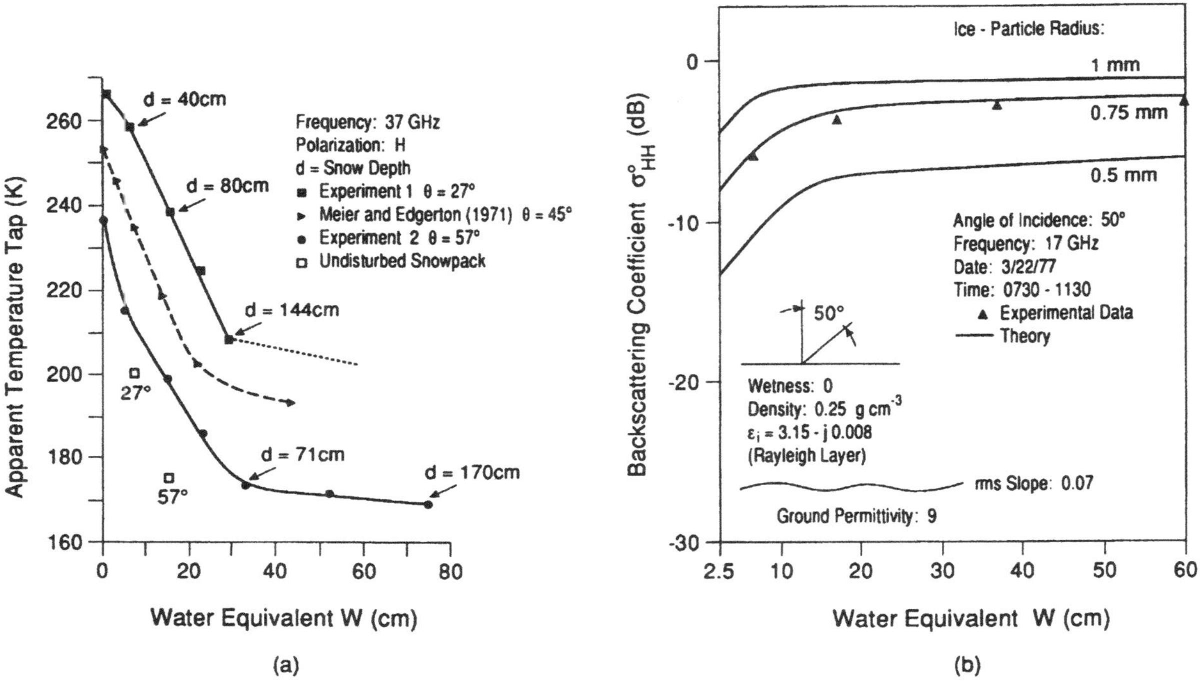

Figure 16. (a) Brightness temperature at 37 GHz, H polarization of dry snow as a function of snow water equivalent (from Ulaby and Stiles [35]). (b) Backscattering of dry snow at 17 GHz vs. water equivalent for various values of ice-particle radius. (From Ulaby et *et al.* [7]).

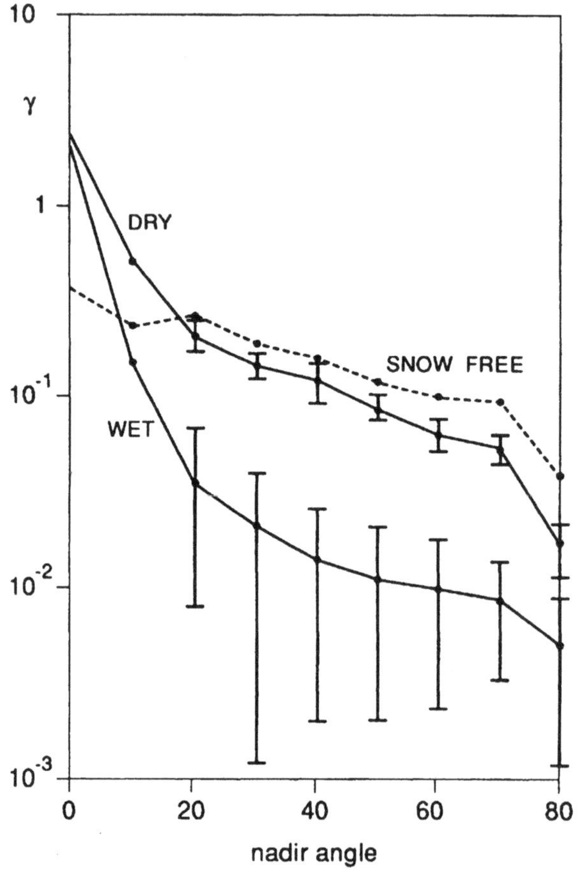

Figure 17. Mean values of backscattering coefficients γ averaged over all wet and dry snow, and over snow-free situations respectively. (From Mätzler and Schanda [42]).

mapping dry snow cover extent and deriving physical properties of dry snow, and a radar operating at a frequency between 5 and 15 GHz (preferably with co- and cross-polarized channels) and incidence angle higher than 20 degrees for imaging wet snow. A dual frequency radar would be more beneficial both for wet snow detection and for supporting microwave radiometry in the determination of the water equivalent of dry snow [36, 40].

3. CONCLUDING REMARKS

The combined use of active and passive sensors in remote sensing of land surfaces is of significant interest in that it can improve the accuracy, the range and the reliability of measurements. In general, the role of imaging radar is to provide unique high resolution imagery with high sensitivity to morphology and a good penetration in arid surfaces. In turn a multifrequency radiometer has a high sensitivity to dielectric properties of natural media and can significantly contribute to monitoring geophysical processes on global scale.

An example of synergism includes use of the complementary properties of active and passive microwave systems to improve the accuracy in soil moisture measurement and in monitoring dry and wet snow cover. In vegetation studies, combined data of emissivity and backscattering are useful in differentiating crop types and in identifying the role played by various plant constituents in the global microwave response and in separating the effects of soil roughness from those of vegetation cover. However, the combined use of high and low-resolution data to study spatial variability of land parameters is not insignificant and requires further investigations. Finally, although emissivity and backscattering, collected in field experiments, are biunivocally related to each other only in particular cases, simultaneous data of emissivity and backscattering can be very useful in improving the understanding of interaction mechanisms between electromagnetic waves and natural media.

Acknowledgements
The author wishes to thank Prof. H. Rott at the University of Innsbruck (Austria), for his revision of the manuscript and very helpful comments and Prof. C. Mätzler at the University of Bern (Switzerland) for his helpful suggestions.

This work has been partially supported by the Italian Space Agency (ASI).

REFERENCES

1. W. H. Peake. Interaction of electromagnetic waves with some natural surfaces. *IRE Trans. Antennas and Propagation*, **AP-7**, S324–S329 (1959).
2. P. Ferrazzoli, G. Luzi, S. Paloscia, P. Pampaloni, G. Schiavon and D. Solimini. Comparison between microwave emissivity and backscatter coefficient of crops. *IEEE Trans. Geosci. Remote Sensing*, **27**, 772–778 (1989).
3. F. T. Ulaby, R. K. Moore and A. K. Fung. *Microwave Remote Sensing. Active and Passive*, Vol. I. Reading, MA, Addison Wesley (1981).
4. L. Tsang, A. J. Blanchard, R. Newton and J. Kong. A simple relation between active and passive remote sensing measurement of earth terrain. *IEEE Trans. Geosci. Remote Sensing*, **GE-20**, 482–485 (1982).
5. J. Kong. Model for microwave remote sensing. In: *Proc. of ESA/NASA Int. Workshop on Passive Microwave Remote Sensing Research Related to Land–Atmosphere Interactions*, Toulouse, January 11–15, 1993 (1993).
6. F. T. Ulaby, R. K. Moore and A. K. Fung. *Microwave Remote Sensing. Active and Passive*, Vol. II. Reading, MA, Addison Wesley (1982).
7. F. T. Ulaby, R. K. Moore and A. K. Fung. *Microwave Remote Sensing. Active and Passive*, Vol. III. Norwood, MA, Artech (1986).
8. F. T. Ulaby, P. P. Batlivala and M. C. Dobson. Microwave backscatter dependence on surface roughness, soil moisture, and soil texture: Part I — bare soil. *IEEE Trans. Geosci. Electr.*, **GE-16**, 286–295 (1978).
9. F. T. Ulaby, G. A. Bradley and M. C. Dobson. Microwave backscatter dependence on surface roughness, soil moisture, and soil texture: Part II — vegetation covered soil. *IEEE Trans. Geosci. Electr.*, **GE-17**, 33–40 (1979).
10. A. Basharinov and A. Shutko. Simulation studies of the SHF radiation characteristics of soils under moist conditions. *NASA Tech. Translation TT*, **F-16** (1975).
11. A. M. Shutko. Microwave radiometry of lands under natural and artificial moistening. *IEEE Trans. Geosci. Remote Sensing*, **GE-20**, 18–26 (1982).
12. T. Schmugge. Remote sensing of soil moisture with microwave radiometers. *Trans. ASAE*, **26**, 748–753 (1983).
13. K. P. Kirdiashev, A. A. Chuklantsev and A. M. Shutko. Microwave radiation of the Earth's surface in the presence of vegetation cover. *Radio Eng. Electron. Phys. Engl. Transl.*, **24**, 256–264 (1979).

14. F. T. Ulaby, M. Razani and C. Dobson. Effects of vegetation cover on the microwave radiometric sensitivity to soil moisture. *IEEE Trans. Geosci. Remote Sensing*, **GE-21**, 51–61 (1983).

15. P. Pampaloni and S. Paloscia. Experimental relationships between microwave emission and vegetation features. *Int. J. Rem. Sensing*, **6** (2), 315–323 (1985).

16. P. Coppo, P. Ferrazzoli, G. Luzi, P. Pampaloni, G. Schiavon and D. Solimini. Microwave remote sensing of land surfaces with airborne active and passive sensors. In: *Proc. of International Geoscience and Remote Sensing Symposium (IGARSS'89)*, pp. 806–809, Vancouver, Canada (1989).

17. P. Ferrazzoli, S. Paloscia, P. Pampaloni, G. Schiavon, D. Solimini and P. Coppo. Sensitivity of microwave measurements to vegetation biomass and soil moisture content: a case study. *IEEE Trans. Geosci. Remote Sensing*, **30**, 750–756 (1992).

18. R. Hüppi, E. Stotzer and E. Schanda. Calibrated microwave signature measurements of soil and wheat. In: *Proc. 3rd Intern. Coll. on Spectral Signatures of Ojects in Remote Sensing*, pp. 351–355, Les Arcs, France. ESA SP-247 (1985).

19. T. Le Toan. Active microwave signatures of soil and crops: significant results of three years of experiments. In: *Proc. of International Geoscience and Remote Sensing Symposium*, Munich, FRG, TP2 3.1–3.5 (1982).

20. R. Bernard, O. Taconet, D. Vidal Majar, J. Thony, M. Vauclin, A. Chapoton, F. Wattrelot and A. Lebrun. Comparison of three '*in situ*' surface soil moisture measurements and application to C-band scatterometer calibration. *IEEEG Trans. Geosci. Remote Sensing*, **GE-22**, 388–394 (1984).

21. F. T. Ulaby, M. C. Dobson and D. R. Brunfeldt. Improvement of moisture estimation accuracy of vegetation-covered soil by combined active/passive microwave remote sensing. *IEEE Trans. Geosci. Remote Sensing*, **GE-21**, 300–307 (1983).

22. Y. Oh, K. Sarabandi and F. T. Ulaby. An empirical model and an inversion technique for radar scattering from bare soil surfaces. *IEEE Trans. Geosci. Remote Sensing*, **GE-30**, 370–382 (1992).

23. F. T. Ulaby, T. F. Bush and P. P. Batlivala. Radar response to vegetation II: 8–18 GHz band. *IEEE Trans. Ant. Prop.*, **AP-23**, 608–618 (1975).

24. F. T. Ulaby and T. F. Bush. Corn growth as monitored by radar. *IEEE Trans. Ant. Prop.*, **AP-24**, 819–828 (1976).

25. F. T. Ulaby, C. T. Allen and G. Eger III. Relating the microwave backscattering coefficient to leaf area index. *Remote Sens. Environ.*, **14**, 113–133 (1984).

26. J. F. Paris. The effect of leaf size on the microwave backscattering by corn. *Remote Sens. Environ.*, **19**, 81–95 (1986).

27. T. Le Toan, H. Laur and E. Mougin. Multitemporal and dual-polarization observations of agricultural vegetation covers by X-band SAR Images. *IEEE Trans. Geosci. Remote Sensing*, **GE-27**, 709–718 (1989).

28. T. Jackson, T. Schmugge and J. Wang. Passive microwave remote sensing of soil moisture under vegetation canopies. *Water Resources Res.*, **18**, 1137–1142 (1982).

29. P. Pampaloni, S. Paloscia and G. Zipoli. Microwave emission of soil and vegetation at X- and Ka-band. In: *Proc. of International Geoscience and Remote Sensing Symposium (IGARSS'83)*, San Francisco, CA (1983).

30. S. Paloscia and P. Pampaloni. Microwave polarization index for monitoring vegetation growth. *IEEE Trans. Geosci. Remote Sensing*, **GE-26** (5), 617–621 (1988).

31. P. Ferrazzoli, L. Guerriero, G. Luzi, S. Paloscia, P. Pampaloni and D. Solimini. Modeling polarization properties of emission from vegetated terrains. *IEEE Trans. Geosci. Remote Sensing* (to be published).

32. P. Pampaloni. Microwave radiometry of soil and vegetation. In: *Proc. of Specialist Meeting on Microwave Radiometry and Remote Sensing Applications, Boulder, CO*, E. Westwater (Ed.), pp. 42–49 (1992).

33. R. Hofer and C. Mätzler. Investigation of snow parameters by radiometry in the 3- to 60-mm wavelength region. *J. Geophys. Res.*, **85**, 453–460 (1980).

34. W. H. Stiles and F. T. Ulaby. The active and passive microwave response to snow parameters, 1: Wetness. *J. Geophys. Res.*, **85**, 1037–1059 (1980).

35. F. T. Ulaby and W. H. Stiles. The active and passive microwave response to snow parameters, 2: Water equivalent. *J. Geophys. Res.*, **85**, 1045–1049 (1980).

36. C. Mätzler, H. Schanda and W. Good. Toward the definition of optimum sensor specification for microwave remote sensing of snow. *IEEE Trans Geosci. Remote Sensing*, **GE-20**, 57–66 (1982).

37. H. Rott and K. Sturm. Microwave signature measurements of antarctic and alpine snow. In: *Proc. of 11th EARSeL Symposium*, pp. 140–151, Graz, Austria (1991).

38. M. T. Hallikainen, F. T. Ulaby and T. E. Van Deventer. Extinction behavior of dry snow in the 18- to 90-GHz range. *IEEE Trans Geosci. Remote Sensing*, **GE-25**, 737–745 (1987).

39. E. Schanda, C. Mätzler and K. Künzi. Microwave remote ensing of snow cover. *Int. J. Rem. Sensing*, **4**, 149–158 (1983).

40. H. Rott. *Private Communication.* (1993).

41. C. Mätzler. Microwave signatures of landscape in winter. Submitted to *Meteorology and Atmospheric Physics' special issue on Physical Retrieval of Hydrological Variables from Spacebased Microwave Measurements'.* (1993).

42. C. Mätzler and E. Schanda. Snow mapping with active microwave sensors. *Int. J. Rem. Sensing*, **5**, 409–422 (1984).

ESA/NASA International Workshop, pp. 117–154
B. J. Choudhury, Y. H. Kerr, E. G. Njoku and P. Pampaloni (Eds)

Theory of microwave scattering from vegetation based on the collective scattering effects of discrete scatterers

LEUNG TSANG,[1] JIN AU KONG,[2] ZHENGXIAO CHEN,[1] KYUNG PAK[1] and CHIH HSU[2]

[1]*University of Washington, Seattle, WA 98195, USA*
[2]*Massachusetts Institute of Technology, Cambridge, MA 02139, USA*

Abstract—In a vegetation canopy, the branches and leaves occur in random clusters rather than in a uniform random distribution. In this paper, we study the collective scattering effects of a conglomeration of particles. The radiative transfer theory is modified by defining the phase matrix and extinction coefficient as, respectively, the bistatic cross section per unit volume of space and the extinction cross section per unit volume of space. The limit of the volume of space is taken such that it is much larger than a wavelength and contains many particles so that the collective scattering effects of the particles within the volume are taken into account. The results between this new version of radiative transfer theory and the classical theory will be compared. We first study the case of point scatterers where the exact solutions of Monte Carlo simulations can be readily computed based on the Foldy–Lax self-consistent multiple scattering equations. We also compare the iterative approach with that of the exact solution. The iterative approach of wave scattering including the collective effects is applied to passive microwave remote sensing of vegetation.

1. INTRODUCTION

Classical vector radiative transfer theory (RT) has been used extensively in studying the electromagnetic scattering effects in microwave remote sensing [1–6]. However, classical radiative transfer theory assumes that the particles scatter independently. This assumption is based on the random phase of scattering by different particles and is valid if the particle positions are independent and the randomness of relative positions is comparable to or larger than a wavelength. Such an assumption can be invalid for microwave scattering by terrain. For example, in dense media with particles closely packed together as in grains in snow and ice, and for scatterers clustered together as in branches and leaves in a vegetation canopy, the randomness of relative positions of scatterers can be less than a wavelength. In such a case, the scatterers scatter collectively. Collective scattering effects include correlated scattering, taking into account the relative phase of scattered waves from the scatterers and their neighbors. The mutual coherent wave interaction between scatterers is also to be included [7–9].

We first review the theoretical approaches and particularly Monte Carlo simulations that have been applied to scattering of waves by random rough surface and discrete scatterers and then discuss how the approach is used in this paper for vegetation scattering.

1.1. Small perturbation method and Kirchhoff approximation for rough surface scattering

These are classical analytic methods for rough surface scattering and are described in references 1 and 10. However, they are limited in regimes of validity and also do not treat composite rough surface. In rough surface scattering, an important geometric characterization is a composite surface which is a rough surface with different scales of roughness. Electromagnetic wave scattering at different frequencies can respond to different roughness scales in a composite rough surface.

1.2. Vector radiative transfer theory (VRT)

The classical radiative transfer theory assumes that the particles scatter independently. The vector radiative transfer theory assumes the incoherent addition of Stokes parameters and generalizes the scalar theory of radiative transfer to include extinction matrix [1, 11, 12] and emission vector [1, 6, 13]. The method has found extensive applications in scattering by vegetation, snow, and ice [2–5].

1.3. Analytic wave theory and approximations

Analytic wave theory starts with Maxwell's equations and takes ensemble averages based on the statistics of the position, sizes, and concentrations of the scatterers. These result in exact equations of Dyson's equation and Bethe-Salpeter equation respectively for the first and second moment of the fields. However, to solve the moment equations, approximations have been made such as the Foldy's approximation, quasicrystalline approximation, coherent potential approximation, ladder approximation, cyclical approximation etc. [1]. Since these approximations start with field (Maxwell's) equations, the effects of correlated scattering from different scatterers and the mutual coherent wave interactions between scatterers can be included in the analytic approximations. In studying coherent wave interaction among scatterers, propagation of wave from one scatterer to another is based on dyadic Green's function which takes into account near field, intermediate field, and far field interactions. For the dense media case, the dense media radiative transfer theory [7-9] has been developed from such an approach.

1.4. Monte Carlo simulations of solutions of Maxwell's equations (MCME)

The interactions of electromagnetic waves with randomly inhomogeneous media are completely governed by Maxwell's equations. Thus, if one can have a numerical solution of Maxwell's equations, the electromagnetic interaction problem is completely solved. However, in numerical solutions of Maxwell's equations for random media, the positions and characteristic of the scatterers are randomly generated according to some prescribed statistics. In a numerical solution of Maxwell's equations there must be a large number of scatterers for volume scattering, and there must be many correlation lengths in rough surface scattering. The process is to be repeated for many statistical ensembles (realizations) and the results are then averaged. Since this is an approach involving numerical solution of Maxwell's equations, we call these solution Monte Carlo simulations of solutions of Maxwell's equations (MCME) to distinguish them from Monte Carlo simulations of solutions of photon transport equation [14, 15]. With the advent of modern computers, and the development of efficient numerical methods in computational electromagnetics, such Monte Carlo simulations have become increasingly possible. There are several advantages of Monte Carlo simulations. Firstly, other methods as listed in sections 1.1, 1.2 and 1.3 all involve approximations that are often hard to justify. In field experiments, there are usually many parameters that are not well characterized. For rough surfaces, there are rms heights and correlation functions. For discrete scatterers, there are sizes, shapes, concentrations, and permittivities of scatterers. It is often possible (though a tedious process) to adjust the parameters to fit the experimental data with the theoretical models. On the other hand, MCME are solutions of Maxwell's equations without adjustable parameters. Thus, it is a challenge for results of analytic approximations to compare well with results of MCME. Secondly, there are physical situations that are so complicated that analytic approximations become too involved and only MCME can give correct solutions. Thirdly, since MCME directly computes fields, it can directly determine the statistics of the fluctuations of amplitudes, intensities, phases, etc. while analytic theory only gives the averaged fields, intensities etc.

Since Monte Carlo simulations play such an important role, it is important that they are calculated correctly. As in numerical solutions of deterministic problems of Maxwell's equations, one needs to have convergence tests with respect to the number of sampling points, energy conservation etc. However, there are at least two additional convergence tests that are unique to random media problems that are not in deterministic problems. Firstly, MCME must obey convergence with respect to sample size. Note that for random media scattering the scattering properties are normalized by area or volume. For rough surface scattering, the bistatic scattering coefficient is scattering cross section per unit area. Thus, the results must converge with respect to surface area or surface length (1-dimensional surface). Suppose the correlation length of rough surface is 1 wavelength, the rms height is 0.5 wavelength, and the incident angle is 10°. Usually a surface length of 30 wavelengths is used. However, it is only sufficiently large provided that one can show that a surface length of more than 30 wavelengths (say 40 wavelengths) will give the same averaged bistatic scattering coefficient. For volume scattering problems, one has to show convergence with respect to the number of scatterers. Thus, if $N = 100$ scatterers seem to be sufficiently large, then one has to show that while keeping the same density of scatterers, more than 100 scatterers (say 200) must give the same averaged electromagnetic scattering result. Secondly, the electromagnetic scattering computations are repeated for many realizations and the results are averaged over realizations. Thus, the computed results must show convergence with respect to the number of realizations.

In the following, we highlight the progress in MCME of rough surface and random discrete scatterers.

1.4.1. MCME of random rough surfaces. MCME of random rough surface based on method of moments have been done since 1978 [16, 17]. Much of the work is restricted to one-dimensional random rough surface with surface length (sample size) of about 25 wavelengths to 40 wavelengths. A banded matrix iterative approach has been used recently that is computationally more efficient than full matrix inversion [18, 19]. The method makes possible the solution of large-scale rough-surface scattering problem with large surface length up to 400 wavelengths with all the coherent wave interactions included. Based on the large surface length, scattering of wave at large incident angles can be studied. The case of composite rough surface with very different roughness scales can be studied. It is shown that the small scale roughness rather than the large scale roughness can dominate the scattering. Recently, the banded matrix iterative approach has been generalized to the sparse-matrix flat-surface iterative approach for 2-dimensional random surface simulations [20]. Results based on MCME generally yield results very different from that of classical methods of (1.1) for cases outside the domain of validity of these classical methods.

1.4.2. MCME of random discrete scatterers. In volume scattering by discrete scatterers, the extinction coefficient and phase function are intrinsic scattering properties of the scattering media and are respectively defined by total cross section per unit volume and bistatic scattering cross section per unit volume. MCME has been performed by using Foldy–Lax self-consistent multiple scattering equations of media containing dielectric spheres that are densely distributed [21]. Convergence with respect to the sample size is demonstrated by showing that the case of 4000 spheres yields the same averaged result as that of 2000 spheres with the same density of scatterers. The MCME computed extinction coefficient agrees very well with that of quasicrystalline approximation with coherent potential of dense media scattering [21]. Recently, the results have also been verified by using a fast computational algorithm known as recursive aggregate T-matrix [22].

In this paper, we study electromagnetic scattering in media with sparse concentration of scatterers with clustering structures. In vegetation, the branches and leaves occur in clusters rather than being distributed randomly uniform in the canopy. The collective scattering effects have been studied with a single scattering approximation. Data matching of scattering from soybeans has also been done with the actual remote sensing case [23]. In this paper, we study the collective scattering effects of a conglomeration of particles in clusters. The vector radiative transfer theory with extinction matrix and emission vector was first developed in the early 1980's [1, 11–13] and it has taken many years for it to be numerically implemented and applied to data matching in vegetation. Thus, the emphasis of this paper is on the fundamentals of collective scattering radiative transfer theory rather than on specific application because the subject of coherent wave multiple scattering from random clusters in vegetation is still at an early stage. Two approaches are used. Firstly, we study Monte Carlo simulations based on point scatterers occurring in clusters. Secondly, we study the collective effects by using a new version of vector radiative transfer theory known as the collective radiative transfer theory. Classically [24], the phase matrix is defined as the product of the number of particles per unit volume and the averaged single particle bistatic cross section. The extinction coefficient is defined as the product of the number of particles per unit volume and the total averaged cross section of a single particle. Both are defined in terms of single particle properties. In Section 2, the radiative transfer theory is modified by redefining the phase matrix and extinction coefficient as, respectively, the bistatic cross section per unit volume of space and the extinction cross section per unit volume of space. The limit of the volume of space is taken such that it is much larger than a wavelength and contains many particles so that the collective and coherent multiple scattering effects of many particles within the volume are taken into account. In Section 3, we test the concept of collective scattering by studying the case of point scatterers. The advantage of using point scatterers is that the exact solutions of Monte Carlo simulations can be readily computed based on the Foldy–Lax self-consistent multiple scattering equations. We compare the case when point scatterers are uniformly distributed and the case when the point scatterers are in clusters. We also describe the iterative approach. It is shown that the first-order solution of the scattering amplitude of the collective scattering of N particles gives valid results for the phase matrix. However, the forward scattering theorem or Foldy's approximation is only applicable if the N particle scattering amplitude is calculated to the second order to obtain the correct extinction rates. It is also shown that the coherent wave gives a delta function in the forward direction in the limit of a large volume and large number of scatterers. Thus, the coherent field has to be subtracted to give a useful definition of phase matrix which describes the scattering of energy away from the forward direction and the extinction coefficient which describes the decay of the wave in the forward direction. In Section 4, we apply the iterative approach to calculate the phase matrix including the collective scattering effects. In Section 5, we perform Monte Carlo simulations by solving wave equations and averaging over many realizations. We compare the results of the exact solution and the iterative approach. It is shown that the iterative approach gives the correct solution for the averaged scattering quantities implying that higher order scattering contributions tend to cancel out on averaging over realizations. In Section 6, the phase matrix for collective scattering is derived and in Section 7, the vector radiative transfer theory with collective scattering is formulated. In Section 8, the results are applied to hypothetical soybeans case and the numerical results are illustrated.

2. RADIATIVE TRANSFER THEORY OF COLLECTIVE SCATTERING BEHAVIOR OF PARTICLES (RTC)

In this section, we define the radiative transfer theory that includes the collective scattering behavior of particles. This distinguishes the new definition from the conventional definition of single particle scattering behavior.

In radiative transfer theory [24], we consider a specific intensity $\bar{I}(\theta, \phi)$ incident upon an elemental volume with cross sectional area A and length ds (Fig. 1). The

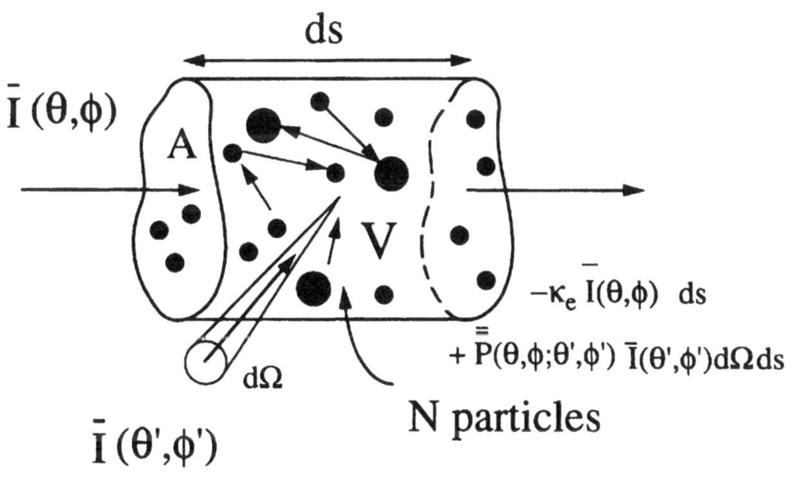

Figure 1. An elemental volume element (V) with arbitrary cross-section from a random medium containing discrete scatterers. The volume element contains many particles. Coherent wave interaction takes place among the particles within volume V. The coherent wave interaction inside V may include near field, intermediate field, and far field ranges. The ensemble average result yield the extinction coefficient and phase matrix in radiative transfer theory, both of which are defined as scattering per unit volume.

cross sectional shape is arbitrary. It is an elemental volume in the development of the radiative transfer equation. It is not small in the sense of the wave equation. Let volume $V = Ads$ obey the following criteria

(i) $V \gg \lambda^3$, where λ is the wavelength;

(ii) there is a large number of particles N, contained in V; the particles are randomly positioned.

Criteria (i) and (ii) are necessary so that the coherent multiple scattering in volume V will have phase fluctuations so that the energy transfer in and out of volume V will become incoherent. Furthermore, if the particles are of different sizes, shapes, and permittivities in volume V, averaging of particle scattering behavior will also be included in the radiative energy transfer within volume V. Criteria (i) and (ii) establish a lower limit of volume V.

Let E_i be the incident field in direction $\hat{k}_i = (\theta_i, \phi_i)$ and E_s be the scattered field from volume V of N particles.

$$E_i = \exp\left(i\vec{k}_i \cdot \vec{r}\right). \tag{1}$$

The scattered field can be decomposed into coherent and incoherent fields.

$$E_s = \langle E_s \rangle + \mathcal{E}_s$$

The scattered field E_s is a statistical field that contains the coherent multiple scattering within volume V and the statistics of E_s will be studied.

Inside volume V, coherent wave interaction among the particles take place. Such coherent wave interaction takes into account near field, intermediate field, and far field ranges. The ensemble averaged results yield extinction coefficients and phase matrices for radiative transfer theory which thus include collective scattering effects. Note that both extinction coefficients and phase matrices are defined as quantities pertaining to scattering per unit volume. Thus, they are not single particle quantities. Nor are they defined as scattering from one particle to another single particle nor to an 'adjacent' particle. The extinction coefficients and phase matrices are also defined in the limit of large V. Since a limit needs to exist for these quantities to be well defined, radiative transfer theory is interpreted as incoherent interaction among large volumes V (Fig. 2). The volume V is arbitrary in shape, and edge effects of the volume for particles at the edges of volume V are not important. Otherwise, the large V limit for extinction coefficients and phase matrices does not exist. Radiative transfer theory is regarded as incoherent wave interaction among volume elements. Near field effects between different volume elements are not important, otherwise the edge effects of volume elements become important and the extinction coefficient and phase matrix would not exist (Note that near field effects are important within the volume element, the effects of which have to be included in coherent wave interaction with the volume element as noted previously). Alternatively, one has to make the volume element large enough until near field interaction between volume elements is not important.

From one elemental volume V to most other elemental volumes, the far field of E_s can be taken. We have

$$E_s = F\left(\hat{k}_s, \hat{k}_i\right) \frac{\exp(ikr)}{r} \tag{3}$$

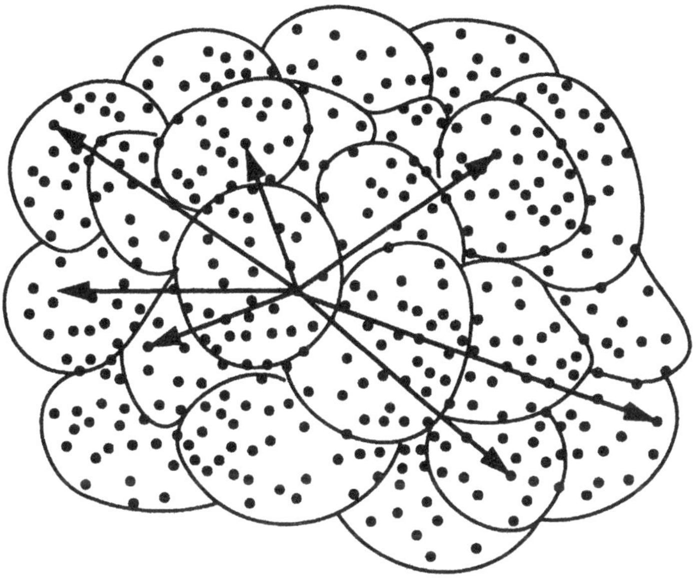

Figure 2. A random medium consisting of many large volume elements defined by arbitrary boundaries as shown. Radiative transfer theory is regarded as incoherent wave interaction among volume elements. Volume elements are arbitrary in shape and are much larger than λ^3. They also contain many particles. Phase matrix is scattering from one volume element to another volume element.

where $\hat{k}_s = (\theta_s, \phi_s)$ is the scattered direction. In (3), $F(\hat{k}_s, \hat{k}_i)$ represents the N particle collective scattering amplitude and includes all the coherent near field interactions among the N particles. It is to be calculated using the Foldy–Lax self-consistent multiple scattering equations of waves. It can be decomposed into coherent and incoherent components.

$$F(\hat{k}_s, \hat{k}_i) = \langle F(\hat{k}_s, \hat{k}_i) \rangle + \mathcal{F}(\hat{k}_s, \hat{k}_i). \tag{4}$$

The optical theorem [1] can be applied to N scatterers which can be viewed as a unit. Thus, for the case of non-absorptive scatterers, we have,

$$\frac{4\pi}{k} \operatorname{Im}\{F(\hat{k}_i, \hat{k}_i)\} = \int_{4\pi} d\Omega_s |F(\hat{k}_s, \hat{k}_i)|^2. \tag{5}$$

The phase function $P(\hat{k}_s, \hat{k}_i)$ and the scattering coefficient κ_s are defined as follows to take into account the collective scattering behavior of N particles.

$$P(\hat{k}_s, \hat{k}_i) = \lim_{V \to \text{large}} \frac{\langle |\mathcal{F}(\hat{k}_s, \hat{k}_i)|^2 \rangle}{V} \tag{6}$$

$$\kappa_s = \int_{4\pi} d\Omega_s P(\hat{k}_s, \hat{k}_i). \tag{7}$$

The limit in Equation (6) is taken for a fixed n_0 where $n_0 = N/V$ is the number of particles per unit volume. The results of (6) and (7) should also be independent of the shape of volume V. Thus, the phase function and the scattering coefficient are defined as per unit volume quantities in the limit of large volume instead of single particle quantities. Note that the limit with large N and large V is taken for convergence. The coherent field is excluded from equations. We will see later that the coherent field approaches a delta function in the forward direction for large V. It also depends on the shape of volume V. Coherent forward scattering does not contribute to the phase function nor the scattering coefficient. For the special case of independent scattering, the definitions of (6) and (7) will also be shown to agree with the conventional theory. The absorption coefficient is then

$$\kappa_{\mathrm{a}} = \frac{W_{\mathrm{a}}}{Q_{\mathrm{i}}} \tag{8}$$

where W_{a} is the power absorbed by the N particles in volume V and Q_{i} is the incident flux. Thus, the absorption coefficient is the absorption cross section per unit volume in the limit of large V. The extinction coefficient is

$$\kappa_{\mathrm{e}} = \kappa_{\mathrm{a}} + \kappa_{\mathrm{s}}. \tag{9}$$

Once the phase matrix and the extinction coefficients are calculated, the radiative transfer equation assumes the following standard form

$$\frac{\mathrm{d}}{\mathrm{d}s} I(\vec{r}, \hat{s}) = -\kappa_{\mathrm{e}} I(\vec{r}, \hat{s}) + \int_{4\pi} \mathrm{d}\Omega' P(\hat{s}, \hat{s}') I(\vec{r}, \hat{s}) \tag{10}$$

where $I(\vec{r}, \hat{s})$ denotes the specific intensity.

To establish the upper limit of V for the definition of the phase function and the scattering coefficient, we note that if V is too large, then the definition of (6) will include the effects described in (10), and is actually the bistatic cross section of the overall medium (rather than the differential volume V in the radiative transfer equation) with multiple coherent and incoherent scattering included. Thus, the upper limit of V is that multiple scattering of incoherent waves will be excluded. Multiple scattering of incoherent waves will take place in the distance scale of the mean free path l_{mfp}

$$l_{\mathrm{mfp}} = \frac{1}{\kappa_{\mathrm{e}}}. \tag{11}$$

Thus, the third criterion of V is an upper limit and is

(iii) $V \ll \left(\frac{1}{\kappa_{\mathrm{e}}}\right)^3$.

The volume V can simultaneously satisfy criteria (i) and (iii) if the mean free path is much larger than the wavelength or the wavenumber is much larger than the extinction coefficient.

$$l_{\mathrm{mfp}} \gg \lambda \tag{12}$$

$$k \gg \kappa_{\mathrm{e}}. \tag{13}$$

The conditions as exhibited in (12) and (13) are actually the conditions for transport-type equations to be valid. When the condition is violated so that $l_{\mathrm{mfp}} \simeq \lambda$ (known as the Ioffe–Regel criterion [25, 26]), strong photon localization may take place and radiative transfer type equations will fail. For remote sensing of geophysical media, conditions of (12) and (13) are usually obeyed. In the following sections, we derive radiative transfer equations with collective particle scattering behavior included (RTC). However, even though (12) and (13) are obeyed, there are still cases when backscattering enhancement (weak photon localization) will take place which will give a maximum of 3 dB difference in the backscattering direction [27, 28]. Backscattering enhancement is a result of cyclical scattering diagrams and is not included in transfer equations which only include ladder diagrams [1, 27, 28]. For the case of independent scattering of N scatterers, $\langle|\mathcal{F}(\hat{k}_s, \hat{k}_i)|^2\rangle = N|f|^2$ so that $P(\hat{k}_s, \hat{k}_i) = n_0|f|^2$, where $n_0 = N/V$ is the number of particles per unit volume and f is the single particle scattering amplitude. Thus, (6) and (7) reduce to the result of conventional radiative transfer theory. In this section, we have introduced the definitions of phase matrix and extinction coefficients. The roles of near field and intermediate field effects are discussed. The size of the elemental volume in radiative transfer is also defined. In the following section, we illustrate the applications of the new definitions to point scatterers that are uniformly distributed.

3. PHASE FUNCTION AND EXTINCTION COEFFICIENT OF POINT SCATTERERS RANDOMLY DISTRIBUTED IN VOLUME V

In this section, we illustrate the phase function and the extinction coefficient for nonabsorptive point scatterers using the definition of Equations (6) and (7). The Foldy–Lax self-consistent multiple equations will be used. The phase function will be separated into coherent and incoherent parts. Then the first-order scattering approximations of the phase functions are derived analytically to show that in order to apply the optical theorem to calculate the scattering coefficient, the collective scattering amplitude has to be carried out to the second-order. The reason for using point scatterers is that Monte Carlo simulations can be performed readily by solving the exact wave equations. The illustrations also give insights on how to perform simulations using the iterative approach of wave scattering instead of using the exact matrix inversion.

Consider an incident plane wave E_i in the direction $\hat{k}_i$ impinging upon a volume V, the size of which obeys the three criteria of Section 2. The volume V contains N nonabsorptive point scatterers located at $\vec{r}_1, \vec{r}_2, \ldots, \vec{r}_N$ (Fig. 3). The point scatterers are uniformly distributed in volume V and have an isotropic scattering amplitude f.

The Foldy–Lax self-consistent multiple scattering equations [1, 29] use the concept that each scatterer j sees a 'final' exciting field E_{ex}^j. Then the scatterer j responds to this 'final' exciting field by giving rise to a scattered field that will be that of the single particle scattering amplitude of f. However, the 'final' exciting field E_{ex}^j obeys multiple scattering equations and has to be solved self-consistently. The multiple scattering equations for the 'final' exciting field are

$$E_{ex}^j(\vec{r}_j) = E_i(\vec{r}_j) + \sum_{\substack{l=1 \\ l\neq j}}^{N} f \frac{\exp\left(ik|\vec{r}_j - \vec{r}_l|\right)}{|\vec{r}_j - \vec{r}_l|} E_{ex}^l(\vec{r}_l) \tag{14}$$

where $j = 1, 2, 3, \ldots, N$. Equation (14) states that the 'final' exciting field E_{ex}^j is the sum of the incident wave and the scattered wave from all other particles except j itself.

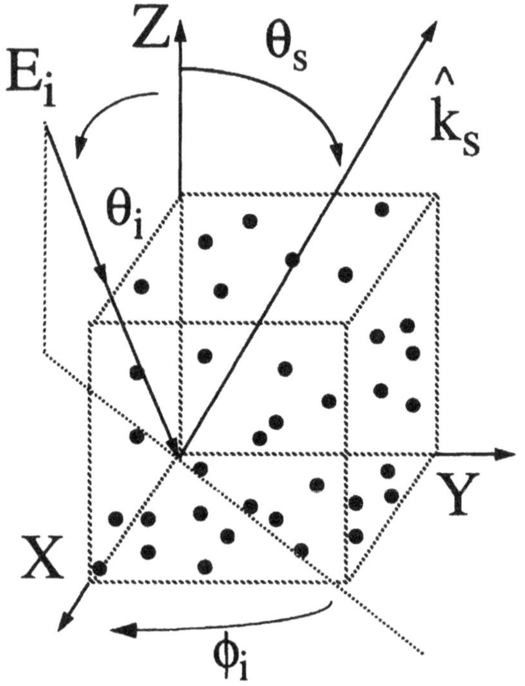

Figure 3. Point particles are randomly and uniformly distributed in a cubic box of size L with volume V.

However, each particle l is also excited by its final exciting field E_{ex}^l. The Foldy–Lax self-consistent multiple scattering equation can be derived from Maxwell's equations and are exact [1]. Since we have an exciting field on both sides of the equations, these are to be solved self-consistently by solving (14) which has N equations and N unknown exciting fields. The exact solution of (14) contains all the multiple scattering effects among N particles. After the exciting field E_{ex}^j, $j = 1, 2, \ldots, N$ is solved, then the total field is given by

$$E(\vec{r}) = E_i + \sum_{l=1}^{N} f \frac{\exp\left(ik|\vec{r} - \vec{r}_l|\right)}{|\vec{r} - \vec{r}_l|} E_{ex}^l(\vec{r}_l). \tag{15}$$

From (3), the scattering amplitude for N particles is

$$F(\hat{k}_s, \hat{k}_i) = \sum_{l=1}^{N} f \exp\left(-i\vec{k}_s \cdot \vec{r}_l\right) E_{ex}^l(\vec{r}_l). \tag{16}$$

When applied to a single scatterer, the optical theorem [1] gives

$$\frac{1}{k}\mathrm{Im}\{f\} = |f|^2. \tag{17}$$

Thus, we shall assume that f is essentially a real number with a small imaginary part given by Equation (5). Thus, $f = f' + ik(f')^2$, where superscript prime denotes the real part. The conventional radiative transfer theory gives the phase function and scattering coefficient as

$$P(\hat{k}_s, \hat{k}_i) = n_0 |f|^2 \tag{18}$$

$$\kappa_s = 4\pi n_0 |f|^2. \tag{19}$$

In the following, we shall apply the definitions of (6) and (7) to derive the new phase functions and scattering coefficients. The equations of (14) can be solved exactly for a reasonable number of scatterers (e.g. $N = 300$). They can also be solved iteratively. Both methods are done for the point scatterer case. However, the iterative approach is more convenient and efficient for more complicated problems. The statistics of scattering amplitude $F(\hat{k}_s, \hat{k}_i)$ will be studied. We assume the single particle probability density function $p(\vec{r}_j)$ and the two-particle joint probability density function $p_2(\vec{r}_i, \vec{r}_j)$ to be, respectively,

$$p(\vec{r}_j) = \frac{1}{V} \tag{20}$$

$$p_2(\vec{r}_i, \vec{r}_j) = \frac{1}{V^2} g(\vec{r}_i - \vec{r}_j). \tag{21}$$

In (21), $g(r)$ is the pair-distribution function which is also proportional to the conditional probability density function. As the particle separation becomes large, their joint probability density function should be independent. Thus

$$\lim_{r \to \infty} g(\vec{r}) = 1. \tag{22}$$

3.1. First-order solution

For the first-order solution, we approximate the 'final' exciting field of (14) by the incident field

$$E_{ex}^j(\vec{r}_j) = \exp(i\vec{k}_i \cdot \vec{r}_j). \tag{23}$$

Then, from (16) the first-order N scattering amplitude is

$$F^{(1)}(\hat{k}_s, \hat{k}_i) = f \sum_{j=1}^{N} \exp(i\vec{k}_d \cdot \vec{r}_j) \tag{24}$$

where $\vec{k}_d = \vec{k}_i - \vec{k}_s$ and the superscript (1) represents the first-order scattering approximation. From (24), we see that the random positions of $\vec{r}_j$ will give random phase fluctuations except in the forward direction of $\vec{k}_d = 0$. Taking the configurational average of (24) gives

$$\begin{aligned} \langle F^{(1)}(\hat{k}_s, \hat{k}_i) \rangle &= f \sum_{j=1}^{N} \int_V d\vec{r}_j \exp(i\vec{k}_d \cdot \vec{r}_j) p(\vec{r}_j) \\ &= \frac{1}{V} f \sum_{j=1}^{N} \int_V d\vec{r}_j \exp(i\vec{k}_d \cdot \vec{r}_j). \end{aligned} \tag{25}$$

Since the variable $\vec{r}_j$ in (25) becomes a dummy integration variable on averaging, the summation over j can be replaced by N. We thus have

$$\langle F^{(1)}(\hat{k}_s, \hat{k}_i)\rangle = n_0 f \int_V d\vec{r}_j \exp\left(i\vec{k}_d \cdot \vec{r}_j\right). \tag{26}$$

Since the volume V is much larger than a wavelength, the coherent field as given by (26) is sharply peaked in the forward direction of $\hat{k}_s = \hat{k}_i$. This concept of strong forward scattering of a coherent wave is only evident in wave theory and is absent from the conventional theory of radiative transfer. Taking the limit of large volume gives

$$\lim_{V\to\infty} \langle F^{(1)}(\hat{k}_s, \hat{k}_i)\rangle = n_0 f (2\pi)^3 \delta(\vec{k}_d) \tag{27}$$

where δ is the Dirac delta function.

Next, we calculate the absolute value of the amplitude squared

$$\left|F^{(1)}(\hat{k}_s, \hat{k}_i)\right|^2 = \sum_{j=1}^{N} \sum_{l=1}^{N} |f|^2 \exp\left(i\vec{k}_d \cdot (\vec{r}_j - \vec{r}_l)\right). \tag{28}$$

The double summation is next separated into $j = l$ and $j \neq l$

$$\left|F^{(1)}(\hat{k}_s, \hat{k}_i)\right|^2 = N|f|^2 + \sum_{j=1}^{N} \sum_{\substack{l=1 \\ j\neq l}}^{N} |f|^2 \exp\left(i\vec{k}_d \cdot (\vec{r}_j - \vec{r}_l)\right). \tag{29}$$

Next, the average of (29) is taken with the second term averaged over the two-particle joint probability density function of (21)

$$\begin{aligned}\left\langle\left|F^{(1)}(\hat{k}_s, \hat{k}_i)\right|^2\right\rangle &= N|f|^2 \\ &+ \sum_{j=1}^{N} \sum_{\substack{l=1 \\ l\neq j}}^{N} \int_V d\vec{r}_j \int_V d\vec{r}_l |f|^2 \exp\left(i\vec{k}_d \cdot (\vec{r}_j - \vec{r}_l)\right) \frac{g(\vec{r}_j - \vec{r}_l)}{V^2}.\end{aligned} \tag{30}$$

The first term represents that of conventional radiative transfer theory and the other terms represent correlation effects. Taking into account that $\vec{r}_j$ and $\vec{r}_l$ become dummy integration variables, (22) gives

$$\begin{aligned}\left\langle\left|F^{(1)}(\hat{k}_s, \hat{k}_i)\right|^2\right\rangle &= N|f|^2 \\ &+ \frac{N(N-1)}{V^2} \int_V d\vec{r}_j \int_V d\vec{r}_l |f|^2 \exp\left(i\vec{k}_d \cdot (\vec{r}_j - \vec{r}_l)\right)\left(g(\vec{r}_j - \vec{r}_l) - 1\right) \\ &+ \frac{N(N-1)}{V^2} \int_V d\vec{r}_j \exp\left(i\vec{k}_d \cdot \vec{r}_j\right) \int_V d\vec{r}_l |f|^2 \exp\left(-i\vec{k}_d \cdot \vec{r}_l\right).\end{aligned} \tag{31}$$

In (31), the last term corresponds to the coherent intensity that is in the forward direction. However, from (24)

$$\frac{1}{k} \mathrm{Im}\{F^{(1)}(\hat{k}_i, \hat{k}_i)\} = \frac{N}{k} \mathrm{Im}\{f\} = N|f|^2. \tag{32}$$

Thus, Equation (32) only contains the first term of (31) and does not contain the pair-distribution function as in (31) nor the sharply peaked forward scattering. Thus, the first order solution by itself does not obey the optical theorem of (5). For the optical scattering theorem to apply, it shall be shown that the $\mathrm{Im}\{F\}$ has to be carried to the second order in scattering. We first calculate the incoherent phase matrix and scattering coefficient. We further divide by V to calculate phase function. This gives

$$\begin{aligned} P(\hat{k}_s, \hat{k}_i) &= \frac{\langle|\mathcal{F}^{(1)}(\hat{k}_s, \hat{k}_i)|^2\rangle}{V} = \frac{\langle|F^{(1)}(\hat{k}_s, \hat{k}_i)|^2\rangle}{V} - \frac{|\langle F^{(1)}(\hat{k}_s, \hat{k}_i)\rangle|^2}{V} \\ &= n_0|f|^2 + n_0\left(n_0 - \frac{1}{V}\right)\int_V \mathrm{d}\vec{r}\,|f|^2 \exp\left(i\vec{k}_d\cdot\vec{r}\right)\left(g(\vec{r}) - 1\right) \\ &\quad - \frac{n_0|f|^2}{V}\int_V \mathrm{d}\vec{r}_j \exp\left(i\vec{k}_d\cdot\vec{r}_j\right)\int_V \mathrm{d}\vec{r}_l \exp\left(-i\vec{k}_d\cdot\vec{r}_l\right) \end{aligned} \tag{33}$$

The $1/V$ in the second term can be dropped. The last term is a result of the slight non-cancellation of the last term of (31) with the coherent intensity. The last term gives a finite contribution in the forward direction. However, in terms of radiative transfer, we should only take into account intensity that, when integrated over directions, will give a finite contribution to power. A function with a sharp peak of finite height and infinitesimal angular width does not contribute to power. Therefore, we next carry out an angular integration of (33) over all scattering directions to calculate the power contribution by each term. We can make use of the integral identity,

$$\int_{4\pi} \mathrm{d}\Omega_s \exp\left(-i\vec{k}_s\cdot\vec{r}\right) = 4\pi\frac{\sin(kr)}{kr} \tag{34}$$

to give

$$\begin{aligned} \int_{4\pi} P(\hat{k}_s, \hat{k}_i)\mathrm{d}\Omega_s &= 4\pi n_0|f|^2 + n_0^2\int_V \mathrm{d}\vec{r}\,|f|^2\left(g(\vec{r}) - 1\right)\frac{4\pi\sin(kr)}{kr}\exp\left(i\vec{k}_i\cdot\vec{r}\right) \\ &\quad - \frac{n_0|f|^2}{V}\int_V \mathrm{d}\vec{r}\exp\left(i\vec{k}_i\cdot\vec{r}\right)\frac{4\pi\sin(kr)}{kr}. \end{aligned} \tag{35}$$

The integral in the last term does not grow as fast as V, so that the last term vanishes for large V. Since it does not contribute to finite power, we shall also ignore it in the phase function which becomes

$$P(\hat{k}_s, \hat{k}_i) = n_0|f|^2 + n_0^2\int_V \mathrm{d}\vec{r}\,|f|^2 \exp\left(i\vec{k}_d\cdot\vec{r}\right)\left(g(\vec{r}) - 1\right). \tag{36}$$

The phase function of collective scattering agrees with the conventional theory when the particle positions are independent, i.e. $g = 1$. Independent particle positions only exist in the idealized case. In real life, because of the finite size of particles, particle positions can never be independent. If a particle has occupied a position, the center of another particle has to be at least a diameter away from the center of the first particle.

We shall next calculate the coherent phase function P_c and also the integration of P_c over directions as follows.

$$P_c(\hat{k}_s, \hat{k}_i) = \frac{1}{V}\left|\langle F(\hat{k}_s, \hat{k}_i)\rangle\right|^2 = \frac{n_0^2|f|^2}{V}\int_V d\vec{r}_j \exp\left(i\vec{k}_d \cdot \vec{r}_j\right)\int_V d\vec{r}_l \exp\left(-i\vec{k}_d \cdot \vec{r}_l\right). \tag{37}$$

The integration of P_c over directions gives the power contained in the coherent wave.

$$\int_{4\pi} d\Omega_s P_c(\hat{k}_s, \hat{k}_i) = n_0^2|f|^2 \int_V d\vec{r} \cos\left(\vec{k}_i \cdot \vec{r}\right) 4\pi \frac{\sin(kr)}{kr}. \tag{38}$$

Then the total power (coherent and incoherent) contained in the wave is the summation of the integration of P and P_c.

$$\begin{aligned}\int_{4\pi} d\Omega_s\left(P(\hat{k}_s, \hat{k}_i) + P_c(\hat{k}_s, \hat{k}_i)\right) &= 4\pi n_0|f|^2 \\ &+ n_0^2\int_V d\vec{r}\,|f|^2 \cos\left(\vec{k}_i \cdot \vec{r}\right)\left(g(\vec{r}) - 1\right)4\pi\frac{\sin(kr)}{kr} \\ &+ n_0^2\int_V d\vec{r}\,|f|^2\cos\left(\vec{k}_i \cdot \vec{r}\right)4\pi\frac{\sin(kr)}{kr}.\end{aligned} \tag{39}$$

Thus the phase function of the coherent wave is a Dirac delta function in the forward direction. Its power as given by (38) is non-convergent with V and also depends on the shape of V. Purely forward scattering does not affect radiative transfer which describes redistribution of radiative energy in different directions. This further justifies the non-inclusion of the coherent wave in the phase matrix.

3.2. *Second-order solution*

In the second-order solution, the exciting field of (14) is approximated by

$$E_{ex}^j(\vec{r}_j) = \exp\left(i\vec{k}_i \cdot \vec{r}_j\right) + \sum_{\substack{l=1 \\ l\neq j}}^{N} f\,\frac{\exp\left(ik|\vec{r}_j - \vec{r}_l|\right)}{|\vec{r}_j - \vec{r}_l|}\exp\left(i\vec{k}_i \cdot \vec{r}_l\right). \tag{40}$$

The forward scattering amplitude is

$$F^{(2)}(\hat{k}_i, \hat{k}_i) = Nf + \sum_{\substack{j\neq l \\ j=1}}^{N} f^2\,\frac{\exp\left(ik|\vec{r}_j - \vec{r}_l|\right)}{|\vec{r}_j - \vec{r}_l|}\exp\left(i\vec{k}_i \cdot (\vec{r}_l - \vec{r}_j)\right). \tag{41}$$

Taking the average and using the property that $g(-\vec{r}) = g(\vec{r})$ gives

$$\begin{aligned}\langle F^{(2)}(\hat{k}_i, \hat{k}_i)\rangle = Nf + \frac{N(N-1)}{V^2}f^2\int_V d\vec{r}_j\int_V d\vec{r}_l\,\frac{\exp\left(ik|\vec{r}_j - \vec{r}_l|\right)}{|\vec{r}_j - \vec{r}_l|} \\ \times\cos\left(\vec{k}_i \cdot (\vec{r}_l - \vec{r}_j)\right)g(\vec{r}_l - \vec{r}_j).\end{aligned} \tag{42}$$

Taking the imaginary part gives the scattering coefficient of (19) for N point scatterers

$$\begin{aligned}\frac{4\pi}{k}\frac{\mathrm{Im}\{\langle F^{(2)}(\hat{k}_i,\hat{k}_i)\rangle\}}{V} &= \frac{4\pi}{k}n_0\mathrm{Im}\{f\} \\ &+ n_0^2 f^2 \int_V \mathrm{d}\vec{r}\,\frac{4\pi\sin(kr)}{kr}\cos\left(\vec{k}_i\cdot\vec{r}\right)\left(g(\vec{r})-1\right) \\ &+ n_0^2 f^2 \int_V \mathrm{d}\vec{r}\,\frac{4\pi\sin(kr)}{kr}\cos\left(\vec{k}_i\cdot\vec{r}\right).\end{aligned} \tag{43}$$

Thus (43) agrees with (39) in view of (17), (19), and the fact that the single particle scattering amplitude f has a small imaginary part. Thus, to apply the optical theorem, the N particle forward scattering amplitude has to be calculated to the second order so that

$$\kappa_e = \lim_{V\to\text{large}}\left[\frac{4\pi\mathrm{Im}\{\langle F^{(2)}(\hat{k}_i,\hat{k}_i)\rangle\}}{kV} - \frac{1}{V}\int_{4\pi}\mathrm{d}\Omega_s\left|\langle F^{(1)}(\hat{k}_s,\hat{k}_i)\rangle\right|^2\right]. \tag{44}$$

In this section, we have used the Foldy–Lax self-consistent multiple scattering equations to derive the phase matrix and extinction coefficients of point scatterers based on the new per volume concept. The iteration approach is used and the importance of carrying to second order solution is shown. The roles of coherent field and incoherent fields are discussed. The coherent field gives rise to a strong forward peak (many dB up) in the forward direction and has to be subtracted from the total integration to give a meaningful definition of phase function. In the next section we apply this per unit volume concept to calculate phase function and extinction coefficients of point scatterers in clusters.

4. PHASE FUNCTION AND EXTINCTION COEFFICIENTS OF POINT SCATTERERS OCCURRING IN CLUSTERS

In this section, we calculate the phase function and extinction coefficient of point scatterers occurring in clusters. It is shown that these quantities depend on two pair functions. Let each cluster be labelled as a primary scatterer and the point scatterers within each cluster be labelled as secondary scatterers. Then the phase function and extinction coefficients depend on g_p, which is the pair distribution function among primary scatterers and on g_s, which is the pair distribution function among secondary scatterers within a primary scatterer.

Consider a volume element V in the radiative transfer as defined in Section 2, the volume containing N_p primary scatterers (clusters) each of which consists of N_s secondary point scatterers (Fig. 4). The N_p clusters are centered at $\vec{r}_\alpha$, $\alpha = 1, 2, \ldots, N_p$ and within each cluster α, the secondary clusters are centered at $\vec{r}_{\alpha j} = 1, 2, \ldots, N_s$. Thus

$$N = N_s N_p \tag{45}$$

is the total number of particles in volume V. Then

$$n_0 = N/V = N_s n_p \tag{46}$$

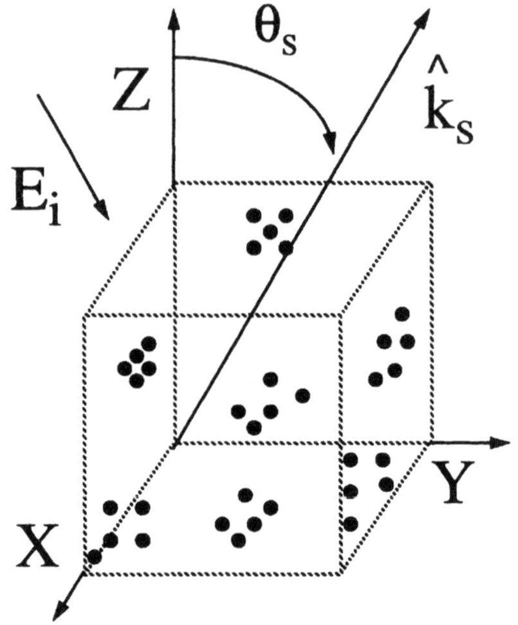

Figure 4. Clustered point particles are randomly distributed in a cubic box of size L with volume V. The clusters are randomly distributed and within each clusters of size l_c, the particles are randomly distributed.

is the number of particles per unit volume and $n_p = N_p/V$ is the number of clusters per unit volume. Note that N and N_p are large numbers in V while N_s does not have to be large. Then, from (24), the first order collective scattering amplitude is

$$F^{(1)}(\hat{k}_s, \hat{k}_i) = f \sum_{\alpha=1}^{N_p} \sum_{j=1}^{N_s} \exp\left(i\vec{k}_d \cdot (\vec{r}_\alpha + \vec{r}_{\alpha j})\right). \tag{47}$$

This can be expressed in terms of the scattering amplitude of the primary scatterers with

$$f_\alpha^P(\hat{k}_s, \hat{k}_i) = f \sum_{j=1}^{N_s} \exp\left(i\vec{k}_d \cdot (\vec{r}_{\alpha j})\right). \tag{48}$$

Thus, the phase function is

$$P(\hat{k}_s, \hat{k}_i) = \frac{\langle |F^{(1)}(\hat{k}_s, \hat{k}_i)|^2 \rangle - |\langle F^{(1)}(\hat{k}_s, \hat{k}_i)\rangle|^2}{V} \tag{49}$$

or

$$\begin{aligned} P(\hat{k}_s, \hat{k}_i) &= n_p \langle |f_\alpha^P(\hat{k}_s, \hat{k}_i)|^2 \rangle \\ &\quad + n_p^2 \langle f_\alpha^P(\hat{k}_s, \hat{k}_i)\rangle \langle f_\beta^P(\hat{k}_s, \hat{k}_i)\rangle \int_{v_p} d\vec{r} \exp\left(i\vec{k}_d \cdot \vec{r}\right)\left(g_p(\vec{r}) - 1\right) \end{aligned} \tag{50}$$

where the joint probability density function of primary scatterers is

$$p_p(\vec{r}_\alpha, \vec{r}_\beta) = \frac{g_p(\vec{r}_\alpha - \vec{r}_\beta)}{V^2} \tag{51}$$

where g_p is the pair distribution function of primary scatterers. Let the primary scatterer size be confined to v_p. Then the first and second moment of primary scatterer scattering amplitude are respectively,

$$\langle f_\alpha^P(\hat{k}_\mathrm{s}, \hat{k}_\mathrm{i})\rangle = \frac{N_\mathrm{s} f}{v_\mathrm{p}} \int_{v_\mathrm{p}} \mathrm{d}\vec{r} \exp\left(i\vec{k}_\mathrm{d}\cdot\vec{r}\right) \tag{52}$$

$$\begin{aligned}\langle |f_\alpha^P(\hat{k}_\mathrm{s}, \hat{k}_\mathrm{i})|^2\rangle = N_\mathrm{s}|f|^2 + \frac{N_\mathrm{s}(N_\mathrm{s}-1)}{v_\mathrm{p}^2} \int_{v_\mathrm{p}} \mathrm{d}\vec{r}_{\alpha j} \int_{v_\mathrm{p}} \mathrm{d}\vec{r}_{\alpha l} |f|^2 \\ \times \exp\left(i\vec{k}_\mathrm{d}\cdot(\vec{r}_{\alpha j} - \vec{r}_{\alpha l})\right) g_\mathrm{s}(\vec{r}_{\alpha j}, \vec{r}_{\alpha l})\end{aligned} \tag{53}$$

where the joint probability density function of secondary scatterers is

$$p_\mathrm{s}(\vec{r}_{\alpha j}, \vec{r}_{\alpha l}) = \frac{g_\mathrm{s}(\vec{r}_{\alpha j}, \vec{r}_{\alpha l})}{{v_\mathrm{p}}^2} \tag{54}$$

with g_s being the pair distribution function of secondary scatterers. For example, if $g_\mathrm{p} = 1$, $g_\mathrm{s} = 1$ and v_p is a cubic volume of l_c^3. Then the phase function is

$$\begin{aligned}P(\hat{k}_\mathrm{s}, \hat{k}_\mathrm{i}) &= n_0|f|^2 + n_0(N_\mathrm{s}-1)|f|^2 \\ &\times \left\{\frac{\sin\left(k_{\mathrm{d}x}\frac{l_\mathrm{c}}{2}\right)}{k_{\mathrm{d}x}\frac{l_\mathrm{c}}{2}}\right\}^2 \left\{\frac{\sin\left(k_{\mathrm{d}y}\frac{l_\mathrm{c}}{2}\right)}{k_{\mathrm{d}y}\frac{l_\mathrm{c}}{2}}\right\}^2 \left\{\frac{\sin\left(k_{\mathrm{d}z}\frac{l_\mathrm{c}}{2}\right)}{k_{\mathrm{d}z}\frac{l_\mathrm{c}}{2}}\right\}^2\end{aligned} \tag{55}$$

where $k_{\mathrm{d}x}$, $k_{\mathrm{d}y}$, and $k_{\mathrm{d}z}$ are respectively the x, y, and z components of $\vec{k}_\mathrm{d}$.

5. MONTE CARLO SIMULATIONS OF POINT SCATTERERS

In this section, we present the Monte Carlo simulation results of two types of configurations as shown in Figs 3 and 4. In Fig. 3, the nonabsorptive particles are uniformly and randomly distributed in a cubic box of size L^3. In Fig. 4, the same number of particles are distributed in clusters in the cubic box of size L^3. The clusters contain N_s particles each. In each cluster the particles are randomly distributed in a cubic volume of size l_c^3. All distance dimensions are in wavelengths (λ). In the Monte Carlo simulations, Equation (14) is solved exactly to find the incoherent phase function as defined by (6). Then the scattering coefficient defined by (7) is found by integrating over all scattered angles. We also find the scattering coefficient and the phase function by approximating the multiple scattering equation to the second order by using Equation (40). The iterative approach is then compared with the exact solution.

The geometry of the two different cases is given in Figs 3 and 4. The cubic box is of volume V and contains many point particles. The length of each side L is 50λ. A scalar plane wave is incident at $\theta_\mathrm{i} = 10°$ and $\phi_\mathrm{i} = 10°$. The incident and the scattering angles are defined as in Figure 3 such that the backscattering direction is $\theta_\mathrm{s} = \theta_\mathrm{i}$ and $\phi_\mathrm{s} = \pi + \phi_\mathrm{i}$. For illustrations, let scattering amplitude $f = (0.008905 + i0.0005)\lambda$. The value of f is typical of a small particle. The real and imaginary parts are chosen to obey the optical theorem of (17) of a single particle. In all cases, the given parameters are fixed. We only change the distribution of particles. The positions of 500 scatterers are generated randomly [30]. For the first case, $N = 500$ center positions are generated randomly within the cubic box (Fig. 3) of side L. This is the uniform and random distribution case. For the second case, $N = 500$ particle positions are generated in two stages (Fig. 4). The 500 particles are of 50 clusters with each cluster containing 10 particles. The cluster volume is a cube with length l_c. Initially, 50 center positions are generated within the cube of length L and they are randomly chosen with the constraint that there is no interpenetration of the cluster cubes l_c. Then in the second stage, we randomly generate 10 particles inside each cluster of volume l_c^3. This is the clustered distribution case. We vary l_c and investigate its effects. For both the uniformly random case and cluster case, the results are calculated for 100 realizations.

In Monte Carlo simulations, it is important to demonstrate convergence with respect to the number of particles (N) in the cubic box and the number of realizations for a fixed particle density n_0. We first present the convergence of extinction coefficients as we increase realizations and the number of particles in Fig. 5 for uniformly random distribution. Initially, we put $N = 63$ particles in a cubic box of size $L = 25\lambda$, so that the particle density is $n_0 = 0.004\lambda^{-3}$ (50 realizations contain 63 particles each while the other 50 realizations contain 62 particles). This is compared to the independent scattering theory ($4\pi n_0|f|^2$). The independent scattering as defined by (19) predicts $\kappa_{ei} = 4.0\times10^{-6}\lambda^{-1}$ where as in the Monte Carlo simulations, $\kappa_e = 3.958\times10^{-6}\lambda^{-1}$ after 100 realizations. To demonstrate convergence with respect to the number of particles,

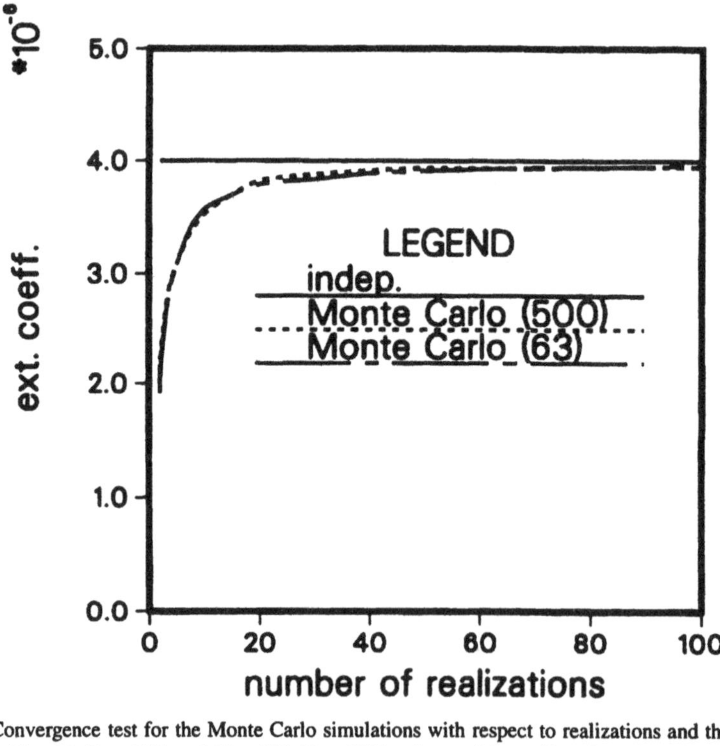

Figure 5. Convergence test for the Monte Carlo simulations with respect to realizations and the number of particles for $N = 63$ ($L = 25\lambda$) and $N = 500$ ($L = 50\lambda$) point particles uniformly and randomly distributed as in Fig. 3. They are compared with the independent scattering result.

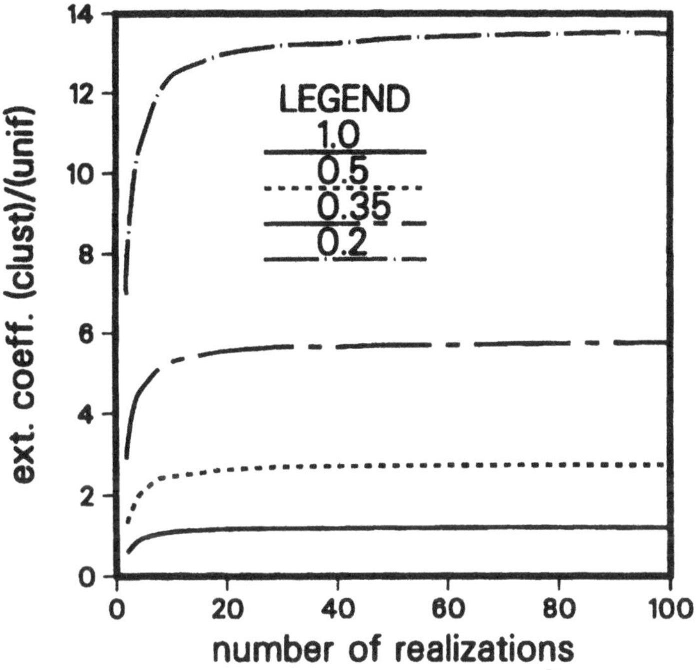

Figure 6. Monte Carlo simulations for the clustered scatterers of Fig. 4. l_c is decreased from 1.0λ to 0.2λ. There are 50 clusters containing 10 scatterers each ($N = 500$, $L = 50\lambda$).

we increase the number of particles to $N = 500$ while n_0 remains at $0.004\lambda^{-3}$. This is realized by increasing the number of particles to 500 and the volume $V = L^3 = (50\lambda)^3$. This new result is also given in Fig. 5, where $\kappa_e = 3.944\times10^{-6}\lambda^{-1}$ after 100 realizations. Figure 5 indicates convergence of the results with respect to the number of realizations and the number of particles. All the Monte Carlo results to be presented from this point on are given for $N = 500$ particles with $L = 50\lambda$, $\theta_i = 10°$, and $\phi_i = 10°$.

Next we consider the second case of clusters of 10 particles in each cluster. Figure 6 shows the convergence of the relative extinction coefficient versus realizations. The relative extinction coefficient (κ_{re}) is the collective extinction coefficient divided by that of independent scattering ($\kappa_{re} = \kappa_e/\kappa_{ei}$, with $\kappa_{ei} = 4.0\times10^{-6}\lambda^{-1}$). There are 50 clusters, each containing 10 particles in all plots given. The cluster cube size l_c is decreased from 1.0λ to 0.2λ. For the cubic box size of 1.0λ there is very little difference between the uniform and clustered distribution as can be observed in the solid curve. However, as the cluster volume is made smaller than a wavelength, κ_{re} can become large. This shows that when considering random media problems such as remote sensing of vegetation, significance of the clustered geometry must be considered. In Figs 7 and 8, respectively, the phase functions of the Monte Carlo results of Fig. 5 ($N = 500$) and the $l_c = 0.2\lambda$ of Fig. 6 are presented. Both the incoherent phase function P_{inc} (dotted line) and the combined phase function $P_{inc} + P_{coherent}$ (solid line) are given. The results are averaged over 100 realizations. In the polar plots, the top half plane gives $\phi_s = 10°$ and θ_s ranges from $0 \rightarrow 180°$ going clockwise. In the bottom half plane $\phi_s = 190°$ and θ_s ranges from $0 \rightarrow 180°$ going counter clockwise (refer to Fig. 3). Comparing Figs 7 and 8,

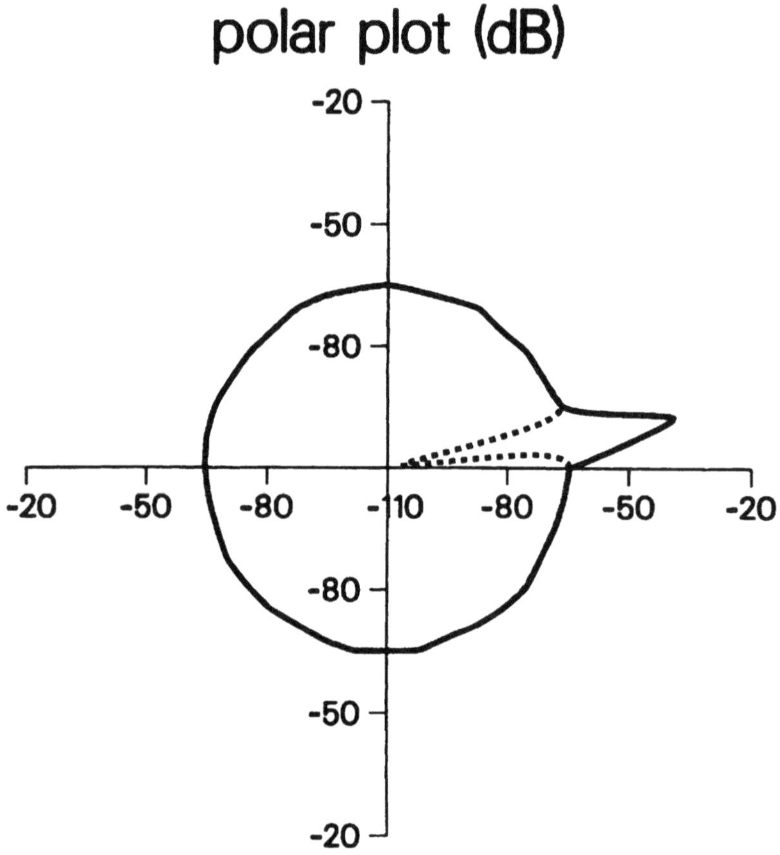

Figure 7. Polar plot (dB) of the phase function of Fig. 5 for $N = 500$ and $L = 50\lambda$ with the geometry given as in Fig. 3. Upper half-plane shows $\phi_s = 10°$ and θ_s is $0° \rightarrow 180°$ going clockwise. Lower half-plane shows $\phi_s = 190°$ and θ_s is $0° \rightarrow 180°$ going counter clockwise. Incident angles are $\theta_i = 10°$ and $\phi_i = 10°$.

the clustered scattering case gives more incoherent contribution than the uniform case, while the opposite is true for coherent contribution. In both figures, the coherent part has a sharp peak of large amplitude (many dB up) in the forward direction ($\theta_s = 170°$, $\phi_s = 10°$). Note that the polar plot is in dB scale, so that the coherent forward peak has much larger amplitude than the other directions. As V approaches infinity, the coherent wave will approach a Dirac delta function. Thus, the coherent peak has to be excluded from the phase matrix and extinction coefficient calculations.

All the Monte Carlo results given so far were obtained by solving the multiple scattering Equation (14) exactly. As shown in the earlier section, we can also carry out the multiple scattering equation by the iterative approach. In order to compare the iterative and the exact method, we compare the incoherent phase function for a single realization using the second-order approximation and the exact method in Fig. 9. In the figure, $l_c = 0.35\lambda$, $N = 500$, $\phi_s = 20°$, and θ_s is varied from 0° to 180°. For one realization, the second-order solution shows differences from the exact solution for most scattering angles. However, when 100 realizations are averaged, the second-order solution agrees with the exact method (Fig. 10). This shows that for a single realization, higher order

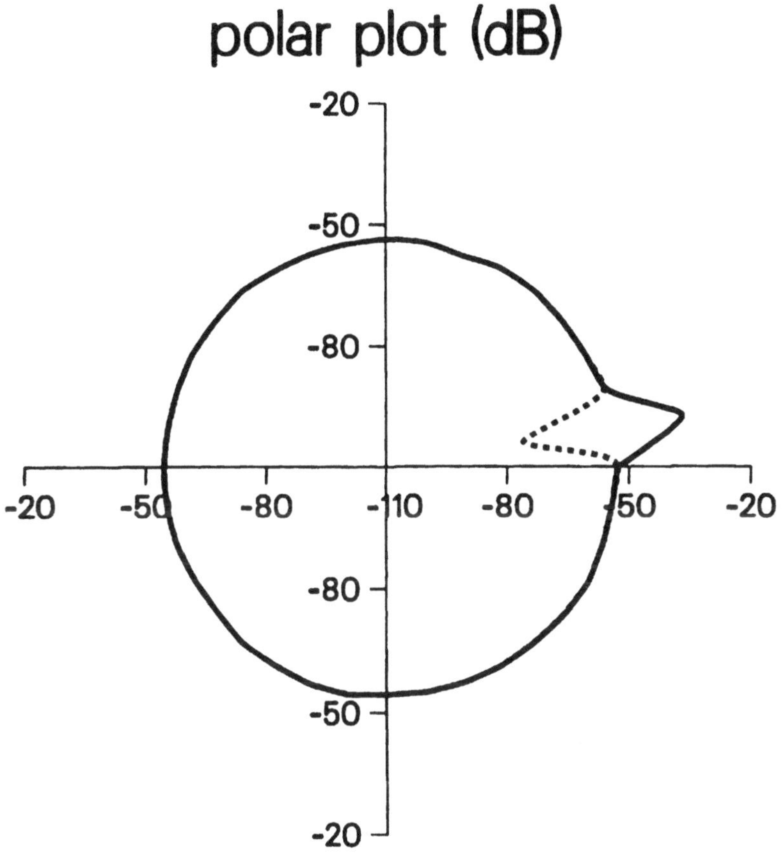

Figure 8. Polar plot (dB) of the phase function of Fig. 6 for $N = 500$, $l_c = 0.2\lambda$ and $L = 50\lambda$ with incident angles at $\theta_i = 10°$ and $\phi_i = 10°$. Upper half-plane shows $\phi_s = 10°$ and θ_s is $0° \rightarrow 180°$ going clockwise. Lower half-plane shows $\phi_s = 190°$ and θ_s is $0° \rightarrow 180°$ going counter clockwise.

scattering effects are important. However, these higher order scattering effects tend to cancel out on averaging over realizations and the second-order theory gives the correct average. This in turn saves significant CPU time while achieving accurate results in Monte Carlo simulations. For one realization containing 500 point particles, the exact solution approach took 1165.68 seconds to solve the multiple scattering equation while for the second-order approximation, only 11.68 seconds were required. The iteration approach can be very useful for more complicated problems of vector electromagnetic scattering by a large number of dielectric scatterers of arbitrary size. In such calculations, the exact solutions become formidable. The iterative approach has been used for vector electromagnetic scattering from a conglomeration of up to 4000 spherical particles [21]. In Fig. 11 the comparison between the analytic collective scattering theory of (55), the independent scattering of (18), and the Monte Carlo simulation of Fig. 10 with 1000 realizations is given. While the Monte Carlo result and the collective scattering theory agree, the independent scattering predicts the wrong result. The dip in the Monte Carlo result is due to subtraction of the coherent part from the total phase function. It is important to note that in comparing Figs 10 and 11, it requires 1000 realizations for the phase functions to converge.

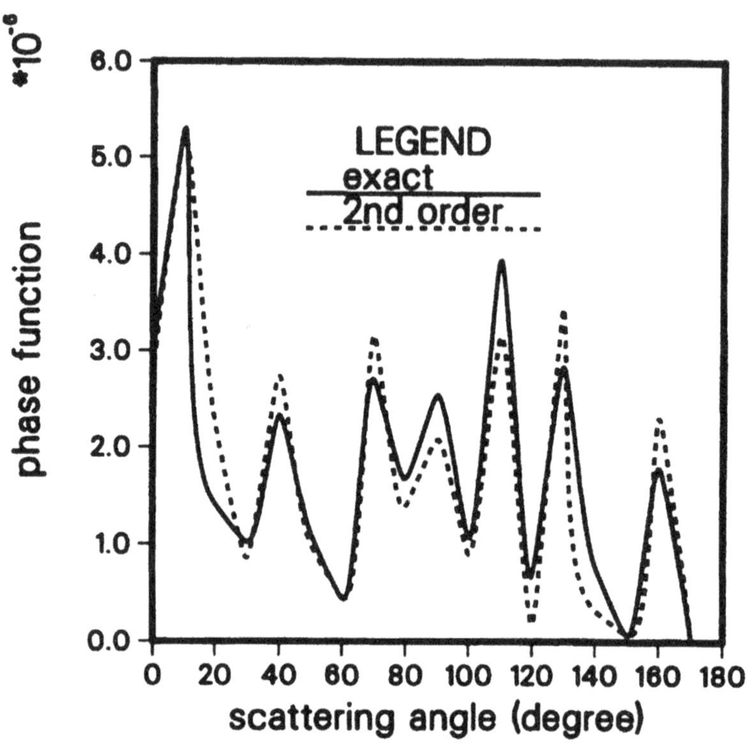

Figure 9. Comparison test between the second-order approach and the exact solution. Phase function of Fig. 6 with $l_c = 0.35\lambda$, $N = 500$ and $L = 50\lambda$ for 1 realization ($\phi_i = 10°$, $\theta_i = 10°$, $\phi_s = 20°$, and θ_s ranges from 0° to 180°).

6. ELECTROMAGNETIC MODELING OF PHASE MATRIX INCLUDING THE EFFECTS OF COLLECTIVE SCATTERING

In this section, we shall extend the model of point scatterers to consider the more general case of the phase matrix of scattering of vector electromagnetic waves by clusters of scatterers including the effects of collective scattering. The basic idea follows that of point scatterers except that in this case it is computationally difficult to solve the exact solution of N scatterers. The effects of the two pair distribution functions g_p and g_s are already indicated. Thus, the first-order iterative solution will be used to calculate the phase matrix. Only the averaged incoherent bistatic intensity will be used in the phase matrix elements as illustrated in previous sections. The second-order solution is difficult to calculate. Thus, we calculate the scattering coefficient by integrating the phase matrix over scattered directions.

Consider a volume V that is large compared with a wavelength containing a large number of N_p primary scatterers (Fig. 12). Each primary scatterer extends over a volume V_p which does not have to be large. The N_p primary scatterers are randomly and uniformly distributed in volume V. Each primary scatterer consists of T types of secondary scatterers with N_1 scatterers of type 1 and N_2 scatterers of type 2, ..., N_T scatterers of type T. (For example, a primary scatterer is a soybean plant, each plant

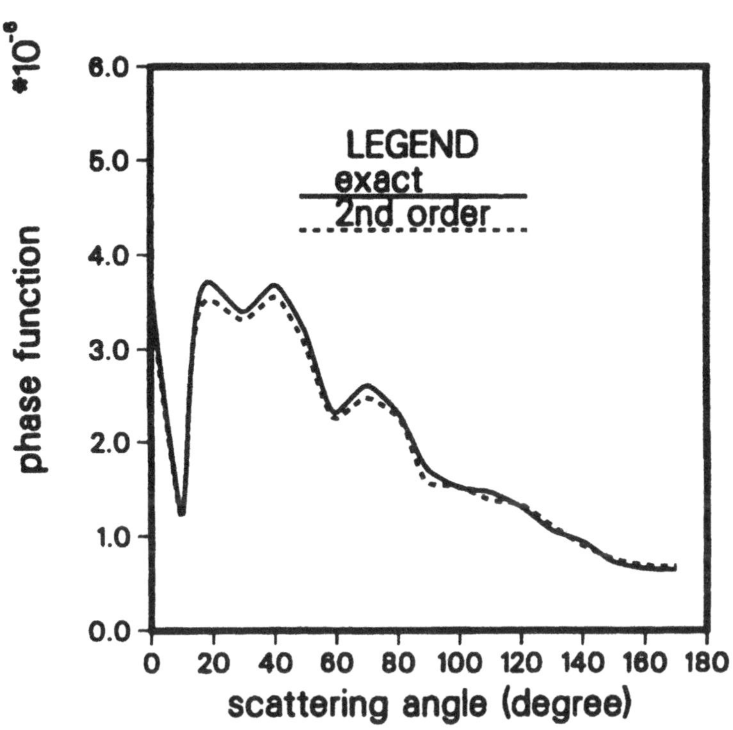

Figure 10. Comparison test between the second-order approach and the exact solution of Fig. 9 is increased to 100 realizations ($\phi_i = 10°$, $\theta_i = 10°$, $\phi_s = 20°$, and θ_s ranges from 0° to 180°).

consists of one stem and N_l leaves, etc.) Let

$$N_s = \sum_{L=1}^{T} N_L \tag{56}$$

be the number of secondary scatterers in a primary scatterer,

$$N = N_s N_p \tag{57}$$

is the number of scatterers in volume V,

$$n_0 = \frac{N}{V} \tag{58}$$

is the number of scatterers per unit volume and

$$n_p = \frac{N_p}{V} \tag{59}$$

is the number of primary scatterers per unit volume. Note that N is large while N_s does not have to be large. Each scatterer has a three label index $\alpha L j$ where $\alpha = 1, \ldots, N_p$ is the index of primary scatterer, $L = 1, \ldots, T$ is the type index of the scatterer and

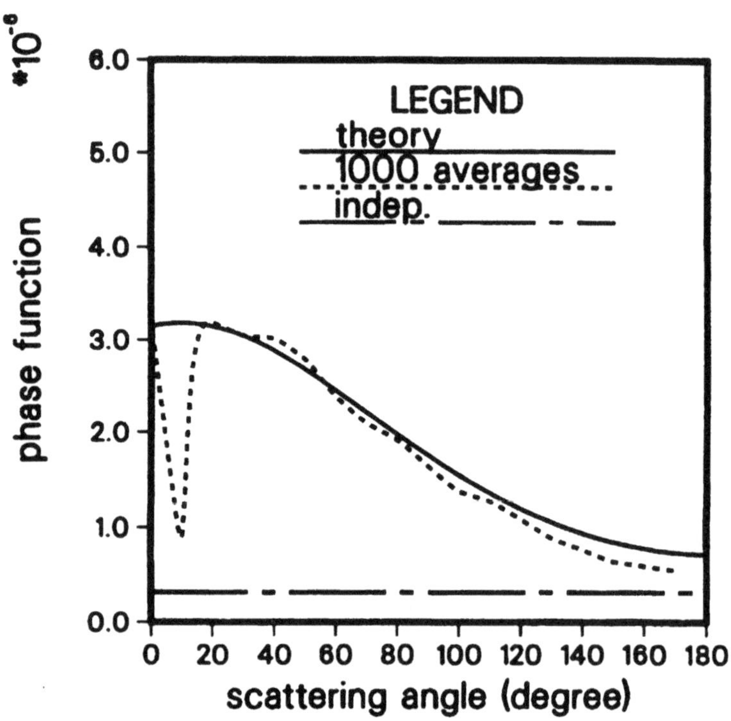

Figure 11. Comparison test between the analytical result from (55), the independent scattering, and the Monte Carlo simulation of Fig. 10 (1000 realizations). The parameters are: $l_c = 0.35\lambda$, $N_s = 10$, $N_p = 50$, $N = 500$, $n_0 = 0.004\lambda^{-3}$ and $L = 50\lambda$ ($\phi_i = 10°$, $\theta_i = 10°$, $\phi_s = 20°$, and θ_s ranges from 0° to 180°).

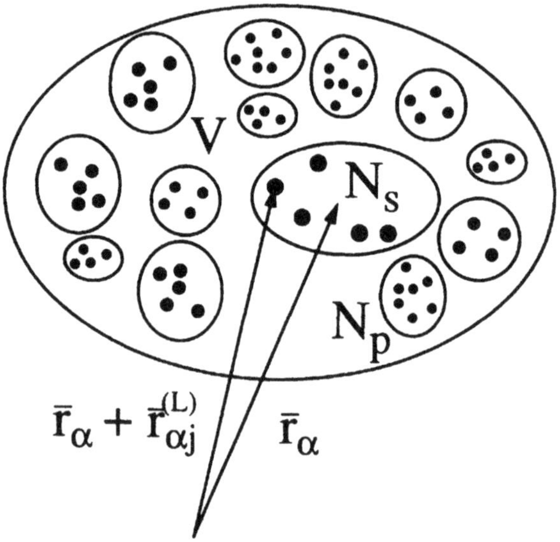

Figure 12. Volume V containing primary scatterers centered at $\vec{r}_\alpha$ and the secondary scatterers centered at $\vec{r}_\alpha + \vec{r}_{\alpha j}^{(L)}$. There are N_p primary scatterers and N_s secondary scatterers.

$j = 1, 2, \ldots, N_L$ is the number index of the secondary scatterer of type L. The position of scatterer αLj is at $\vec{r}_\alpha + \vec{r}_{\alpha j}^{(L)}$. Thus, $\vec{r}_\alpha$ is the location of the center of the primary scatterer and $\vec{r}_{\alpha j}^{(L)}$ is the location of the center of the secondary scatterer relative to the primary scatterer. Let $\overline{\overline{f}}_{\alpha j}^{(L)}(\hat{k}_s, \hat{k}_i, s_{\alpha j}^{(L)})$ denote the scattering dyad of the αLj scatterer if it is located at the origin. The argument $s_{\alpha j}^{(L)}$ denotes properties such as the size and orientation of the scatterer αLj. It will later be averaged so that the scattering properties can be averaged over size and orientation.

Consider an incident plane wave with polarization $\hat{e}_i$ incident on the volume V in direction $\hat{k}_i$. Then the solution of the scattered field from N scatterers can be expressed as

$$\overline{E}_s(\vec{r}) = \overline{\overline{F}}(\hat{k}_s, \hat{k}_i) \cdot \hat{e}_i \frac{\exp(ikr)}{r} \tag{60}$$

where $\overline{\overline{F}}(\hat{k}_s, \hat{k}_i)$ is the scattering dyad of the N scatterers in volume V. Thus, we define

$$\overline{\overline{\mathcal{F}}}(\hat{k}_s, \hat{k}_i) = \overline{\overline{F}}(\hat{k}_s, \hat{k}_i) - \langle \overline{\overline{F}}(\hat{k}_s, \hat{k}_i) \rangle \tag{61}$$

as the scattering dyad of the incoherent wave. The phase matrix elements are in terms of the average of the product of the elements of the incoherent scattering dyad in the limit of large V as discussed in previous sections. Thus, we shall calculate

$$\begin{aligned} \lim_{V \to \text{large}} \frac{1}{V} \langle \overline{\overline{\mathcal{F}}}(\hat{k}_s, \hat{k}_i) \otimes \overline{\overline{\mathcal{F}}}^*(\hat{k}_s, \hat{k}_i) \rangle = \lim_{V \to \text{large}} \frac{1}{V} \Big\{ &\langle \overline{\overline{F}}(\hat{k}_s, \hat{k}_i) \otimes \overline{\overline{F}}^*(\hat{k}_s, \hat{k}_i) \rangle \\ &- \langle \overline{\overline{F}}(\hat{k}_s, \hat{k}_i) \rangle \otimes \langle \overline{\overline{F}}^*(\hat{k}_s, \hat{k}_i) \rangle \Big\} \end{aligned} \tag{62}$$

where $\otimes$ denotes direct product. In the first-order iteration approach, we have the first-order solution of the scattering dyad as given by

$$\overline{\overline{F}}^{(1)}(\hat{k}_s, \hat{k}_i) = \sum_{\alpha=1}^{N_p} \sum_{L=1}^{T} \sum_{j=1}^{N_L} \overline{\overline{f}}_{\alpha j}^{(L)}(\hat{k}_s, \hat{k}_i, s_{\alpha j}^{(L)}) \exp\left(i\vec{k}_d \cdot (\vec{r}_\alpha + \vec{r}_{\alpha j}^{(L)})\right) \tag{63}$$

where $\vec{k}_d = (\hat{k}_i - \hat{k}_s)k$. To simplify the treatment, we also define the scattering dyad of a primary scatterer centered at the origin

$$\overline{\overline{f}}_{\alpha}^{(P)}(\hat{k}_s, \hat{k}_i) = \sum_{L=1}^{T} \sum_{j=1}^{N_L} \overline{\overline{f}}_{\alpha j}^{(L)}(\hat{k}_s, \hat{k}_i, s_{\alpha j}^{(L)}) \exp\left(i\vec{k}_d \cdot \vec{r}_{\alpha j}^{(L)}\right) \tag{64}$$

so that

$$\overline{\overline{F}}^{(1)}(\hat{k}_s, \hat{k}_i) = \sum_{\alpha=1}^{N_p} \overline{\overline{f}}_{\alpha}^{(P)}(\hat{k}_s, \hat{k}_i) \exp\left(i\vec{k}_d \cdot \vec{r}_\alpha\right). \tag{65}$$

The superscript (P) is used to denote primary scatterers which consist of the scattering from a conglomeration of secondary scatterers grouped together in a cluster. By taking the average of (63), we can calculate the scattering dyad of the coherent wave. For a sparse concentration of scatterers, it is reasonable to assume that the probability density function

$$p(\vec{r}_\alpha, \vec{r}_{\alpha j}^{(L)}, s_{\alpha j}^{(L)}) = p(\vec{r}_\alpha) p(\vec{r}_{\alpha j}^{(L)}, s_{\alpha j}^{(L)}) \tag{66}$$

and the single primary scatterer probability density function $p(\vec{r}_\alpha)$ are uniformly and randomly distributed

$$p(\vec{r}_\alpha) = \frac{1}{V}. \tag{67}$$

Taking the average of (64) and (65) and using (66) and (67) as the probability density function gives

$$\left\langle \overline{\overline{F}}^{(1)}(\hat{k}_s, \hat{k}_i) \right\rangle = \frac{N_p}{V} \int_V d\vec{r}_\alpha \left\langle \overline{\overline{f}}_\alpha^{(P)}(\hat{k}_s, \hat{k}_i) \right\rangle \exp\left(i\vec{k}_d \cdot \vec{r}_\alpha\right) \tag{68}$$

where

$$\begin{aligned} &\left\langle \overline{\overline{f}}_\alpha^{(P)}(\hat{k}_s, \hat{k}_i) \right\rangle \\ &= \sum_{L=1}^{T} N_L \int_{V_p} d\vec{r}_{\alpha j}^{\,(L)} \int_{V_p} ds_{\alpha j}^{(L)} p(\vec{r}_{\alpha j}^{\,(L)}, s_{\alpha j}^{(L)}) \overline{\overline{f}}_{\alpha j}^{(L)}(\hat{k}_s, \hat{k}_i, s_{\alpha j}^{(L)}) \exp\left(i\vec{k}_d \cdot \vec{r}_{\alpha j}^{\,(L)}\right). \end{aligned} \tag{69}$$

Thus, Equation (68) states that the average N particle scattering amplitude is the averaged primary scatterer scattering dyad that is further averaged by their position as weighted by the phase factor. Equation (69) states that the average primary scatterer scattering dyad is the cluster of secondary scatterer scattering dyads weighted by their sizes and orientation and further averaged over their position relative to the center of the primary scatterer as weighted by the phase factor. Since the volume V is large compared with wavelength, the coherent wave scattering dyad $\langle \overline{\overline{F}}^{(1)}(\hat{k}_s, \hat{k}_i) \rangle$ as given by (68) is strongly peaked in the forward direction $\vec{k}_d = 0$. From (59) and (68), we have

$$\begin{aligned} \left\langle \overline{\overline{F}}^{(1)}(\hat{k}_s, \hat{k}_i) \right\rangle \otimes \left\langle \overline{\overline{F}}^{(1)*}(\hat{k}_s, \hat{k}_i) \right\rangle &= n_p^2 \int_V d\vec{r}_\alpha \int_V d\vec{r}_\beta \left\langle \overline{\overline{f}}_\alpha^{(P)}(\hat{k}_s, \hat{k}_i) \right\rangle \\ &\otimes \left\langle \overline{\overline{f}}_\beta^{(P)*}(\hat{k}_s, \hat{k}_i) \right\rangle \exp\left(i\vec{k}_d \cdot (\vec{r}_\alpha - \vec{r}_\beta)\right). \end{aligned} \tag{70}$$

To obtain the bistatic scattering intensity, we form the direct product of (65) with itself to obtain a double summation $\sum_\alpha \sum_\beta$. The double summation is decomposed into terms of $\beta = \alpha$ and $\beta \neq \alpha$, so that

$$\begin{aligned} \left\langle \overline{\overline{F}}^{(1)}(\hat{k}_s, \hat{k}_i) \right\rangle \otimes \left\langle \overline{\overline{F}}^{(1)*}(\hat{k}_s, \hat{k}_i) \right\rangle &= \sum_{\alpha=1}^{N_p} \overline{\overline{f}}_\alpha^{(P)}(\hat{k}_s, \hat{k}_i) \otimes \overline{\overline{f}}_\alpha^{(P)*}(\hat{k}_s, \hat{k}_i) \\ &+ \sum_{\substack{\alpha=1 \\ \alpha\neq\beta}}^{N_p} \sum_{\beta=1}^{N_p} \overline{\overline{f}}_\alpha^{(P)}(\hat{k}_s, \hat{k}_i) \otimes \overline{\overline{f}}_\beta^{(P)*}(\hat{k}_s, \hat{k}_i) \exp\left(i\vec{k}_d \cdot (\vec{r}_\alpha - \vec{r}_\beta)\right). \end{aligned} \tag{71}$$

We assume the following joint probability density function. For $\alpha \neq \beta$

$$\begin{aligned} p\left(\vec{r}_\alpha, \vec{r}_{\alpha j_1}^{\,(L_1)}, s_{\alpha j_1}^{(L_1)};\ \vec{r}_\beta, \vec{r}_{\beta j_2}^{\,(L_2)}, s_{\beta j_2}^{(L_2)}\right) &= g_p(\vec{r}_\alpha - \vec{r}_\beta) p(\vec{r}_\alpha) p(\vec{r}_\beta) \\ &\times p(\vec{r}_{\alpha j_1}^{\,(L_1)}, s_{\alpha j_1}^{(L_1)}) p(\vec{r}_{\beta j_2}^{\,(L_2)}, s_{\beta j_2}^{(L_2)}) \end{aligned} \tag{72}$$

where $g_{\mathrm{p}}(\vec{r}_\alpha - \vec{r}_\beta)$ is the pair-distribution function for two different primary scatterers. As $|\vec{r}_\alpha - \vec{r}_\beta|$ becomes much larger than the primary scatterer size V_{p}, the pair function g_{p} asymptotically approaches unity. Thus, it is conventional to write $g_{\mathrm{p}} = 1 + (g_{\mathrm{p}} - 1)$. Using (72) in (71) gives

$$\begin{aligned}
&\left\langle \overline{\overline{F}}^{(1)}(\hat{k}_{\mathrm{s}}, \hat{k}_{\mathrm{i}}) \otimes \overline{\overline{F}}^{(1)*}(\hat{k}_{\mathrm{s}}, \hat{k}_{\mathrm{i}}) \right\rangle \\
&= N_{\mathrm{p}} \left\langle \overline{\overline{f}}_{\alpha}^{(P)}(\hat{k}_{\mathrm{s}}, \hat{k}_{\mathrm{i}}) \otimes \overline{\overline{f}}_{\alpha}^{(P)*}(\hat{k}_{\mathrm{s}}, \hat{k}_{\mathrm{i}}) \right\rangle \\
&\quad + \frac{N_{\mathrm{p}}(N_{\mathrm{p}} - 1)}{V^2} \int_V \mathrm{d}\vec{r}_\alpha \int_V \mathrm{d}\vec{r}_\beta \left[g_{\mathrm{p}}(\vec{r}_\alpha - \vec{r}_\beta) - 1 \right] \\
&\quad \times \exp\left(i\vec{k}_{\mathrm{d}} \cdot (\vec{r}_\alpha - \vec{r}_\beta) \right) \left\langle \overline{\overline{f}}_{\alpha}^{(P)}(\hat{k}_{\mathrm{s}}, \hat{k}_{\mathrm{i}}) \right\rangle \otimes \left\langle \overline{\overline{f}}_{\beta}^{(P)*}(\hat{k}_{\mathrm{s}}, \hat{k}_{\mathrm{i}}) \right\rangle \\
&\quad + \frac{N_{\mathrm{p}}(N_{\mathrm{p}} - 1)}{V^2} \int_V \mathrm{d}\vec{r}_\alpha \int_V \mathrm{d}\vec{r}_\beta \exp\left(i\vec{k}_{\mathrm{d}} \cdot (\vec{r}_\alpha - \vec{r}_\beta) \right) \\
&\quad \times \left\langle \overline{\overline{f}}_{\alpha}^{(P)}(\hat{k}_{\mathrm{s}}, \hat{k}_{\mathrm{i}}) \right\rangle \otimes \left\langle \overline{\overline{f}}_{\beta}^{(P)*}(\hat{k}_{\mathrm{s}}, \hat{k}_{\mathrm{i}}) \right\rangle
\end{aligned} \tag{73}$$

where $\langle \overline{\overline{f}}_{\alpha}^{(P)}(\hat{k}_{\mathrm{s}}, \hat{k}_{\mathrm{i}}) \otimes \overline{\overline{f}}_{\alpha}^{(P)*}(\hat{k}_{\mathrm{s}}, \hat{k}_{\mathrm{i}}) \rangle$ is the bistatic intensity from the same primary scatterer and, in view of (64), involves four summations $\sum_{L_1} \sum_{j_1} \sum_{L_2} \sum_{j_2}$. We decompose these into three types of terms:

(i) $L_1 = L_2$, $j_1 = j_2$,

(ii) $L_1 = L_2$, $j_1 \neq j_2$, and

(iii) $L_1 \neq L_2$.

We further express the joint probability density function of secondary scatterers of (α, L_1, j_1) and (α, L_2, j_2) of the same primary scatterer in terms of pair-distribution function g_{s}

$$\begin{aligned}
p\left(\vec{r}_{\alpha j_1}^{(L_1)}, s_{\alpha j_1}^{(L_1)}; \vec{r}_{\alpha j_2}^{(L_2)}, s_{\alpha j_2}^{(L_2)}\right) &= g_{\mathrm{s}}\left(\vec{r}_{\alpha j_1}^{(L_1)}, s_{\alpha j_1}^{(L_1)}; \vec{r}_{\alpha j_2}^{(L_2)}, s_{\alpha j_2}^{(L_2)}\right) \\
&\quad \times p\left(\vec{r}_{\alpha j_1}^{(L_1)}, s_{\alpha j_1}^{(L_1)}\right) p\left(\vec{r}_{\alpha j_2}^{(L_2)}, s_{\alpha j_2}^{(L_2)}\right)
\end{aligned} \tag{74}$$

where g_{s} also asymptotically approaches unity as the separation between the two secondary scatterers becomes large. Substituting (74) into (73), and noting $g_{\mathrm{s}} = (g_{\mathrm{s}} - 1) + 1$, and together with the decomposition gives five terms as follows.

$$\begin{aligned}
&\left\langle \overline{\overline{f}}_{\alpha}^{(P)}(\hat{k}_{\mathrm{s}}, \hat{k}_{\mathrm{i}}) \otimes \overline{\overline{f}}_{\alpha}^{(P)}(\hat{k}_{\mathrm{s}}, \hat{k}_{\mathrm{i}}) \right\rangle \\
&= \sum_{L=1}^{T} N_L \left\langle \overline{\overline{f}}_{\alpha j}^{(L)}(\hat{k}_{\mathrm{s}}, \hat{k}_{\mathrm{i}}, s_{\alpha j}^{(L)}) \otimes \overline{\overline{f}}_{\alpha j}^{(L)*}(\hat{k}_{\mathrm{s}}, \hat{k}_{\mathrm{i}}, s_{\alpha j}^{(L)}) \right\rangle \\
&\quad + \sum_{L=1}^{T} N_L(N_L - 1) \left\langle \overline{\overline{f}}_{\alpha j_1}^{(L)}(\hat{k}_{\mathrm{s}}, \hat{k}_{\mathrm{i}}, s_{\alpha j_1}^{(L)}) \exp\left(i\vec{k}_{\mathrm{d}} \cdot \vec{r}_{\alpha j_1}^{(L)} \right) \right\rangle \\
&\quad \otimes \left\langle \overline{\overline{f}}_{\alpha j_2}^{(L)*}(\hat{k}_{\mathrm{s}}, \hat{k}_{\mathrm{i}}, s_{\alpha j_2}^{(L)}) \exp\left(-i\vec{k}_{\mathrm{d}} \cdot \vec{r}_{\alpha j_2}^{(L)} \right) \right\rangle
\end{aligned}$$

$$
\begin{aligned}
&+\sum_{L=1}^{T} N_L(N_L-1)\int d\vec{r}_{\alpha j_1}^{(L)}\int ds_{\alpha j_1}^{(L)}\int d\vec{r}_{\alpha j_2}^{(L)}\int ds_{\alpha j_2}^{(L)}\\
&\times\left(g_s\left(\vec{r}_{\alpha j_1}^{(L)},s_{\alpha j_1}^{(L)},\vec{r}_{\alpha j_2}^{(L)},s_{\alpha j_2}^{(L)}\right)-1\right)p\left(\vec{r}_{\alpha j_1}^{(L)},s_{\alpha j_1}^{(L)}\right)p\left(\vec{r}_{\alpha j_2}^{(L)},s_{\alpha j_2}^{(L)}\right)\\
&\times\overline{\overline{f}}_{\alpha j_1}^{(L)}\left(\hat{k}_s,\hat{k}_i,s_{\alpha j_1}^{(L)}\right)\exp\left(i\vec{k}_d\cdot\vec{r}_{\alpha j_1}^{(L)}\right)\\
&\otimes\overline{\overline{f}}_{\alpha j_2}^{(L)*}\left(\hat{k}_s,\hat{k}_i,s_{\alpha j_2}^{(L)}\right)\exp\left(-i\vec{k}_d\cdot\vec{r}_{\alpha j_2}^{(L)}\right)\\
&+\sum_{\substack{L_1=1\\L_1\neq L_2}}^{T}\sum_{L_2=1}^{T}N_{L_1}N_{L_2}\left\langle\overline{\overline{f}}_{\alpha j_1}^{L_1}\left(\hat{k}_s,\hat{k}_i,s_{\alpha j_1}^{(L_1)}\right)\exp\left(i\vec{k}_d\cdot\vec{r}_{\alpha j_1}^{(L_1)}\right)\right\rangle\\
&\otimes\left\langle\overline{\overline{f}}_{\alpha j_2}^{L_2*}\left(\hat{k}_s,\hat{k}_i,s_{\alpha j_2}^{(L_2)}\right)\exp\left(-i\vec{k}_d\cdot\vec{r}_{\alpha j_2}^{(L_2)}\right)\right\rangle\\
&+\sum_{\substack{L_1=1\\L_1\neq L_2}}^{T}\sum_{L_2=1}^{T}N_{L_1}N_{L_2}\int d\vec{r}_{\alpha j_1}^{(L_1)}\int ds_{\alpha j_1}^{(L_1)}\int d\vec{r}_{\alpha j_2}^{(L_2)}\int ds_{\alpha j_2}^{(L_2)}\\
&\times\left(g_s\left(\vec{r}_{\alpha j_1}^{(L_1)},s_{\alpha j_1}^{(L_1)},\vec{r}_{\alpha j_2}^{(L_2)},s_{\alpha j_2}^{(L_2)}\right)-1\right)p\left(\vec{r}_{\alpha j_1}^{(L_1)},s_{\alpha j_1}^{(L_1)}\right)p\left(\vec{r}_{\alpha j_2}^{(L_2)},s_{\alpha j_2}^{(L_2)}\right)\\
&\times\overline{\overline{f}}_{\alpha j_1}^{(L_1)}\left(\hat{k}_s,\hat{k}_i,s_{\alpha j_1}^{(L_1)}\right)\exp\left(i\vec{k}_d\cdot\vec{r}_{\alpha j_1}^{(L_1)}\right)\\
&\otimes\overline{\overline{f}}_{\alpha j_2}^{(L_2)*}\left(\hat{k}_s,\hat{k}_i,s_{\alpha j_2}^{(L_2)}\right)\exp\left(-i\vec{k}_d\cdot\vec{r}_{\alpha j_2}^{(L_2)}\right).
\end{aligned}
\tag{75}
$$

Using Equations (70) and (73) in (62), the bistatic intensity of the incoherent wave can be calculated. We note that in the limit of large V and N_p, the last term in (73) will cancel that of (70). Hence, we have

$$
\begin{aligned}
&\lim_{V\to\text{large}}\frac{1}{V}\left\langle\overline{\overline{\mathcal{F}}}^{(1)}(\hat{k}_s,\hat{k}_i)\otimes\overline{\overline{\mathcal{F}}}^{(1)*}(\hat{k}_s,\hat{k}_i)\right\rangle\\
&=n_p\left\langle\overline{\overline{f}}_{\alpha}^{(P)}(\hat{k}_s,\hat{k}_i)\otimes\overline{\overline{f}}_{\alpha}^{(P)*}(\hat{k}_s,\hat{k}_i)\right\rangle\\
&\quad+n_p^2\left\langle\overline{\overline{f}}_{\alpha}^{(P)}(\hat{k}_s,\hat{k}_i)\right\rangle\otimes\left\langle\overline{\overline{f}}_{\beta}^{(P)*}(\hat{k}_s,\hat{k}_i)\right\rangle\\
&\quad\times\int d\vec{r}\left(g_p(\vec{r})-1\right)\exp\left(i\vec{k}_d\cdot\vec{r}\right).
\end{aligned}
\tag{76}
$$

Once the extinction coefficient and the phase matrix are calculated, these are readily incorporated into the collective radiative transfer equations (RTC) as shown in the next section. Numerical solutions are also readily calculated.

7. VECTOR RADIATIVE TRANSFER EQUATION INCLUDING THE EFFECTS OF COLLECTIVE SCATTERING

The vector radiative transfer equation is of the following form,

$$
\begin{aligned}
\cos\theta\,\frac{d\overline{I}(\theta,\phi,z)}{dz}&=-\overline{\overline{\kappa}}_e(\theta,\phi)\cdot\overline{I}(\theta,\phi,z)\\
&\quad+\int_0^{2\pi}d\phi'\int_0^{\pi}d\theta'\sin\theta'\,\overline{\overline{P}}(\theta,\phi;\,\theta',\phi')\overline{I}(\theta',\phi',z)
\end{aligned}
\tag{77}
$$

where $\overline{I}(\theta, \phi, z)$ is the 4 × 1 Stokes vector, $\overline{\overline{\kappa}}_e(\theta, \phi)$ is the extinction matrix, and $\overline{\overline{P}}(\theta, \phi; \theta', \phi')$ is the phase matrix. The phase matrix elements are now defined as covariance of the elements of the incoherent scattering dyad of N scatterers in the limit of large V. For example,

$$P_{11}(\theta, \phi;\ \theta', \phi') = \lim_{V \to \text{large}} \frac{1}{V} \langle \mathcal{F}_{vv}(\hat{k}, \hat{k}') \mathcal{F}^*_{vv}(\hat{k}, \hat{k}') \rangle \tag{78a}$$

$$P_{12}(\theta, \phi;\ \theta', \phi') = \lim_{V \to \text{large}} \frac{1}{V} \langle \mathcal{F}_{vh}(\hat{k}, \hat{k}') \mathcal{F}^*_{vh}(\hat{k}, \hat{k}') \rangle \tag{78b}$$

$$P_{13}(\theta, \phi;\ \theta', \phi') = \lim_{V \to \text{large}} \frac{1}{V} \mathrm{Re}\{\langle \mathcal{F}_{vv}(\hat{k}, \hat{k}') \mathcal{F}^*_{vh}(\hat{k}, \hat{k}') \rangle\} \tag{78c}$$

with similar definitions for other phase matrix elements with 1 to 1 correspondence with that of conventional independent scattering [1]. In (77) and (78), $\hat{k}' = (\theta', \phi')$ and $\hat{k} = (\theta, \phi)$ represent, respectively, the incident and scattered directions. The extinction matrix elements in the past have been calculated by using Foldy's approximation or the forward scattering theorem. However, as shown in the previous section, this requires a higher order approximation (in this case, at least a second-order approximation) to the N particle scattering dyad $\overline{\overline{F}}$, which is not easy to calculate in this case. Thus, we use the alternative of first calculating the absorption coefficient and then adding the absorption coefficient to the scattering coefficient. The absorption coefficient is calculated by assuming independent scattering. Thus,

$$\kappa_{\mathrm{a}\beta}(\theta, \phi) = n_{\mathrm{p}} \sum_{L=1}^{T} N_L \int \mathrm{d}s_{\alpha j}^{(L)} p(s_{\alpha j}^{(L)}) \sigma_{\mathrm{a}\beta}(s_{\alpha j}^{(L)}, \hat{k}) \tag{79}$$

where $\beta = \mathrm{v}, \mathrm{h}$ and $\sigma_{\mathrm{a}\beta}(s_{\alpha j}^{(L)}, \hat{k})$ is the absorption coefficient of the secondary scatterer $s_{\alpha j}^{(L)}$ of polarization β with incident direction $\hat{k}$. An average over orientation and size is as indicated in Equation (79). The extinction coefficients for vertically and horizontally polarized waves are, respectively,

$$\begin{aligned}\kappa_{\mathrm{ev}}(\theta, \phi) = \kappa_{\mathrm{av}}(\theta, \phi) + \int_0^{2\pi} \mathrm{d}\phi' \int_0^{\pi} \mathrm{d}\theta' \sin\theta' \big(P_{11}(\theta, \phi;\ \theta', \phi') \\ + P_{21}(\theta, \phi;\ \theta', \phi')\big)\end{aligned} \tag{80}$$

$$\begin{aligned}\kappa_{\mathrm{eh}}(\theta, \phi) = \kappa_{\mathrm{ah}}(\theta, \phi) + \int_0^{2\pi} \mathrm{d}\phi' \int_0^{\pi} \mathrm{d}\theta' \sin\theta' \big(P_{12}(\theta, \phi;\ \theta', \phi') \\ + P_{22}(\theta, \phi;\ \theta', \phi')\big).\end{aligned} \tag{81}$$

8. APPLICATION TO PASSIVE REMOTE SENSING

The purpose of this paper is to discuss the fundamentals of collective scattering and how they can be included in the framework of radiative transfer theory of collective scattering (RTC). The results in the previous section indicate that two additional parameters are important in coherent collective scattering. They are respectively g_p, the pair distribution function between clusters and g_s, the pair distribution function of secondary scatterers within a cluster. How these two quantities are determined is a problem of geometry. Thus, the problem of geometrical characterization of the vegetation canopy is an important one for collective scattering. For example, the idea of fractals and fractal growth [33] may be used to simplify the geometrical description. Nevertheless, in this section we attempt to illustrate the results of collective scattering for a soybean canopy. However, there are gross simplifications of geometry such as pair distribution functions, number of leaves in a cluster etc. Further refinement of the geometrical description has to be made before the collective theory can be actually applied to data matching.

In this section, we apply the formalism to passive remote sensing of soybean plants. In terms of the notations of clusters of scatterers, each soybean plant is a primary scatterer (Fig. 13). The soybean plants are of height d and there are N_a plants per unit area. Thus,

$$n_p = \frac{N_a}{d}. \tag{82}$$

We choose the center of the primary scatterer to be at $(0,0,0)$. Each soybean plant consists of two types of secondary scatterers: 1 stem and N_l leaves. Thus, $N_1 = 1$ and $N_2 = N_l$. Each leaf consists of a cylinder (branch) with a disk attached at the end. The stem is represented by a vertical cylinder with scattering dyad $\overline{\overline{f}}^{(s)}$. The stem is centered at $(0,0,-d/2)$. Note that the position of the secondary scatterer $\vec{r}_{\alpha j}^{(L)}$ is defined with respect to the center of the primary scatterer. Thus, $\vec{r}^{(1)} = -d/2\hat{z}$, and for type 1 secondary scatterer

$$\overline{\overline{f}}^{(1)}(\hat{k}_s, \hat{k}_i, s^{(1)}) \exp\left(i\vec{k}_d \cdot \vec{r}^{(1)}\right) = \overline{\overline{f}}^{(s)}(\hat{k}_s, \hat{k}_i) \exp\left(ik_{dz}\frac{d}{2}\right) \tag{83}$$

where $\vec{k}_d = k_{dx}\hat{x} + k_{dy}\hat{y} + k_{dz}\hat{z}$. For type 2 scatterer, let the branch be of length L_b and the axis be in direction $\hat{z}_b$, where

$$\hat{z}_b = \sin\beta\cos\alpha\hat{x} + \sin\beta\sin\alpha\hat{y} + \cos\beta\hat{z}. \tag{84}$$

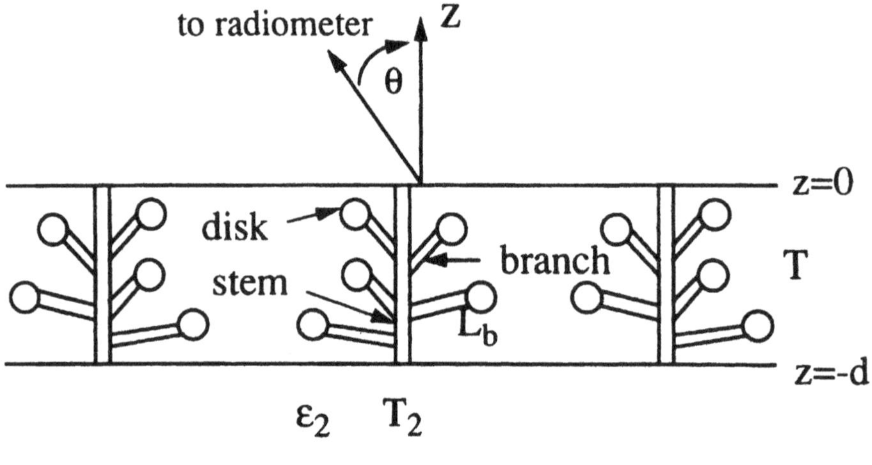

Figure 13. Sketch of a soybean canopy.

For the disk, the axis is along direction $\hat{z}_d$ with

$$\hat{z}_d = \sin\beta_d \cos\alpha_d \hat{x} + \sin\beta_d \sin\alpha_d \hat{y} + \cos\beta_d \hat{z}. \tag{85}$$

The stem/leaf junction is at $z_{aj}\hat{z}$ where z_{aj} is a random variable and $-d \leqslant z_{aj} \leqslant 0$. The center of the branch is at $L_b/2\hat{z}_b$ with respect to the stem/leaf junction and the center of the disk is at $L_b\hat{z}_b$ with respect to the stem/leaf junction. Hence, the scattering dyad of the primary scatterer is

$$\begin{aligned}\overline{\overline{f}}_\alpha^{(P)}(\hat{k}_s, \hat{k}_i) &= \overline{\overline{f}}^{(s)}(\hat{k}_s, \hat{k}_i) \exp\left(-ik_{dz}\frac{d}{2}\right) \\ &+ \sum_{j=1}^{N_l} \left[\overline{\overline{f}}^{(b)}(\hat{k}_s, \hat{k}_i, s_{bj}) \exp\left(i\vec{k}_d \cdot \hat{z}_{bj}\frac{L_b}{2}\right)\right. \\ &\left. + \overline{\overline{f}}^{(d)}(\hat{k}_s, \hat{k}_i, s_{dj}) \exp\left(i\vec{k}_d \cdot \hat{z}_{bj} L_b\right)\right] \exp(ik_{dz} z_{aj})\end{aligned} \tag{86}$$

where $\overline{\overline{f}}^{(b)}$ and $\overline{\overline{f}}^{(d)}$ are the scattering dyads of the branch and the disk, respectively, with both centered at the origin, and s_b and s_d denote the orientation dependence of the scattering dyads. Let the position and orientation distributions be given by the probability density function as follows

$$p(z_a) = \frac{1}{d} \tag{87}$$

$$p(\alpha, \beta;\ \alpha_d, \beta_d) = \frac{p(\beta)p(\beta_d)}{4\pi^2} \tag{88}$$

where $0 \leqslant \alpha \leqslant 2\pi$, and $0 \leqslant \alpha_d \leqslant 2\pi$. Hence, using (86)–(88) in (69), we have the averaged scattering dyad of the primary scatterer

$$\left\langle \overline{\overline{f}}_\alpha^{(P)}(\hat{k}_s, \hat{k}_i) \right\rangle = \overline{\overline{f}}^{(s)}(\hat{k}_s, \hat{k}_i) e^{-ik_{dz}\frac{d}{2}} + N_l \left\langle \overline{\overline{f}}^{(l)}(\hat{k}_s, \hat{k}_i) e^{ik_{dz} z_a} \right\rangle \tag{89}$$

where $\overline{\overline{f}}^{(l)}$ denotes the leaf scattering dyad with weighted average

$$\begin{aligned}&\left\langle \overline{\overline{f}}^{(l)}(\hat{k}_s, \hat{k}_i) \exp\left(ik_{dz} z_a\right) \right\rangle \\ &= \frac{\exp\left(-ik_{dz}\frac{d}{2}\right)}{4\pi^2} \frac{\sin\left(k_{dz}\frac{d}{2}\right)}{k_{dz}\frac{d}{2}} \int_0^{2\pi} \mathrm{d}\alpha \int_0^{2\pi} \mathrm{d}\alpha_d \int_0^{\pi} \mathrm{d}\beta \int_0^{\pi} \mathrm{d}\beta_d \\ &\times p(\beta)p(\beta_d)\left[\overline{\overline{f}}^{(b)}(\hat{k}_s, \hat{k}_i, s_b) \exp\left(i\vec{k}_d \cdot \hat{z}_b \frac{L_b}{2}\right)\right. \\ &\left. + \overline{\overline{f}}^{(d)}(\hat{k}_s, \hat{k}_i, s_d) \exp\left(i\vec{k}_d \cdot \hat{z}_b L_b\right)\right].\end{aligned} \tag{90}$$

Using (86)–(88) in (75), the covariance of the primary scatterer is

$$\left\langle \overline{\overline{f}}_\alpha^{(P)}(\hat{k}_s, \hat{k}_i) \otimes \overline{\overline{f}}_\alpha^{(P)*}(\hat{k}_s, \hat{k}_i) \right\rangle$$

$$
\begin{aligned}
&= \overline{\overline{f}}^{(s)}(\hat{k}_s, \hat{k}_i) \otimes \overline{\overline{f}}^{(s)*}(\hat{k}_s, \hat{k}_i) \\
&+ \frac{N_l}{4\pi^2} \int_0^{2\pi} d\alpha \int_0^{2\pi} d\alpha_d \int_0^{\pi} d\beta \int_0^{\pi} d\beta_d \, p(\beta) p(\beta_d) \\
&\times \left[\overline{\overline{f}}^{(b)}(\hat{k}_s, \hat{k}_i, s_b) \exp\left(i\vec{k}_d \cdot \hat{z}_b \frac{L_b}{2}\right) + \overline{\overline{f}}^{(d)}(\hat{k}_s, \hat{k}_i, s_d) \exp\left(i\vec{k}_d \cdot \hat{z}_b L_b\right) \right] \\
&\otimes \left[\overline{\overline{f}}^{(b)*}(\hat{k}_s, \hat{k}_i, s_b) \exp\left(-i\vec{k}_d \cdot \hat{z}_b \frac{L_b}{2}\right) + \overline{\overline{f}}^{(d)*}(\hat{k}_s, \hat{k}_i, s_d) \exp\left(-i\vec{k}_d \cdot \hat{z}_b L_b\right) \right] \\
&+ N_l(N_l - 1) \left\langle \overline{\overline{f}}^{(l)}(\hat{k}_s, \hat{k}_i) \exp(ik_{dz} z_a) \right\rangle \otimes \left\langle \overline{\overline{f}}^{(l)*}(\hat{k}_s, \hat{k}_i) \exp(-ik_{dz} z_a) \right\rangle \\
&+ N_l \overline{\overline{f}}^{(s)}(\hat{k}_s, \hat{k}_i) \exp\left(ik_{dz} \frac{d}{2}\right) \otimes \left\langle \overline{\overline{f}}^{(l)*}(\hat{k}_s, \hat{k}_i) \exp(-ik_{dz} z_a) \right\rangle \\
&+ N_l \left\langle \overline{\overline{f}}^{(l)}(\hat{k}_s, \hat{k}_i) \exp(ik_{dz} z_a) \right\rangle \otimes \overline{\overline{f}}^{(s)*}(\hat{k}_s, \hat{k}_i) \exp\left(-ik_{dz} \frac{d}{2}\right).
\end{aligned} \tag{91}
$$

By using (90) and (91), the phase matrix and extinction coefficient including collective scattering effects can be readily calculated.

For passive remote sensing (Fig. 13), assuming statistical azimuthal symmetry as indicated by the orientation distributions of (88), the third and fourth Stokes parameters are equal to zero and the Stokes parameters are independent of angle ϕ. Thus, we have

$$
\begin{aligned}
\cos\theta \frac{dI_v(\theta, z)}{dz} &= \kappa_{av}(\theta) CT - \kappa_{ev}(\theta) I_v(\theta, z) \\
&+ \int_0^{\pi} d\theta' \sin\theta' \left(p_{11}(\theta, \theta') I_v(\theta', z) + p_{12}(\theta, \theta') I_h(\theta', z) \right)
\end{aligned} \tag{92}
$$

$$
\begin{aligned}
\cos\theta \frac{dI_h(\theta, z)}{dz} &= \kappa_{ah}(\theta) CT - \kappa_{eh}(\theta) I_h(\theta, z) \\
&+ \int_0^{\pi} d\theta' \sin\theta' \left(p_{21}(\theta, \theta') I_v(\theta', z) + p_{22}(\theta, \theta') I_h(\theta', z) \right)
\end{aligned} \tag{93}
$$

where

$$
p_{ij}(\theta, \theta') = \int_0^{2\pi} d\phi' P_{ij}(\theta, \phi;\ \theta', \phi') \tag{94}
$$

$i, j = 1, 2$, $C = K_b/\lambda^2$, K_b is the Boltzmann's constant, λ is the wavelength, and T is the physical temperature of the soybean. The radiative transfer equations are subject to the boundary conditions [1]. For $0 < \theta < \pi/2$

$$
I_\beta(\pi - \theta,\ z = 0) = 0 \tag{95}
$$

$$
I_\beta(\theta, z = -d) = r_\beta(\theta) I_\beta(\pi - \theta,\ z = -d) + CT_2(1 - r_\beta(\theta)) \tag{96}
$$

where $\beta = $ v, h, and θ is now defined with respect to the azimuth direction (Fig. 13). $r_\beta(\theta)$ is the Fresnel reflectivity at the soil interface and T_2 is the physical temperature of the soil. While we have assumed that the rough surface is assumed to be flat, a rough surface boundary can be inserted. Rough surface effects are important at L-band. However, from the point of view of illustration, we have decided to suppress other effects so as to clearly exhibit the effects of collective volume scattering in passive remote sensing.

After the parameters are calculated by solving the radiative transfer equations, the brightness temperatures are given for $0 < \theta < \pi/2$ by,

$$T_{B\beta}(\theta) = \frac{I_\beta(\theta,\ z=0)}{C}. \tag{97}$$

In the following, we illustrate the collective scattering effects by solving the radiative transfer equation numerically. By using the method of quadratures, the continuous angle θ is discretized into $2N$ intervals. The integral–differential equations are converted into a system of ordinary differential equations [1]. The calculations are done for microwave thermal emission of the soybean canopy at 1.4 GHz (L-band) and 5.0 GHz (C-band). Scattering from the stem and branches is neglected by setting $\overline{\overline{f}}^{(s)} = \overline{\overline{f}}^{(b)} = 0$ in Equations (83) to (91). Only scattering from the disks is considered. However, the clustering effects still exist because the disks are attached to the branches and stems in a manner that is indicated in Fig. 13. The physical optics approximation is used to calculate scattering by a dielectric disk [2].

We choose quadrature number $2N = 16$ in solving the radiative equation. The soil volumetric moisture m_v, which has a large influence at 1.4 GHz and 5.0 GHz is set at 0.2 (wet soil). The soil complex relative dielectric constant is then $5.7 + i1.25$ [32]. The leaf moisture is assumed to be 0.3. The complex relative dielectric constants for a leaf are, respectively, $20 + i6.6$ at 1.4 GHz and $16.9 + i5.7$ at 5.0 GHz [32]. The disks are assumed to be circular with a radius of 2.8 cm and a thickness of 0.012 cm. The orientations of the branches and the disks are such that for branches

$$p(\beta) = \frac{\sin\beta}{\cos\beta_1 - \cos\beta_2} \qquad \beta_1 \leqslant \beta \leqslant \beta_2 \tag{98}$$

and for disks

$$p(\beta_d) = \frac{\sin\beta_d}{\cos\beta_{1d} - \cos\beta_{2d}} \qquad \beta_{1d} \leqslant \beta \leqslant \beta_{2d}. \tag{99}$$

The $\sin\beta$ factor accounts for the smallness of the solid angle at small β. We used $\beta_1 = 70°$, $\beta_2 = 90°$, $\beta_{1d} = 0°$, $\beta_{2d} = 90°$, The physical temperatures are $T = T_2 = 300$ K.

Figures 14 and 15 show the differences between independent scattering theory and collective scattering theory. The parameters considered in this case are as follows: The height of canopy L is 70 cm. The branch length L_b is 2 cm. The number of soybean plants is $N_A = 3.6/\text{m}^2$. The number of leaves is 10/cm along the stem. Figure 14 shows both vertical and horizontal brightness temperatures at 1.4 GHz for different observation angles. Figure 15 shows both vertical and horizontal brightness temperatures at 5.0 GHz for different observation angles. The leaf fractional volume, which equals (number of plants per unit area) × (number of leaves per unit length along the stem) × (volume of a single disk), is 0.001 in this case. The vertical brightness temperature generally increases with observation angle and the horizontal brightness temperature generally decreases with observation angle because of air-soil boundary reflectivity. Compared with independent scattering theory, the collective scattering theory decreases the vertical

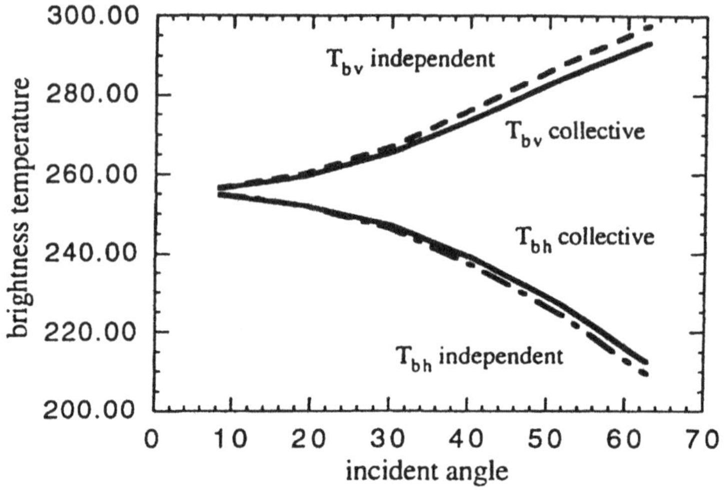

Figure 14. Comparison of both vertical and horizontal brightness temperatures as computed by independent scattering theory (dotted lines) and collective scattering theory (solid lines) at 1.4 GHz (L-band).

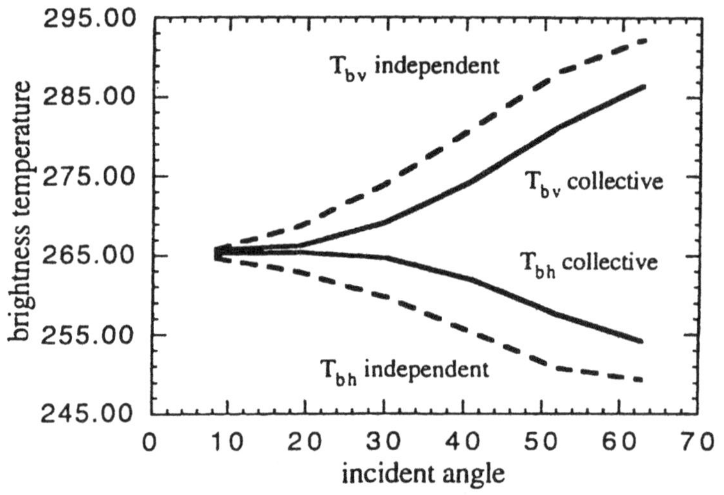

Figure 15. Comparison of both vertical and horizontal brightness temperatures as computed by independent scattering theory (dotted lines) and collective scattering theory (solid lines) at 5.0 GHz (C-band).

brightness temperatures and increases the horizontal brightness temperatures so that the difference between the vertical and horizontal brightness temperatures decreases. This is because the clustering property gives more scattering than independent scattering. The collective scattering theory introduces significantly more scattering (mostly in the forward direction). Scattering redistributes energy between the two polarizations and decreases the polarization contrast of brightness temperatures. At C-band, the difference drop between vertical and horizontal brightness temperatures is more significant than that at L-band. The collective scattering increases scattering much more at C-band than

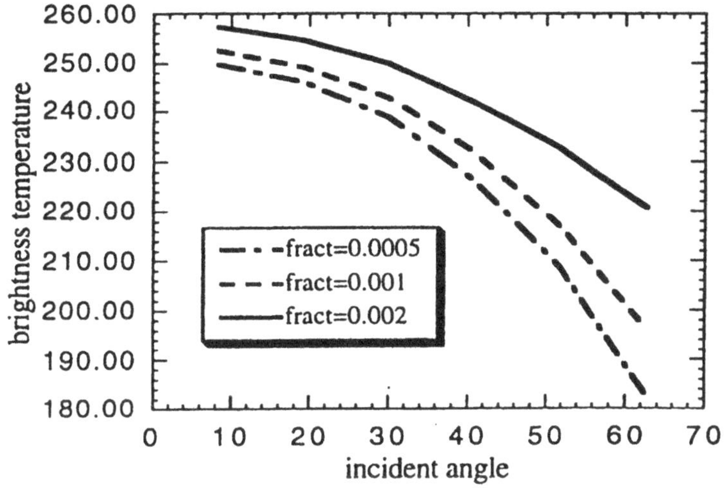

Figure 16. Horizontal brightness temperatures for different fractional volumes at 1.4 GHz (L-band).

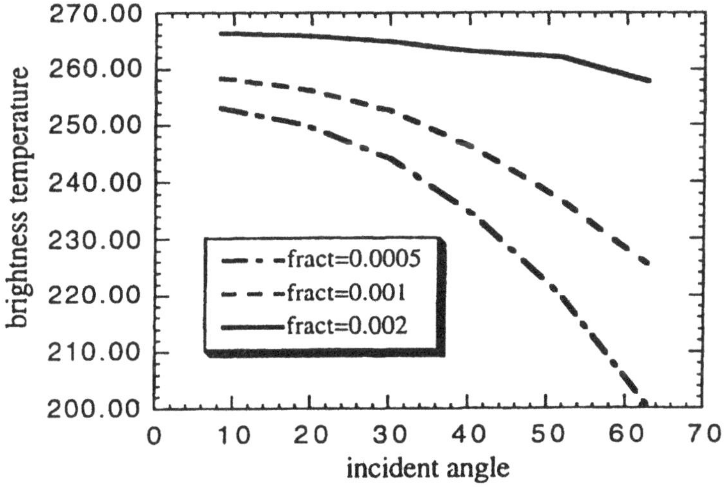

Figure 17. Horizontal brightness temperatures for different fractional volumes at 5.0 GHz (C-band).

at L-band. The increase in extinction rates (from independent scattering to collective scattering) is much bigger than that at L-band.

Figures 16 and 17 show the horizontal brightness temperatures for different fractional volumes at L- and C-bands. The parameters considered in this case are as follows: The height of canopy L is 35 cm. The branch length L_b is 5 cm. The number of soybean plants is $N_A = 7.2/\text{m}^2$. The number of leaves per stem is 2.5, 5 and 10 for three different curves in each figure and this will give fractional volumes of 0.0005, 0.001 and 0.002, respectively. L-band has a lower brightness temperature than C-band. At an observation angle of 52°, L-band gives $T_{bh} = 208$ K while C-band gives $T_{bh} = 220$ K for fractional volume = 0.0005. This is because the absorption rate at L-band is significantly lower. For the above case, the absorption rate at L-band is 8.0×10^{-4}/cm for horizontal polarization while at C-band, the absorption rate is 2.1×10^{-3}/cm. For a fractional volume as low as 0.0005, the brightness temperature at L-band is only slightly higher than that of bare soil, which is 205 K. The vegetation plays a smaller role at L-band. Increasing the fractional volume will increase the brightness temperature at C-band more dramatically. At fractional volume = 0.002, L-band gives $T_{bh} = 232$ K, which is 24 K higher than that at fractional volume = 0.0005. C-band gives $T_{bh} = 262$ K, which is 42 K higher than that at fractional volume = 0.0005. The absorption rates are 3.2×10^{-3}/cm and 8.2×10^{-3}/cm for L- and C-band, respectively, at fractional volume 0.002.

In collective radiative transfer theory the computation of extinction coefficient and phase matrix have to be weighted by pair distribution functions and are computationally more intensive than the classical definitions. However, the solution techniques of radiative transfer equation remain unchanged and the CPU of this part is unaffected. Also a more efficient fractal geometrical description [33–36] in the future may save CPU for the computation of extinction coefficients and phase matrices.

9. CONCLUSIONS

In this paper, we have studied the effects of coherent collective scattering with applications to microwave remote sensing of vegetation. In vegetation, branches and leaves occur in clusters and their collective scattering effects can be important. The collective scattering effects are integrated into a collective radiative transfer theory (RTC). This is done by recognizing that the extinction coefficient and phase matrix in radiative transfer theory are scattering cross sections per unit volume and thus are intrinsic properties of the random media. By taking the limit of large volume with volume much larger than wavelengths and the volumes containing a large number of scatterers, the collective coherent scattering effects of scatterers can be incorporated into the definition of extinction coefficients and phase matrices and subsequently into the collective radiative transfer theory. In the derivation, it is shown that two pair functions g_p and g_s directly affect the solutions. The quantity g_p describes the pair distribution function between primary scatterers (clusters) and the quantity g_s describes the pair distribution function of secondary scatterers within a primary scatterer. The paper also uses the example of point scatterers to illustrate how the iterative approach can be used to solve the Foldy–Lax self-consistent multiple scattering equations and why the very strong coherent forward peak (many dB up) has to be subtracted in order to have a well defined phase matrix and extinction coefficient. An important advantage of using the point scatterer model is that exact solutions of the Foldy–Lax multiple scattering equations can be readily computed for making comparisons with iterative approach. The paper also discusses the importance of Monte Carlo simulations of Maxwell's equations (MCME) which has attracted considerable interest recently in random media and rough surfaces. This is because the solution of electromagnetic fields in microwave remote sensing, given the statistics and joint statistics of the geometry of the random media, are completely and exactly governed by Maxwell's equations. Thus MCME provide the authoritative check without adjustable parameters on the validity of approximate analytic methods. The paper also applies the RTC to passive remote sensing of vegetation. However, the emphasis is more on numerical illustration of the method of RTC rather than on direct application and data matching because the study of collective scattering effects in vegetation is still in its infancy. It is also pointed out that a discrete scatterer model has many parameters. Thus, the problem of geometrical characterization of vegetation canopy is an important problem. One possibility is to use fractals with corresponding primary and secondary scatterers and pair distribution functions.

REFERENCES

1. L. Tsang, J. A. Kong and R. T. Shin. *Theory of Microwave Remote Sensing*. Wiley Interscience, New York (1985).
2. F. T. Ulaby and C. Elachi (Ed.). *Radar Polarimetry for Geoscience Applications*. Artech House, Norwood, Mass. (1990).
3. J. A. Kong (Ed.). *Polarimetric Remote Sensing, Progress in Electromagnetic Research*. Vol. 3, Elsevier, New York (1990).
4. L. Tsang and K. H. Ding. Polarimetric signatures of a layer of random nonspherical discrete scatterers overlying a homogeneous half space based on first and second order vector radiative transfer theory. *IEEE Trans. Geosci. Remote Sensing*, **29**, 242–253 (1991).
5. M. Karam and A. K. Fung. Electromagnetic scattering from a layer of finite length, randomly oriented dielectric circular cylinders over a rough interface with application to vegetation. *Int. J. Rem. Sensing*, **9**, 1109–1134 (1988).
6. L. Tsang. Polarimetric passive remote sensing of rough surfaces and discrete scatterers. *J. Electr. Waves Appl.*, **5**, (1), 41–57 (1991).
7. L. Tsang. Passive remote sensing of dense nontenuous media. *J. Electr. Waves Appl.*, **1** (2), 159–173 (1987).
8. B. Wen, L. Tsang, D. P. Winebrenner and A. Ishimaru. Dense medium radiative transfer theory: comparison with experiment and application to microwave remote sensing and polarimetry. *IEEE Trans. Geosci. Remote Sensing*, **GE-28**, 46–59 (1990).
9. L. Tsang. Dense media radiative transfer theory for dense discrete random media with spherical particles of multiple sizes and permittivities. In: *Progress in Electromagnetics Research*, Vol. 6, *Dielectric Properties of Heterogeneous Materials*, Chapter 5, pp. 181–230, Elsevier, New York (1992).
10. F. T. Ulaby, R. K. Moore and A. K. Fung. *Microwave Remote Sensing, Active and Passive*, Vols 1, 2, and 3. Article House, Lexington, Mass (1981, 1981 and 1986).
11. A. Ishimaru and C. W. Yeh. Matrix representations of the vector radiative transfer equations for randomly distributed nonspherical particles. *J. Opt. Soc. Am. A*, **1**, 359–364 (1984).
12. L. Tsang, J. A. Kong and R. T. Shin. Radiative transfer theory for active and remote sensing of a layer of nonspherical particles. *Radio Science*, **19**, 629–642
13. L. Tsang. Thermal emission of nonspherical particles. *Radio Science*, **19**, 966–974 (1984).
14. G. I. Manchuck, G. A. Mikhailou, M. A. Nagaralieu, R. A. Darbinjin, B. A. Kargin and B. S. Elepou. *The Monte Carlo Methods in Atmosphere Optics*. Springer-Verlag, New York (1980).
15. H. T. Chuah and H. S. Tan. A microwave emission model for vegetative medium using the Monte Carlo method, Part I and II. *J. Electr. Waves Appl.*, **6**, (7), 799–852 (1992).
16. R. M. Axline and A. K. Fung. Numerical computation of scattering from a perfectly conducting rough surface. *IEEE Trans. Ant. Prop.*, **AP-26**, 482–488 (1978).
17. J. S. Chen and A. Ishimaru. Numerical simulation of the second order Kirchhoff approximation from very rough surfaces and study of backscattering enhancement. *J. Acoust. Soc. Am. A.*, **88**, 1846–1850 (1990).
18. L. Tsang, C. H. Chan and H. Sangani. Banded matrix iterative approach to Monte Carlo simulations of scattering of waves by large-scale random rough surface problems: TM case. *Electronics Letters*, **29**, 166–167 (1993).

19. L. Tsang, C. H. Chan, H. Sangani, A. Ishimaru and P. Phu. A banded matrix iterative approach to Monte Carlo simulations of large scale random rough surface scattering: TE case. *J. Electr. Waves Appl.*, **7** (9), 1185–1200 (1993).

20. L. Tsang, C. H. Chan and K. Pak. Monte Carlo simulations of a two-dimensional random rough surface using the sparse-matrix flat-surface iterative approach. *Electronics Letters*, **29** (13), 1153–1154 (1993).

21. L. Tsang, C. E. Mandt and K. H. Ding. Monte Carlo simulations of the extinction rate of dense media with randomly distributed dielectric spheres based on solution of Maxwell's equations. *Optics Letters*, **17** (5), 314–316 (1992).

22. C. C. Lu, Y. M. Wang, W. C. Chew and L. Tsang. The application of recursive aggregate T-matrix algorithm in the Monte Carlo simulations of the extinction rate of random particles. In: *Proceedings of IEEE Antenas and Propagation Symposium*, Ann Arbor, Michigan (1993).

23. S. H. Yueh, J. A. Kong, J. K. Jao, R. T. Shin and T. LeToan. Branching model for vegetation. *IEEE Trans. Geosci. Remote Sensing*, **30**, 390–402 (1992).

24. S. Chandrasekhar. *Radiative Transfer*. Dover, New York (1960).

25. S. John. Localization of light. *Physics Today*, **44**, 32–40 (May 1991).

26. P. Sheng (Ed.). *Scattering and Localization of Classical Waves in Random Media*. World Scientific, Singapore (1990).

27. Y. Kuga and A. Ishimaru. Retroreflectance from a dense distribution of spherical particles. *J. Opt. Soc. Am. A.*, **1**, 831–835 (1984).

28. L. Tsang and A. Ishimaru. Backscattering enhancement of random discrete scatterers. *J. Opt. Soc. Am. A.*, **1**, 836–839 (1984).

29. M. Lax. Multiple scattering of waves. *Rev. Mod. Phys.*, **23**, 287–310 (1951).

30. K. H. Ding, C. Mandt, L. Tsang and J. A. Kong. Monte Carlo simulations of pair distribution function of dense discrete random media with multiple sizes of particles. *J. Electr. Waves Appl.*, **6**, 1050–1030 (1992).

31. R. W. Newton and J. W. Rouse. Microwave radiometer measurements of soil moisture content. *IEEE Trans. Ant. Prop.*, **AP-28**, (5), 680–686 (Sept 1980).

32. M. W. Whitt. *Microwave Scattering From Periodic Row-Structured Vegetation*. PhD Dissertation, Department of Electrical Engineering, University of Michigan, Ann Arbor (1991).

33. B. B. Mandelbrot. *The Fractal Geometry of Nature*. W. H. Freeman, San Francisco (1983).

34. T. Vicsek. *Fractal Growth Phenomena*. Second edition. World Scientific, Singapore (1992).

35. H. A. Lauwerier. *Fractals: Endlessly Repeated Geometrical Figures*. Princeton University Press, Princeton, New Jork (1991).

36. P. Prusinkiewicz and A. Lindenmayer. *The Algorithmic Beauty of Plants*. Springer-Verlag, New York (1990).

ESA/NASA International Workshop, pp. 155–191
B. J. Choudhury, Y. H. Kerr, E. G. Njoku and P. Pampaloni (Eds)

Synergism of optical and microwave observations for land surface studies

BHASKAR J. CHOUDHURY

Code 974, Hydrological Sciences Branch, NASA/Goddard Space Flight Center, Greenbelt, MD 20771, USA

Abstract—There is much to be gained from synergistic use of optical (visible to infrared) and microwave observations, from atmospheric corrections to and extraction of land surface parameters from passive microwave observations. It is desirable and sometimes it might be necessary to correct microwave observations for atmospheric effects before extracting surface parameters. Optical data could provide information about surface and air temperatures, precipitable water vapor in the atmosphere, cloud top temperature and its liquid water content, which could be used for atmospheric correction. Visible and near-infrared reflectances can be used to correct for the effect of vegetation on microwave emissivities at low frequencies for estimating soil moisture. It is, however, recognized that while water content of vegetation is the key parameter needed for correcting microwave observations, visible and near-infrared reflectances are determined by the chlorophyll concentration and cellular structure of leaves. Microwave emissivities at high frequencies and visible reflectance might provide complementary information about vegetation characteristics, which is explored both by model simulations and analysis of satellite data. Concurrent optical and microwave field measurements are highly desirable to verify some of the modelling studies. Better understanding needs to be developed regarding the effects of spatial heterogeneity on coarse resolution satellite observations. High correlation between visible reflectance and microwave reflectivity could be used to obtain effective surface temperature.

1. INTRODUCTION

Thermal emission at various microwave frequencies can provide information about such land surface characteristics as surface-layer soil moisture, seasonal inundation of rivers and swamps (wetlands), fractional vegetation cover and its water content and snow cover [1, 2]. Estimation of these surface characteristics from microwave observations necessarily involves models relating radiative characteristics (emissivity or reflectivity) to physical characteristics (soil moisture, vegetation water content, etc.), and in general several physical characteristics with differing sensitivities determine the radiative characteristic at any frequency of observation. This non-uniqueness of the relation between radiative characteristics and a particular physical characteristic can lead to uncertainties in estimating surface variables from microwave observations. Synergism of microwave and optical data could aid in a better estimation of the surface variables. Such synergism can also be of considerable help in correcting satellite observations for atmospheric effects, so as to calculate surface emissivity or reflectivity from the brightness temperatures.

This paper discusses synergistic use of optical (visible to infrared) and microwave observations for atmospheric corrections and for relating microwave emissivities to surface variables (soil moisture, vegetation parameters and the effective surface temperature).

While field observations to quantify land surface characteristics are immensely valuable for development, clarification and validation of basic concepts and methodologies, satellite data are indispensable for regional and global studies. The disparity of spatial scales involved in going from typical field observations (c. 10–100 m) to satellite data (c. 20–100 km) presents the major difficulty in directly transferring the methodologies developed and tested over fields to the satellite observations. This disparity of spatial scales might be addressed through large-scale field studies (spatial scales 20–100 km), aircraft observations from different altitudes and model simulations, but nevertheless remains an outstanding problem. Thus, the interpretation of satellite data given in this paper must be considered tentative, requiring validation through large-scale field studies or other approaches.

2. ATMOSPHERIC CORRECTION

Extraction of land surface parameters from satellite microwave observations may require atmospheric correction, particularly when the observations are at high microwave frequencies (say higher than 10 GHz). Since atmospheric effects are being dealt with in detail elsewhere within this volume, the emphasis in this section will be on synergism with optical (visible to infrared) measurements for calculating surface reflectivity or emissivity for studies related to vegetation and soil moisture from satellite observed brightness temperatures, focusing on 19.3 and 37 GHz observations by the Special Sensor Microwave Imager (SSM/I) on board the F-8 satellite of the US Defense Mapping Satellite Programme (DMSP-F8). Examples of the calculated reflectivity are given at the end of this section and further discussed in the next section.

The brightness temperature at polarization p (horizontal or vertical), T_s^p, observed from a satellite is schematically illustrated in Fig. 1, which can be described by the following equation:

$$T_s^p = T_u + T_B^p t_a \tag{1}$$

where T_u is upwelling atmospheric emission, t_a is atmospheric transmission coefficient and T_B^p is the brightness temperature which would be measured close to the ground, given by:

$$T_B^p = R_p T_d + (1 - R_p) T_g \tag{2}$$

where T_d is the downwelling atmospheric emission, T_g is effective surface temperature [3] and R_p is the surface reflectivity, from which the emissivity ε can be obtained as $\varepsilon = 1 - R_p$. The surface reflectivity can be calculated from satellite observed brightness temperatures using Equations (1) and (2) as:

$$R_p = \left(T_g - \frac{T_s^p - T_u}{t_a} \right) \frac{1}{T_g - T_d}. \tag{3}$$

The major atmospheric variables determining T_u, T_d and t_a under non precipitating conditions are oxygen, precipitable water vapor and cloud characteristics (most importantly being the cloud height and liquid water content) and fractional cloud cover in the field of view. The transmission coefficient t_a can be obtained from optical thickness τ as:

$$t_a = \exp(-\tau/\mu) \tag{4}$$

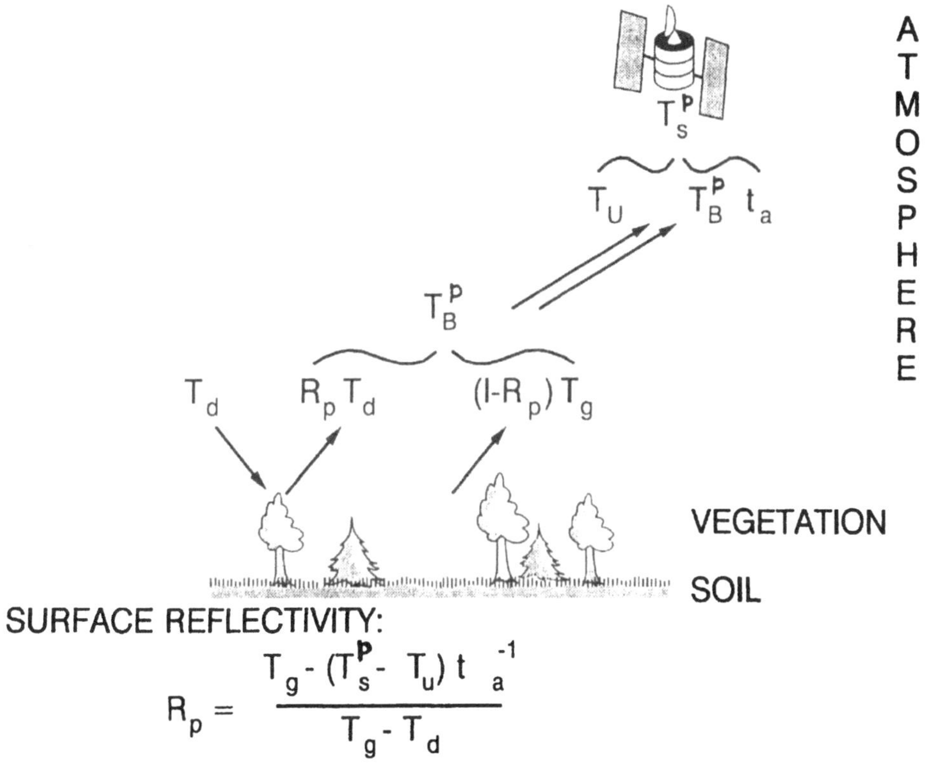

Figure 1. Schematic illustration of radiative transfer components associated with satellite observed brightness temperatures. The symbols are defined in the text.

where μ is cosine of the incidence angle of radiometer and τ is given by:

$$\tau = \int_0^\infty \kappa(z)\,\mathrm{d}z \tag{5}$$

where $\kappa(z)$ is the sum of absorption coefficients due to oxygen, water vapor and cloud as a function of air pressure, varying with height z.

Figure 2 shows the absorption coefficients of oxygen and water vapor as a function of the frequency of observation [4], suggesting the need for atmospheric correction to the satellite data above 15 GHz if surface reflectivity is to be obtained from these data. Calculations of τ using upper air sounding data for eight selected locations from temperate and tropical climates away from oceans (see caption) gave the following relations for 19.3 (Fig. 3a) and 37 GHz (Fig. 3b):

$$\text{for}\quad 19.3\ \text{GHz:}\qquad \tau = 0.011 + 0.0026\,V \tag{6a}$$

with $r^2 = 0.998$ and standard error of estimate (SEE) of 0.0015.

$$\text{and for}\quad 37\ \text{GHz:}\qquad \tau = 0.037 + 0.0021\,V \tag{6b}$$

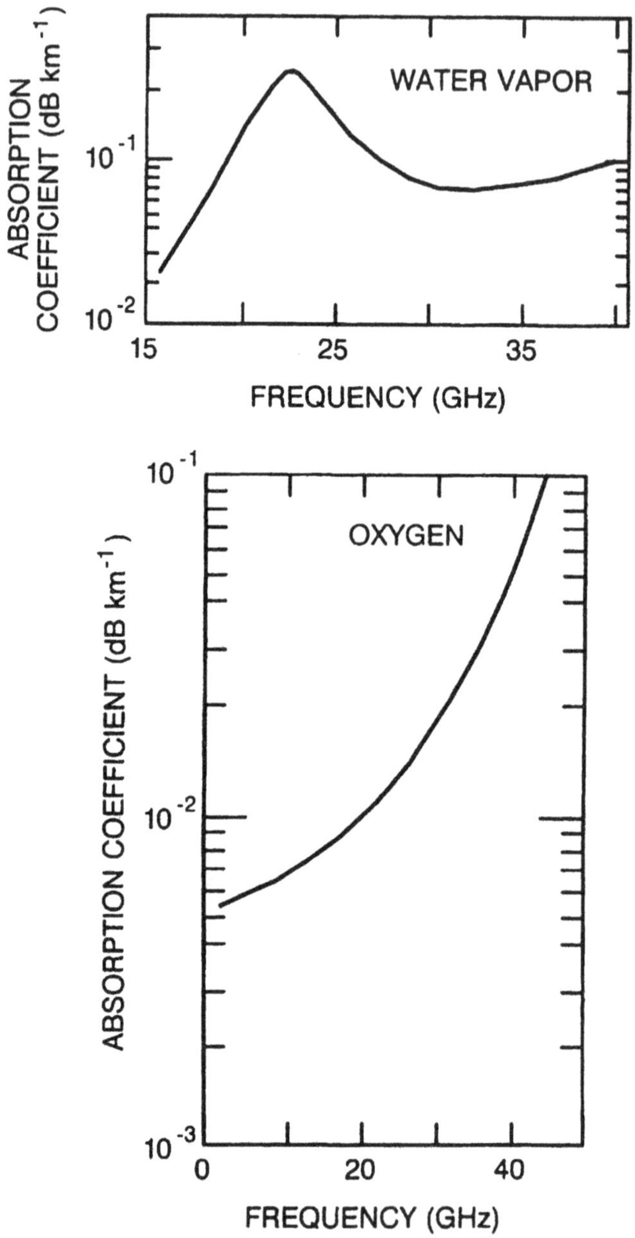

Figure 2. Absorption coefficients of oxygen and water vapor as a function of frequency [4].

with $r^2 = 0.98$ and standard error of estimate (SEE) of 0.0044,
and where V is precipitable water vapor in the atmosphere (mm) and the first term on the right hand side of Equation (6) is due to oxygen absorption. Larger scatter in Fig. 3(b) for 37 GHz compared to that appearing in Fig. 3(a) for 19 GHz is primarily due to higher oxygen absorption at 37 GHz, which is more sensitive to variations of the

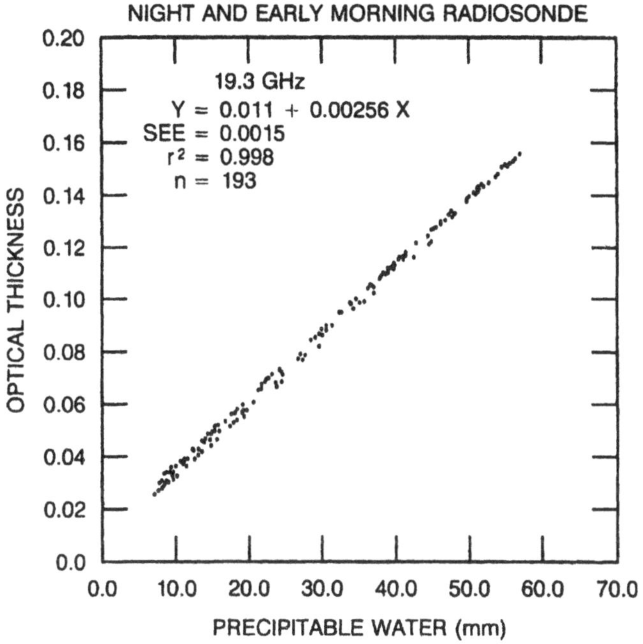

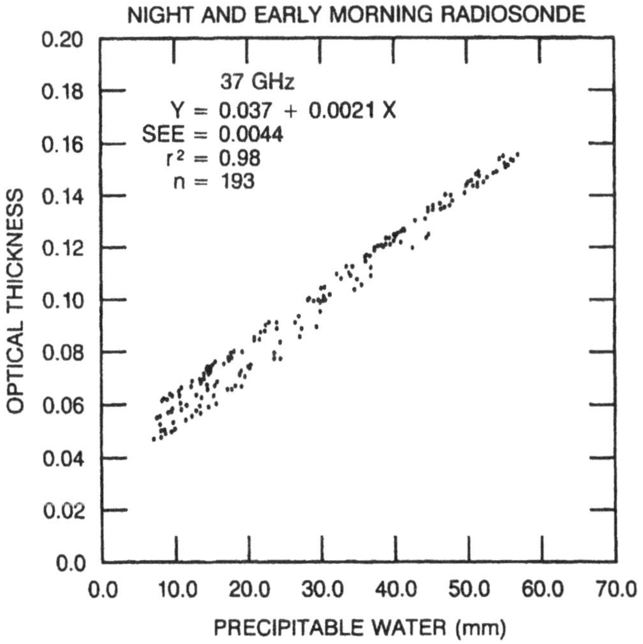

Figure 3. Calculated optical thickness as a function of precipitable water based on night and early morning upper air sounding observations for (a) 19.3 GHz, and (b) 37 GHz frequencies of SSM/I. The station name (number of soundings) used in the calculation are: Washington, DC, USA (24), Longview, USA (23), Nashville, USA (25), Manaus, Brazil (23), Vilhena, Brazil (12), Chiang Mai, Thailand (32), Upington, South Africa (31) and Neuquen, Argentina (23).

surface air pressure and also temperature profile. Similar equations have been obtained previously from analysis of actual observations [5, 6]. It is seen from Equation (6) that atmospheric correction to the satellite observed brightness temperatures under clear sky conditions will require a knowledge of precipitable water at the spatial and temporal scales of the satellite data. In this context one should note that while precipitable water can be derived from upper air sounding (radiosonde) data, these soundings are generally not coincident with and spatially representative of the satellite data. There is a severe lack of sounding data for many parts of the world.

Spatial and temporal distributions of precipitable water vapor needed for atmospheric correction may be obtained by synergism with optical data. Observations by TOVS (Tiros Operational Vertical Sounder on board NOAA satellites) have been used operationally to produce global daily data set for V at $2.5° \times 2.5°$ resolution. The uncertainty in these V values has been estimated to be about 30% [7]. This uncertainty and spatial resolution of V values from TOVS might be suitable for regional and global studies using the satellite microwave data, but may not be appropriate for precise calculations at pixel scales of the microwave data (c. 20–50 km). Surface observations of vapor pressure may also be used to estimate V from empirical equations [8, 9], but these estimates may not be spatially representative for satellite observations. Also, daily observations of surface vapor pressure are available at much more widely separated locations as compared to monthly mean values. Thus, the surface observations of vapor pressure may not be temporally representative of the atmospheric conditions corresponding to the satellite data. High spatial resolution sounders like Atmospheric Infrared Sounder, Advanced Microwave Sounding Unit and Microwave Humidity Sounder (AIRS/AMSU/MHS) planned to be on board the Earth Observing System (EOS) are expected to provide the precipitable water to supplement observations by the Multifrequency Imaging Microwave Radiometer (MIMR) during the EOS era (c. 2000).

The absorption coefficient of water clouds as a function of frequency at different cloud temperatures is shown in Fig. 4 [4]. The absorption coefficient increases with increasing frequency or decreasing cloud temperature, and it is about a factor 3.5 larger at 37 GHz compared to that at 19 GHz. For cloudy conditions at San Nicolas Island (33.3°N, 119.6°W) during July, Westwater *et al.* [5] found:

$$\text{for } 20.60 \text{ GHz:} \quad \tau = 0.018 + 0.0042\,V + 0.06\,L_c \tag{7a}$$

$$\text{for } 31.65 \text{ GHz:} \quad \tau = 0.032 + 0.0018\,V + 0.14\,L_c \tag{7b}$$

where L_c is cloud liquid water content (mm). The coefficients for cloud liquid water would approximately be 0.06 and 0.2, respectively, for 19 and 37 GHz for atmospheric conditions similar to those over the San Nicolas Island, although recognizing that these coefficients depend upon the cloud temperature (cf. Fig. 4). For colder mid-latitudes temperate conditions (air temperature c. 273 K) the coefficients could be 0.09 and 0.3, respectively, for 19 and 37 GHz. We see from Equation (6) that precipitable water vapor has similar effect at 19 and 37 GHz observations, while from Equation (7) we see that cloud liquid water will affect 37 GHz more than 19 GHz. A knowledge of cloud temperature is required together with its liquid water content to correct microwave data for cloudy atmospheres. The effect of cloud on the microwave data will generally be lower for hot tropical conditions (higher cloud temperatures) than for cold temperate conditions under similar liquid water content.

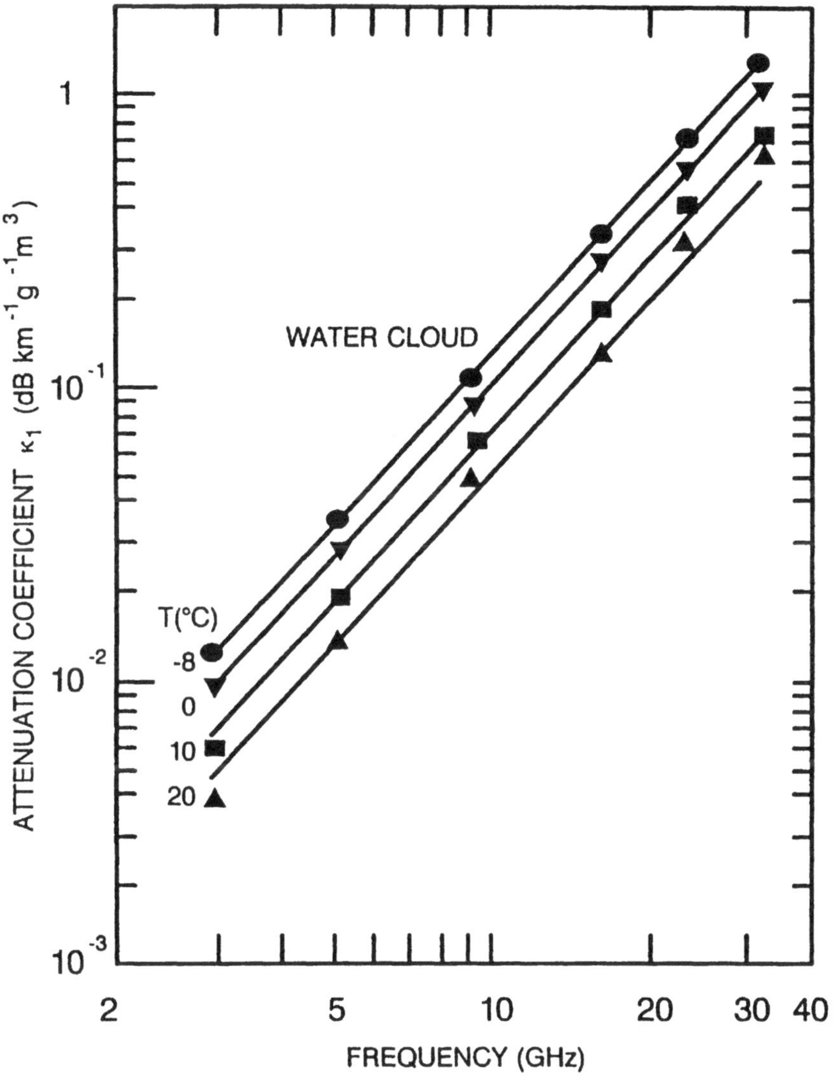

Figure 4. Attenuation coefficients of cloud at different temperatures as a function of frequency [4].

The optical thickness for clouds under non-raining conditions can be calculated fairly accurately when cloud temperature and liquid water content are prescribed [4]. Data sets of cloud fraction and cloud top pressure (temperature) have been produced operationally using the TOVS observations, although there are some biases in these data [10]. These cloud characteristics have also been produced using observations by Nimbus-7 Temperature Humidity Infrared Radiometer (THIR) and Total Ozone Mapping Spectrometer (TOMS) [10]. Attempts to improve and validate these global data sets are continuing.

Very little data is currently available for cloud liquid water content L_c over land at the spatial resolution of satellite observations, although much research has been done

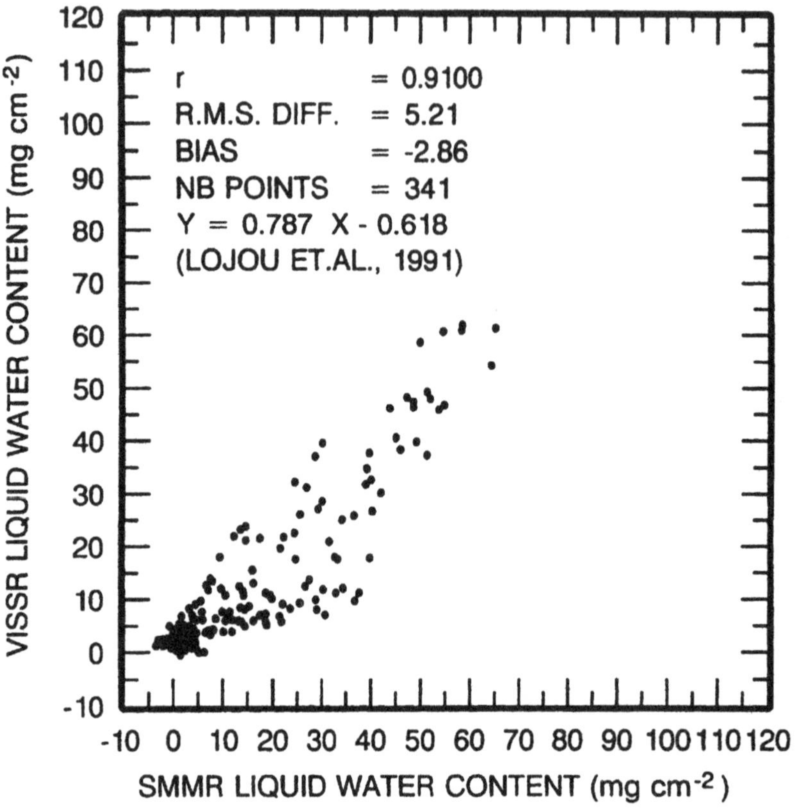

Figure 5. Comparison of cloud liquid water content over ocean derived from visible reflectance and microwave brightness temperatures. Microwave method has been tested extensively for estimation of cloud liquid water over ocean. This comparison and theoretical studies suggest that optical method might be applicable over land for cloud liquid water estimation.

to extract this information over oceans [11–13]. Stephens [14] showed from radiative transfer calculations that cloud optical thickness and reflectance for visible radiation (0.3–0.7 μm) is primarily determined by the liquid water content. Lojou *et al.* [15] found that the liquid water content determined from visible reflectance agrees well with that determined from microwave observations over ocean (Fig. 5). The feasibility of estimating L_c from visible reflectance is encouraging for possible applications over land surface, although the methodology needs to be verified. The cloud optical thickness τ_c at 0.6 μm, determined from satellite observed radiances through a radiative transfer model is part of the ISCCP data set [16], from which L_c could be obtained as [17]:

$$L_c = \frac{4}{3}\frac{r\tau_c}{Q} \tag{8}$$

where Q (= 2.12) is the extinction efficiency for visible radiation and cloud drop radius r has been assumed to be 0.01 mm. It is pertinent to note that cloud drop size is a variable parameter [18], which leads to uncertainties in the calculated L_c values from Equation (8). Nevertheless, synergism with optical measurements from satellites can provide valuable information about precipitable water vapor and cloud liquid water for calculating τ and hence t_a. An analysis of the Optical Line Scanner (OLS; on board the DMSP satellites) data for cloud liquid water content to supplement SSM/I observations could be valuable. The liquid water content of non-raining clouds is generally less than 0.3 mm, although this value could be higher than 1 mm for some very dense clouds [11–13, 15]. For a humid tropical atmosphere with air temperature of 300 K and precipitable water vapor of 40 mm, the satellite observed brightness temperatures will be higher by about 1 and 2 K, respectively, for 19 and 37 GHz SSM/I data under cloudy conditions with 0.3 mm liquid water over a dense vegetation ($R_p = 0.06$) as compared to that for clear sky conditions, while these increases in brightness temperatures are about 2 and 6 K respectively for a higher surface reflectivity ($R_p = 0.16$; horizontal polarization over deserts); the increases in the brightness temperatures are about 20% higher when the precipitable water vapor is decreased to 20 mm. These calculations show that it is more important to correct 37 GHz observations than 19 GHz for effects of clouds.

Calculation of T_u and T_d requires distribution of air pressure, temperature, water vapor and cloud characteristics as a function of height [4]. The difference between these two brightness temperatures is generally less than a few degrees [11]. Equation for T_d is:

$$T_d = \int_0^\infty \kappa(z)T(z)\exp\left(-\int_0^z \kappa(z')\,\mathrm{d}z'/\mu\right)\mathrm{d}z/\mu \tag{9}$$

where $T(z)$ is air temperature at height z and other symbols are as defined for Equations (4) and (5). Equation (9) can be easily evaluated for an isothermal air at temperature T_e to obtain

$$T_d = T_e(1 - t_a) \tag{10}$$

where t_a is defined by Equation (4). Although air is not isothermal, one can obtain Equation (10) for reasonable approximations for variations of air pressure $P(z)$, temperature $T(z)$ and vapor density $\phi(z)$ as:

$$P(z) = P_a\exp\left(-\frac{z}{\alpha}\right) \tag{11a}$$

$$T(z) = T_a - \Gamma z \tag{11b}$$

$$\phi(z) = \phi_a\exp\left(-\frac{z}{\beta}\right) \tag{11c}$$

where P_a, T_a and ϕ_a are the atmospheric surface-layer values for the corresponding variables, Γ is the lapse rate and α and β are damping coefficients. Inserting Equation (11) in (9), one can show by Taylor series expansion that the equation for T_d is as given by Equation (10) with T_e expressed as:

$$T_e = T_a - \Gamma\beta[1 - 3\tau/(4\mu)]/(1 + \beta/\alpha) \tag{12}$$

where τ and μ are as defined in Equation (4). Reasonable values of the parameters appearing in Equation (12) are: $\Gamma = 4$–$7\ \mathrm{K\,km^{-1}}$, $\alpha = 7.5$ km and $\beta = 1.5$–3.0 km, and, as a first approximation, one can use T_a for T_e in Equation (10) with a small adjustment (T_e about 10 K lower than T_a). With the upper air sounding data acquired during night and early-morning hours used to derive Equation (6) it was found that the following equation for T_e gave a fairly accurate estimates of T_d (Fig. 6):

$$\text{for 19.3 GHz:} \quad T_e = T_a - (8 + 0.06\,V) \tag{13a}$$

$$\text{for 37.0 GHz:} \quad T_e = T_a - (18 - 0.12\,V) \tag{13b}$$

where V is the precipitable water (mm). The presence of lapse rate in Equation (12) show that day and night time microwave observations would have slightly different values for T_e.

Darnell *et al.* [19] found good agreement between surface observations of air temperature T_a over United States and the surface layer air temperatures (surface to 85 kPa pressure level) derived from TOVS observations during one year period at one location, while Choudhury *et al.* [20] found the TOVS temperatures to be somewhat lower than surface air observations over northern Africa due probably to different times of observations and the effective atmospheric layer for which TOVS temperatures are representative (the TOVS temperatures might be representative of a temperature in-between T_a and T_e). The uncertainties in the surface layer air temperature derived from TOVS is about 2.5 K [7]. Estimates of T_a and V derived from TOVS observations could provide a global data set for atmospheric correction to 19 and 37 GHz brightness temperature observations by SSM/I for regional and global studies. Certainly, for an accurate atmospheric correction to the SSM/I data at pixel scales one would require detailed spatially representative atmospheric parameters.

Although observations needed for effective surface temperature T_g are not discussed above (see Njoku, and Schmugge in this volume), it is worth noting that the difference between T_g and T_a will not generally exceed 5 K for night and early morning observations (ascending orbits of SSM/I). The T_g for daytime satellite observations (descending orbits of SSM/I) could be 5–15 K higher than T_a, and one may use heat balance models to estimate T_g in the absence of direct observations. Infrared temperature observations by the Advanced Very High Resolution Radiometer (AVHRR, on board NOAA satellites) may also provide a good estimator for T_g for SSM/I data under clear sky conditions when adjusted for different times of observations during a day by these two sensors. Furthermore, while infrared temperatures when corrected for emissivity can provide the surface temperature, T_g is the effective temperature of a surface layer whose thickness depends upon soil moisture and the microwave frequency. This distinction between infrared temperature and T_g is particularly important for deserts and semi-arid regions because of low soil moisture and strong soil temperature gradient. Since T_g and R_p appear as a product (cf. Equation (2)), it is desirable to calculate both of these surface characteristics simultaneously. Synergism with optical data could be used to estimate T_g (discussed below).

Notwithstanding the present uncertainties in the derived surface and air temperatures, precipitable water and cloud liquid water content from optical observations by different space-borne sensors, it is clear that optical data can provide valuable information needed for atmospheric correction to microwave observations for estimating surface reflectivity. These optical data from satellites can provide spatially representative values globally at fairly high temporal resolution (e.g. twice daily), which cannot be obtained otherwise. Since both air temperature and cloud cover can have strong diurnal variations, it is preferable to have the optical data coincident with the microwave observations.

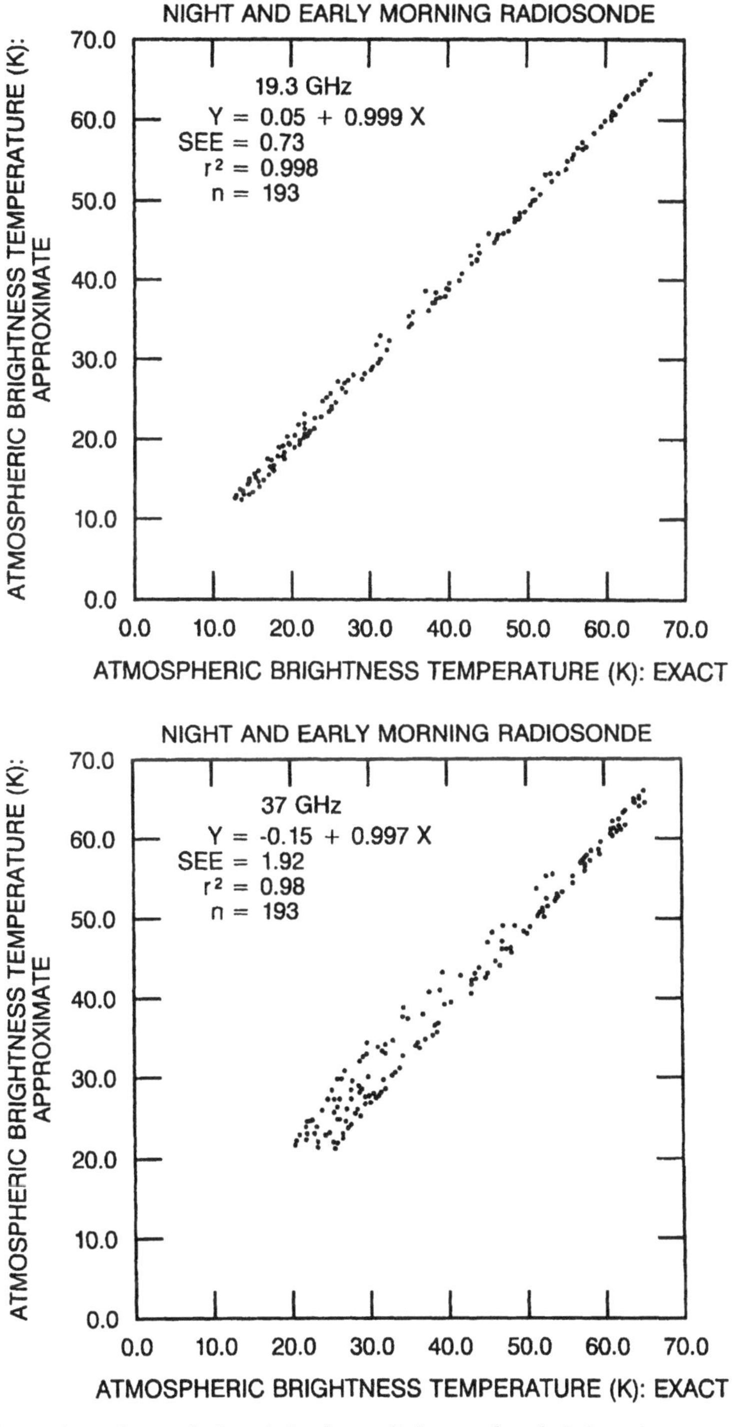

Figure 6. Comparison of atmospheric emission from radiative transfer calculation using upper air sounding data (cf. Fig. 3) and from parameterized equations (Equations (6) and (13)). (a) 19.3 GHz, and (b) 37 GHz.

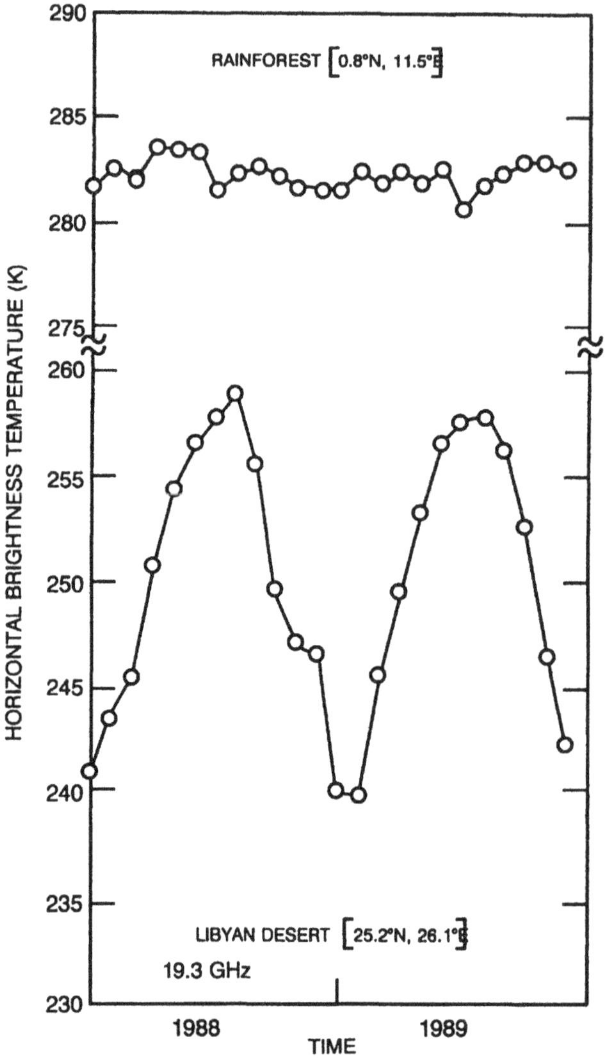

Figure 7(a). Temporal variations of brightness temperatures over Libyan desert and rainforest during 1988–1989 as observed at 19.3 GHz frequency of SSM/I for horizontal polarization. The brightness temperatures have been gridded into 0.25° × 0.25° cells by determining the beam center, and an average value of nine pixels (3 × 3 matrix) centered on the location have been computed.

Temporal variations of horizontally polarized brightness temperature observed by the SSM/I (ascending orbits) at 19 GHz over a sand desert (25.2°N, 26.1°E) and rainforest (0.8°N, 11.5°E) are shown in Fig. 7(a), while these variations for vertically polarized brightness temperature are shown in Fig. 7(b). These brightness temperatures may not directly provide a realistic assessment of spatial and temporal changes of such surface characteristics as vegetation or soil moisture; one needs to consider both atmospheric effect and surface temperature variations for such an assessment. With lower precipitable water vapor over the desert (c. 15 mm) compared to that over the rainforest (c. 45 mm),

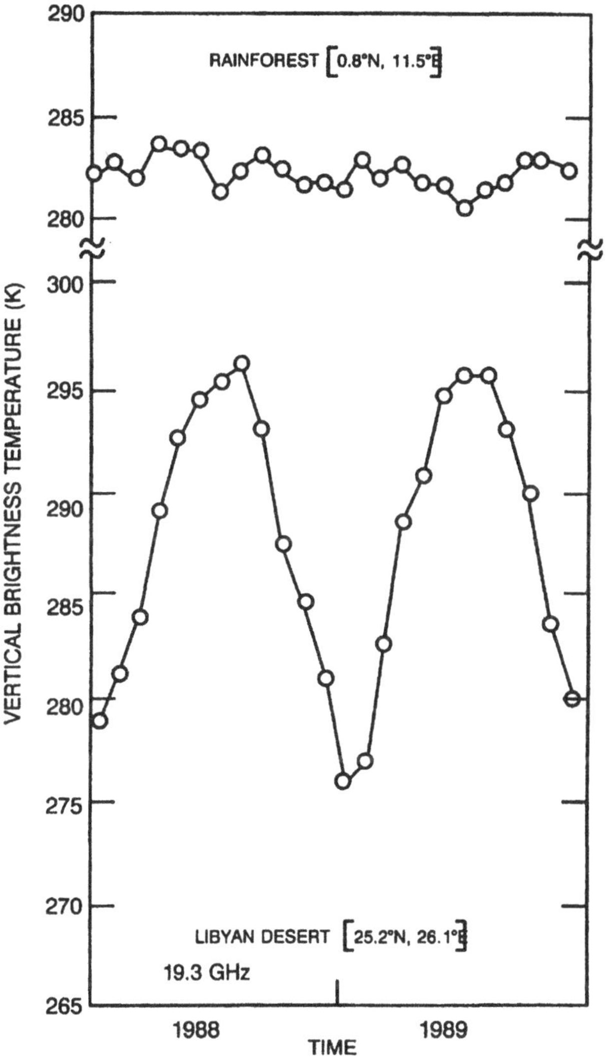

Figure 7(b). Temporal variations of brightness temperatures over Libyan desert and rainforest during 1988–1989 as observed at 19.3 GHz frequency of SSM/I for vertical polarization. The brightness temperatures have been gridded into 0.25° × 0.25° cells by determining the beam center, and an average value of nine pixels (3 × 3 matrix) centered on the location have been computed.

the atmospheric emission is lower and transmission coefficient is higher for the desert location. Also, the seasonal variations of air and surface temperatures for the desert are significantly more pronounced compared to those for the rainforest. Mean monthly surface observations (Monthly Climatic Data of the World, NOAA) of air temperature and vapor pressure at Kufra (24.2°N, 23.3°E) and Dakhla (25.5°N, 29°E) were used for the sand desert, while these observations at Mitzic (0.8°N, 11.5°E) were used for rainforest to calculate atmospheric emission and transmission coefficient. The observed daily mean air temperatures were adjusted for the early morning time of observations

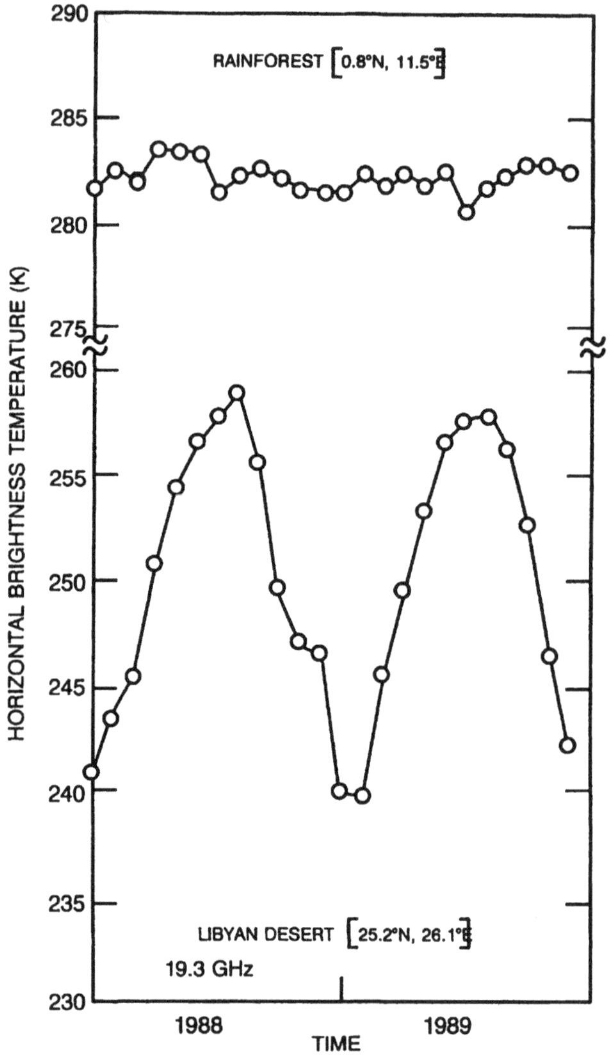

Figure 8(a). Temporal variation of the calculated surface reflectivity for Libyan desert and rainforest from the data in Fig. 7 for horizontal polarization.

by the SSM/I and effective surface temperatures were computed using a heat balance equation [21]. The surface reflectivity calculated from Equations (3), (4), (6) and (13) assuming clear sky conditions are shown in Figs 8(a) and 8(b), respectively, for horizontal and vertical polarizations. It is seen that temporal variations of the reflectivity for desert are markedly different from that observed for the corresponding brightness temperatures. One can also see that while the vertical reflectivity for the desert is always lower than that for rainforest, the corresponding brightness temperatures are higher during most months for desert compared to rainforest due to atmospheric effects and surface temperature dependence associated with the brightness temperatures. These examples illustrate the need for calculating the reflectivities by correcting the satellite

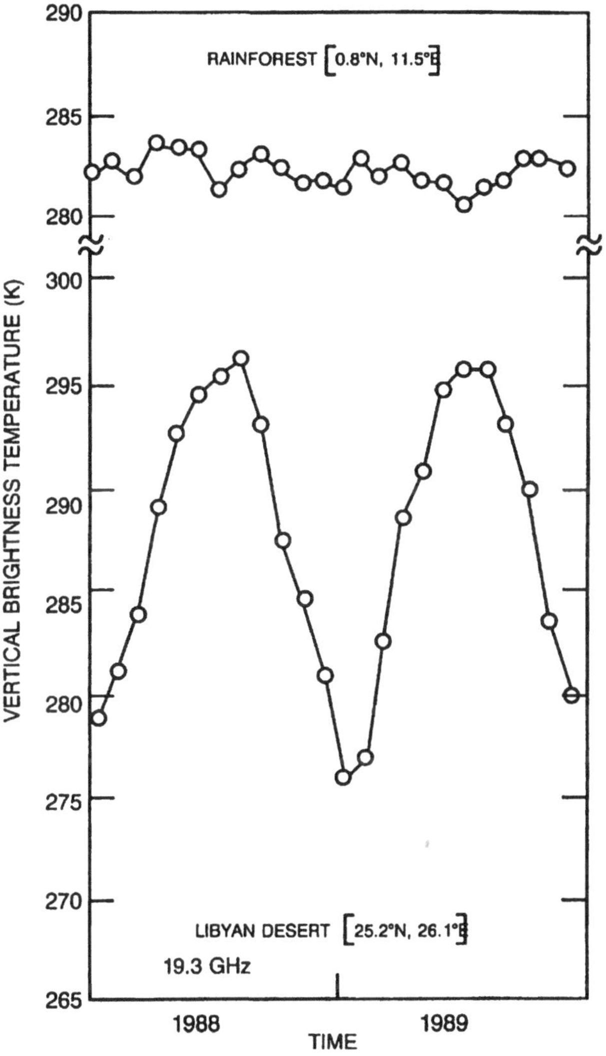

Figure 8(b). Temporal variation of the calculated surface reflectivity for Libyan desert and rainforest from the data in Fig. 7 for vertical polarization.

observed brightness temperatures for variations of surface temperature and atmospheric effects for a realistic analysis of spatial and temporal variations of vegetation or soil moisture (addressed below). The calculated surface reflectivity for varied land surface types is discussed below.

3. SURFACE CHARACTERISTICS: SOIL MOISTURE, VEGETATION PARAMETERS & EFFECTIVE SURFACE TEMPERATURE

Since remote sensing of soil moisture, vegetation parameters and effective temperature using microwave observations are being reviewed separately (see papers by Shutko,

Wang and Choudhury, Kerr, and Njoku), the emphasis in this section will be on synergism with optical measurements.

3.1. Soil moisture

From theoretical considerations as well as field, aircraft and satellite data, it has been demonstrated that low frequency (1–10 GHz) microwave observations are most suitable for estimating soil moisture, and at these frequencies the atmospheric effect is small under most conditions. The basic physics underlying remote sensing of soil moisture is that dielectric constant of soil, which determines the microwave emissivity, increases with soil moisture [22]. The basic relation between the emissivity and soil moisture, however, is modified by two primary factors: surface roughness and vegetation. A simple expression for the emissivity ε ($= 1 - R_p$) showing the dependencies on volumetric soil moisture (θ), surface roughness height h and vegetation optical thickness τ_v covering fractional land area f_v is:

$$\varepsilon = 1 + \left[\varepsilon_0(\theta) - 1\right] \exp(-h)\left[f_v \exp(-2\tau_v) + (1 - f_v)\right] \tag{14}$$

where $\varepsilon_0(\theta)$ is the emissivity of a smooth bare soil, given approximately as follows for L-band (1.4 GHz or 21 cm wavelength) and C-band (5 GHz or 6 cm wavelength) nadir observations over sandy loam soils with moisture in the surface 25 mm or 5 mm layer [23]:

$$\varepsilon_0(\theta) = 0.99 - 1.38\ \theta \tag{15a}$$

$$\varepsilon_0(\theta) = 0.99 - 1.37\ \theta. \tag{15b}$$

Analysis of many observations over grasses and crops have shown that the nadir optical thickness τ_v is related to vegetation water content W (kg m^{-2}) as [24]:

$$\tau_v = b'W \tag{16}$$

where the constant b' is approximately proportional to the microwave frequency; for L-band is $b' = 0.1$.

It is clear from the above discussion that estimation of soil moisture from microwave observations requires information about surface roughness and vegetation. Although surface roughness could vary from one agricultural field to another (most often prior to crop emergence), the range of such variations is not generally large for spatial resolutions of satellite data (c. 20 km); major exceptions would be large topographic variations (mountainous areas). The vegetation characteristics, on the other hand, vary substantially with time and space. Thus, the need for vegetation data is more crucial than surface roughness data for estimating soil moisture, which points to synergism with optical remote sensing. As a first approximation, the emissivity of a smooth bare soil (which is a direct indicator of soil moisture) can be obtained from the emissivity ε determined from satellite observations, as

$$\varepsilon_0(\theta) = 1 + [\varepsilon - 1]\left[f_v \exp(-2\tau_v) + (1 - f_v)\right]^{-1}. \tag{17}$$

Theis *et al.* [25] showed that aircraft observations of L-band brightness temperature, infrared temperature and visible and near-infrared reflectances over a range of agricultural

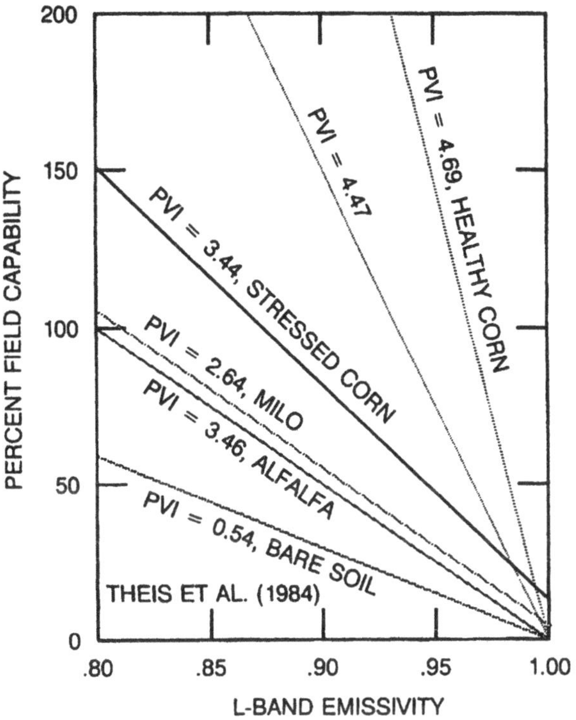

Figure 9. Variation of L-band (1.4 GHz) emissivity with soil moisture (expressed as percent field capacity) for different crop biomass as represented by the perpendicular vegetation index (PVI) derived from visible and near-infrared observations. Theis *et al.* [25] used PVI to account for vegetation effect on the L-band emissivity for estimating soil moisture.

fields were very effective for estimation of soil moisture. The emissivity was calculated by dividing the brightness temperatures by the infrared temperature. The reflectances were used to compute perpendicular vegetation index (PVI), as an indicator of vegetation biomass. Figure 9 shows that the range of L-band emissivity due to variations of soil moisture decreases as PVI increases. The PVI was used to correct the emissivities for vegetation effect. A predictive equation for soil moisture was developed based on emissivity and PVI, such that the predicted soil moisture accounted for 75% of the variance for the observed soil moisture.

Owe *et al.* [26] developed a predictive equation for soil moisture in terms of emissivity at 18 GHz from SMMR (Scanning Multichannel Microwave Radiometer on board the Nimbus-7 satellite) observations (normalizing the brightness temperatures by air temperatures) and normalized difference vegetation index (NDVI) calculated from visible and near-infrared reflectances observed by AVHRR. Choudhury and Golus [27] developed a predictive equation for antecedent precipitation index in terms of brightness temperatures at 6.6 GHz from SMMR observations and NDVI. Spatial resolution of SMMR 6.6 and 18 GHz data are, respectively, about 150 and 50 km. Both of these studies were performed over the US Southern Great Plains. Owe *et al.* [26] found that the predicted soil moisture accounted for 71% of the variance of the soil moisture as indicated by the antecedent precipitation index (API) for three validation sites of roughly

150 km area (Fig. 10). Choudhury and Golus [27] found the accounted variance of 76% for 10 independent validation areas of 150 km size. The higher accounted variance in Choudhury and Golus might be expected due to the use of lower microwave frequency in the predictive equation, although neither 6.6 nor 18 GHz are optimum frequencies for soil moisture estimation [28].

A major difficulty in analyzing SMMR data for soil moisture has been the coarse spatial resolution of these data, which had necessitated modeled soil moisture (API) to surrogate for the observed soil moisture. Owe *et al.* [29] evaluated the relation between SMMR 6.6 GHz emissivities and simulated soil moisture over 150 km area in southern Africa for a three year period. The soil moisture model used in the study was validated against observed soil moisture at several locations within the study area, and observed rainfall at 12 locations within the area was used to calculate soil moisture. The emissivity was estimated by dividing horizontally polarized brightness temperatures by air temperature. The effect of vegetation on the emissivity was quantified using NDVI. The relation between 6.6 GHz transmissivity of vegetation T and NDVI is shown in Fig. 11. This study suggested the following relation for estimating the emissivity of a

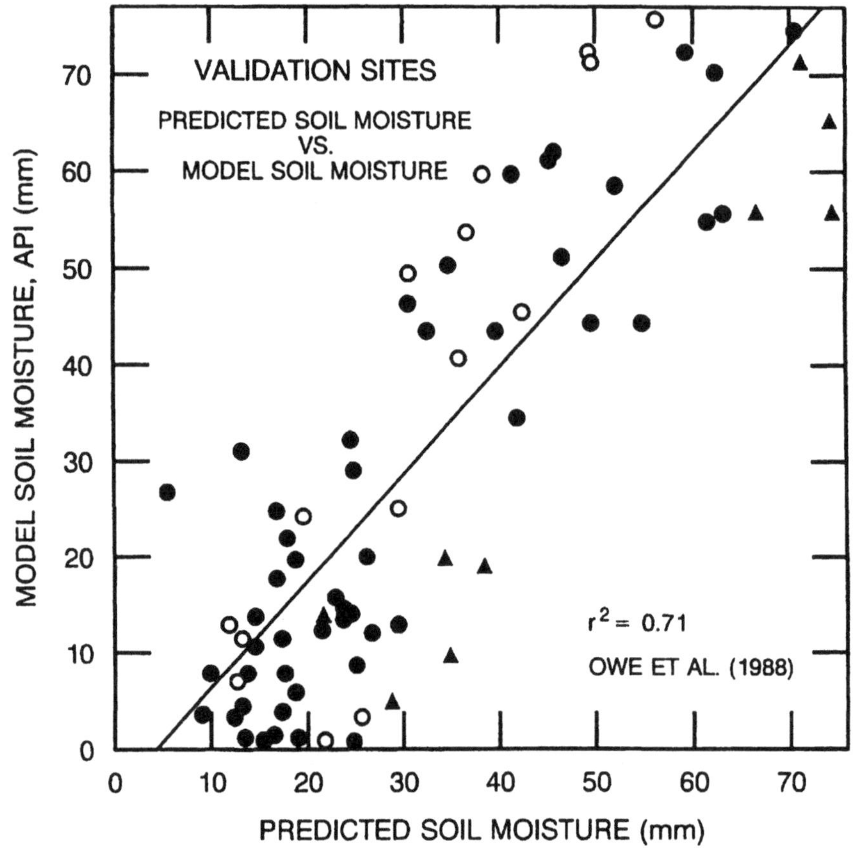

Figure 10. Scatter plot of simulated soil moisture (antecedent precipitation index, API) and soil moisture derived from 18 GHz emissivity and normalized difference vegetation index (NDVI) derived from visible and near-infrared observations by advanced very high resolution radiometer (AVHRR) on board NOAA satellites.

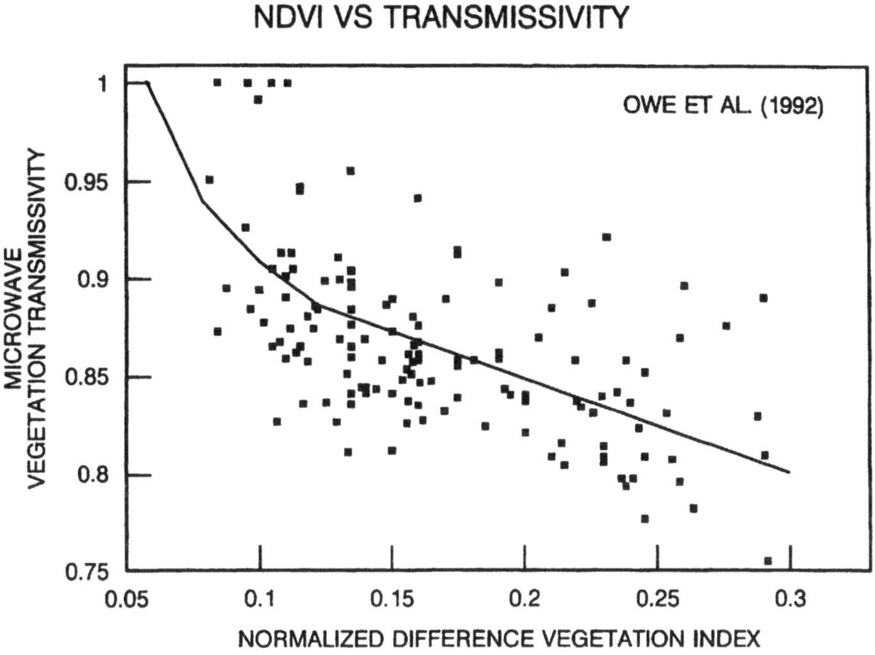

Figure 11. Calculated relation between the transmissivity of vegetation at 6.6 GHz of SMMR and NDVI (calculated from AVHRR observations).

smooth bare soil in terms of SMMR 6.6 GHz emissivity ε and NDVI:

$$\varepsilon_0(\theta) = 1 + 1.35[\varepsilon - 1]T^{-2} \tag{18}$$

where

$$T = 0.945 - 0.48\ \text{NDVI}, \qquad \text{for NDVI} > 0.123$$
$$T = 0.827 + 0.945 \exp(-17.1\ \text{NDVI}), \qquad \text{for NDVI} < 0.123$$

and the numerical factor 1.35 in Equation (18) has been used as an adjustment for surface roughness effect. The Equation (18) was developed for a sparsely vegetated area (low NDVI; less than 0.3) and it may not be accurate for densely vegetated areas (high NDVI).

Although only few studies so far have combined optical and microwave data for estimating soil moisture, both theory and above examples show the usefulness of such synergism, particularly for regional studies using satellite data where ground measurements for vegetation are not likely to be available. It is pertinent to note that optical data may not always be a good indicator of the effect of vegetation on microwave observations, since optical data are indirectly related to the vegetation water content. Also, as discussed below, microwave and optical measurements respond to different vegetation parameters and therefore to a large degree they provide independent measurements for vegetation. High frequency microwave observations could provide an alternative data set to correct for vegetation effect on low frequency observations.

3.2. *Vegetation*

Optical remote sensing has made a significant contribution towards remote sensing of vegetation over the last three decades [30, 31]. This contribution has been through detailed field observations and development of theoretical methods for interpreting these observations. These studies have been supplemented by aircraft observations and global monitoring by the TIROS and later the NOAA series of satellites. Although the temporal span of global observations from satellites in the optical (visible and near-infrared) and microwave regions are comparable, far fewer field studies for vegetation have been done using microwave observations (see Kerr in this volume). Nevertheless, available data and modelling studies suggest that synergism of microwave and optical measurements could provide additional biophysical information. Some key aspects of visible (0.6–0.7 μm) and near-infrared (0.7–1 μm) reflectances are briefly summarized below before discussing synergism.

Extensive field observations over agricultural crops and grasses and radiative transfer modelling for homogeneous green leaf canopies have provided a good understanding of the factors determining visible and near-infrared reflectances of green canopies [30, 31]. Chlorophyll and other pigments within green leaves absorb 80–90% of incident visible radiation (for photosynthesis), while leaf mesophyll scatters most (c. 90–95%) of near-infrared radiation due to refractive index discontinuities within the leaf [32]. Thus, in the interaction of shortwave radiation with leaves, absorption dominates in the visible region and scattering dominates in the near-infrared region. As a result of this difference in the interaction process, the visible reflectance of green canopies decreases, while near-infrared reflectance increases with increasing ground cover or leaf area index [33]. Soil reflectance then determines the overall surface reflectance. These basic features of visible and near-infrared reflectances for a cotton canopy are illustrated in Fig. 12 [34]. The angular distribution of leaves and the angle of incident radiation (direction of sun) are among the factors determining the angular distribution of the reflectances; a rather prominent feature of this angular variation is substantially high reflectance measured opposite to sun (the 'hot spot' effect). The visible reflectance of partially vegetated soil changes relatively more with changing solar zenith angle than near-infrared reflectance due primarily to the importance of shadows in determining the visible reflectance. A number of indices, based on visible and near-infrared reflectances, have been developed for inferring vegetation characteristics of crops [35]. Observations show that while visible reflectance almost always decreases with increasing vegetation cover, the near-infrared reflectance does not always increase for arid land vegetation and forests because of shadows and non-green components in the canopy, for example bark and senesced leaves [2]. Recent studies have focused on understanding the reflectances of forests and spatially heterogeneous canopies [31, 36–38].

The interaction of microwave and visible/near-infrared radiation with leaves may be compared as follows: (1) The cellular structure of leaves, which plays a rather significant role in determining the near-infrared reflectance, is of much less importance in determining the interaction of microwave radiation because the wavelength of radiation in the microwave region (c. 10–200 mm) is much larger than the thickness of most leaves (c. 0.1–0.3 mm); microwave radiation cannot 'see' the cellular structure of leaves as near-infrared radiation can, and (2) While concentration of chlorophyll and other pigments largely determine the visible reflectance of leaves, water content of leaves (or the dielectric constant) is the primary factor determining reflection and

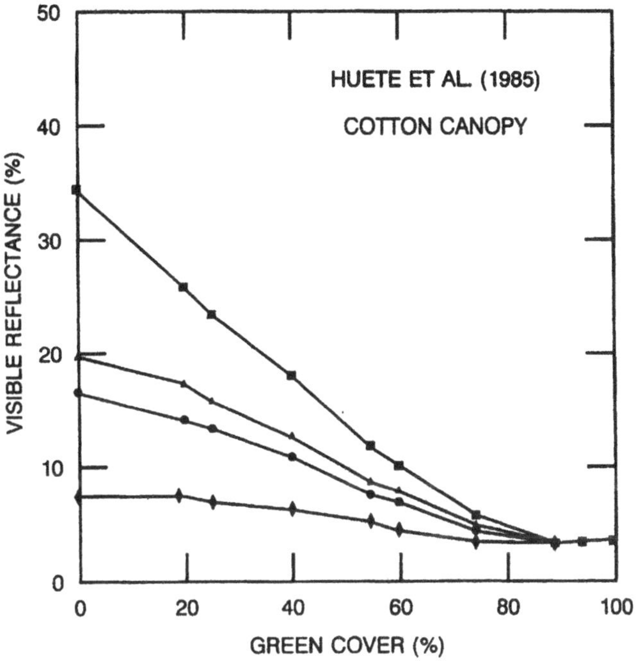

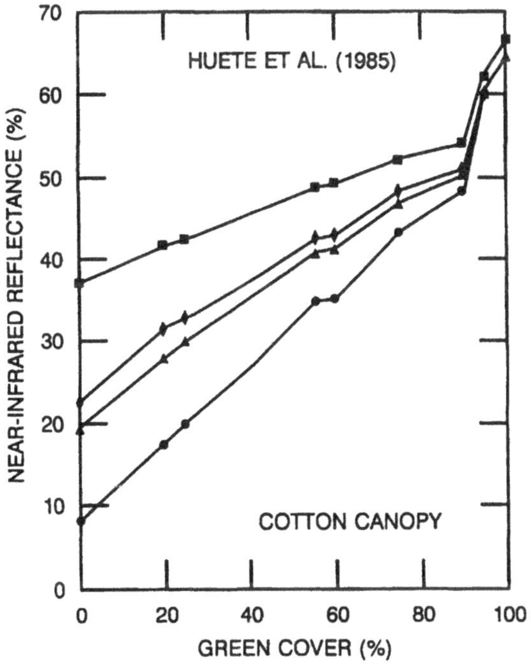

Figure 12. Field observations of reflectance of a cotton canopy as a function of fractional vegetation cover for different soil reflectances as could be determined from the reflectances at zero fractional cover, (a) visible reflectance, and (b) near-infrared reflectance. Each curve is for different soil reflectance as can be determined from the reflectance values for zero cover.

transmission of microwave radiation. Changes in the leaf water content due to stress may also affect visible and near-infrared reflectances, but indirectly through changes in pigment concentration and cellular structure [32, 39, 40]. The reflectance of a canopy can also change under soil water stress due to changes in the canopy geometry, such as leaf folding etc. [41]. We have little theoretical or experimental knowledge of the comparative effects of woody structure on microwave and optical measurements, but there is no obvious reason to expect that this effect will be identical for both measurements.

It is pertinent to note some relations among biophysical parameters determining spectral reflectances and microwave emission [32, 42]. The absorbance of leaves for visible radiation generally increases as leaves unfold and mature due to pigment accumulation, and with leaf maturation its thickness and water content also generally increase leading to an increased microwave absorption [43]. The absorbance for visible radiation generally decreases with the onset of senescence due to chlorophyll deterioration; the water content of leaves begins to decrease after senescence. However, temporal changes in the water content and pigment concentration are not generally synchronous. While water content of leaves changes diurnally in response to transpiration, comparable changes do not occur in the pigment concentration. Changes in some soil physical characteristics (e.g. soil moisture and surface roughness) can have significantly higher impact on the microwave emissivity, while changes in other soil characteristics (e.g. soil reflectance) can directly impact on the overall soil-canopy reflectance without any noticeable impact on the microwave emissivity. Associated with these similarities and differences in the relations among biophysical parameters, there could be similarities and differences in the spatial and temporal variations of reflectance and microwave measurements.

Detailed field studies to evaluate similarities and differences of microwave and optical measurements acquired concurrently are not yet available. Radiative transfer models might be used to gain some understanding of the expected relation between visible reflectance and microwave reflectivity of green leaf canopies both for homogeneous and partially vegetated soils [44, 45]. The visible reflectance of a homogeneous canopy of uniformly oriented leaves with leaf area index L is given by:

$$\rho = \rho_s \exp\left(-\frac{\zeta L}{\mu}\right) + \rho_v\left[1 - \exp\left(-\frac{\zeta L}{\mu}\right)\right] \tag{19}$$

where ρ_s and ρ_v are, respectively, reflectivity of soil and dense vegetation (randomly oriented leaves), with

$$\rho_v = \frac{\omega}{8\mu_0\zeta} \tag{20}$$

and,

$$\zeta = 0.5\left(1 + \frac{\mu}{\mu_0}\right) \tag{21}$$

where ω is the single scattering albedo of a leaf ($= 1 -$ absorbance), μ and μ_0 are cosine of zenith angle of observation and sun, respectively. Typical ranges for ρ_v and ζ will, respectively, be 3–5% and 0.8–1.3. Soil reflectance can vary over a substantial range, say 3–40%, with desert soils being at the higher end of the reflectance and forest soils being at the lower end.

Microwave reflectivity for polarization p at high frequencies might be expressed as:

$$R_p = R_p^s \exp\left(-\frac{\xi L}{\mu}\right) + R_v\left[1 - \exp\left(-\frac{\xi L}{\mu}\right)\right] \tag{22}$$

where R_p^s and R_v are the reflectivity of soil and dense vegetation (randomly oriented leaves), with

$$R_v = \frac{1-\chi}{1+\chi} \tag{23}$$

$$\chi = \left(1 - \frac{r_l}{1-t_l}\right)^{1/2} \tag{24}$$

$$\xi = \left[(1-t_l)(1-r_l-t_l)\right]^{1/2} \tag{25}$$

and r_l and t_l are, respectively, reflectivity (average of horizontal and vertical polarizations) and transmissivity of a leaf, which can be calculated knowing the dielectric constant. From r_l and t_l values given in Mätzler and Sume [43] one finds the typical range for R_v is 6–14% (5–11%) and for ξ is 0.40–0.65 (0.4–0.6) for 35 GHz (21 GHz) observations. These estimated canopy reflectivities appear reasonable [46], but the estimated attenuation coefficients ξ are lower than the values derived from observations [46] due to ignoring water content of stems in the above model. A direct comparison of model predictions with observations has not yet been made.

The leaf area index of a canopy L is a highly variable parameter both spatially and temporally, and it is expected to be one of the major determinants of the spatial and temporal variations of the microwave reflectance and emissivity, although intermittent changes in both reflectance and emissivity can occur due to soil moisture variation following rainfalls (the magnitude of the variation will depend mainly upon the leaf area index). The radiative transfer models presented above, while simplistic, suggest similar functional dependence for visible reflectance and microwave reflectivity on the leaf area index. By manipulating Equations (19) and (22) one can obtain the following relation between these two measurements:

$$\frac{\Delta R}{\Delta R_s} = \left[\frac{\rho - \rho_v}{\rho_s - \rho_v}\right]^{(\xi/\zeta)} \tag{26}$$

where ΔR and ΔR_s are the difference of horizontal and vertical reflectivity for, respectively, vegetated and bare soils. A linear relation between these two measurements may result when the exponent ratio ξ/ζ is close to one, otherwise the relation will be non-linear.

The expected relation between ΔR and ρ for partially vegetated surfaces has been explored by Choudhury [44]. If visible reflectance is assumed to near-nadir measurements with sun close to zenith (so that shadows are negligible), then ρ for a partially vegetated surface with dense clumped foliage can be written as:

$$\rho = f_v\rho_v + (1 - f_v)\rho_s \tag{27}$$

where f_v is fractional vegetation cover and other symbols are as defined with Equation (19). The ΔR for such a surface can be expressed as:

$$\Delta R = \Delta R_s(1-f_v)^{(\eta+1)} + \Delta R_v\left[1 - (1-f_v)^{(\eta+1)}\right] \tag{28a}$$

where ΔR_s and ΔR_v are reflectivity differences, respectively, for soil and dense vegetation and

$$\eta = \frac{\tan\varphi}{b} \tag{28b}$$

where φ is the zenith angle of the radiometer (50° for SSM/I) and b is a parameter determined by the clump shape. Equations (27) and (28a) can be combined by eliminating f_v to obtain the following relation between ΔR and ρ:

$$\frac{\Delta R - \Delta R_v}{\Delta R_s - \Delta R_v} = \left[\frac{\rho - \rho_v}{\rho_s - \rho_v}\right]^{(\eta+1)} \tag{29}$$

which does not differ substantially from Equation (26) with $\Delta R_v = 0$. However, the conceptualization used to obtain Equation (26) is much different from that for Equation (29): a uniform canopy with variable leaf area index for Equation (26), while partial cover with clumped foliage in the case of Equation (29). These relations remain to be validated against field data.

In the absence of concurrent field observations of optical and microwave reflectivity, satellite observations by AVHRR and SSM/I are presented below to illustrate the relationship between these two measurements. Temporal variations of ΔR for 37 GHz and visible reflectance ρ for sand desert and two seasonally vegetated locations are shown in Fig. 13 for the 1988–1989 period. The visible reflectances are derived from the AVHRR–Global Vegetation Index (GVI) weekly composites data, which have been made monthly composites by restricting the scan angle to be within 30° from nadir to further minimize cloud contamination and bi-directional effect. The radiances have been normalized by the cosine of solar zenith angle and nominal corrections for sensor degradation and atmospheric effects (ozone absorption, Rayleigh scattering, and aerosol scattering based upon 25 km visibility) have been made to obtain the reflectances [47–49].

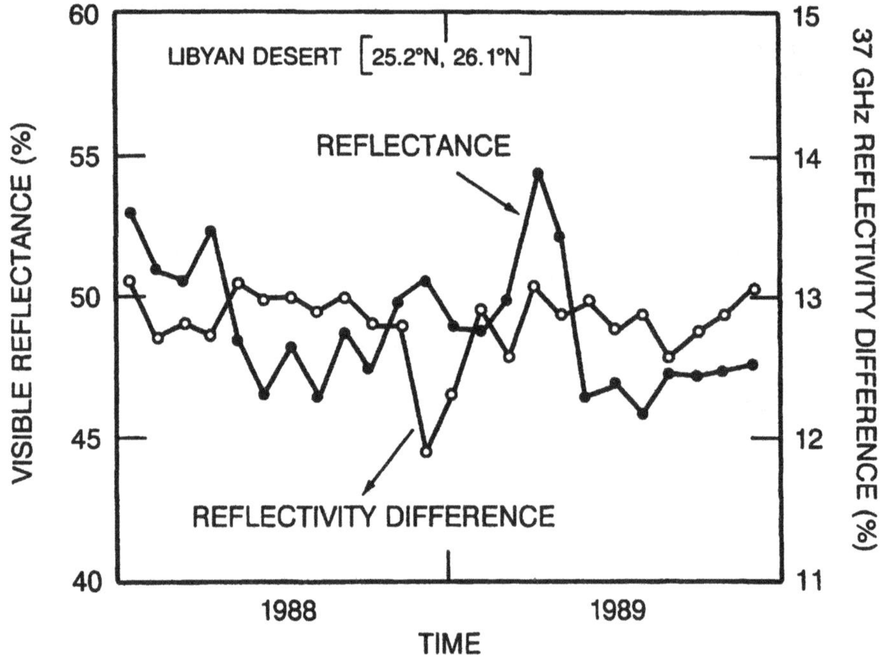

Figure 13(a). Temporal variations (January 1988 to December 1989) of visible reflectance and 37 GHz reflectivity difference (difference of horizontal and vertical polarizations) for Libyan desert. The visible reflectances are derived from NOAA–Global Vegetation Index (GVI) product based on AVHRR observations with nominal corrections for sensor degradation and atmospheric effects [47–49].

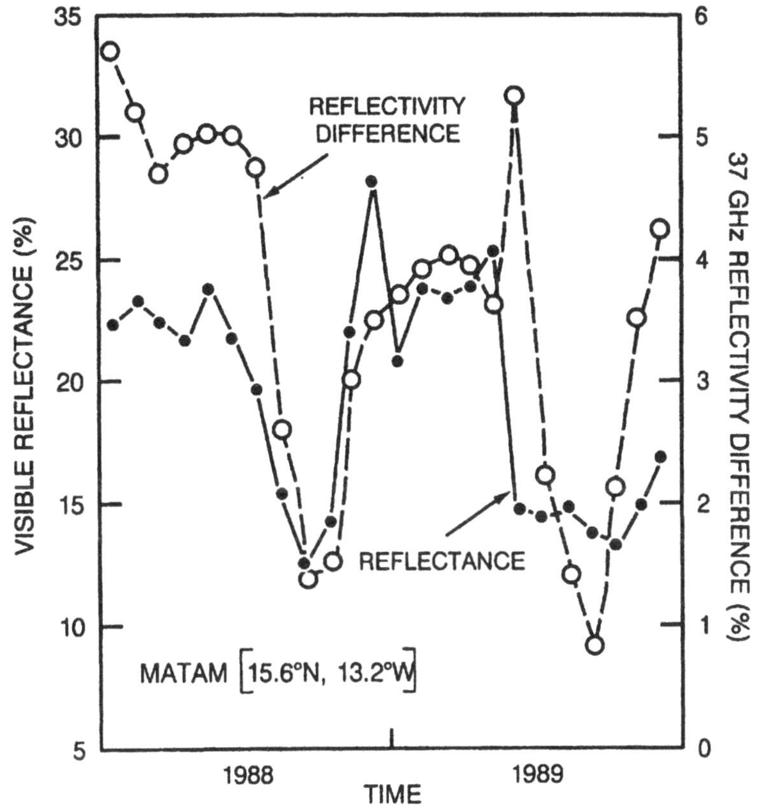

Figure 13(b). Temporal variations (January 1988 to December 1989) of visible reflectance and 37 GHz reflectivity difference (difference of horizontal and vertical polarizations) for Matam, Senegal. The visible reflectances are derived from NOAA–Global Vegetation Index (GVI) product based on AVHRR observations with nominal corrections for sensor degradation and atmospheric effects [47–49].

These reflectances have been averaged over $0.25° \times 0.25°$ cells. The SSM/I data processing and the calculation of the microwave reflectivities are discussed in detail in [21]. The atmospheric correction to the microwave data was based on the assumption of clear sky conditions. For each location, an average value of nine $0.25° \times 0.25°$ pixels (3×3 matrix) centered on the location has been calculated.

Systematic patterns of temporal variation are not seen for either of these two measurements over the desert (Fig. 13a). The temporal patterns of these two measurements are seen to be in good correspondence for the seasonally vegetated areas (Figs 13b and 13c). These temporal variations over the seasonally vegetated areas could be interpreted through vegetation phenology resulting from rainfall patterns [50]: both ΔR and ρ decrease with vegetation growth and they increase as leaves senesce and dry. Correlation between ΔR and ρ seen in this figures appears to be mediated through the correlation of the relevant biophysical parameters determining these measurements, as noted above.

Scatter plot of ΔR and ρ observed during July and August of 1987 for 204 globally distributed locations is shown in Fig. 14 for 37 GHz. The data processing aspects are

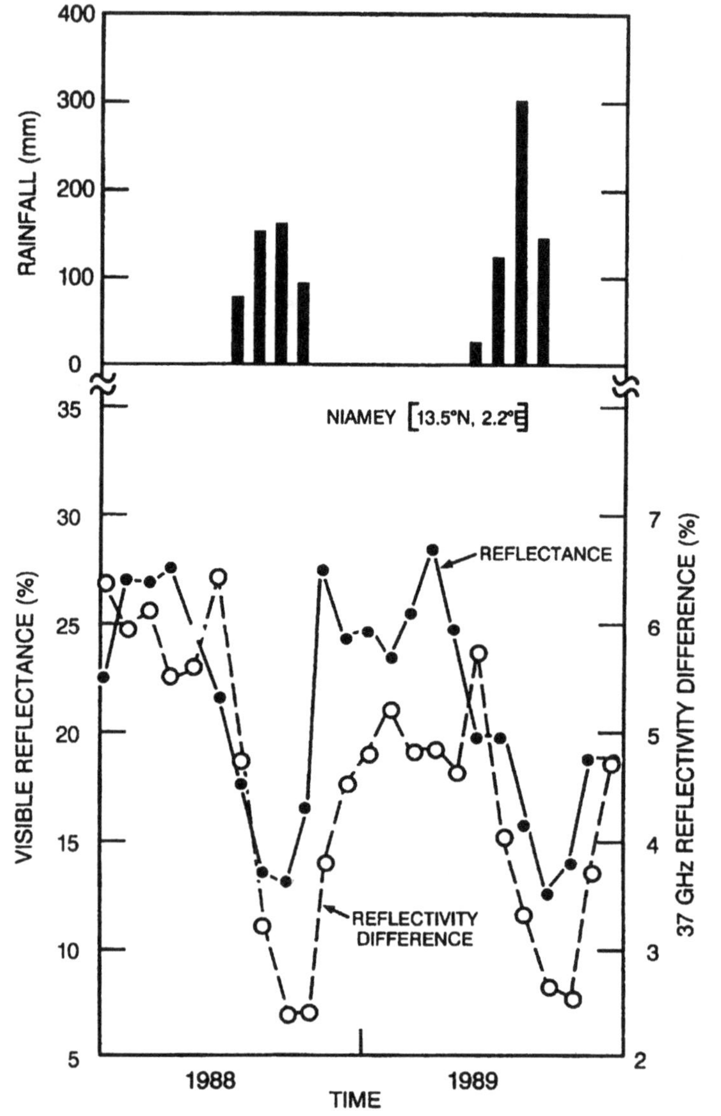

Figure 13(c). Temporal variations (January 1988 to December 1989) of visible reflectance and 37 GHz reflectivity difference (difference of horizontal and vertical polarizations) for Niamey, Niger. Rainfall values for Niamey are also shown. The visible reflectances are derived from NOAA–Global Vegetation Index (GVI) product based on AVHRR observations with nominal corrections for sensor degradation and atmospheric effects [47–49].

discussed above as for Fig. 13. The locations for these data, shown in Fig. 15, were selected subjectively with some consideration: (1) To minimize the effects of exposed water (away from coast, major lakes, rivers, swamps) and rough terrain (mountainous areas with little vegetation), since both exposed water and surface roughness can affect these measurements in a way not related to vegetation (e.g. exposed water can increase ΔR and decrease ρ, and the effect of surface roughness can be to decrease ΔR without

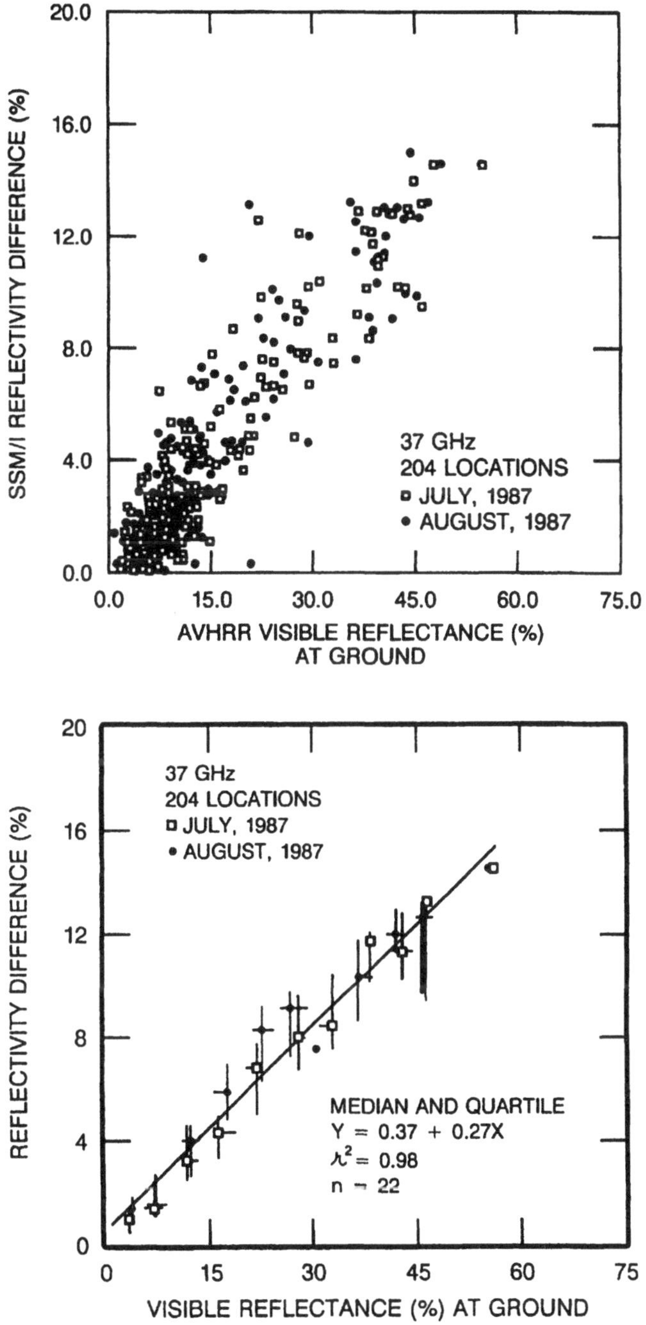

Figure 14. (a) Scatter plot visible reflectance from NOAA-GVI product based on AVHRR data (cf. Fig. 13) and 37 GHz reflectivity difference for 204 globally distributed locations (see Fig. 15). Observations for each location are for July and August, 1987. Both the microwave reflectivity and visible reflectance are average of nine pixels centered on the location, where the pixel values have been gridded into 0.25° × 0.25° cells. (b) Median and quartile range for the data shown in (a).

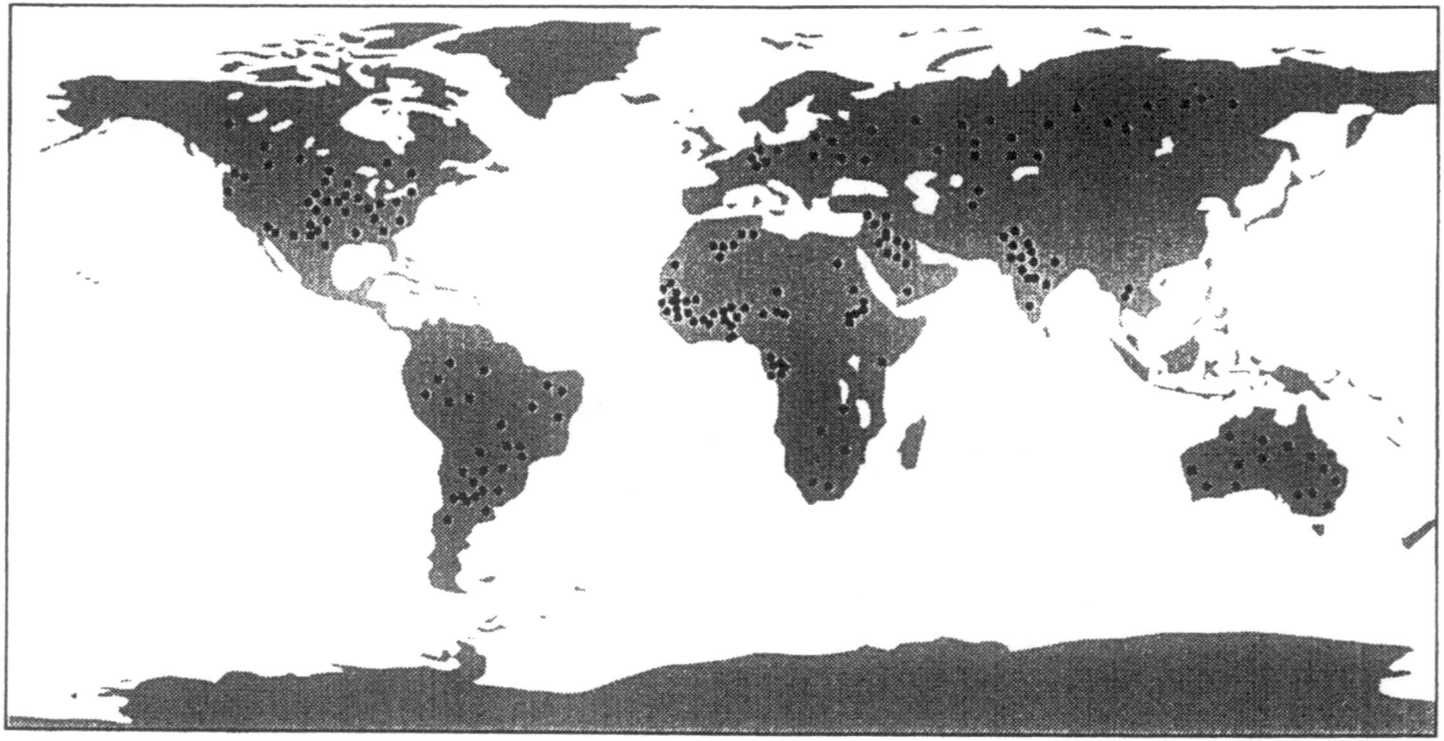

Figure 15. Location of data values used in Fig. 14. These locations have been selected subjectively as discussed in the text. The locations are a little schematic because of the map scale.

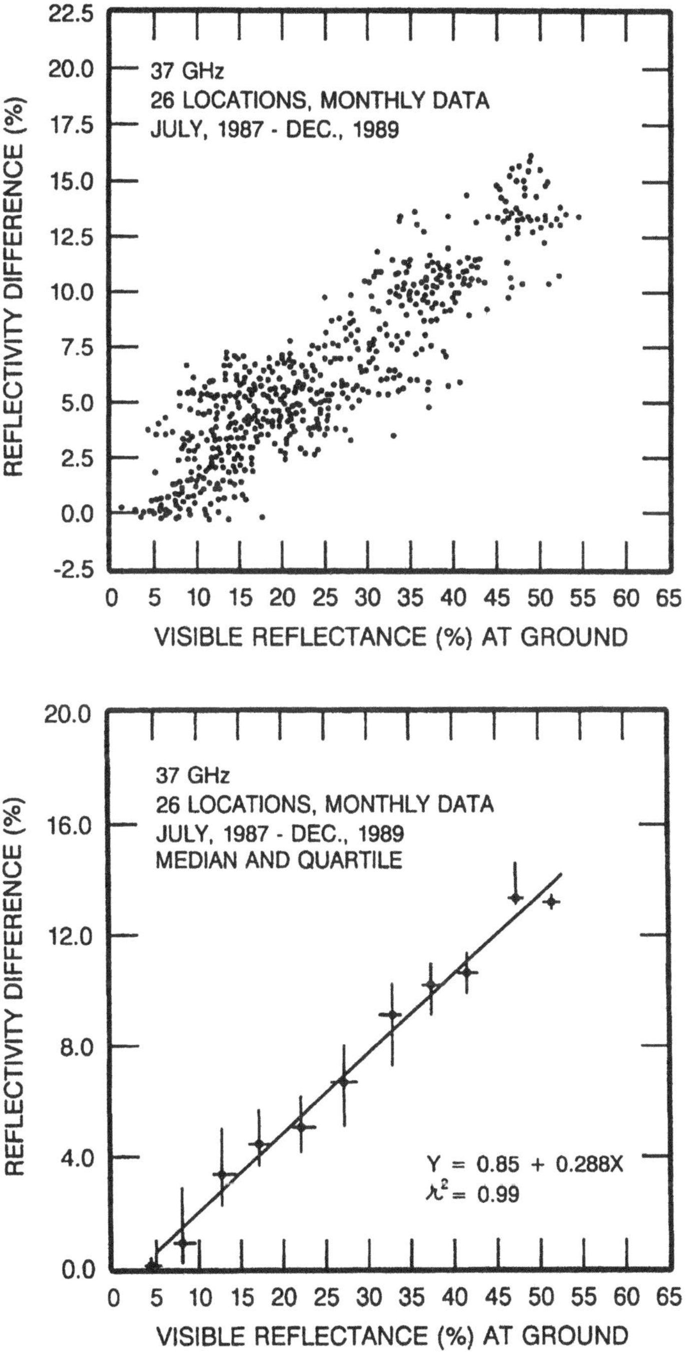

Figure 16. (a) Scatter plot of 37 GHz reflectivity difference and visible reflectance for 26 selected locations from Africa, Australia and Saudi Arabia for a 30 month period (July 1987 to December 1989). The SSM/I data for December 1987 for all locations are interpolated values between November 1987 and January 1988. (b) Median and quartile range for the data shown in (a).

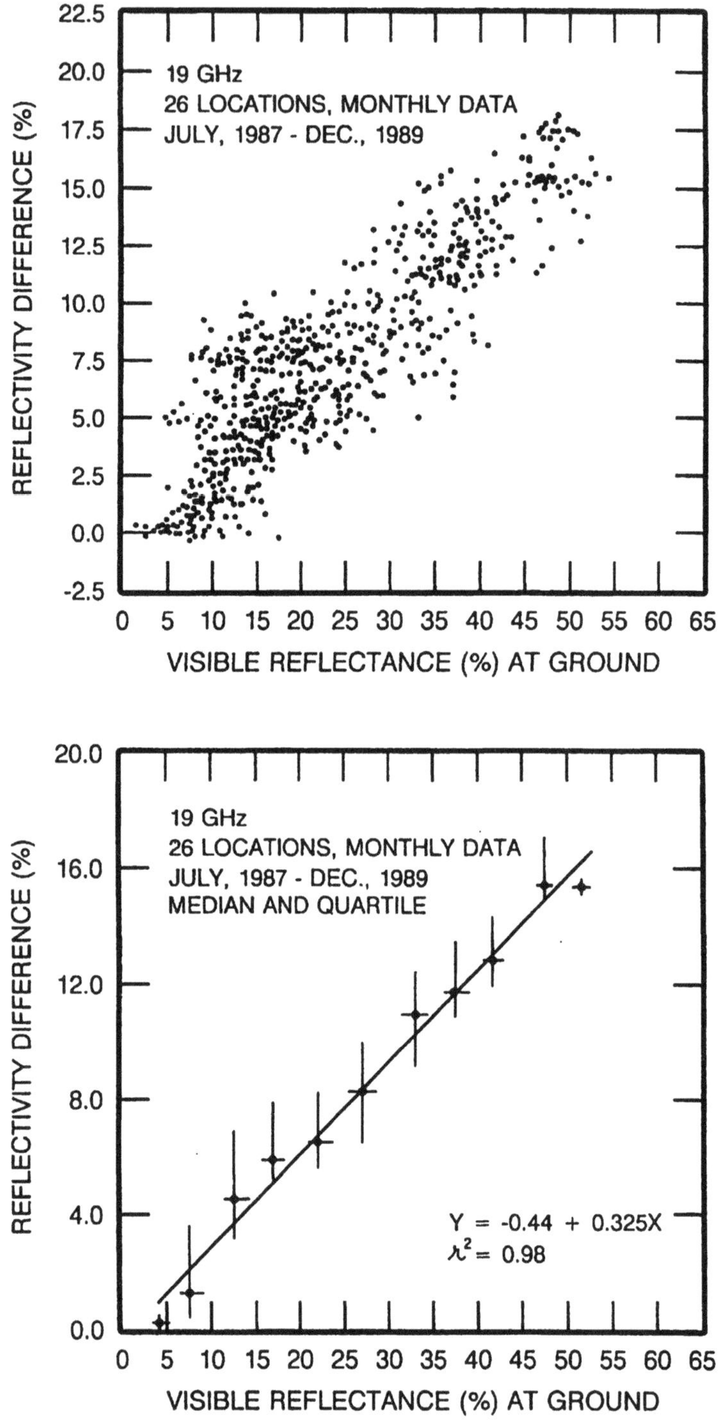

Figure 17. (a) Similar to Fig. 16(a) except for 19.3 GHz, and (b) similar to Fig. 16(b) except for 19.3 GHz.

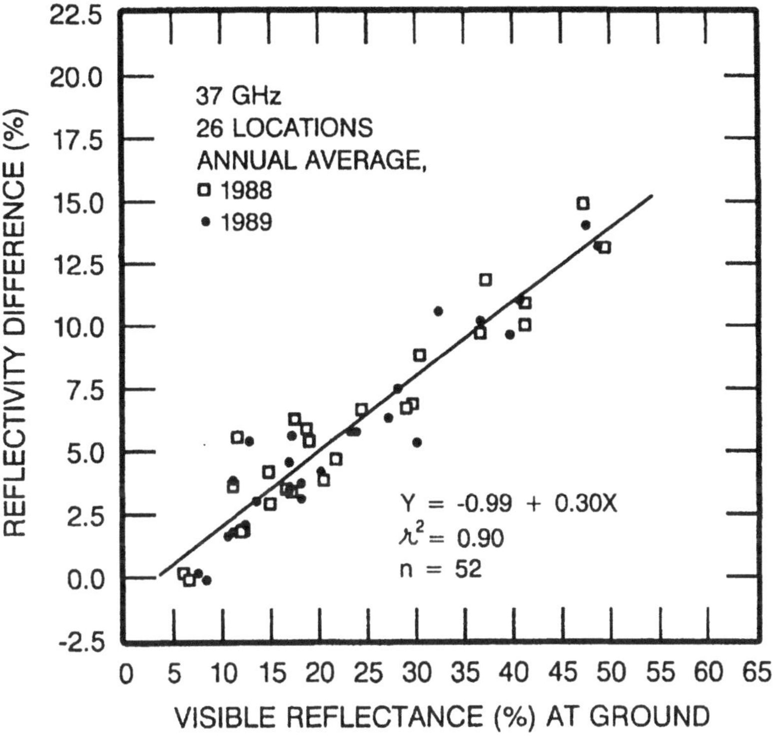

Figure 18. Annual average values for 1988 and 1989 for 26 locations for which the monthly data are shown in Fig. 16.

significantly affecting ρ), (2) availability of meteorologic data for air temperature and vapor pressure together with good estimates or observations for the diurnal air temperature range to account for surface temperature variations and atmospheric effects, and (3) to be representative of major vegetation types of the world (crops, grasslands, savanna, woodland, temperate deciduous forest, boreal forest, rainforest and deserts). Despite scatter in the data values in Fig. 14, a direct relation appears to exist between the ΔR and ρ measurements. To further assess the relation between these two measurements, the data from July 1987 to December 1989 (30 months for each location) for 26 selected locations in Africa, Australia and Saudi Arabia are shown in Fig. 16 for 37 GHz and Fig. 17 for 19 GHz. The annual average values for 1988 and 1989 for these 26 locations are shown Figs 18 and 19. We again see a direct relation between these two measurements in Figs 16–19 with scatter similar to that appearing in Fig. 14. We also note that the scatter for 19 GHz ΔR vs. ρ is similar to that for 37 GHz ΔR vs. ρ. Some of these scatter points could be originating from errors in the atmospheric corrections to these data, possible inclusion of roughness and exposed water within the pixels, and changes in soil reflectance. It has not yet been possible to quantify the magnitude of error associated with each of the possible causes.

The radiative transfer models presented above are simplistic, but they do suggest a biophysical basis for the observed correlations between ΔR and ρ. Detailed field observations are imperative to provide a firm basis for interpreting these correlations.

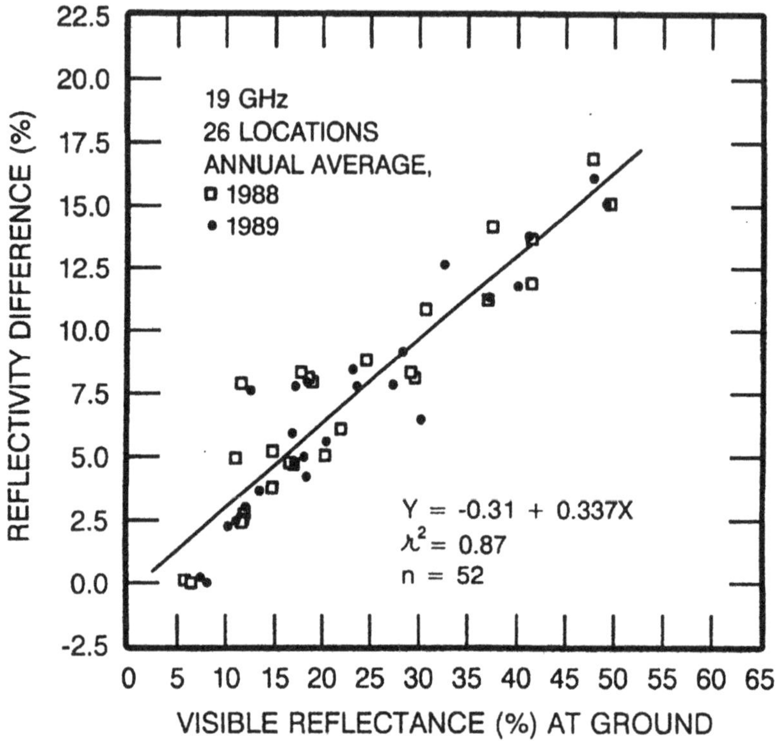

Figure 19. Annual average values for 1988 and 1989 for 26 locations for which the monthly data are shown in Fig. 17.

The above discussion also suggests synergistic use of these measurements for a better understanding of the biophysical process, since the onset of soil water stress and of senescence, for example, are expected to be distinguishable through their differing effects on these measurements. We need to understand the effect of woody structure on these measurements. Since there is no obvious reason to expect that the effect of woody structure will be identical on these measurements, their combined use might be able to distinguish plant communities with different structure. Thus, combined optical and microwave measurements are expected to provide a better understanding of the structure and functioning of plant communities.

3.3. Effective surface temperature

Calculation of the surface reflectivities and their relations with soil moisture and vegetation characteristics were discussed above. The results in Fig. 8(b) and observations of Mätzler [46] suggest that vertical reflectivity varies over a rather narrow range in going from bare to vegetated surfaces. Corresponding to Fig. 14, the relationship of visible reflectance ρ and vertical reflectivity R_v at 19.3 GHz is shown in Fig. 20. The variations of the vertical reflectivity for dense vegetation appear to be higher than expected, but it is not clear to what extent this variation has been introduced by errors in the calculation.

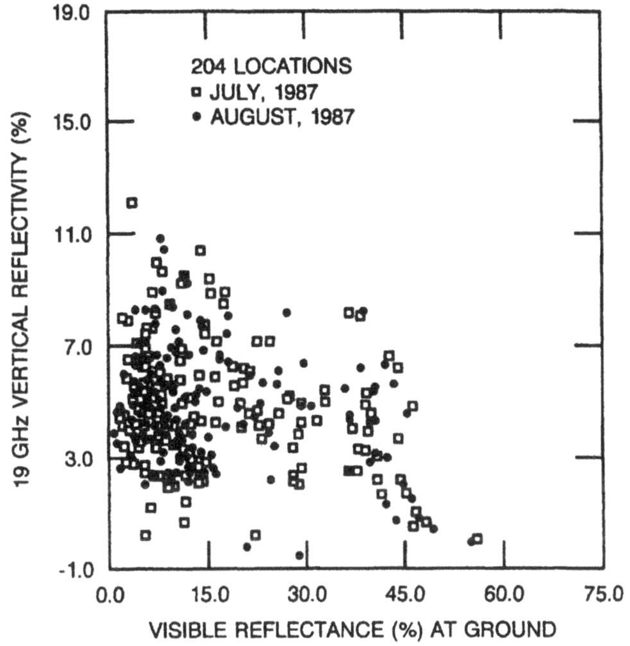

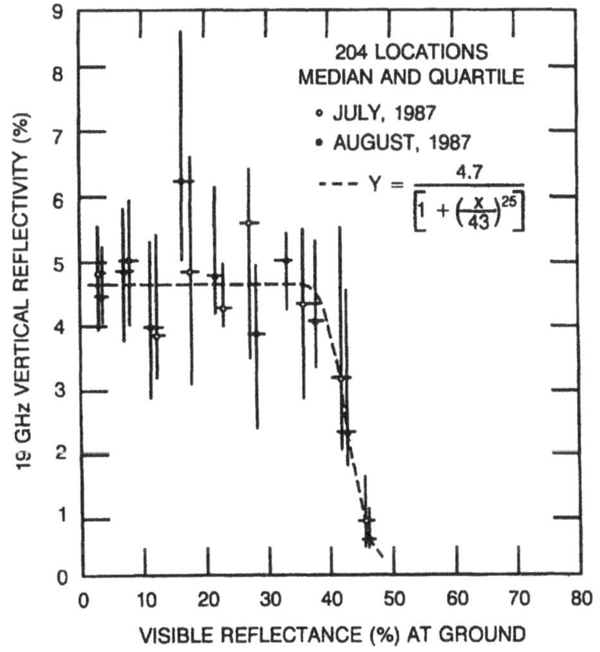

Figure 20. (a) Scatter plot of vertical reflectivity at 19.3 GHz and visible reflectance for 204 globally distributed locations (Fig. 15) for July and August, 1987, (b) Median and quartile range for the data in figure (a), and fitted equation to the data.

It is seen that within an uncertainty of about 2% one can express R_v as:

$$R_v = \frac{4.7}{1 + (\rho/43)^{25}}. \tag{30}$$

With this information about R_v, one can rearrange Equation (3) for vertically polarized observations to calculate the effective surface temperature as:

$$T_g = \frac{T_s^v - T_u - t_a R_v T_d}{t_a(1 - R_v)}. \tag{31}$$

If we approximate the atmospheric emissions (T_u and T_d) as:

$$T_u = T_d = T_g(1 - t_a) \tag{32}$$

then T_g could be obtained from the following equation:

$$T_g = \frac{T_s^v}{1 - R_v t_a^2} \tag{33}$$

Now, because the product ($R_v t_a^2$) is much smaller than 1, we can approximate Equation (33) as:

$$T_g = T_s^v(1 + R_v t_a^2) \tag{34}$$

which shows that 1% error in R_v corresponds approximately to 3 K error in estimating T_g.

The above approach for estimating T_g is physically based, and the scatter in Fig. 20 shows that the error in estimating T_g could be 12 K (Fig. 20b suggests the rms error to be less than 6 K, although no quantitative estimate has yet been made). These errors in T_g are larger than those obtained through regression models, namely 2–3 K, for a study over the United States [51], while the above analysis is based on a global data set without model calibration. The data for atmospheric correction could be obtained through synergism with the optical measurements, as has been discussed above. The above analysis needs to be extended to understand the variations of the vertical reflectivity, quantify the uncertainties in the calculated T_g and validate against surface observations.

4. SUMMARY AND CONCLUSIONS

Extraction of land surface parameters such as soil moisture, vegetation characteristics and effective surface temperature from satellite observations could be performed more accurately by correcting the satellite data for atmospheric effects and adjusting for changes in surface temperature. Optical data could be of much help both for atmospheric correction and for adjustment of changes of surface temperature by providing information for surface and air temperatures, atmospheric water vapor content, cloud top temperature and its liquid water content. Estimation of soil moisture from microwave emissivity requires correction for the effect of vegetation, and optical data could be of help for such correction. Optical and microwave measurements of vegetation pertain to different biophysical parameters (e.g. pigments and cellular structure of leaves vs. its water content) and thus concurrent observations could enhance our capability to monitor ontogenic and stress related changes of plant communities. Synergism with optical data might also provide an estimate for the effective surface temperature.

5. ACKNOWLEDGEMENTS

Data processing assistance was provided by Nizam Ahmed. Work was financially supported by the NASA Headquarters.

REFERENCES

1. B. J. Choudhury. Passive microwave remote sensing contribution to hydrological variables. *Surveys Geophysics*, **12**, 63–84 (1991).
2. B. J. Choudhury. Multispectral satellite data in the context of land surface heat balance. *Rev. Geophysics*, **29**, 217–236 (1991).
3. B. J. Choudhury, T. J. Schmugge and T. Mo. A parameterization of effective soil temperature for microwave emission. *J. Geophys. Research*, **87**, 1301–1304 (1982).
4. F. T. Ulaby, R. K. Moore and A. K. Fung. *Microwave Remote Sensing: Active and Passive. Vol 1*. Addison-Wesley, Massachusetts (1981).
5. E. R. Westwater, J. B. Snider and M. J. Falls. Ground-based radiometric observations of atmospheric emission and attenuation at 20.6, 31.65 and 90.0 GHz: A comparison of measurements and theory. *IEEE Transactions on Antennas and Propagation*, **38**, 1569–1579 (1990).
6. C. Mätzler. Ground-based observations of atmospheric radiation at 5, 10, 21, 35, and 94 GHz. *Radio Science*, **27**, 403–415 (1992).
7. K. B. Kidwell. *NOAA Polar Orbiter Data Users Guide*. National Oceanic and Atmospheric Administration, Washington DC (1984).
8. S. B. Idso. Atmospheric attenuation of solar radiation. *J. Atmosph. Sci.*, **26**, 1088–1095 (1969).
9. J. L. Monteith. An empirical method for estimating long-wave exchanges in the British Isles. *Quarterly J. Roy. Meteorol. Soc.*, **87**, 171–179 (1961).
10. L. L. Stowe, H. Y. Yeh, C. G. Wellemeyer, H. L. Kyle and the Nimbus-7 Cloud Data Processing Team. Nimbus-7 global cloud climatology. Part II. First year results. *J. Climate*, **2**, 671–709 (1989).
11. N. C. Grody, A. Gruber and W. C. Shen. Atmospheric water content over the tropical Pacific derived from the Nimbus-6 scanning microwave spectrometer. *J. Appl. Meteorol.*, **19**, 986–996 (1980).
12. C. Prabhakara, I. Wang, A. T. C. Chang and P. Gloersen. A statistical examination of Nimbus-7 SMMR data and remote sensing of sea surface temperature, liquid water content in the atmosphere and surface wind speed. *J. Climate Appl. Meteorol.*, **22**, 2023–2037 (1983).
13. J. C. Alishouse, J. B. Snider, E. R. Westwater, C. T. Swift, C. S. Ruf, S. A. Snider, J. Vongsathorn and R. A. Ferraro. Determination of cloud liquid water content using the SSM/I. *IEEE Trans. Geosci. Remote Sensing*, **28**, 817–821 (1990).
14. G. L. Stephens. Radiation profiles in extended water clouds. II. Parameterization schemes. *J. Atmosph. Sci.*, **35**, 2123–2132 (1978).
15. J.-Y. Lojou, R. Frouin and R. Bernard. Comparison of Nimbus-7 SMMR and GOES-1 VISSR atmospheric liquid water content. *J. Appl. Meteorol.*, **30**, 187–198 (1991).
16. W. B. Rossow, L. C. Garder, P.-J. Lu and A. Walker. *International Satellite Cloud Climatology Project (ISCCP) Documentation of Cloud Data.* WMO/TD-No. 266, World Meteorological Organization, Geneva (1988).
17. W. B. Rossow, L. C. Garder and A. A. Lacis. Global, seasonal cloud variations from satellite radiance measurements. Part 1: Sensitivity of analysis. *J. Climate*, **2**, 419–458 (1989).
18. B. A. Baum, B. A. Wielicki and P. Minnis. Cloud-property retrieval using merged HIRS and AVHRR data. *J. Appl. Meteorol.*, **31**, 351–369 (1992).
19. W. L. Darnell, S. K. Gupta and W. F. Staylor. Downward longwave surface radiation from sun-synchronous satellite data: Validation of methodology. *J. Climate Appl. Meteorol.*, **25**, 1012–1021 (1986).
20. B. J. Choudhury, E. R. Major, E. A. Smith and F. Becker. Atmospheric effects on SMMR and SSM/I 37 GHz polarization difference over the Sahel. *International J. Remote Sensing*, **13**, 3443–3463 (1992).

21. B. J. Choudhury. Reflectivities of selected land surface types at 19 and 37 GHz from SSM/I observations. *Rem. Sens. Environ.*, **46**, 1–27 (1993).

22. T. Schmugge, T. J. Jackson, W. P. Kustas and J. R. Wang. Passive microwave remote sensing of soil moisture: Results from HAPEX, FIFE and MONSOON 90. *ISPRS J. Photogrammetry Remote Sensing*, **47**, 127–143 (1992).

23. J. R. Wang. Microwave emission from smooth bare fields and soil moisture sampling depth. *IEEE Trans. Geosci. Remote Sensing*, **GE-25**, 616–622 (1987).

24. T. J. Jackson and T. J. Schmugge. Vegetation effect on the microwave emission of soils. *Rem. Sens. Environ.*, **36**, 203–212 (1991).

25. S. W. Theis, B. J. Blanchard and R. W. Newton. Utilization of vegetation indices to improve microwave soil moisture estimates over agricultural lands. *IEEE Trans. Geosci. Remote Sensing*, **GE-22**, 490–496 (1984).

26. M. Owe, A. Chang and R. E. Golus. Estimating soil moisture from satellite microwave measurements and satellite derived vegetation index. *Rem. Sens. Environ.*, **24**, 331–345 (1988).

27. B. J. Choudhury and R. E. Golus. Estimating soil wetness using satellite data. *International J. Remote Sensing*, **9**, 1251–1257 (1988).

28. J. R. Wang. Effect of vegetation on soil moisture sensing observed from orbiting microwave radiometers. *Rem. Sens. Environ.*, **17**, 141–151 (1985).

29. M. Owe, A. A. van de Griend and A. T. C. Chang. Surface moisture and satellite microwave observations in semiarid southern Africa. *Water Resources Research*, **28**, 829–839 (1992).

30. G. Asrar (Ed.). *Theory and Applications of Optical Remote Sensing*. John Wiley, New York (1989).

31. R. B. Myneni and J. Ross (Eds). *Photon–Vegetation Interactions*. Springer-Verlag, New York (1991).

32. E. A. Walter-Shea and J. M. Norman. Leaf optical properties. In: *Photon-Vegetation Interactions*, R. B. Myneni and J. Ross (Eds), pp. 229–251, Springer-Verlag, New York (1991).

33. P. J. Curran and E. J. Milton. The relationships between the chlorophyll concentration, LAI and reflectance of a simple vegetation canopy. *International J. Remote Sensing*, **4**, 247–255 (1983).

34. A. R. Huete, R. D. Jackson and D. F. Post. Spectral response of a plant canopy with different soil background. *Rem. Sens. Environ.*, **17**, 37–53 (1985).

35. R. D. Jackson. Spectral indices in n-space. *Rem. Sens. Environ.*, **13**, 409–421 (1983).

36. T. Nilson. A timber reflectance model. *Soviet J. Remote Sensing*, **8**, 467–482 (1990).

37. R. B. Myneni, G. Asrar, D. Tanre and B. J. Choudhury. Remote sensing of solar radiation absorbed and reflected by vegetated land surfaces. *IEEE Trans. Geosci. Remote Sensing*, **30**, 302–314 (1992).

38. X. Li and A. H. Strahler. Geometric-optical bidirectional modeling of the discrete crown vegetation canopy: Effect of crown shape and mutual shadowing. *IEEE Trans. Geosci. Remote Sensing*, **30**, 276–292 (1992).

39. W. D. Bowman. The relationship between leaf water status, gas exchange, and spectral reflectance of cotton leaves. *Rem. Sens. Environ.*, **30**, 249–255 (1989).

40. W. J. Ripple. Spectral reflectance relationships to leaf water stress. *Photogrammetric Engineering Remote Sensing*, **52**, 1669–1675 (1986).

41. R. D. Jackson and C. E. Ezra. Spectral response of cotton to suddenly induced water stress. *International J. Remote Sensing*, **6**, 177–185 (1985).

42. H. W. Gausman, W. A. Allen, D. E. Escober, R. R. Rodriguez and R. Cardenas. Age effects of cotton leaves on light reflectance, transmittance and absorptance and on water content and thickness. *Agronomy Journal*, **63**, 465–469 (1971).

43. C. Mätzler and A. Sume. Microwave radiometry of leaves. In: *Microwave Radiometry and Remote Sensing Applications*, P. Pampaloni (Ed.), pp. 133–148, VSP, Utrecht (1989).

44. B. J. Choudhury. A comparative analysis of satellite-observed visible reflectance and 37 GHz polarization difference to assess land surface change over the Sahel zone, 1982–1986. *Climatic Change*, **17**, 193–208 (1990).

45. B. J. Choudhury. Multispectral satellite observations for arid land studies. *ISPRS J. Photogrammetry Remote Sensing*, **47**, 101–126 (1992).

46. C. Mätzler. Seasonal evolution of microwave radiation from an oat field. *Rem. Sens. Environ.*, **31**, 161–173 (1991).

47. D. Tanre, C. Deroo, P. Duhaut, M. Herman, J. J. Morcrette, J. Perbos and P. Y. Deschamps. *Simulation of the Satellite Signal in the Solar Spectrum (5S), User's Guide*. Laboratoire d'Optique Atmospherique, USTL, Villeneuve d'Ascq, France (1986).

48. B. N. Holben, Y. Kaufman and J. D. Kendall. NOAA-11 AVHRR visible and near-IR inflight calibration. *International J. Remote Sensing*, **11**, 1511–1519 (1990).

49. R. W. Saunders. The determination of broad band albedo from AVHRR visible and near-infrared radiances. *International J. Remote Sensing*, **11**, 49–67 (1990).

50. H. N. Le Houerou. *The Grazing Land Ecosystems of the African Sahel*. Springer-Verlag, New York (1989).

51. M. J. McFarland, R. L. Miller and C. M. U. Neale. Land surface temperature derived from the SSM/I passive microwave brightness temperatures. *IEEE Trans. Geosci. Remote Sensing*, **28**, 839–845 (1990).

ESA/NASA International Workshop, pp. 193–204
B. J. Choudhury, Y. H. Kerr, E. G. Njoku and P. Pampaloni (Eds)

Radiometry at infrared and microwave frequencies

T. SCHMUGGE

USDA/ARS Hydrology Lab., Beltsville, MD 20705 USA

Abstract—Measurements of thermal radiation at infrared and microwave wavelengths yields much information about the land surface. In the infrared the primary use is for surface temperature determination as the emissivity is usually close to one. It is fortuitous that the peak in the thermal emission occurs in an atmospheric transmission window. There are variations in emissivity of some minerals and soils in the 7 to 14 μm region which can be interpreted for mineral identification. In the microwave there is more interest in the information which can be inferred from variations in the surface emissivity; most of these relate to the presence or lack of water in the liquid state. This arises from the effect of the electric dipole of the water molecule on the dielectric properties of water. The dielectric constant of water changes from about 80 at around 1 GHz to less than 5 at 100 GHz and higher. For ice the transition occurs at much lower frequencies, $\approx$ 10 kHz. At infrared wavelengths the molecular structure and vibrations have a more significant effect. For example the emissivity of quartz sand can be as low as 0.6 due to the SiO_2 restrahlen bands. The emissivity for vegetative canopies has been found to be close to one with little spectral variation. The effect of these dielectric constant variations on microwave and infrared observations is presented, along with the effect of the atmosphere on these observations.

1. INTRODUCTION

The intensity of the thermal radiation at microwave and thermal infrared wavelengths is described by the Planck black body relationship given as a function of frequency ν in Equation (1a) and as a function of wavelength λ in Equation (1b):

$$B_\nu(T) = \frac{2h\nu^3/c^2}{\exp(h\nu/kT) - 1} \tag{1a}$$

$$B_\lambda(T) = \frac{2hc^2/\lambda^5}{\exp(hc/\lambda kT) - 1} \tag{1b}$$

where h is Planck's constant (6.63×10^{-34} joule sec), c is the speed of light and k is Boltzmann's constant (1.38×10^{-23} joule/K). The units are $\mathrm{w\,m^{-2}\,ster^{-1}}$ per Hz for Equation (1a) or $\mathrm{w\,m^{-2}\,ster^{-1}}$ per meter for Equation (1b). Equation (1a) is plotted in Fig. 1 for several temperatures. In this figure the reflected solar radiation is also plotted for albedos of 0.1 and 1 to show that near wavelengths where the terrestrial thermal radiation peaks, i.e. at $\lambda \approx 10$ μm, the reflected solar radiation is several orders of magnitude weaker. The wavelength for the maximum in the thermal emission curve is given by the Wien displacement law where $\lambda_{max}T \approx 2.898 \times 10^{-3}$ mK. Thus for a temperature of 300 K, $\lambda_{max} \approx 9.7$ μm. Thus measurements around this wave-

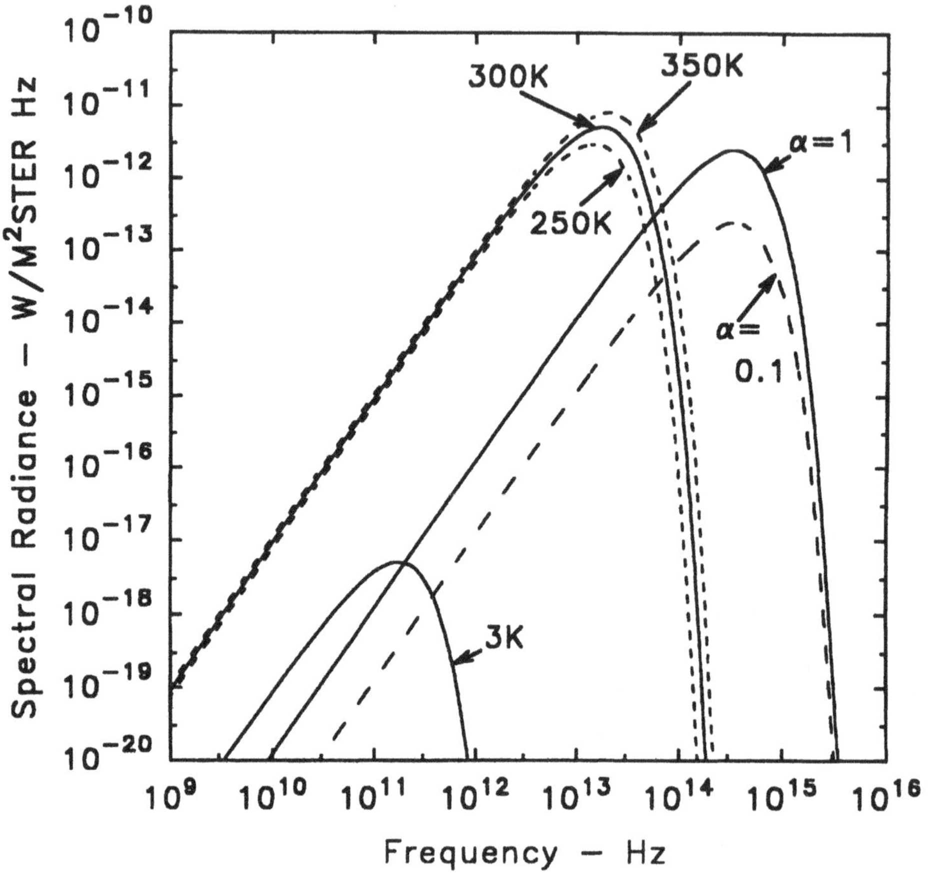

Figure 1. The spectral radiance of a black body according to Equation (1a) for a typical range of terrestrial temperatures, the cosmic background (3 K) and the reflected solar irradiance with indicated values of the albedo (α). Adapted from Schanda [1].

length should yield information on surface temperature without any contamination by reflected solar radiation. The cosmic background radiation at $T = 3$ K is also shown for reference.

Here we consider the range in wavelength/frequency from about $\lambda = 10$ cm or $n = 3$ GHz in the microwave to $\lambda = 10$ μm or 3×10^4 GHz in the thermal infrared. Thus we are talking about 4 orders of magnitude difference in wavelength or frequency. It is interesting to note that in the infrared we always speak in terms of wavelength while in the microwave frequency and wavelength are used interchangeably. I think this results from the techniques used to quantify the difference in the waves: in the infrared wavelengths can be measured but not frequency while in the microwave both frequency and wavelength can be measured. It should be noted that the convention in the far-infrared to millimeter band is to use spatial frequency, $\nu = 1/\lambda$, in cm^{-1}. We will use this convention here. Also it is good to keep in mind that it is the frequency that remains constant when a wave passes from one medium to another.

At visible and infrared wavelengths Equation (1a) is usually represented as

$$B_\nu(T) = \frac{\varepsilon C_1 \nu^3}{\exp\left(C_2\nu/T\right) - 1} \tag{2}$$

where C_1 is 1.19096×10^{-8} W/(m^2 sr cm^{-4}), C_2 is 1.4388 cm K and ν is expressed in wavenumber (cm^{-1}) and ε is the emissivity. For a perfect emitter or black body $\varepsilon = 1$, and for real surfaces $\varepsilon < 1$.

Since the peak in thermal radiation for terrestrial surface is at about 10 μm the intensity in the microwave at these temperatures is very low. Also because frequency is so low ($h\nu/kT \ll 1$), we can make use of the Rayleigh–Jeans approximation and expand Equation (1a) to obtain one which is linear in the temperature radiance relationship:

$$L = 2kT(\nu/c)^2 = 2kT(1/\lambda)^2. \tag{3}$$

Thus we usually refer to the intensity of the black body radiation in the microwave region in terms of a brightness temperature T_B. This approximation is good to better than 1% if ν in GHz is less than T in K, i.e. for the entire microwave portion of the spectrum when viewing terrestrial surfaces. Measurement of the thermally emitted radiation at various wavelengths from the earth's surface can yield much useful information about parameters such as surface soil moisture and temperature. These two parameters yield very important information about the land-atmosphere interaction. The partitioning of the net radiation into latent and sensible heating components is determined by the moisture available in the soil for the evapotranspiration process. The value of the surface temperature is the result of this partitioning. By available moisture we mean that which can be exchanged with the atmosphere either through direct evaporation from the soil or by transpiration from the plant drawing from the entire root zone. Thus, if sufficient moisture is available the surface temperature will be close to that of the air. If not the surface temperature will rise above that of the air.

To estimate surface temperatures, radiation at wavelengths around 10 μm is used because the peak intensity of Equation (1) occurs in this region for terrestrial temperatures ($\sim$ 300 K) and in addition the atmosphere is relatively transparent. Therefore the variations in the intensity of radiation are mainly related to surface temperature variations. However it must be kept in mind that variations in emissivity of 1% yield brightness temperature variations of 0.3 K to 0.6 K depending on sky radiance. To estimate surface soil moisture, radiation at much longer wavelengths (10's of cm) is used. Changes in the intensity of the emitted radiation at these wavelengths are primarily due to the variation of the emissivity with the moisture content of the soil. This effect results from the large dielectric contrast between water and dry soils. This difference is demonstrated in Fig. 2 where infrared (10–12 μm) and microwave (21 cm) brightness temperatures are presented. These data are from a NASA aircraft flight over the Salton Sea and the Imperial Valley in southern California at an altitude of 600 m [2]. They show about a 10 K variation in the infrared due to surface temperature and about an 80 K variation in the microwave brightness temperature over land due to soil moisture variations. The percentages indicate the ground measurements of soil moisture in the top 5 cm of the soil for the numbered fields. The wide range of moisture values results from irrigation. The low microwave emissivity, ≈ 0.4, for a water surface is also apparent.

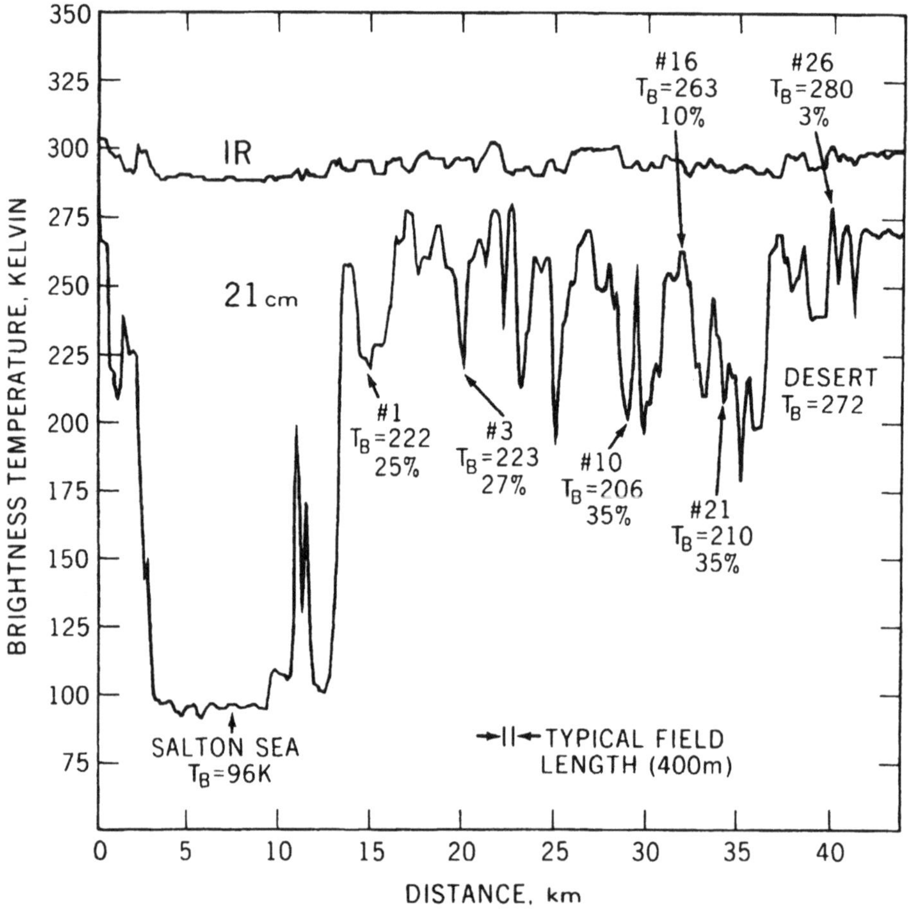

Figure 2. Thermal infrared (10 to 12 μm) and microwave (21 cm) brightness temperatures versus distance from a 600 m altitude flight above the north end of the Imperial Valley, California. The 21 cm brightness temperatures and measured soil moistures are indicated for several irrigated fields [2].

In order to understand possible synergism between measurements of thermal radiation infrared and microwave wavelengths we need to understand better the emissivity in these two wavelength regions. Since there are many other papers at this workshop discussing microwave emissive properties of the land surface, I will concentrate on the properties of infrared observations of the surface and in particular the emissivity of terrestrial materials or surfaces.

2. THERMAL INFRARED

In Fig. 3, we have plotted Equation (2) from 5 to 20 μm for temperatures 280, 290, and 300 K, i.e. near the low range of terrestrial temperatures. At these temperatures the peak emission occurs in the 8 to 10 μm range of wavelength. In this figure we have also plotted the atmospheric transmission calculated with the Lowtran-7 path radiance

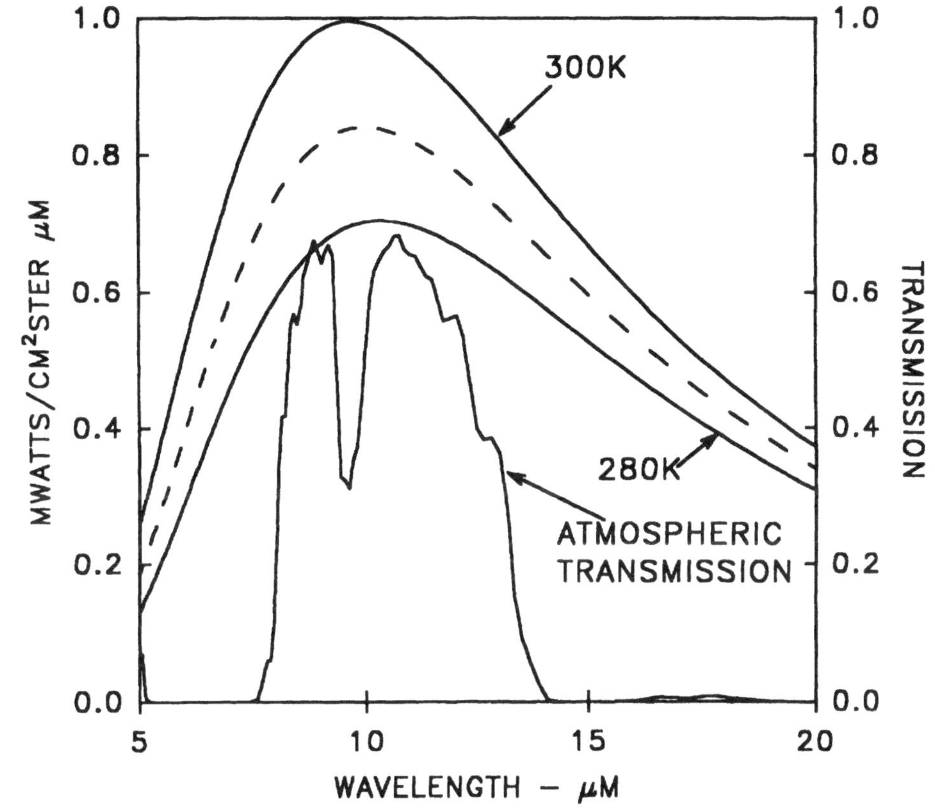

Figure 3. The spectral radiance of a black body according to Equation (2) for typical terrestrial temperatures and the atmospheric transmission for the US standard mid-latitude summer atmosphere.

model [3] for the US standard mid-latitude summer atmosphere, assuming the radiometer is at satellite altitude. While the atmosphere is relatively transparent in the 8 to 12 μm range compared to adjacent wavelengths, there is still significant attenuation. As seen in Fig. 3 there is only 60 to 70% transmission with a major dip at about 9.5 μm due to ozone absorption. With the exception of this dip, water vapor is the dominant absorber in the 8 to 12 μm window. This is due mainly to what is called the water vapor continuum since only in the 8–9 μm range are there any relatively strong absorption lines. Thus the magnitude of the atmospheric effect will depend on the water vapor content of the intervening atmosphere. This unknown or uncertain atmospheric contribution is one of the problems for the remote sensing of surface temperature at infrared wavelengths.

The radiation values given in Fig. 3 are those which would be observed right at the surface; the relation between the radiation, L_i, seen by a radiometer on a satellite or aircraft and the surface temperature is:

$$L_i = \left[\varepsilon_i B_i(T) + \left(1 - \varepsilon_i\right) L_{\text{atm}}(\text{dn})\right] \tau_i + L_{\text{atm}}(\text{up}) \tag{4}$$

where the subscript i indicates the integral of these quantities over the band width for channel i of the radiometer, B_i is the Planck function given by Equation (1), the L_{atm}'s are the upward and downward components of the atmospheric radiation and τ is the atmospheric transmission. The values of L_{atm} and τ can be calculated using a model for atmospheric path radiance such as Lowtran-7, however it would be preferable to eliminate the atmospheric effects using the multispectral information.

Several approaches have been developed for eliminating atmospheric effects in the estimation of sea surface temperature from space using multichannel thermal data. The technique used with the AVHRR (Advanced Very High Resolution Radiometer) data from the NOAA series of polar orbiting satellites involves the differential water vapor absorption in the 10 to 13 μm window with the 11.5 to 13 μm portion being more strongly absorbed, the so called split window technique [4, 5]. However this assumes that the surface emissivity is constant over this spectral band, which is not the case for land surfaces [5, 6, 7]. There have been a number of recent papers on the use of split window techniques over land with the AVHRR data [8, 9]. To understand this possibility we need to look at the infrared emissivities for typical land surfaces.

3. INFRARED EMISSIVITY

There has been much use of the spectral variation of infrared emissivities for geological purposes [10]. For the purposes of this workshop we are more interested in the emissivities of soils and vegetation. This discussion will based on recent papers by Salisbury and d'Aria [11, 12] which present many emissivity spectra in the 8 to 14 μm range for soils and vegetation. These results were based on laboratory and field measurements made with an interferometer. To extract the emissivity from the radiance measurements, the assumption is made that the emissivity is unity somewhere in the wave band observed. A blackbody curve is fitted to that portion of the spectra to obtain the temperature. The emissivity is then determined by dividing the observed radiance by the black body radiance at that temperature [13].

However to do this with field measurements it is necessary to take into account the reflected sky radiation as indicated in Fig. 4 where the true and apparent emissivity are plotted as functions of wavelength for a case with significant cloud cover and thus downwelling radiance. This example shows the importance of measuring the downwelling sky radiance in emittance measurements made outside. This factor can also be important for microwave emission measurements. The important result from Fig. 4 is the close correspondence between the emissivity measurements and the hemispherical reflectance results. The latter are much easier to make and can be made in a laboratory, yielding a value for one minus the emissivity. We will present results for a number of other materials in this form. Note that for the quartz sand shown here the emissivity is less than 0.6 at $\lambda = 9.2$ μm.

An example of the emissivity variation of soils is given in Fig. 5 where spectral variation for two soils from the southwest of France are presented. The curves were determined in the laboratory at University of Strasbourg by Nerry *et al.* [14]. They show the pronounced effect of quartz (SiO_2) on the emissivity spectra for the two soils with emissivity being less than 0.9 for $\lambda < 9.5$ μm. Note that in the 10 to 12 μm range the emissivity is high ~ 0.95 and relatively constant. The response functions for the 6 channels of the TIMS (Thermal Infrared Multichannel Scanner) radiometer are also

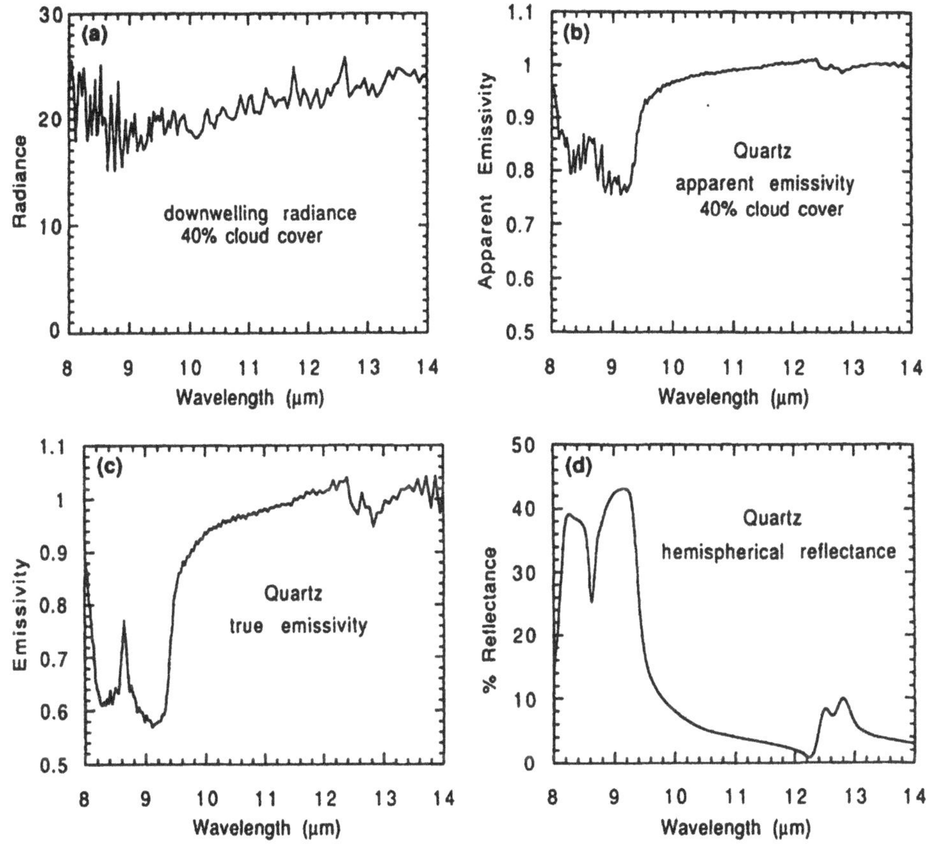

Figure 4. Results of a field measurement of emissivity for a quartz sand compared to laboratory measurements of the directional hemispherical reflectance. a) Downwelling sky radiance for a humid tropical atmosphere with 40% cloud cover. b) Apparent emissivity with the reflected downwelling sky radiance included. c) True emissivity of the quartz sand after removal of the reflected downwelling sky radiance. d) Laboratory spectrum of directional hemispherical reflectance of the quartz sand [11].

shown. Emissivity values derived from TIMS data obtained on an aircraft platform are shown in Fig. 6. These data are normalized to the calculated emissivity for channel 5 and for the hot spot show good agreement with laboratory measurements. The behavior of the cool spot is not understood at the present time. These results indicate that it is possible to extract these emissivity variations with remotely sensed data.

Results for a large number of soils are given in the paper by Salisbury and D'Aria [12]. For almost all of their soils the emissivity in the 10 to 12 μm band is higher than in the 8 to 9.2 μm band. The effect of quartz on the soils' emissivity depends on particle size, with the effect being greater for coarser soils, i.e. those with larger particle sizes. Since soil organic matter is highly absorbing in the thermal infrared with a relatively flat spectrum it reduces the spectral contrast of the quartz restrahlen bands. A similar but smaller effect arises from the presence of soil moisture because of the flat absorption spectrum of water. This is shown in Fig. 7 for an entisol with 94% sand at 3 soil moisture levels. It is seen that the effect of the moisture is to increase the emissivity in the 8 to 9.2 μm band.

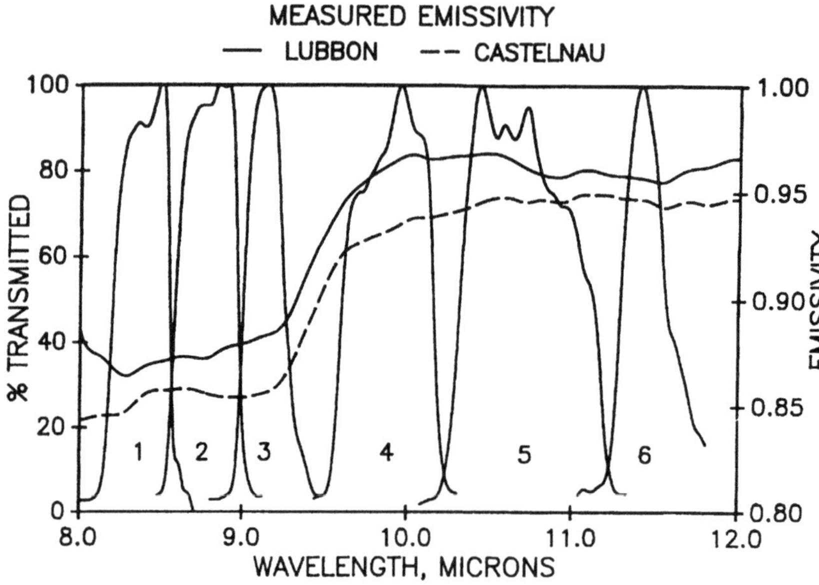

Figure 5. Laboratory measurements of the emissivity for two soils from the HAPEX area [14] with the filter functions for the six TIMS channels.

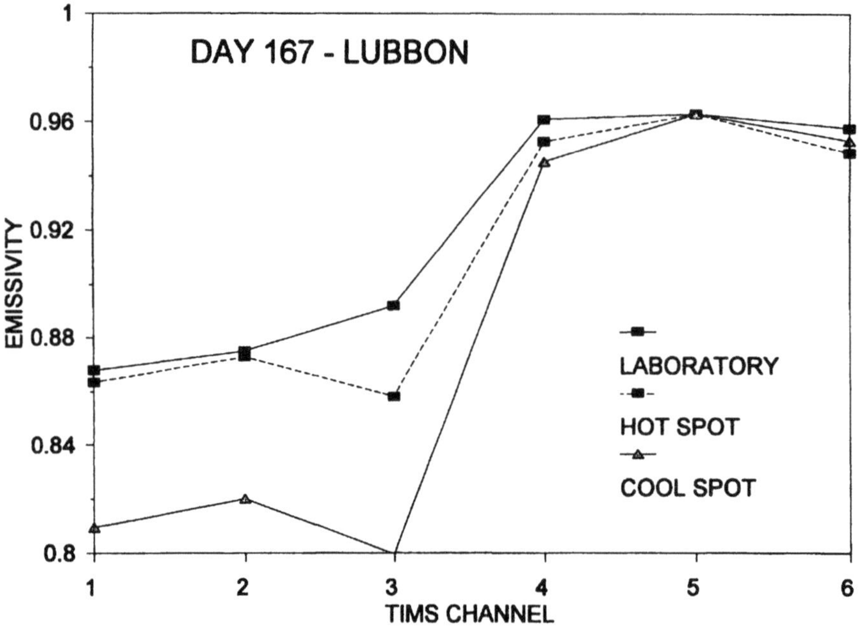

Figure 6. Comparison of the aircraft emissivity values with laboratory measurements for a bare soil field at Lubbon. The aircraft values are for hot and cool spots in the field derived from data obtained with the TIMS sensor at an altitude of 1500 m [16].

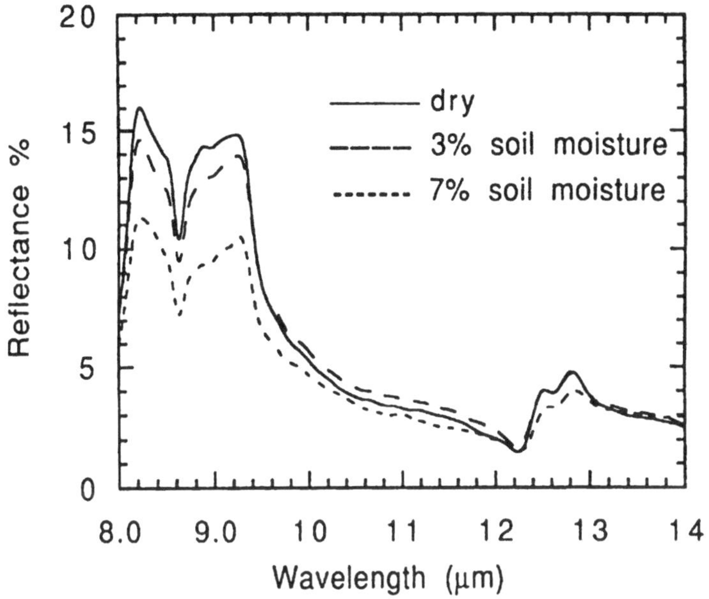

Figure 7. Effect of increasing gravimetric soil moisture on the hemispherical reflectance spectrum for a quartz sand from theeastern shore of Maryland [12].

In addition to minerals and soils Salisbury and D'Aria [11] present spectra for a number of terrestrial materials. The reflectance spectra for a few relevant materials are given in Fig. 8. These include various vegetative components such as green and senescent foliage. For the green foliage the reflectances are generally less than 5%, emissivity > 0.95, while for the senescent vegetation reflectance is generally higher. As expected the soil litter has low reflectance.

When these components are combined to form a vegetative canopy the results become more complicated because of scattering within the canopy. The net result is simpler because multiple scattering by the leaf surfaces with their low reflectance yields a canopy emissivity very close to one with little spectral variation. This was observed for the prairie grass in FIFE by Palluconi *et al.* [15] who found an emissivity of 0.99 ± 0.01, with no spectral variation. In HAPEX–MOBILHY Schmugge *et al.* [16] found little or no spectral variation over a coniferous forest or an oat field using TIMS data from an aircraft platform. This is shown in Fig. 9. The values for the bare field are those which were used to obtained the emissivity values shown if Fig. 6. These results for vegetation are in substantial agreement with modeling studies of Norman *et al.* [17] who derived a value of 0.99 for the emissivity of the grass canopy in FIFE.

DISCUSSION

The results of Salisbury and D'Aria [12] indicate that a detailed analysis of emissivity spectra for soils may yield information on the soil texture (particle sizes), organic content and/or moisture content. While these properties are detectable in laboratory spectra it remains to be seen if they will be observable from a remote platform such

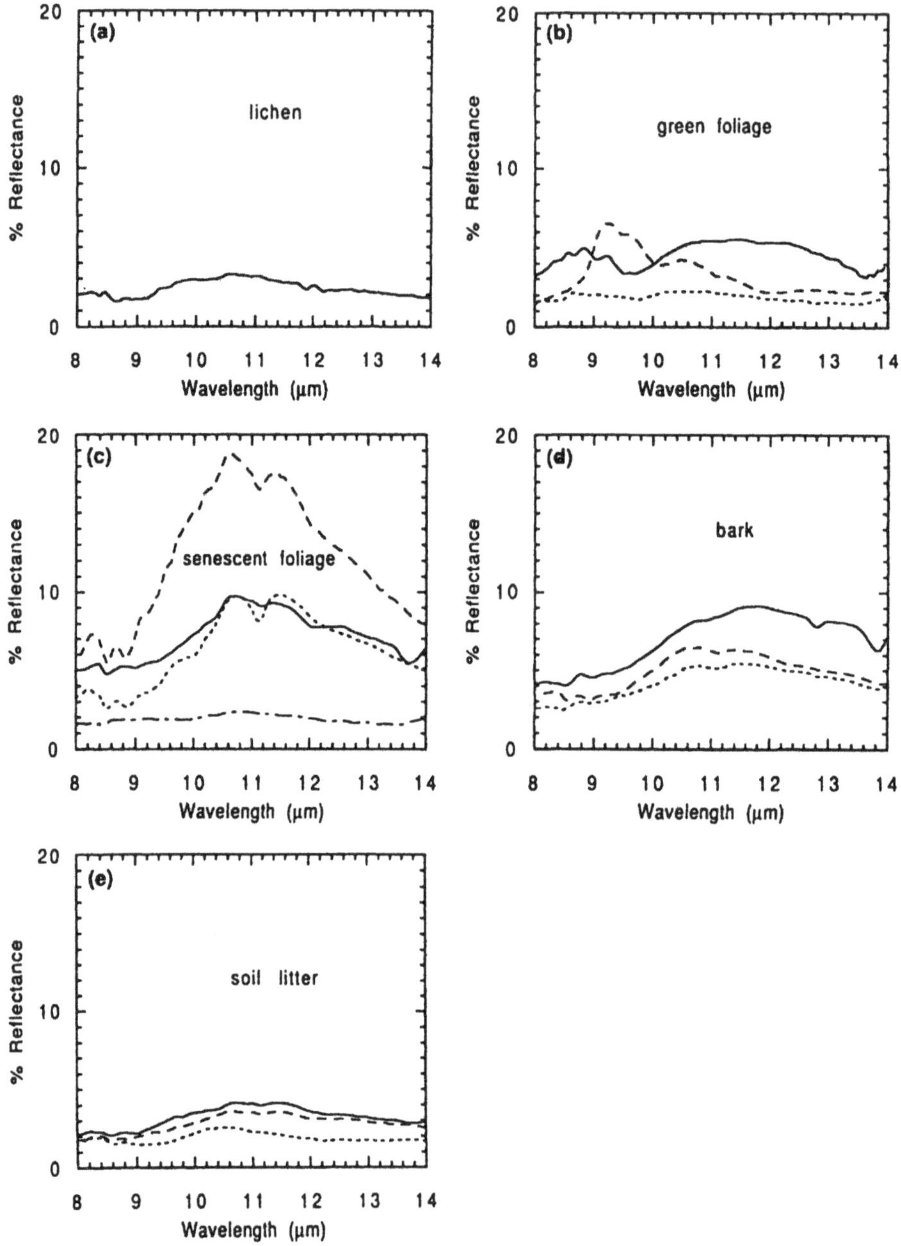

Figure 8. Directional hemispherical reflectance spectra for several types of vegetative cover: a) tree lichen from Maryland; b) green foliage: red oak leaves — solid, indiangrass — dashed, and white pine needles — dotted; c) senescent foliage: beech leaves — dashed, red oak leaves — solid, rye grass — dotted, and white pine needles — dash-dot; d) bark samples: red oak — solid, yellow poplar — dashed, and white pine — dotted; e) decaying soil litter: deciduous leaf fragments, and twigs — solid, decaying wood — dashed, and decaying pine needles — dotted [11].

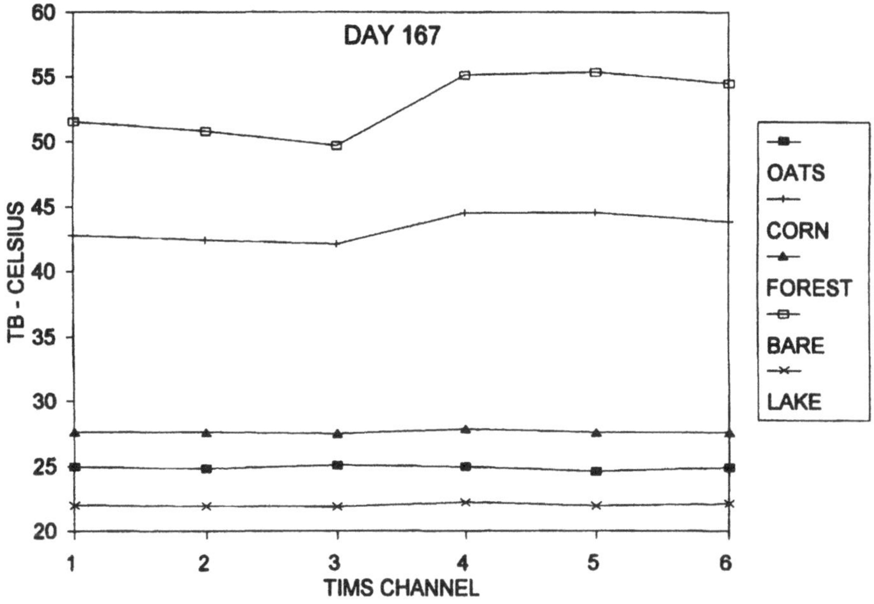

Figure 9. TIMS brightness temperatures for several surfaces at the HAPEX–MOBILHY central site near Lubbon on 16 June 1986 (day 167). These values represent the upwelling brightness at the surface and have not been corrected for reflected downwelling atmospheric radiance [16].

as the aircraft instrument with a small number of bands, e.g. the TIMS or the Advance Spaceborne Thermal Emission and Reflection (ASTER) Radiometer to be flown on the EOS-AM spacecraft in 1998 [18]. The variation of bare soil emissivity with wavelength λ indicates that care must be taken in using broad band infrared radiometer results to infer surface temperature. The results for vegetated canopies, i.e. unity emissivity with little or no spectral variation, are very encouraging for the possible use of split window approaches for eliminating atmospheric effects in thermal infrared determinations of surface temperatures of vegetation.

REFERENCES

1. E. Schanda. *Physical Fundamentals of Remote Sensing.* Springer-Verlag Publishers, Berlin (1986).
2. T. Schmugge, T. Wilheit, W. Webster and P. Gloersen. *Remote Sensing of Soil Moisture with Microwave Radiometers II.* NASA Tech. Note TN D-8321 (1976).
3. F. X. Kniezys, E. P. Shettle, L. W. Abreu, J. H. Chetwynd, G. P. Anderson, W. O. Gallery, J. E. A. Selby and S. A. Clough. *User Guide to Lowtran-7,* Air Force Geophysics Laboratory Report. No. AFGL-TR-88-0177, Hanscom AFB, Massachusetts 01731 (1988).
4. E. P. McClain, W. G. Pichel, C. C. Walton, Z. Ahmed and J. Sutton. Multichannel improvements to satellite derived global sea surface temperatures. *Adv. Space Res.*, **2**, 23–47 (1983).
5. H. Grassl. Extraction of surface temperature from satellite data. In: *Applications of Remote Sensing to Agrometeorology*, F. Toselli (Ed.), pp. 199–220 (1989).
7. F. Becker. The impact of spectral emissivity on the measurements of land surface temperature from a satellite. *Int. J. Rem. Sensing*, **8**, 1509–1522 (1987).
8. Y. H. Kerr, J.-P. Lagouarde and J. Imbernon. Accurate land surface temperature retrieval from AVHRR data with use of an improve split window algorithm. *Remote Sens. Environ.*, **41**, 197–209 (1992).

9. F. Becker and Z.-L. Li. Towards a local split window over land surfaces. *Int. J. Rem. Sensing*, **11**, 369–393 (1990).
10. M. J. Bartholomew, A. B. Kahle and G. Hoover. Infrared spectroscopy (2.3–20 μm) for the geological interpretation of remotely sensed multispectral thermal infrared data. *Int. J. Rem. Sensing*, **10**, 529–544 (1989).
11. J. W. Salisbury and D. M. D'Aria. Emissivity of terrestrial materials in the 8–14 μm atmospheric window. *Remote Sens. Environ.*, **42**, 83–106 (1992).
12. J. W. Salisbury and D. M. D'Aria. Infrared (8–14 μm) remote sensing of soil particle size. *Remote Sens. Environ.*, **42**, 157–165 (1992).
13. A. B. Kahle and R. E. Alley. Separation of temperature and emittance in remotely sensed radiance measurements. *Remote Sens. Environ.*, **42**, 107–112 (1992).
14. F. Nerry, J. Labed and M. Ph. Stoll, Essai de détermination absolue des signatures spectrales de sols nus dans l'infrarouge thermique, en émission et en réflexion, In: *Proc. 4th Int'l Colloquium on Spectral Signatures in Remote Sensing*, Aussois France, 18–22 January 1988, ESA SP-287, pp. 185–188, (1988).
15. F. Palluconi, A. B. Kahle, G. Hoover and J. E. Conel. The spectral emissivity of prairie and pasture grasses at Konza Prairie, Kansas. In: *Symp. on FIFE, Amer. Met. Soc.*, pp. 77–78 (1990).
16. T. J. Schmugge, F. Becker and Z.-L. Li. Spectral emissivity variations observed in airborne surface temperature measurements. *Remote Sens. Environ.*, **35**, 95–104 (1991).
17. J. M. Norman, J. Chen and N. Goel. Thermal emissivity and infrared temperature dependency of plant canopy architecture and view angle, In: *Proc. IEEE Int Geoscience and Remote Sensing Symp.* (IGARSS'90), 2–24 May 1990, College Park, Maryland., Vol. III, pp. 1747–1750 (1990).
18. G. Asrar and D. J. Dokken. *EOS Reference Handbook — 1993*. NASA Earth Science Support Office, Washington, DC, 20024 USA (1993).

ESA/NASA International Workshop, pp. 205–224
B. J. Choudhury, Y. H. Kerr, E. G. Njoku and P. Pampaloni (Eds)

Synergism of active and passive microwave data for estimating bare soil surface moisture

SASAN S. SAATCHI,[1] ENI G. NJOKU[1] and URS WEGMÜLLER[2]

[1] *Jet Propulsion Laboratory, California Institute of Technology, 4800 Oak Grove Drive, Pasadena, California 91109, USA*
[2] *Institute of Applied Physics, University of Bern, Switzerland*

Abstract—In this paper, a unique set of active and passive microwave measurements have been combined in conjunction with backscatter and emission models of bare soil surfaces to develop algorithms for soil moisture estimation. A multi-frequency ground-based radiometer–scatterometer system working at frequencies between 1 GHz and 12 GHz at incidence angles ranging from 0° to 70° has been used to examine the effect of soil moisture and surface roughness on microwave scattering and emission. The results of the experiment for a limited range of soil roughness and moisture shows that the ratio of the backscattering coefficient over the surface reflectivity is independent of the soil moisture and varies almost linearly with respect to the surface rms height times the propagation constant.

To examine this effect further, the rough soil surface has been modeled using the classical approaches of the small perturbation, physical optics, and geometrical optics approximations. The bistatic scattering coefficients of the surface are then used in conjunction with principles of reciprocity and conservation of energy and integrated over the hemispherical solid angle to obtain the reflectivity of the surface. The ratio of the backscattering coefficient over reflectivity is computed and its sensitivity to soil moisture and roughness variations is examined. It has been found that over a large range of surface roughness this ratio is independent of soil moisture. A retrieval algorithm based on the estimation of the surface rms height from this ratio and the correction of the emissivity data for roughness effects has been developed. Active and passive microwave measurements of bare soil surfaces have been used in conjunction with this algorithm to estimate surface soil moisture. The estimated soil moisture values when compared with the *in situ* measurements have shown reasonable agreement.

1. INTRODUCTION

Active and passive microwave remote sensing techniques have been applied effectively to the problem of estimating the surface soil moisture in a variety of environmental conditions. Research to date has shown that both types of sensors are sensitive to the surface roughness and the vegetation cover. In estimating the soil moisture, the effect of the vegetation and roughness are often corrected either by acquiring multi-configuration (frequency and polarization) data or by adjusting the surface parameters in order to match the model predictions to the measured data. Due to the limitations in the use of multi-configuration spaceborne systems and the frequent lack of *a priori* knowledge of the surface characteristics for parameter adjustments, synergistic use of active and passive sensors may improve the estimation of soil moisture [1].

In this paper, we examine the synergism of radar and microwave radiometer measurements of soil moisture for bare or low vegetated surfaces. Both types of sensors, with limited configurations, have the potential of being on orbiting platforms in future [2]. Studies focused on soil moisture estimation, however, have usually dealt with only one instrument at a time [3–7]. The lack of efforts to study the synergism of the two techniques stems from 1) the difficulty of making simultaneous measurements with radar and radiometer systems and 2) the independent development of soil surface electromagnetic models for active and passive microwave remote sensing. In most studies to date, backscattering models have been physically based and are derived by approximating the exact surface scattering formulations [6, 7]; whereas, the emissivity models are usually empirical or semi-empirical with non-physical parameters [6, 7, 9]. Thus, regardless of the accuracy of each model, inter-comparisons of the models and their sensitivities to the same surface parameters are difficult unless the models are derived from the same physical principles and approximations.

In order to overcome these difficulties, we have used microwave scattering models for rough surfaces to simulate both radar and radiometric measurements. By necessity, the surface scattering models are approximations and are only valid for a limited range of frequencies and/or roughness parameters [7, 8]. To compare these models with experimental data and to validate their range of applicability requires precise measurements of surface parameters. Due to difficulties in measuring roughness parameters, scattering models are often used with little knowledge of surface roughness. In particular, in the radiometric case, the roughness parameters are usually treated as adjustable parameters for estimating soil moisture [3, 5, 9]. However, recent radiometric measurements have shown that there is a need to accurately model the surface roughness at various frequencies in order to improve the sensitivity of the observed brightness temperature to soil moisture [10]. Similar conclusions can be drawn in the case of radar measurements of the soil moisture. In a recent study, a polarimetric scatterometer system has been used to examine the accuracy of the surface scattering models and to develop inversion algorithms for soil moisture [11]. The results of this study show that the precise estimation of the surface roughness is the key to correctly infering the soil moisture information from the radar data. However, the implementation of the proposed algorithm requires a multi-polarization radar system [11]. In many current spaceborne systems, the configuration of the sensor is limited to one polarization and one frequency (e.g. ERS1, JERS1, ESTAR, etc.). Thus, it is important to develop inversion algorithms which can make use of single configuration data from multiple sensors.

In this work, we first describe the field experiment and the characteristics of the sampled and monitored soil surfaces in Section 2. Then, we employ classical rough surface scattering theories to model both the active and passive observations over bare soil surfaces. The bistatic scattering coefficients obtained from the small perturbation method (SPM), physical optics (PO), and geometrical optics (GO) approximations are used in conjunction with the energy conservation and reciprocity principle to obtain the surface emissivity. In Section 3, the accuracy and validity conditions for backscattering and emissivity modes are discussed. In Section 4, we combine models and measurement results to obtain a synergistic approach for measuring the surface soil moisture. The synergistic algorithm is then implemented on the measured data and the results are summarized in Section 5.

2. EXPERIMENT

This study was motivated by the availability of simultaneous backscattering and emissivity measurements of soil surfaces from the same platform. The experimental data were acquired using a combined radiometer-scatterometer unit (RASAM). The system consists of a 7-frequency Dicke radiometer with horizontal and vertical polarization channels and a 1–12 GHz noise scatterometer with 4 polarization channels (HH, VH, HV, VV) [12, 13]. Measurements were made from 0° to 70° incidence angles in steps of 10 degrees from nadir by both systems. In the case of the radiometer, the measured signals were converted to brightness temperature by using calibration data (hot load at 80 °C and cold load at sky temperature). The accuracy of the radiometer in measuring the brightness temperature was between 1–2 K (Kelvin). The scatterometer system was also calibrated by using external targets [13].

The instrument was installed on a cherry picker approximately 8 m above the soil surface. Some of the surfaces were chosen in winterwheat fields which had less than 20 g/m^2 of wet biomass. The measurements were conducted during the entire winter season and the fields were not cultivated during this period. The effect of vegetation on the measurements was assumed negligible since the microwave signatures were dominated by the soil characteristics. To examine the effect of soil characteristics on the measured data, we considered cases where the contributions of soil dielectric constant (moisture) and roughness could be separated. Thus, the experiment was focused on two scenarios: 1) the effect of soil moisture, which was measured during the freezing and thawing condition of soil surface, and 2) the effect of surface roughness which was measured when the surface characteristics changed due to erosion and rain [12].

In the first scenario, the soil surface froze to a depth of about 5 cm and its volumetric liquid moisture content was similar to a dry soil condition. During the thawing period the surface was wet and the liquid moisture of the top layer increased. The surface roughness and the soil texture remained constant during this cycle and therefore the only significant change was due to the liquid moisture variation of the top 5 cm of the soil surface. Table 1 shows the data obtained from the *in situ* measurements for this case. The dielectric constants of the surface were obtained from the measured volumetric soil moisture using a semi-empirical model developed by Dobson *et al.* [15] and by using the measured bulk density of 1.19 g/cm^3 and texture of 37% sand, 45% silt and 18% clay in the model. For the frozen soil, an equivalent liquid soil moisture of 0.05 cm^3/cm^3 was assumed to simulate the frozen soil dielectric constant.

Average volumetric moisture measurements of the top 5 cm of the soil surface may not be appropriate for analysis of data from all frequency channels. As the frequency increases, the penetration depth of microwave energy in the soil decreases such that,

Table 1.
Dielectric constant of soil during freezing and thawing cycle. The parameter ms stands for the volumetric soil moisture content (cm^3/cm^3) and ε' and ε'' are the real and imaginary parts of the dielectric constant, respectively

	Frozen soil ($ms = 0.05$)		Thawed soil ($ms = 0.38$)	
f (GHz)	ε'	ε''	ε'	ε''
2.5	3.9	0.1	21.3	2.2
3.1	4.0	0.5	22.7	5.7
4.6	3.9	0.5	21.0	7.4
7.2	3.7	0.6	17.8	9.1
10.2	3.7	0.4	18.0	6.9
11.0	3.5	0.7	13.8	9.5

rms height = 1.23 cm, correlation length = 3.92 cm.

Table 2.
Soil surface parameters during surface roughness variation. The parameter s is the rms height of the surface, l is the correlation length, and the surface slope m is defined as s/l

Surface	T (K)	s (cm)	l (cm)	$m = s/l$	ms (cm^3/cm^3)
F-1	275	0.38	1.35	0.28	0.21, 0.28
F-2	278	0.87	3.81	0.23	0.33
F-3	278	1.23	3.92	0.31	0.05, 0.38
F-4	282	1.70	4.05	0.42	0.38

for example, at 10 GHz, and for a wet soil, only moisture in the top 1 cm of the soil may affect the microwave measurements. On the other hand, for frozen soil, at larger wavelengths, the penetration depth may be greater than 5 cm. However, the 5 cm values were used in the absence of more detailed information on the soil profile, and the possible introduction of error was noted.

In the second scenario, the same field was observed over a three month period. During this period, the surface roughness changed slowly duc to rain and erosion. We have focused on data acquired on four different days in which the soil moisture had approximately the same value and only the roughness parameters changed. The data obtained from the *in situ* measurements are listed in Table 2. For both scenarios, the measured soil temperature and surface roughness parameters were recorded. Soil thermometric temperature was measured at various depths and the surface temperature was determined using an infrared radiometer [12]. From all these measurements, an effective soil temperature was calculated. In the data analysis, this temperature was used to calculate the surface emissivity from the radiometer brightness temperature measurements. Since the temperature and soil moisture profiles are correlated to brightness temperature measurements, the use of an effective soil temperature may introduce an error in the emissivity calculations at various frequencies.

For the roughness parameters, a laser profilometer was used to measure the surface rms height from several 1 m long height profiles. The laser profiler had a horizontal resolution of 0.5 mm and a vertical accuracy of 0.25 mm. It was found that in all cases the autocorrelation function of the surface height was much closer to an exponential function than a Gaussian. Hence, an exponential autocorrelation function was used to obtain the correlation length. In Table 2, we have shown four field situations that fit the requirement of significantly different roughness and similar moisture content. These cases were those used below with modeling and data analysis over the frequency range of RASAM system.

3. THEORY

In this section, the backscattering and emission from a random rough surface are formulated by using classical electromagnetic surface scattering models. The classical models are approximate models which are valid for different degrees of roughness from 'smooth' to 'very rough'. The criterion for degree of roughness was first introduced by Rayleigh [16]. Rayleigh's criterion for degrees of roughness is somewhat arbitrary. It assumes that if the phase difference between two rays scattered from separate points on the surface is less than $\pi/2$ radians, then the surface is 'smooth'; otherwise it is 'rough'.

This was mainly due to the fact that when the phase difference is small compared to π, the waves will be in phase and constructively interfere and will scatter mainly in the specular direction (smooth surface). When the phase difference is large, the waves destructively interfere, leading to scattered energy in all directions with little contribution in the specular direction [16]. Rayleigh's assumption, being a first order classifier of surface smoothness, illustrates the important point that the roughness is not an intrinsic property of a surface but depends on the frequency and angle of the incident wave. However, for modeling microwave scattering of natural surfaces where the rms height is of the order of the wavelength, a more stringent condition for roughness such as the Fraunhofer criterion has been used [7].

Classical rough surface scattering models are described according to the scale of roughness relative to the wavelength. Here, we consider three scattering models, namely, the small perturbation model (SPM) and the Kirchhoff model under the physical optics (PO), and geometrical optics (GO) approximations. The detailed derivation of each model and its region of applicability is discussed elsewhere in the literature [7, 17, 18]. Here, we concentrate on the model results and their uses in computing surface emission.

Since the main objective is the comparison between scatterometer and radiometer measurements of soil surfaces, only expressions that can indirectly relate the backscattering coefficient to the surface emission are described. This is done by deriving expressions for surface emissivity from the surface scattering cross sections. The brightness temperature $T_{\mathrm{p}}(\theta)$, observed by a radiometer system may be described as:

$$T_{\mathrm{p}}(\theta) = \varepsilon_{\mathrm{p}}(\theta) T_0 = \left[1 - \Gamma_{\mathrm{p}}(\theta)\right] T_0 \tag{1}$$

where $\varepsilon_{\mathrm{p}}(\theta)$ is the surface emissivity, $\Gamma_{\mathrm{p}}(\theta)$ is the surface reflectivity (p being either H or V polarization and θ being the look direction), and T_0 is the physical temperature or the equivalent temperature of the surface [19]. Equation (1) defines the emissivity of the surface according to energy conservation and reciprocity principles. This equation assumes that the amount of energy which is not scattered from the surface is absorbed. Under thermal equilibrium absorbed energy is the same as energy emitted from the surface. The reflectivity of a random surface, can be expressed as the total average fraction of incident energy scattered from the surface [7] and it can be expressed as:

$$\Gamma_{\mathrm{p}}(\theta) = \frac{1}{4\pi \cos\theta} \int_0^{2\pi} \int_0^{\pi/2} \left[\sigma_{\mathrm{qq}}(\theta, \theta_s, \phi_s) + \sigma_{\mathrm{pq}}(\theta, \theta_s, \phi_s)\right] \sin\theta_s \, \mathrm{d}\theta_s \, \mathrm{d}\phi_s \tag{2}$$

where $\sigma_{pq}(\theta, \theta_s, \phi_s)$ is the bistatic scattering coefficient. The surface reflectivity can be further decomposed into coherent and incoherent contributions as a result of integrating the coherent and incoherent components of the scattering coefficients [20]. These components are given by

$$\Gamma_{\mathrm{p}}(\theta) = \Gamma_{\mathrm{p}}^{\mathrm{coh}}(\theta) + \Gamma_{\mathrm{p}}^{\mathrm{inc}}(\theta) \tag{3a}$$

$$\Gamma_{\mathrm{p}}^{\mathrm{coh}}(\theta) = |R_{\mathrm{p}}|^2 \exp\left[-4k^2 s^2 \cos^2\theta\right] \tag{3b}$$

$$\Gamma_{\mathrm{p}}^{\mathrm{inc}}(\theta) = \frac{1}{4\pi \cos\theta} \int_0^{2\pi} \int_0^{\pi/2} \left[\sigma_{\mathrm{pp}}^{\mathrm{inc}}(\theta, \theta_s, \phi_s) + \sigma_{\mathrm{qp}}^{\mathrm{inc}}(\theta, \theta_s, \phi_s)\right] \sin\theta_s \, \mathrm{d}\theta_s \, \mathrm{d}\phi_s \tag{3c}$$

where $k = 2\pi/\lambda$ is the propagation constant at the operating wavelength λ, s is the rms height of the surface, and R_p is the Fresnel reflection coefficient of the surface for polarization $p(\mathrm{H}, \mathrm{V})$. Equation (3b) is obtained by integrating the coherent scattering coefficients over the hemispherical solid angle. In Equation (3c), $\sigma_{qp}^{inc}(\theta, \theta_s, \phi_s)$ is the bistatic scattering coefficient of the surface from the incident direction θ and polarization p to the scattered direction (θ_s, ϕ_s) and polarization q. Since (3c) is a general expression, the accuracy of calculating the surface reflectivity and ultimately the emissivity depends on the accuracy of the model used to specify the scattering coefficients. Unlike the coherent contribution, no analytical expression can be found for the incoherent part due to the numerical expression necessary to compute Equation (3c). In most studies, the coherent part of the reflectivity alone is used to match experimental data by adjusting a parameter that can be interpreted as an effective rms height which is smaller than the measured height [9].

In Fig. 1, the coherent and the incoherent (diffuse) fields are shown schematically. The scattered coherent field is always predominantly in the specular direction and there is no backscattered energy. The strength of the scattered field is dependent on the dielectric constant of the surface. As the surface becomes rougher, the scattered field becomes more diffuse and the amount of energy radiated in the backscatter direction increases [16]. For a very rough surface, as shown in Fig. 1, the coherent field completely diminishes and there is only a diffuse or incoherent field. Thus, in the radar case, the backscatter energy is due to the incoherent field. In the radiometric case, since the energy is integrated over all directions, both coherent and incoherent fields contribute to the surface reflectivity. In what follows, we examine the backscattered energy and

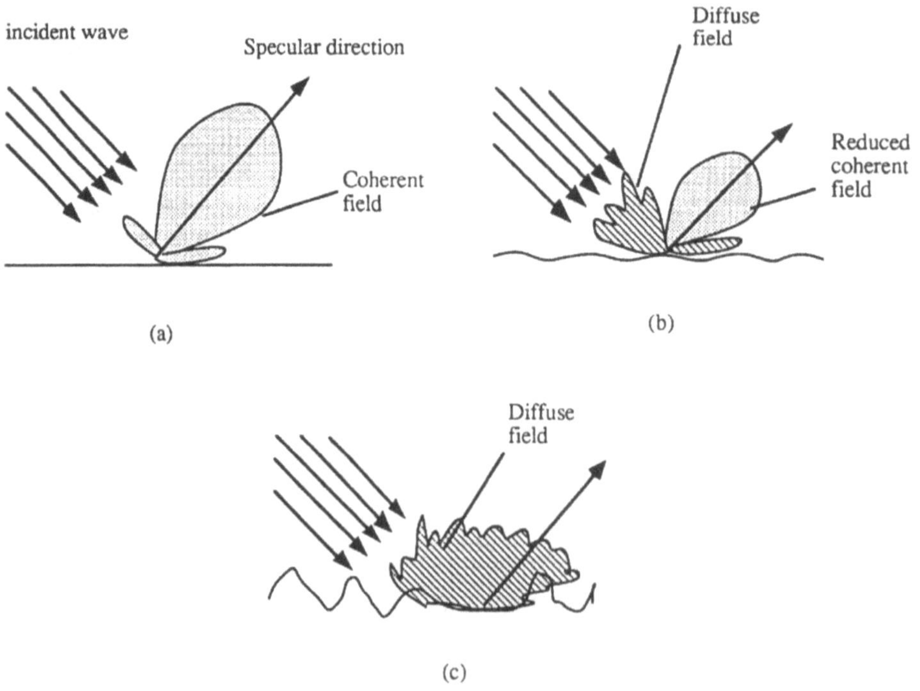

Figure 1. Schematic of changes in scattered energy distribution as surface roughness increases, (a) smooth surface, (b) slightly rough surface, and (c) very rough surface.

the surface reflectivity as calculated by using the SPM, PO, and GO approximations. Here, we have limited ourselves to only the HH polarized backscattering coefficient and H polarized emissivity.

3.1. Slightly rough surface (SPM)

The small perturbation model (SPM) is used for slightly rough surfaces when both the rms height and the correlation length of the surface are small compared to the wavelength. In this approximation, the bistatic scattering cross section is determined by the surface dielectric constant and the two-point correlation function of the roughness structure. The accuracy of this method depends on the order of the terms retained in obtaining the scattered field. The perturbation parameter is ks, with s being the root mean square of the surface height (rms height) and k being the wave number ($k = 2\pi/\lambda$, where λ is the wavelength). In this study we have included only the first order terms. Thus, the theory does not take into account the multiple scattering and shadowing effects.

For the simulations, we have considered a bare soil surface with exponential correlation function which will be used later for comparisons with the experimental data. The validity region of the first order SPM for non-Gaussian correlation functions is $ks \leqslant 0.3$, and $kl \leqslant 6.0$, where l is the correlation length of the surface [6, 17, 22]. In Fig. 2, the backscattering coefficient and the emissivity are shown for a slightly rough surface with respect to the incidence angle and ks for different values of ms (with ms being the soil moisture). Figures 2(a) and 2(b) show the angular response of the SPM model for the backscattering and emissivity and the comparison with the measured data over field F-1. As was mentioned earlier, only the HH channel of scatterometer and H channel of radiometer data have been used for the comparisons. In both figures, the measured data are shown to follow the model simulations quite closely. In the backscattered case, the model simulation underestimates measured data, in particular near nadir and grazing angles where errors due to the antenna pattern corrections for coherent field and multiple scattering effects are enhanced [6]. In the case of emissivity the discrepancy increases towards the grazing angles mainly because of errors in calculating the bistatic scattering coefficients, and the experimental errors.

According to the Equations (3), the emissivity of slightly rough surfaces is due to both the coherent and incoherent components of the scattered fields, whereas, the backscattering is only due to the incoherent or diffuse component of the scattered field. Figures 2(c) and 2(d) show that as the surface roughness increases, as indicated by increasing ks, the backscattering or the incoherent component increases and the surface reflectivity remains approximately constant (slightly decreases). The surface reflectivity, for slightly rough surfaces, is mainly dominated by the coherent component. Furthermore, the changes in soil moisture have opposite effects on the backscattering and emissivity. In the former case (Fig. 2c) the backscattering increases as the surface becomes more wet. In the latter case (Fig. 2d) the emissivity decreases for wet surfaces. The 40° incidence angle has been selected in order to show the effect of roughness and soil moisture simultaneously. This angle is also close to the look direction of many spaceborne sensors such as ERS1 and SMMR.

The effect of correlation length appears to be less important than the rms height in the radiometric case. Saatchi and Wegmüller [14] have shown that for a wide range of surface roughness, the emissivity is almost independent of the correlation length.

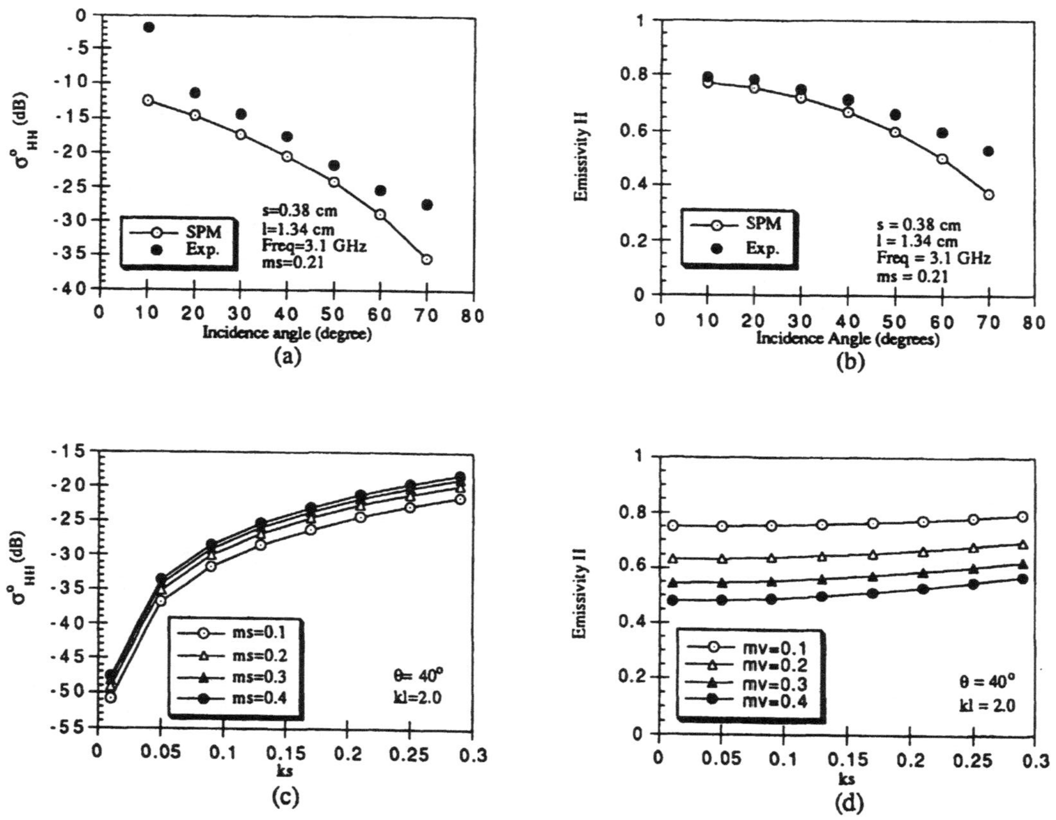

Figure 2. SPM simulations for backscattering HH polarization and emissivity H polarization. Comparison of angular response with data: a) backscattering, b) emissivity. Surface rms height and moisture effects at 40° incidence angle: c) backscattering, d) emissivity.

Recently, a similar study by Oh *et al.* [11] has shown, through a series of measurements, that the backscattering coefficient is also more sensitive to the rms height than the correlation length.

3.2. Moderately rough surface (PO)

The physical optics (PO) approximation known as the Kirchhoff approximation has a validity range that is appropriate for modeling moderately rough surfaces. It is widely used for rough surface scattering because of its simplicity. In particular, for the radiometric case, this theory in conjunction with Equations (3) has shown reasonable agreement with measured data [23]. In this approximation, every point on the surface is treated as though it is part of an infinite plane parallel to the local surface tangent. The theory is exact for an infinite smooth plane surface and is approximate for a rough surface. Because of the tangent plane approximation, the theory fails to predict the polarization dependence of the scattered field correctly. The region of validity of the PO approximation is considered to include surfaces with large correlation length and small rms height, such that the rms slope of the surface s/l is less than 0.25. However, this is considered a broad criterion. A narrower criterion states that the local radius of curvature of the surface should be large compared to the wavelength [21]. This validity condition also embeds the fact that the PO approximation is progressively less accurate as the angle of incidence increases and at larger departures of scattering angle from the specular direction [21]. The theory is mostly used at high frequencies and for moderately rough surfaces, due to these underlying assumptions [7, 17].

To implement the PO model, the formulation given by Ogilvy [17] for the bistatic scattering cross section is used. In this formulation, the multiple scattering and shadowing effects are not included. The model, therefore, is less accurate at high incidence angles where both effects are important. To illustrate the dependence of the backscattering coefficient on surface parameters under the PO approximation, we have plotted the HH polarized backscattering coefficient versus incidence angle. Figure 3a shows the result of the model simulations for field F-3 (Table 2) with two types of surface moisture conditions corresponding to frozen and thawed soil conditions in winter. In both cases, the agreement with measurements is good except near nadir and close to the grazing angle. In the case of emissivity (Fig. 3b), there is good agreement between model and measured data within 50° from nadir.

Note that for moderately rough surfaces, the coherent scattered field is generally less than the incoherent part [14]. As the roughness increases, the incoherent term dominates the coherent component. In Figs 3(c) and 3(d), the backscattering and emissivity, respectively, are plotted in terms of ks. The range of ks chosen for this simulation overlaps with the SPM approximation. The validity condition of the PO approximation for non-Gaussian surfaces is known to be $kl > 6$ and $ks < 1$ [17]. In this range, the surface reflectivity and the backscattering coefficient increase with ks. However, if one runs the simulation for slightly larger values of ks (i.e. $1 < ks < 2$), the reflectivity starts decreasing till the model becomes completely invalid for slopes larger than 0.25. This effect is due to the strong dependence of the PO approximation on the surface slope. However within the validity range of the model and at certain incidence angles one can show that the emissivity is almost independent of the correlation length [14].

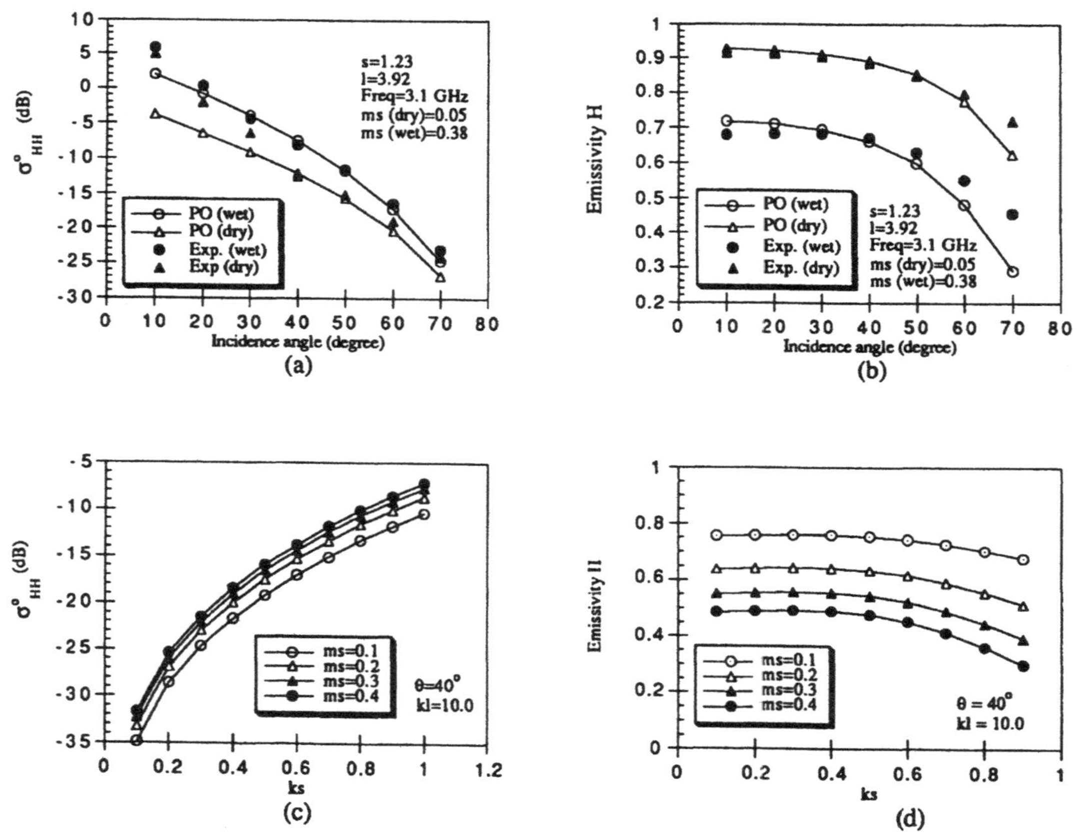

Figure 3. PO simulations for backscattering HH polarization and emissivity H polarization. Comparison of angular response with data for freeze and thaw conditions: a) backscattering, b) emissivity. Surface rms height and moisture effects at 40° incidence angle: c) backscattering, d) emissivity.

3.3. Very rough surface (GO)

For very rough surfaces, the geometrical optics approximation (GO) is most appropriate for evaluating the scattering cross-section of the surface. In this approximation, which is also referred to as the Kirchhoff model under the stationary phase approximation, it is assumed that only the specular points or the facets on the surface contribute to the scattered field. The coherent field, therefore, will be negligible and the total scattered field is due to the diffuse or incoherent field. The bistatic scattering coefficients and consequently the surface reflectivity are proportional to the probability distribution of the specular points on the surface [22]. In addition, at higher incidence angles, part of the surface may be shadowed by other parts. In the formulation of the problem for GO approximation, we have included the shadowing function [7].

The GO model is valid when the rms height of the surface is very large compared to the wavelength ($2ks\cos\theta > 3.0$) [7]. The accuracy of this approximation also deteriorates as the incidence angle increases. One distinctive feature of this theory is the lack of dependence upon frequency [17, 22]. In addition, the surface slope in the GO model can be chosen to be small but surface can never approach a plane.

Using this approximation to simulate the surface of the field F-4 (Table 2), good agreement is found between the GO model results and the scatterometer and radiometer measurements. In Fig. 4(a), the backscattering coefficient for the dry case deviates from measurements less than 1 dB at most angles except at near nadir and near grazing angles. The difference between simulated and measured emissivity however, is much larger (0.1 at 70°). In addition to the model inaccuracy at high incidence angles, scattering in the sparse 8 cm thick vegetation layer above the soil surface could also be playing a role at the relatively high frequency (10.2 GHz).

To examine the relation between backscattering and the roughness parameter ks, we have plotted the model simulations in Fig. 4(c). It is obvious that as ks increases the backscattering increases; however, it reaches a saturation level for $ks > 5$. As one moves closer to nadir the saturation occurs at smaller values of roughness. This saturation, which is a limitation on the use of the GO model for very rough surfaces, is mainly due to multiple scattering effects and the use of a shadowing function which increases the error in the forward scattering amplitude where multiple scattering is important. The decrease in reflectivity from a smooth surface to a rough surface is due to the increase in the absorption of the incoming energy by the surface.

The effect of soil moisture in the GO model is similar to the SPM and PO models. The results for all three models indicate that the the soil moisture variations do not translate linearly to backscattering coefficient. In all cases the backscattering approaches saturation as the surface moisture increases, whereas the emissivity does not. This, however, does not cause any limitations on the use of the models for emissivity calculations because the soil also saturates at values above 40% of volumetric moisture.

From the model simulations and comparisons with experimental data one can conclude that the accuracy of the classical rough surface models, used in this study, depends on many factors. For example, the angle of incidence and scattering, which in turn determine whether shadowing and multiple scattering are important, and the surface roughness parameters and statistics which dictate whether the exponential or Gaussian correlation function is appropriate. In addition, the uncertainties in using the right parameters in the model such as the top layer soil moisture, soil bulk density, soil texture, and the mixing model for converting the soil moisture to permittivity, are potential

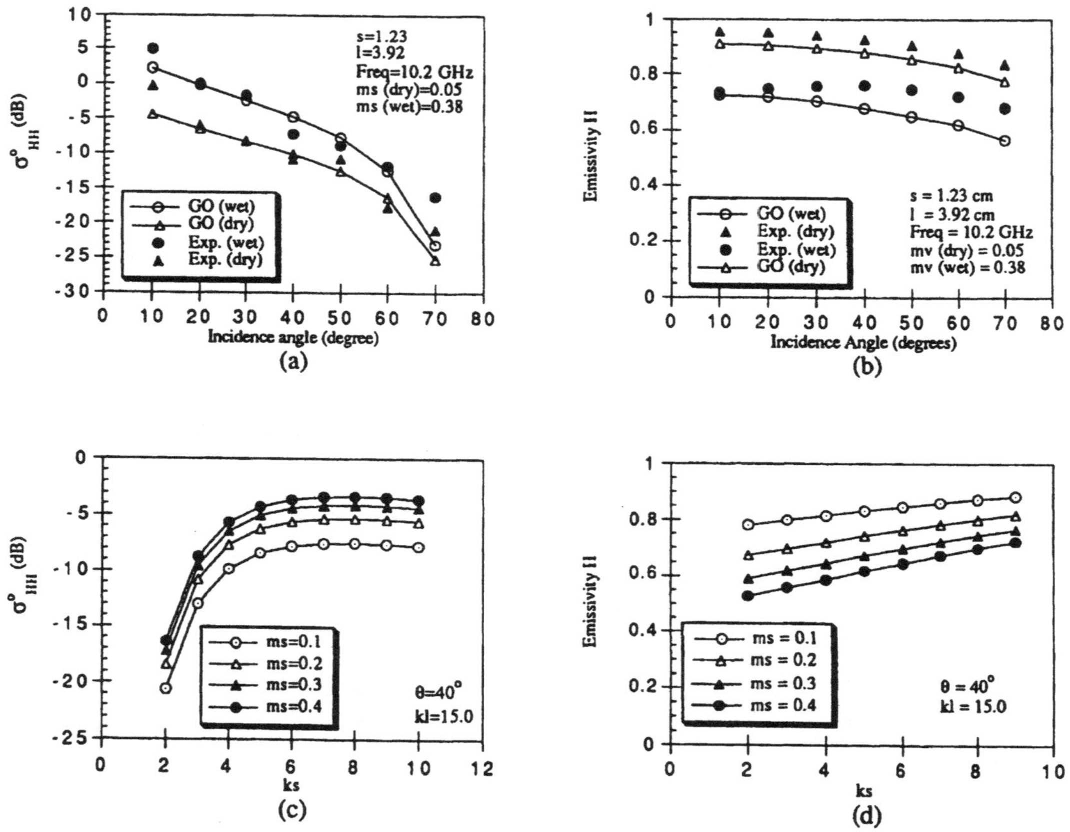

Figure 4. GO simulations for backscattering HH polarization and emissivity H polarization. Comparison of angular response with data for freeze and thaw conditions: a) backscattering, b) emissivity. Surface rms height and moisture effects at 40° incidence angle: c) backscattering, d) emissivity.

sources of errors. For example, the assumption of the sensitivity of the microwave measurements to the top 5 cm soil moisture data does not hold at high frequencies (> 5 GHz) because the skin depth in the soil is less than 5 cm, while at long wavelengths (< 2 GHz), and for frozen or dry soil the skin depth is greater than 5 cm. Nevertheless, when the assumptions for each model are satisfied, the comparisons of model simulations with measurements are in most cases favorable. This is very encouraging since the simulated values were obtained without the need for adjusting arbitrary parameters.

All three models correctly predict the surface emissivity at H polarization for angles within 50° from nadir. By comparing the sensitivity of the backscattering and emissivity to surface parameters, we find that indeed the emissivity is less sensitive to the surface roughness parameters, and varies mainly with the surface moisture (see Figs 2d, 3d, and 4d). However, the surface roughness is more important in determining the backscattering coefficient than the soil moisture (see Figs 2c, 3c, and 4c). Remarkably, the surface features which determine the coherent and incoherent scattering patterns, when averaged over the scattering angles, will have little effect on emissivity. On the other hand, the surface dielectric constant or moisture, which affect the Fresnel reflection coefficients, have an important impact on determining the emissivity. Note that even though the roughness has little effect on emissivity, when multiplied by the surface temperature it may cause a large error on the surface brightness temperature. Thus, the effects of roughness need to be corrected and one may conclude that backscattering and emissivity measurements from bare soil surfaces can be used in a complementary fashion to estimate surface parameters. In the next section, we use this property in estimating the surface soil moisture by the synergism of active and passive microwave data.

4. SYNERGISM

With the results of the above study one can make use of backscattering and emissivity data to estimate the surface soil moisture. Similar studies have been conducted in the past to enhance the capability of remotely estimating the soil moisture over bare and vegetation covered land [25, 26]. For example, in the study conducted by Ulaby *et al.* [25], L-band radiometer and C-band radar data at angles less than 30° have been used to estimate the soil moisture by correcting for the vegetation effects through an optimization technique. Theis *et al.* [26] also used a combination of L- and C-band radar and radiometers and a set of empirical relations to find the best linear fit to the radiometric data. However, most of the recent studies have shown that a more rigorous treatment of the problem in terms of modeling and data analysis is needed for a successful synergistic approach [10, 27].

In our study, we have established a relationship between the backscattering coefficient and emissivity by using the same basic principles. By formulating the problem according to Equation (3) and the analysis of the experimental data, we have shown that in calculating the surface emissivity the terms relating to the roughness parameters (which in turn determine the scattering pattern in all directions) are averaged over the hemisphere above the surface. The dependence of emissivity on the variations of these parameters is consequently much smaller than the backscattering coefficient. Therefore, the ratio of the backscattering coefficient over the surface reflectivity (unity minus emissivity) for the same polarization and incidence angle should be strongly related to the roughness of the surface. Indeed, by using a subset of the data obtained over the fields in Table 2, we obtain this result in Fig. 5. In this figure, we have used the 40° incidence angle data at 3.1, 7.2, and 10.2 GHz over all four fields and all moisture conditions. The backscattering coefficient has been also normalized by $\cos\theta$ (θ being the incidence angle). A linear regression equation has been applied to the data and the best fit ($r^2 = 0.92$) is also shown in this plot. Analytical proof of this result is very difficult for all cases due to the numerical integration in Equation (3). However, for the SPM model it can be readily shown that by taking the ratio of the backscattering coefficient and the reflectivity (coherent term), the Fresnel reflection coefficient common in both expressions cancel, and the quotient is dependent only on surface roughness [14]. Note that the 40° incidence angle has been used for two reasons: 1) it is in the range of angles where the best agreement between model and measurements has been obtained, and 2) it is also close to the range of incidence angles of aircraft or spacecraft radar and radiometer measurements.

To investigate this result further, we have plotted the same ratio for the SPM, PO, and GO model simulations in Fig. 6. Although the range of validity of the models do not necessarily cover the range of measured surface parameters, the results in Fig. 6 indicate that there is not a unique function that can define this ratio for a wide range of ks values. Figure 7 shows the ratio in terms of the volumetric soil moisture and it reveals that σ^0_{HH}/Γ_H is almost invariant of soil moisture. Strictly speaking, the invariance property is valid for moderately rough surfaces. As the roughness increases the GO approximation becomes less accurate and the ratio becomes dependent on soil moisture. Therefore, by combining the radar and radiometer measurements, one can estimate the surface rms height from this ratio.

On developing an algorithm based on the above results, first we need to model this ratio as a function of the surface rms height. According to the model simulations, this function is nonlinear. However, the experimental data shows that one can approximate this function as a regression line. Here, for the sake of simplicity, we also assume

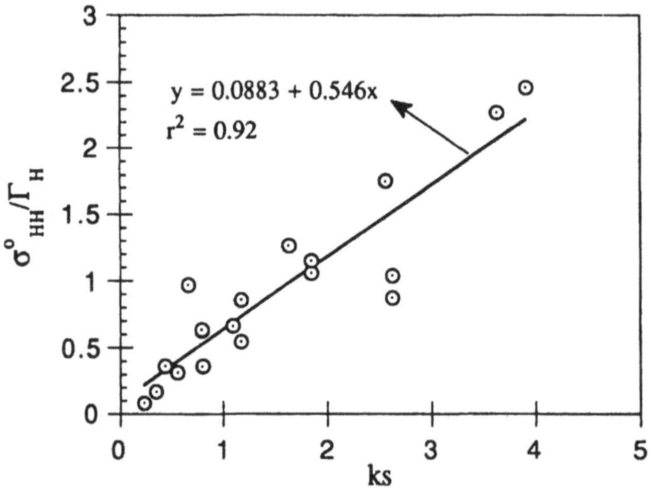

Figure 5. Ratio of backscattering coefficient (HH) normalized by cos(40°) over surface reflectivity (H) versus ks. Circles are the data points and the solid line is the best fit as a result of regression analysis.

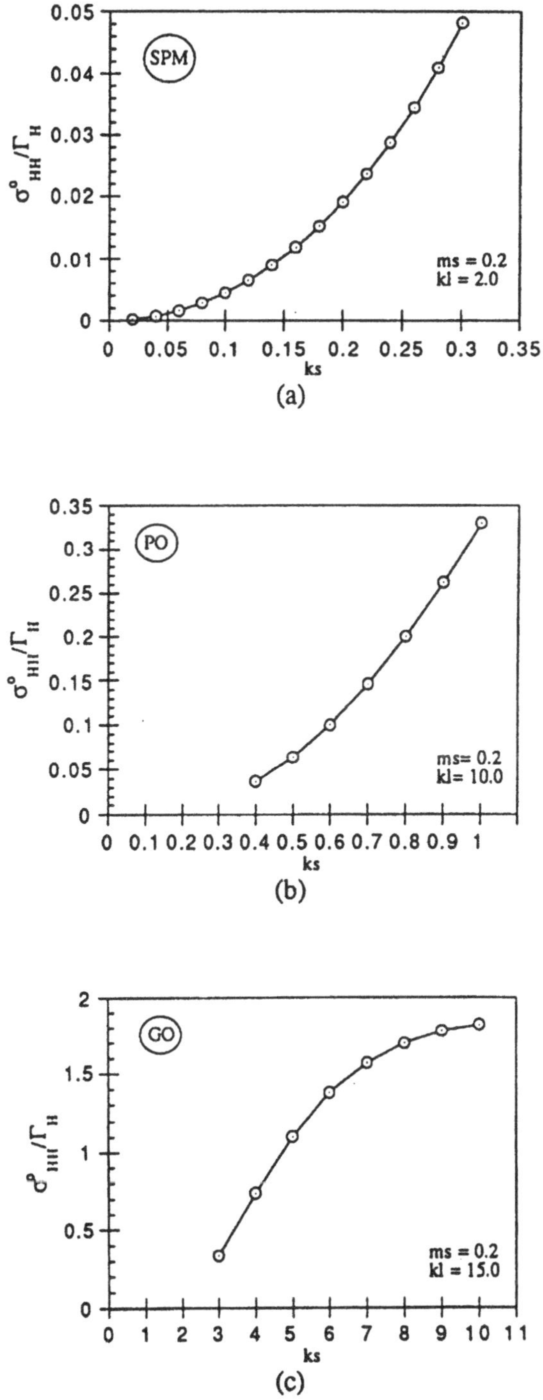

Figure 6. Model simulations of the ratio of the normalized backscattering (HH) over reflectivity (H) for a) SPM model, b) PO model, and c) GO model.

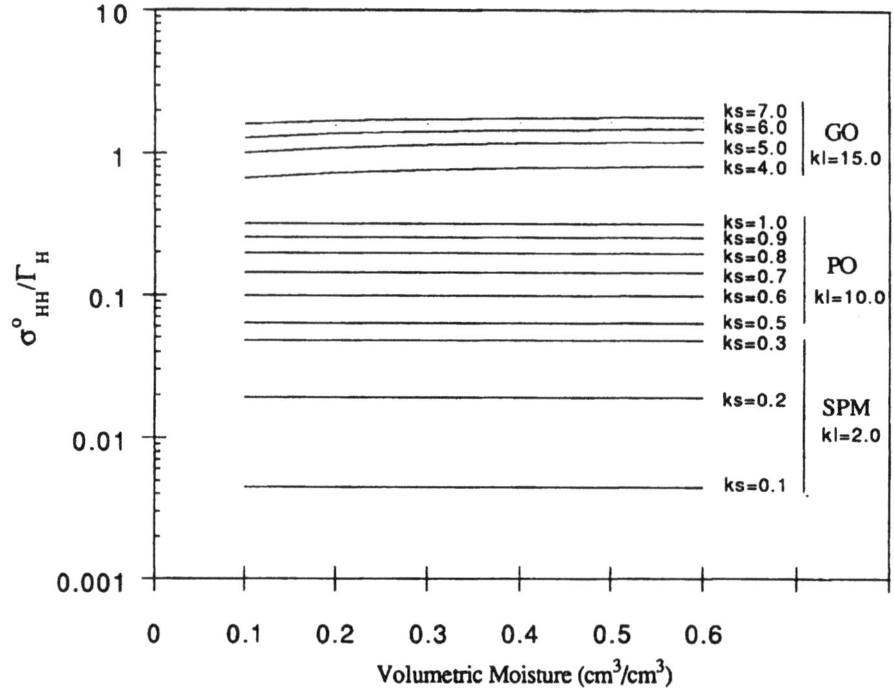

Figure 7. The ratio of backscattering over surface reflectivity versus the soil moisture variations.

Table 3.
Measurements and model simulations of the backscattering over reflectivity ratio used in Fig. 8

	ks	Ratio-Exp.	Ratio-Model
1	0.24000	0.081250	0.039600
2	0.36000	0.16773	0.082600
3	0.45000	0.36174	0.15100
4	0.57000	0.31275	0.26300
5	0.67000	0.96521	0.42700
6	0.80000	0.63455	0.57300
7	0.80000	0.62416	0.53300
8	0.81000	0.35957	0.53000
9	1.1000	0.66005	1.0300
10	1.1800	0.54368	1.0300
11	1.1800	0.85430	0.92100
12	1.6400	1.2641	1.4200
13	1.8500	1.1481	1.4100
14	1.8500	1.0550	1.2400
15	2.5600	1.7536	1.8900
16	2.6200	1.0357	
17	2.6200	0.87113	0.82000
18	3.6300	2.2719	2.1800
19	3.9100	2.4606	2.1800

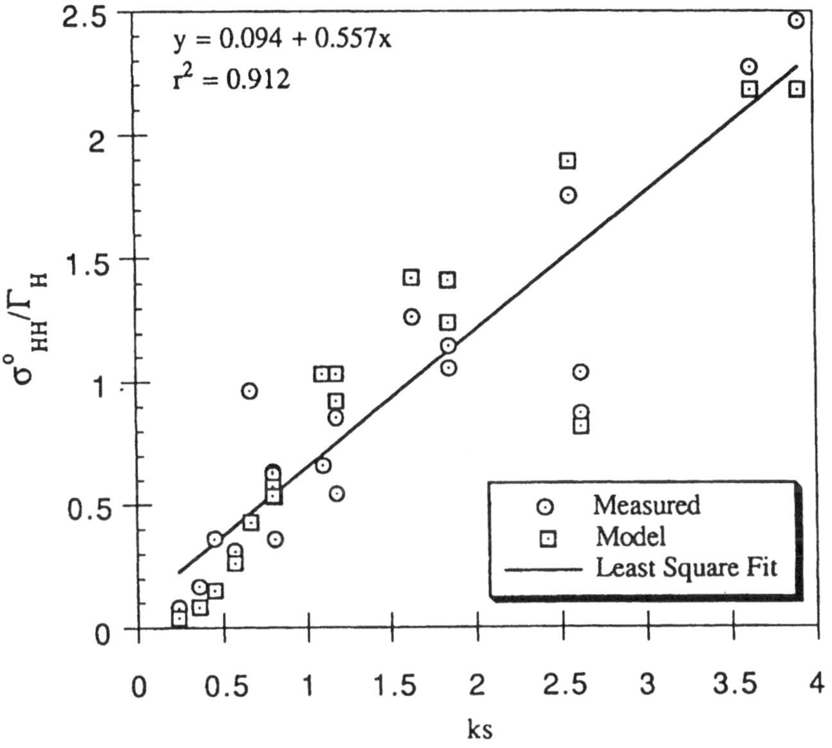

Figure 8. Ratio of backscattering over emissivity from measurements and model simulations. The line is a least-square fit to both the model and data.

that the ratio of backscattering to reflectivity is a linear function of surface rms height. However, to derive the equation for the regression line, we use both measurement and model simulation results. In Fig. 8, we have plotted the calculated values of the ratio from measurements and model simulations for the same surface parameters. These values are also shown in Table 3. Then, we have derived a least squares fit to all the points plotted in Fig. 8. The resulting line can be also interpreted as a minimization of the difference between the model and measurement values. This line is given by

$$\frac{\sigma^0_{\mathrm{HH}}}{\Gamma_{\mathrm{H}}} = 0.094 + 0.557\,ks. \tag{4}$$

In the above expression the backscattering coefficient has been normalized by cos(40°).

Knowing an estimate of ks from Equation (4), the surface emission can be corrected for the roughness effects. Thus, the next step is to use an analytical expression to relate the emission to surface permittivity or moisture. Since we are interested in only a single emissivity measurement from the radiometer, we use the semi-empirical expression derived by Saatchi and Wegmüller [14] for H polarized emissivity. The expression is given as follows:

$$e_{\mathrm{H}} = 1 - |R_{\mathrm{H}}(\theta)|^2 \exp(-4k^2s^2\cos^2\theta) - \sin\theta|R_{\mathrm{H}}(0)|^2\left[1 - \exp(-2ks\cos\theta)\right] \tag{5}$$

where $R_H(\theta)$ and $R_H(0)$ are the Fresnel reflection coefficients at incidence angles θ and 0 respectively. This expression readily summarizes the results obtained by model simulations in the previous section. On the right side of the equation, we have two terms which describe the coherent and incoherent contributions to the emissivity. The coherent term has often been used to model the reflectivity of the surface [5, 9]. The incoherent contribution, which is the second term on the far right of Equation (5), has been added in order to take into account the incoherent contribution from very rough surfaces ($ks > 5$) within the angular range of 20–50°. As the coherent contribution decreases by increasing the roughness, the incoherent contribution increases till the total emissivity saturates. The validity condition of Equation (5) covers the regions of SPM and PO approximations. However, it is more stringent than the GO approximation.

By using Equations (4) and (5) and a set of radar and radiometer measurements at H polarization, we can retrieve the surface permittivity and moisture. In order to test this retrieval algorithm, we choose a subset of RASAM measurements that have not been used in Figs 5 and 8. By forming the ratio of backscattering over reflectivity from the data and Equation (4), we estimate ks. Then, we substitute ks in Equation (4) and invert the equation to obtain the surface permittivity which is assumed to be a real number. By using the mixing model [15], we convert the permittivity to soil moisture. The result depicted in Fig. 9, shows that this technique works quite well. The rms error associated with the estimated soil moisture is approximately 0.05 cm^3/cm^3 . Knowing the sources of errors both in model simulations and measured data, such as the effect of correlation length on the ratio and the definition and measurement of soil moisture, it is possible to modify the algorithm for better estimation of the surface soil moisture. Note that the inversion technique is based on simultaneous radar and radiometer measurements at the same frequency. Choosing data from two different frequencies will change the dependence of the ratio to ks and soil moisture, unless the frequencies are close to each other.

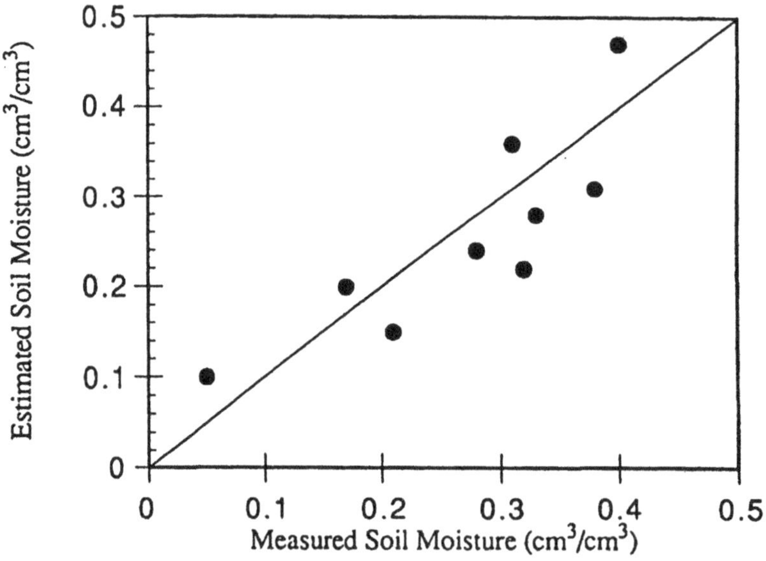

Figure 9. Estimated soil moisture plotted against measured soil moisture. The rms error between estimated and measured values is almost 0.05 cm^3/cm^3.

5. CONCLUSION

In this study, the properties of the rough surface scattering models, the SPM, PO, and GO approximations for simulating the backscattering and emissivity of bare soil surfaces have been examined. Results of the simulations have been compared with experimental data obtained by a radiometer-scatterometer system (RASAM) mounted on a tower. The comparisons have shown reasonable agreement, in particular for angles within 50° from nadir. The results have also verified the fact that the emissivity for a rough surface is less sensitive to roughness parameters than the backscattering coefficients. The emissivity is shown to be almost independent of the correlation length of the surface. The validity conditions for each model, and the sources of errors in simulating the emissivity for the measured data, have been discussed.

The results of the model simulations and the measured data were used to obtain the ratio of the HH polarized backscattering coefficient over H polarized reflectivity. This ratio, regardless of the values of the correlation length, can be approximated by a linear function with respect to the normalized surface rms height ks. This function has been obtained by the best linear fit to the data and the corresponding simulated values. As a result of this ratio, a synergistic inversion procedure has been developed to estimate the soil moisture. In this procedure, a single polarization (HH) band of the radar data and a single band of the radiometer data (H-polarized emissivity or brightness temperature) were used. The ratio and the semi-empirical model developed earlier [14] were employed in this algorithm to correct the emissivity data for the rms height of the surface and to estimate the soil moisture. The accuracy of the synergistic inversion approach has only been tested by using the data from the RASAM system. The error in estimating the soil moisture is approximately 0.05 cm^3/cm^3. The preceding analysis and discussion illustrate the use of radar and radiometer system with minimum configurations as a soil moisture mapper. However, the application of this technique on other data sets from either airborne or spaceborne systems, where resolution is also of concern, will verify and improve the accuracy of the synergistic algorithm.

REFERENCES

1. E. T. Engman. Applications of microwave remote sensing of soil moisture for water resources and agriculture. *Remote Sens. Environ.*, **35**, 213–226 (1991).
2. NASA. *From Pattern to Process: The Strategy of the Earth Observing System.* EOS Science Steering Committee Report, Vol. II.
3. T. Mo, T. J. Schmugge and J. R. Wang. Calculation of microwave brightness temperature of rough soil surfaces. *IEEE Trans. Geosci. Remote Sensing*, **GE-25**, 47–54 (1987).
4. T. J. Jackson, T. J. Schmugge and J. R. Wang. Passive microwave sensing of soil moisture under vegetation canopies. *Water Resources Res.*, **18**, 1137–1142 (1982).
5. J. R. Wang, J. C. Shiue, T. J. Schmugge and E. T. Engman. Mapping soil moisture with L-band radiometric measurements. *Remote Sens. Environ.*, **27**, 305–312 (1989).
6. F. T. Ulaby, R. K. Moore and A. K. Fung. *Microwave Remote Sensing: Active and Passive.* Vol. 3. Dedham, MA, Artech House (1986).
7. F. T. Ulaby, R. K. Moore and A. K. Fung. *Microwave Remote Sensing: Active and Passive.* Vol. 2. Reading, MA, Addison–Wesley (1982).
8. P. Coppo, G. Luzi, S. Paloscia and P. Pampaloni. Effect of soil roughness on microwave emission: comparison between experimental data and models. In: *Proceeding of IGARSS'91*, (Helsinki, Finland), pp. 1167–1170 (1991).

9. B. J. Choudhury, T. J. Schmugge, R. W. Newton and A. Chang. Effect of surface roughness on the microwave emission from soils. *J. Geophys. Res.*, **81**, 3660–3666 (1979).

10. T. J. Schmugge. Remote sensing of soil moisture: Recent advances. *IEEE Trans. Geosci. Remote Sensing*, **GE-21**, 336–344 (1983).

11. Y. Oh, K. Sarabandi and T. Ulaby. An empirical mode and an inversion technique for radar scattering from bare soil soil surfaces. *IEEE Trans. Geosci. Remote Sensing*, **GE-30**, 370–381 (1992).

12. U. Wegmüller. The effect of freezing and thawing on microwave signatures of bare soil. *Remote Sens. Environ.*, **33**, 123–135 (1990).

13. R. Hüppi. *Rasam: a radiometer scatterometerto measure microwave signatures of soil, vegetation and snow. Dissertation.* Institut für Angewandte Physik, Universität Bern, Bern (December 1987).

14. S. S. Saatchi and U. Wegmüller. Modeling and measurement of microwave emission from bare soil surfaces. Submitted to *IEEE Trans. Geosci. Remote Sensing* (1992).

15. C. Dobson, F. Ulaby, M. Hallikainen and M. El-Rayes. Microwave dielectric behavior of wet soil. Part II: Dielectric mixing models. *IEEE Trans. Geosci. Remote Sensing*, **GE-23**, 35–43 (1985).

16. J. W. (Lord) Rayleigh. *The Theory of Sound.* MacMillan, New York (1945).

17. J. A. Ogilvy. *Theory of Wave Scattering from Random Rough Surfaces.* Bristol, England, Adam Hilger (1991).

18. P. Beckman and A. Spizzichino. *The Scattering of Electromagnetic Waves from Rough Surfaces.* New York, Macmillan (1963).

19. W. H. Peake. Interaction of electromagnetic waves with some natural surfaces. *IRE Trans. AP*, **7**, S324–S329 (1951).

20. L. Tsang and R. W. Newton. Microwave emission from soils with rough surfaces. *J. Geophys. Res.*, **87**, 9017–9024 (1982).

21. A. K. Fung and H. J. Eom. An approximate model for backscattering and emission from land and sea. In: *IEEE International Geoscience and Remote Sensing Symposium*, Vol. 1, pp. 620–628 (1981).

22. M. Vieno-Vespisito. *Scattering and Diffraction in Physical Optics.* John Wiley, New York (1991).

23. T. Mo and T. J. Schmugge. A parametrization of the effect of surface roughness on microwave emission. *IEEE Trans. Geosci. Remote Sensing*, **GE-25**, 481–486 (1987).

24. C. T. Swift. Passive microwave remote sensing of the ocean — a review. In: *Passive Radiometry of the Ocean*, J. F. R. Gower (Ed.), pp. 25–54, D. Reidel, Publishing Company, The Netherlands (1980).

25. F. T. Ulaby, M. C. Dobson and D. R. Brunfeldt. Improvement of moisture estimation accuracy of vegetation-covered soil by combined active/passive microwave remote sensing. *IEEE Trans. Geosci. Remote Sensing*, **GE-21**, 300–307 (1983).

26. S. W. Theis, B. J. Blanchard and A. J. Blanchard. Utilization of active microwave roughness measurements to improve passive microwave soil moisture estimates over bare soils. *IEEE Trans. Geosci. Remote Sensing*, **GE-24**, 334–339 (1986).

27. H. Haefner and P. Pampaloni. Water resources. *Int. J. Rem. Sensing*, **13**, 1277–1303 (1992).

Snow

ESA/NASA International Workshop, pp. 227–243
B. J. Choudhury, Y. H. Kerr, E. G. Njoku and P. Pampaloni (Eds)

Intercomparison of snow retrieval algorithms by means of spaceborne microwave radiometry

HELMUT ROTT and THOMAS NAGLER

Institut für Meteorologie und Geophysik, Universität Innsbruck, A-6020 Innsbruck, Austria

Abstract—The physical basis of snow retrievals by means of microwave radiometry is reviewed. Effects of atmospheric attenuation and emission on the microwave radiation measured at satellite altitude are discussed and examples for the modification of microwave radiation of the earth surface due to atmospheric effects are shown. Algorithms are presented which enable the inversion of satellite measurements in respect to areal extent, water equivalent, and depth of the snow cover. Intercomparisons of different algorithms for snow mapping show considerable differences of the retrieved snow cover products. Reasons for these differences include (i) regional variations in snow morphology and land cover type, which affect the relation between the emitted radiation and the depth and water equivalent of the snow cover; (ii) the use of different radiometer channels for the retrievals; (iii) differences in the multitemporal treatment of the data. Existing algorithms have been developed and verified at regional scales, but an accurately verified global snow retrieval algorithm is still missing. Such an algorithm needs to include factors accounting for regional diversities of snow morphology and land cover type, in particular forest cover. Enhanced efforts in this respect are needed to further utilize the great potential of microwave radiometry for monitoring the extent and mass of the global snow cover, a crucial element for climate research and for the global water and energy cycle.

1. INTRODUCTION

Consistent data sets on areal extent and properties of the global snow cover are of great significance for climate research and for studies of the global energy and water cycle. Synoptic information on the snow cover is required as input to global circulation models. Spaceborne microwave radiometers are promising tools for global snow cover monitoring with high temporal repetition rate, because of the potential of microwaves to observe the earth surface through clouds and to provide information on the internal properties of the snowpack. These capabilities are not only of interest for global snow cover surveys, but also for snow monitoring at regional scales, including applications in hydrological monitoring and forecasting, agricultural management, and regional climatology.

Algorithms for inverting spaceborne microwave measurements in respect to snowpack properties, including the areal extent, the depth, and the water equivalent, have been developed by various research groups. The primary sensors for these studies have been the Scanning Multichannel Microwave Radiometer (SMMR) on Nimbus-7 and the Special Sensor Microwave/Imager (SSM/I) on DMSP (US Defence Metereological Satellite Program) satellites. Snow cover products, derived by different algorithms, may show considerable differences even if the same satellite data set is used for the snow retrieval. The purpose of this paper is to explain the reasons for these differences. In order to achieve this goal, the basic principles of snow retrievals using microwave radiometric data are briefly reviewed, algorithms for deriving extent, depth, and water equivalent of snow are described, and snow cover products derived by means of different algorithms are quantitatively compared.

We hope that this discussion will contribute to the further refinement of snow retrieval algorithms, with the final goal to come to an accurate, generally accepted algorithm for deriving global snow extent and mass. Concluding from the present knowledge, such an algorithm will have to account for regional diversities of snow morphology and land cover, and consequently will not be a simple algorithm.

2. BRIGHTNESS TEMPERATURE OF SNOW COVERED TERRAIN

In order to explain the possibilities for inverting microwave measurements of snow in respect to physical properties of the snowpack such as the water equivalent, the main aspects of the microwave interaction with snow are briefly reviewed. The propagation of microwave radiation in a snowpack at a given frequency can be described by the radiative transfer equation [1], which can be written using the Rayleigh–Jeans approximation:

$$\mu \frac{\mathrm{d}T_{\mathrm{B}}(\tau,\mu)}{\mathrm{d}\tau} = -T_{\mathrm{B}}(\tau,\mu) + J(\tau,\mu) \qquad (1)$$

where μ is the cosine of the zenith angle, τ is the optical depth, and T_{B} is the brightness temperature. The source function J accounts for emission and scattering:

$$J(\tau,\mu) = \left[1-\omega_0\right]T(\tau,\mu) + \frac{\omega_0}{2}\int_{-1}^{1} P(\mu,\mu')T_B(\tau,\mu)\,\mathrm{d}\mu' \qquad (2)$$

where T is the thermodynamic temperature, P is the normalized scattering phase function, and μ' is the cosine of the scattering angle. The single scattering albedo ω_0 is related to the scattering coefficient k_{s} and absorption coefficient k_{a} of the volume by

$$\omega_0 = \frac{k_{\mathrm{s}}}{k_{\mathrm{a}} + k_{\mathrm{s}}}\,. \qquad (3)$$

Solutions for the radiative transfer for dry snow, assuming sparsely distributed Rayleigh scatterers, show dominating effects of the ratio (a/λ) (grain diameter/wavelength) on the scattering albedo and brightness temperature [2]. Assuming grain diameter of 1 mm and snow density of 300 kg m^{-3}, k_{s} approaches the same magnitude as k_{a} at a frequency ν of about 15 GHz ($\lambda = 1.5$ cm), and increases rapidly at frequencies above 20 GHz [3]. Recent improvements of radiative transfer models for snow account for correlations between the densely populated scatterers [4, 5]. Numerical results of dense medium radiative transfer in a medium with small particles show that attenuation rates and scattering albedo are less than those predicted by the assumption of independent scattering.

Hallikainen *et al.* [6] measured the extinction properties of several dry snow types in the 18 to 90 GHz range and came up with the following empirical expressions for the extinction coefficient k_{e}:

for 18–60 GHz, in decibels per meter: $k_{\mathrm{e}} = 0.0018\ f^{2.8}a^{2.0}$ (4)

for 90 GHz, in decibels per meter: $k_{\mathrm{e}} = 300\ a^{1.9}$ (5)

where a is the observed particle diameter in millimeters and f is the frequency in GHz. The extinction coefficient is the sum of the absorption coefficient and scattering coefficient: $k_e = k_a + k_s$. Assuming $a = 0.5$ mm, which is a value for dry winter snow in mid-latitudes, we calculate the following penetration depths d_p: 2.43 m at 19 GHz, 0.45 m at 37 GHz, and 0.05 m at 85 GHz. Slightly higher values for d_p were measured for dry winter snow in the Alps [7, 8].

At frequencies below 15 GHz the absorption losses are dominating over the scattering losses at the snow grains, but reflections at internal boundaries between layers with different snow density become more effective than scattering at grains. For strongly layered semi-infinite snow this may result in T_B-differences of more than 50 K between horizontal and vertical polarization at the Brewster angle [9].

The presence of liquid water in snow strongly increases the dielectric losses, and the emitted signal is originating from a thin layer at the top of the snowpack. The penetration depth of wet snow with liquid water content of 5% by volume is of the order of one wavelength only [3, 8].

Figure 1 shows the average emission behaviour of 3 distinct snow types, based on long-term measurements carried out by scientists of the University of Bern in the Swiss Alps, where during winter comparatively deep snow builds up [8, 10]. The emissivity ε at polarization p (vertical, v, or horizontal, h) was calculated from the $T_{B,p}$ measurements

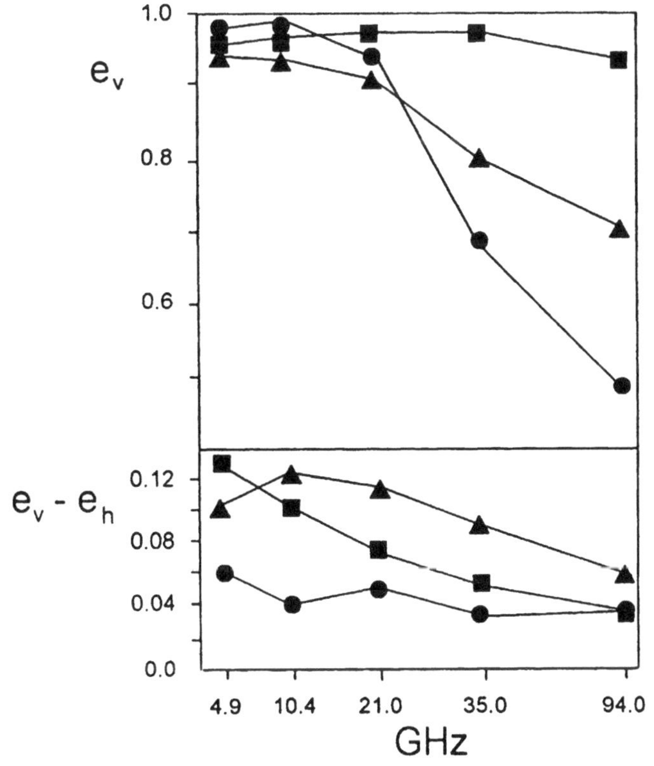

Figure 1. Emissivities at vertical polarization e_v at 50° nadir angle versus microwave frequency, and polarization difference $e_v - e_h$, for 3 different snow types. From measurement in the Swiss Alps by Mätzler [8, 10]. ■ wet snow, ▲ dry winter snow, ● refrozen crust on top of wet snow.

Table 1.
The main parameters influencing the emissivity ε of snow lying on ground without or with sparse vegetation, for the frequency range of 5 to 100 GHz

Parameter	Effect	Frequency range primarily affected
Liquid water content LWC	Increase of ε with increasing LWC	5–100 GHz
For dry snow:		
Snow water equivalent SWE	Decrease of ε with increasing SWE	15–40 GHz, $f(\nu)$
Grain size GS	Decrease of ε with increasing GS	15–100 GHz, $f(\nu)$
Layering	Increase of $(\varepsilon_v - \varepsilon_h)$ for layered snow	5–15 GHz, $f(\nu)$
Soil properties	Contributing to total emission	5–40 GHz, $f(\nu)$

according to:

$$\varepsilon_p = \frac{T_{B,p} - T_B^{\downarrow}}{T_s - T_B^{\downarrow}} \tag{6}$$

where T_s is the thermodynamic temperature of the snow, and $T_B^{\downarrow}$ is the downwelling brightness temperature of the atmosphere at the earth surface. The emissivities in Fig. 1 were derived from measurements at an incidence angle $\Theta = 50°$, which is close to the incidence angle of satellite-borne microwave radiometers and also close to the Brewster angle of dry snow, resulting in maximum polarization differences.

The emissivity of winter snow in Fig. 1 is clearly decreasing in the frequency range of 18 to 90 GHz due to increased scattering in the snow volume. At frequencies $\nu \leqslant 10$ GHz the main part of the emitted radiation in case of dry snow originates from the soil below the snowpack. The polarization difference $(\varepsilon_v - \varepsilon_h)$ in the 10 to 20 GHz range results from reflections at internal boundaries [8, 10]. Crusts of refrozen snow cause significant increase of scattering at frequencies above about 30 GHz because of the larger grain size compared to dry winter snow. This demonstrates a basic problem of inverting microwave radiometer measurements in respect to snow water equivalent SWE. Increase of grain size has a similar effect on the signatures as increase of the scattering layer. The polarization differences hold a potential for separating refrozen snow from dry winter snow, because $(\varepsilon_v - \varepsilon_h)$ of refrozen snow is smaller.

Wet snow shows high emissivities at all frequencies. This causes problems for discriminating wet snow versus snow-free surfaces, because snow-free terrain frequently also shows high emissivities [8, 10, 16]. In regions where snow-free areas reveal small values of $(\varepsilon_v - \varepsilon_h)$, this parameter enables the detection of wet snow areas by means of microwave radiometry [11]. Wet snow with even just a thin frozen crust on top can be detected at high microwave frequencies, because the emissivity is clearly reduced due to scattering at the coarse grains [8]. Table 1 summarizes the main parameters of relevance for the microwave radiation emitted by snow-covered soil.

3. ATMOSPHERIC EFFECTS ON THE BRIGHTNESS TEMPERATURE AT SATELLITE ALTITUDE

The brightness temperatures measured at satellite altitude $T_B(H)$ are affected by attenuation and emission of gases, clouds, and hydrometeors, as well as by scattering at cloud and hydrometeor particles. Figure 2 shows examples for the modification of T_B due to

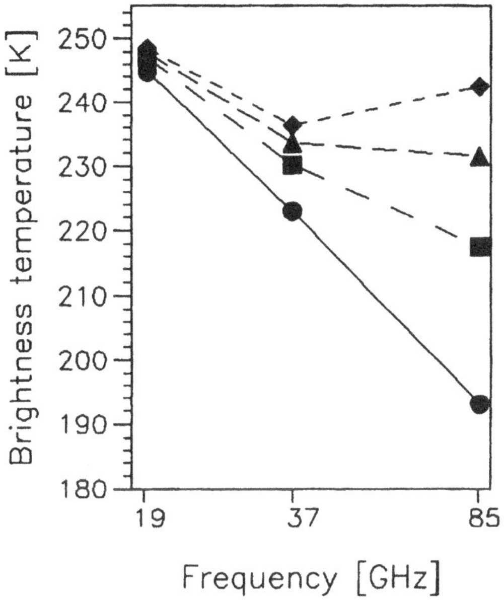

Figure 2. Calculated brightness temperature at satellite altitude $H = 830$ km at 19, 37, and 85 GHz for an incidence angle of 50°, assuming the emissivities (e_v in Fig. 1) of dry winter snow and surface temperature of 272 K. The propagation conditions are: ● — fully transparent; ■ — cloud free, RH = 60%; ▲ — altostratus, RH = 80%, V = 2.4–2.9 km, W = 0.15 g m^{-3}; ♦ — stratus, RH = 80%, V = 0.5–1.0 km, W = 0.20 g m^{-3}. (RH = relative humidity, V = cloud vertical extent, W = cloud water content.)

the atmosphere, calculated at 3 frequencies corresponding to channels of the satellite-borne Special Sensor Microwave Imager (SSM/I). The calculations were carried out for a non-scattering atmosphere in local thermodynamic equilibrium for a given frequency according to [12, 13]:

$$T_B(H) = t(0,H)\left[\varepsilon T_s + (1-\varepsilon)T_B^{\downarrow} + (1-\varepsilon)t(\infty,0)T_c\right] + T_B^{\uparrow} \tag{7}$$

where ε is the surface emissivity, T_s is the surface temperature, and T_c is the brightness temperature of the cosmic background. $T_B^{\uparrow}$ and $T_B^{\downarrow}$, the upward and downward radiating brightness temperatures due to the atmospheric contribution, were calculated from the atmospheric temperature profile $T(z)$ and the vertical derivative of the transmittance t:

$$T_B^{\uparrow} = \int_0^H T(z)\frac{\partial t(z,H)}{\partial z}\,\mathrm{d}z \tag{8}$$

$$T_B^{\downarrow} = \int_0^{\infty} T(z)\frac{\partial t(0,z)}{\partial z}\,\mathrm{d}z. \tag{9}$$

The transmittance t between two altitudes (z_1, z_2) is given by:

$$t(z_1, z_2) = \exp\left[-\int_{z_1}^{z_2} k_a(z)\,\mathrm{d}z\right]. \tag{10}$$

For the calculations shown in Fig. 2 surface temperature $T_s = 272$ K was assumed, and vertically polarized emissivites e_v for the dry winter snow case shown in Fig. 1. The atmospheric radiative transfer calculations were based on average temperature and humidity profiles of January 1988 at the synoptic station Berlin–Tempelhof in Germany.

The calculations show clearly the decrease of the T_B-differences between the different frequencies resulting from the decrease of atmospheric transmittance t with increasing frequency. The following values were calculated for $t(0, H)$ at the incidence angle $\Theta = 50°$ for the 3 atmospheric conditions shown in Fig. 2:

		19	37	85 GHz	Symbol in Fig. 2
Case (1)	$t(0, H) =$	0.96	0.93	0.82	■
Case (2)	$t(0, H) =$	0.94	0.87	0.70	▲
Case (3)	$t(0, H) =$	0.93	0.84	0.60	◆

In cloudy conditions, as assumed in cases 2 and 3, the surface contribution to $T_B(H)$ is significantly reduced at 85 GHz. For deriving properties of the earth surface such as parameters of the snowpack, it is necessary to apply corrections for atmospheric effects or to exclude the cases with significant propagation losses at 85 GHz due to precipitation and dense clouds. In case of optically thin clouds, snow covered areas can be clearly detected at 85 GHz even without atmospheric corrections, because the reduced emissivity due the scattering at the snow grains is a dominating signal.

The calculations shown above do not include scattering at cloud droplets and hydrometeors, which reflect the cold cosmic background towards the downward looking satellite radiometer. At frequencies of about 20 GHz scattering at hydrometeors becomes effective in reducing $T_B(H)$ observed from the satellite over land surfaces for rain rates of several mm/h [14]. The effects of scattering on hydrometeors are increasing with microwave frequency. For this reason rainfall areas show spectral decrease of $T_B(H)$, which is similar to the spectral decrease of dry snow areas. This is taken into account in snow classification procedures.

4. RETRIEVING SNOW COVER FROM SPACEBORNE MICROWAVE RADIOMETERS

The Scanning Multichannel Microwave Radiometer (SMMR), which operated on board of Nimbus-7 between November 1978 and June 1987, and the Special Sensor Microwave/Imager (SSM/I) operating on board of DMSP satellites since June 1987, are the primary data sources for microwave derived snow cover products. The SMMR snow retrievals are usually based on T_B-data of the 18 GHz and 37 GHz channels (v or/and h polarization), or on 37 GHz data as the only source. The cell size of the gridded T_B-data is 60 km at 18 GHz and 30 km at 37 GHz, the incidence angle at the earth surface is 50.3°. Because of the swath width of 800 km and the alternate day duty cycle, 5 consecutive days were required for global coverage by SMMR. SSM/I scans the earth surface at the frequencies 19.3, 37.0, and 85.5 GHz in v and h polarization, and at 22.2 GHz in v polarization; the incidence angle at the surface is 53.3°, the effective fields of view range from 69 km × 43 km (19 GHz) to 15 km × 13 km (85 GHz) [15]. The swath width of 1394 km provides complete daily coverage at latitudes > 42°.

Algorithms have been developed for deriving the following snow parameters from SMMR data [16]: the areal extent of the snowpack, the snow water equivalent SWE and/or snow depth SD, and the extent of melting/freezing snow areas. In Canada an algorithm has been developed recently for mapping melting snow areas in the Prairies, based on 37 GHz (v − h) polarization differences [11]. Most of the algorithms were derived empirically, correlating the satellite data with ground based snow measurements, in some cases also with airborne snow surveys. For meteorology, climate research, and hydrology the snow mass on the ground (SWE) is of more interest than SD. However, measurements of SWE over extended areas with suitable spatial density, as required for development of microwave snow monitoring algorithms, have been available only for few regions.

The most frequently applied algorithms for retrieving SWE and extent of dry snow from microwave radiometry, are based on the relation [12, 17]:

$$\mathrm{SWE} = A_1 + A_2 \frac{T_B(\nu_1) - T_B(\nu_2)}{\nu_2 - \nu_1} \text{ [mm]}, \qquad \text{for SWE} > 0 \tag{11}$$

where A_1 and A_2 are empirically derived coefficients.

The condition for snow covered areas is:

$$\text{if } \left[T_B(\nu_1) - T_B(\nu_2)\right] > -\frac{A_1}{A_2}\left(\nu_2 - \nu_1\right), \qquad \text{then snow.} \tag{12}$$

In case of SMMR, $\nu_1 = 18$ GHz and $\nu_2 = 37$ GHz are used in most cases. In Finland good results for SWE determination were also obtained with $\nu_1 = 10.7$ GHz [17]; a disadvantage of this channel is the lower spatial resolution. In case of SSM/I, $\nu_1 = 19$ GHz and $\nu_2 = 37$ GHz are used. In the next paragraph we present an algorithm, which uses additionally the 85 GHz channel.

The coefficient A_2, which represents the sensitivity of the brightness temperature gradient $(\Delta T_B/\Delta\nu)$ in respect to SWE, shows considerable regional variations, depending mainly on the morphology of the snowpack and the land cover type. Forests, masking the radiance emitted from the snow cover, reduce the response of $(\Delta T_B/\Delta\nu)$ in respect to SWE. These regional variations can be considered in the inversion calculations, for example by including information on the forest cover [18]. Interannual variations of A_2 in a given region result primarily from variations in snowpack stratigraphy and grain size due to weather events and snow metamorphosis [17]. These effects are more difficult to compensate; one possibility is the use of anchor stations, representative for a region, to adjust the coefficient seasonally. Table 2 shows examples for the coefficients A_1 and A_2 in 4 large regions, derived empirically by correlating satellite T_B's and ground data of SWE, based on data sets of several years.

Table 2.
Coefficients A_1 and A_2 for SWE retrieval, using 18 GHz and 37 GHz channels in case of SMMR, and 19 GHz and 37 GHz in case of SSM/I

Region	A_1 [mm]	A_2 [(mm GHz)/K]	Data source	Ref.
(1) Southern Finland	−98.0	191.9	SMMR TCT	[17]
(2) Northern Finland	−108.7	165.3	SMMR TCT	[17]
(3) Eastern Germany	−6.0	54	SMMR CELL	[12]
(4) Canadian Prairies	−20.7	49.3	SSM/I	[19]

Vertically polarized T_B's were applied for the regions (1), (2), and (4), and the mean of h- and v-polarized T_B's for region (3). Regional differences in the slope A_2 are primarily related to differences in the forest cover; 77% of southern Finland and 47% of northern Finland are covered by forest [17]. Forests cover 28% of the study region in Germany, but these forests are in general not dense and include a large percentage of deciduous trees. The offset in ΔT_B for the snow-free case (given by $-A_1/A_2$) is significantly larger in Finland than in the two other regions; this can, at least partly, also be explained by the forest cover [20].

The standard error of estimate for SWE from SMMR for region (3) was 18 mm, the maximum value of SWE measured on ground was 70 mm [12]. Hallikainen and Jolma [17] report a standard error of 27 mm for SWE $\leqslant$ 100 mm and of 35 mm for all SWE data of southern Finland, and corresponding values of 23 mm and 45 mm for northern Finland. If the two large areas in Finland are further subdivided, the relation between ΔT_B and SWE is clearly improved. Further improvements can be achieved, if the interannual variation of snowpack morphology can be taken into account. Errors in the relation between ground-based and satellite-derived SWE result from errors in the spatial values of SWE, which are derived from point measurements on the ground, as well as from errors in the retrieval of SWE from the satellite microwave data.

Some of the differences between the coefficients in Table 2 result from different absolute calibration of the various data sources. SSM/I shows calibration differences of several K compared to the corresponding channels of SMMR; the differences vary from one channel to the other, resulting in slightly smaller gradients [($T_B(18) - T_B(37)$] for SMMR [13, 21]. Calibration differences were also found between the different SMMR products (TCT — Temperature Cell Tapes, CELL — Cell All Tapes) [13, 22].

A simple algorithm has been developed by Chang *et al.* [23] for mapping the extent and depth of the global snow cover:

$$\mathrm{SD} = A\left[T_B(\nu_1) - T_B(\nu_2)\right] \quad [\mathrm{cm}]. \tag{13}$$

The SMMR channels 18 GHz, horizontal polarization, and 37 GHz, horizontal polarization, and the coefficient $A = 1.59$ [cm/K] are used for global snow mapping [23]. The first global snow cover maps produced from SMMR data were based on the same SMMR channels, but used a slightly different coefficient A and applied a multitemporal condition, selecting the minimum value of $[T_B(\nu_1) - T_B(\nu_2)]$ for each 5-day period to reduce the effects of melting snow and atmospheric transmittance [16]. As evident from the complex relationship between T_B and SD or SWE, global maps of SD derived from microwave radiometer measurements using just one global coefficient may show large errors in certain regions. For improvments of global snow cover products it will be necessary to account for the spatial variations of snow morphology and land cover.

5. SSM/I SNOW MAPPING ALGORITHM

An algorithm for mapping the areal extent of the snow cover and the snow depth SD, which was developed based on SSM/I data over central and eastern Europe, using the 19, 37, and 85 GHz vertically polarized channels [13, 24], is described in this chapter. SSM/I data in original scan geometry (Sensor Data Records) of the period January 13, 1988, to March 31, 1988, were available for the investigations, covering the region 5°E to 35°E and 42°N to 62°N. Three-hourly meteorological observations and daily SD measurements at the synoptic stations of the whole region were used as correlative data for algorithm development. Because only few SWE measurements were available, our algorithm derives SD, but not SWE. The performance of the algorithm was tested over North America, based on SSM/I data of the same period. Because of the deterioration of the radiometric quality of the 85 GHz channels of SSM/I on the DMSP F8 satellite, we could not test the algorithm with data of the following winters. After launch of another SSM/I sensor on DMSP F11 in 1991, we hope to obtain data for further investigations.

Figure 3 shows an example of a SSM/I derived snow cover map of the study area in Europe. Only few pixels show values of SD larger than 15 cm. This is in agreement with the ground based observations. During winter 1987/88 the snow cover was shallow throughout the investigation area, the majority of the ground based measurements did not exceed SD of 10 cm. The sensitivity of the various SSM/I channels to SD is illustrated in Fig. 4, showing the emission behaviour of 3 classes of SD for dry snow and of snow-free areas, based on brightness temperatures measured at the satellite $T_B(H)$. The mean values and the standard deviations of $T_B(H)$ are shown for each class. The T_B data were grouped into the 4 classes according to the ground data.

The snow-free areas, based on a limited data set from several days in January 1988, show little spectral differences and small standard deviations. For all 3 snow cover classes, the mean T_B is clearly decreasing and the variance is increasing with increasing

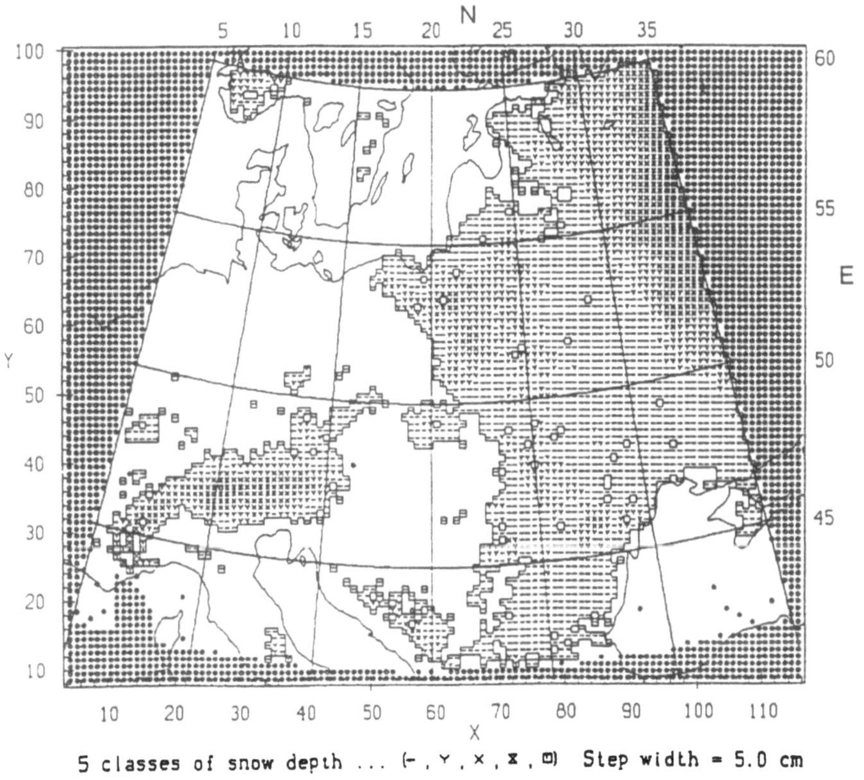

Figure 3. Snow cover map of central and eastern Europe derived from SSM/I data of the days 3 to 5 February 1988. 5 classes of snow depth: 0.5–5, 6–10, 11–15, 16–20, > 20 cm.

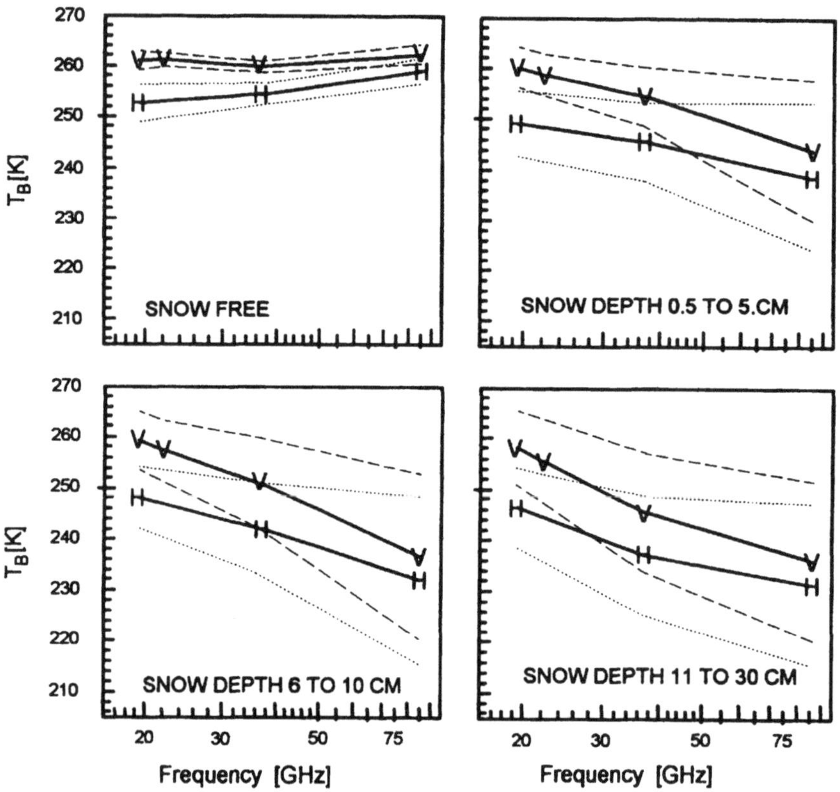

Figure 4. Mean values and standard deviations of brightness temperatures at 19, 22, 37, and 85 GHz, vertical (V) and horizontal (H) polarization, from SSM/I measurements over Europe in winter 1988. (A) Snow-free surfaces close to the snow boundary; (B) snow with depths 0.5 to 5 cm; (C) snow with depths 6 to 10 cm; (D) snow with depths 11 to 30 cm.

frequency, the polarization differences show a slight decrease with frequency. This corresponds to the typical behaviour of dry winter snow (see Fig. 1 and [8]). The spectral difference $[T_B(19) - T_B(37)]$ increases from the shallow to the deeper snow classes. $[T_B(37) - T_B(85)]$ reaches the highest value at 7 cm SD [24], because the penetration depth at 85 GHz is of the order of few centimeters only. For this reason the 85 GHz data are mainly of interest for monitoring shallow snow.

The pronounced increase of the standard deviation of $T_B(H)$ from 19 to 85 GHz for the snow covered areas has several reasons. One reason is the increase of atmospheric transmittance t with frequency, causing larger temporal and spatial variations of t at 85 GHz than at the lower frequencies. Another reason is temporal and spatial variation of snow morphology. In particular, the effects of grain size are strongly increasing between 19 and 85 GHz [6]. Finally, there is the problem of the limited representativeness of the *in situ* point measurements for snow studies over extended areas, which causes significant problems for the development of inversion algorithms from satellite measurements. Algorithm development was based on correlations of satellite measurements with *in situ* data of SD, where a single SD measurement on ground was typically representing an area of a few hundred km^2.

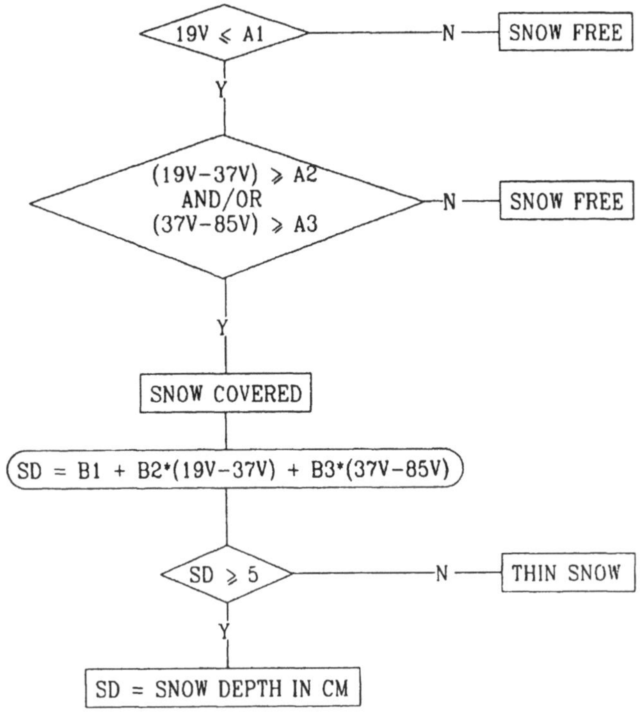

Figure 5. Decision tree for snow classification and calculation of snow depth using SSM/I data.

The algorithm for automatic mapping of extent and depth of the snow cover is descibed in Fig. 5. The following thresholds $A1$, $A2$, $A3$ and coefficients $B1$, $B2$, $B3$ were derived from the data set over Europe: $A1 = 266$ K, $A2 = 4$ K, $A3 = 3$ K, $B1 = -2.41$ cm, $B2 = 1.2$ cm/K, $B3 = -0.16$ cm/K. The first node of the decision tree eliminates snow-free surfaces, including rainfall areas, using a threshold of 266 K for $T_B(19V)$. The second node uses the spectral difference $[T_B(19V) - T_B(37V)]$ and/or $[T_B(37V) - T_B(85V)]$ to separate snow-covered from snow-free areas with $T_B(19V) < A1$ (cold or wet surfaces). The 85 GHz data were found very useful for detecting areas with thin snow cover. For the calculation of SD the main weight is in the spectral difference $[T_B(19V) - T_B(37V)]$. The difference $[T_B(37V) - T_B(85V)]$ shows weak negative correlation with SD for the whole data set, because of saturation in the range of 5 to 10 cm snow depth. For SD less than the saturation depth the correlation between $[T_B(37V) - T_B(85V)]$ and SD is positive; above this depth, the correlation is weakly negative. For this reason, and because of the weak sensitivity of $[T_B(19V) - T_B(37V)]$ in regard to thin snow, we propose to include all pixels classified as $D \leqslant 5$ cm in one class ('thin snow') without further discrimination of SD.

Figure 6 illustrates the importance of the two spectral gradients for the snow classification over Europe. Each vertical bar shows the percentage of the snow covered area classified by both spectral differences (black), classified by $[T_B(19V) - T_B(37V)]$ only (×-hatched), and classified by $[T_B(37V) - T_B(85V)]$ only (striped), based on SSM/I data of 3 consecutive days. 65% of the snow covered areas are identified by both

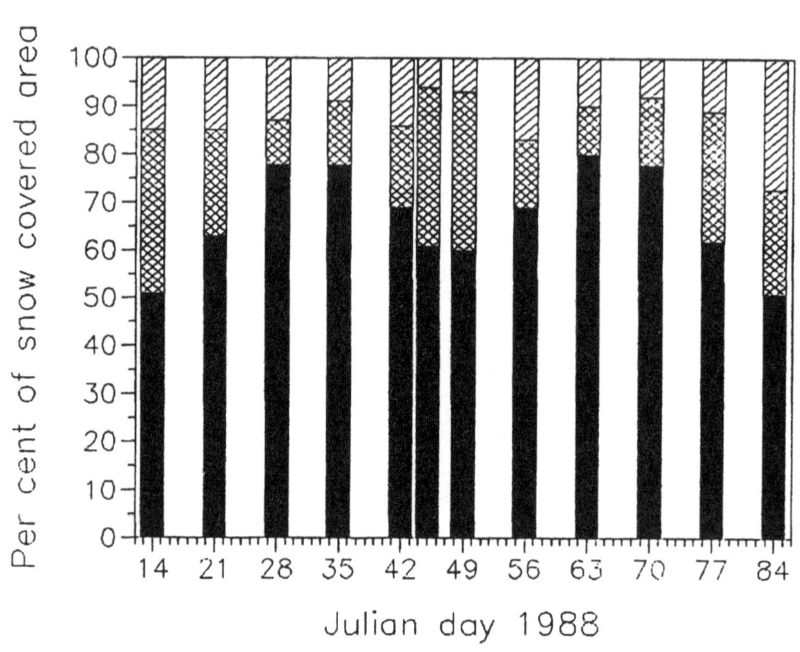

Figure 6. Assessment of the different contributions of the SSM/I snow cover algorithm to the total snow covered area during winter 1988 in Europe. Black: percentage of snow covered area at the given date classified by $[T_B(19V) - T_B(37V)]$ and $[T_B(37V) - T_B(85V)]$; ×-hatched: classified by $[T_B(19V) - T_B(37V)]$ only; striped: classified by $[T_B(37V) - T_B(85V)]$ only.

gradients, 14% are classified by $[T_B(37V) - T_B(85V)]$ only; these are in general areas with shallow snow, but may also be areas with very thin frozen crust on top of wet snow, which reduces T_B at 85 GHz, but not at lower frequencies. The areas, which are classified by $[T_B(19V) - T_B(37V)]$ only, are generally covered by SD > 15 cm, including cases when T_B (85 GHz) measured at the satellite is increased due to the atmospheric contribution (e.g. in case of low clouds). Comparisons with *in situ* snow measurements confirm the improvements by the use of 3 frequencies for snow classifications [13].

6. INTERCOMPARISONS OF SNOW MAPPING ALGORITHMS

Previous studies on SMMR snow mapping demonstrated already the importance of multitemporal data handling for microwave based snow classification, in order to reduce effects of temporary melting and of variations of atmospheric transmittance [16]. This is of particular importance if 85 GHz data are used, because of the reduced atmospheric transparency at this frequency. We investigated various procedures of multitemporal data treatment, using SSM/I data of three consecutive days for generation of each snow cover map. The SSM/I T_B's from the individual swaths (in the Sensor Data Record format) of 3 days were binned to a $25 \times 25\ \text{km}^2$ grid in polar stereographic projection, true at 60° N with 3 different binning procedures:

Procedure (1): Select the T_B's for the overflight when $\left[T_B(37V) - T_B(85V)\right] = MAX$.

Procedure (2): Select the mean T_B's of all passes.

Procedure (3): Select the mean value of the T_B's of two or more passes, for which the pixel is classified as snow covered (iteratively).

The snow classifications were carried out after the T_B's had been selected according to the multitemporal procedures. Procedure (1) gives the maximum snow extent of a three day period, and selects also in general the case when the atmosphere is most transparent at 85 GHz. However, it is sensitive to data errors, and may also lead to misclassifications of precipitation areas in respect to snow, because not all precipitation areas are eliminated by the first condition of the decision tree. Procedure (1) results in overestimation of the snow extent during the period of dry snow. Averaged over the investigation period in winter 1988, the snow extent by procedure (1) was 11% (of the total area) higher than with procedure (2). Procedure (2) reduces the atmospheric influences and sensor effects, but reduces also the spectral differences, if the snow cover is melting during part of the overflights. For this reason, snow extent and depth are in general slightly underestimated, in particular in areas with melting snow. Considering the potential for reducing temporal fluctuations due to the sensor and the atmosphere, as well as eliminating partly melting areas, procedure (3) shows the best results. The average snow extent with procedure (3) was higher by 2% of the total area than with procedure (2). The closest overall agreement with *in situ* measurements of snow extent and depth is obtained with procedure (3) [13]. Due to the low snow depth we had only few areas with melting snow staying over periods of several days. This reduced the possibilities for studying long term melting snow.

We compared our algorithm with snow classification algorithms of Chang *et al.* [23] and of Fiore and Grody [25, 26]. Chang's algorithm was developed with SMMR data and uses the difference $[T_B(18H) - T_B(37H)]$ as specified in Equation (13); an area is classified as snow covered if $SD > 2.5$ cm. In the first classification step, Fiore and Grody identify scatterers (precipitation, snow, desert), using a weighted spectral difference of T_B(19V and 22V) minus $T_B(85V)$. Their data base are antenna temperatures T_A's, which differ from the brightness temperatures of the Sensor Data Records (SDR's), which we used [27]. They use the following condition to separate snow cover from the other scattering media [25]:

$$\text{snow,} \quad \text{if} \quad \left[T_A(22V) < 257\ \text{K} \quad \text{and} \quad T_A(22V) < \left(155 + 0.49 \times T_A(85V)\right)\ \text{K}\right]. \quad (14)$$

Figure 7 compares the number of pixels classified as snow covered by the 3 different algorithms based on the SSM/I data set of winter 1988 over the investigation area in Europe. Because of the polar stereographic projection used (Fig. 3), the equivalent area of one pixel varies between 466 km^2 and 625 km^2, depending on latitude. Our classification, which uses multitemporal procedure (3) provides closest agreement with the snow extent from ground observations. Chang's algorithm results in some underestimation of snow extent, because the thin snow cover is not identified by the 18 and 37 GHz channels.

Because we had only SDR-T_B's available for the intercomparison we had to change the thresholds of the classification condition of Equation (14). For the snow classification 'F' in Fig. 7 we used the following thresholds according to N. Grody (personal

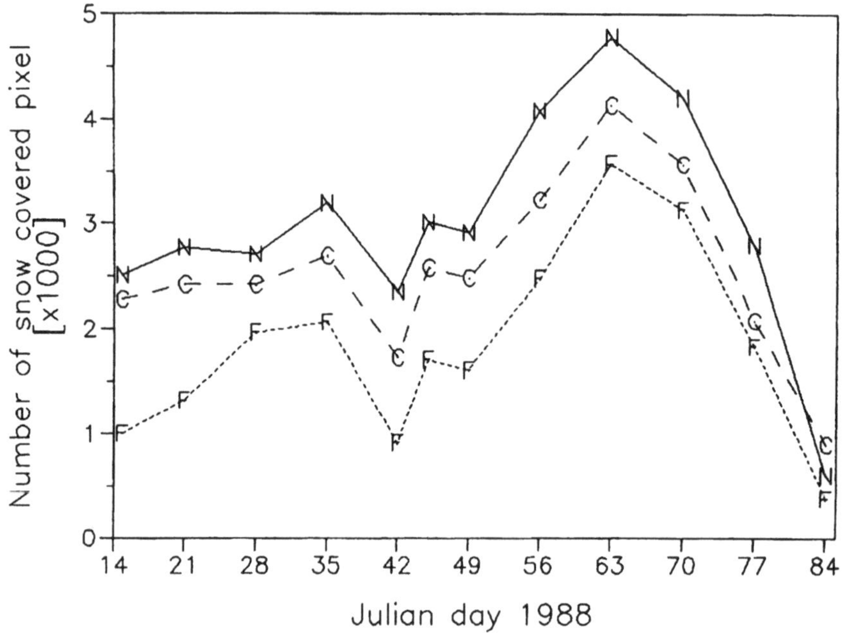

Figure 7. Snow extent over the investigation area in central and eastern Europe during the period 15 January to 25 March 1988, in number of snow covered pixels, derived from SSM/I data applying 3 different algorithms. N — our algorithm using multitemporal procedure (3); C — Chang algorithm with mean T_B of a 3 day period; F — algorithm of Fiore and Grody.

communication): $T_B(22V) < 264$ K and $T_B(22V) < 176 + 0.49 \times T_B(85V)$ K. Due to corrections for feedhorn spillover and cross-polarization coupling, the T_B's are higher than the antenna temperatures [27]. Further differences of the thresholds are due to the fact that Fiore and Grody convolved the 85 GHz TA's to the 60 km resolution of the 22 GHz data, whereas we used the full resolution. According to Grody [26] the use of the 22 GHz channel reduces the mis-classification of rain over wet ground as snow compared to the use of the 19 GHz channel. The algorithm of Fiore and Grody underestimates the snow extent in regions of thin snow cover. A possible reason for this is the reduction of the spectral difference T_B(19V and 22V) $- T_B(85V)$ due to atmospheric effects, resulting in mis-classification of thin snow as snow-free ground.

The mean snow depth, $\langle SD \rangle$ averaged over the pixels classified as snow covered, is compared in Fig. 8 for 3 different classification procedures. On the average, $\langle SD \rangle$ by Chang's algorithm is 4 cm larger than with our algorithm, using multitemporal procedure (3). This has two reasons: (i) very thin snow is not detected; (ii) the coefficient relating the spectral difference (18–37) to SD is higher than in our algorithm. $\langle SD \rangle$ using our multitemporal procedure (2) is lower by 1.5 cm compared to using procedure (3). The difference is highest in March, when the snow cover is receding. This shows that procedure (3) helps to reduce the effects of melting. A similar trend was found for the snow extent and mean snow depth $\langle SD \rangle$ over North America, retrieved by the 3 algorithms for the period January to March 1988. We obtained $\langle SD \rangle = 21.8$ cm with the Chang algorithm, $\langle SD \rangle = 16.8$ cm with our procedure (3), and $\langle SD \rangle = 15.4$ cm with our procedure (2).

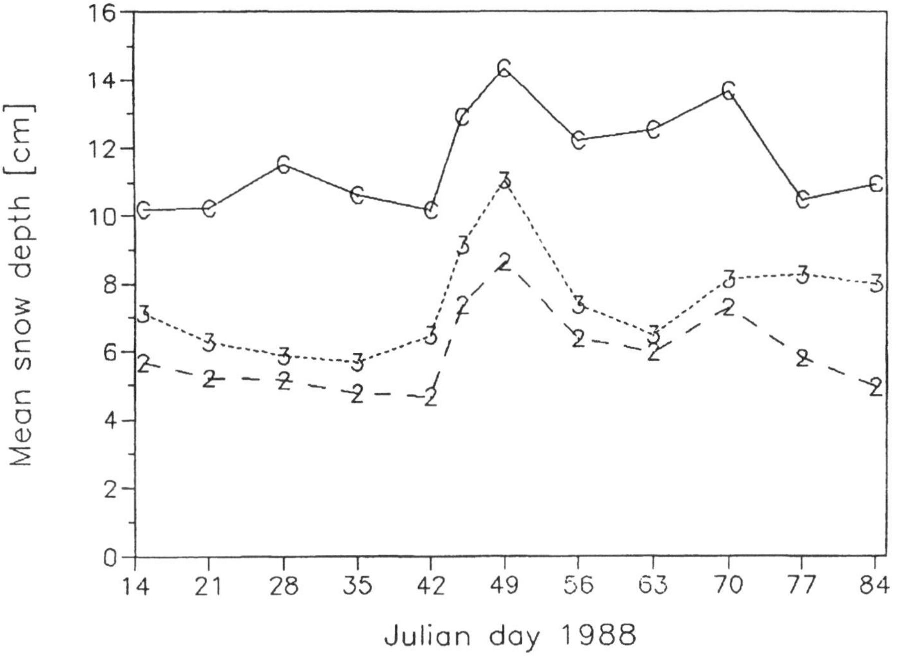

Figure 8. Snow depth averaged over the areas classified as snow covered, based on the same SSM/I data set used for Fig. 7. Algorithms applied: C — Chang; (3) our algorithm with multitemporal procedure (3); (2) — our algorithm with multitemporal procedure (2).

7. CONCLUSIONS

Spaceborne imaging microwave radiometers hold a unique potential for monitoring the areal extent, depth, and water equivalent of the snow cover at global and regional scales with high temporal repetition rates. Application of different algorithms, which have been developed for inversion of microwave measurements in respect to the snow cover, show differences in the derived snow cover products. The main reasons for these differences are:

(1) Most of the algorithms are tuned to regional snow conditions. If applied to regions with different snow morphology and land cover, the performance deteriorates. Because algorithm development is generally based on correlations between ground based and satellite measurements, the performance depends also on the quality of the correlative data sets.

(2) Different channels and data products (brightness temperatures, antenna temperatures) of the same sensor are applied in various algorithms. Most of the snow retrieval algorithms are based on measurements in two channels: 37 GHz and 18 GHz (SMMR) or 19 GHz (SSM/I). Some minor differences result from application of different polarizations (vertical, horizontal). Recently developed algorithms apply also 85 GHz data (based on SSM/I), which are more sensitive to thin snow, but are more affected by the atmosphere.

(3) The multitemporal treatment of the data is of great significance. Appropriate procedures help to reduce the effects of melting snow and of atmospheric prop-

agation attenuation and emission. As an example, SSM/I images the surface at high latitudes at least 6 times during 3 days, providing good opportunity for multitemporal analysis if one snow map is derived for a 3 day period.

Looking at the intercomparison of the various snow retrieval algorithms, we believe that the development of a generally agreed and accurately verified algorithm for deriving the extent and mass of the global snow cover should be considered with high priority. Such an algorithm will have to include parameters or coefficients accounting for the spatial patterns of snow morphology and land cover (in particular forest cover). Regional algorithms, which have been developed for various parts of Europe and North America, are an appropriate starting base to come to a globally applicable, but regionally deversified algorithm. So far, verifications of microwave derived snow cover products are lacking over extended areas such as Siberia and the boreal forests. Another field for further research is the detection of wet snow. A promising technique has been developed for the Canadian Prairies, but studies in other regions are needed to learn about the general applicability.

Consistent data sets on the snow cover, as required for climate monitoring, have to rely on accurately calibrated satellite sensors. In this respect, generation of absolutely calibrated microwave data sets, including recalibration of SMMR, has to be given high priority.

Acknowledgements
This work has been supported by the Austrian Science Funds, Project P 8476. The SSM/I data were made available by the World Data Center-A/National Snow and Ice Data Center, University of Colorado, Boulder.

REFERENCES

1. S. Chandrasekhar. *Radiative Transfer*. Dover, New York (1960).
2. A. W. England. Thermal microwave emission form a scattering layer. *J. Geophys. Res.*, **80** (32), 4484–4496 (1975).
3. H. Rott, G. Domik, C. Mätzler, H. Miller and K. G. Lenhart. *Study on Use and Characteristics of SAR for Land Snow and Ice Applications*. Final Reoprt, ESA Contract Nr 5441/83/D/IM(SC), Inst. Met. and Geophys., Univ. Innsbruck (1985).
4. A. K. Fung and M. F. Chen. Modelling of microwave emission and scattering from snow and soil. *Adv. Space Res.*, **9** (1), 297–306 (1989).
5. B. Wen, L. Tsang, D. P. Winebrenner and A. Ishimaru. Dense medium radiative transfer theory: comparison with experiment and application to microwave remote sensing and polarimetry. *IEEE Trans. Geosc. Rem. Sensing*, **28**, 46–59 (1990).
6. M. T. Hallikainen, F. T. Ulaby and T. E. van Deventer. Extinction behaviour of dry snow in the 18- to 90-GHz range. *IEEE Trans. Geosci. Rem. Sens.*, **25**, 737–745 (1987).
7. R. Hofer and C. Mätzler. Investigations on snow parameters by radiometry in the 3- to 60-mm wavelength region. *J. Geophys. Res.*, **85** (C1), 453–460 (1980).
8. C. Mätzler. Application of the interaction of microwaves with the natural snow cover. *Remote Sensing Reviews*, **2**, 159–387 (1987).
9. H. Rott, K. Sturm and H. Miller. Active and passive microwave signatures of Antarctic firn by means of field measurements and satellite data. *Annals of Glaciology*, **17**, 337–342 (1993).
10. C. Mätzler and R. Hüppi. Review of signature studies for microwave remote sensing of snowpacks. *Adv. Space Res.*, **9** (1), 253–265 (1989).

11. A. Walker and B. Goodison. Discrimination of wet snowcover using passive microwave satellite data. *Annals of Glaciology*, **17**, 307–311 (1993).
12. J. Aschbacher. *Land Surface Studies and Atmospheric Effects by Satellite Microwave Radiometry*. PhD Thesis, Univ. Innsbruck (1989).
13. T. Nagler. *Verfahren der Analyse der Schneebedeckung aus Messungen des Special Sensor Microwave/Imager (SSM/I)*. Diploma Thesis, Univ. Innsbruck (1991).
14. R. Wu and J. A. Weinman. Microwave radiances from precipitating clouds containing aspherical ice, combined phase, and liquid hydrometeors. *J. Geophys. Res.*, **89** (D5), 7170–7178 (1984).
15. J. E. Hollinger, J. E. Pierce and G. A. Poe. SSM/I instrument evaluation. *IEEE Trans. Geosci. Rem. Sens.*, **28**, 781–790 (1990).
16. K. Künzi, S. Patil and H. Rott. Snow-cover parameters retrieved from Nimbus-7 Scanning Multichannel Microwave Radiometer (SMMR) data. *IEEE Trans. Geosci. Rem. Sens.*, **20**, 452–467 (1982).
17. M. T. Hallikainen and P. A. Jolma. Comparison of algorithms for retrieval of snow water equivalent from Nimbus-7 SMMR data in Finland. *IEEE Trans. Geosci. Rem. Sens.*, **30**, 124–131 (1992).
18. J. L. Foster, A. T. C. Chang, D. K. Hall and A. Rango. Derivation of snow water equivalent in boreal forests using microwave radiometers. *Arctic*, **44**, 147–152 (1991).
19. B. E. Goodison. *Report to the SSM/I Products Working Team*. Unpublished (1992).
20. M. T. Hallikainen and P. A. Jolma. Retrieval of snow water equivalent in Finland by satellite microwave radiometry. *IEEE Trans. Geosci. Rem. Sens.*, **24**, 855–862 (1986).
21. H. Rott and K. Sturm. Microwave signature measurements of Antarctic and Alpine snow. In: *Proc. of 11th EARSeL Symposium, Graz, Austria*, pp. 140–151 (1991).
22. E. A. Francis. *Calibration of the Nimbus-7 Scanning Multichannel Microwave Radiometer (SMMR), 1979–1984*. Master Thesis, Oregon State Univ. (1987).
23. A. T. C. Chang, J. L. Foster and D. K. Hall. Nimbus-7 SMMR derived global snow cover patterns. *Annals of Glaciology*, **9**, 39–44 (1987).
24. T. Nagler and H. Rott. Development and intercomparison of snow mapping algorithms based on SSM/I data. *IEEE Proc. IGARSS'92*, 812–814 (1992).
25. J. V. Fiore and N. C. Grody. Classification of snow cover and precipitation using SSM/I measurements: case studies. *Int. J. Remote Sens.*, **13**, 3349–3361 (1992).
26. N. C. Grody. Classification of snow cover and precipitation using the Special Sensor Microwave Imager. *J. Geophys. Res.*, **96** D4, 7423–7435 (1991).
27. J. Hollinger. *DMSP Special Sensor Microwave/Imager Calibration/Validation, Final Report of the DMSP SSM/I Cal/Val Team*. Vol. I, Naval Research Lab., Washington DC (1989).

ESA/NASA International Workshop, pp. 245–262
B. J. Choudhury, Y. H. Kerr, E. G. Njoku and P. Pampaloni (Eds)

Canadian development and use of snow cover information from passive microwave satellite data

B. E. GOODISON and A. E. WALKER

Canadian Climate Centre, Atmospheric Environment Service,
4905 Dufferin Street, Downsview, Ontario, Canada M3H 5T4

Abstract—This paper summarizes the Canadian Climate Centre (CCC) program on algorithm development and validation to derive snow water equivalent (SWE) from satellite passive microwave data for hydrological and climatological applications. A vertically polarized brightness temperature gradient ratio (GTV) algorithm provided the most consistent and compatible SWE over the Canadian prairies when validated using Nimbus-7 Scanning Multichannel Microwave Radiometer (SMMR) and DMSP Special Sensor Microwave Imager (SSM/I) data. Since 1988/89, near real-time SWE maps have been evaluated by users in operational hydrological forecasting. Their recommendations have spurred the development of a wet snow indicator to discriminate between areas of wet snow and snow-free land for both SSM/I and SMMR data. Examples of the marked improvement in snow cover products after using this indicator are given. It is shown to be critical in the accurate development and assessment of time series of snow cover for the Prairie region for 1979-1986, for the study of variability and change. Problems and new initiatives related to the underestimation of SWE in forested regions and the validation of algorithms in data-sparse areas are discussed. Future research thrusts of the CCC program and its links to EOS/CRYSYS, including recommendations to the international community to achieve improved snow cover products using passive microwave data, are given.

1. INTRODUCTION

The study of regional and global water cycles, including the study of future change, requires a good understanding of processes combined with reliable, accurate and consistent data for all components of the energy and water cycle. For much of Canada, snow cover (extent, water equivalent) is arguably the most important storage component of the hydrological cycle. At the same time, it plays an interactive role in the regional and global climate system, thus influencing future potential change in the hydrological system. Snow cover, like other cryospheric components, is also felt to be a sensitive integrator of basic climate elements and variations in snow cover may potentially be an effective indicator of regional and global change. Systematic observation and analysis of snow cover is one contribution in monitoring the variability of the climate system and in the identification of significant change.

The development of a consistent, compatible data series for spatial and temporal analyses of snow cover is critical for such hydrological and climatological investigations. In many regions of Canada, there is little or no conventional snow cover information, limiting our ability to define or model hydrological and climatological processes accurately in space and time [1]. Remotely sensed information, however, does offer a potential complementary data source.

The availability of visible and near-infrared images of the earth since the 1960's has allowed the development of techniques for the mapping of the areal extent of snow cover. However, information on the water equivalent, depth and wetness of the snowpack are essential for energy and water balance investigations at all scales. Microwave radiometry, with its generally all-weather capability, offers the promise of augmenting and complementing conventional information by providing greater spatial coverage frequently. It is a remote sensing technique ideally suited to Canadian conditions and needs. The focus of the Canadian Climate Centre (CCC) research program has been to develop, validate and apply passive microwave satellite data to determine snow water equivalent, extent and snowpack state for both real-time operational use in hydrological, climatological and agricultural applications and for long term monitoring in the study of climate variability and the detection of change. The objective of this paper is to summarize the development and current status of the Canadian regional snow water equivalent (SWE) algorithms and their successful use in Western Canada for producing near realtime products for water resource applications and climatological analyses. Problem areas are discussed in the context of on-going algorithm development.

2. ALGORITHM DEVELOPMENT

Early investigations [2, 3, 4] using Nimbus-7 SMMR (Scanning Multichannel Microwave Radiometer) and Nimbus-5 and -6 ESMR (Electrically Scanning Microwave Radiometer) demonstrated the potential for utilizing passive microwave data to monitor snow cover parameters, especially areal extent, water equivalent, depth and onset of melt. The Canadian Climate Centre (CCC) snow cover/passive microwave research effort was initiated in 1981 as part of a multi-agency international snow cover research study to support the activities of operational water resource agencies in the Canadian Prairies [5].

An intensive field campaign conducted in February 1982 in southern Saskatchewan provided the basis for the derivation of the CCC regional snow cover algorithms. Figure 1 outlines the Canadian Prairie region where algorithm testing, validation and operational application have been conducted using Nimbus-7 SMMR and DMSP SSM/I (Special Sensor Microwave/Imager) satellite data. The region is largely agricultural, with winter land cover characterized by fallow, stubble, cultivated stubble, pasture and natural prairie fields. There are a few isolated small forest stands dotted across the open prairie; north of the treeline, continuous boreal forest mixed with lakes and wetlands dominates the landscape. Although the open prairie region may be considered as relatively 'uniform' in a broad physiographic context, the mixture of agricultural land cover types in winter and the propensity for the snow to be readily redistributed by the wind results in a heterogeneous snow cover from a small basin hydrological perspective.

Airborne microwave measurements (18 GHz and 37 GHz, H and V polarizations) were collected along an 1100 km flight line network (Fig. 1) using NASA's multifrequency microwave radiometer (MFMR). Coincident airborne gamma radiation measurements (US National Weather Service) provided the 'areal ground true' SWE [6, 7]. The gamma and airborne microwave data were averaged over 20–25 km segments, a resolution comparable to the 25–30 km resolution of the SMMR 37 GHz satellite data. Intensive ground snow surveys, stratified by land cover, were conducted along

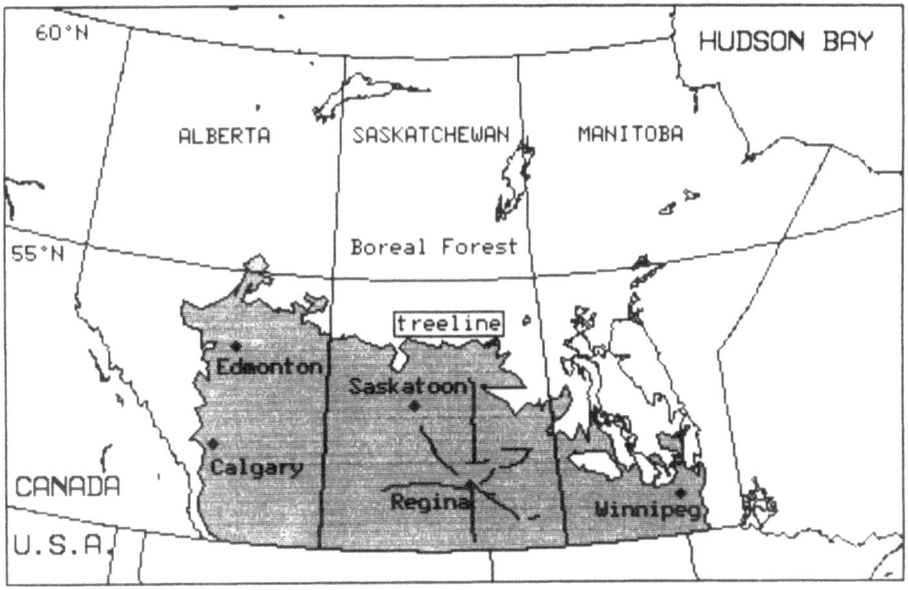

Figure 1. Western Canadian study area. Non-forested prairie region is shaded. Airborne flight lines in southern Saskatchewan are shown.

the flight lines for validation of the gamma and microwave measurements. A more complete discussion of the experimental programme is provided in [7].

Using the gamma data as 'ground true', a series of empirical airborne-based algorithms were derived for determining snow water equivalent from passive microwave measurements. Since there were some areas of melt, or at least with a 'wet' surface snow, as identified by field and airborne personnel, the flight segments were separated into dry and wet cases. Figure 2 presents the graphical comparison between gamma measurements of SWE and the 18 and 37 GHz vertically polarized microwave brightness temperature gradient (GTV) [8] for dry and 'wet' line segments. Statistically, the correlation between GTV and SWE is not significant when the data are grouped together. However, there is a highly significant relationship ($r^2 = 0.89$) for the dry snow cases. Almost identical results occur when SWE is regressed against 37 V brightness temperatures ($r^2 = 0.84$ for dry snow only). Figure 3 shows the relationship between 37 H and SWE for all cases; there is considerably more scatter in the measurements, probably attributable to the different snowpack structures encountered throughout the study area (snow; snow with crusts or ice layers; ice on the ground). Based on these results, two SWE algorithms were selected for use — one using only 37 V, the other employing the vertical brightness temperature gradient (37 V−18 V)/19. Recent European results have also shown that vertically polarized data provide the most significant relationship to snow water equivalent [9], while other researchers have preferred to use horizontally polarized data for algorithm derivation [10].

The application of the airborne-based snow water equivalent algorithms to satellite microwave measurements for testing and validation purposes initially involved SMMR data. Selected maps of snow water equivalent were produced for the Canadian Prairie

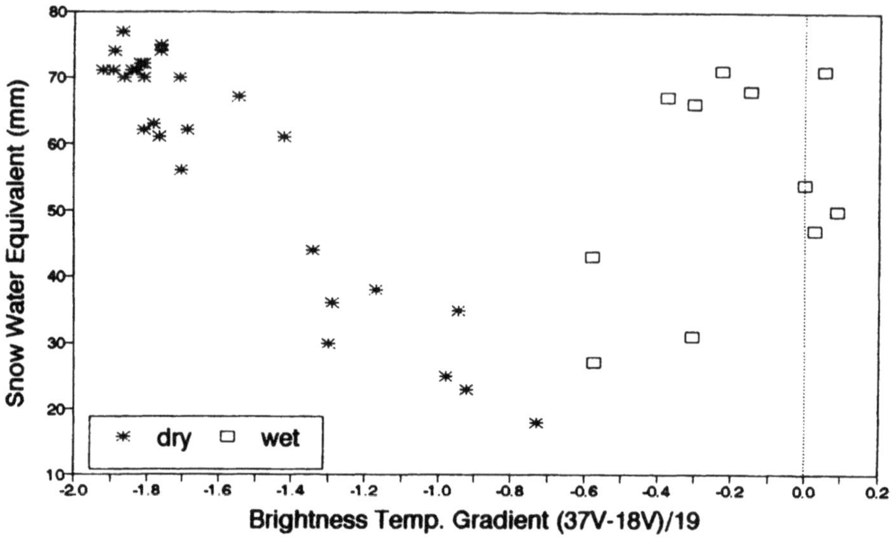

Figure 2. Relationship between NWS airborne gamma SWE and MFMR airborne radiometer GTV for all flight segments. Dry snow and wet snow cases are identified separately.

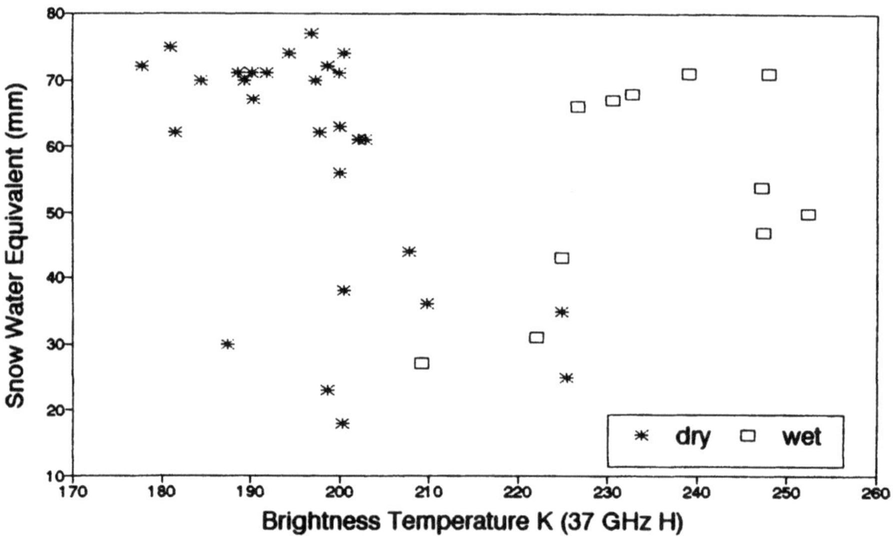

Figure 3. Relationship between NWS airborne gamma SWE and 37 GHz horizontally polarized data from MFMR airborne radiometer. Dry snow and wet snow cases are identified separately.

region primarily representing maximum snow cover conditions for 1979 to 1986. Figure 4 compares the NWS airborne gamma measurements of February 16–17, 1982 with the distribution of SWE derived from SMMR CELL-ALL data for the nighttime orbit on February 15, 1982, using the GTV algorithm. No adjustments have been made to the airborne-based coefficients when applying them to the satellite data. The SWE estimates

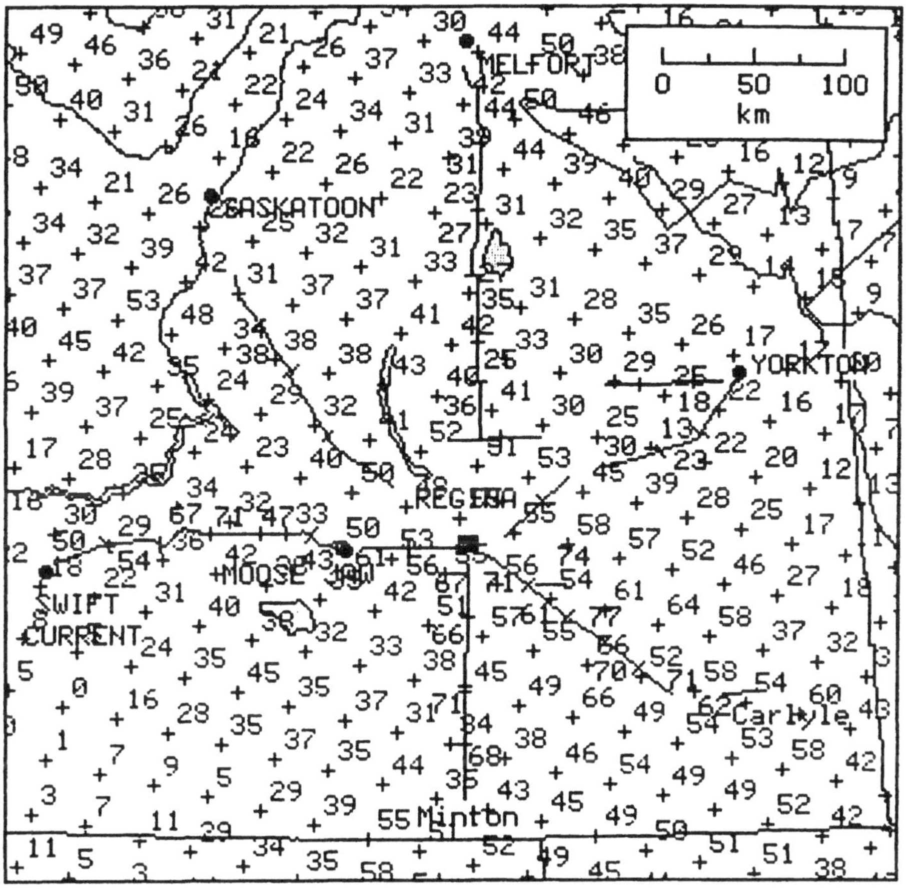

Figure 4. Distribution of SWE on February 15, 1982 derived from SMMR CELL-ALL data using GTV airborne algorithm. Light grey numbers are SMMR pixel estimates; black numbers are gamma estimates of SWE for flight line segments.

south of the treeline are very good (within 10 mm) compared to the airborne gamma measurements and to conventional ground survey measurements [7]. The only significant departure between the two estimates of SWE is in southwestern Saskatchewan, south and west of Regina, where patchy wet snow was reported. This resulted in lower SWE estimates from SMMR than would be expected from a dry snowpack (Fig. 2). It should be noted that this algorithm has been found to be less reliable in forested areas, as the derived SWE values are too low. The implications of both these problems will be discussed further in the context of real-time operations.

Additional comparisons with conventional ground snow survey measurements in other parts of the prairie region showed the microwave SWE estimates to be generally within 10–20 mm of ground measurements [11]. Given the discrepancies between point and areally averaged measurements, this difference was deemed very acceptable. These comparisons also confirmed that, in the prairie region, the GTV algorithm provided the most consistent spatial snow cover data as compared to conventional snow surveys or special airborne gamma missions.

3. ALGORITHM VALIDATION USING NEAR REAL-TIME SSM/I DATA

The usefulness of Nimbus-7 SMMR data for monitoring snow cover in the Canadian Prairie region on a regular basis was found to be limited by a relatively narrow 800 km swath width, an alternate day acquisition cycle, and non real-time availability. Complete coverage of the region was only possible by combining several SMMR orbits, which could possibly span 3 or 5 calendar days, and due to the research nature of the satellite, the time delay in data availability was generally weeks or months after satellite viewing. The launch of the SSM/I on the U.S. Defense Meteorological Satellite Program (DMSP) F-8 satellite in June 1987 overcame these limitations. The 1400 km swath width of the SSM/I combined with twice daily orbital passes (ascending and descending) provides complete coverage of the Canadian Prairie region on a daily basis. The operational nature of the SSM/I also provides the capability of data access in near real-time, and hence, of monitoring the spatial and temporal changes in snow water equivalent conditions over the region on a regular basis throughout the winter season. The accessibility of SSM/I data also resulted in greater opportunities for testing and validating the airborne snow water equivalent algorithms which were modified for use with the SSM/I 37 and 19 GHz frequencies (e.g. GTV=(37 V−19 V)/18). Repeated comparison between SSM/I SWE values and SWE data acquired from ground surveys and airborne gamma measurements have shown that the regional patterns of snow distribution agree very well, and average values are within 10–20 mm of each other [12].

Access to SSM/I data in near real-time via the Civilian Navy/NOAA Oceanographic Data Distribution System (CNODDS) has provided the capability to generate snow products within hours after data acquisition. Since the 1988/89 winter season, maps depicting current SWE conditions for the Canadian Prairie region have been produced on a weekly basis using near real-time SSM/I data. These maps are disseminated by facsimile machine to a number of water resources agencies who use the information to monitor snow conditions, plan for ground surveys, and prepare runoff forecasts. This program [13] is cooperative in nature and has served to enhance the validation process of the SSM/I SWE algorithms. The users who receive the maps are an important source of validation by providing feedback on the representativeness of the SWE information presented on the maps and identifying problem areas which may require further research. After four winter seasons in operation, this program has successfully demonstrated a useful application of passive microwave derived snow cover information for hydrological analyses.

4. IDENTIFICATION OF PROBLEM AREAS AND ALGORITHM RESEARCH

The initial development of SWE algorithms in Canada was driven by hydrological applications and the need for reliable information on winter snow cover conditions in order to augment conventional ground measurements which may be too sparse or too costly to gather in some regions. The operational SSM/I SWE map program emphasized the need to generate a product which was as accurate and representative as possible since the users required reliable information for their decision-making process. Problems in algorithm performance identified during the validation process of the operational SWE maps have become the focus of research into algorithm improvement, including (i) wet snow; (ii) forest cover (vegetation) effects; (iii) algorithm validation in areas with little or no conventional ground information; (iv) seasonal changes in snowpack structure; and (v) limitations of a coarse pixel resolution (25 km) for hydrological applications. These problems are not unique to the CCC algorithms. The following sections describe the CCC algorithm research which address the problems of greatest concern in the Canadian context.

4.1. Wet snow

The presence of liquid water within a snowpack alters the microwave emission characteristics and severely limits the quantitative determination of snow cover parameters. Under wet snow conditions, the SSM/I SWE algorithm will produce 'zero' values which erroneously indicate that the area is snow-free land. In the Canadian Prairie region, periods with temperatures near and above 0 °C during the winter months have occurred often in recent years, resulting in more frequent wet (melting) snow conditions over extensive areas. With access to SSM/I on a daily basis in near real-time, it has been demonstrated that during periods of winter melt, the changes in the SSM/I SWE values can be quite dramatic over short time periods due to changes in weather; the snow cover in affected regions may 'disappear' and 'reappear' on successive orbit passes [14]. Figures 5(a) and (b) illustrate such an event, using SSM/I data from February 1 and February 3, 1991. During the 48 hour period between the two orbits, the occurrence of mild temperatures resulted in an extensive area of the snow cover over the region 'disappearing' due to the presence of wet snow conditions. Although the use of nighttime orbit passes may be one way of mitigating the wet snow problem (i.e. high probability of temperatures below freezing), recent winter melt events over the Canadian prairie region have shown that wet snow conditions can persist throughout the overnight period if the temperatures remain mild (i.e. above or close to 0 °C).

With users depending on a reliable representation of the spatial extent of snow cover as well as quantitative SWE information, the need to overcome the limitation of the SSM/I SWE algorithm under wet snow conditions became a high priority research effort. The access to near real-time SSM/I data played an important role in this effort, by making it possible to identify and monitor wet snow conditions, and analyse changes in SSM/I brightness temperatures. This led to the development of a 'wet snow indicator' which discriminates between areas of wet snow and actual snow-free land based on the difference between 37 GHz vertical and horizontal polarization brightness temperatures. The derivation of this indicator is described in [15]. If a SSM/I SWE value is calculated as zero, a polarization difference of 10 K or greater signifies wet snow, while a difference of less than 10 K indicates snow-free ground. The basis of this discrimination is illustrated in Fig. 6 using the 37 GHz polarization differences for dry snow, wet snow and snow-free land corresponding to the melt event depicted in Figs 5(a) and (b).

The use of the SSM/I wet snow indicator is demonstrated in Fig. 7 using the same February 3, 1991 SSM/I orbit as displayed in Fig. 5(b). The area of snow cover which 'disappeared' between February 1 and 3, 1991 is confirmed as being wet snow through the use of the indicator. Although it is not possible to derive an accurate snow water equivalent value under wet snow conditions, the use of the wet snow indicator provides a more representative depiction of the spatial extent and condition of the snow cover, thus enhancing the accuracy of the SSM/I SWE map product and indirectly improving the performance of the SWE algorithm. Testing and validation of the indicator is an ongoing process, which has so far been limited to the Canadian Prairie region; it may also be

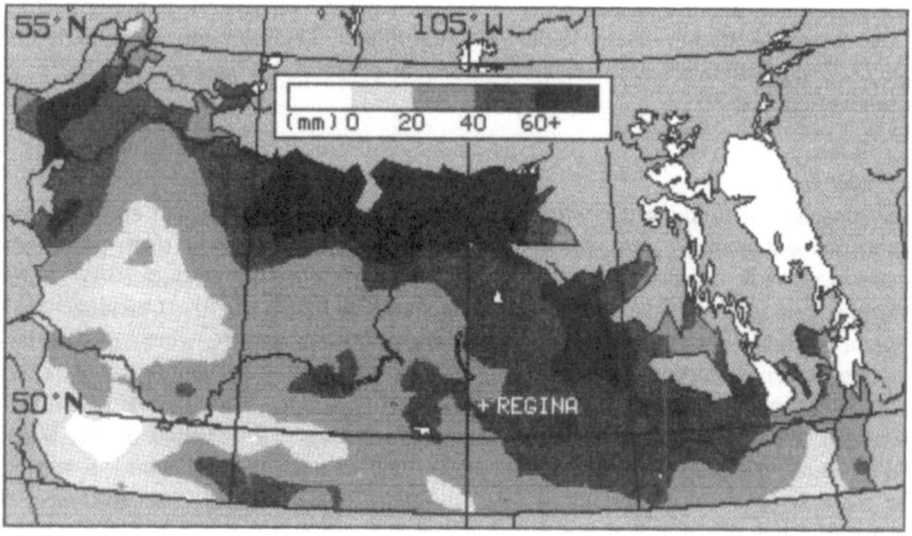

Figure 5a. Snow water equivalent (in mm) over the southern Canadian Prairies, derived from SSM/I data for February 1, 1991 using the GTV algorithm.

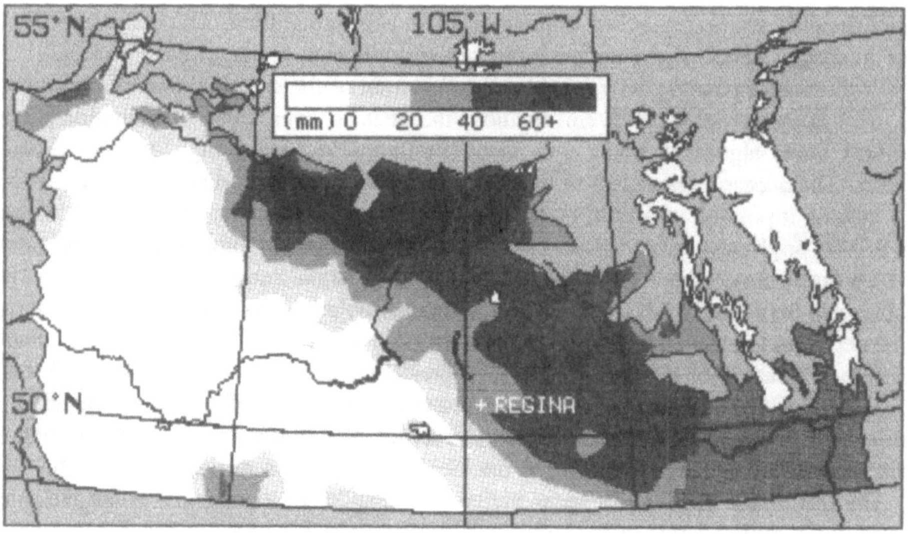

Figure 5b. Snow water equivalent (in mm) over the southern Canadian Prairies, derived from SSM/I data for February 3, 1991 using the GTV algorithm. Within a 48-hour period the snow cover has apparently 'disappeared' over a large portion of the study area, due to the effect of wet snow on the SSM/I SWE retrievals.

applicable to other regions of similar landscape characteristics. A similar indicator for Nimbus-7 SMMR data has recently been derived and is currently being evaluated.

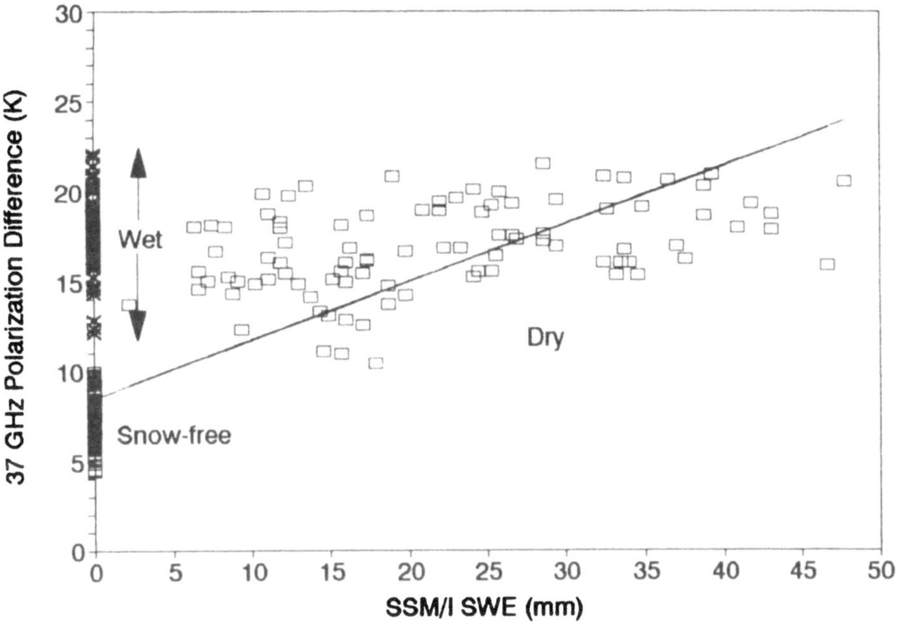

Figure 6. SSM/I 37 GHz polarization differences corresponding to wet snow, dry snow and snow-free land from the February 1991 melt event illustrated in Figs 5(a) and (b). Wet snow and snow-free land are clearly distinguishable using the 10 K discriminant value.

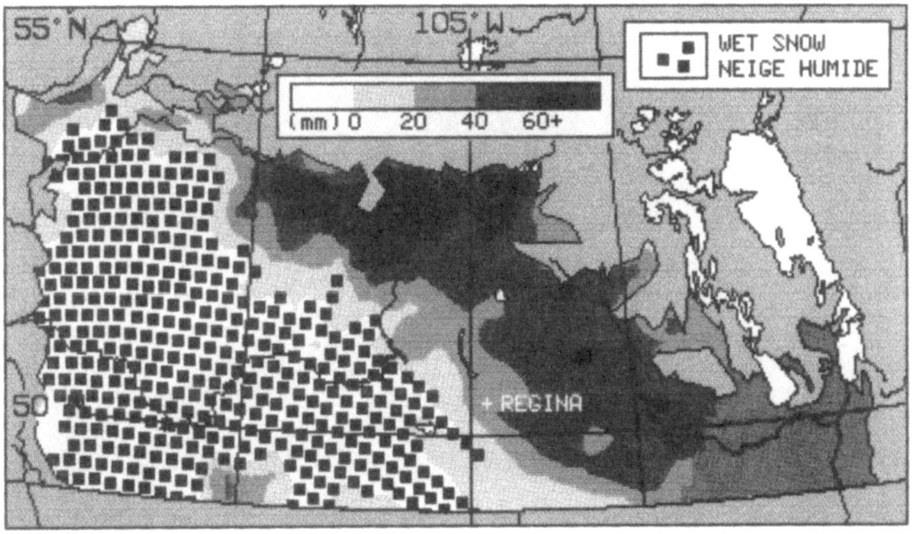

Figure 7. Wet snow areas defined using the SSM/I wet snow indicator for February 3, 1991.

4.2. Forest effects

As illustrated in Fig. 1, a large part of the Canadian Prairie region is covered by boreal forest. The accuracy of SWE retrievals from passive microwave data is affected by the type of land cover present and the terrain conditions. A vegetation cover will attenuate the emission from the underlying snowpack and contribute its own emission to the signal received at the satellite. This results in a decreased microwave sensitivity to SWE in vegetated areas, which can cause underestimations in derived SWE by as much as 50% as compared to open agricultural areas [16].

As expected, the SSM/I SWE retrievals derived from the airborne-based algorithms are generally much lower than comparative ground measurements within the Canadian Prairie boreal forest region [17]. The influence of the forest cover on SWE retrievals from the SSM/I GTV algorithm is demonstrated in Fig. 8. At the boundary (treeline) between open prairie and forest, the SWE values decrease from 20 mm over the prairie region to zero over the forest where the vegetation cover is most dense. Further north the SSM/I SWE retrievals increase again as the vegetation cover becomes less dense.

Within the Canadian Prairie boreal forest region, few conventional snow measurements are available due to the remoteness of the region and the sparse station network. Thus, there is a strong interest among water resource and meteorological agencies in other potential sources of snow cover information for the region. Development of a representative SWE algorithm for the boreal forest region has therefore become a high priority. Research efforts are currently being focussed on deriving a series of regional algorithms which will compensate for the effects of varying forest cover density and type. Forest cover data-sets containing information on cover density and type compiled from ground surveys and other remote sensing data sources are essential to support this research.

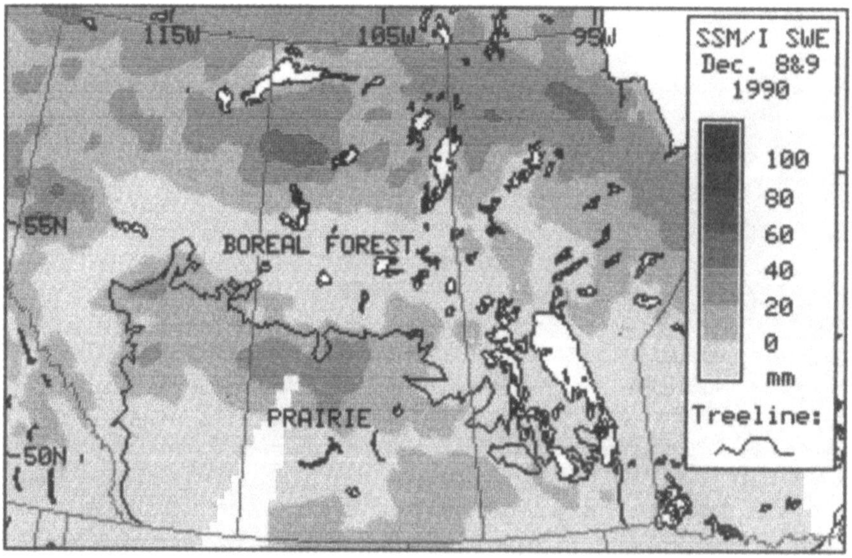

Figure 8. Snow water equivalent over the Canadian Prairies derived from SSM/I data for December 8–9, 1990 using the GTV algorithm. The influence of the forest cover on the SWE retrievals is apparent with the decrease in SWE at the treeline boundary.

4.3. Algorithm validation in data-sparse regions

One of the greatest potential contributions of satellite data is to provide reliable information for data-sparse regions of the earth. Research has been initiated to derive and assess the reliability of snow cover information derived from SSM/I satellite data for sub-arctic and high arctic regions of northern Canada, where there is very limited spatial and temporal information on the regional snow cover. Initially, the GTV and 37 V prairie algorithms have been applied to SSM/I orbital data over the Mackenzie Basin portion of the Northwest Territories (NWT), where there are some ground-based snow courses available for comparison [18]. This part of the NWT is covered by transitional forest with densities ranging from 15% to 75%. As noted, the presence of significant forest vegetation of varying cover density can complicate the retrieval of representative snow cover information from passive microwave data.

Figure 9 presents an example of SWE derived from SSM/I data for March 27, 1990 for the western portion of the NWT using the 37V open prairie algorithm. There appears to be high spatial variability of SWE over this area with values ranging from less than 50 mm to greater than 100 mm; however, the effects of a varying density of forest cover on these observed spatial patterns must be considered in any assessment of the derived SWE information. The difference in SWE between SSM/I and conventional snow course measurements is shown in Fig. 9; the area of high SWE (80–120 mm) north of Great Slave Lake exhibited the best agreement, with the SSM/I values being within 10 mm of the snow course measurements. Although both algorithms tended

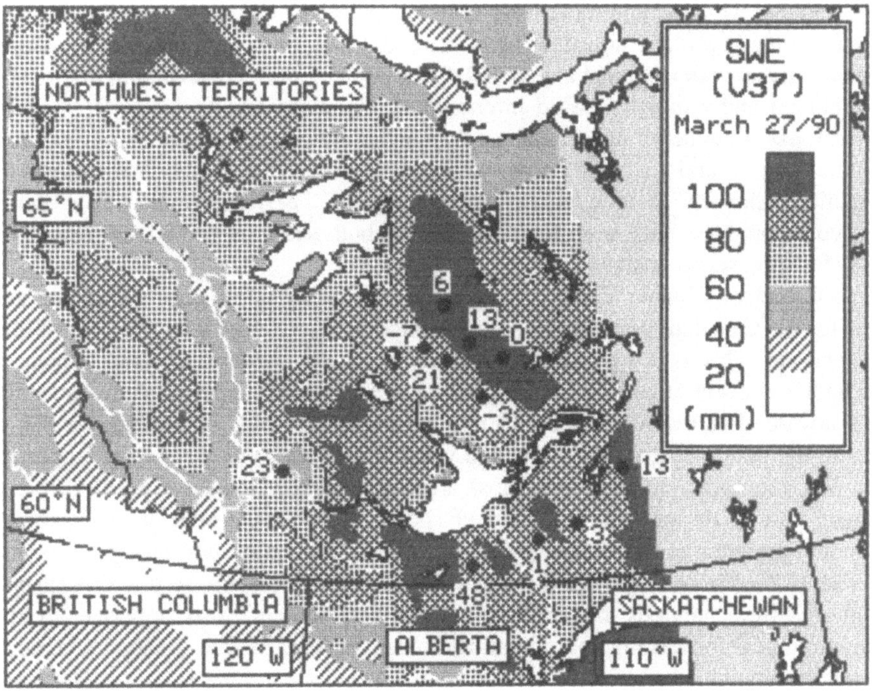

Figure 9. Snow water equivalent over the western portion of NWT derived from SSM/I data for March 27, 1990 using the 37 V algorithm. Values represent the difference in SWE (mm) between conventional snow course measurements and the SSM/I retrievals.

to underestimate the SWE throughout the region, the single frequency 37 V algorithm yielded slightly higher values.

The results from 1990 are very encouraging, but there are limited ground observations to compare with the satellite retrievals. For a satellite method to be effective, SWE retrievals must be consistent during a winter period as well as from year to year. Subsequent comparisons were done in 1991 and 1992. In 1992 the SSM/I SWE estimates were 40–60 mm lower than the ground measurements over the entire area. Potential reasons for this are currently under investigation. Was the temporal accumulation pattern different from other years? Were there more melt periods, with variable areas of wet snow having an influence on SWE retrievals? Was the snowpack structure substantially different from other years? Are the ground-based measurements consistent from year to year? All of these could contribute to problems in accurate snow cover determination. This emphasizes the need to validate algorithms under all conditions over a period of years in many different regions. If the information is to be used for hydrological and climatological applications, more than an individual case study is required in order to demonstrate the utility of remote sensing.

5. CLIMATE VARIABILITY AND CHANGE

To assess future global change, monitoring of the climate system through observation and analysis of seasonal and interannual fluctuations of climate variables, such as snow cover, is necessary. Snow extent is important, but snow water equivalent or snow mass is the variable which would more truly reflect variability in the energy and water cycle. Identifying the important 'signal' to be monitored, however, is not easy [19]. Ideally, the routine analysis of the regional and global distribution of snow cover should involve both conventional and remotely sensed information. Goodison and Walker [19] show that conventional snow depth measurements from synoptic weather stations are very useful in addressing several temporal snow cover questions, but there is limited information on spatial variability of snow cover over the region. The passive microwave derived snow cover products have been found to provide both temporal and spatial information useful for time series analyses.

The Canadian Climate Centre is currently completing an atlas of bi-weekly snow cover for the Canadian prairie region (as depicted in Fig. 1) for the 1979–1986 SMMR time period. This is a first step in creating a much longer time series which will incorporate conventional data from earlier years and SSM/I derived snow cover from 1987 onward. Figures 10 and 11 illustrate the marked contrast in snow cover that can occur in the prairie region in consecutive years during the same week. The importance of having the ability to map wet snow areas is clearly demonstrated. In 1981, the southern part of the prairies is devoid of snow. In 1982, the area of wet snow was originally depicted as snow-free [19], but in fact the entire prairie region is snow covered. Although the SWE of the wet snow area is not known on this day (it might be inferred from previous days or conventional data sources), it is important to know for hydrological forecasting and numerical modelling that the area is snow covered and potentially melting.

Passive microwave derived snow cover information offers the great advantage of assessing changes in areal snow water equivalent on a daily basis during the winter and from year to year [19], given the limitations noted previously. This allows the creation

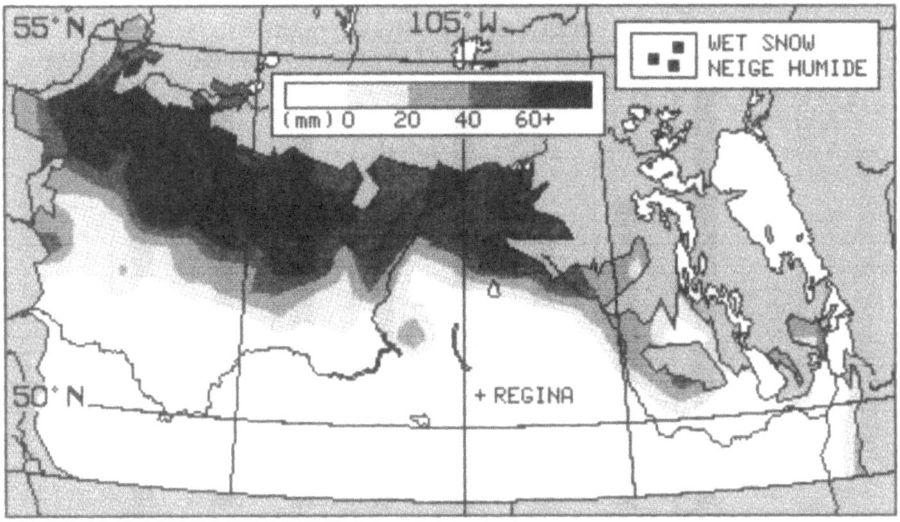

Figure 10. Distribution of snow water equivalent over the Canadian Prairies for March 1, 1981 derived from Nimbus-7 data, including the mapping of areas of wet snow.

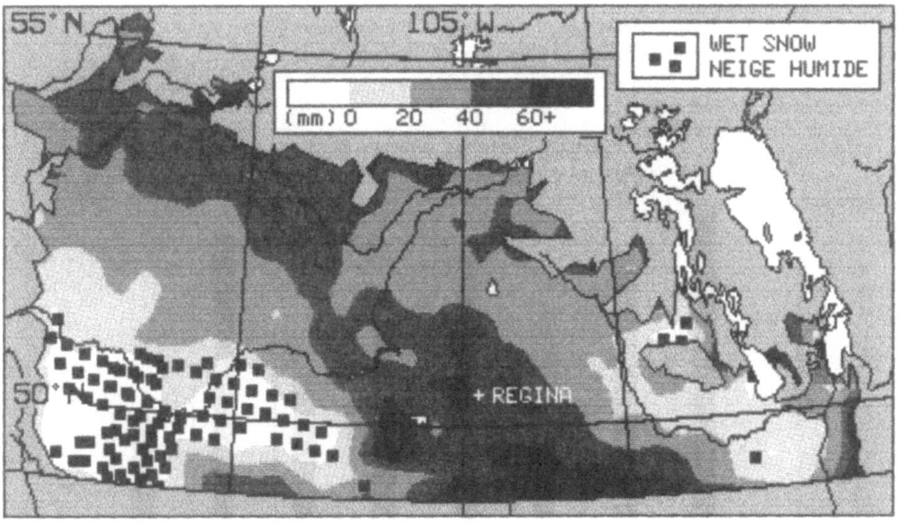

Figure 11. Distribution of snow water equivalent over the Canadian Prairies for March 1, 1982 derived from Nimbus-7 data, including the mapping of areas of wet snow.

of snow cover time series which can then be used in the study of climate variability, GCM model validation or hydrological forecasting. Figures 12 and 13 compare the snow covered area, as derived from SMMR data, for the first weeks of February and March for 1979 to 1986 for the prairie region. Percent snow cover, including and excluding areas of wet snow, are shown. It is very noticeable that there were areas of melting snow on 12 of the 16 weeks sampled. Without the ability to map the areas of

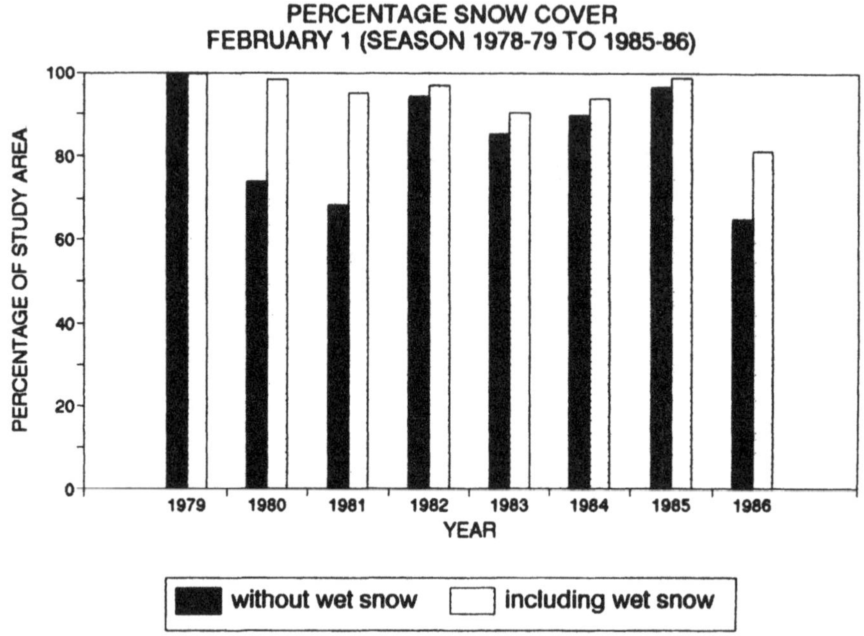

Figure 12. Snow cover as a percentage of total area of the Canadian Prairie region for the first week of February, 1979 to 1986, with and without consideration of areas of wet snow.

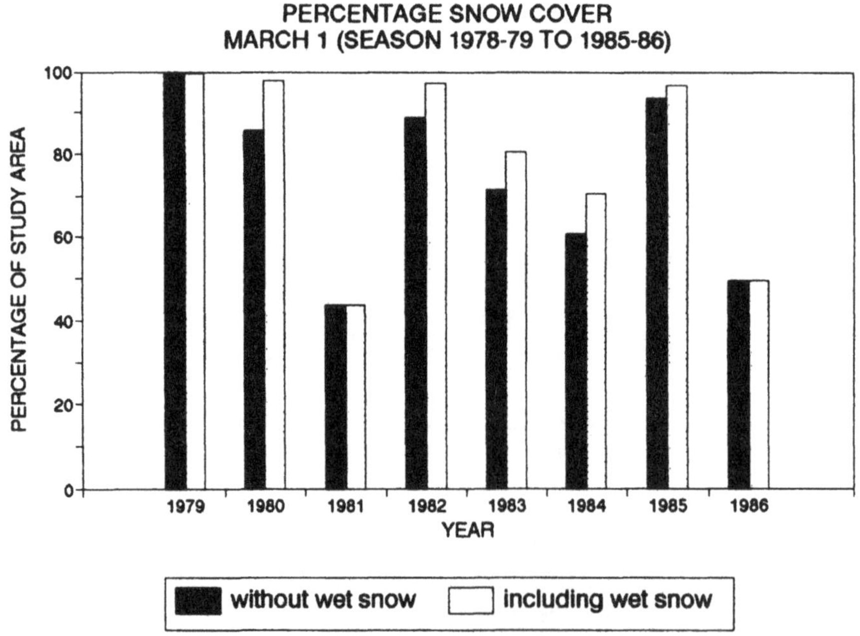

Figure 13. Snow cover as a percentage of total area of the Canadian Prairie region for the first week of March, 1979 to 1986, with and without consideration of areas of wet snow.

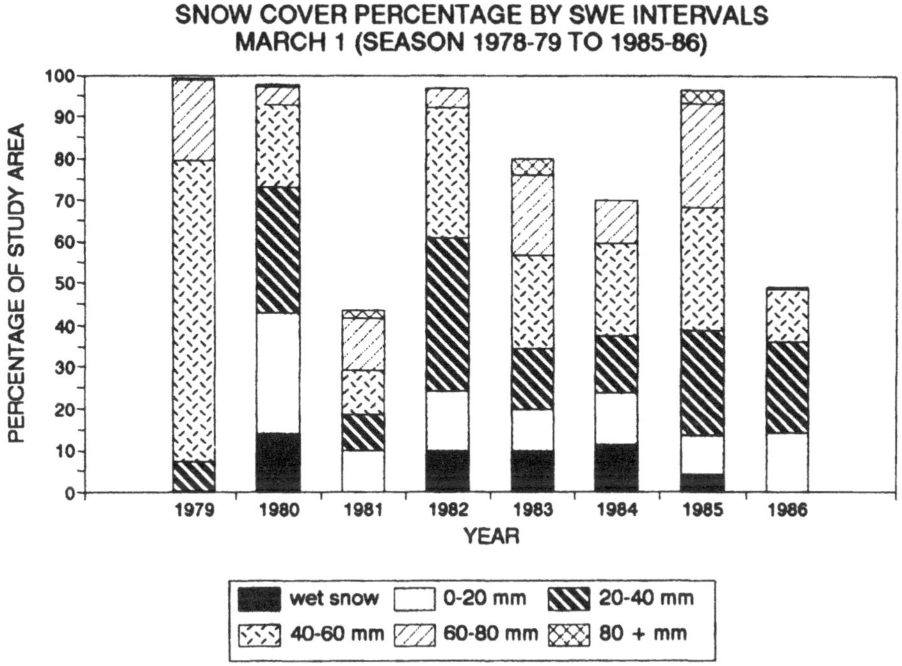

Figure 14. Snow cover as a percentage of total snow covered area for 20 mm intervals of SWE, including the area of wet snow, for the first week of March, 1979 to 1986.

wet snow the estimate of total snow covered area could be seriously underestimated, even in mid-winter when one might normally assume that there is little or no melt occurring. For this period no conclusion on the variability in snow covered area should be made, since the SMMR record alone is too short to provide a meaningful analysis.

A substantial contribution of passive microwave data is the ability to map snow water equivalent and monitor the change from year to year. Figure 14 compares the area covered by different amounts of snow over the prairie region, including the area covered by wet snow for which no specific SWE value is determined. The satellite record is the only data source which allows a time series of SWE to be prepared on a weekly basis (daily is possible for the SSM/I record). As the SWE algorithms are further validated and improved, such time series will become feasible for many other regions of the world, especially data-sparse areas, and a more accurate assessment of variability and change will be possible.

6. CONCLUSIONS AND RECOMMENDED FUTURE THRUSTS

The research of the Canadian Climate Centre has focussed on the development, validation and implementation of snow cover products derived from satellite passive microwave data for both near real-time use in hydrology and numerical modelling and the assessment of climate variability. Passive microwave derived snow cover information offers a potentially invaluable data source for such studies. However, for the snow cover information derived from passive microwave data to be used to its fullest, the following thrusts have been identified and are part of the CCC research plan:

- continued validation of the current airborne algorithms for open prairie areas through operational application by users;
- development of regional SWE and snow cover algorithms which reflect varying land cover and snowpack conditions;
- assess the transferability of regional algorithms;
- develop improved algorithms for shallow snow and early autumn periods when the ground is 'warm' or unfrozen;
- develop and refine methods for validation of algorithms where little or no conventional data exist;
- creation of a time series of SWE (and other snow cover parameters) based on conventional (pre-SMMR) data, remotely sensed information combined with conventional data where possible, and only remotely sensed information, such as in data-sparse areas;
- continue to foster the technology transfer of this satellite derived information to hydrological, climatological and agricultural users, including the GCM and numerical modelling groups.

To achieve these goals there are three additional requirements which we urge the international community to support:

- development of an accurate gridded set of brightness temperatures to provide a common data base for global analysis, such as the EASE-GRID proposed by the National Snow and Ice Data Center (NSIDC);
- development of a compatible SMMR/SSM/I time series of brightness temperatures for the globe (or at least the Northern Hemisphere), corrected for sensor biases and geo-location errors;
- higher resolution passive microwave sensors, including consideration of individual smallsat missions.

The Canadian Climate Centre initiatives are a major component of EOS/CRYSYS (Use of a Cryospheric System to Monitor Global Change in Canada), an accepted multidisciplinary project within the NASA Earth Observing System (EOS) Program. The main objective of CRYSYS is the development of methods to extract information on cryospheric parameters from satellite measurements which can be used as the basis of improving our knowledge of interactions between the cryosphere, the atmosphere and the oceans and the prediction of impacts arising from global change. The research summarized above provides a strong foundation for achieving the goals of the snow component within CRYSYS and those of the Canadian Climate Centre with respect to climate variability and change detection. The availability of data from the EOS Multifrequency Imaging Microwave Radiometer (MIMR) to complement that from SMMR and SSM/I will be a major advance. To have operational algorithms capable of use with MIMR data upon launch is an objective of the CCC research effort.

Acknowledgements
The authors would like to acknowledge the continuing support of our co-operators, particularly Mr David Graham (Alberta Environment), Mr Alf Warkentin (Manitoba Natural Resources), and Mr Brian Latham (Indian and Northern Affairs Canada) who continue to provide valuable input to our algorithm research. Ph.D. Associates Inc. have played an instrumental role in implementing the real-time data acquisition and product delivery to our cooperators, and in data processing for the support of algorithm development and validation. Special thanks are extended to Michael Davey (Canadian Climate Centre) for his help in data manipulation and analysis.

REFERENCES

1. B. E. Goodison, J. Glynn, D. Harvey and J. Slater. Snow surveying in Canada: a perspective. *Canadian Water Resources J.*, **12**, 27–42 (1987).
2. J. L. Foster, A. Rango, D. K. Hall, A. T. C. Chang, L. J. Allison and B. C. Diesen III. Snowpack monitoring in North America and Eurasia using passive microwave satellite data. *Remote Sens. Environ.*, **10**, 285–298 (1980).
3. W. H. Stiles and F. T. Ulaby. The active and passive microwave response to snow parameters. *J. Geophys. Res.*, **85** (C2), 1037–1044 (1980).
4. A. Rango, A. T. C. Chang and J. Foster. The utilization of spaceborne radiometers for monitoring of snowpack properties. *Nordic Hydrology*, **10**, 25–40 (1979).
5. B. E. Goodison, A. Banga and R. A. Halliday. Canada–United States Prairie snow cover runoff study. *Canadian Water Resources J.*, **9**, 99–107 (1984).
6. T. R. Carroll, J. E. Glynn and B. E. Goodison. A comparison of US and Canadian gamma radiation snow water equivalent measurements. In: *Proc. Western Snow Conference, 51st Annual Meeting*, pp. 27–37, Vancouver, Washington, April 19–21, 1983 (1983).
7. B. E. Goodison, I. Rubinstein, F. W. Thirkettle and E. J. Langham. Determination of snow water equivalent on the Canadian prairies using microwave radiometry. In: *Modelling Snowmelt Induced Processes (Proceedings of the Budapest Symposium, July 1986). IAHS Publ. no. 155*, pp. 163–173 (1986).
8. K. F. Kunzi, S. Patil and H. Rott. Snow cover parameters. Retrieval from NIMBUS-7 Scanning Multichannel Microwave Radiometer (SMMR) data. *IEEE Trans. Geosci. and Rem. Sens.*, **GE-20**, 452–467 (1982).
9. M. T. Hallikainen and P. A. Jolma. Comparison of algorithms for retrieval of snow water equivalent from Nimbus-7 SMMR data in Finland. *IEEE Trans. Geosci. Rem. Sens.*, **30**, 124–131 (1992).
10. A. T. C. Chang, J. L. Foster, D. K. Hall, A. Rango and B. K. Hartline. Snow water equivalent estimation by microwave radiometry. *Cold Regions Sci. Technol.*, **5**, 259–267 (1982).
11. B. E. Goodison, I. Rubinstein and F. W. Thirkettle. Snow cover determination on the Canadian Prairies using microwave radiometry. In: *Abstracts Volume of 20th Annual Congress*, pp. 59–60, Canadian Meteorological and Oceanographic Society, Regina, Saskatchewan, 3–6 June, 1986 (1986).
12. B. E. Goodison. Determination of areal snow water equivalent on the Canadian prairies using passive microwave satellite data. In: *Proceedings, IGARSS'89*, pp. 1243–1246, Vancouver, July 1989 (1989).
13. F. Thirkettle, A. Walker, B. Goodison and D. Graham. Canadian Prairie snow cover maps from near real-time passive microwave data: from satellite data to user information. In: *Proceedings of the 14th Canadian Symposium on Remote Sensing*, pp. 172–177, Calgary, Alberta, May 1991 (1991).
14. B. E. Goodison, A. E. Walker and F. W. Thirkettle. Determination of snow cover on the Canadian Prairies using passive microwave data. In: *Proceedings of the International Symposium on Remote Sensing and Water Resources*, pp. 127–136, Enschede, The Netherlands, August 1990 (1990).
15. A. E. Walker and B. E. Goodison. Discrimination of a wet snowcover using passive microwave satellite data. *Annals of Glaciology*, **17**, 307–311 (1993).

16. M. Hallikainen and P. Jolma. Development of algorithms to retrieve the water equivalent of snow cover from satellite microwave radiometer data. In: *Proceedings of IGARSS'86*, pp. 611–616, Zurich, Switzerland, September 1986 (1986).
17. A. E. Walker and B. E. Goodison. Monitoring snow cover in the boreal forest region of the Canadian Prairies using passive microwave data. In: *Proceedings of the 14th Canadian Symposium on Remote Sensing*, pp. 296–299, Calgary, Alberta, May 1991 (1991).
18. A. E. Walker and B. E. Goodison. Passive microwave derived snow cover for hydrological and climatological applications in northern regions In: *Proceedings, Ninth International Symposium, Northern Research Basins*, Whitehorse, Yukon Territory, August 17–21, 1992, pp. 575–588 (1993).
19. B. E. Goodison and A. E. Walker. Use of snow cover derived from satellite passive microwave data as an indicator of climate change. *Annals of Glaciology*, **17**, 137–142 (1993).

ESA/NASA International Workshop, pp. 263–272
B. J. Choudhury, Y. H. Kerr, E. G. Njoku and P. Pampaloni (Eds)

Application of SSM/I data for snow cover and climate research

RICHARD L. ARMSTRONG

World Data Center-A for Glaciology, NSIDC/CIRES C.B. 449, University of Colorado, Boulder, CO 80309 USA

Abstract— Snow cover is an important variable for climatic and hydrologic models due to its effects on surface albedo, energy, and moisture budgets. Only satellite observations can provide a comprehensive hemispheric-scale record of the short-term, as well as interannual, variation in snow cover. Passive microwave sensors provide an excellent method to monitor temporal and spatial variations in large-scale snow cover parameters, avoiding problems of cloud cover and polar night.

The National Snow and Ice Data Center (NSIDC) at the University of Colorado is developing a data system to produce, archive, and distribute validated hemispheric-scale snow cover products derived from the DMSP Special Sensor Microwave Imager (SSM/I). Currently work focusses on providing the global change researcher with an SSM/I data structure and format which is easier to use than swath format, yet maximizes the radiometric integrity of the original temperature values, provides high spatial and temporal precision, and involves no averaging of the original swath data. In October of 1992, NSIDC distributed a prototype Version 0.0 of the Equal Area SSM/I Earth Grid (EASE-Grid) on CD-ROM for initial evaluation. The main intent of the Version 0.0 prototype is to solicit input on the interpolation, data storage format, earth grid scheme and the projections of the proposed EASE-Grids. Feedback from the user community will determine the eventual structure and format to be used by NSIDC for the distribution of gridded SSM/I data.

1. BACKGROUND

The first American space-based passive microwave sensors were NASA instruments, including the Electronically Scanning Microwave Radiometer (ESMR), flown aboard NIMBUS-5 and -6, and the Scanning Multichannel Microwave Radiometer (SMMR) on NIMBUS-7. These two instruments operated from 1972 to 1987. Based on this experience, the Department of Defense procured several Special Sensor Microwave Imager (SSM/I) instruments to enhance its capability to determine terrestrial, oceanic, and atmospheric conditions. The SSM/I instruments were built by Hughes Aircraft Company under the direction of the Air Force Space Division (USAF SD) and the Naval Space Systems Activity (NSSA) and represent a joint Air Force/Navy operational program to obtain synoptic maps of critical atmospheric, oceanographic, and selected land parameters on a global scale. The first SSM/I to become operational was launched aboard Defense Meteorological Satellite Program (DMSP) F8 in June 1987. This was the first of seven SSM/I's scheduled for launch over the next two decades. More recently, an SSM/I was carried on DMSP F10, launched in December 1990 and F11 was launched in November 1991.

The SSM/I operates at four frequencies (19.4, 22.2, 37.0, and 85.5 GHz) with orthogonal (horizontal and vertical) polarizations measured at each frequency except 22 GHz, which has only a vertical polarization channel. Its spatial resolution is frequency-dependent and ranges from about 50 km at 19 GHz to better than 15 km at 85 GHz. The radiometric characteristics of the SSM/I channels were initially selected to achieve specific objectives, principally in the areas of ocean and atmospheric research. For example, a channel with a center frequency of 22 GHz was included in order to obtain estimates of atmospheric water vapor. The 19 and 37 GHZ channels were used to achieve the maximum transfer of scientific experience and algorithms from the ESMR and SMMR instruments. The 85 GHz channels were chosen in order to more effectively measure rain and cloud properties. The focus on efforts to derive land surface products from SMMR and SSM/I has been a more recent development.

The SSM/I is carried on a spacecraft which is in a circular sun-synchronous, near-polar orbit at an altitude of 833 km with an inclination of 98.8° and an orbit period of 102.0 minutes resulting in 14.1 full orbit revolutions per day. With a swath width of almost 1400 km, the SSM/I provides near global coverage every day. Each day there are small diamond-shaped areas near the equator which are not 'seen' by the sensor on that particular day but which will be seen in the next 2 to 4 days. Also, surrounding both the north and south poles are small circular sectors of 2.4° latitude that are never seen, since the plane of the SSM/I orbit is tilted with respect to the Earth's axis of rotation. A detailed description of the sensor and its radiometric performance can be found in Hollinger *et al.* [1]. The details of sensor calibration and validation as well as algorithm analysis can be found in a later report coordinated by Hollinger [2].

In recent years it has been recognized that microwave techniques permit unique observations of land surface parameters critical to the study of the Earth's hydrologic cycle, such as soil moisture, vegetation, precipitation, and snow. Increasing amounts of global, multiyear data have been, and will be, acquired from spaceborne radiometers such as the SMMR on the Nimbus-7 satellite and the SSM/I on the DMSP satellites, and there is now an awareness of the need for quantitative estimation of land surface parameters over large spatial and temporal scales for environmental monitoring and as boundary conditions for validation of regional and global climate models [3, 4]. These developments have led to increased interest in the potential of microwave radiometry for global observations of land surfaces.

2. SNOW COVER PRODUCTS DERIVED FROM PASSIVE MICROWAVE DATA

Snow cover is an important variable for climate and hydrologic models due to its effects on energy and moisture budgets. A primary factor controlling the amount of solar radiation absorbed at the earth surface is the extent of snow cover due to its high albedo. Any decrease in snow cover resulting from a warming trend results in increased absorption of solar radiation and additional heat to melt additional snow. This results in the classic positive temperature-albedo feedback mechanism which is included in nearly all climate models. In addition to the albedo effect, snow cover represents a significant heat sink during the warming period of the seasonal cycle due to a relatively high latent heat of fusion. As a result, the seasonal snow cover provides a major source of thermal inertia within the total climate system as it takes in and releases large quantities of energy with little or no fluctuation in temperature. Therefore, snow cover is an important variable for global climate monitoring and change detection.

During the past two decades much important information on hemispheric-scale snow extent has been provided by satellite remote sensing in the visible wavelengths [5, 6]. One major problem with the application of visible wavelength data is the need for clear sky conditions. Persistent cloud cover in winter may, in some instances, limit data to only a few days per month. In addition, visible-band data do not provide the opportunity to extract snow water equivalent or depth information.

Passive microwave remote sensing offers several advantages over visible-band data. It is currently the best method to monitor temporal and spatial variations in snow cover on hemispheric to global scales. Because measurements are not in the visible portion of the electromagnetic spectrum and because of the ability to penetrate clouds (19 and 37 GHz channels) passive microwave remote sensing allows data collection during darkness and in nearly all weather conditions. In addition, passive microwave provides the capability to compute snow water equivalent and to detect melt [7–9]. These additional variables comprise important input to energy budget, hydrologic, and global circulation models and enhance the capability to evaluate climate change [10].

Microwave energy emitted from the surface of the earth is currently measured by the Defense Meteorological Satellite Program (DMSP) Special Sensor Microwave Imager (SSM/I) [11]. Data are global and frequent, at least daily for locations above 43° latitude, and once every two to three days for those areas where snow might be expected at latitudes below 43°. When snow covers the ground, some of the microwave energy emitted by the underlying soil is scattered by the snow grains. When moving from snow-free to snow-covered land surfaces, a sharp decrease in emissivity provides a nearly unambiguous indicator of the presence of snow. Empirical studies have demonstrated that the amount of scattering is well correlated with snow amount (number of grains) and specific wavelength. Based on this relationship, algorithms have been developed which indicate the presence of snow and compute snow water equivalent, or snow depth given an assumed density [12–22].

There are also disadvantages to the passive microwave data. The resolution is coarse when compared to that typically available from visible-band sensors. Resolution depends on the channel frequency and is approximately 30 km for those channels used in most snow cover algorithms. In addition, the application of passive microwave is currently limited to dry snow. Once liquid water is present on the ice grains, the snow surface becomes a strong emitter and thus may be indistinguishable from bare ground. This dramatic signal change serves as a clear indicator of the onset of melt, but as long as the snow is wet, water equivalent cannot be easily determined using SSM/I data alone. However, recent work [23] describes a method to discriminate areas of wet snow from areas of snow-free land. If the snow refreezes, it again become 'visible' to the microwave sensor.

Another problem results from the fact there may be objects which are primarily emitters which extend above the snow cover, such as dense coniferous forests. The increased emission from such surface features will tend to reduce the total amount of scattering integrated across the pixel and thus may indicate a lesser amount of snow cover than is actually present [24–26].

Finally, there are complications which result from the snow structure itself. Microwave scattering not only increases with increasing number of grains, but also with increasing grain size [27]. Thus, without regard for grain size variation, any increased scattering would imply an increase in snow amount. For example, according to the initial results presented by Chang and others [12] a change in mean grain diameter of 0.4 mm (e.g. from 0.6 mm to 1.0 mm) would produce a brightness temperature (the product of emissivity and physical temperature) change of more than 50 K. As a result of this 0.4 mm mean diameter change, estimated snow water equivalent could vary by a factor of 5 (0.1–0.5 m), given typical algorithm sensitivity. Such a range of water equivalent values covers a large portion of the total sensitivity scale of typical snow cover algorithms. Thus initial theory [12] suggested that without detailed stratigraphic mean grain size data, accurate snow water equivalent estimates from microwave remote sensing would be difficult, if not impossible, to achieve. In reality, such detailed snow structure information for areas the size of a SSM/I pixel would simply be impossible to obtain. However, algorithms are typically applied without consideration for the variable grain sizes found in the natural snow cover, and some success has been reported by each investigator [7, 13, 16, 18, 19, 21].

Thus, in spite of the strong dependency on grain size, results have been reasonable under a wider range of conditions than might have been expected considering theory alone. The exact reason for this is not currently well understood. In addition, the basic assumptions in the scattering theory applied to modeling the natural snow cover are being re-evaluated. In a dense medium with an appreciable fractional volume of scatterers, the assumption of independent scatters used in conventional radiative transfer theory is not valid. This has been verified by controlled laboratory experiments [28] and has been studied theoretically [29, 30]. Recently a dense medium radiative transfer theory which accounts for correlated scattering has been developed [31]. Dense medium theory predicts a less steep drop in brightness temperature with increasing grain size as compared with the conventional radiative transfer calculations.

Although not as sensitive as theory might indicate, testing has shown that algorithms which assume a single grain size will eventually produce unreasonable values when the mean grain size of the actual snow cover exceeds about 2.0 mm in diameter [32, 33]. As long as the snow cover remains sub-freezing, only one process, temperature-gradient or kinetic metamorphism, will cause the mean grain size of a layer to exceed a diameter of 2.0 mm [34, 35]. A method which will identify conditions where mean grain sizes larger than 2.0 mm are expected to prevail is under development [36]. This method is based on relationships between average snow depth for a given region and current depth and temperature data. Given this information, a technique to adjust the microwave algorithms for grain size can be developed.

3. SNOW ALGORITHM VALIDATION

The National Snow and Ice Data Center (NSIDC) is developing a prototype global snow depth model which provides daily snow extent and depth at a 25 km grid resolution [37]. Time-series data generated by the model will provide important input to the study of global change with respect to surface energy budgets and the hydrologic cycle as well as provide validation data sets for remote sensing algorithms. Specific data generated for each grid point include calculated average and maximum snow depth, age of the total snow cover and time since last snowfall, along with appropriate data source and confidence flags. The model represents the integration of both surface and satellite

(SSM/I) observations. Surface measurements used in current model development are from the WMO/GTS synoptic data collection network.

In geographic locations where measurements from surface reporting stations are available with a sufficiently high density (at least one data point within 100 km of a given grid point) these data represent the primary model input and also support algorithm validation. A fixed data base consists of latitude, longitude, and elevation for each grid point as well as a list of the 10 closest reporting stations. For each model run, the reports from the 5 closest surface stations are obtained. Each grid point value is meant to reflect the best estimate of the average snow depth within the area immediately surrounding that grid point. In order to achieve this, an inverse-distance weighting scheme is applied to populate a 25 km global grid. In regions where there is an average of less than one reporting station within 100 km of a grid point, the model applies data from SSM/I snow cover algorithms.

Preliminary testing at NSIDC has demonstrated the potential to map snow extent with an accuracy of about one to two pixels on the hemispheric scale. However, algorithms developed using SMMR data may not be directly applied to SSM/I data due to instrument calibration differences. Such application frequently causes algorithms to indicate the presence of shallow snow where no snow is present. As a result of this lack of compatibility, the algorithms being tested are separated according to whether they were developed for SMMR or SSM/I.

At this stage in our study it appears that no single algorithm is appropriate for hemispheric-scale retrieval of water equivalent. This is partly due to the fact that nearly all algorithms developed for the retrieval of snow cover information have been derived and validated using snow cover data from local to regional scale studies. For global scale studies, algorithms must be modified and adjusted in order to minimize the influence of other physical properties within the target area.

Difficulties arise in attempts to validate passive microwave snow cover algorithm output using point surface measurements due to the large pixel size, approximately 1000 km^2, for the SSM/I channels used in the algorithms. In order to better match the data from single measurement points to the large area integration represented by the SSM/I data, a surface measurement point to grid cell interpolation technique has been developed based on distance weighting criteria. The interpolation scheme is supported by a digital elevation data base (5-arc minute resolution, NOAA ETOPO5) in order that the influence of a measurement site on given grid cell will decay with changes in elevation as well as horizontal distance.

In addition to the WMO data set currently in use, NSIDC is in the process of archiving selected snow cover data sets from the former USSR which are considered important to the study of global climate change and which will also serve as validation data for remote sensing. The extensive snow cover data sets represent a range of spatial resolutions from relatively high density measurements over a small area (e.g. the Valdai Research Station north of Moscow) to the 3000 to 5000 stations of the Hydrometeorological Service which are spread throughout the former USSR. These data are currently available in hard copy form only and therefore are being digitized both at NSIDC and at the Geographical Institute in Moscow. The transfer of these data to NSIDC, as well as the necessary digitization and formatting, will be an ongoing process over at least the next year. The first data sets to be digitized will be the post-1987 period to coincide with the SSM/I.

One specific Soviet data set of interest to remote sensing comprises measurements of snow depth, density, and snow structure at approximately 200 m intervals along transects that are typically 1.0 to 2.0 km in length. These data represent uniquely valuable information for the validation of passive microwave snow cover retrieval algorithms because the transects represent a resolution between conventional point measurements and the large area integration of the passive microwave footprint.

4. GLOBAL SSM/I DATA SET

NSIDC is developing a data system to produce, archive, and distribute validated hemispheric-scale snow cover products derived from the DMSP Special Sensor Microwave Imager (SSM/I). In addition, NSIDC anticipates the opportunity to distribute snow products derived from a full global SMMR time series in a format compatible with SSM/I as a result of work underway at the Jet Propulsion Laboratory to recalibrate and produce a new Level 1b (brightness temperatures) SMMR data set [38].

In order to proceed with the development of the SSM/I and SMMR data systems, it was first necessary to provide a data and image format which would allow standardized comparisons of geophysical products derived from SSM/I data. Therefore, work at NSIDC currently focusses on providing the general user of global change data with a data structure (interpolation from swath coordinates to an earth grid, projections, and actual earth grid scheme) which is easier to use than swath format, yet maximizes the radiometric integrity of the original temperature values, provides high spatial and temporal precision, and involves no averaging of the original swath data.

Since 1989 NSIDC has distributed DMSP SSM/I brightness temperature grids for the polar regions on CD-ROM [11]. This is an example of a time-averaged data set. The technique used in the production of this data set reduces temporal precision by averaging all satellite passes over a given grid cell for the 24 hour period and original radiometric values are degraded by the use of the 'drop in the bucket' method of interpolation (the earth grid cell which contains the center of the satellite observation footprint is given the whole weight of that particular observation). In contrast, the basic purpose of the Equal Area SSM/I Earth Grid (EASE-Grid) is to provide a format which is between swath data (one file per orbit) and an averaged (time and space) daily or multi-day product with its inherent reduction in precision. In October of 1992, NSIDC distributed the prototype version of the EASE-Grid to the user community for initial evaluation. The response to the proposed format of the EASE-Grid, described below, has generally been positive with no major problems in the area of global scale studies.

Based on the recommendations of the SSM/I Products Working Team (SPWT), the point of departure for the EASE-Grid prototype format was the methodology of Galantowicz and England [39] which is based on earlier work by Stogryn [40] and by Poe [41]. The SPWT, is a multi-agency and multi-disciplinary working group formed through guidance from NASA Headquarters in 1989 which focusses on the problems associated with extracting land surface (primarily vegetation, soil, and snow cover) information from SSM/I. SPWT activities are coordinated by NSIDC.

The method used in the EASE-Grid to resample SSM/I swath data to points on the earth grid retains the spatial characteristics of the original data. The process involves interpolation to increase (16 times) the density of antenna temperature measurements in the satellite swath reference frame (sample interval of 25.0 km for 19, 22, and 37 GHz and 12.5 km for 85 GHz). This process uses actual antenna pattern to create the oversampled array and the net effect is as if the additional samples had been made by the satellite radiometer itself — i.e. the beam patterns and spatial resolutions of the interpolated data approximate those of the original samples [39]. The brightness temperature for a given earth grid cell is obtained from the oversampled array by the nearest neighbor method, which is considered appropriate when moving from higher to lower resolution (grid density).

The EASE-Grid data sets will provide daily brightness temperatures on CD-ROM for all channels, separated into ascending and descending files, including the time (UTC) for each sample location. At latitudes of less than 55° all data are binned (0–2 satellite passes per day). At latitudes greater than 55° where more than two passes per day are possible, data selected for the grid are from that pass which is closest to the local time equivalent of the equator crossing time. The current version of EASE-Grid provides all channels gridded to 25 km, as well as the 85 GHz channel at 12.5 km.

The EASE-Grid uses a conventional (i, j) array format for the sample data. The capability to convert between the array coordinates (i, j) and the latitude and longitude of a given earth grid cell is included on the CD-ROM. Therefore, no special software tools are required to view the data as the (i, j) format is compatible with all standard image processing packages. The integers of the earth grid coordinates are the centers of the cell. Cell center points are coincident (beam centered). The grid organization is row/column and the data are band sequential (i.e. not interleaved) with one file (array) for each channel, ascending and descending, for each projection, and a time file. Data precision is 0.1 K for brightness temperatures, 0.1 hours for time, and 0.01 degrees for latitude/longitude.

The EASE-Grid was also created to support direct visual and digital comparison of data sets from other sensor systems. Although the specific interpolation scheme used to resample from satellite sampling intervals to earth-gridded cells will be unique to each sensor, the fundamental projection and gridding code of the EASE-Grid provides the basis for a standard inter-use method.

5. CONCLUSIONS

Snow cover extent represents a primary control over energy exchange at the Earth's surface and a major component of the surface hydrologic regime. At the annual maximum, snow may cover more then 33 percent of the Earth's surface. Long-term records of snow cover variability are essential for detecting the greenhouse signal, quantifying the sensitivity of the climate system to internal feedbacks, and adapting to expected changes. Conventional surface measurements are effective for assessing local temporal variability, but are of limited use for spatial variability. During the past two decades much important information on hemispheric-scale snow extent has been provided by satellite remote sensing in the visible wavelengths. Problems include lack of data during hours of darkness and the need for clear sky conditions. Passive microwave remote sensing offers several advantages over visible-band data. It is currently the best method to monitor temporal and spatial variations in snow cover on hemispheric to global scales. Passive microwave remote sensing allows data collection in nearly all weather conditions and during darkness, and also provides the potential to compute snow water equivalent and to detect melt.

To support the use of SSM/I data in snow cover research, as well as other environmental studies, NSIDC has developed an optimal swath to earth-grid binning procedure

(EASE-Grid) and has created an archiving and distribution system for SSM/I data and products. Irregularly distributed surface data have been interpolated to the EASE-Grid format, weighted according to distance and elevation. Preliminary results from comparison of algorithm output with gridded surface data indicates that most algorithms provide snow covered area to an accuracy of about one to two grid cells (25 to 50 km), as long as the snow is subfreezing. However, with respect to snow depth and water equivalent, accuracy can be highly variable. On a regional scale accuracy may be high, on the order of 5–10 mm water equivalent in some cases. However, when the same algorithm is applied on a global scale, errors can be significant, even within areas of similar terrain. In locations where surface measurements represent a spatial density comparable to that of the SSM/I grid resolution, direct quantitative comparisons can be made between the surface measurements and the algorithm output.

In the future, NSIDC will expand the surface station data set to support algorithm validation by including regions of higher density measurements in the United States and Canada. The image analysis tools which have been developed will be applied to longer time-series analysis. We will determine whether or not the difference between the algorithm output and the validation data are systematic, and if so, can the regions showing consistent errors be correlated with specific conditions identifiable using ancillary data (e.g. topography, vegetation, snow structure, temperature)? Using data from the SSM/I F11 satellite we will evaluate algorithms requiring the 85 GHz channels. It is expected that the increased scattering produced at this frequency will enhance the capability to identify areas of shallow snow cover. We will also analyze the feasibility of applying polarization difference algorithms in order to better discriminate between wet and dry snow. NSIDC will continue to develop the concept of a global snow cover data set based on passive microwave remote sensing, but which is produced through continuous interaction with a dynamic surface changes data base which would include station snow depth and water equivalent measurements.

REFERENCES

1. J. Hollinger, R. Lo, G. Poe, R. Savage and J. Pierce. *Special Sensor Microwave/Imager User's Guide.* El Segundo, CA, Space and Communications Group, Hughes Aircraft Company, p. 120 (1987).
2. J. Hollinger. *DMSP Special Sensor Microwave/Imager Calibration/Validation.* Naval Research Laboratory, Washington, DC, Cal/Val Team Final Report, Vols I and II (1991).
3. F. Becker, H. J. Bolle and P. R. Rowntree. *The International Satellite Land Surface Climatology Project (ISLSCP).* ISLSCP Report #10, ISLSCP Secretariat, Berlin, FRG (1988).
4. GEWEX. *Scientific Plan for the Global Energy and Water Cycle Experiment.* World Meteorological Organization, Geneva, Switzerland, Report #WCRP-40 (WMO/TD #376) (1990).
5. D. A. Robinson, F. T. Keimig and K. F. Dewey. Recent variations in northern hemisphere snowcover. In: *Proceedings of the 15th Annual Climate Diagnostics Workshop*, pp. 219–224 (1991).
6. K. F. Dewey. Satellite-derived maps of snow cover frequency for the northern hemisphere. *J. Climate Appl. Meteor.*, **26** (9), 1210–1229 (1987).
7. K. F. Kunzi, S. Patil and H. Rott. Snow-cover parameters retrieved from Nimbus-7 Scanning Multichannel Microwave Radiometer (SMMR) data. *IEEE Trans. Geosci. and Remote Sensing*, **GE-20**, (4), 452–467 (1982).
8. J. L. Foster, D. K. Hall, A. T. C. Chang and A. Rango. An overview of passive microwave snow research and results. *Reviews of Geophysics and Space Physics*, **22**, 195–208 (1984).
9. H. Rott. *Remote Sensing of Snow.* International Association of Hydrological Sciences, Publication No. 166, pp. 279–290 (1987).

10. B. E. Goodison and A. E. Walker. Use of snow cover derived from satellite passive microwave data as an indicator of climate changes. *Annals Glaciology*, **17**, 137–142 (1993).

11. R. Weaver, C. Morris and R. G. Barry. Passive microwave data for snow and ice research: Planned products from the DMSP SSM/I system. *EOS*, **68** (39), 769–777 (1987).

12. A. T. C. Chang, J. L. Foster, D. K. Hall, A. Rango and B. K. Hartline. Snow water equivalent determination by microwave radiometry. In: *Goddard Space Flight Center, NASA Technical Memorandum 82074*, p. 18 (1981).

13. A. T. C. Chang, J. L. Foster, P. Gloersen, W. J. Campbell, E. G. Josberger, A. Rango and Z. F. Danes. Estimating snowpack parameters in the Colorado River basin. In: *Large-Scale Effects of Seasonal Snow Cover*, B. E. Goodison *et al.* (Eds), pp. 343–352, IAHS Publication 166 (1987).

14. A. T. C. Chang, J. L. Foster and D. K. Hall. Nimbus-7 SMMR derived global snow cover parameters. *Annals of Glaciology*, **9**, 39–44 (1987).

15. K. F. Kunzi, S. Patil and H. Rott. Snow-cover parameters retrieved from Nimbus-7 Scanning Multichannel Microwave Radiometer (SMMR) data. *IEEE Trans. Geosci. and Remote Sensing*, **GE-20** (4), 452–467 (1982).

16. M. T. Hallikainen and P. A. Jolma. Retrieval of the water equivalent of snow cover in Finland by satellite microwave radiometry. *IEEE Trans. Geosci. Remote Sensing*, **GE-24** (6), 855–862 (1986).

17. M. J. McFarland, G. D. Wilke and P. H. Harder. Nimbus-7 SMMR investigation of snowpack properties in the Northern Great Plains for the winter of 1978–1979. *IEEE Trans. Geosci. Remote Sensing*, **GE-25** (1), 39–46 (1987).

18. B. E. Goodison. Determination of areal snow water equivalent on the Canadian prairies using passive microwave satellite data. *IGARSS '89 Proceedings*, **3**, 1243–1246 (1989).

19. J. Aschbacher. *Land Surface Studies and Atmospheric Effects by Satellite Microwave Radiometry*. (Unpublished PhD dissertation), p. 202, University of Innsbruck, Austria (1989).

20. E. G. Josberger, C. Ling, W. J. Campbell, P. Gloersen, A. T. C. Chang and A. Rango. Correlations of Scanning Multichannel Microwave Radiometer (SMMR) observations with snowpack properties of the Upper Colorado River Basin for Water Year 1986. *IGARSS'89 Proceedings*, **3**, 1239–1242 (1989).

21. H. Rott, T. Nagler and J. Aschbacher. Algorithm development for monitoring global snow cover by spaceborne microwave radiometry. In: *Proceedings of the 11th EARSeL Symposium*, Graz, Austria, July 1991 (1991).

22. N. C. Grody. Classification of snow cover and precipitation using the special sensor microwave imager. *J. Geophysi. Res.*, **96** (D4), 7423–7435 (1991).

23. A. E. Walker and B. E. Goodison. Discrimination of a wet snow cover using passive microwave satellite data. *Annals of Glaciology*, **17**, 307–311 (1993).

24. J. L. Foster, A. T. C. Chang, D. K. Hall and A. Rango. Derivation of snow water equivalent in boreal forests using microwave radiometry. *Arctic*, **44** (1), 147–152 (1991).

25. A. T. C. Chang, J. L. Foster and D. K. Hall. Effect of vegetation cover on microwave snow water equivalent estimates. In: *Proceedings of International Symposium of Remote Sensing and Water Resources*, pp. 137–146, Enschede, The Netherlands (1990).

26. M. T. Hallikainen, P. A. Jolma and J. M. Hyppa. Satellite microwave radiometry of forest and surface types in Finland. *IEEE Transactions on Geoscience and Remote Sensing*, **26**, 622–628 (1988).

27. A. T. C. Chang, P. Gloersen, T. Schmugge, T. T. Wilheit and H. J. Zwally. Microwave emission from snow and glacier ice. *Journal of Glaciology*, **16**, 23–29 (1976).

28. A. Ishimaru and Y. Fuga. Attenuation constant of a coherent field in a dense distribution of particles. *Journal of Optical Society of America*, **72**, 1317–1320 (1982).

29. L. Tsang, J. A. Kong and R. T. Shin. *Theory of Microwave Remote Sensing*. Wiley-Interscience, New York (1987).

30. B. Wen, L. Tsang, D. P. Winebrenner and A. Ishimaru. Dense medium radiative transfer theory: comparison with experiment and application to microwave remote sensing and polarimetry. *IEEE Trans. Geosci. and Remote Sensing*, **28**, 46–59 (1990).

31. L. Tsang. Passive remote sensing of dense non-tenuous media. *J. of Electr. Waves and Appl.*, **1**, 159–173 (1987).

32. R. L. Armstrong. *Monitoring Snow Grain Size for Passive Microwave Studies*. EOS Transactions, American Geophysical Union **71**, No. 43, p. 1333 (1990).

33. D. K. Hall, A. T. C. Chang and J. L. Foster. Detection of the depth-hoar layer in the snow-pack of the arctic coastal plain of Alaska, USA, using satellite data. *Journal of Glaciology*, **32** (110), 87–94 (1986).

34. R. L. Armstrong. *Metamorphism in a Subfreezing Seasonal Snow Cover*. (PhD thesis), p. 174. University of Colorado, Boulder (1985).

35. S. C. Colbeck. The layered character of snow covers. *Reviews of Geophysics*, **29** (1), 81–96 (1991).

36. R. L. Armstrong, A. Rango, A. T. C. Chang and E. G. Josberger. Snow depths and grain size relationships with relevance for passive microwave studies. *Annals of Glaciology*, **17**, 171–176 (1993).

37. R. L. Armstrong and M. Hardman. Monitoring global snow cover. In: *Proceedings of IGARSS'91*, Helsinki, Finland, Vol. IV, pp. 1947–1949 (1991).

38. E. Njoku. Personal communication. (1992).

39. J. F. Galantowicz and A. W. England. *The Michigan Earth Grid: Description, Registration Method for SSM/I Data and Derivative Map Projections*. Radiation Laboratory, Department of Electrical Engineering and Computer Science, University of Michigan, Ann Arbor, 48109-2122, Technical Report 027396-2-T, p. 19 (1991).

40. A. Stogryn. Estimates of brightness temperatures from scanning radiometer data. *IEEE Trans. Antennas Propagation*, **AP-26**, 720–726 (1978).

41. G. A. Poe. Optimum interpolation of imaging microwave radiometer data. *IEEE Trans., Geosci. Remote Sensing*, **GE-28**, 800–810 (1990).

ESA/NASA International Workshop, pp. 273–283
B. J. Choudhury, Y. H. Kerr, E. G. Njoku and P. Pampaloni (Eds)

Size parameters of snow grains in scattering and emission models

YA-QIU JIN

Department of Electronic Engineering, Fudan University, Shanghai, China 200433

Abstract—By using a computer method of morphologic mapping to the sectional image of snow sample, a size distribution of spherical ice crystals is obtained. Following the calculation of the effective propagation constant of size distributed, dense scatterers, the effective radius of snow grains in scattering model is derived. By solving the vector radiative transfer equations for multi-layer random media embedded by different dense scatterers and coupling boundary conditions in discrete ordinate and eigen-analysis method, the polarized brightness temperatures of dry and wet snow are numerically calculated and are compared with experimental data. Numerical simulation and the functional dependence on various parameters are discussed.

1. INTRODUCTION

In microwave remote sensing, snow is usually modelled as a layer of strongly fluctuating, continuous random medium characterized by a correlation function, or a layer of random, densely distributed spherical scatterers. The correlation length, which corresponds to the averaged sectional size of grains, or the radius of one-sized spherical scatterers have to be properly chosen to enable the theoretical results to match the data. It is known that scattering coherence from particles must be taken into account due to the appreciable fractional volume of snow grains. Since the albedo is governed by the size and dielectric properties of snow grains, the numerical results of radiative transfer calculation are very sensitive to the correlation length or the size of scatterers. At low frequency, scattering is proportional to particle size to the sixth power so that the larger particles dominate the scattering although they are fewer in number. If one goes with the mean radius of ice grain or one measures the correlation length in the random medium sense, one will only obtain a size that is close to the average size and severely underestimate the scattering. At high frequencies, Mie scattering effects are important and size distribution can give rise to a variety of the frequency dependence. Therefore, to obtain the size distribution and the effective radius for scattering is very important. Snow crystals are generally nonspherical and multiple sized. To extract the proper parameters from the ground truth data is essential to understand the remote sensing data and numerical results of radiative transfer. On the other hand, in the case of wet snow, the imaginary part of dielectric constant of liquid-coating grains due to melting may significantly increase the brightness temperature. The thermal emission of wet snow is significantly affected by the fractional volume of liquid water in the wet surface layer. Therefore, a comprehensive model of multi-layer random media embedded by different and dense scatterers with size distribution should be developed.

In this paper, by using the digitalized photographs of cross-sectional slices of a snow pile, a computer algorithm of morphologic mapping by various circles is developed. A size distribution of equivalent spherical particles is then obtained. Following the calculation of the effective propagation constant of size distributed, dense spherical scatterers derived by Tsang *et al.* [1] the effective one-sized radius of snow grains in scattering and emission model is obtained. This effective radius is closely related with the scattering coherence due to appreciable fractional volume of scatterers, and is significantly different from both arithmetic and volumetric means. It is also different from the mode size which is assumed to have the maximum number in the distribution. By making use of a model with a surface layer embedded by liquid-coated spheres for melting wet snow, a comprehensive model of two-layer random media embedded by different scatterers, respectively, is developed. The modified vector radiative transfer equations of dense scatterers (DMRT) for two-layer random media and four coupling boundary conditions are solved by discrete ordinate and eigen-analysis method. Numerical results of polarized brightness temperatures are obtained and compared with experimental data of dry and wet snow remote sensing. The numerical simulation and functional dependence on various parameters are quantitatively discussed.

2. SIZE DISTRIBUTION AND EFFECTIVE RADIUS

Figure 1 is a digitalized sectional image of snow sample, where the black background is air and the white part is ice crystals. we developed a computer method of morphologic mapping by scanning and filling in the white part with different spheres. Then we replaced the sectional picture of Fig. 1 by the size distributed equivalent spheres. Figure 2 shows a mapping process. Employing the modified Gamma distribution function [1],

$$n(a) = Aa^p \exp(-Ba^q), \quad \mathrm{mm}^{-3}\mathrm{mm}^{-1} \tag{1}$$

Figure 1. Digitalized image of snow sample.

Figure 2. At morphologic mapping.

we try to find the proper choice of the parameters in Equation (1) to match the size distribution as shown in Fig. 2. In Equation (1), a is the radius of spherical particles; $n(a)\,\mathrm{d}a$ gives the particle number per unit volume with radii between a and $a + \mathrm{d}a$; the parameters A and B are determined by the fractional volume of particles f_{v} and the mode radius a_{c} as follows

$$f_{\mathrm{v}} = \frac{4\pi}{3}\frac{A}{q}B^{\frac{-(p+4)}{q}}\Gamma\left(\frac{p+4}{q}\right) \tag{2a}$$

$$a_{\mathrm{c}} = \left(\frac{p}{Bq}\right)^{\frac{1}{q}} \tag{2b}$$

where Γ is the Gamma function. Since the two-dimensional section as Fig. 1 can not always cut all particles across their diameters, the number of small particles would be overestimated in this approach. By averaging the results from several samples, we may choose an appropriate mode size a_{c} which has maximum number in the distribution. Then, ignoring the overestimated number of smaller ($< a_{\mathrm{c}}$) particles and using Equation (1) to match the distribution of larger ($\geqslant a_{\mathrm{c}}$) particles, the size distribution of snow grains of Fig. 1 may be obtained as shown in Fig. 3, where solid circles are the sample data.

For an electromagnetic wave through a medium of dense spherical scatterers, the effective propagation constant has been derived by using a non-linear approximated Dyson equation and employing the Percus–Yevick pair distribution function at low frequency [1]. The effective propagation constant takes account of scattering coherence of dense scatterers. For multiple component, spherical scatterers (radius a_i, dielectric constant ε_i, fractional volume f_i, and particle number per unit volume n_i, $i = 1, \ldots, n_0$),

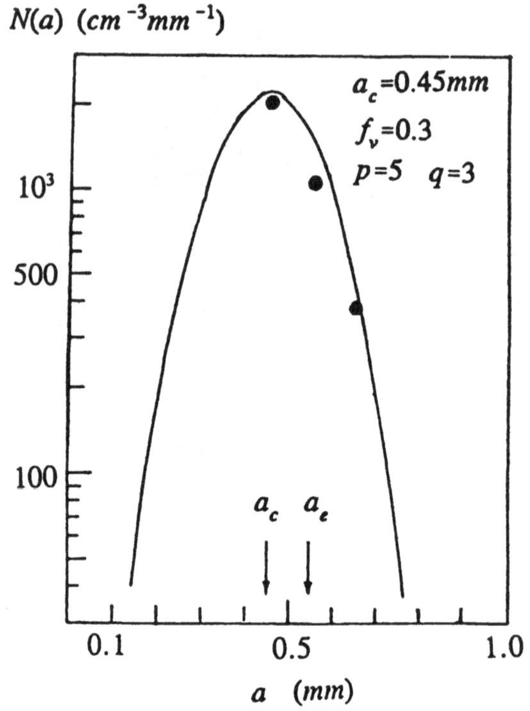

Figure 3. Size distribution of snow grains.

the effective propagation constant may be written as

$$K^2 = k_0^2 + \frac{3\,K_0^2}{D}\sum_{i=1}^{n_0} f_i y_i \left[1 + i\frac{2\,K_0^3}{3D}\left(a_i^3 y_i + \sum_{j=1}^{n_0} a_j^3 y_j \sqrt{\frac{n_j}{n_i}}\,H_{ij}\right)\right] \tag{3}$$

where

$$K_0^2 = k_0^2 + \frac{3\,K_0^2}{D}\sum_{i=1}^{n_0} f_i y_i \tag{4a}$$

$$y_i = \frac{k_{si}^2 - k_0^2}{3\,K_0^2 + \left(k_{si}^2 - k_0^2\right)}, \qquad k_{si}^2 = k_0^2\,\frac{\varepsilon_i}{\varepsilon_0} \tag{4b}$$

$$D = 1 - \sum_{i=1}^{n_0} f_i y_i \tag{4c}$$

$$H_{ij} = \sqrt{n_i n_j}\int \mathrm{d}\vec{r}\,\left[g_{ij}(\vec{r}) - 1\right] \tag{4d}$$

Here, g_{ij} is the pair distribution function and is calculated by using the Percus–Yevick distribution function; H_{ij} may be numerically calculated by matrices computation [1].

If all particles are supposed to have the same radius and dielectric constant, Equation (3) may be written as

$$K^2 = k_0^2 + \frac{f_v z_0}{1 + (1 - f_v) z_0 / (3 K_0^2)} \times \left[1 + i \frac{2}{9} K_0 \frac{z_0}{1 + (1 - f_v) z_0 / (3 K_0^2)} a^3 \frac{(1 - f_v)^4}{(1 + 2 f_v)^2} \right] \quad (5a)$$

$$\equiv k_0^2 + f_v c \quad (5b)$$

where $z_0 = k_s^2 - k_0^2$. Comparing Equations (3) and (5a), we know that if the model of one-sized scatterers with the same ε_s is used, such effective radius a_e, which causes the same scattering effect, should be chosen as

$$a_e^3 = \frac{(1 + 2 f_v)^2}{f_v (1 - f_v)^4} \left[\sum_{i=1}^{n_0} f_i a_i^3 + \sum_{i=1}^{n_0} f_i \sum_{j=1}^{n_0} a_j^3 \sqrt{\frac{n_j}{n_i}} H_{ij} \right] . \quad (6)$$

Numerical results show that a_e is significantly different from the arithmetic mean a_0 $(= \sum n_i a_i / n_0)$, volumetric mean, and the mode radius a_c. The effective a_e tends to the larger size. This is due to the dominant scattering effect caused by larger scatterers, especially. In Fig. 3 the values of a_e and a_c are given in an example. The difference between a_e and a_0, a_c is correlated with the size distribution and fractional volume. Figures 4 and 5 show some examples of these dependencies.

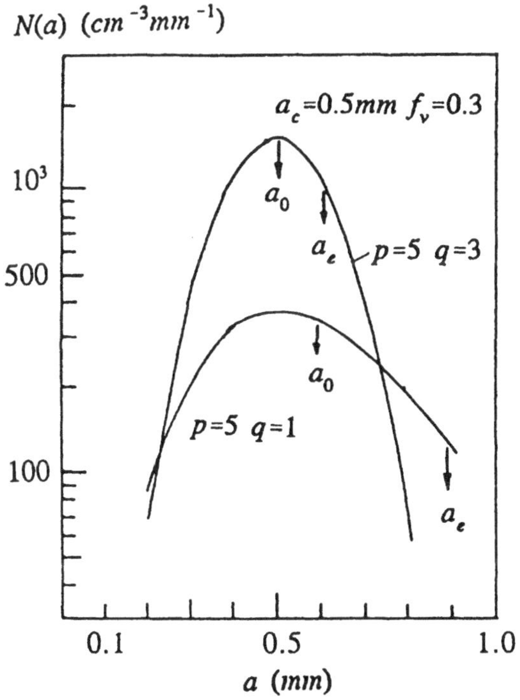

Figure 4. Size distribution with different parameters.

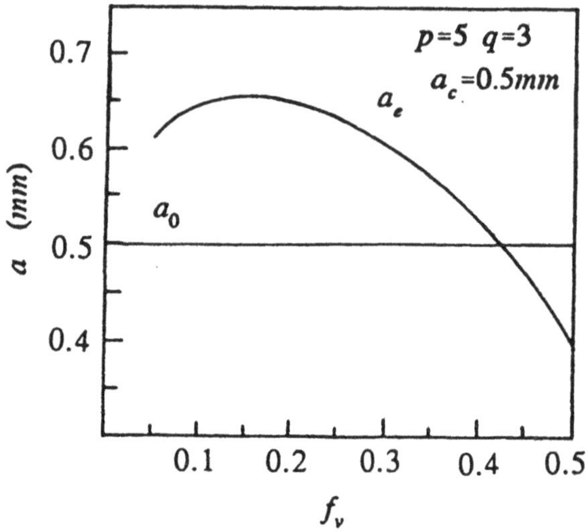

Figure 5. The effective radius vs. fractional volume.

We use the effective dielectric constant ε_{ws} of ice spheres with liquid water coating [2] for the scatterers in wet snow layer. It may be written as

$$\frac{\varepsilon_{ws}-\varepsilon_0}{\varepsilon_{ws}+2\varepsilon_0}=\frac{(\varepsilon_2-\varepsilon_0)(\varepsilon_1+2\varepsilon_2)+S(\varepsilon_1-\varepsilon_2)(\varepsilon_0+2\varepsilon_2)}{(\varepsilon_2+2\varepsilon_0)(\varepsilon_1+2\varepsilon_2)+2S(\varepsilon_2-\varepsilon_0)(\varepsilon_1-\varepsilon_2)} \tag{7}$$

where ε_1, ε_2 are the ice and water dielectric constants, respectively; $S=a_1/a_2$, where a_1, a_2, the inner and outer radii of the coating sphere, may be related with the fractional volume of liquid water f_w and total fractional volume f_t by $S=a_1/a_2=(1-f_w/f_t)^{1/3}$. Numerical results show that the imaginary part of K of wet snow is especially different from that of dry snow.

3. NUMERICAL RESULTS OF DMRT EQUATIONS FOR MULTI-LAYER RANDOM MEDIA

The model of two-layer random media embedded by different scatterers with size distribution is shown in Fig. 6. The surface layer, region 1, can be seen as a melting or refrozen layer of snow grains for wet and dry snow remote sensing. By using the Gaussian quadrature, the DMRT equation in region m ($m=1,2$) is written as

$$\cos\theta\frac{d}{dz}\vec{I}_m(\theta,z)=-\kappa_{em}\vec{I}_m(\theta,z)+\kappa_{am}C\vec{T}_m+\sum_{j=-N}^{N}a_j\kappa_{sm}\overleftrightarrow{P}(\theta,\theta_j)\cdot\vec{I}_m(\theta_j,z) \tag{8}$$

where N is quadrature number; and the extinction, scattering and absorption coefficients, κ_{em}, κ_{sm}, κ_{am} are calculated by the effective propagation constant K_m of Equation (3), and they are written, respectively, as

$$\kappa_{em}=2K_m'' \tag{9a}$$

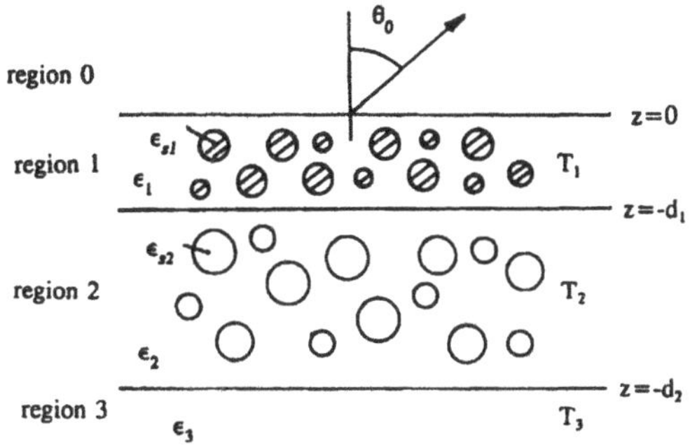

Figure 6. Geometry of two-layer random media.

$$\kappa_{sm} = \frac{2}{9} f_v a_e^3 |c|^2 \frac{(1 - f_{vm})^4}{(1 + 2f_{vm})^2} \tag{9b}$$

and $\kappa_{am} = \kappa_{em} - \kappa_{sm}$, $m = 1, 2$. The coupling boundary conditions are written as $(0 \leqslant \theta_1,\ \theta_2 \leqslant \pi/2)$

$$\vec{I}_1(\pi - \theta_1, z = 0) = \overleftrightarrow{R}_{10}(\theta_1)\, \vec{I}_1(\theta_1, z = 0) \tag{10a}$$

$$\vec{I}_1(\theta_1, z = -d_1) = \overleftrightarrow{R}_{12}(\theta_1)\, \vec{I}_1(\pi - \theta_1, z = -d_1) + \overleftrightarrow{t}_{21}(\theta_2)\, \vec{I}_2(\theta_2, z = -d_1) \tag{10b}$$

$$\vec{I}_2(\pi - \theta_2, z = -d_1) = \overleftrightarrow{R}_{21}(\theta_2)\, \vec{I}_2(\theta_2, z = -d_1) + \overleftrightarrow{t}_{12}(\theta_1)\, \vec{I}_1(\pi - \theta_1, z = -d_1) \tag{10c}$$

$$\vec{I}_2(\theta_2, z = -d_2) = \overleftrightarrow{R}_{23}(\theta_2)\, \vec{I}_2(\pi - \theta_2, z = -d_2) + \overleftrightarrow{t}_{23}(\theta_2)\, C\vec{T}_3 \tag{10d}$$

where R_{ab}, t_{ab} are the reflectivity and transmitivity matrices from the region a to region b; θ_1, θ_2 obey the Snell's law. Actually, the main difference between the effective propagation constants K_1, K_2 in the case of wet snow comes from their imaginary parts. Thus, we may have R_{21} $(= R_{12}) \approx 0$, and $\theta_1 = \theta_2 \equiv \theta$, $N_1 = N_2 \equiv N$. In the following numerical calculations, we take $N = 8$.

By using discrete ordinate and eigen-analysis method [3, 4] to solve Equations (8) and (10a-d), the polarized brightness temperatures are obtained as

$$\vec{T}_B(\theta_0) = \overleftrightarrow{t}_{10}(\theta) \cdot \left[\vec{T}_1 + \left(\overleftrightarrow{E}_1 + \overleftrightarrow{Q}_1\right) \cdot \vec{x}_1 + \left(\overleftrightarrow{E}_1 - \overleftrightarrow{Q}_1\right) \cdot \overleftrightarrow{U}_1) \cdot \overleftrightarrow{U}_1(d_1) \cdot \vec{y}_1 \right] \tag{11}$$

where $\overleftrightarrow{E}_1$, $\overleftrightarrow{Q}_1$ are the matrices with the eigenvalue and eigenvector solutions of DMRT Equation (8) in region 1; $U_1(z)$ is a diagonal matrix with the elements $\exp(-\alpha_{1i} z)$, here α_{1i} denotes the i-th eigenvalue in region 1. The coefficients vector $\vec{x}_1$, $\vec{y}_1$ are determined by the coupling boundary conditions Equations (10a-d).

Figure 7 shows the functional dependence of vertically polarized brightness temperature (T_{Bv}) of dry snow on the layer depth (d) at 36 GHz, and comparison with the data of T_{Bv} versus different depths measured by our radiometer in Jingyutang Remote Sensing Site of Chinese Academy of Sciences in February 1991. At 36 GHz, the larger particles with diameter larger than 1 mm have Mie scattering behaviour so that Equation (6) may not be valid. The approach to DMRT equation for Mie scattering

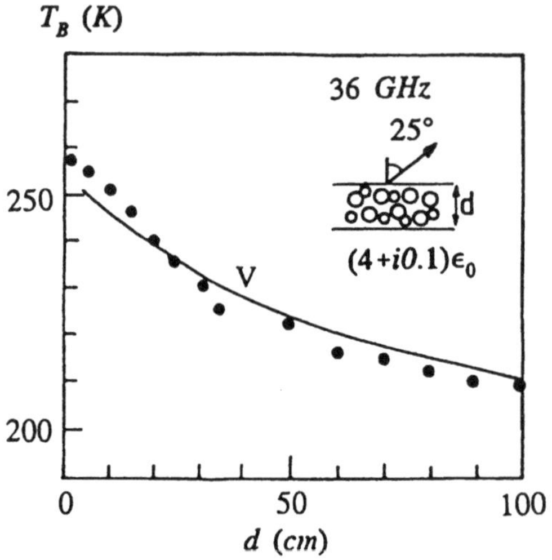

Figure 7. 36 GHz vertically polarized T_B vs. layer depth: $T_1 = 265$ K, $T_2 = 270$ K, $f_v = 0.25$, $a_c = 0.5$ mm, $p = 5$, $q = 3$.

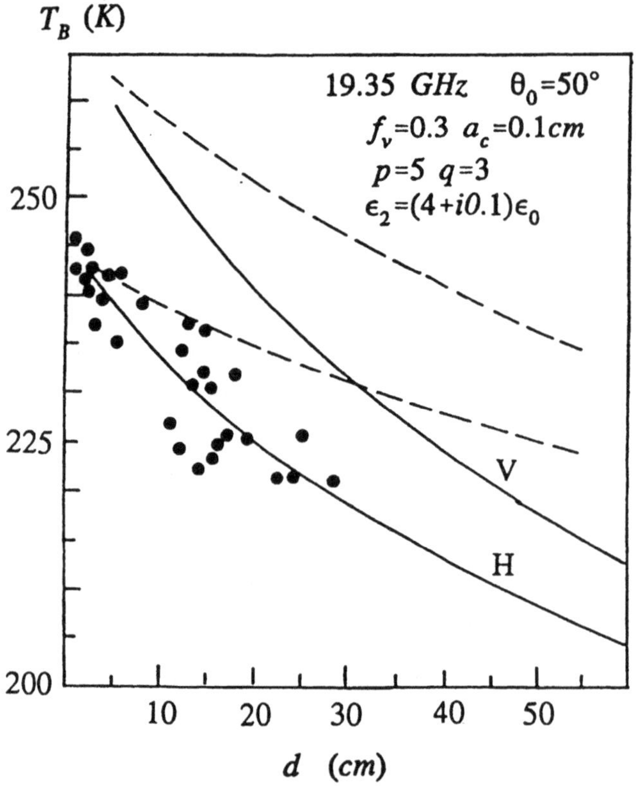

Figure 8. 19.35 GHz T_B vs. layer depth: $T_1 = T_2 = 270$ K.

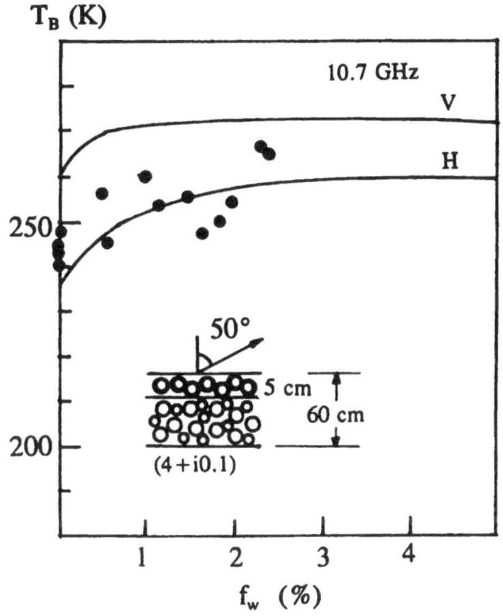

Figure 9. Polarized T_B of wet snow vs. fractional volume of liquid water f_w: $T_1 = 273$ K, $T_2 = T_3 = 270$ K, $f_t = 0.25$, $a_c = 0.1$ cm, $p = 5$, $q = 3$.

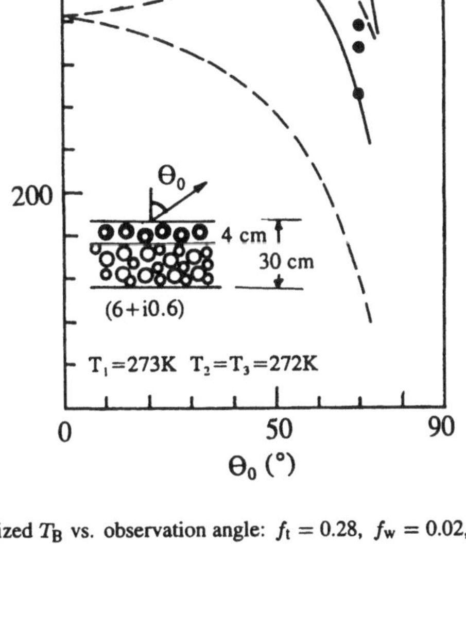

Figure 10. 10.7 GHz polarized T_B vs. observation angle: $f_t = 0.28$, $f_w = 0.02$, $a_c = 0.1$ cm, $p = 5$ $q = 3$.

remains to be done. In this figure, the parameter a_c is especially chosen to match the data. Figure 8 shows the same dependence at 19.35 GHz and comparison with the data of space-borne remote sensing [5]. The dashed lines in the figure were obtained by the model of one-sized scatterers with the arithmetic mean a_0. It can be seen that the brightness temperatures are very sensitive to the size parameters, especially at higher frequencies such as 19.35 GHz. By using our two-layer model, the brightness temperature versus fractional volume of liquid water of wet snow are numerically calculated and compared with the experimental data [6] as shown in Fig. 9. The existence of liquid water in the surface layer ($d_1 = 5$ cm) significantly increases the brightness temperature. As the fractional volume f_w continuously increases, the brightness temperatures would be gradually saturated. As another example of wet snow, Fig. 10 shows the angular dependence of brightness temperature and comparison with data [7]. The dashed lines are obtained by one-layer model of dry snow with the depth $d = 30$ cm. If the total layer ($d = 30$ cm) was assumed to be embedded by water-coated particles, too much melting would overestimate the emissivity.

Figure 11 shows some simulation data of dry and wet snow, and soil surface without snow cover at 18 and 37 GHz. The dry snow data from right to left are obtained for the depth $d = 5, 10, 20, \ldots, 100$ cm, respectively. Wet snow data correspond to $d_2 = 20$, 30, 50 cm, $d_1 = 5$ cm, and $f_w = 0.02$. The soil surface data from up to down are obtained for surface volumetric wetness = 0, 0.05, 0.1, 0.15, respectively. It can be seen that for dry snow $T_{B18} > T_{B37}$; for wet snow $T_{B18} \approx T_{B37}$; and for soil surface

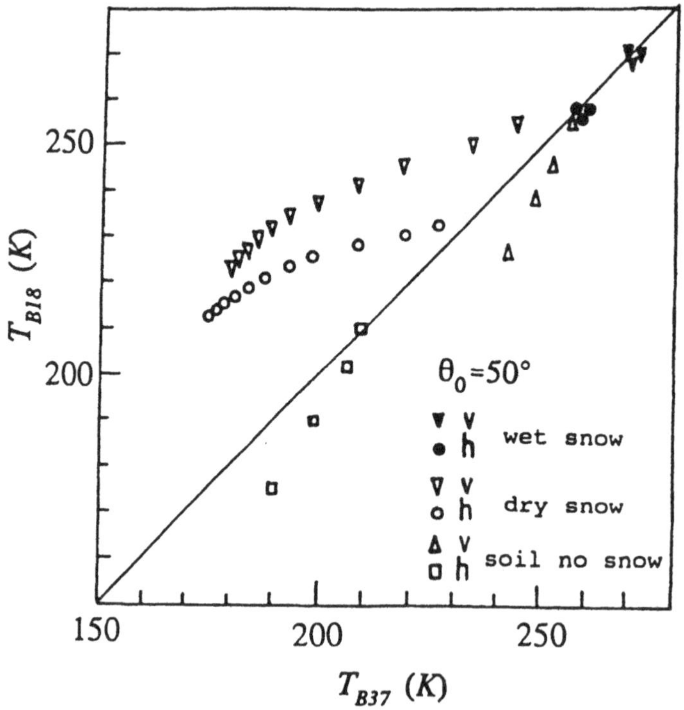

Figure 11. Simulated data of 18, 37 GHz polarized T_B dry snow: $T_1 = 260$ K, $T_2 = 263$ K, $f_v = 0.3$, $a_c = 0.85$ mm; wet snow: $T_1 = 273$ K, $T_2 = 260$ K, $T_3 = 263$ K, $f_t = 0.3$, $a_c = 0.85$ mm; $\varepsilon_3 = (4 + i0.1)\varepsilon_0$, $p = 5$, $q = 3$.

$T_{B18} < T_{B37}$, especially for wet soil. These simulation data are consistent with the observation of space-borne remote sensing for snow [8], and should be applicable to snow parameter retrieval.

4. CONCLUSIONS

In this paper, the radiative transfer of two-layer random media embedded by multiple components, dense spherical scatterers with a size distribution is discussed. By developing a computer method of morphologic mapping, an example of a size distribution of spherical crystals from a snow sample is obtained. Following the calculation of the effective propagation constant in DMRT theory, the effective radius of snow grains in the scattering model is derived. This effective radius is significantly different from both arithmetic and volumetric means. By using Gaussian quadrature and discrete ordinate and eigen-analysis approach, the DMRT equations of two-layer random media and four coupling boundary conditions are solved. Numerical results of polarized brightness temperatures and the functional dependence on frequency, observation angle, size distribution, layer depth, liquid water content, layer structure, etc. are discussed, and are compared with experimental data of snow remote sensing. The sensitivities of size parameters and dielectric properties of snow grains are tested. This approach and numerical simulation should be applicable to snow parameter retrieval.

Acknowledgement
This work was supported by the National Natural Science Foundation of China and UK Royal Society K. C. Wong Fellowship.

REFERENCES

1. L. Tsang, K. H. Ding and B. Wen. In: *Polarimetric Remote Sensing*. J. A. Kong (Ed.), pp. 75–141. Elsevier (1990).
2. C. F. Bohren and D. R. Huffman. *Absorption and Scattering of Light by Small Particles*. John Wiley & Sons, New York (1983).
3. L. Tsang and J. A. Kong. Thermal microwave emission from a three layer random medium with three-dimensional variations. *IEEE Trans. Geosci. Rem. Sens.*, **GE-18**, 212–216 (1980).
4. Y. Q. Jin and R. G. Isaacs. Simulation and statistical retrieval for inhomogeneous and non-isothermal atmospheric presipitation. *J. Quant. Spectrosc. Radiat. Transfer*, **37**, 461–469 (1987).
5. A. Rango, A. Chang and J. L. Foster. Application of a snowmelt-runoff model using Landsat data. *Nodic Hydrology*, **10**, 25–40 (1979).
6. W. H. Stiles and F. T. Ulaby. The active and passive microwave response to snow parameters, I. wetness. *J. Geophys. Res.*, **85**, 1037–1044 (1980).
7. A. K. Fung and H. J. Eom. Emission from a Rayleigh layer with irregular boundaries. *J. Quant. Spectrosc. Radiat. Transfer*, **26**, 397–409 (1981).
8. K. F. Kunzi, S. Patel and H. Rott. Snow-cover parameters retrieved from Nimbus-7 scattering multichannel microwave radiometer (SMMR) data. *IEEE Trans. Geosci. Rem. Sens.*, **20**, 452–467 (1982).

ESA/NASA International Workshop, pp. 285–294
B. J. Choudhury, Y. H. Kerr, E. G. Njoku and P. Pampaloni (Eds)

Monitoring snow cover on the Tibetan Plateau using passive microwave satellite data

DAVID A. ROBINSON and THERESE E. SPIES

Department of Geography, Rutgers University, New Brunswick, NJ 08903, USA

Abstract— Output from satellite-borne passive microwave sensors has increasingly been recognized as an important means of monitoring snow cover. However, employing a single global algorithm to extract a snow signal from multichannel microwave data does not accurately monitor snow across all landscapes. One such region where difficulties occur is the Tibetan Plateau, where global algorithms overestimate the extent of cover. Here, adjustments we have made to the Chang *et al.* [1] global algorithm are discussed. Given the linear fit between snow depth and brightness temperature in the algorithm, estimated snow depths for individual 1° × 1° cells are reduced in 10 mm increments. This continues until the number of regional cells estimated to be snowfree agree with a detailed analysis of snow cover using shortwave (visible and near-infrared) satellite imagery. In 15 Plateau test cases, an average of 80 mm of snow is subtracted from original microwave estimates of cell snow depth over the more level portions of the Plateau and 60 mm over more mountainous areas to reach agreement with shortwave values. Possible reasons for the original overestimation are discussed. The microwave offsets are applied to over 500 snow charts derived from Scanning Multichannel Microwave Radiometer data using the Chang *et al.* algorithm between 1979–87. These data show that the average annual (Sep–Aug) percent coverage of snow on the Plateau is 35%, with the most snow observed in the 1982/83 season (39%) and the least in 1982/83 (26%). The mean is 10% larger than less accurate estimates obtained from lower resolution hemispheric snow charts that are derived from shortwave satellite data.

1. INTRODUCTION

The Tibetan Plateau, with an average elevation of approximately 5000 meters, covers an area of about two million square kilometers in western China. The combination of altitude and area makes this a unique climatologic region, and influences atmospheric dynamics in locations adjacent to as well as farther from the Plateau [2–5]. Given its high elevation and lower middle latitude location, large quantities of solar radiation reach the surface of the Plateau throughout the year. The major variable influencing the distribution of this radiation once it reaches the surface is snow cover. A fresh snow cover as shallow as 0.1 m may increase the albedo of the essentially treeless landscape to 0.80, this being about 0.60 higher than when the surface is snowfree. The melt of snow increases soil moisture, resulting in large quantities of absorbed solar energy leaving the surface and atmospheric boundary layer as latent heat. Thus, between the reflective and latent losses of energy, sensible heating on the Plateau is greatly reduced by the presence of snow.

Despite the importance empirical and modelling investigations ascribe to Plateau snow cover, knowledge of the kinematics of Plateau snow remains less than adequate. Ground station observations of snow are virtually absent in the western half of the

Plateau. The limited number of stations in the eastern half tend to be in inhabited locations where the climate is likely to be most tolerable, thus may not be representative of regional conditions [6]. Satellite observations of surface conditions on the Plateau have been made using shortwave imagery since the 1960s. However, National Oceanic and Atmospheric Administration (NOAA) weekly snow charts [7], the only shortwave (visible and near-infrared) product to continuously report the extent of snow cover over the Northern Hemisphere during this period, did not monitor snow on the Plateau in a consistent manner until the late 1970s [8–9]. Even since then, cloudiness frequently obscures portions of the region for a week or more at a time, limiting the timeliness of these reports. Furthermore, no information on snow water equivalent or depth is directly attainable from shortwave analyses.

Plateau snow conditions may also be monitored by measuring microwave radiation emitted from the surface. Sensors on-board polar orbiting satellites measure this radiation in several wavelengths, and empirical or theoretical algorithms are applied to microwave brightness temperatures to estimate snow extent and mass (expressed as water equivalent or depth). A microwave approach provides information on snow extent and mass regardless of sky conditions, making for a potentially invaluable means of monitoring regional and global snow conditions. However, to date, only generic algorithms derived for the global study of snow have been used to estimate Plateau snow, and these overestimate the extent of cover [10]. Most of the cover estimated from microwave data is quite shallow. This is verified by available station observations, and suggests that global microwave algorithms might be successfully adjusted to better estimate Plateau snow cover. This manuscript describes such an effort. The global algorithm of Chang *et al.* [1] is adjusted to better represent Plateau snow extent by comparing microwave estimates with detailed shortwave satellite analyses of regional cover. Concentration is on snow extent, as the absence of ground truth over most of the Plateau makes estimates of depth difficult to evaluate. Finally, Plateau snow cover will be examined from 1978 to 1987 using the adjusted microwave charts and the results compared with estimates of Plateau cover from NOAA snow charts that are derived from shortwave satellite data.

2. DATA

2.1. Microwave

Microwave data used in this investigation were measured by the Scanning Multichannel Microwave Radiometer (SMMR) on-board the Nimbus-7 satellite. This instrument flew from November 1978 to August 1987. Data were collected on a one day on one day off basis, requiring five days (or three days on) to obtain global coverage. Chang *et al.* [1] developed a theoretical algorithm that uses the difference in brightness temperatures of 18 and 37 GHz SMMR night time returns to derive a snow depth/brightness temperature relationship for a uniform snow field:

$$\mathrm{SD} = 15.9 \times (T_{18\mathrm{H}} - T_{37\mathrm{H}})\ \mathrm{mm} \tag{1}$$

where SD is snow depth (mm), and $T_{18\mathrm{H}}$ and $T_{37\mathrm{H}}$ are the horizontally polarized brightness temperatures for the SMMR 18 GHz and 37 GHz radiometers. Resolutions at these freqencies are 55 and 27 km, respectively. This algorithm is based on a regression relationship derived from theoretically calculated brightness temperatures for a layer of snow of varying depth (up to 1.0 m) and density above frozen soil. This generic algorithm assumes a snow density of 300 kg m^{-3} and a mean snow-grain radius of 0.3 mm. It was tuned for a uniform snow field with moderate snow depth over low elevation regions.

Algorithm results for the five-day periods closest to the middle of the months of November 1978–79, 1984–86 and January and March 1979–80, 1985–87 are used in the case studies discussed in Sections 3 and 4. Data are mapped to a 0.5° × 0.5° grid, with snow of $\geqslant$ 30 mm reported in 10 mm increments. The original grid is averaged to a 1° × 1° grid to conform with the shortwave data discussed in the next section.

2.2. *Shortwave*

Detailed charts of snow cover over western China are constructed using shortwave Defense Meteorological Satellite Program (DMSP) imagery. Image resolution is 2.8 km and the spectral range of the DMSP sensor is 0.4–1.1 μm. Snow is identified on shortwave imagery by recognizing characteristic textured surface features and brightness.

Images from five consecutive days are used to document mid-month cover for the intervals listed in the previous section. The five days permit most of the surface to be observed, although clouds occasionally persist over a portion of the region for the full observation period. Charted cover is digitized for 1° × 1° grid cells. Where snow is present, a cell is determined to have a brightness value ranging from 1 to 3, depending on the spatial coverage and depth of the snow cover. In this lightly vegetated region, the brightness increases as the cover becomes more spatially complete and deep. For comparison with microwave values, all cells charted as class 1 (low brightness), 2, or 3 (high brightness) cover are classified as snow cells, even though less than half of the surface is snow covered for all class 1 cells and even some class 2 cells. At least two trained observers charted Plateau surface conditions for each of the 15 study cases. A close agreement between the analysts occurred in every case, and for this study, the shortwave charts are considered accurate for mapping the extent of snow cover.

3. METHODOLOGY

The microwave and shortwave products are analyzed using GRASS (Geographic Resources Analysis Support System), a raster-based geographic information system developed by the US Army Corps of Engineers Construction Engineering Research Laboratory. This permits rapid, consistent and repeatable analyses of the two spatially referenced data sets. Analyses include data filtering, map overlaying and map algebra.

Digital terrain analysis is also performed using GRASS to objectively define the high altitude study region. National Geophysical Data Center digital elevation model data with a grid cell resolution of 0.08° × 0.08° are averaged for one degree cells throughout western China. All one degree cells having an average elevation exceeding 3500 meters are used in the study. These high elevation cells are divided into two subregions to determine whether relief has any impact on the microwave values. The division is based on an analysis of the diversity of elevation within a grid cell. The low diversity cells will be referred to as the Plateau plains and the high diversity cells as the mountainous Plateau. The shaded cells in Fig. 1(a) represent the entire study region. The plains

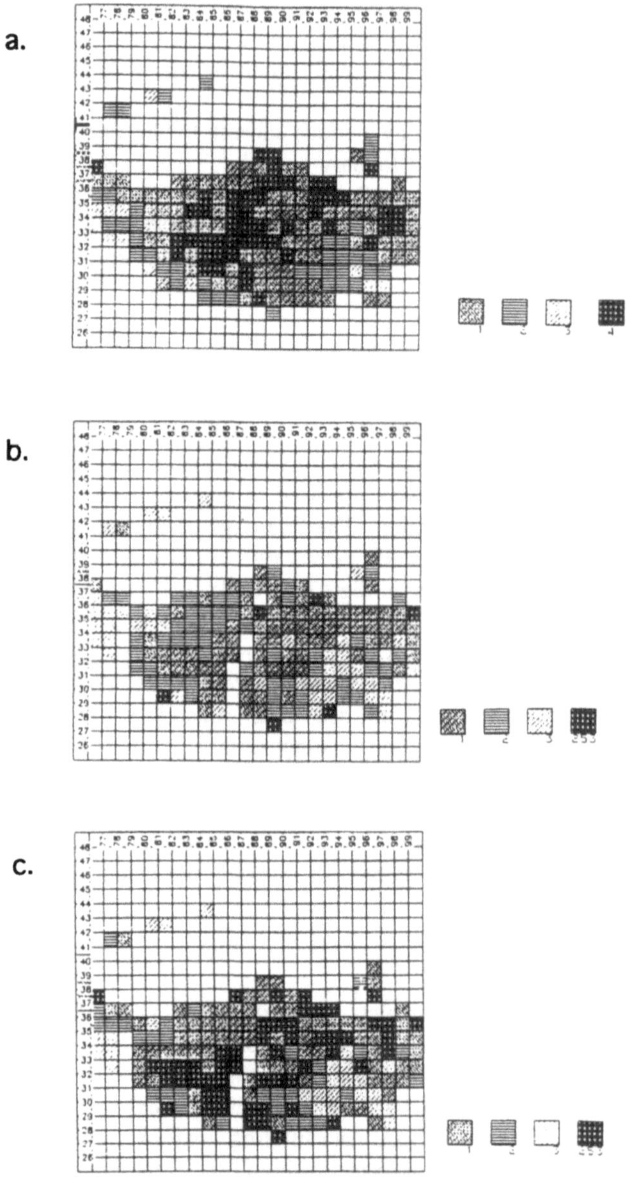

Figure 1. Shortwave (DMSP) and microwave charts of snow cover for mid January 1987 over the Plateau study region. Latitude (°N) and longitude (°E) are shown on the left and top edges of the charts, respectively. The shortwave chart (a) shows the distribution of snow cover brightness classes 1–3 and those cells observed to be snowfree (4). Chart (b) is the unadjusted microwave depiction of snow cover. Snow depth categories include (1) 30–100 mm, (2) 110–200 mm, and (3) >200 mm. Category 253 represents snowfree cells. The empty cells in this chart and chart (c) that are shaded in (1) are missing microwave data. Chart (c) depicts the distribution of snow cover after the mean offset of −80 mm is applied to the Plateau plain cells and the −60 mm offset is applied to the mountainous Plateau cells. The cell categories may be interpreted as in chart (b).

dominate the Plateau west of 90°E, occupying 101 of the 119 cells. The mountainous cells in the west include portions of the Himalayan, Kunlun Shan and Tian Shan ranges. East of 90°, 65 of the 83 cells are mountainous, but with less well defined ranges than in the west.

An initial comparison of the original microwave snow output with the shortwave charts indicates a significant microwave overestimation of the extent of snow cover over the Plateau subregions. This is exemplified in Fig. 1, where the shortwave chart (a) reports snow cover in 74% of the cells, while the generic Chang microwave analysis (b) finds a 97% coverage of snow, with only six of the 202 cells snowfree.

The linear fit between snow depth and brightness temperature in the Chang algorithm, and the shallow snow depths (<100 mm) reported by the microwave analysis in most cells shown to be snowfree in the shortwave charts suggests that by applying an offset to Equation (1), microwave estimates of snow coverage over the Plateau can be adjusted to better match detailed shortwave observations. In the case shown in Fig. 1, offsets of −80 mm over the Plateau plain and −60 mm over the mountainous Plateau result in the best match between the adjusted microwave (72% cover) (Fig. 1c) and shortwave estimates. Results of this analysis for the 15 study cases are reported in the next section.

4. ALGORITHM ADJUSTMENTS

The microwave offset needed to achieve the closest spatial agreement with a shortwave chart, ranges between −40 mm and −160 mm over the Plateau plain (Table 1). Offsets range from 0 to −160 mm over the mountainous Plateau (Table 2). The mean offset for the 15 cases is −80 mm over the Plateau plain and −60 mm over the mountainous Plateau. There is no apparent relationship between the size of the offset and the study

Table 1.
Results of the 15 case studies of snow extent on the Plateau plain. Included is the percent coverage of snow as determined from the shortwave (SW), unadjusted microwave (uaMW) and best-fit adjusted microwave (aMW) analyses. The offset (mm) that resulted in the best fit is also listed

Month/year	SW	uaMW	Offset	aMW
Jan 79	20.9	84.9	−160	20.1
Jan 80	36.8	89.5	−120	40.5
Jan 85	68.0	87.8	−40	71.0
Jan 86	83.2	93.7	−50	83.5
Jan 87	71.8	99.1	−80	70.6
Mar 79	53.5	83.9	−80	50.0
Mar 80	25.0	81.5	−110	29.3
Mar 85	45.1	71.3	−50	46.0
Mar 86	65.1	87.0	−60	71.1
Mar 87	68.1	92.1	−70	67.6
Nov 78	59.1	79.6	−60	56.0
Nov 79	17.6	81.6	−150	15.0
Nov 84	60.6	85.5	−50	63.5
Nov 85	79.5	91.9	−60	80.0
Nov 86	72.7	92.2	−60	75.4
Mean	55.1	86.8	− 80	56.0

Table 2.
Same as Table 1, except for the mountainous Plateau

Month/year	SW	uaMW	Offset	aMW
Jan 79	39.8	86.4	−160	40.7
Jan 80	51.8	93.5	−120	50.7
Jan 85	66.0	81.3	−40	67.5
Jan 86	95.2	92.4	0	92.4
Jan 87	78.3	94.0	−50	78.3
Mar 79	56.6	91.5	−70	54.9
Mar 80	47.0	81.7	−100	45.1
Mar 85	37.0	70.0	−50	43.8
Mar 86	78.0	78.1	0	78.1
Mar 87	78.0	80.5	0	80.5
Nov 78	70.3	81.1	−40	71.6
Nov 79	30.0	89.7	−130	33.3
Nov 84	69.0	76.0	−40	67.1
Nov 85	92.8	89.0	0	89.0
Nov 86	72.0	82.3	−40	73.4
Mean	64.1	84.5	−60	64.4

month. However, larger offsets are generally required when shortwave charts indicate a low extent of snow. In all cases, not only is a close agreement achieved with respect to percent snow coverage, but the spatial distribution of the cover agrees well between the shortwave and adjusted microwave analyses.

5. PLATEAU SNOW COVER: 1978–1987

The mean offsets for the two Plateau subregions have been applied to all microwave charts generated between December 1978 and August 1987. On average, the Plateau is 31% snow covered in fall, 59% in winter, 40% in spring, and 9% in summer (Table 3). The annual average cover is 35%. The average coverages over the two subregions are within a few percent of each other in all but the winter, when the mountainous Plateau

Table 3.
Seasonal and annual percent snow cover over the Plateau. Winter (Dec–Feb), spring (Mar–May), summer (Jun–Aug), fall (Sep–Nov), annual (Sep–Aug)

	Fall	Winter	Spring	Summer	Annual
1978/79	–	63	40	11	–
1979/80	35	59	39	7	36
1980/81	29	55	40	7	33
1981/82	25	59	43	10	35
1982/83	30	63	50	11	39
1983/84	36	61	34	9	35
1984/85	24	44	30	8	26
1985/86	36	72	38	7	38
1986/87	34	56	44	9	37
Mean	31	59	40	9	35

has 8% more cover than the Plateau plain. Coverage estimates are biased towards the high side, as many cells on the Plateau that are categorized as snow covered according to the shortwave or the adjusted microwave analyses are not fully covered. This situation is not balanced by partial cover in those cells reported to be snowfree.

A year-to-year survey of snow extent indicates annual (Sep–Aug) cover varying from 39% in 1982/83 to 26% in 1984/85. Fall cover was at a maximum in 1985, winter cover in 1985/86 and spring cover in 1983. Minima for these three seasons were set in the 1984/85 snow season. Figures 2 and 3 show the annual and seasonal departures from

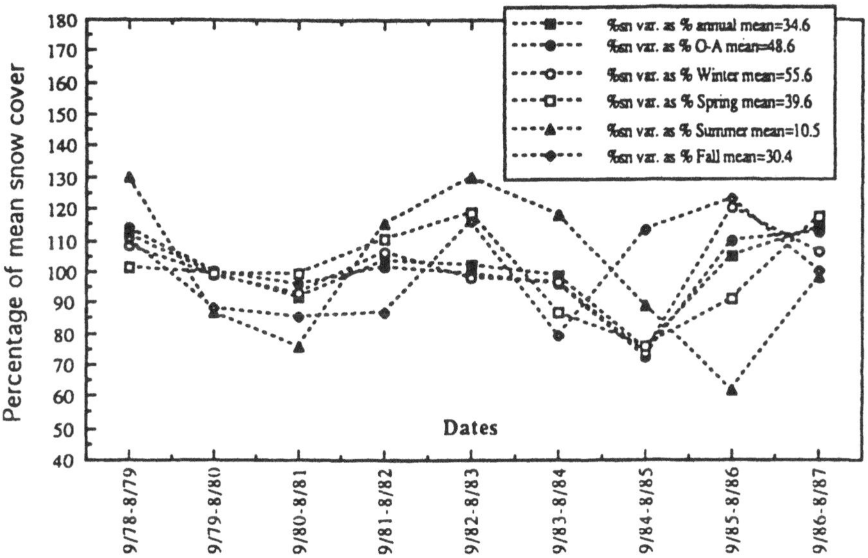

Figure 2. Departures of snow cover on the Plateau plain expressed as the percentage of the seasonal or annual mean.

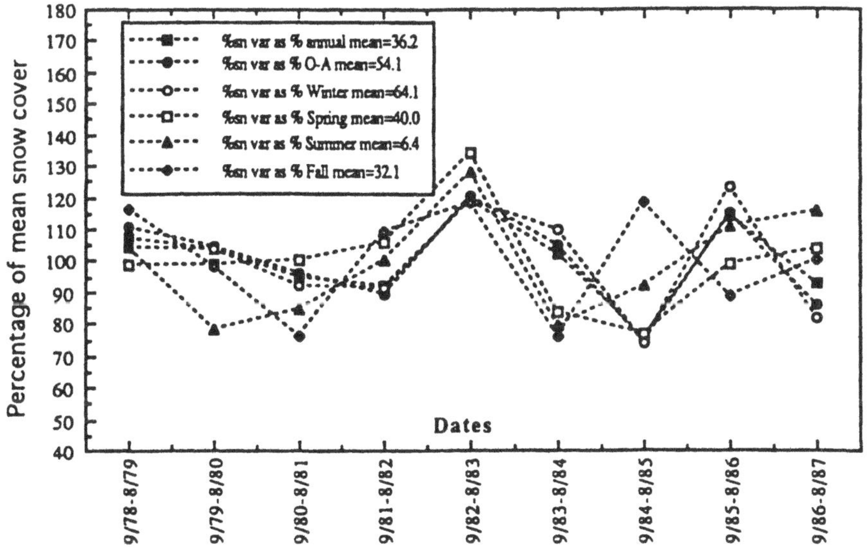

Figure 3. Same as Fig. 2, except for the mountainous Plateau.

normal for the Plateau plain and mountainous Plateau subregions. They further illustrate the year-to-year variations in snow extent, and suggest that there is a somewhat greater persistence of anomalies from one season to another over the mountainous subregion than on the Plateau plain. This is observed best in 1982/83.

Seasonal microwave estimates are larger than ones obtained from the NOAA shortwave charts. The annual NOAA value of 25% snow coverage is 10% lower than the microwave average, and fall (25%), winter (38%) and spring (25%) NOAA estimates are from 6 to 21% below microwave averages. Summer values differ by 1% (NOAA 10%). Comparisons of NOAA and microwave estimates with DMSP derived values for the 15 test cases show an absolute error of 23% for NOAA and 17% for microwave estimates (microwave values are computed using mean offsets). In 12 of the 15 cases NOAA areas are lower than DMSP values, while only 9 microwave cases are lower. When taking into account the sign of the difference, the microwave mean of the 15 cases is only 1% above the DMSP value while the NOAA mean is 23% below the DMSP value. The higher spatial and temporal resolution of the microwave analysis makes it more accurate than the NOAA product. The temporal factor is primarily a result of the frequent cloud cover over the Plateau preventing shortwave analyses, yet not interfering with microwave evaluations.

6. DISCUSSION

The overestimation of snow cover over the Tibetan Plateau when applying global snow algorithms to passive microwave data is probably due to the fact that (1) the assumed snow density and mean snow-grain radius in the algorithm are different from those in western China, and (2) the algorithm was tuned for a uniform snow field with moderate snow depth in low elevation regions. Elsewhere around the Northern Hemisphere microwave analyses tend to underestimate snow extent. A comparison of mean monthly snow extent over the entire Northern Hemisphere finds values derived using the Chang microwave algorithm to be between 80 and 90% of areas calculated from NOAA shortwave charts in winter and spring, 20 to 40% of the NOAA estimates in summer, and 40 to 70% of NOAA areas in fall [11]. At this scale the NOAA values are considered to be more accurate than the microwave estimates due to wet and shallow snow along the periphery of low elevation continental snowpacks remaining undetected in the standard microwave analyses.

A number of other efforts are underway to better understand the microwave problems that lead to errors in snow monitoring. These include regional investigations of snow extent and volume oriented towards the development of additional region-specific algorithms. Landscapes of interest include mountains [12–13], forest [14–15], tundra [16] and prairie [17]. Some studies include data from the 85 GHz channel on the Special Sensor Microwave Imager currently on-board DMSP satellites. Results indicate that adding this channel improves the recognition of shallow snow cover [18].

7. CONCLUSIONS

With snow cover having an important role in the energy and hydrologic budgets both locally and beyond, it is critical that accurate information on the extent and mass of the Tibetan snowpack be available. In this investigation, we modified the generic microwave algorithm of Chang *et al.* [1] in order to improve the monitoring of snow extent on the Plateau. A lack of ground truth prevented an adequate assessment of snow mass estimates. Analyses of over 500 microwave snow charts covering the 1979–87 period found that on average the Plateau is 31% snow covered in fall, 59% in winter, 40% in spring, and 9% in summer. The annual (Sep–Aug) average cover is 35%, with the most snow observed in the 1982/83 season (39%) and the least in 1982/83 (26%). These values are from 6% to 21% higher in the fall, winter and spring than values derived from shortwave-derived NOAA snow charts. The microwave estimates are more accurate, having a higher spatial and temporal resolution than the NOAA analyses. The temporal factor is primarily a result of the frequent cloud cover over the Plateau preventing shortwave analyses, yet not interfering with microwave evaluations.

The improvements demonstrated in this study suggest that region-specific algorithms can lead to a more accurate assessment of snow kinematics. Additional work is needed, and in some cases in progress, to develop algorithms for different snow conditions and physiographic regions.

Acknowledgments

Thanks to A. Chang at NASA for the original microwave snow data and to the National Snow and Ice Data Center, University of Colorado for loan of the DMSP imagery. We are also grateful to A. Chang, B. Goodison, R. Armstrong and G. Kukla for their valuable input at various stages of the project. Particular appreciation is extended to Li Peiji and Cao Meisheng of the Lanzhou Institute of Glaciology and Geocryology for their efforts in many aspects of the investigation. This work is supported by NOAA under grant NA90AA-D-AC518, and by the National Science Foundation through grants from the International (INT-8713878) and Geography and Regional Science (SES-9011869) programs.

REFERENCES

1. A. T. C. Chang, J. L. Foster and D. K. Hall. Nimbus-7 derived global snow cover parameters. *Annals of Glaciology*, **9**, 39–45 (1987).
2. J. Namias. Influence of abnormal heat sources and sinks on atmospheric behavior. In: *Proceedings of the International Symposium on Numerical Weather Prediction*, pp. 615–627 (1962).
3. H. He, J. W. McGinnis, Z. Song and M. Yanai. Onset of the Asian summer monsoon in 1979 and the effect of the Tibetan Plateau. *Mont. Weat. Rev.*, **115**, 1966–1995 (1987).
4. M. Domros and P. Gongbing. *The Climate of China*. Springer-Verlag, Berlin (1988).
5. T. P. Barnett, L. Dumenil, U. Schlese, E. Roeckner and M. Latif. The effect of Eurasian snow cover on regional and global climate variations. *J. Atmos. Sci.*, **46**, 661–685 (1989).
6. P. Li. Seasonal snow resources and their fluctuations in China. In: *Large Scale Effects of Seasonal Snow Cover*, Proceedings of the Vancouver Symposium, IAHS Publ. 166, pp. 93–104 (1987).
7. M. Matson, C. F. Ropelewski and M. S. Varnadore. *An Atlas of Satellite-derived Northern Hemispheric Snow Cover Frequency*. NOAA/NESDIS/NWS, Washington (1986).
8. G. Kukla and D. A. Robinson. Accuracy of snow and ice monitoring. In: *Snow Watch 1980, Glaciological Data*, Report GD-5, pp. 91–97 (1981).
9. C. F. Ropelewski, A. Robock and M. Matson. Comments on 'An apparent relationship between Eurasian spring snow cover and the advance period of the Indian summer monsoon'. *J. Climate and Appl. Meteor.*, **23**, 341–342 (1984).
10. D. A. Robinson, K. Kunzi, H. Rott and G. Kukla. Comparative utility of microwave and shortwave satellite data for all-weather charting of snow cover. *Nature*, **312**, 434–435 (1984).

11. D. A. Robinson. Monitoring northern hemisphere snow cover. In: *Snow Watch '92: Detection Strategies for Snow and Ice. Glaciological Data Report*, **GD-25**, pp. 1–25 (1993).
12. A. T. C. Chang, J. L. Foster and A. Rango. Utilization of surface cover composition to improve the microwave determination of snow water equivalent in a mountain basin. *Int. J. Rem. Sensing*, **12**, 2311–2319 (1991).
13. R. Armstrong. Personal communication (1993).
14. D. K. Hall, J. L. Foster and A. T. C. Chang. Measurement and modeling of microwave emission from forested snowfields in Michigan. *Nordic Hydrology*, **13**, 129–138 (1982).
15. M. T. Hallikainen and P. A. Jolma. Retrieval of the water equivalent of snow cover in Finland by satellite microwave radiometry. In: *IEEE Transactions on Geoscience and Remote Sensing*, **GE-24**, pp. 855–862 (1986).
16. D. K. Hall, A. T. C. Chang and J. L. Foster. Detection of the depth hoar layer in the snowpack of the Arctic coastal plain of Alaska using satellite data. *Journal of Glaciology*, **32**, 87–94 (1986).
17. B. E. Goodison, A. E. Walker and F. W. Thirkettle. Determination of snow cover equivalent on the Canadian prairies using near real-time passive microwave data. In: *Proceedings of the Workshop on Applications of Remote Sensing in Hydrology*, pp. 297–316 (1990).
18. T. Nagler and H. Rott. *Intercomparison of Snow Mapping Algorithms over Europe using SSM/I Data.* Interim Report to the SSM/I Products Working Team (1991).

ESA/NASA International Workshop, pp. 295–314
B. J. Choudhury, Y. H. Kerr, E. G. Njoku and P. Pampaloni (Eds)

Passive microwave remote sensing of snow: volume scattering in snow and parametric inversion of snow parameters with an artificial neural network

LEUNG TSANG, DANIEL T. DAVIS, RICHARD WEST, ZHENGXIAO CHEN, JENQ-NENG HWANG and DALE P. WINEBRENNER

Department of Electrical Engineering, University of Washington, Seattle, Washington 98195, USA

Abstract—Passive microwave remote sensing of snow is affected by volume scattering which can be a result of layering or scattering by ice grains. Because of the smallness of ice grains, layering plays a dominant role at low microwave frequencies. In this paper, we study the brightness temperature of Antarctic firn at C-band with a centimeter layering model. Comparisons are made with ground-based experimental data using ground truth determined density layering profiles. Scattering by ice grains in snow is important at frequencies above 18 GHz. We study the volume scattering effects with the dense media radiative transfer theory. We also perform the inversion of snow parameters from passive microwave remote sensing measurements using a neural network trained with a dense media multiple scattering model. Inversion of five parameters is simultaneously performed from five brightness temperatures. The five brightness temperatures are 18 GHz vertical and horizontal, 22 GHz vertical, and 37 GHz vertical and horizontal polarization. The five parameters are: mean-grain size of ice particles in snow, snow density, snow temperature, snow depth, and the real part of the ground permittivity. A constrained iterative scheme is used with error analysis illustration. The neural network constrained iterative inversion algorithm is also illustrated by performing synthetic mapping of a terrain where all five parameters are simultaneously inverted. The reconstructed contours are in good agreement with the original ones.

1. INTRODUCTION

Passive microwave remote sensing of snow has attracted investigators for many years [1–3]. Measurements have indicated that volume scattering plays an important role in determining the brightness temperature of the snow medium. Volume scattering can take two different forms. First, the individual ice grains can scatter. Second, the layering structure of snow can reflect electromagnetic waves. Grain sizes are usually in the range of 0.1 to 2 mm in diameter. They can cause significant scattering for microwaves at frequencies above 15 GHz. The model of radiative transfer theory using independent scattering and a Mie scattering phase function has been used for many years [4–6]. It has been shown that such a model overestimates scattering [6–8]. In recent years, a dense medium radiative transfer model has been developed which takes into account the dense medium scattering effects [6, 9, 10]. For frequencies below 10 GHz, the smallness of the grain size in snow when compared with wavelength does not produce a significant amount of volume scattering because scattering is proportional to grain size to the sixth power. Thus the brightness temperature darkening at frequencies below 10 GHz can be mostly attributed to layering structure.

In passive microwave remote sensing of snow, an objective is to retrieve snow parameters from brightness temperature measurements. The important geophysical parameters include snow depth, snow temperature and the soil permittivity below the snow. However, brightness temperatures are also affected by other snow parameters such as snow density, grain size etc. Thus, the inversion problem in snow consists of simultaneous multi-parameter inversion from multi-frequency and dual-polarization brightness temperature measurements. With the use of artificial neural network [11–13], we have developed efficient inversion algorithms [14–17]. The neural network is first trained with the electromagnetic scattering model. The training time can be large. However, once the network is trained, the parameters can be retrieved quickly from the brightness temperatures.

In the first part of this paper, we study the brightness temperature of Antarctic firn at C-band with the layered medium model. Previous work in layering assumes that the layering is a continuous Gaussian random process [6, 18, 19]. However, to generate significant scattering, this model requires layering correlation lengths of 1 to 2 mm [18]. Recent ground truth measurements indicate layers in the Antarctic firn that are on the order of centimeters thick [20–22]. At the interfaces of the layers, the snow density changes abruptly over intervals small compared to layer thickness. This feature of the layers makes them effective reflectors at centimeter wavelengths. Thus, we studied the C-band signature with a different random process similar to the telegraph process where the layered medium is represented by discrete centimeter scale layering. The brightness temperature is calculated by a model based on the fluctuation–dissipation approach for the emission of a stratified medium with temperature variation [6, 23, 24]. Monte Carlo simulations of the brightness temperatures are performed by randomly generating the permittivity of the profile and then solving the brightness temperature with the layered medium model. The calculated brightness temperatures are compared between the previous model using a Gaussian process with a millimeter scale correlation length, and this recent model based on discrete centimeter-scale layering. Results were also compared with a data set collected by Helmut Rott [22]. Rott measured the density of Antarctic firn at 5 cm intervals for several different locations. This paper makes comparisons with results from the Vestestraumen site (74.0°S, 13.8°W).

Various techniques for solving inverse problems in remote sensing have been proposed in the last few decades [25, 26]. The well-known technique in inversion is the Backus–Gilbert inversion technique [25, 26]. In this technique, the approximation of single scattering is used so that the scattering measurements are linearly related to the medium parameters. However, in microwave remote sensing problems, especially for dry snow, multiple scattering can be a dominant effect. The relation between remote sensing measurements and the medium parameters are highly nonlinear.

In this paper, we apply the constrained iterative inversion technique [15, 27] to invert snow parameters for a layered snow medium from brightness temperatures using a neural network trained with the dense media radiative transfer theory. The parameters are snow depth, mean grain size of ice particles in snow, snow density, and snow temperature. From the snow density and snow depth, the snow water equivalent can be derived. The performance of the constrained iterative inversion technique [15, 27] is also compared with the performance of the conventional explicit inversion technique [14].

2. BRIGHTNESS TEMPERATURE OF ANTARCTIC FIRN AT C-BAND

In Fig. 1, the snow density profile at Vestestraumen of Antarctic firn is shown. The varying snow density gives rise to a snow permittivity profile. To calculate the brightness temperature of a layered medium, we consider a layered medium with permittivity ε_n and temperature T_n for layer n. By using fluctuation dissipation theory, and the wave approach based on the layered medium Green's function, the brightness temperature of such a medium can be expressed in terms of recurrence relations and propagation matrix [6, Chapter 2]. This approach is based on the wave approach so that reflection from each layer is treated coherently. The approach has been applied to passive remote sensing of soil moisture. The analytic expressions from which brightnes temperatures are calculated can be found in the references [6, 23, 24].

In the past, a random medium model based on a continuous Gaussian random process has been used to account for the brightness temperature in Antarctic firn [18, 19]. In this model, the permittivity ε is assumed to be a Gaussian random process with mean permittivity ε_m. The correlation function at two points z and z' is assumed to be

$$\langle \varepsilon(z)\varepsilon(z') \rangle = \delta\varepsilon_m^2 \mathrm{e}^{-\left(\frac{|z-z'|}{l_z}\right)} \tag{1}$$

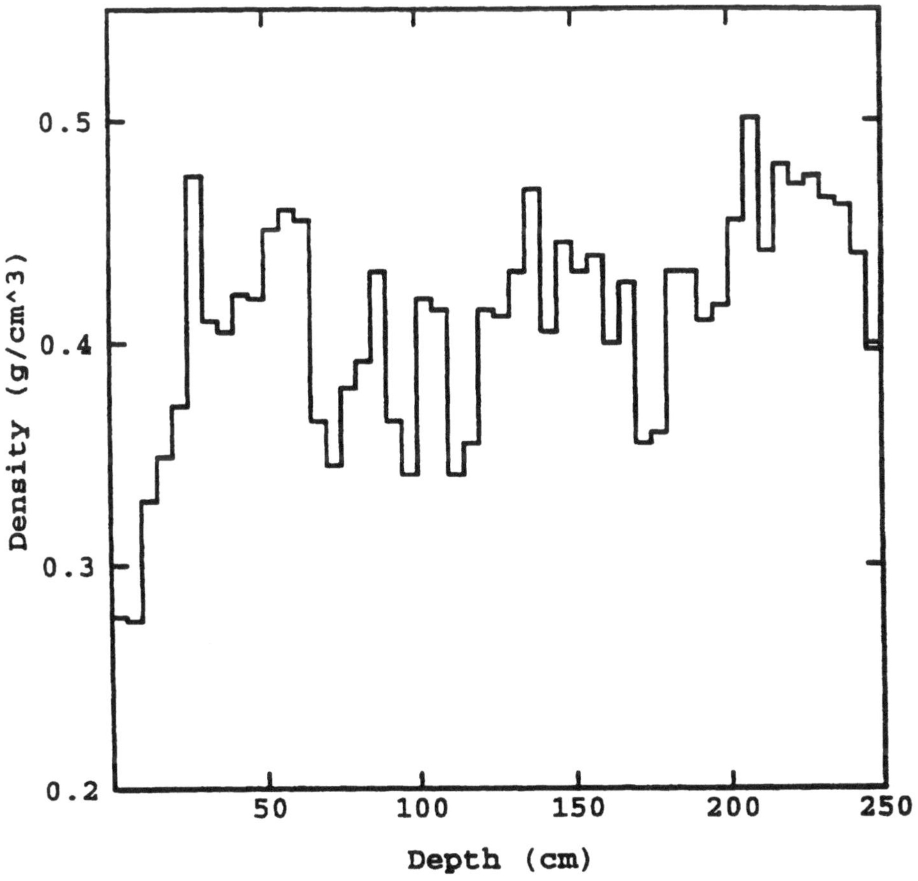

Figure 1. The snow density profile measured by H. Rott at Vestestraumen [22]. The profile was obtained by measuring the average density in consecutive 5 cm layers.

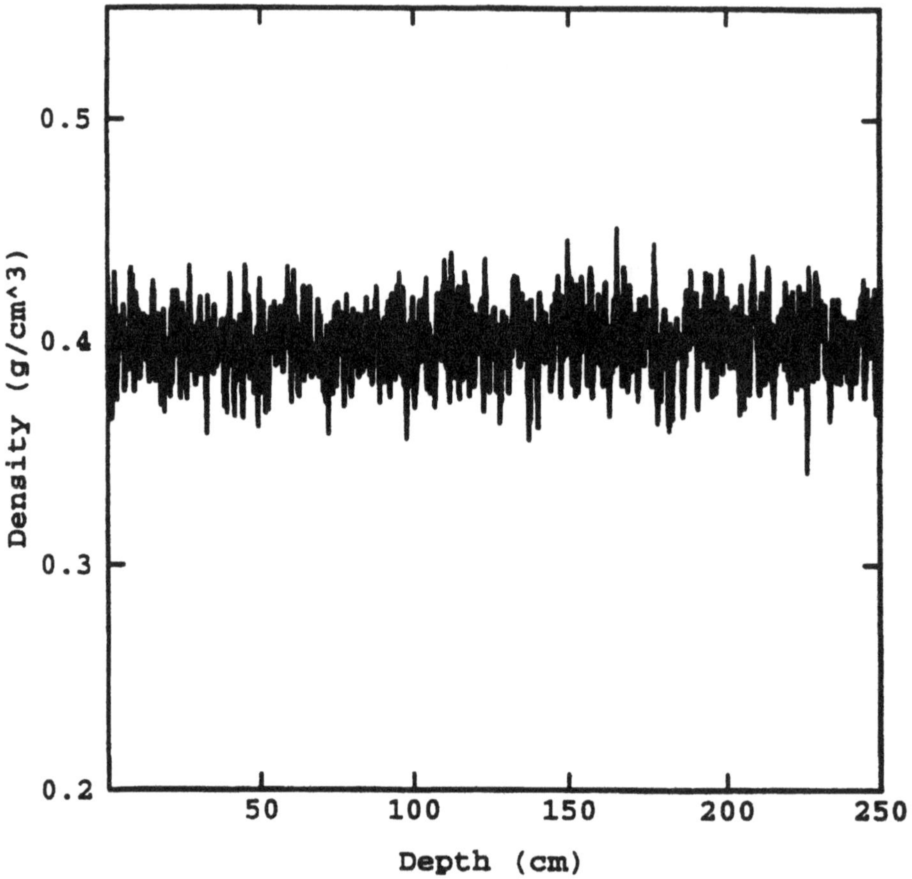

Figure 2. A single realization of a continuous Gaussian random profile with a mean density of 0.4 g/cm^3, a correlation length of 2 mm, and a standard deviation in density of 0.0156 g/cm^3.

where δ is the variance and l_z is the correlation length. Brightness temperatures have been calculated based on the single scattering Born approximation [18] and also the multiple scattering radiative transfer solution [19]. In the following, we calculate the brightness temperature of such a random process rigorously by Monte Carlo simulations. First the snow density profile $\rho(z)$ is assumed to be a Gaussian random process with given standard deviation and correlation length. The process for one realization is randomly generated on the computer. In Fig. 2, we show a simulated density profile using a correlation length of 2 mm and a standard deviation of 0.0156 g/cm^3. The continuous profile of each realization is generated down to a depth of 21 meters, and discretized into 32768 layers. This gives a layer thickness of 0.6 mm. The permittivity of each layer is calculated using the following empirical mixing formulas for dry snow [28].

$$\varepsilon' = 1 + \frac{1.60\rho}{1 - 0.35\rho} \tag{2}$$

$$\varepsilon'' = \varepsilon''_{\text{ice}}\left(0.52\rho + 0.62\rho^2\right). \tag{3}$$

In these equations, $\varepsilon = \varepsilon' + i\varepsilon''$ is the complex relative dielectric constant, and ε''_{ice} is the imaginary part of the relative dielectric constant of pure ice.

The brightness temperature of such a profile is then calculated using the layered medium model. The layered medium model also accounts for a temperature profile. The temperature profile is assumed to have the form of $T_0 + T_h e^{\gamma z}$ [6, Chapter 4, p. 260]. T_0 is the mean annual temperature, T_h is the difference between the surface temperature and the mean annual temperature, and γ is the exponential decay rate. This form of temperature profile is characteristic of Antarctic firn in the summer. Figure 3 shows the brightnes temperatures of two different realizations which have the same correlation length and variance as the realization shown in Fig. 2. Since the layered medium is based on a coherent appproach, the results have to be averaged over many realizations. In a real life situation, there can be built-in incoherence. For example the interfaces of the layered structure can be rough with rms height larger than 1/16 of a wavelength

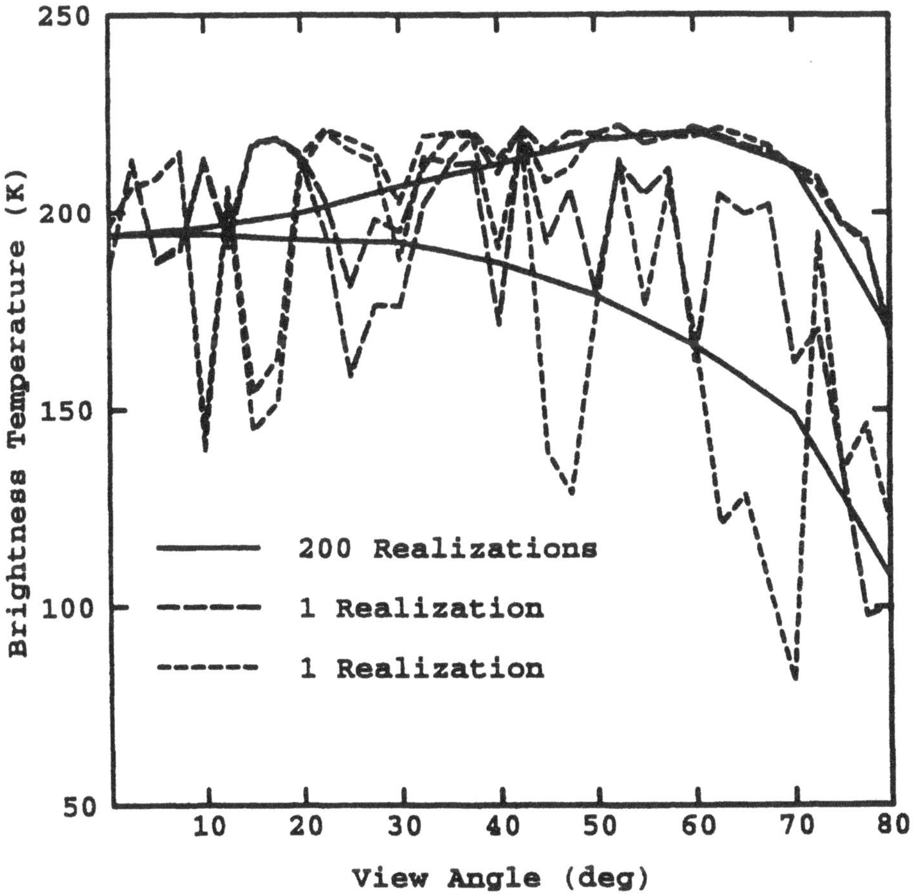

Figure 3. Computed brightness temperatures using fluctuation dissipation theory and a discretized continuous Gaussian random density profile with a mean density of 0.4 g/cm^3, a correlation length of 2 mm, and a standard deviation in density of 0.0156 g/cm^3. The temperature profile is: $T(z) = 222 + 34\exp(0.81z)$. ($z$ is 0 at the top and negative in the medium.) Both profiles are carried down to a depth of 21 meters, and discretized into 32768 layers, giving a layer thickness of 0.6 mm. Both vertically and horizontally polarized brightness temperatures are shown, with $T_v > T_h$.

so that the reflection is rendered incoherent. In Fig. 3, we also show the brightness temperature averaged over 200 realizations. Note that all the coherent oscillations have been smoothed out in the average. However, in order to produce a sigificant amount of decrease in brightnes temperature due to reflection, a small correlation length of 2 mm is required which does not agree with the ground truth measurements of the density profile as shown in Fig. 1 which indicates the presence of centimeter-scale layering.

To incorporate ground truth data on the density profile, we apply the layered medium model to a theoretical profile statistically similar to the measured profile. The measured profile in Fig. 1 shows a background trend of the form $1 - e^{az}$ ($\delta \leqslant 0$ for snow medium) which saturates below about 1 meter of depth. A theoretical profile similar to the measured profile can be generated using two random variables, and a background exponential trend. Layer thickness is assumed to be exponentially distributed. The

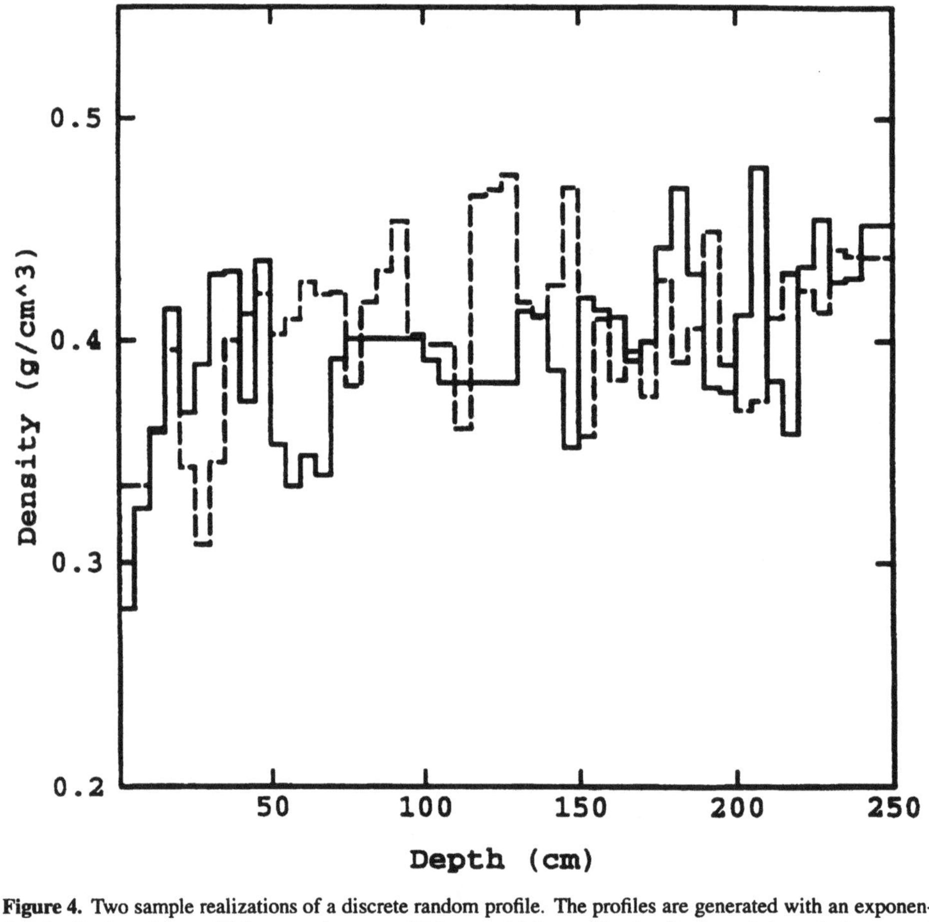

Figure 4. Two sample realizations of a discrete random profile. The profiles are generated with an exponential distribution of layer thickness and a mean thickness of 4.1 cm. The density in each layer is the sum of a deterministic part given by 0.25 (g/cm^3) + 0.16(1 − exp(z/(20 cm))) (g/cm^3), and a normally distributed random variable with zero mean and a standard deviation of 0.04 g/cm^3. (z is 0 at the top and negative in the medium.) For display purposes, the resulting profile is then averaged over 5 cm intervals, as was Rott's measured profile (see Fig. 1). Calculation of brightness temperatures uses the original profile without the 5 cm averaging.

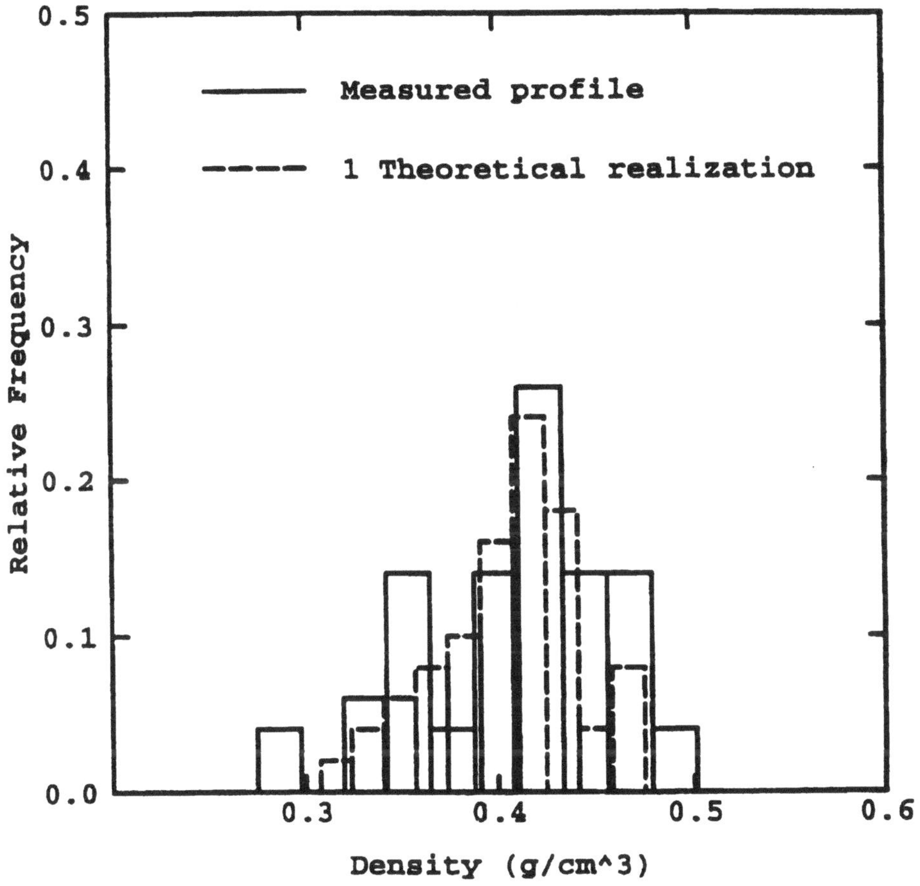

Figure 5. A histogram of Rott's measured density profile superimposed on a histogram of one realization of a density profile using the discrete random profile described in Fig. 4.

layer-to-layer density variations are assumed to be normally distributed, and then superimposed on top of the background exponential profile. The profile is carried down to 10 meters depth. Below this depth, layered scattering is expected to drop off as physical processes smooth out the layer interfaces. The temperature profile is again assumed to have the form of $T_0 + T_h e^{\gamma z}$. Since Rott's measurements [22] were made in the Antarctic summer, the temperature distribution decays from a high temperature at the surface down to the mean annual temperature at depths greater than 10 m. Figure 4 shows two sample theoretical profiles generated with a mean layer thickness of 4.1 cm, and a standard deviation in density of 0.04 g/cm^3. Figure 5 shows the corresponding histogram of the densities in a theoretical profile along with the histogram of the actual densities. The background exponential profile was set to provide the correct density level in the theoretical profile. Using the layered medium model, the theoretical brightness temperatures for a range of observation angles are computed and averaged over many realizations. Figure 6 shows the results when 600 such realizations are averaged together. The measured brightness temperatures [22] are also shown in Fig. 6, along with the results which would be expected from a uniform density profile and the same

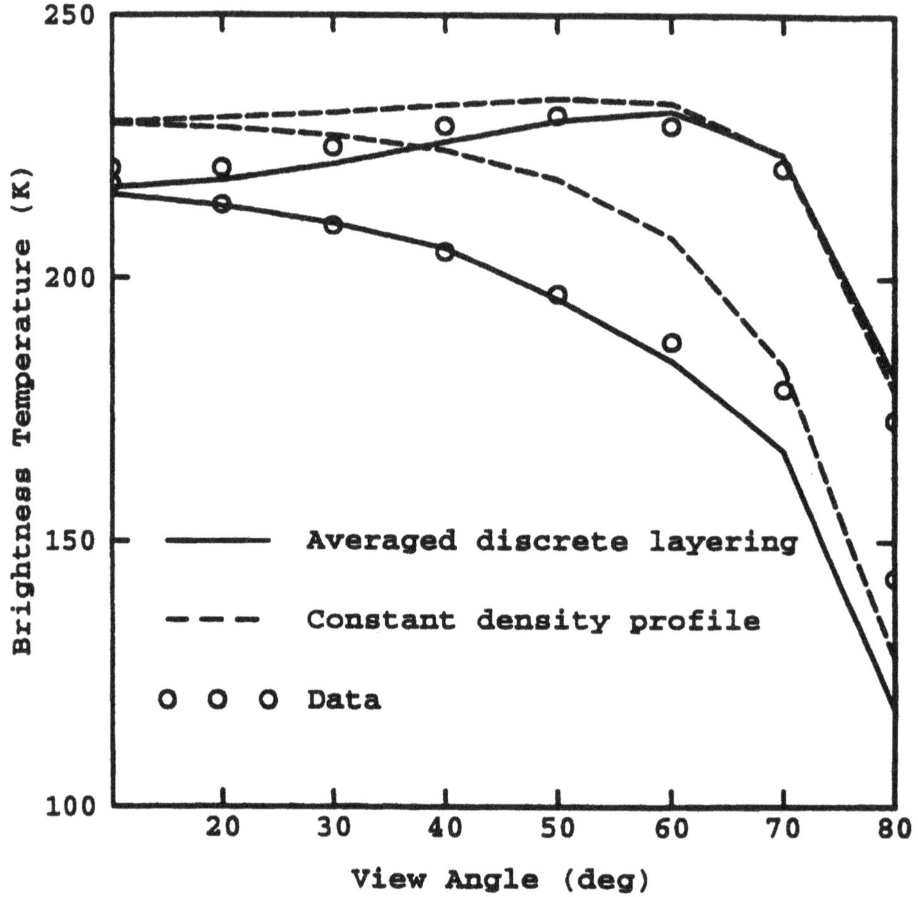

Figure 6. A comparison of Rott's measured brightness temperatures with theoretical results using the discrete random profile described in Fig. 4. Average layer thickness = 4.1 cm, standard deviation of density variation = 0.04 g/cm^3, and a background trend given by: 0.25 (g/cm^3)+0.16(1 − exp(z/(20 cm))) (g/cm^3). Temperature profile: $T(z) = 234 + 24\exp(0.81z)$. ($z$ is 0 at the top and negative in the medium.) Both profiles are carried down to a depth of 10 meters. Vertically and horizontally polarized brightness temperatures are shown, with $T_v > T_h$ at all view angles.

temperature profile. The layered scattering model gives good agreement with the measured results out to an observation angle of 50 deg. Note that the discrete layering used by the theory (Fig. 4) is statistically similar to the observed profile (Fig. 1). No ad-hoc correction factors are required. The half space results show less polarization contrast and higher temperatures, indicating that layer scattering is an important phenomenon at C-band.

3. RETRIEVAL OF SNOW PARAMETERS WITH AN ARTIFICIAL NEURAL NETWORK

An objective of passive microwave remote sensing of snow is the mapping and retrieval of useful parameters, such as physical temperature and snow−water equivalent, from the remote sensing measurements [1]. However, remote sensing measurements are also dependent on a large number of unknown physical parameters such as snow mean grain size, snow density, snow layering, etc... Thus, the inversion problem consists of multi-parametric inversion from multi-frequency and polarimetric remote sensing measurements.

To study volume scattering in snow, we use the dense medium radiative transfer theory [9, 10]. The dense medium radiative transfer theory accounts for dense medium scattering effects and is applied to snow because the ice particles in snow are densely packed. The theory agrees well with controlled laboratory experiments [29] and numerical simulations of Maxwell's equations [30]. The dense media theory also includes multiple scattering effects so that the relation between the brightness temperatures and the snow parameters are nonlinear.

To have a more flexible technique for inverse parameter retrieval using a neural network, constrained iterative inversion is used [27, 31]. The iterative inversion of the network generates the parameters from a trained forward mapping, $\mathbf{\Phi}$, that gives a network output as close as possible to a desired measurement vector. By taking advantage of the duality between the weights and the input activation values in minimizing the mean squared error, E, the iterative gradient descent algorithm can again be applied to obtain the desired input vector. The idea is similar to the back-propagation algorithm, where the error signals are propagated back through the network to adjust the weights so as to decrease the output error.

When the conventional explicit inversion technique is applied to the present 5-measurement-5-parameter problem, parameters are inverted point by point assuming each point is independent. Inversion is not accurate because of the non-unique correspondence between the measurements and the parameters. However, measurements obtained from geophysical terrain are supposed to be related such that no large variation in physical parameters can exist between one point and its neighbors. The constrained iterative inversion algorithm exploits this relation and starts from known points (ground truth) to accomplish the inversion for the whole region. We illustrate the algorithm in this paper by performing synthetic mapping, where we draw contours of each of the five parameters for the whole region of interest. Then we calculate the 5-channel brightness temperatures for each point of the region by using the dense media theory. The contours are then reconstructed by using the constrained iterative inversion techniques with the trained neural network. It is shown that the reconstructed contours agree well with the original ones.

4. DENSE MEDIA RADIATIVE TRANSFER EQUATION FOR PASSIVE REMOTE SENSING

Consider thermal emission from a layer of dielectric particles of permittivity ε_s embedded in a background medium of permittivity ε overlayering a homogeneous half space of dielectric medium of permittivity ε_2. (See Fig. 7.) Note for the case of snow, the background permittivity ε is ε_0 of free space and the particle permittivity ε_s will be that of ice. The particle sizes obey a size distribution $n(a)$ which is the number of particles per unit volume with radii between a and $a + \mathrm{d}a$. The medium is of uniform temperature T. Then the dense media radiative equation for passive remote sensing assumes

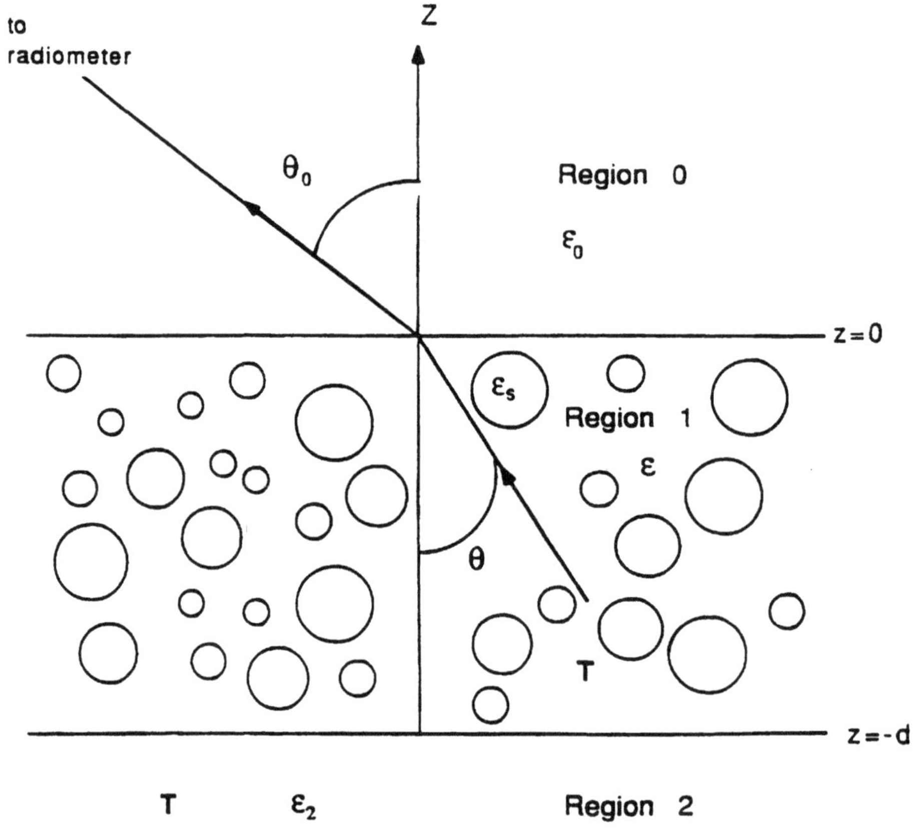

Figure 7. Passive remote sensing with observation of brightness temperatures of a layer of dielectric particles embedded in a background medium of permittivity ε overlayering a homogeneous half space of dielectric medium of permittivity ε_2. The particles obey Rayleigh size distribution. The temperatures of the layer and the low half space are T. The observation is in the direction of θ_0.

the following matrix form [9, 10]. For $0 \leqslant \theta \leqslant \pi$:

$$\cos\theta \frac{\partial}{\partial z}\bar{I}(z,\theta) = -\kappa_e \bar{I}(z,\theta) + \kappa_e\big(1-\widetilde{\omega}\big)CT\begin{bmatrix}1\\1\end{bmatrix} + \frac{3}{8}\kappa_e\widetilde{\omega}\int_0^{\pi} \mathrm{d}\theta' \sin\theta' \Big(\bar{\bar{P}}(\theta,\theta')\cdot\bar{I}(z,\theta')\Big) \tag{4}$$

where

$$\bar{I}(z,\theta) = \begin{bmatrix} I_v(z,\theta) \\ I_h(z,\theta)\end{bmatrix} \tag{5}$$

and I_v, I_h are the vertical and horizontal specific intensities, respectively. Also in (4), $C = (K_b K'^2)/(\lambda^2 k^2)$, K_b is Boltzmann's constant, λ is free space wavelength, K' is the real part of the effective wave number in region 1, k is the free space wave number and $\widetilde{\omega}$ is the albedo. The phase matrix $\bar{\bar{P}}$, the extinction rate κ_e and the albedo are calculated from dense medium theory and are different from the conventional radiative transfer theory [9, 10].

The boundary conditions for (4) are, for $0 \leqslant \theta \leqslant (\pi/2)$, $\beta = \mathrm{v, h}$ at $z = 0$

$$I_\beta(z, \pi - \theta) = r_\beta(\theta) I_\beta(z, \theta) \tag{6}$$

and at $z = -d$

$$I_\beta(z, \pi - \theta) = r_{12\beta}(\theta) I_\beta(z, \theta) + \big(1 - r_{12\beta}(\theta)\big) C T_2 \tag{7}$$

where r_β ($\beta = \mathrm{v, h}$) are the Fresnel reflectivities at the top boundary and $r_{12\beta}$ ($\beta = \mathrm{v, h}$) are the Fresnel reflectivities at the bottom boundary. The effective propagation constant for the snow medium has been taken into account in calculating both r_β and $r_{12\beta}$.

After (4) is solved subject to the boundary conditions of (6) and (7), the brightness temperatures in the direction θ_0, where $\theta_0 = \sin^{-1}(K' \sin\theta / k)$ is related to θ by Snell's law, for vertical and horizontal polarizations, are given by

$$\begin{bmatrix} T_{\mathrm{Bv}}(\theta_0) \\ T_{\mathrm{Bh}}(\theta_0) \end{bmatrix} = \frac{1}{C} \begin{bmatrix} (1 - r_{\mathrm{v}}(\theta)) I_{\mathrm{v}}(0, \theta) \\ (1 - r_{\mathrm{h}}(\theta)) I_{\mathrm{h}}(0, \theta) \end{bmatrix}. \tag{8}$$

For the numerical simulation and training of the neural network, we have used a Rayleigh size distribution with number density $n(a)$

$$n(a) = \frac{\pi f a}{16 \langle a \rangle^5} \exp\left(- \frac{\pi a^2}{4 \langle a \rangle^2} \right) \tag{9}$$

where $\langle a \rangle$ is the mean radius and f is the fractional volume of all the ice particles. The advantage of the Rayleigh size distribution is that there are only two parameters, f and $\langle a \rangle$.

Thus, the medium input physical parameters of the dense medium radiative transfer equation using the Rayleigh size distribution are:

(i) Complex relative complex constant of the particles ε_s: in this paper, it is assumed to be a known parameter. We choose $\varepsilon_s = 3.2 + i0.001$.

(ii) Fractional volume occupied by all the particles f: the snow density $\rho = f \times 0.917\ \mathrm{g/cm^3}$.

(iii) Mean radius $\langle a \rangle$.

(iv) Physical temperature T.

(v) Real part of the complex relative dielectric constant of the lower half space $\mathrm{Re}(\varepsilon_2)$.

(vi) Imaginary part of the complex permittivity of the lower half space $\mathrm{Im}(\varepsilon_2)$: we have assumed this to be a known parameter and have chosen $\mathrm{Im}(\varepsilon_2) = 0.0024$. This only has a small effect on the brightness temperature.

(vii) Snow depth d.

From these parameters, the brightness temperatures of vertical and horizontal polarization at specified frequencies and observation angles of 53° can be calculated by the above procedure (see (4)–(9)). We choose 7 values for each of the four parameters: ρ, $\langle a \rangle$, T and d, and 5 values for the last parameter, $\mathrm{Re}(\varepsilon_2)$. There are thus 12 005 input–output pairs generated. The range of the parameters are: $0.07\ \mathrm{g/cm^3} \leqslant \rho \leqslant 0.7\ \mathrm{g/cm^3}$; $0.01\ \mathrm{cm} \leqslant \langle a \rangle \leqslant 0.064\ \mathrm{cm}$; $207\ \mathrm{K} \leqslant T \leqslant 270\ \mathrm{K}$; $10\ \mathrm{cm} \leqslant d \leqslant 100\ \mathrm{cm}$; $3 \leqslant \mathrm{Re}(\varepsilon_2) \leqslant 7$. The snow water equivalent can be expressed in terms of snow density and snow depth. In the neural network input–output training and inversion, all the five parameters have their respective ranges normalized to lie between 0.05 and 0.95.

5. FORWARD MODELS FOR INVERSE PROBLEMS

Inverse problems admit of two lines of attack: creating a forward model $\hat{m} = \hat{\phi}(p)$ of the process $m = \phi(p)$ which then must be manipulated to yield an inverse [27], or creating an explicit inverse $\hat{p} = \hat{\phi}^{-1}(m)$ of the physical process [14], where p is the set of physical parameters, and m is the resulting set of measurements. An explicit inverse suffers from many-to-one or ill-conditioned problems when more than one parameter set could account for a particular measurement set. Forward models, on the other hand, can accurately model a causal relationship, since they map each parameter set to their resulting measurement set. With a method of inverting the forward model, we can find a possible multiplicity of solutions from which we can select according to other information or additional constraints we wish to impose.

The inversion of a forward model takes the form of a search in the input space of the model for an input which produces the desired output. With gradient information relating the input to some performance criteria, the search of the input space can proceed as a directed search, usually taken in the direction of this gradient.

For example, the passive remote sensing of snow is modelled by the dense media theory scattering model. These scattering models are implicit functions; we do not have a simple closed form equation from parameters to measurements, but must instead rely on numerical methods to calculate the solutions. As such, it is unfeasible to calculate gradients of the output measurement error with respect to the input parameters. We are left with no way to perform a directed search in the parameter space. One solution is to train an explicit inverse $\hat{p} = \hat{\phi}^{-1}(m)$, with the measurements as input and parameters as output. This suffers under familiar many-to-one problems.

A better solution is to train a data-driven model $\hat{m} = \hat{\phi}(p)$ to represent our implicit function, the scattering model $m = \phi(p)$, and as long as we can take gradients of the output measurement error with respect to the parameters, we can perform a directed search to obtain our inverted parameter value.

6. A DATA DRIVEN MODEL: THE MULTILAYER PERCEPTRON

Multilayer perceptrons, or MLPs, are feed forward neural networks which have one or more layers of hidden neurons between the input and output layers. They have a simple layer structure, where successive layers of neurons are fully interconnected, with connection weights controlling the strength of the connections. They can be trained to represent arbitrary input–output relationships. Once trained, the gradient of the input with respect to each output can be calculated, allowing a directed search in the input space with respect to criteria based on the output values.

6.1. Training multilayer perceptrons

The training phase of a multilayer perceptron uses the back propagation learning rule, an iterative gradient descent algorithm designed to minimize the mean squared error, E, between the desired target vectors, $\mathbf{t}$, and the actual output vectors, $\mathbf{z}$, given the input vector $\mathbf{y}$. The updating of the weights, $\mathbf{w}$, can be described as [11]:

$$\mathbf{w} \Longleftarrow \mathbf{w} + \Delta\mathbf{w} \Longleftarrow -\eta \frac{\partial E}{\partial \mathbf{w}} \tag{10}$$

where

$$E = E(\{\mathbf{w}\}, \{\mathbf{y}\}) = \frac{1}{2}\|\mathbf{t} - \mathbf{z}\|^2 \tag{11}$$

and the η term in (10) is a constant adjusting the size of the update terms. The θ values are updated analogously to the weights, $\mathbf{w}$. This is sometimes known as the recursive update rule. As long as the training vectors are randomly presented to the MLP, the running average of the gradient changes should roughly approximate the gradient found for the error function over the whole training set.

The gradient descent algorithm can be further refined to include a momentum term. The weights are updated according to the gradient of the current error and the previous weight change, where the previous weight change is scaled according to a momentum coefficient, μ.

$$\mathbf{w}_{n+1} \Longleftarrow \mathbf{w}_n + \Delta\mathbf{w}_n \Longleftarrow \mathbf{w}_n - \eta\frac{\partial E}{\partial \mathbf{w}} + \mu\Delta\mathbf{w}_{n-1}. \tag{12}$$

6.2. Multilayer perceptron inversion

In the backpropagation algorithm, error vectors are formed at the output of each layer of neurons. These error vectors are the gradient of the error function E with respect to the neuron output values. By running the backpropagation algorithm back to the input layer $\mathbf{y}$, we can find the gradient of the error function with respect to the input neurons [12]. With this gradient information in hand, we can iteratively update the values of the input neurons to minimize the output error. This process is known as MLP inversion, as it performs a gradient descent search in the input space to find the point with the lowest output error for a given desired vector, essentially providing the inverse function of the MLP.

The inversion of a network will generate the input vector, $\mathbf{y}$, that can produce a desired output vector. By taking advantage of the duality between the weights and the input activation values in minimizing the mean squared error, E, the iterative gradient descent algorithm can again be applied to obtain the desired input vector.

$$\mathbf{y} \Longleftarrow \mathbf{y} + \Delta\mathbf{y} \Longleftarrow \mathbf{y} - \eta\frac{\partial E}{\partial \mathbf{y}}. \tag{13}$$

As in backpropagation, the inversion process uses an η value to scale the search step size. The stopping criteria are set by output error criteria, and a maximum number of iterations.

7. APPLYING MLPS TO A REMOTE SENSING SNOW MODEL

We explored the performance of an explicit inverse and a forward model trained using data produced by our snow model.

7.1. Backpropagation training

Of the 12 005 input–output pairs created with the dense media theory scattering model, 11 005 were used for training, and the remaining 1000 for testing. The values were preprocessed to ranges appropriate to a MLP, a range between 0 and 1. For each of

Table 1.
Comparison of RMS error after training

Method	Testing RMS error	Training RMS error
Explicit inverse	0.310452	0.308993
Forward model	0.012229	0.012334

the snow parameters, their full range was linearly scaled to lie between 0.05 and 0.95, leaving some room for extrapolation. The measurement temperatures were all linearly scaled so that the temperature range 0K to 275K was mapped to the full 0 to 1 range.

We trained two 2-layer MLPs: a forward model $\hat{m} = \hat{\phi}(p)$, training from the snow parameters to the predicted satellite measurements, and an explicit inverse $\hat{p} = \hat{\phi}^{-1}(m)$, training from the predicted satellite brightness temperature measurements to the snow parameters. Each MLP had one hidden layer with 30 neurons, and was trained by backpropagation with momentum (see Section 6.1) with an η value of 0.1 and a λ value of 0.9 for 1000 iterations of the training set.

The input-output pairs were randomly ordered, and the MLPs connection weights in $\mathbf{w}(l+1)$ were initialized randomly by a uniform distribution, from $[-5/N_l, 5/N_l]$, where N_l is the number of neurons in the lth layer.

Each network was then evaluated against the training data and testing data for training accuracy. The results are presented in Table 1, along with a comparison of their average training errors. Note that the errors are calculated from the output errors: measurement error for the forward model, and snow parameter error for the explicit inverse. Note that the explicit inverse had a much higher error rate. This was caused by the highly nonlinear, many-to-one nature of the inverse scattering problem. The forward model did not have these problems, and trained to a small error.

7.2. Testing on a synthetic landscape

We have created an artificial landscape and defined the snow parameters at each point on our landscape, within the range of values over which the network was trained. The landscape was a 16 by 16 grid of snow parameter values, with the associated satellite measurements calculated by the scattering model mapping. The satellite measurements were then used by each method to reconstruct the parameters of our synthetic landscape.

We assumed that there were certain locations in the map for which the snow parameters were known a priori. By using a weighted potential of these points, weighting them according to the inverse of the distance from the ground truth points, we found initial guesses for the snow parameter values across the landscape, and used these initial guesses in the inversion of the forward model. Inversion of the forward model used five initial points of ground truth, at the center of the plot (8,8), and at (4,4), (4,13), (13,4), and (13,13), to supply initial guesses for each point.

Applying topological constraints into our reconstruction method may give us a good chance of constraining our predicted surface sufficiently to eliminate snow parameters which could have caused a particular measurement, but can not be smoothly fitted into the contours of parameters.

Figures 8, 9, and 10 show the original synthetic landscape, and the landscapes for the reconstructions using the explicit inverse and inversion on the forward model, respectively. For the purposes of comparison, all five parameters are shown normalized to the ranges described at the end of Section 4.

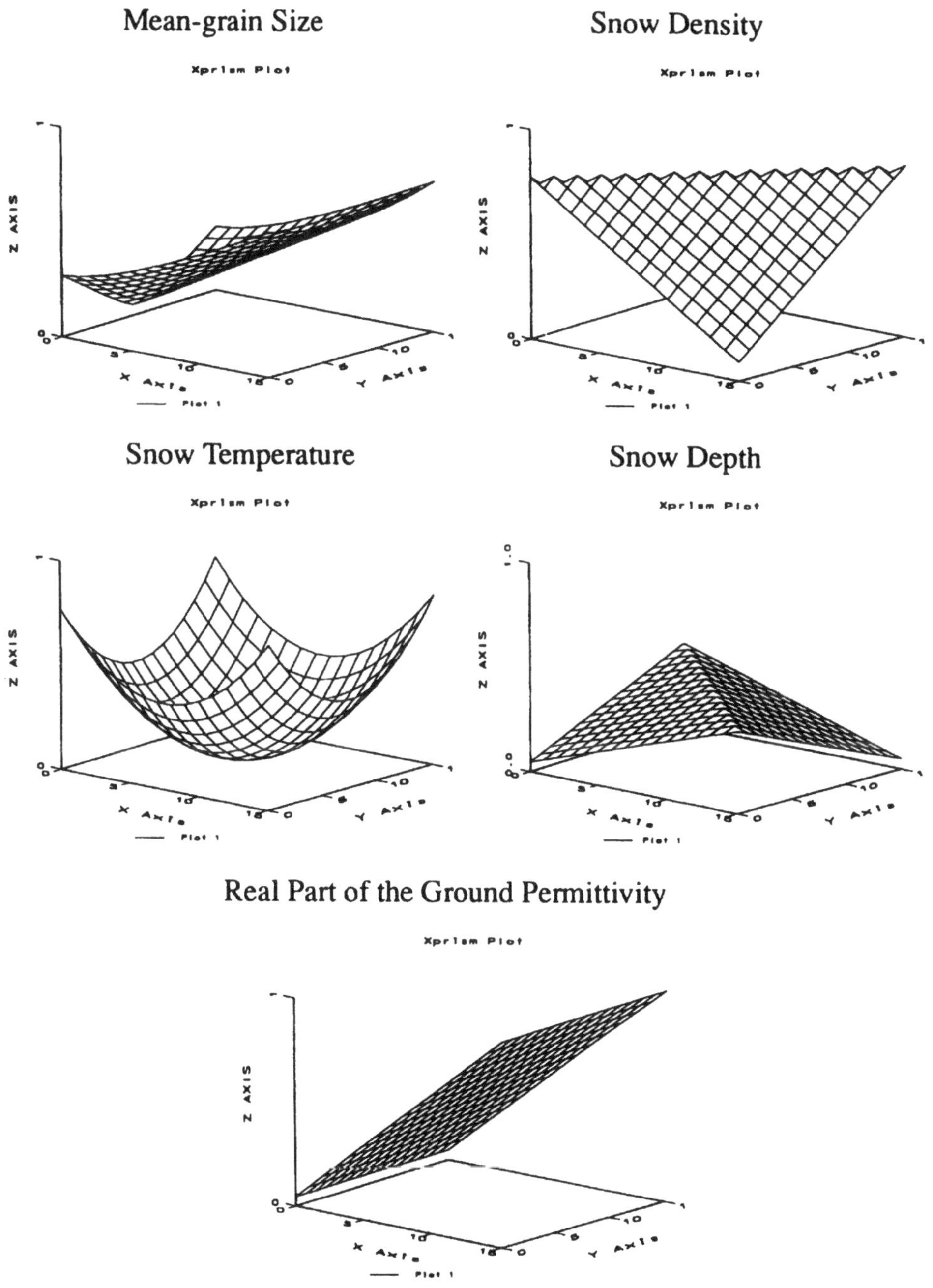

Figure 8. Original synthetic contours of snow parameters.The range of the parameters are: $0.07\ \mathrm{g/cm^3} \leqslant \rho \leqslant 0.7\ \mathrm{g/cm^3}$; $0.01\ \mathrm{cm} \leqslant \langle a \rangle \leqslant 0.064\ \mathrm{cm}$; $207\ \mathrm{K} \leqslant T \leqslant 270\ \mathrm{K}$; $10\ \mathrm{cm} \leqslant d \leqslant 100\ \mathrm{cm}$; $3 \leqslant \mathrm{Re}(\varepsilon_2) \leqslant 7$. The snow water equivalent can be expressed in terms of snow density and snow depth. All the five parameters have their respective ranges normalized to lie between 0.05 and 0.95.

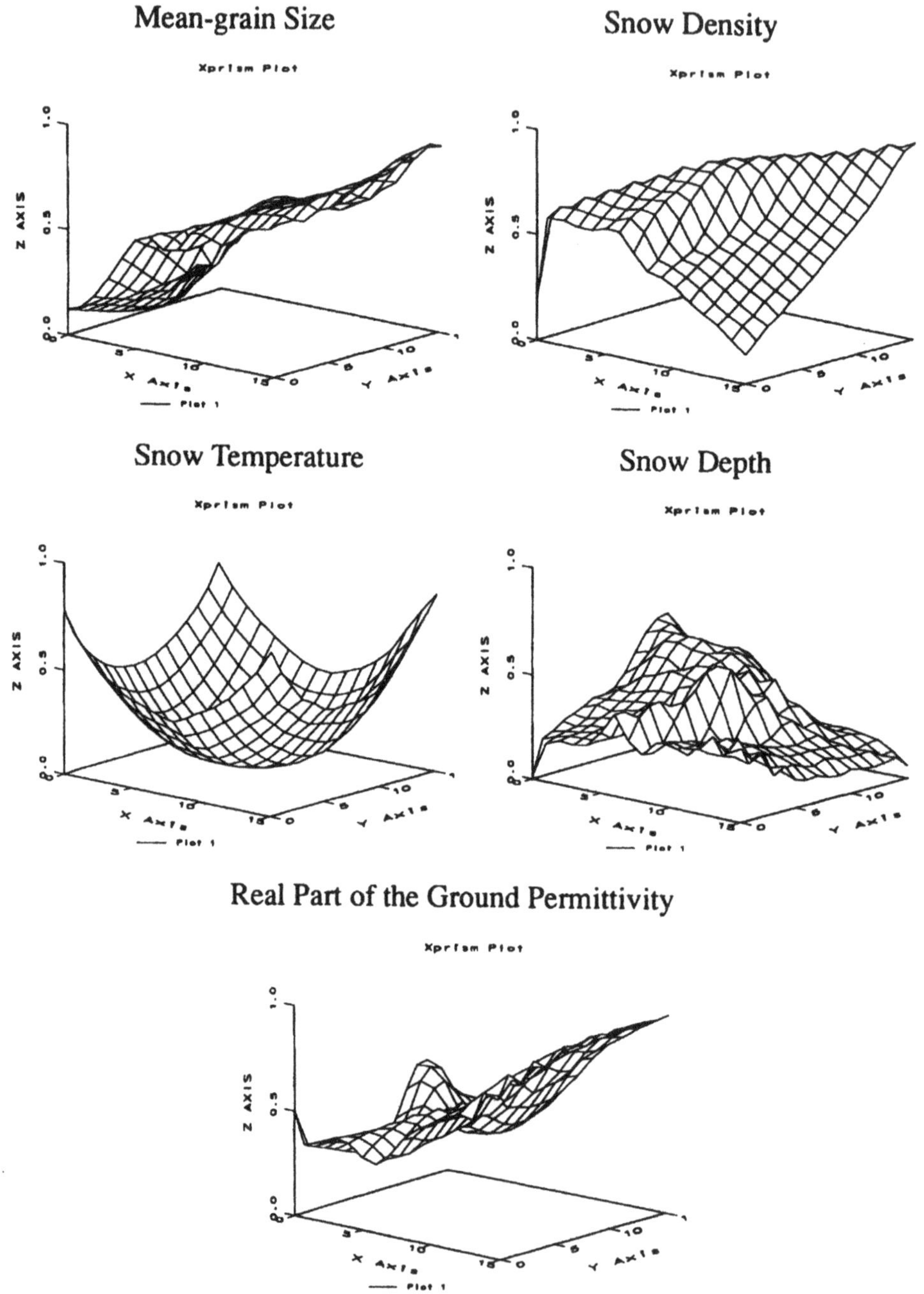

Figure 9. Contours reconstructed by explicit inverse. The range of the parameters are: $0.07\,\mathrm{g/cm^3} \leqslant \rho \leqslant 0.7\,\mathrm{g/cm^3}$; $0.01\,\mathrm{cm} \leqslant \langle a \rangle \leqslant 0.064\,\mathrm{cm}$; $207\,\mathrm{K} \leqslant T \leqslant 270\,\mathrm{K}$; $10\,\mathrm{cm} \leqslant d \leqslant 100\,\mathrm{cm}$; $3 \leqslant \mathrm{Re}(\varepsilon_2) \leqslant 7$. The snow water equivalent can be expressed in terms of snow density and snow depth. All the five parameters have their respective ranges normalized to lie between 0.05 and 0.95.

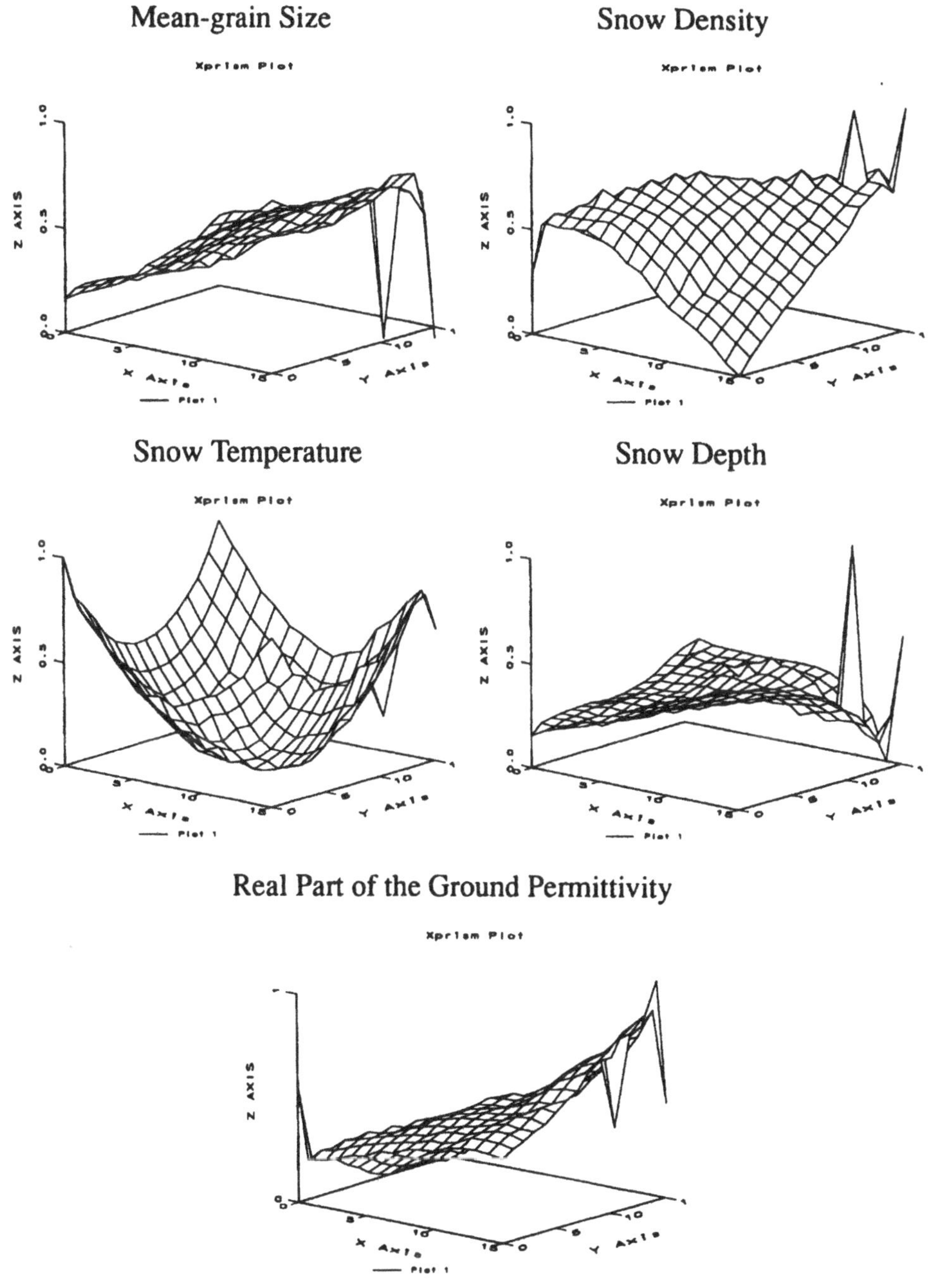

Figure 10. Contours reconstructed by iterative inversion. The range of the parameters are: $0.07\,\mathrm{g/cm^3} \leqslant \rho \leqslant 0.7\,\mathrm{g/cm^3}$; $0.01\ \mathrm{cm} \leqslant \langle a \rangle \leqslant 0.064\ \mathrm{cm}$; $207\,\mathrm{K} \leqslant T \leqslant 270\ \mathrm{K}$; $10\ \mathrm{cm} \leqslant d \leqslant 100\ \mathrm{cm}$; $3 \leqslant \mathrm{Re}(\varepsilon_2) \leqslant 7$. The snow water equivalent can be expressed in terms of snow density and snow depth. All the five parameters have their respective ranges normalized to lie between 0.05 and 0.95.

We find that both methods were extremely accurate in predicting snow temperature, with the explicit inverse performing slightly better. For all other parameters, inversion of the forward model gives a discernable performance enhancement. Except for a couple of outliers in the landscape reconstruction, the landscapes produced by inversion of the forward model are clearly smoother. The outliers give clear discontinuities in the reconstruction, and as such we have no doubt that they may easily be smoothed from the reconstruction. Given the slightly superior performance for the explicit inverse on snow temperature, it would probably make sense to use the prediction of the explicit inverse as a fixed parameter in the iterative inversion method, thus using both models on the parameters which they predict best.

8. CONCLUSION

In this paper, we have proposed using a data-driven model, the multilayer perceptron, to facilitate solving the inverse problem for a snow scattering model. The MLP is first trained to match the input-output relationship of the scattering model. With the trained MLP, we are able to take the gradients of the snow parameters with respect to the satellite measurements, and so we have the means to perform a directed search of the input space to locate a set of snow parameters which could account for any given set of satellite measurements. The process of searching the input space of a MLP using the gradient of the output error is known as iterative inversion.

Iterative inversion of a trained forward model has further advantages. It allows use of many different cost functions, and the imposition of many different constraints. For example, it allows incorporation of ground truth data and topological constraints into the reconstruction of a grid of points with known topological structure, and we demonstrate this ability in our performance comparison between this method and the method of training an explicit inverse to reconstruct a landscape of snow parameter values from their associated satellite measurements. Iterative inversion of a trained forward model is shown superior to using an explicit inverse to perform the inverse problem.

The ability to incorporate additional information into the iterative inversion process gives this method real power. The use of data-driven models also gives the opportunity to incorporate ground truth data into the training of the model, giving us a synthesis of theoretical and measured data in the trained forward model [16].

REFERENCES

1. A. T. C. Chang, B. J. Choudhury and P. Gloersen. Microwave brightness of polar firn as measured by Nimbus-5 and -6 ESMR. *J. Glaciology*, **25** (91) (1980).
2. K. F. Kunzi, S. Patil and H. Rott. Snow-cover parameters retrieved from Nimbus-7 Scanning Multi-Channel Microwave Radiometer (SMMR) data. *IEEE Trans. Geosci. Remote Sensing*, **20**, 453–467 (1982).
3. A. T. C. Chang, J. L. Foster and D. K. Hall. Nimbus-7 derived global snow-cover parameters. *Annals Glaciology*, **9**, 39–44 (1987).
4. T. C. Chang, P. Gloersen, T. Schmugge, T. T. Wilheit and H. J. Zwally. Microwave emission from snow and glacier ice. *J. Glaciology*, **16**, 23–29 (1976).
5. J. A. Kong, R. Shin, J. C. Shiue and L. Tsang. Theory and experiment for passive remote sensing of snow-packs. *J. Geophys. Res.*, **84**, 5669–5673 (1979).

6. L. Tsang, J. Kong and R. Shin. *Theory of Microwave Remote Sensing*. New York, Wiley-Interscience (1985).
7. H. J. Zwally. Microwave emissivity and the accumulation rate of polar firn. *J. Glaciology*, **18**, 195–215 (1977).
8. J. R. Wang, A. T. C. Chang and A. K. Sharma. On the estimation of snow depth from microwave radiometer measurements. *IEEE Trans. Geosci. Remote Sensing*, **30** (4), 785–792 (1992).
9. L. Tsang. Passive remote sensing of dense nontenuous media. *J. Electr. Waves Appl.*, **1** (2), 159–173 (1987).
10. L. Tsang. Dense media radiative transfer theory for dense discrete random media with spherical particles of multiple sizes and permittivities. Ch. 5. In: *Progress in Electromagnetic Research*, Vol. 6, Elsevier, New York (1992).
11. D. E. Rumelhart, G. E. Hinton and R. J. Williams. Learning internal representations by error propagation. In: *Parallel Distributed Processing (PDP): Exploration in the Microstructure of Cognition*, Vol. 1, Chapter 8, pp. 318–362, MIT Press, Cambridge, Massachusetts (1986).
12. A. Linden and J. Kindermann. Inversion of multilayer nets. In: *Proc. Int'l Joint Conf. on Neural Networks*, Vol. II, pp. 425–430, Washington DC (1989).
13. R. P. Lippmann. Introduction to computing with neural nets. *IEEE ASSP*, **4** (2), 4–22 (1987).
14. L. Tsang, Z. Chen, Seho Oh, R. J. Marks II and A. T. C. Chang. Inversion of snow parameters from passive microwave remote sensing measurements by a neural network trained with a multiple scattering model. *IEEE Trans. Geosci. Remote Sensing*, **30**, 1015–1024 (1992).
15. D. T. Davis, Z. Chen, L. Tsang, J. N. Hwang and A. T. C. Chang. Inversion of snow parameters by neural network with iterative inversion. *IEEE Trans. Geosci. Remote Sensing*, **31**, 842–852 (1993).
16. J. N. Hwang, D. T. Davis and L. Tsang. A locally tuned neural network for ground truth incorporation. In: *Int'l Geoscience and Remote Sensing Symposium, Houston*, Vol. 2, pp. 1064–1066 (1992).
17. A. T. C. Chang and L. Tsang. A neural network approach to inversion of snow water equivalent from passive microwave measurements. *Nordic Hydrology*, **23**, 173–182 (1992).
18. A. S. Gurvich, V. L. Kalinin and D. T. Matveyer. Influence of the internal structure of glaciers on their thermal radio emission. *Atm. Oceanic Physics*, USSR, **9**, 713–717 (1973).
19. L. Tsang and J. A. Kong. The brightness temperature of a half space random media with non-uniform temperature profile. *Radio Science*, **10** (12), 1025–1033 (1975).
20. P. J. Stephenson. Some considerations of snow metamorphism in the antarctic ice sheet in the light of ice crystal studies. In: *International Conference on Low Temperature Science-Conference on Physics of Snow and Ice Proceedings*, Vol. 1, part 2, pp. 725–740 (1967).
21. R. Alley, J. Bolzan and I. Whillans. Polar firn densification and grain growth. *Annals Glaciology*, **3**, 7–11 (1982).
22. H. Rott, K. Sturm and H. Miller. Active and passive microwave signatures of antarctic firn by means of field measurements and satellite data. *Annals Glaciology*, **17**, 337–343 (1993).
23. L. Tsang, E. Njoku and J. A. Kong. Microwave thermal emission from a stratified medium with non-uniform temperature distribution. *J. Appl. Physics*, **46**, 5127–5133 (1975).
24. E. G. Njoku and J. A. Kong. Theory of passive sensing of near-surface soil moisture. *J. Geophys. Res.*, **82**, 3108–3118 (1977).
25. A. Ishimaru. *Wave Propagation and Scattering in Random Media*. Vol. 2, Academic Press (1978).
26. G. Backus and F. Gilbert. Uniqueness in the inversion of inaccurate gross earth data. *Phil. Trans. Roy. Soc. London* Ser. A266, 123–192 (1970).
27. J. N. Hwang, and C. H. Chan. Iterative constrained inversion and its applications. In: *24th. Conf. on Information Systems and Sciences*, pp. 754–759, Princeton (1990).
28. C. Mätzler. Applications of the interaction of microwaves with the natural snow cover. *Remote Sensing Reviews*, **2**, 259–392 (1987).
29. A. Ishimaru and Y. Kuga. Attenuation constant of coherent field in a dense distribution of particles. *J. Opt. Soc. Am.*, **72**, 1317–1320 (1982).

30. L. Tsang, C. Mandt and K. H. Ding. Monte Carlo simulations of extinction rate of dense media with randomly distributed spheres based on solution of Maxwell's equations. *Optics Letters*, **17** (5), 314–316 (1992).

31. J. N. Hwang, C. H. Chan and R. J. Marks II. Frequency selective surface design based on iterative inversion of neural network. In: *Int'l Joints Conf. on Neural Networks*, pp. 39–44, San Diego (1990).

Vegetation

ESA/NASA International Workshop, pp. 317–344
B. J. Choudhury, Y. H. Kerr, E. G. Njoku and P. Pampaloni (Eds)

Vegetation models and observations
A review

YANN H. KERR[1] and JEAN PIERRE WIGNERON[2]

[1]*LERTS, UMR CNES-CNRS, 18 Avenue Edouard Belin, 31055 Toulouse Cedex, France*
[2]*INRA, Bioclimatologie, Domaine St Paul, Montfavet, France*

Abstract—The aim of this paper is to give an overview of the vegetation canopy modelling in passive microwaves. A very concise overview of existing observations is also given together with some possible recommendations concerning the planning of future field experiments. Passive microwaves have been used, over land surfaces, primarily to assess surface soil moisture. Yet, vegetation canopies do have an influence on the measured signal and have to be taken into account in any soil moisture retrieval algorithm. It has also been found that the microwave signal gave information such as biomass or structure on the canopy itself, information that might not always be obtainable with measurements at other wavelengths (i.e. optical and thermal bands).

Several approaches have been developed to model vegetation canopies. They can be separated mainly into 3 types. The first type is somewhat simplistic and considers vegetation as a uniform layer whose characteristics (opacity, albedo, ...) are constant and computed with simple formulas or empirical relationships. The two other methods differ in the way the canopy is defined. They are somewhat more complex. The second approach still considers the medium as continuous but defines it through correlation lengths. The third method uses a discrete characterization of the medium (lengths and orientations of the elements such as stalks and leaves). The paper presents the different approaches and describes the state of the art. The inversion problem and algorithm development are then discussed, i.e. the state of the art, the philosophy driving the research, and the planned future activities in direct and inverse modelling.

1. INTRODUCTION

Passive microwaves have long been used for a variety of applications, mainly over oceans and for characterizing the atmosphere. Over land surfaces, applications have been somewhat less, mainly due to the discrepancy between the spatial resolution of current satellite borne sensors and the spatial variability of surface parameters. Moreover, due to the availability of data collected in the visible and near infrared with finer resolution, studies over land performed with passive microwaves have been mainly restricted to surface moisture estimation. This is due to the poor spatial resolution attainable with passive microwaves versus the typical high resolution optical data and the fact that soil moisture is directly measured by microwave sensors only. Consequently vegetation assessment with use of passive microwaves has not yet reached a high degree of maturity.

It is apparent however that these considerations have to be reconsidered for the following reasons: at low frequencies, passive microwave radiometry is much less affected by the atmosphere than shortwave radiometry; recent progress in radiometer design has brought the availability of very stable and sensitive instruments (see for example the Special Sensor Microwave Imager SSM/I) and future polar platforms will carry such an instrument having also low frequency channels and improved spatial resolutions (Multi-frequency Imaging Microwave Radiometer MIMR). Finally the somewhat coarse spatial resolution coupled with a high temporal resolution makes the data of very high interest for global or mesoscale climate models.

As far as vegetation is considered, passive microwaves are of high potential interest simply because the signal is directly linked to the integrated water content of the layer. It can be related to vegetation parameters such as biomass, leaf area index (LAI), and water stress, and eventually might give an estimate of water availability in the root extraction zone. However the complexity of the task must not be overlooked, and results are not yet at hand, as we will see later.

Vegetation modelling for microwaves started in the seventies with semi-empirical, physically based approaches. Linked to the more sophisticated modelling carried out in the active microwave domain, more sophisticated approaches have been developed. Three types of approaches will thus be described. The first one, often labeled 'simple' approach, considers vegetation as a uniform layer and is easy to use. It consists in estimating the vegetation opacity through the use of physically based semi-empirical relationships between vegetation integrated water content, canopy structure, frequency and the opacity. Though simplistic, this approach gives satisfactory results at low frequencies and has been adapted to higher frequencies by coupling an optical approach. The advantage of such an approach is that the input parameters are relevant to vegetation characteristics and rather simple. Consequently inversion algorithms deliver directly useful parameters.

The second and third types of approach are more sophisticated. They are most often based on models developed in the active microwaves. They take into account volume absorption and scattering by the discrete scatterers of the canopy (trunks, stalks, branches, leaves). So as to solve the radiative transfer equations one needs to compute the scattering and absorption with the phase matrix and the extinction and absorption coefficients. For this, two approaches are possible: either to consider the canopy as a continuous medium the dielectric perturbations are characterized by two parameters, the correlation lengths in the vertical direction and the horizontal plane (second approach). Or, as in the third approach, the canopy is considered as a discrete medium characterized by the scatterers (elliptical or cylindrical). The continuous medium approach offers the advantage of being somewhat simpler, especially as far as the characterization of the canopy is concerned. However, for the inversion procedure, the output parameters (correlation lengths) are difficult to relate to actual vegetation characteristics. The discrete approach is much more satisfying since it actually takes into account the vegetation structure and characteristics. But, in this latter case, the number of parameters is large and the model requires very sophisticated knowledge of the canopy. Inversion is very complex.

The general idea is thus, at the moment, to use a discrete approach to model vegetation canopies and, through the use of sensitivity analysis, to assess the most relevant canopy parameters. From this, simplified models could be established, allowing the derivation of tractable inversion algorithms. Also, it can be used to establish relationships between relevant vegetation characteristics and opacity and single scattering albedo as a function of frequency. These relationships could then be used in simpler models.

1. MODELS

1.1. Background

The first point to be well stressed is that the measured signal is a complex convolution of the influence of the atmosphere, the canopy itself and the soil underneath the canopy. While in the shortwave part of the spectrum only the cloudless atmosphere is semi-transparent, in the microwave spectrum, the atmosphere (including clouds) and the canopy are transparent at low frequencies. In order to quantify grossly, it can be said that at 85 GHz the atmosphere is still sufficiently transparent for the sensor to 'see' the surface except in the case of rain and precipitating clouds. In a more quantitative way, Fig. 1 shows the atmospheric opacity as a function of frequency for several values of integrated vapour and liquid water contents [1]. Optical radiometry gives information on the envelop of the canopy, while microwave radiometry gives information on an integrated layer of the vegetation, the depth of the integrated layer being linked to the frequency. The measured signal is therefore much richer in the microwaves but more complex to analyse. Generally speaking, the higher the frequency the smaller the penetration depth will be. The signal emanating from the soil will be obscured by vegetation depending upon frequency and vegetation integrated water content as shown on Fig. 2 [2]. One can see that at C band (6 cm), for instance, vegetation almost masks the signal from the soil when its integrated water content is higher than 4 kg/m^2.

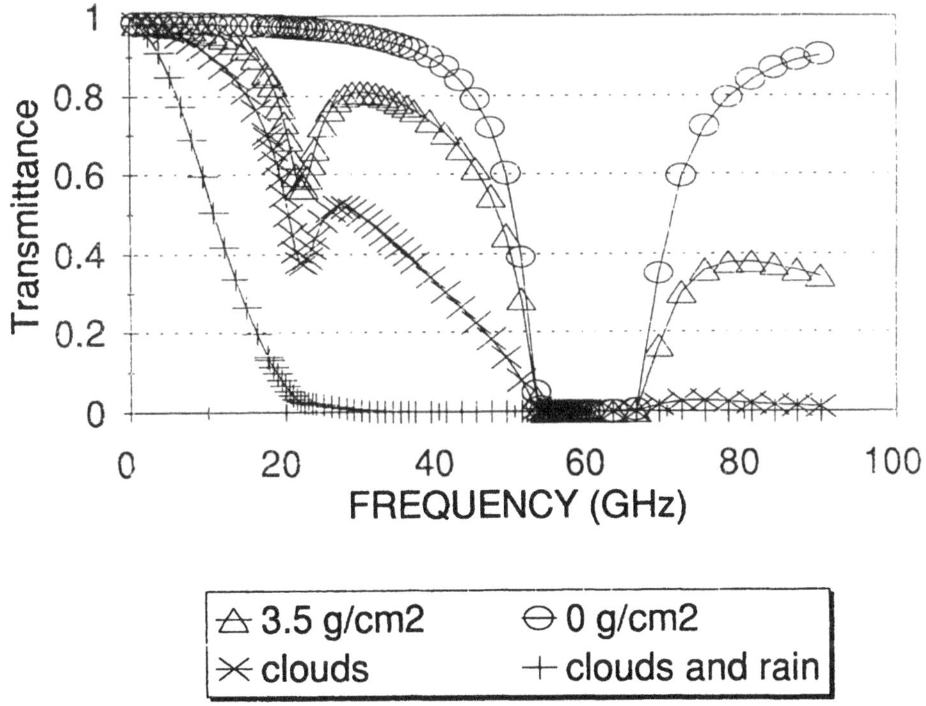

Figure 1. Influence of the atmosphere on passive microwaves as a function of frequency. The calculations were done using the model described in [1] for a) an atmosphere without any water, b) an atmosphere with an integrated water content of 3.5 g/cm^2 and a 15°N atmosphere profile, c) same atmosphere as in b) and with a cumulus congestus cloud (ILWC = 0.134 g/cm^2), d) same as c) with also a rain of 25 mm/hr.

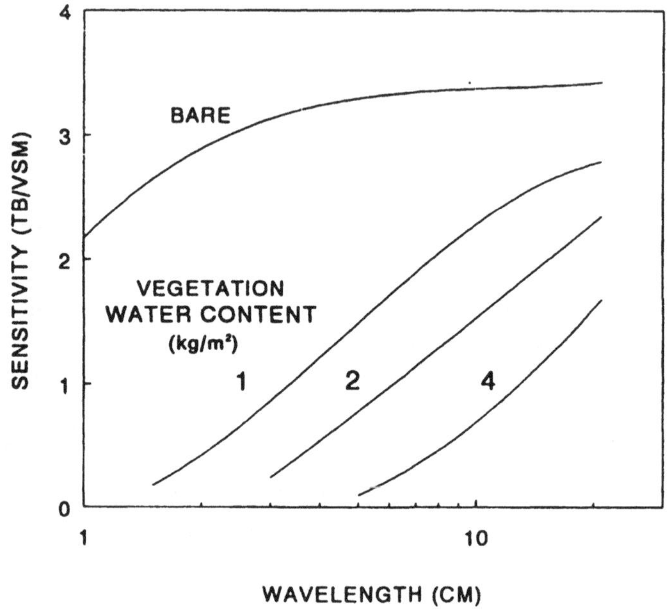

Figure 2. Sensitivity of brightness temperature to soil moisture as a function of wavelength for bare and vegetated conditions (H polarization, 10° view angle). From Jackson, (1993) [2].

The brightness temperature T_{Bp} measured by a sensor at polarization p, consists of three components: the upwelling atmospheric contribution T_{au}; the surface emission T_{bp} attenuated by the atmosphere (τ_{au}); and the sum of the cosmic background T_{sky} attenuated by the atmosphere on the downward path (τ_{tad}) and the atmospheric downwelling atmospheric emission T_{ad} reflected at the surface (r_{sp}) and attenuated along the upward path by the atmosphere (τ_{au}). Combining these components gives the radiative transfer equation (see for example [1]):

$$T_{Bp} = T_{au} + T_{bp}\exp(-\tau_{au}) + \big(T_{ad} + T_{sky}\exp(-\tau_{ad})\big)\exp(-\tau_{au})r_{sp} \tag{1}$$

where T_{bp} is the term of interest for surface studies, though, strictly speaking, the surface characteristics also have an influence on the third term of Equation (1). Using simplified radiative transfer equations, the surface contribution (i.e. brightness temperature of the target only) can be expressed as [1]:

$$\begin{aligned} T_{bp} = {} & \varepsilon_{sp}T_e\exp(-\tau_c/\mu) + (1-\omega_p)T_c\big[1-\exp(-\tau_c/\mu)\big] \\ & + (1-\varepsilon_{sp})(1-\omega_p)T_c\big[1-\exp(-\tau_c/\mu)\big]\exp(-\tau_c/\mu) \end{aligned} \tag{2}$$

where ε_{sp} is the soil emissivity, T_e is the soil effective temperature, τ_c is the canopy opacity, μ is the cosine of the viewing angle, ω_p is the single scattering albedo, T_c is the canopy effective temperature and the subscript p denotes the polarization (h or v).

It is then apparent from Equation (2) that, to describe the surface contribution, a number of parameters must be quantified. Moreover, as the vegetation rarely covers 100% of the footprint, the relative contributions of bare soil and vegetation covered soil have to be quantified. In the following we will consider only the vegetation canopy and assume that the underlying soil and the atmosphere are well known and modelled. It can nevertheless be noted that to compute the soil emissivity, it is necessary to quantify the equivalent temperature, the surface soil moisture and roughness, and the soil texture [1].

To characterize the vegetation contribution, it is necessary to quantify the vegetation fractional cover, vegetation type and water content, and vegetation equivalent temperature so as to assess the opacity and single scattering albedo. Consequently, the first statement that comes to mind is that the number of unknowns is large and retrieval of vegetation parameters does not appear to be a simple matter.

Generally speaking, vegetation models fall into two big categories and two main approaches can be identified:

i) Either the aim is to retrieve soil moisture, and then vegetation is considered as a perturbating factor (similar to the atmosphere), and characterized with a simplistic approach (i.e. water content, general structure);

ii) or the aim is to estimate vegetation parameters, which leads to more sophisticated models. Those sophisticated models result mainly from work done with scatterometers and adapted to emission models. They rely on an as accurate as possible characterization of the canopy geometry and vegetation components dielectric properties.

The first approach assimilates the canopy to a uniform layer with very little description on the canopy structure, when the second approach requires knowledge of the vegetation geometry (branches and leaves sizes, orientation, number of elements per unit volume, etc.). In what follows we will give an overview of the most used models for vegetation.

1.2. Semi-empirical models

Semi-empirical models are qualified here by the term 'simple' since they are most often straightforward to use and require limited information on the vegetation. It must nevertheless be noted that 'simple' models are physically based. In the latter case the main assumption is that multiple scattering is neglected. Most of the 'simple' models were originally developed for assessing the contribution of the vegetation layer in order to retrieve soil moisture. Consequently they are mostly valid at low frequencies. They frequently consider vegetation as an absorbing layer where scattering is low or negligible. In particular for radiative transfer equations, the phase matrix, which will be introduced in the following section about 'physical models', is neglected. Therefore analytical formulations can be derived. The general philosophy consists in assuming that the canopy is a homogeneous layer with an effective complex dielectric constant ε_c.

1.2.1. Dielectric properties of vegetation: an example. In order to assess the vegetation contribution, one approach consists in computing the dielectric properties of vegetation. Vegetation can be considered as a mixing of air, dry matter and water (either free or bounded). Ulaby and El-Rayes [3] proposed a model for computing vegetation dielectric properties. After assessing from experiments, the relaxation frequency of bound water (0.18 GHz) in comparison to that of free water (18 GHz) at 295 K, and the relationship between vegetation water content and the proportion of bound water, the dielectric constants ε_v of vegetation can be computed:

$$\varepsilon_v = \varepsilon_r + \varepsilon_{bw} v_{bw} + \varepsilon_{fw} v_{fw} \tag{3}$$

where v_{bw} and v_{fw} are respectively the volume fractions of bound and free water, ε_{bw} and ε_{fw} the dielectric constants of bound and free water, ε_r the dielectric constant of dry matter. It is then possible to estimate the value of ε_v as a function of gravimetric water content that can be related to the integrated water content once the vegetation height d and volume fraction vol_{fv} are known. Also, formulation of ε_v as a function of frequency and dry matter content has been recently developed for higher frequencies by Mätzler and Sume [4]. Dry matter content is estimated from the average leaf thickness and its water equivalent thickness. Figure 3 depicts the results obtained at 21, 35, and 94 GHz. The method is based upon a linear fit and the frequency dependence is obtained by assuming the medium obeys a Debye relaxation spectrum and that the 35 GHz data were the best reference to fit the curves. They found thus the expression for the complex dielectric constant of a leaf as given by:

$$\varepsilon_v = \varepsilon_{inf} + \left[(\varepsilon_s - \varepsilon_{inf})/(1 - i f/f_0)\right] \tag{4}$$

with:

$$\varepsilon_{inf} = 2.3 + 3.8\,\mathrm{md} \tag{5}$$

$$\varepsilon_s = 41 - 40\,\mathrm{md} \tag{6}$$

$$f_0 = 15\,\mathrm{GHz}$$

$$\mathrm{md} = d - \left[(M - M_d)/S\rho\right]/d$$

where d is the leaf mean thickness, S the leaf area, and ρ the leaf density (varies between 0.95 and 1 g/cm^3).

It is then possible to derive the canopy dielectric constant ε_c through the expression:

$$\varepsilon_c^\beta = 1 + \mathrm{vol}_{fv}(\varepsilon_v^\beta - 1). \tag{7}$$

β is here a constant. For $\beta = 1$, Equation (7) is known as the linear model, while for $\beta = 0.5$ as the refractive model and for $\beta = 0.33$ as the cubic model [5].

In a recent paper, Schmugge and Jackson [6] tested the above formulation to calculate the bulk dielectric constant of canopies. The canopy opacity was then derived and compared to field measurements. Figure 4 shows the resulting values of the vegetation dielectric constant. They correspond to an integrated water content of 2.4 (solid lines) and 3.2 (dashed lines) kg/m^2. The authors found that the square root dielectric mixing model was better than the linear one which tended to overestimate the losses in the canopy (Fig. 5). They also found that the model, though simple, gave a reasonable accuracy at relatively low frequencies and high water content, but might be questionable above 20 GHz. For low vegetation water contents, the expression is not valid [4]. Several other dielectric mixing models have been described [5, 7], but validations are yet to be investigated.

(a)

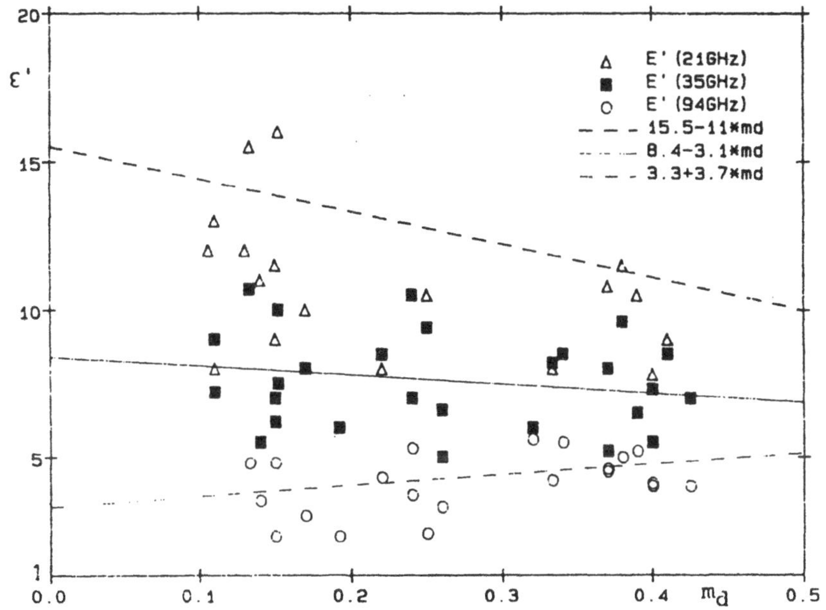

(b)

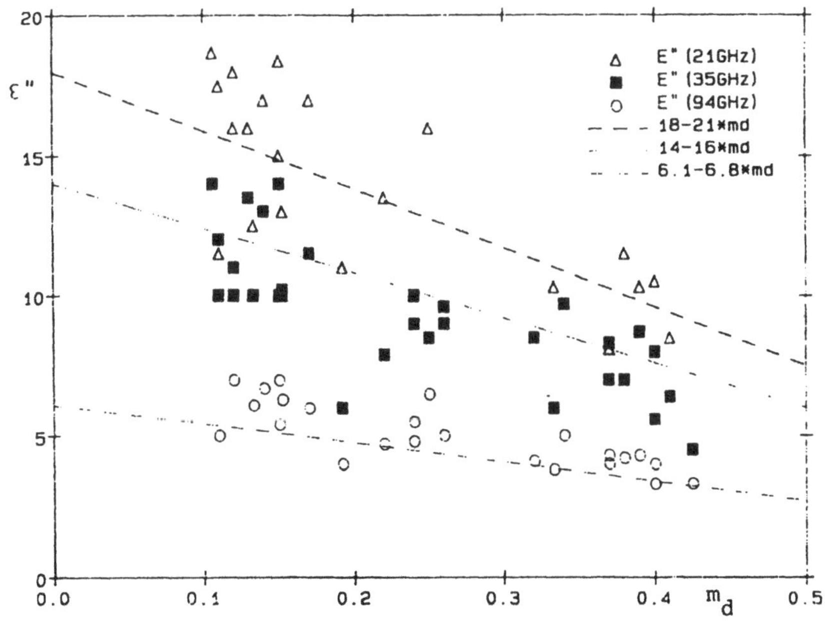

Figure 3. Measured values of the real (a) and imaginary part (b) of the dielectric constant of a leaf at 21, 35, and 94 GHz. The solid line is the 35 GHz linear fit and the dashed lines follow from the Debye equation as given in Equations (4) to (6). From Mätzler and Sume, 1989 [4].

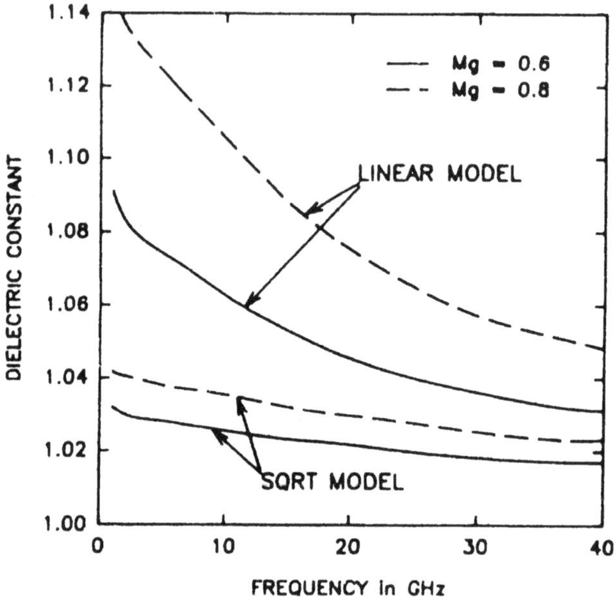

(a)

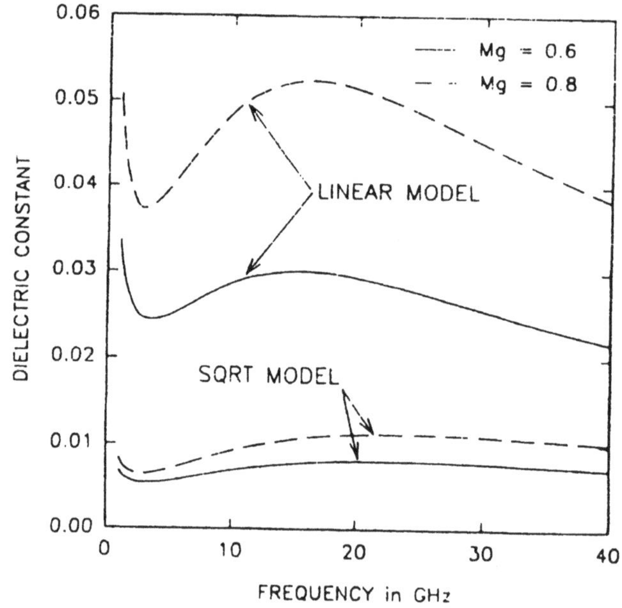

(b)

Figure 4. Real (a) and imaginary (b) part of the dielectric constant for vegetation canopy as computed with Equation (7) with $\beta = 1$ (linear model) and 0.5 (square root model), and assuming a vegetation height of 1 m, and a vegetation filling fraction of 4×10^3. The solid lines are for a gravimetric water content of 0.6 g/cm^3, and the dashed lines for a gravimetric water content of 0.8 g/cm^3. From Schmugge and Jackson, 1992 [6].

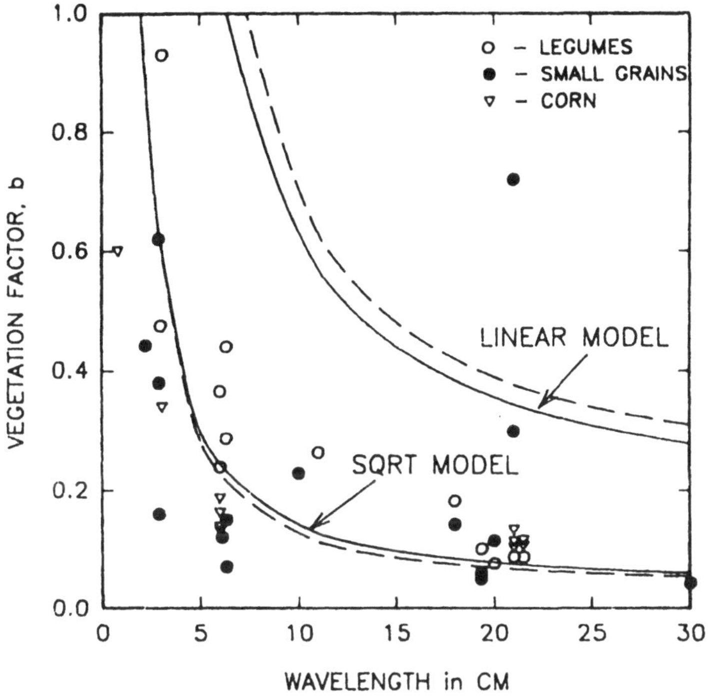

Figure 5. A comparison of the values of the coefficient b of Equation (11) and those derived from the linear and square root dielectric mixing models for the canopy. The solid curves are for WC values of 2.4 (solid lines) and 3.2 (dashed lines) kg/m^2. From Schmugge and Jackson, 1992 [6].

1.2.2. Canopy opacity. It is possible to obtain the canopy opacity directly from the canopy dielectric constant:

$$\tau_c = 4\pi h/\lambda \, \mathrm{Im}\left(\sqrt{\varepsilon_c}\right) \tag{8}$$

where h is the canopy height and ε_c is given by Equation (7). Good agreement was found with measurements [6], though the validity above 20 GHz is still to be assessed. It must be noted that the canopy water content is taken into account in the formulation of ε_c.

Other approaches have also been investigated. To characterize vegetation in a simple way, one needs to express the opacity and single scattering albedo of the vegetation. To achieve this purpose one may perform measurements over vegetation and derive empirical laws, or use models, or any mixture of the two. Several attempts were made to compare theory with ground measurements (see for example [8, 9]). In [8] the measurements were made at 2.7 and 5.1 GHz. It was found that the single scattering albedo ω_p varied between 0.05 and 0.13 but frequency and polarization dependence could not be addressed with the range of measurements available. The dependency on crop geometry with time was supposed to be negligible for a given crop. However other authors reported measurements giving much higher values for the single scattering albedo, as high as 0.5 [4], suggesting that more efforts should be performed in the modelling of ω_p.

As far as opacity is concerned, canopy geometry appeared to have a strong influence [10, 11]. In [9] the measurements were performed at 10 and 37.5 GHz. It was found that ω_p was always less than 0.1 and that the opacity τ_c^* could be expressed as a function of wavelength λ and integrated plant water content WC (in kg/m^2) according to:

$$\tau_c^* = \left(k/\sqrt{\lambda}\right) \ln(1 + \mathrm{WC}) \tag{9}$$

where k is a constant for a given canopy.

However, the most used expression for τ_c^* is derived from the expression first suggested by Kirdyashev *et al.* [12]. The canopy opacity can be derived from the vegetation integrated water content WC through a linear relationship:

$$\tau_c^* = A\,\mathrm{WC}\,\varepsilon''_{sw} f \tag{10}$$

where A is a coefficient related to the canopy structure [12], WC is the height integrated vegetation water content per unit area (kg/m^2), ε''_{sw} is the imaginary part of the dielectric constant of saline water in the vegetation, and f is the frequency of observation expressed in GHz. As for a given canopy the coefficients are constant, it is possible to write [13]:

$$\tau_c^* = b\,\mathrm{WC}. \tag{11}$$

This formulation of the canopy opacity gives very good results at low frequencies. It was found through an exhaustive analysis of published data [4], that b is not very

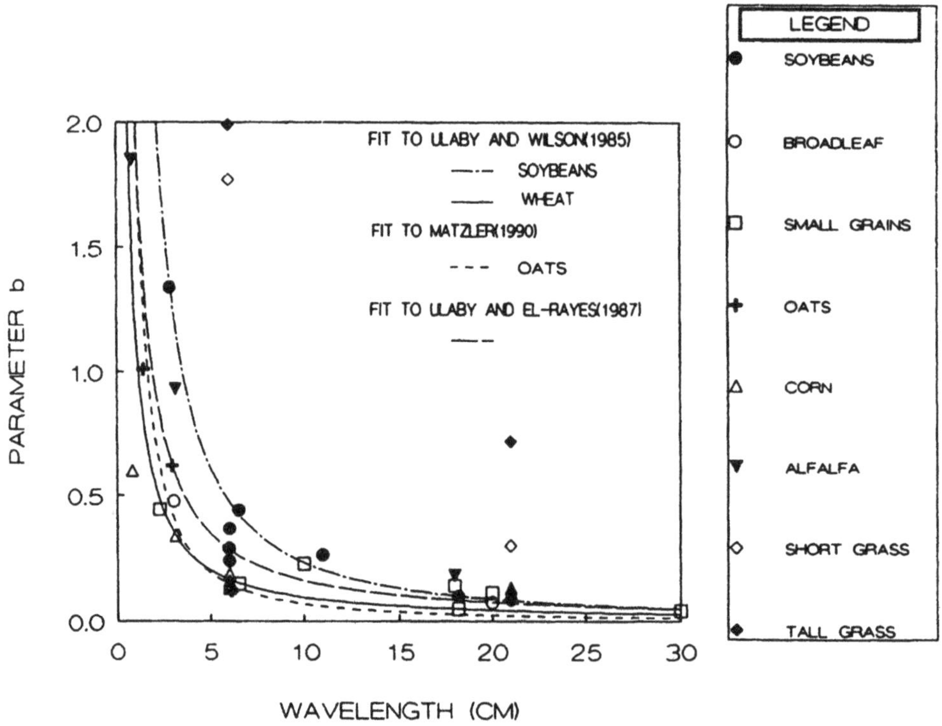

Figure 6. Same as Fig. 5 but with fits to several datasets. From Jackson and Schmugge, 1992, [6].

sensitive to vegetation type at low frequency. Vegetation can be divided into three categories: leaf dominated (e.g. soybeans, cotton, alfalfa), stem dominated (i.e. corn, wheat) and grasses. Consequently, b can be assessed through a classification process based on visible and near infrared data. At low frequencies, b is simply proportional to frequency as can be seen on Fig. 6.

Such a formulation is very well adapted to remote sensing applications, but is only valid at low frequencies (typically under 10–15 GHz) as it relies on assumptions only valid at low frequencies [15]. It seemed useful to investigate how to adapt the formulation to higher frequencies using geometrical optics [15]. It allows to derive an expression of the canopy opacity through the calculation of the absorption coefficient of the type:

$$\tau_c^* = A f \mathrm{WC} \varepsilon''_{\mathrm{sw}} t_p \qquad (12)$$

which is exactly similar to the expression (10) but for t_p corresponding to the transmittivity of a single leaf at polarization p. It can be noted that the coefficient A is the same as in Equation (10).

With a similar approach the scattering coefficient can also be derived, allowing calculation of the single scattering albedo ω. It was found that ω had a very weak frequency dependence. Table 1 gives some of the measurements available in the literature. However most of the references are rather old and it is now considered that single scattering albedo might in fact reach much higher values (0.5 at 35 GHz as reported in [4]).

In [16] a model to simulate brightness temperatures at 37 GHz is presented. This model treats the vegetation as a uniform layer of green vegetation (leaves) containing woody material. The radiative transfer parameters are related to biophysical characteristics of the scene. The model solves the radiative transfer equations by a two point Gauss quadrature method (multiple scattering within the canopy). The model is used to simulate brightness temperatures that are compared with measured values. It is found that polarization difference at 37 GHz is strongly related to biophysical parameters of the canopy, and linearly related to reflectance in the visible and near infrared (see Fig. 7).

1.3. Physical models

In order to better understand the behavior of vegetation, it appears necessary to develop 'physical' models. Those models take into account multiple scattering effects, which are very significant in a vegetation layer when frequency increases, whereas in the 'simple model' scattering effects are usually considered as low.

The 'physical' models were essentially developed in the active microwaves. They can be separated in two main classes: the coherent models, where phase and amplitude of the electromagnetic field are computed; and incoherent models, where only the amplitude is taken into consideration. The vegetation canopy can be characterized with two methods: either as a continuous medium, or as a discrete medium where all the vegetation scatterers (leaves, stems, branches, trunks, ...) have to be accurately described in terms of dimensions, shape, and distribution of orientation.

1.3.1. Coherent approach. Usually in the coherent approach, wave equations are used to derive the bistatic scattering coefficient, which is related to emission through the energy conservation law [17]. A schematic diagram of the different possible ways to tackle the coherent approach is given in Fig. 8. The mean field in the medium is calculated based

Table 1.
Measured/estimated values of the single scattering albedo as reported in the literature

Authors, vegetation type	F (GHz)	WC (kg/m^2)	LAI	ω^*
Mo *et al.*, 1982 [42]				
corn	1.42	1.34	–	0.04
soybeans	1.42	1.54	–	0.07
grass	1.42	0.66	–	0.04
Ulaby *et al.*, 1983 [60]				
corn	1.4	82%	–	0.04
corn	5.0	82%	–	0.05
Brunfelt and Ulaby, 1984 [8]				
wheat	2.7V	–	–	0.075
wheat	2.7H	–	–	0.118
wheat	5.1V	–	–	0.065
wheat	5.1H	–	–	0.081
corn	2.7V	–	–	0.049
corn	2.7H	–	–	0.082
corn	5.1V	–	–	0.092
corn	5.1H	–	–	0.085
soybeans	2.7V	–	–	0.103
soybeans	2.7H	–	–	0.124
soybeans	5.1V	–	–	0.124
soybeans	5.1H	–	–	0.127
Pampaloni and Paloscia, 1986 [9]				
corn	9.7	2.0	2.0	0.08
corn	37.5	2.0	2.0	0.03
corn	9.7	5.3	6.0	0.08
corn	37.5	5.3	6.0	0.06
alfalfa	9.7	1.6	4.0	0.05
alfalfa	37.5	1.6	4.0	0.05
alfalfa	9.7	0.5	1.5	0.04
alfalfa	37.5	0.5	1.5	0.04

on approximations: mainly the Born approximation and the renormalization bilocal approximation. In the first order Born approximation the mean field is the transmitted field inside a medium characterized by a constant dielectric permittivity [18]. This approach neglects scattering effects. The first order Born approximation was also used to study the effects of the anisotropy of the medium structure on the microwave emissivity of a vegetation canopy [19]. The first order renormalization bilocal approximation assumes that the mean field satisfies the Dyson equation, and takes into account both absorption and scattering [20]. Also for discrete media, the mean field equation can be derived based on Foldy's approximation [21]. The bistatic scattering coefficients are then deduced from the correlation function of the scattered field through the distorted Born approximation [5].

To solve the complex wave equations analytically, hypothesis of weak dielectric fluctuations has to be made (through Born and renormalization), which is not met for a vegetation layer. Also the calculation of emissivity using the energy conservation law

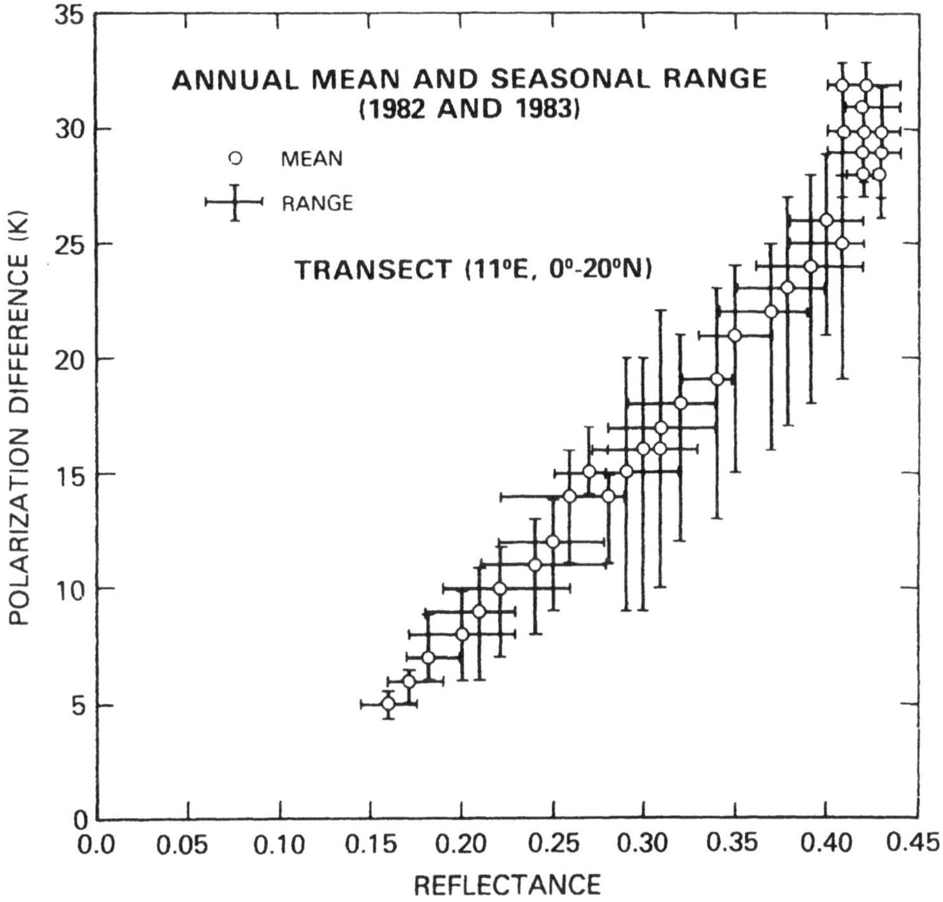

Figure 7. Relationship between the average reflectance of channel 1 and 2 of the AVHRR (580–680 nm and 725–1100 nm) and the SMMR Polarization difference at 37 GHz ($TB_v - TB_h$) for various locations along the transect over Africa ranging from 0° to 20°N at the 11° longitude over a two year period. From Choudhury *et al.*, 1990 [16].

[17] is not well-suited for layered media. In particular a constant thermal profile must be considered inside the medium. In the field of microwave modelling, the trend is now to pursue field calculations as far as possible to take into account coherent phase interactions of scattered radiations. Then the incoherent approach is used to take into account multiple scattering effects. The study of Tsang in this book is a good example of this scheme [22].

1.3.2. Incoherent approach. In the incoherent approach, phase correlations between radiations are neglected; only intensity is taken into account. The main criteria for rejecting coherent interactions are that the average distance between vegetation scatterers is large in comparison with the wavelength and they are randomly distributed. It can be noted that recently, radiative transfer equations have been modified, since this criterion is not met in a vegetation canopy, where the branches and leaves occur mainly in random clusters rather than in a uniform random distribution [22].

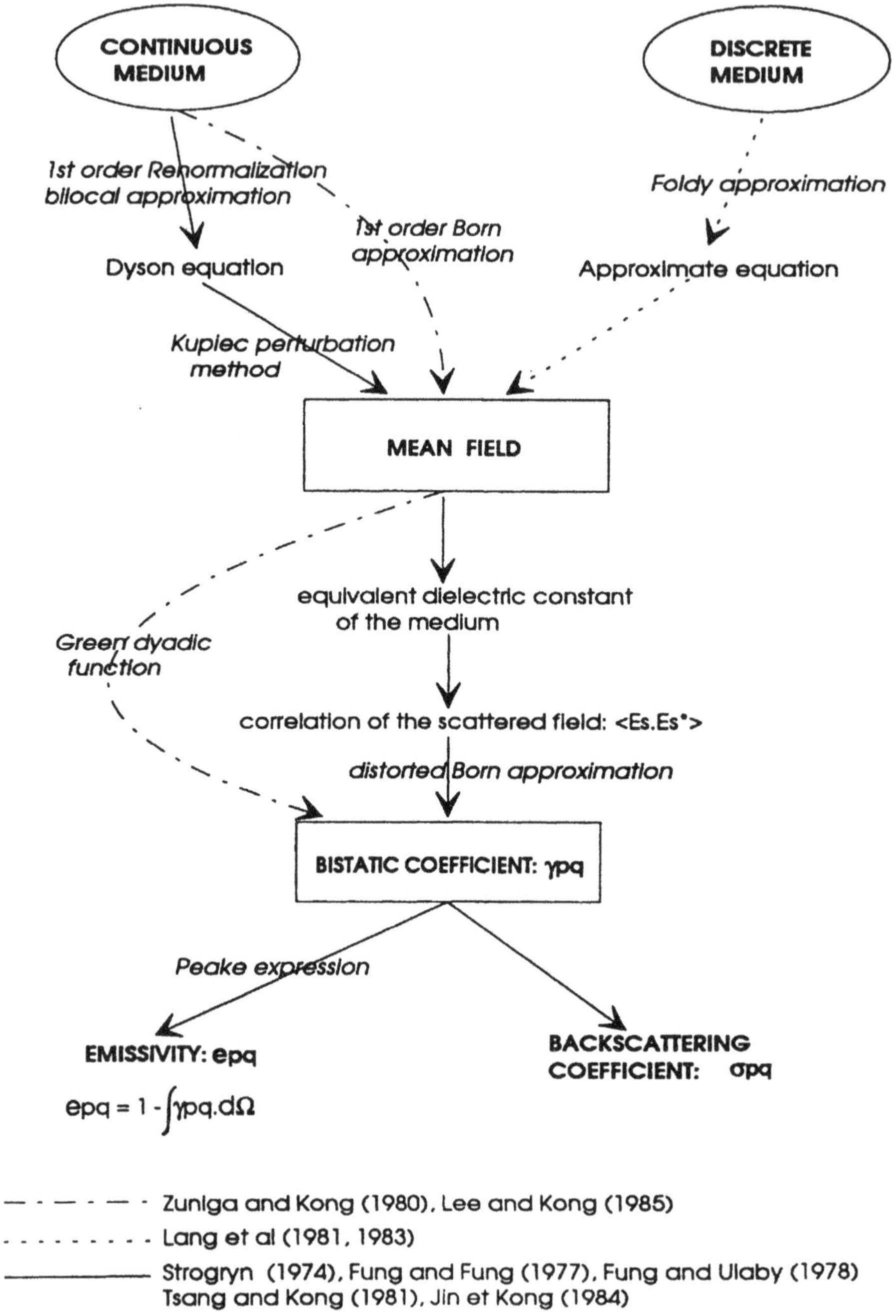

Figure 8. Schematic diagram of the coherent approach for computing volume scattering.

A diagram depicting the possible ways to deal with the incoherent approach is given in Fig. 9. The first step of the incoherent approach is to derive the phase matrix P (based partly on coherent calculations) that describes the redistribution of scattered

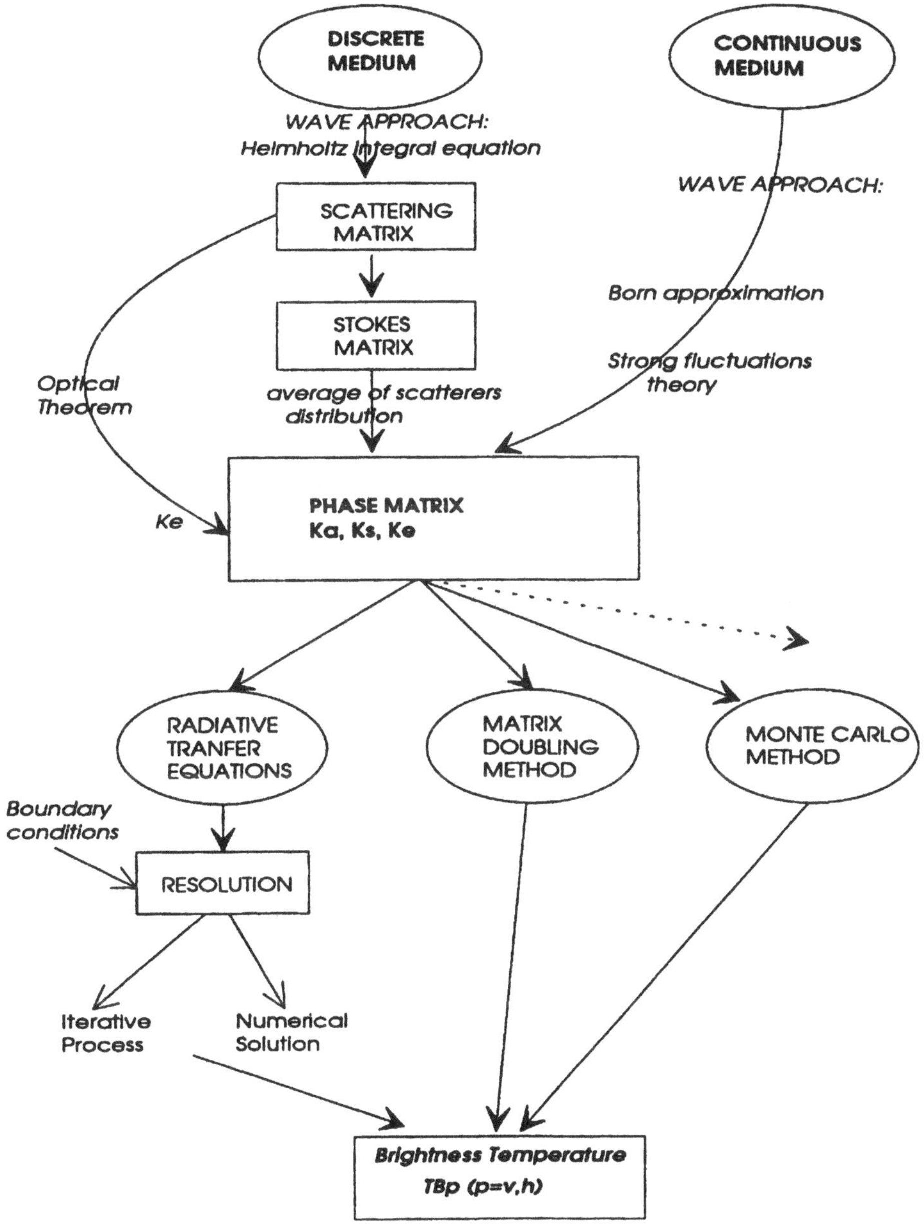

Figure 9. Schematic diagram of the incoherent approach for computing volume scattering.

energy away from the forward direction. The phase matrix P, the scattering coefficient k_s simply deduced from P by integration over the scattered directions, and the absorption coefficient k_a, are included in different descriptions of incoherent radiative exchanges.

Incoherent radiative exchanges are usually described through radiative transfer equations [23]. This approach is efficient to model multiple scattering effects between vegetation scatterers, which have strong dielectric contrasts with the air [24]. In passive microwave theory, polarimetric effects are often neglected and the radiative transfer equations are solved using numerical methods [25]. To take into account polarimetric and multiple scattering effects, iterative solutions up to first and second order (which accounts for multiple scattering effects within the canopy), have been developed both in active and passive microwave remote sensing [24, 26, 27].

Beside radiative transfer equations, incoherent transfer can be described using Matrix Doubling theory [28] or Monte Carlo simulations [22].

1.3.3. Description of the medium. The vegetation layer can be considered as either a continuous or a discrete medium. In the discrete approach all the different vegetation scatterers are accurately characterized in terms of dielectric permittivity, dimensions, shape, and distribution of orientations. To obtain analytical representation of scattering properties, the scatterers are described by simple geometric objects: mainly discs (for leaves), cylinders (for branches and trunks) [26] or spheroids. These different scatterers are distributed in one or several layers, with plane interfaces. The scattering matrix and the Stokes matrix of discrete scatterers are derived through the Helmholtz integral equation. The phase matrix and the scattering coefficient are then deduced from the Stokes matrix by averaging over the scatterers distribution (size and orientation) [26]. The extinction coefficient can be derived from the forward scattering theorem [29]. The discrete approach is well-suited to perform sensibility analysis of vegetation microwave emission and to specify optimal configurations of the microwave sensor to observe land surface features [24, 26, 28, 30]. Figures 10 and 11 show a comparison between ground measurements and the discrete model output either as a function of angle or as a function of frequency. The striking feature is the excellent fit between the two, but it must be noted that the soil moisture used in the model was derived from a best fit estimate.

In the continuous model, scattering is described by statistical dielectric fluctuations inside the canopy. The autocorrelation function of the vegetation permittivity is parameterized by two correlation lengths l_z and l_ρ, describing dielectric fluctuations respectively in the vertical direction and in the horizontal plane [25]. Based on this scheme, a strong fluctuation theory has then been developed to take into account strong dielectric contrasts between vegetation scatterers and the air, inside the canopy. Recently Jin [31], calculated the phase matrix of strongly fluctuating continuous media, which is useful in radiative transfer theory. Figure 12 shows a comparison between ground measurements and the continuous model output throughout the growing season. The continuous medium structure is characterized by very few parameters (two correlation lengths, the vegetation volume fraction, and the height of canopy) in comparison with the discrete approach. So it is a pertinent approach to simulate vegetation emission in relation with large scale satellite observations, and with inversion problems [7, 32]. Figure 13 shows the retrieval of vegetation volume fraction using the continuous approach throughout most of a growing season for soybeans. It must be noted that at the L band (1.4 GHz), the retrieval algorithm significantly overestimates the vegetation volume fraction after day of year (DOY) 240. This is mainly due to a calibration problem on the L band of the radiometer at the end of the experiment.

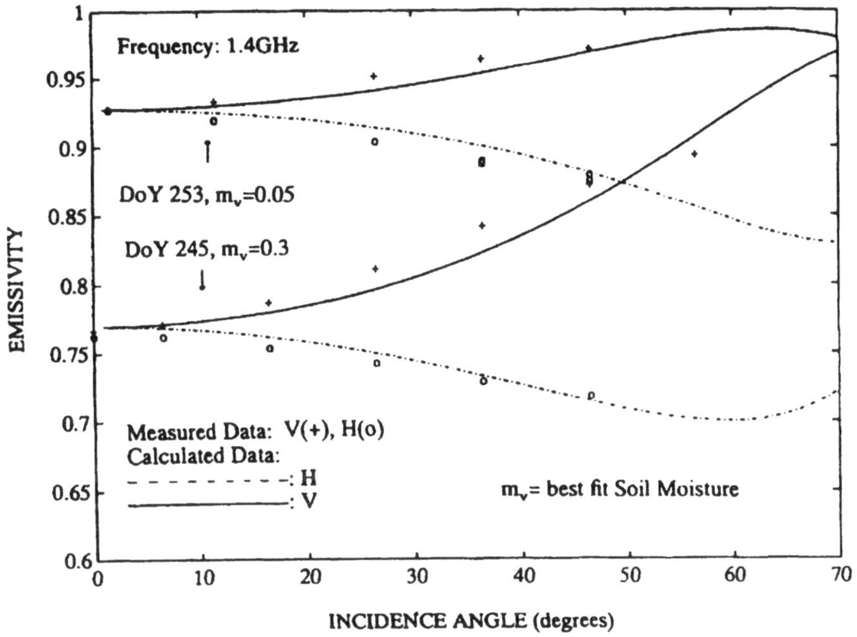

(a)

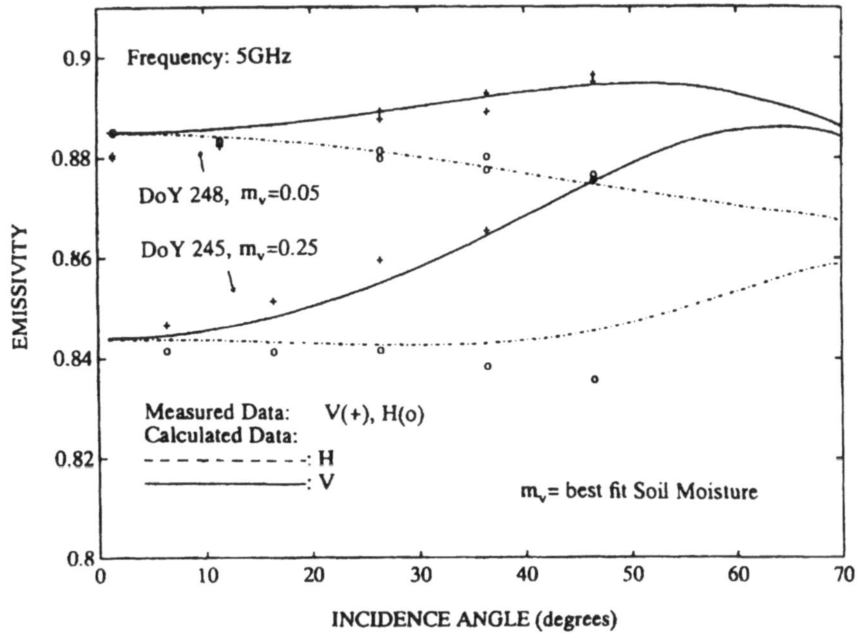

(b)

Figure 10. Comparison between measured (soybean) and modelled emissivity as a function of viewing angle at 1.4 GHz (a) and 5 GHz (b). The discrete model was used. From Wigneron *et al.*, 1993 [30].

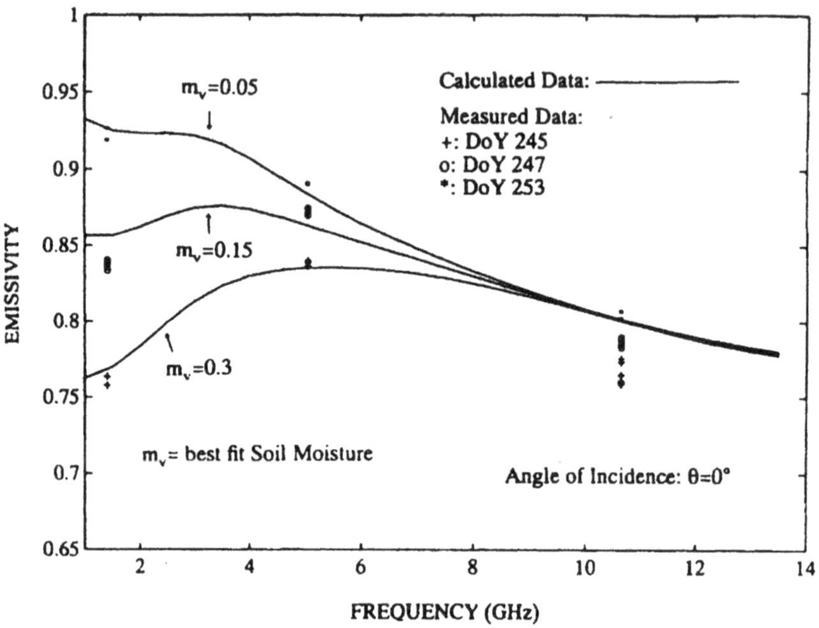

(a)

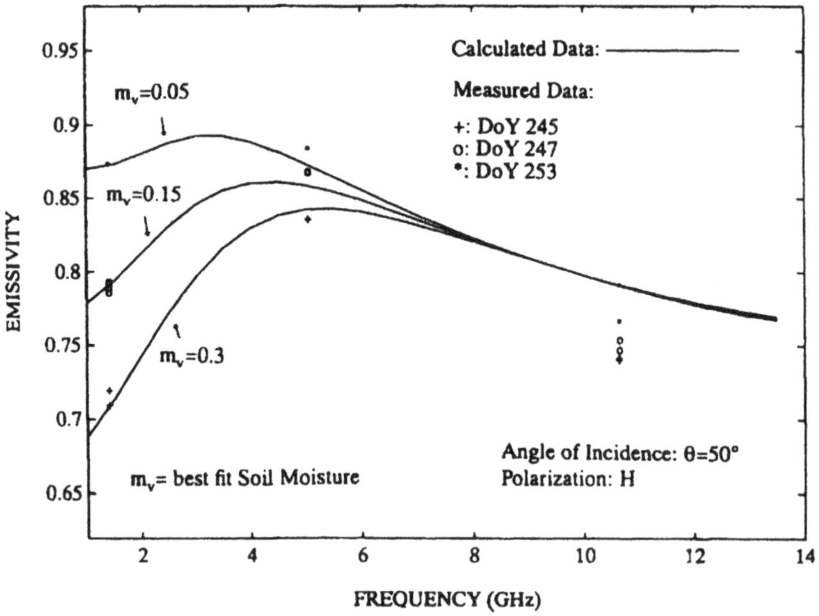

(b)

Figure 11. Comparison between measured (soybean) and modelled emissivity as a function of frequency for an incidence angle of 0° (a) and 50° (b) at H polarization. The discrete model was used. From Wigneron *et al.*, 1993 [30].

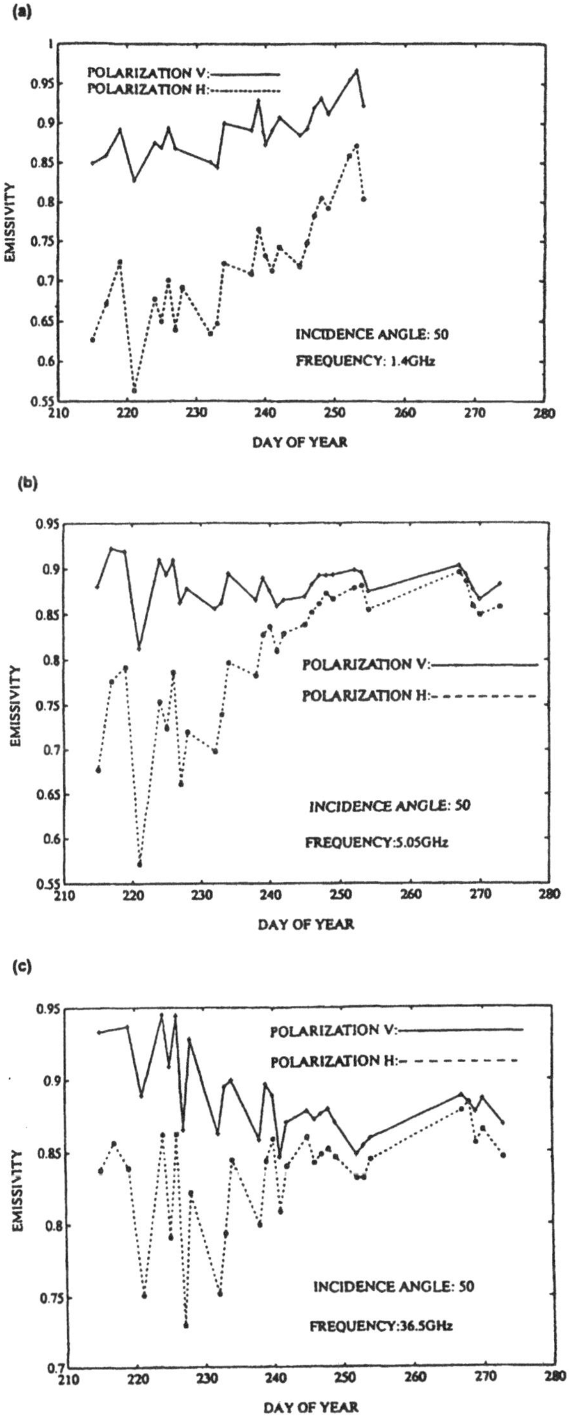

Figure 12. Temporal evolution of measured and computed (continuous model) brightness temperature at an incidence angle of 50° and for 1.4 (a), 5.05 (b), and 36.5 (c) GHz.

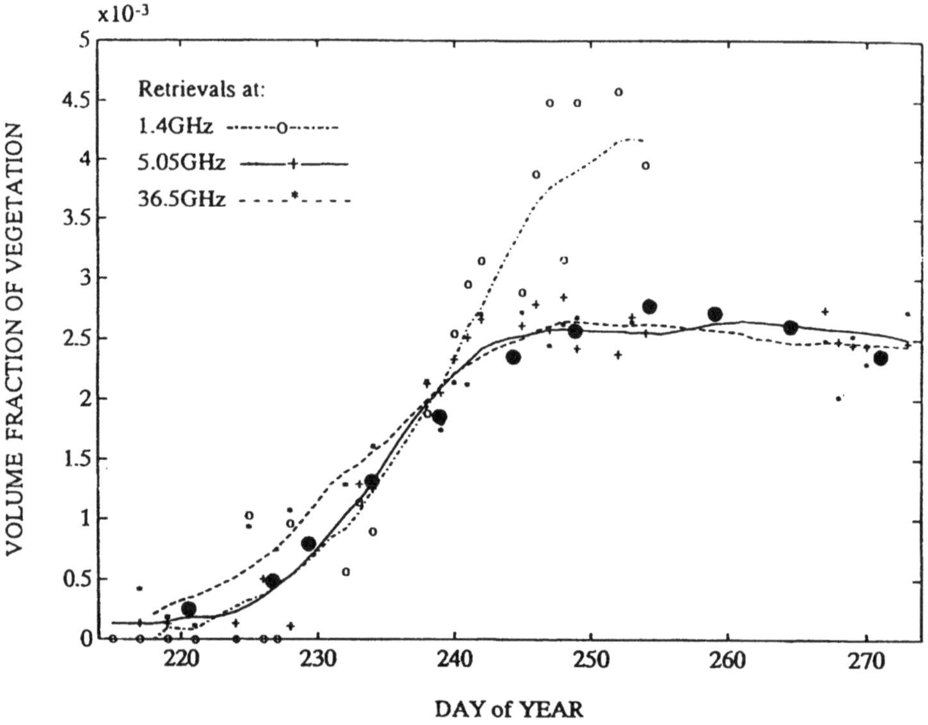

Figure 13. Temporal evolution of the retrieved vegetation volume fraction using the continuous approach at 1.4, 5.05, and 36.5 GHz. The points correspond to the retrieved values and the lines to the retrieved values smoothed using a running average. The large dots correspond to the measurements. From Wigneron *et al.*, 1993 [32].

2. OBSERVATIONS

To develop and test models, it is necessary to perform observations. To be useful, measurements need to be performed with a well adapted methodology. They can be separated into two main groups: measurements done over well controlled sites, and large scale measurements mainly for remote sensing applications. It is also necessary to perform measurements over the whole range of possible observation conditions (frequency, polarization, viewing angles) and for a large variety of surfaces (various crops, natural vegetation, forests, types of soils and soil moisture). It seems thus necessary to optimize design of field experiments so as to cover as much as possible the range of possibilities and fill possible gaps [33]. It seems also useful to build in the design of large scale experiments the methodology to address scaling issues, and the relevant ground data to allow the validation of satellite data retrieval algorithms. This approach was carried out for the HAPEX SAHEL experiment that took place in Niger during the summer of 1992 [34, 35]. A complete description of the various field experiments is too large and would be outside the scope of this paper. However, it seemed useful to give a number of references to which the interested reader could refer to. We have thus endeavoured to give very basic information on what has been done and existing instruments. From our experience it seemed also useful to give some possible recommendations as to the best way to optimise the use of ground data and plan future experiments.

2.1. Ground based experiments

There are some instruments available throughout the world, either for purely ground based measurements (i.e. on a fixed platform at a height from 2 to 20 m), or airborne (at a height from 150 to several hundreds of meters). Some have a limited range of frequencies, some are more exhaustive. To name but a few, there are the instruments made by the Bern group (PAMIR) that cover the range 5–94 GHz and are described in [36] and [37]. The Italian group has instruments ranging from 1.4 to 37 GHz [9]. In the former USSR, several experiments were performed with radiometers functioning in the range 0.6 to 10 GHz [12, 38]. In the United States, various instruments were used in the range 1–37 GHz [39–41]. Most of the published data concern data acquired between 1.4 and 5 GHz [42–44]. In China, the Changchum Institute of Geography has designed a radiometer working in the range 1–36 GHz [7], and in Hungary, results obtained with a radiometer at L-band have been published [45]. In France a radiometer covering the range 1.4–90 GHz has recently been made available. Its main characteristic is that it is a possible MIMR simulator.

The above list is, of course, not exhaustive. It simply shows that the range 1–40 GHz is somewhat well covered with instruments of various characteristics. Consequently, one could assume that data sets covering a wide range of viewing conditions and surface types are available.

Unfortunately, this is not always the case for different reasons:

* Most often, these instruments are used in a given location (controlled site) and over a given type of vegetation (generally agricultural crops).
* Even though measurements might be performed at different frequencies, they are often only analyzed for a limited set of the available range.
* In many cases, the data sets are not available, and/or the experimental conditions not well documented.

It seems important, if we want to foster the research in passive microwaves and the development of algorithms to make several efforts in order to:

* ensure that existing and future data sets are made available and well documented, so that anyone could make full use of them,
* ensure that the instruments' characteristics are well documented, and, whenever possible, intercalibrate existing instruments,
* standardize a minimum set of measurements to be performed systematically for characterizing canopies and environmental conditions, together with measurement protocols/instruments,
* ensure that the 1–90 GHz range of frequencies is covered, and in particular, the frequencies available on existing and future space-borne instruments,
* ensure that measurements are performed over the largest range of vegetation types possible, and in particular, over natural vegetation and for different climatic environments. To achieve this, development of small, fully portable instruments such as MARMOTTE [46] seems to be a sensible approach. Finally the representativity of the targets should be carefully assessed.

2.2. Large scale field experiments

Ground based experiments are particularly useful for modelling purposes, since all the relevant surface parameters can be assessed over well controlled areas. However, if inverse algorithms are to be developed, it is necessary to make a step toward larger resolution elements. For this purpose several large scale field experiments have been performed such as HAPEX MOBILHY in 1986, FIFE (1987 and 1989), MONSOON 1990 [47]. It must however be noted that the main goal of the passive microwave part of these experiments was to assess soil moisture. More recently, the HAPEX SAHEL experiment [34] was partly designed to address the use of passive microwaves over vegetation with particular emphasis on scaling issues and mixed pixels [35]. The scope was to test inverse algorithms, analyze how the retrievals could be used for Global Circulation Models (GCM), and possible validations of the inverse algorithms.

Generally speaking, measurements done during large scale field experiments are made available to the scientific community and are very often quality checked. However, some recommendations can nevertheless be made considering past experience. Here again it seems necessary to standardize ground measurements (protocols and instruments), to generate community level processed satellite data and endeavour to gather in a data base all measurements made at other wavelengths (visible, near infrared, far infrared, active microwaves) as well as mesoscale or GCMs results. The objective is to generate data sets enabling researchers to address scaling issues, synergism and algorithm development and validation.

It seems necessary also to make full use of planned large scale experiments to sample as many ecosystems as possible, and eventually add specific sub-experiments to cover possible gaps in the measurement space (frequencies, vegetation type/ecosystem, polarizations, etc.).

3. DISCUSSION AND GOALS FOR THE NEAR FUTURE

Simple models have now been developed and work efficiently at low frequencies, as has been described in the first part of the modelling chapter. These models are rather coarse and somewhat simplistic, but they seem to be adapted to satellite scale measurements, even though usually they are not valid at frequencies above 10 to 20 GHz. Improvements are necessary for use at frequencies such as 37 GHz, and are currently under study. It is important to continue efforts in the physical modelling of vegetation in order to better understand the frequency and polarization dependence of several parameters such as the single scattering albedo.

For discrete models, we are currently dealing with fairly sophisticated descriptions of the canopy. Consequently, such models are usually intractable for practical use. They are nevertheless a very powerful tool to understand the behavior of vegetation canopies in the passive microwave domain. It can also be envisioned that with such models, coupled with generic vegetation models and vegetation growth models, the complexity of the description of the canopy could be reduced, making discrete models tractable issues.

Inversion of continuous models is much simpler since these models rely on a simplified description of the canopy. However, they are still complex and some parameters resulting from the inversion (the correlation lengths) are yet to be related to 'measurable' biophysical parameters.

In all the above the problem of 'mixed pixels' such as those seen by spaceborne sensors is also to be carefully studied before one can start using the physical models for retrieval of vegetation parameters.

Applications of passive microwaves for assessing vegetation parameters have been tested. Several parameters such as, for instance, plant water content [9, 32], plant water stress [48], leaf characteristics [30], plant growth monitoring [49], forest and surface types [50–52] have been estimated with use of passive microwaves. However, one can say that for the time being, all the questions are not yet answered and that, even though promising, the above mentioned results are most often preliminary.

Several issues do remain to be solved: the pixel size in relation to spatial variability, the contribution of the soil and the atmosphere, and the mixed pixels.

Considering the number of surface variables to be addressed, it seems very important to address possible addition of information to be able to distinguish more accurately the vegetation signal from background (soil and atmosphere). This can be performed using the synergism in the general sense or, in other words, synergism with other sensors, within the time/space domain, and with models.

3.1. Synergism with other sensors

The most obvious synergism lies with the use of visible and near infrared sensors since they can give information on the surface cover at a much higher resolution (spatial). They can also deliver relevant information on some of the canopy characteristics and fractional cover. Far infrared sensors also give relevant information, though it must be considered with much care since:

* synchronous measurements are necessary due to the very high temporal variability of the surface temperature,
* far infrared gives information on skin temperature while passive microwaves give information integrated over the penetration depth, which can lead to variable biases.

In the microwave part of the spectrum, apart from the obvious use of several microwave frequencies to deconvolute the different contributions to the signal linked to their different sensitivities for given variables, one can also use active microwaves [37, 53]. For instance scatterometer data such as the data collected by wind scatterometers might be of significant help. Those sensors make backscattering coefficient (σ^0) measurements at several incidence angles. By monitoring the slope and intercept of σ^0 as a function of view angle, one can monitor spatial and temporal evolution of surface roughness, and biomass. The resolution of satellite borne scatterometers is coarse (50 km in the case of Windscatt on ERS-1), but well adapted to the type of resolutions available for passive microwave sensors such as the Scanning Multichannel Microwave Radiometer (SMMR) or SSM/I. Studies are under way at the moment on the possible use of wind scatterometer data, and it was found that the slope and intercept of σ^0 is quite dependent on the incidence angle of the central beam, which reduces the amount of useful data. Moreover the contribution of soil moisture on the backscatter signal is more significant than that predicted by previous studies.

3.2. Synergism in the time–space domain

This method relies on the fact that the different contributions to the signal have different spatial and temporal signatures. For example, with time, soil texture is constant, soil roughness varies very slowly, soil moisture varies within days, and soil temperature within hours. Vegetation has a temporal frequency of the order of the week. Atmosphere has a temporal frequency ranging from days/weeks (integrated water content), to hours (clouds) and finally minutes (rain).

On the spatial scale, atmosphere has a low to medium variability (integrated water content to clouds/rain) while surface characteristics have a medium to high variability.

Considering that with microwave sensors of the type of MIMR we can have access to data quite often (the so-called all-weather capacity), one can see that it should be possible to use the spatio-temporal synergism to help in distinguishing the various atmospheric, and eventually, soil contributions. Such an approach would thus greatly enhance more classical synergistic approaches involving only other sensors/wavelengths.

3.3. Other synergism

Finally, one can mention the possible use, even though the method is not yet mature, of vegetation/crop growth models [54]. Once the vegetation type is known and with, as an input, the climatic conditions, it is possible to model vegetation growth. From such vegetation models, it is possible to derive vegetation parameters such as, for example, woody material or leaf area index. The results are regularly updated with satellite measurements, with methods similar to those used in weather forecasting. Such outputs could then be used as inputs for physical microwave models, enabling the inversion algorithms to be tractable. The latter would then give more detailed information on vegetation canopies. These vegetation growth models exist now for a limited number of agricultural crops. Models for natural vegetation are yet to be developed and validated.

Another approach is to use synergism with mesoscale or global climate models. These models can either deliver information useful for data correction (atmosphere), or give brackets for surface conditions (temperature, moisture). They can also be used for validation purposes.

4. CONCLUSION

The use of passive microwaves for vegetation studies is not yet fully mature. This can partly be explained by the lack of well adapted space-borne instruments, and partly to the complexity of the problem.

However, in the near term, the MIMR instrument will give us an opportunity to have an instrument whose frequencies are well adapted, and the spatial and temporal resolutions satisfactory. It seems thus highly necessary to be ready for the EOS/METOP era.

Significant progress has been made during the last few years. Simple models, coarse but efficient, have been developed and give satisfactory results at low frequencies and large scales. They are now being improved to be valid at higher frequencies. More complex models (physical) are also being developed and give good results for the direct modelling at the local scale. Efforts are now being made to inverse those algorithms and relate the outputs to useful vegetation biophysical parameters.

The scope for the future is probably now to:

* keep on the modelling effort and corresponding field experiments. Document and make available existing data sets/instruments,
* derive from physical models simple laws giving the opacity and single scattering albedo as functions of frequency, polarization, vegetation type, as well as exhaustive sensitivity analysis,
* address the scaling issues, both spatial and temporal,
* address the mixed pixel issue,
* make more detailed studies on the different types of synergism, and how to use them in inversion algorithms,
* create generic vegetation models to be able to consider vegetation as an ecosystem rather than a species,
* interact strongly with the GCMs and mesoscale modelling communities,
* generate inverse algorithms, and test them. Establish validation procedures adapted to the time and space resolution of space-borne passive microwave sensors.

Finally, it seems timely that the models are considered as a whole, from surface to sensor. Up to now, most of the studies cope with one aspect: vegetation, or soil, or atmosphere. If we want to have in the near future reliable algorithms for vegetation canopies, it seems that the different models for each 'layer' should be concatenated and the inversion problem considered as a whole.

REFERENCES

1. Y. H. Kerr and E. G. Njoku. A semi-empirical model for interpreting microwave emission from semiarid land surfaces as seen from space. *IEEE Trans. Geosci. Remote Sensing*, **28**, 3, 384–393 (1990).
2. T. J. Jackson. Measuring surface soil moisture using passive microwave remote sensing. *Hydrological Processes*, **7**, 139–152 (1993).
3. F. T. Ulaby and M. A. El-Rayes. Microwave dielectric spectrum of vegetation Part II: Dual-dispersion model. *IEEE Trans. Geosci. Remote Sensing*, **GE-25** (4), 550–557 (1987).
4. C. Mätzler and A. Sume. Microwave Radiometry of Leaves. In: *Microwave Radiometry and Remote Sensing Applications*, P. Pampaloni (Ed.), pp. 133–148, VSP Utrecht, The Netherlands (1989).
5. F. T. Ulaby, R. K. Moore and A. K. Fung. *Microwave Remote Sensing: Active and Passive*. Vol. 3, *From Theory to Applications*, Artech House, Dedham MA (1986).
6. T. J. Schmugge and T. J. Jackson. A dielectric model of the vegetation effects on the microwave emission from soils. *IEEE Trans. Geosci. Remote Sensing*, **30** (4), 757–760 (1992).
7. Y. Q. Jin, J. Zhang and R. Zhao. Numerical modelling of radiative transfer and multifrequency measurement of thermal emission from crop canopies. *Int. J. Rem. Sensing*, **13** (15), 2801–2812 (1992).
8. D. R. Brunfelt and F. T. Ulaby. Measured microwave emission and scattering in vegetation canopies. *IEEE Trans. Geosci. Remote Sensing*, **GE-22** (6), 520–524 (1984).
9. P. Pampaloni and S. Paloscia. Microwave emission and plant water content: A comparison between field measurements and theory. *IEEE Trans. Geosci. Remote Sensing*, **GE-24** (6), 900–905 (1986).
10. F. T. Ulaby and E. A. Wilson. Microwave attenuation properties of vegetation canopies. *IEEE Trans. Geosci. Remote Sensing*, **23**, 746–753 (1985).

11. F. T. Ulaby, A. Tavakoli and T. B. Senior. Microwave propagation constant for a vegetation canopy with vertical stalks. *IEEE Trans. Geosci. Remote Sens.*, **25**, 714–725 (1987).
12. K. P. Kirdyashev, A. A. Chukhlantsev and A. M. Shutko. Microwave radiation of the Earth's surface in the presence of vegetation cover. *Radio Eng. Electron. Phys.*, **24**, 256–264 (1979).
13. T. J. Jackson, T. J. Schmugge and J. R. Wang. Passive microwave remote sensing of soil moisture under vegetation canopies. *Water Resources Res.*, **18**, 1137–1142 (1982).
14. T. J. Jackson and T. J. Schmugge. Vegetation effects on the microwave emission of soils. *Remote Sens. Environ.*, **36**, 203–212 (1991).
15. U. Wegmüller and C. Mätzler. Models to relate canopy opacity to vegetation parameters. In: *Microwave Radiometry and Remote Sensing Applications*, R. Westwater (Ed.), pp. 79–86, United States Department of Commerce, NOAA Boulder, CO (1992).
16. B. J. Choudhury, J. R. Wang, A. Y. Hsu and Y. L. Chien. Simulated and observed 37 GHz emission over Africa. *Int. J. Rem. Sensing*, **11** (10), 1837–1868 (1990).
17. W. H. Peake. Interaction of electromagnetic waves with some natural surfaces. *IEEE Trans. Antennas Propagation*, **AP-7**, 324–329 (1959).
18. M. Zuniga and J. A. Kong. Active remote sensing of random media. *J. Appl. Phys.*, **51**, 74–79 (1980).
19. J. K. Lee and J. A. Kong. Passive microwave remote sensing of an anisotropic random-medium layer. *IEEE Trans. Geosci. Remote Sensing*, **GE-23**, 924–932 (1985).
20. L. Tsang and J. A. Kong. Scattering of electromagnetic waves from random media with strong permittivity fluctuations. *Radio Science*, **16**, 303–320 (1981).
21. R. H. Lang. Electromagnetic backscattering from a sparse distribution of lossy dielectric scatterers. *Radio Science*, **16**, 15–30 (1981).
22. L. Tsang, J. A. Kong, Z. Chen, K. Pak and C. Hsu. Theory of microwave scattering from vegetation based on the collective scattering effects of discrete scatterers. In: *ESA/NASA International Workshop*, B J. Choudhury, Y. H. Kerr, E. G. Njoku and P. Pampaloni (Eds), pp. 117–154, VSP, Utrecht (1994).
23. S. Chandrasekhar. *Radiative Transfer*. Dover Publications, Inc., New York (1960).
24. F. T. Ulaby, K. Sarabandi, K. Mcdonald, M. Whitt and M. C. Dobson. Michigan microwave canopy scattering model (MIMICS). *Int. J. Rem. Sensing*, **12**, 1223–1253 (1990).
25. L. Tsang and J. A. Kong. Thermal microwave emission from a three-layer random medium with three dimensional variations. *IEEE Trans. Geosci. Remote Sensing*, **18**, 212–216 (1980).
26. M. A. Karam, A. K. Fung, R. H. Lang and N. S. Chauhan. A microwave scattering model for layered vegetation. *IEEE Trans. Geosci. Remote Sensing*, **30**, 767–784 (1992).
27. L. Tsang. Polarimetric passive microwave remote sensing of random discrete scatterers and rough surface. *J. Electr. Waves Appl.*, **5**, 41–57 (1991).
28. P. Ferrazzoli, L. Guerriero, S. Paloscia, P. Pampaloni and D. Solimini. Modeling polarization properties of emission from soil covered with vegetation. *IEEE Trans. Geosci. Remote Sensing*, **30**, 157–165 (1992).
29. L. Tsang, J. A. Kong and T. S. Shin. *Theory of Microwave Remote Sensing*. Wiley-Interscience Publications, New York (1985).
30. J.-P. Wigneron, J.-C. Calvet, Y. Kerr, A. Chanzy and A. Lopes. Microwave emission of vegetation: sensitivity to leaf characteristics. *IEEE Trans. Geosci. Remote Sensing*, **31** (3), 716–726 (1993).
31. Y. Q. Jin. The radiative transfer equation for strongly-fuctuating continuous random media. *J. Quant. Spectrosc. Radiat. Transfer*, **42**, 529–537 (1989).
32. J.-P. Wigneron, Y. Kerr, A. Chanzy and Y. Q. Jin. Inversion of surface parameters from passive microwave measurements over a soybean field. *Remote Sens. Environ*, **46**, 61–72 (1993).
33. Y. H. Kerr, T. J. Schmugge, E. G. Njoku, H. Suess *et al.* Report of the land surface working group, San Miniato Workshop, San Miniato, Italy, September 1990.
34. J. P. Goutorbe, T. Lebel, A. Tinga, P. Bessemoulin, J. Brower, A. J. Dolman, E. T. Engman, M. Gash, J. H. C. Hoepffner, P. Kabat, Y. H. Kerr, B. Monteny, S. D. Prince, F. Said, P. J. Sellers and J. Wallace. Hapex Sahel: a large scale study of land atmosphere interactions in the semi-arid tropics, submitted to *Annales Geophysicae* (1992).

35. S. D. Prince, Y. H. Kerr, J. P. Goutorbe, T. Lebel, A. Tinga, J. Brouwer, A. J. Dolman, E. T. Engman, J. C. H. Gash, M. Hoepffner, P. Kabat, B. Monteny, F. Said, P. J. Sellers and J. Wallace. The Hydrology-Atmosphere Pilot Experiment in the Sahel (HAPEX SAHEL): Remote sensing, Submitted to *Remote Sens. Environ.* (1993).

36. C. Mätzler. Seasonal evolution of microwave radiation from an oat field. *Remote Sens. Environ.*, **31**, 161–173 (1990).

37. U. Wegmüller. Remote sensing signature studies on agricultural fields with ground based radiometry and scatterometry, PhD thesis, Applied Physics, University of Bern (1990).

38. A. M. Shutko. *Microwave Radiometry of Water Surfaces and Grounds*, Nauka, Moscow (1986).

39. J. Wang, J. Shiue, E. Engman, *et al. Remote Measurements of Soil Moisture by Microwave Radiometers at BARC Test Site*, NASA TM 80720 (1980).

40. J. Wang, P. E. O'Neill, E. Engman *et al. Remote Measurements of Soil Moisture by Microwave Radiometers at BARC Test Site II*, NASA TM 83954 (1982).

41. J. Wang, T. Jackson, E. Engman, W. Gould, J. Fuchs, W. Glazer, P. O'Neill, T. Schmugge and J. McMurtrey. *Microwave Radiometer Experiment of Soil Moisture Sensing at BARC Test Site During Summer 1981*, NASA TM 86056 (1984).

42. T. Mo, B. J. Choudhury, T. J. Schmugge, J. R. Wang and T. J. Jackson. A model for microwave emission from vegetation covered fields. *J. Geophys. Res.*, **87**, 11229–11237 (1982).

43. T. Mo, T. J. Schmugge and J. R. Wang. Calculations of the microwave brightness temperature of rough soil surfaces: bare fields. *IEEE Trans. Geosci. Remote Sensing*, **GE-25**, 47–54 (1987).

44. T. J. Jackson and P. E. O'Neill. Attenuation of soil emissivity by corn and soybeans at 1.4 and 5 GHz. *IEEE Trans. Geosci. Remote Sensing*, **GE-28**, 978–980 (1990).

45. A. D. Vyas, A. J. Trivedi, O. P. N. Calla, S. S. Rana, S. B. Sharma and A. B. Vora. Experimental data for separation of vegetation and soil and estimation of soil moisture using passive microwaves. *Int. J. Rem. Sensing*, **11** (8), 1421–1438 (1990).

46. Y. H. Kerr. MARMOTTE: A portable field microwave radiometer. In: *Specialists Meeting on Microwave Radiometry and Remote Sensing Applications*, R. Westwater (Ed.), pp. 169–173 (1992).

47. T. J. Schmugge, T. J. Jackson, W. P. Kustas and J. R. Wang. Passive microwave remote sensing of soil moisture: results from HAPEX, FIFE and MONSOON 90. *ISPRS J. Photogramm. Remote Sensing*, **47**, 127–143 (1992).

48. S. Paloscia and P. Pampaloni. Microwave remote sensing of plant water stress. *Remote Sens. Environ.*, **16** (3), 249–255 (1984).

49. S. Paloscia and P. Pampaloni. Microwave polarization index for monitoring vegetation growth. *IEEE Trans. Geosci. Remote Sensing*, **26** (5), 617–621 (1988).

50. M. Hallikainen and P. Jolma. Classification of forest and surface types by satellite microwave radiometry. *IGARSS'86, Zürich, 8–11 September 1986*, ESA SP-254, 853–858 (1986).

51. J. C. Calvet, J.-P. Wigneron, E. Mougin, Y. H. Kerr and J. L. S. Brito. Use of passive microwaves over the Amazon Forest, *IGARSS'92, Houston Texas, USA* (1992).

52. M. T. Hallikainen, P. A. Jolma and J. M. Hyppä. Satellite microwave radiometry of forest and surface types in Finland. *IEEE Trans. Geosci. Remote Sensing*, **26** (5), 622–628 (1988).

53. P. Ferrazzoli, S. Paloscia, P. Pampaloni, G. Schiavon, D. Solimini and P. Coppo. Sensitivity of microwave measurements to vegetation biomass and soil moisture content: A case study. *IEEE Trans. Geosci. Remote Sensing*, **30** (4), 750–756 (1992).

54. J. P. Wigneron. Modélisation de l'émission micro-onde d'un couvert végétal – Mise en relation de la mesure hyperfréquence passive avec les échanges énergétiques et hydriques d'une culture de soja. PhD Thesis, Université Paul Sabatier, Toulouse, France (1993).

55. A. K. Fung and H. S. Fung. Application of first order renormalization method to scattering from a vegetation-like half space. *IEEE Trans. Geosci. Electr.*, **15**, 189–195 (1977).

56. A. K. Fung and F. T. Ulaby. A scatter model for leafy vegetation. *IEEE Trans. Geosci. Electr.*, **16**, 281–285 (1978).

57. Y. Q. Jin and J. A. Kong. Strong fluctuation theory for electromagnetic wave scattering by a layer of random discrete scatterers. *J. Appl. Phys.*, **55**, 1364–1369 (1984).

58. R. H. Lang and J. S. Sidhu. Electromagnetic backscattering from a layer of vegetation: a discrete approach. *IEEE Trans. Geosci. Remote Sensing*, **21**, 62–71 (1983).
59. A. Stogryn. Electromagnetic scattering by random dielectric constant fluctuations in a bounded medium. *Radio Science*, **9**, 509–518 (1974).
60. F. T. Ulaby, M. Razani and M. C. Dobson. Effects of vegetation cover on the microwave radiometric sensitivity to soil moisture. *IEEE Trans. Geosci. Remote Sensing*, **GE-21**, 51–61 (1983).

ESA/NASA International Workshop, pp. 345–355
B. J. Choudhury, Y. H. Kerr, E. G. Njoku and P. Pampaloni (Eds)

Progress in multi-frequency radiometry of natural objects

CHRISTIAN MÄTZLER and URS WEGMÜLLER

Institute of Applied Physics, University of Bern, Sidlerstrasse 5, CH-3012 Bern, Switzerland

Abstract—We report on our recent progress made in multi-frequency microwave radiometry by concentrating on vegetation experiments made at a test site near Bern. For this purpose we used the 5-frequency radiometer system PAMIR which covers the entire range of the planned Multi-channel Imaging Microwave Radiometer (MIMR).

Characteristic properties of vegetation are a near-black body behaviour with a small degree of polarization. Exceptions occur when anisotropic orientations dominate. Temperature can also play an important role. Our experience is based on two types of experiments: (1) observation of radiation from the surface of the test site and (2) observation of transmitted and reflected thermal radiation from layers of natural objects. Examples of both types are presented.

1. INTRODUCTION

The terrestrial remote sensing group at the Institute of Applied Physics of the University of Bern has been involved in microwave radiometry, especially for signature studies, for more than 20 years. Our research has been concentrated on experimental studies with surface-based equipment at several test sites, mainly for the investigation of snow, soil, vegetation and sea ice. Multi-frequency aspects and the synergism of microwave radiometry with active microwave sensors helped us in an improved understanding of microwave interaction with the terrestrial surface.

Now we would like to share our experience on the one hand with those who like to test and validate their theoretical models, and on the other hand with those who like to apply remote sensing data to the extraction of geophysical information. The purpose of this paper is to introduce the type of information we have gained during the last few years. Here we will focus on data obtained from a set of five microwave radiometers, operating over the frequency range from 5 to 100 GHz. This system is called PAMIR. Since 1987 PAMIR has been operated at the agricultural test site 'Tannacker' near Bern, Switzerland. There the majority of experiments were dedicated to transmission, emission and reflection measurements of vegetation. The test site is shown in the photographs of Fig. 1.

For the sake of completeness we should point out that we got additional experience with another multi-frequency instrument covering the frequency range from 2 to 12 GHz. It is a radiometer-scatterometer system, called RASAM [1–3]. The present paper, however, is concerned with experiments made with PAMIR. We will highlight two examples of recent experiments: (1) microwave emission spectra of a grass field under frozen condition, and (2) microwave transmission spectra of a sprucefir twig under various conditions.

Figure 1. Test site Tannacker in Moosseedorf, some 10 km north of Bern (570 m above sea level, 7°29′E/47°00′N). The PAMIR is visible as a white box to the left of the farmhouse (top), and the bottom picture is a close view of the radiometer setup.

2. PAMIR, AN *IN-SITU* MIMR SIMULATOR

PAMIR is the acronym for 'Passive and Active Microwave and Infrared Radiometer'. This sensor system consists of five microwave radiometers (Table 1), an X-band scatterometer and a thermal infrared radiometer. The radiometers were built and first operated by Roland Hofer [4] for signature research on alpine snowpacks. These measurements continued until 1987 at the alpine test site, Weissfluhjoch [5]. Then PAMIR was relocated to the agricultural test site 'Tannacker' at Moosseedorf near Bern [6]. There, the scatterometer and the infrared radiometer were removed.

The PAMIR frequency range covers that of the Multi-channel Imaging Microwave Radiometer (MIMR) to be flown on future earth observation satellites. Therefore the observations made by PAMIR can be used to simulate spectra of future MIMR data.

Table 1.
Specifications of the PAMIR microwave radiometers

Radiometer number	1	2	3	4	5
Center frequency, GHz	4.9	10.4	21	35	94
Bandwidth, GHz	0.2	1.2	0.8	0.8	2.0
Sensitivity, K (at 1-s integr. time)	0.1	0.1	0.1	0.1	0.2
Viewing angle	all between nadir and zenith				
Polarization	any linear polarization				
3-dB beam width of horn antenna, deg	10	9	9	9	9

3. RECENT SURFACE OBSERVATIONS

At the test site, Tannacker, two types of PAMIR measurements were made. The first type consisted of radiometric measurements of the natural surface in front of the instruments. In this section we report on these observations. The range of incidence angles covered 50° to 80° for the surface and 120° to 180° for sky measurements. (With the present configuration using the trailer setup we cannot use incidence angles below 50°.) These data were used to determine spectra of emissivities (or reflectivities, according to Equation (1)) at both vertical and horizontal linear polarization, and of atmospheric zenith opacities. Together with the collected temperature information these data can be used to estimate the brightness temperatures at satellite level.

The complete PAMIR dataset was compiled as a comprehensive catalogue [7] with more than 100 different situations. They consist, respectively, of bare soil in frozen and unfrozen condition, of oat and barley canopies and of grass with and without snowcovers of various types. The microwave data are complemented by the relevant ground-truth information.

As an example of such an observation we show the data of Measurement 108 of December 14, 1992 in Figs 2 and 3. The object is short grass with hoar frost. Figure 2 shows the angular plots of all ten reflectivities r_{p}, where

$$r_{\mathrm{p}} = 1 - e_{\mathrm{p}}; \qquad \mathrm{p} = \mathrm{v}, \mathrm{h} \tag{1}$$

and e_{p} is the p-polarized emissivity of the given frequency. Figure 3 is the spectral plot of r_{v} and r_{h} at an incidence angle of 60°. The ground-truth information for the time

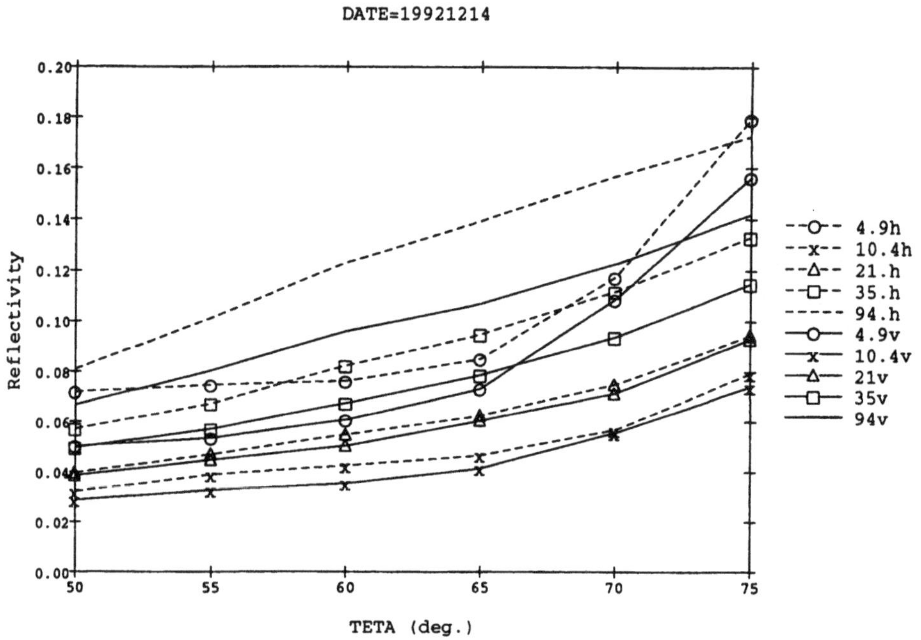

Figure 2. Reflectivities at all PAMIR channels versus incidence angle. The object is a grass field with hoar frost at freezing temperature.

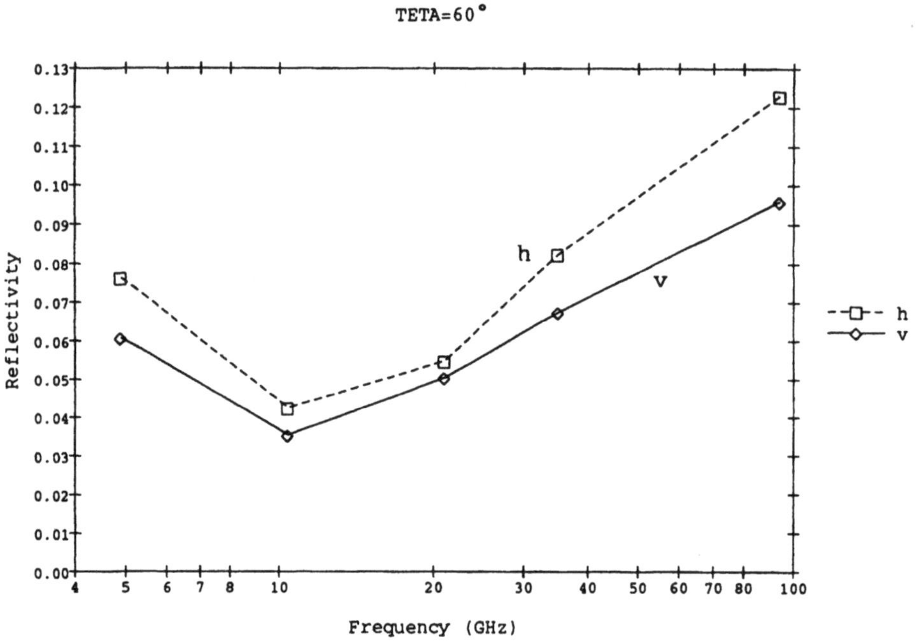

Figure 3. Reflectivities at vertical (solid line) and horizontal (dashed line) polarization versus frequency for same situation as Figure 2.

Table 2.
Ground-truth data for PAMIR measurements of December 14, 1992

Date	92–12–14
1 Time (h)	8.90
2 CI	0.00
3 T_{air} (C)	−2.40
4 T_{bb} (C)	−3.40
5 T_{veg} (C)	−4.20
6 T_{sol} (C)	−0.80
7 T (C)	−3.00
8 ρ (g/cc)	1.29
9 m_g (%)	31.30
10 m_d (%)	19.10
11 H_{veg} (cm)	7.00
12 d_w (kg/m^2)	1.10
13 E'sol	5.20
14 T_{snow} (C)	
15 H_{snow} (cm)	0.10
16 Water eq. (mm)	0.10
17 Comment	Grass with hoar frost

of the microwave measurements is shown in Table 2. It gives date, time, a cloudiness index (CI) [8], temperatures, soil density (ρ), gravimetric soil moisture (m_g), dry matter fraction of the vegetation (m_d), height of vegetation (H_{veg}), water content of vegetation canopy (d_w), soil dielectric constant at 700 MHz and, finally, three snow parameters (if present). The indicated snow height of 1 mm does not represent an actual height, but it indicates the presence of hoar frost. The water content (d_w) is defined as the plant-water mass (frozen and unfrozen) per unit area of the ground surface.

The low value, 5.2, of the soil dielectric permittivity (measured from the surface to a depth of 1 cm) and the freezing temperatures indicate that the water of the soil was at least partly frozen. This is also true for the grass canopy.

Figure 2 shows a slow increase of all reflectivities with increasing incidence angle, and the polarization difference $r_h - r_v$ is small. The lack of a Brewster effect indicates that the object cannot be modelled by a flat surface. The grass canopy plays an important role even in the partly frozen condition (Table 2). Also the ice crystals of the hoar frost show an influence by their Rayleigh scattering. This can be seen in Fig. 3 where we observe an increase of r_v and r_h at frequencies increasing from 10 to 94 GHz. Grass covers at the same temperature but without hoar frost would not show such an increase. Over the 10 to 100 GHz range the reflectivities would be close to 0.05. The enhanced reflectivities at 4.9 GHz as compared to 10.4 GHz indicate that at the lowest frequency the grass canopy is partly transparent showing the influence of the larger reflectivity at the soil surface. A presentation and discussion of passive microwave signatures of landscapes in winter on the basis of the PAMIR catalogue can be found in [9].

The brightness temperatures T_p (p = v, h) near the surface follow from

$$T_p = e_p T + r_p T_{sky} \tag{2}$$

Table 3.
Sky brightness temperatures for the PAMIR data of December 14, 1992

f (GHz)	T_{sky} (°K)
4.9	7.2
10.4	13.3
21	27.5
35	27.5
94	53.2

and the sky brightness temperature T_{sky} is given by

$$T_{sky} = T_a(1 - e^{-\tau}) + T_{ex}e^{-\tau} \tag{3}$$

where T is the surface temperature, T_a is the effective air temperature, τ the effective opacity of the atmosphere and T_{ex} is the extraterrestrial brightness, usually equal to the 3 K cosmic background temperature.

For the data of December 14, 1992, the T_{sky} values at a zenith angle of 60° are given in Table 3 where τ is twice the zenith opacity.

The sky brightness temperatures at 21, 35, and 94 GHz strongly depend on the total contents of atmospheric water vapor and liquid water. With the algorithm of [8] the data of our example (Table 3) lead to a water vapor content of 6.6 kg/m^2 and to an upper limit of the liquid water content of 30 g/m^2. In the PAMIR catalogue the sky data are directly given in terms of zenith opacities.

4. TRANSMISSION AND REFLECTION MEASUREMENTS ON SEMI-TRANSPARENT OBJECTS

The second type of measurements made with PAMIR served for the investigation of wave propagation in layered media. The results are valuable on the one hand for the measurement of physical properties — such as the dielectric permittivity — and on the other hand for testing radiative transfer models. Both types of studies are required for the development of physical algorithms for the inversion of MIMR data. The measurement technique is simple; it consists of four radiometric measurements for transmissivity and reflectivity measurements. The methods were described by Sume *et al.* [10] and by Mätzler and Sume [11], and they are shown in Fig. 4.

For isothermal situations in which the Dicke temperature, the object temperature T and the black-body temperature T_{bb} are the same the transmissivity is determined by

$$t_p = \frac{T - T_{xp}}{T - T_{sky}} \tag{4}$$

and the reflectivity by

$$r_p = \frac{T - T_{yp}}{T - T_{sky}} \tag{5}$$

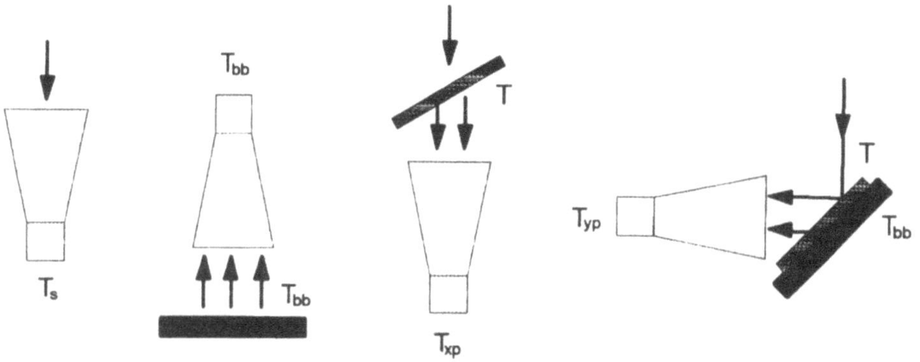

Figure 4. Four situations for the measurement of transmissivity and reflectivity, from left to right: sky calibration, black-body calibration, transmissivity measurement of test object, and reflectivity measurement with test objects above a black body.

where T is the physical object temperature, T_{sky} is the sky brightness temperature and T_{xp} and T_{yp} are the brightness temperatures of the object at polarization p in the last two configurations of Fig. 4.

Now Equation (1) must be changed in order to account for a transparent object:

$$e_p + r_p + t_p = 1. \qquad (6)$$

We should remember that the emissivity e_p is equal to the absorptivity a_p. In this sense Equation (6) represents the energy conservation. With Equations (4), (5) and (6) we can determine t_p, r_p and e_p (or a_p). In this way PAMIR was used for radiometric measurements of broad leaves [11]. The results were used to determine the dielectric properties of the leaf material.

Here we show another example: Twigs of spruce fir (*Picea excelsa*) were fixed on slabs of styropor. Figure 5 shows sample number 3 (RT3). This sample was measured at several freezing temperatures between −2 and −10°C. The result for $T = -2$°C is shown in Fig. 6. In order to compare the transmissivity measurement made at vertical incidence ($\Theta = 0$) with the reflectivity measurements made at $\Theta = 50°$ we averaged

$$r = (r_v + r_h)/2$$

and we assumed that r is independent of Θ. We observe that r is small, less than 0.05, which is much smaller than the not-transmitted fraction $1 - t$ of the incident radiation. Therefore the absorptivity a (or emissivity) dominates the interaction. All other reflectivity measurements of spruce fir were equally small so we will focus on the transmissivity, or $1 - t$, in the following.

An interesting result is the comparison of transmission measurements made with a drying sample of spruce fir (RT1) at temperatures between 8° and 10°C and the freezing sample (RT3). With increasing loss of water, the transmissivity of the drying sample increased. For the freezing sample the water content remained almost constant. Nevertheless its transmissivity increased with decreasing temperature. Furthermore, the transmissivity spectra of both samples are quite similar. The results are shown in Fig. 7.

Figure 5. Spruce fir twigs, sample RT3 used for the freezing experiment, Dec. 29, 1992 to Jan. 4, 1993. Top: total sample, size 50 × 60 cm, bottom: enlarged view.

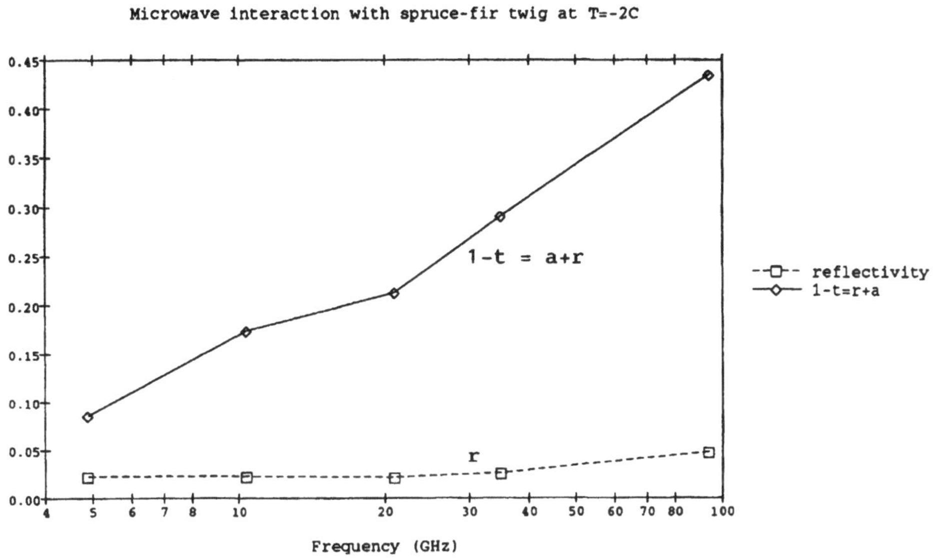

Figure 6. Reflectivity (dashed line) and $1 - t$ (solid line) versus frequency of spruce fir twig shown in Fig. 5 at $T = -2°C$.

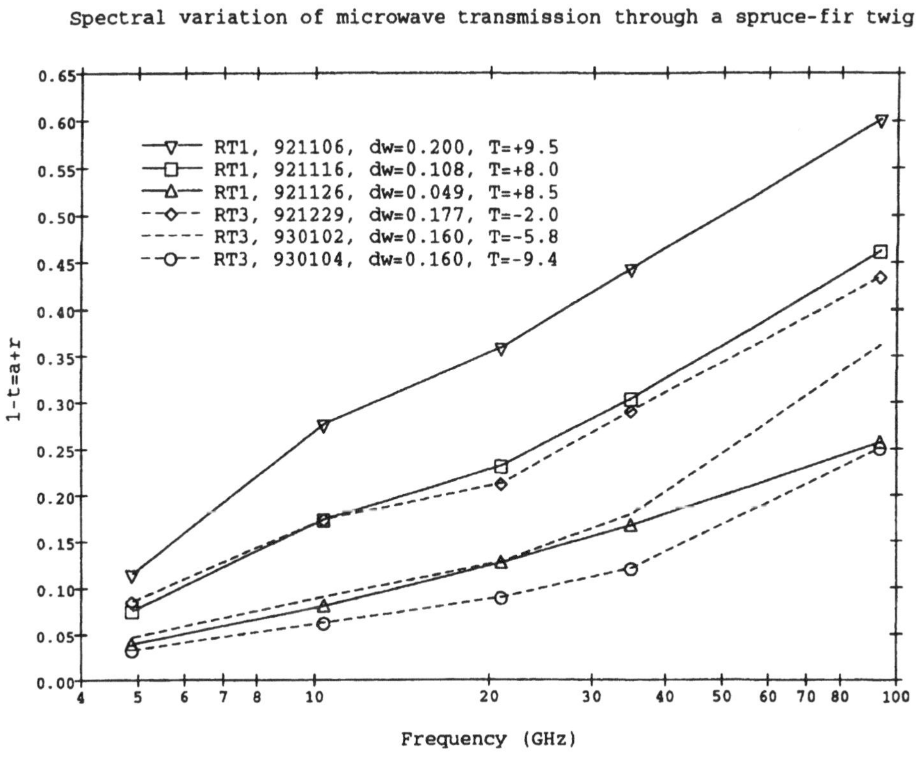

Figure 7. Comparison of $1 - t$ versus frequency for drying (RT1) and freezing (RT3) twigs of spruce fir.

We express the water content of the samples by the water mass, d_w, per unit area in kg/m^2. Numerically, this quantity is equal to the water equivalent thickness in mm. The close similarity of the $1 - t$ spectra of drying and of freezing twigs suggests that the reduction of liquid water is similar in both cases. Most of the absorption is due to liquid water, whereas absorption in ice is negligible. From the similarity of curves 2 and 4 we estimate that at $T = -2\,°C$ about 56% of the water is still in the liquid phase. This percentage follows from the ratio of the two d_w values. Similarly, at a temperature between $-6°$ and $-9\,°C$ about 30% of the water remains in the liquid phase.

These new results are of interest in physiological studies of the water status in plants. On the other hand, the results are also useful for the modelling of microwave signatures in forested areas. The temperature decrease can increase the transparency of forest trees in winter conditions. We will investigate if this effect can influence the retrieval of snowpack properties from MIMR data.

5. CONCLUSIONS

A PAMIR data catalogue 1989 to 1992 was completed and can now be used for microwave modelling of vegetation-covered ground and for the development of algorithms for MIMR data inversion. A similar data catalogue [12] of active and passive microwave data obtained with the 2 to 12 GHz radiometer and scatterometer RASAM [2, 3] is also available.

In addition to these surface observations a new type of microwave measurement gave valuable new insight into the behaviour of wave propagation in complex media, the transmission and reflection measurements made on slabs of vegetation material. In addition to the small objects it is possible to study the transmission through the entire canopy. In this way interesting results of leaves, twigs and trees were found. In future experiments we will improve the technique by reducing the shadowing effect produced by the radiometer box and by increasing the mobility of the instruments.

REFERENCES

1. R. Hüppi and E. Schanda. L- to X-band scatter and emission measurements of vegetation. In: *Proc. IGARSS '86 Symp., ESA SP-254 Vol. II*, pp. 1113–1118, Zürich, Switzerland (1986).
2. R. Hüppi. RASAM: a radiometer-scatterometer to measure microwave signatures of soil, vegetation and snow. PhD Thesis, University of Bern, Switzerland, December 1987 (1987).
3. U. Wegmüller. *Remote Sensing Signature Studies on Agricultural Fields with Ground-Based Radiometry and Scatterometry*. PhD Thesis, Faculty of Natural Sciences, University of Bern (1990).
4. E. Schanda and R. Hofer. Microwave multispectral investigations of snow. In: *Proc. 11th Internat. Symp. on Remote Sensing of Environment*, pp. 601–607, Ann Arbor, MI (1977).
5. C. Mätzler. Applications of the interaction of microwaves with the natural snow cover. *Remote Sens. Rev.*, **2**, 259–392 (1987).
6. C. Mätzler. Seasonal evolution of microwave radiation from an oat field. *Remote Sens. Environ.*, **31**, 161–173 (1990).
7. C. Mätzler. *Passive Microwave Signature Catalog 1989–1992*. Report of the Institute of Applied Physics, Bern, Dec. 1992 (1992).
8. C. Mätzler. Ground-based observations of atmospheric radiation at 5, 10, 21, 35, and 94 GHz. *Radio Science*, **27**, 403–415 (1992).
9. C. Mätzler. Microwave signatures of landscapes in winter. Submitted to *Meteorology and Atmospheric Physics* (1993).

10. A. Sume, C. Mätzler, R. Hüppi and E. Schanda. *Microwave Radiometer and Scatterometer Measurements of Vegetation*. Försvarets Forskningsanstalt (FAO) Rapport C 30494–3.2, Linköping, Sweden (1988).
11. C. Mätzler and A. Sume. Microwave radiometry of leaves. In: *Microwave Radiometry and Remote Sensing Applications*, P. Pampaloni (Ed.), pp. 133–148, VSP, Utrecht, The Netherlands (1989).
12. U. Wegmüller and C. Mätzler. *Active and Passive Microwave Signature Catalog (2-12 GHz)*. Report of the Institute of Applied Physics, Bern (1993).

ESA/NASA International Workshop, pp. 357–374
B. J. Choudhury, Y. H. Kerr, E. G. Njoku and P. Pampaloni (Eds)

Microwave emission from vegetation

SIMONETTA PALOSCIA

C.N.R.–I.R.O.E., Via Panciatichi 64, 50127 Firenze, Italy

Abstract—In this paper we summarize the results obtained in the past years using aircraft and ground-based observations on natural surfaces in a frequency range between 1.4 GHz and 37 GHz, at vertical and horizontal polarizations. It is shown that microwave emission is sensitive to vegetation biomass and that an interesting parameter for describing the vegetation growth is the polarization index measured at X-band and incidence angle higher than 40 degrees. The relation between brightness temperature and plant biophysical parameters is discussed by means of a simplified model based on radiative transfer theory.

1. INTRODUCTION

Remote sensing techniques, mainly by means of satellites, offer the possibility of observing the Earth's surface from space systematically during the year and over several years' time. This fact allows a practically continuous monitoring of evolution in time and space of phenomena at a large scale, such as desertification, deforestation, erosion processes and so on, which can therefore be examined thoroughly in order to look for the real causes, both natural or man-made, and to identify the eventual consequences.

In recent years, due to the rise in concentration of atmospheric carbon dioxide, the monitoring of vegetation on both regional and global scales has become more and more important for a deeper understanding of land surface heat and carbon balance and for quantitative evaluation of land degradation. Global satellite observations, when properly understood and validated, are expected to provide a key data source for the analyses of the biosphere. The main spectral bands used for this purpose have been in the visible and infrared domain and it has been pointed out that these types of observations can provide both spatial and temporal patterns of vegetation. Nevertheless, microwaves are also significant for their characteristics of weather independence and high sensitivity to water present in the observed surfaces. The latter characteristic is due to the large difference existing between the dielectric constant of water and that of dry matter. Many experimental and theoretical investigations pointed out the actual possibility of measuring the soil moisture content of a soil layer of several centimeters by using wavelengths in the order of 20–30 cm [e.g. 1–3]. However, analyses of field and satellite data of microwave emission at millimetric and centimetric wavelengths suggest that these data might provide potentially new information about vegetation as well.

Field observations and simulations have shown that microwave emission at centimetric and millimetric wavelengths from land surface is sensitive to vegetation features [4–7]. In general, dry bare soils show higher microwave emission than vegetated canopies, whilst on agricultural fields emission of moist bare soils may be lower or fairly similar to that of crops, depending primarily on the soil surface roughness. In both cases the use of multiband/multipolarization measurements may be very useful in separating bare from vegetated soils and for estimating vegetation parameters. Bidimensional diagrams obtained from 10 GHz and 36 GHz emissions have shown in fact the possibility of identifying different clusters representing bare soil and at least two main classes of crops: broad leaves with orientation mainly horizontal (i.e. sugarbeet, sunflower, corn) and small-randomly oriented leaves (i.e. wheat, alfalfa) [8]. In general, microwave emission from a canopy cover includes contributions from soil and vegetation layers; depending on the frequency and observation angle one contribution or the other is dominant. In order to minimize the soil effect and to investigate a layer of vegetation deep enough, the best frequency for studying vegetation features seems to be around 6–18 GHz. However, if the observation angle is in the order of 40 degrees from nadir or higher, even the L-band (1.4 GHz) emission was found to be sensitive to crop biomass [9]. On the other hand, a fairly good sensitivity for vegetation monitoring has been found up to 37 GHz, as has been suggested by some preliminary analyses of 37 GHz satellite data carried out by Choudhury and other colleagues [10–12]. Spaceborne microwave radiometer observations were obtained by the multichannel microwave radiometer (SMMR) on board the Nimbus-7 satellite and are currently being obtained by the special sensor microwave imager (SMM/I) on board DMSP satellite, with a spatial resolution of about 25 km or more. At these spatial resolutions, a single pixel of the image may contain many vegetation types including agricultural crops. The main drawback, which in fact has hampered and still hampers the use of microwave radiometers from space, is the poor ground resolution which is in the order of kilometers depending on the wavelength. The poor ground resolution limits a detailed analysis of the territory, mainly in rather heterogeneous areas, however it does not hinder and may favour the study of large scale processes. On the other hand, more detailed analysis of territory can be performed by means of aircraft sensors.

In this paper we summarize the results obtained with aircraft and ground-based observations on bare and vegetated fields at different microwave frequencies from 1.4 GHz to 37 GHz, in vertical and horizontal polarizations. The emissivity of various crop types has been correlated to some ground measured biophysical parameters, which mostly affect microwave emission in the tested frequency range. An interesting microwave parameter, able to describe the plant growth, was found to be the polarization index measured at X-band (10 GHz) and incidence angle higher than 40 degrees from nadir. Finally, experimental results have been interpreted using a simplified radiative transfer model of the soil–vegetation system.

2. EQUIPMENT AND EXPERIMENTAL RESULTS

Microwave radiation depends on the thermodynamic temperature of the emitting medium and on its emissivity. The latter contains major information on the physical conditions of the surface and depends on frequency, incidence angle, and polarization of the observing system. Also, the polarization difference is an important parameter in retrieving geophysical parameters from the observed brightness temperature at a fixed off-nadir incidence angle.

Experimental data used in this paper were collected by means of the MSP (Multiband Sensor Package), operating at I.R.O.E. from aircraft and ground platforms.

The MSP consists of three channel microwave radiometers 1.4 GHz (V or H polarization), 9.6 GHz (V and H), 36.6 GHz (V and H) and of a thermal infrared sensor operating in the 8–14 μm band. Microwave radiometers have an accuracy (repeatibility) of ± 1 K and a sensitivity of about ± 0.5 K. The thermal infrared sensor is a commercial type with a sensitivity and an accuracy of 0.5 K. The sensors were operated at a continuous variable incidence angle (0°–60°) from truck (10° step) and at two fixed directions (10° and 40°) from aircraft [13].

The measured quantities considered in this paper are the brightness temperature T_b and the polarization index PI:

$$\mathrm{PI} = 2 \times \frac{T_{bv} - T_{bh}}{T_{bv} + T_{bh}}$$

where T_{bv} and T_{bh} are vertical and horizontal polarization brightness temperatures. These parameters are available both from aircraft and truck-mounted sensors. Since a thermal infrared radiometer is a fixed part of the MSP, the microwave brightness temperature can be normalized to the infrared temperature in order to obtain an approximation of the emissivity, i.e. the normalized temperature T_n.

During the experiments, the following measurements of the main parameters of soil and vegetation have been carried out: gravimetric and volumetric soil moisture at different depths standard deviation of the surface heights H_{std} and correlation length of surface roughness computed from surface profiles obtained by means of a metallic needle profilometer. Besides the soil characteristics, the following crop parameters were measured: leaf area index LAI (m^2/m^2), plant water content PWC (kg/m^2), number of leaf, leaf and stem dimensions, vegetation temperature by means of the thermal infrared sensor.

An example of the expected levels of microwave emission, which can be observed in monitoring natural surfaces, is shown in Fig. 1, where the values of (a) mean normalized temperature

$$T_{nm} = \frac{T_{nv} + T_{nh}}{2}$$

and (b) polarization index PI, measured in June 1986 along a flight line crossing a river on an agricultural area in Northern Italy from the airborne MSP at 36 GHz (incidence angle $\theta = 40°$), are represented. On the Po river (left side of the diagram) T_{nm} values are quite low (≈ 0.5), while PI values are fairly high (≈ 0.15, corresponding to a difference between V and H polarizations higher than 35 K). Over agricultural fields T_{nm} values are fairly high (≈ 0.90–0.98), while PI values are close to zero. Additionally, the fluctuations in the T_{nm} values over agricultural fields are due to the presence of vegetation: the T_{nm} values are higher (0.98–0.99) over dry bare fields (stubble) than over vegetated fields (sugarbeet) which show an average value of about 0.91–0.92. On the contrary, PI values are slightly positive ($\delta T_b \approx 3$ K) on field of wheat stubble and close to zero on sugarbeet. This behaviour is quite understandable because polarized emission from soil is attenuated by vegetation (which, in turn, emits unpolarized or weakly polarized radiation).

2.1. Crop emission

The measurements of T_n and PI during the growth cycle of some crops observed in different environmental conditions have shown that microwave parameters are affected by the features of vegetation cover. In particular, both parameters seem rather sensitive

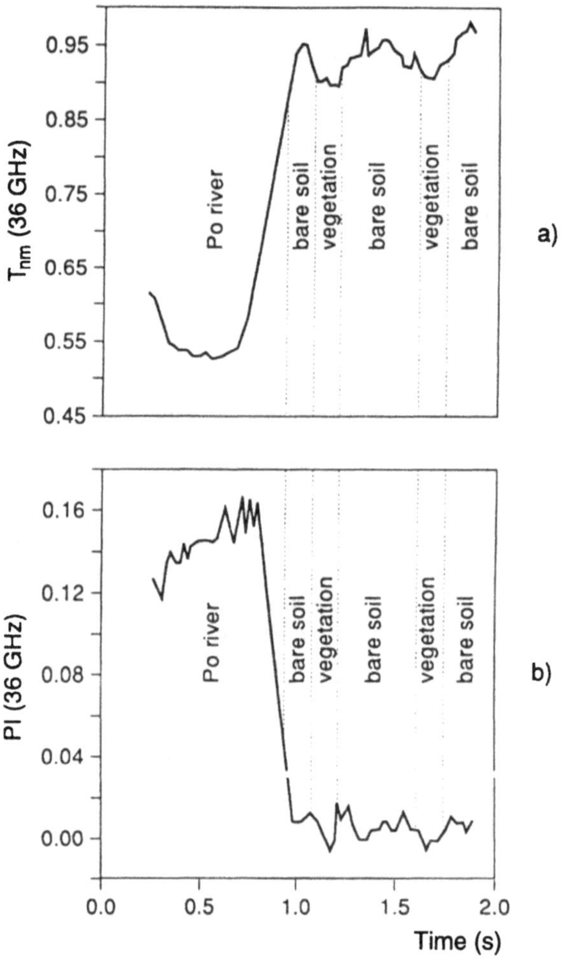

Figure 1. The mean normalized temperature T_{nm} (a) and the polarization index PI (b), measured by aircraft at 36 GHz and 40° incidence angle (θ), represented as a function of time (seconds) during a flight over the Po river and several agricultural fields carried out in June 1986 in Northern Italy.

to the total vegetation biomass and to the dimensions of plant constituents in terms of wavelengths.

A plant parameter, which well represents the stage of development of the canopy and is strictly related to the net productivity and, at least during the growth phase, to the biomass and the water content of plants, is the leaf area index LAI (m^2/m^2). In Fig. 2 aircraft data of T_n (H polarization) at 1.4 GHz (a), 10 GHz (b), and 36 GHz (c) is plotted against LAI of alfalfa (A), sugarbeet (S), sunflower (F), and corn (C). From these diagrams, we see that T_n changes with plant growth according to the observation frequency and crop type [9]. More specifically, at 1.4 GHz T_n generally increases when LAI increases, whereas at 10 GHz T_n values tend to increase as vegetation grows in plants which have small and rather randomly oriented leaves (i.e. wheat and alfalfa), and decrease in plants with broad and almost horizontal leaves (corn and sugarbeet). At 36 GHz a general, though slight, decreasing trend with LAI can be appreciated.

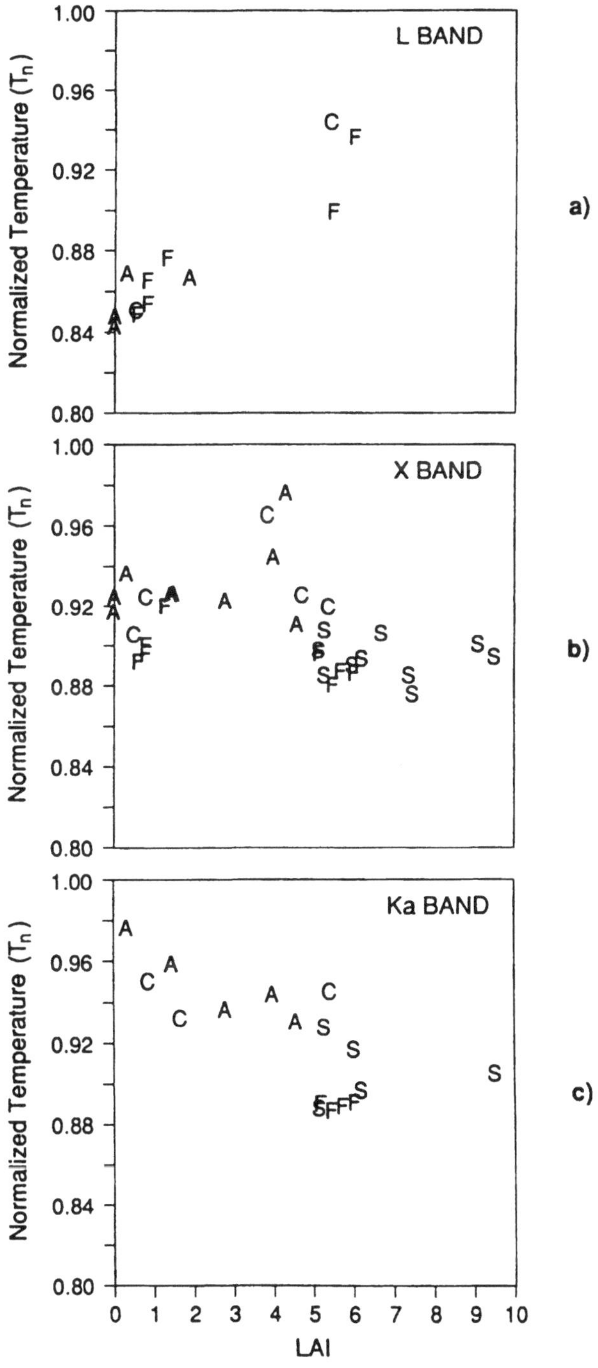

Figure 2. T_n measured in H polarization ($\theta = 40°$) at L-band (1.4 GHz) (a), X-band (10 GHz) (b), and Ka-band (36 GHz) (c) as a function of the leaf area index (LAI) of different crop types: alfalfa (A), sugarbeet (S), corn (C), sunflower (F).

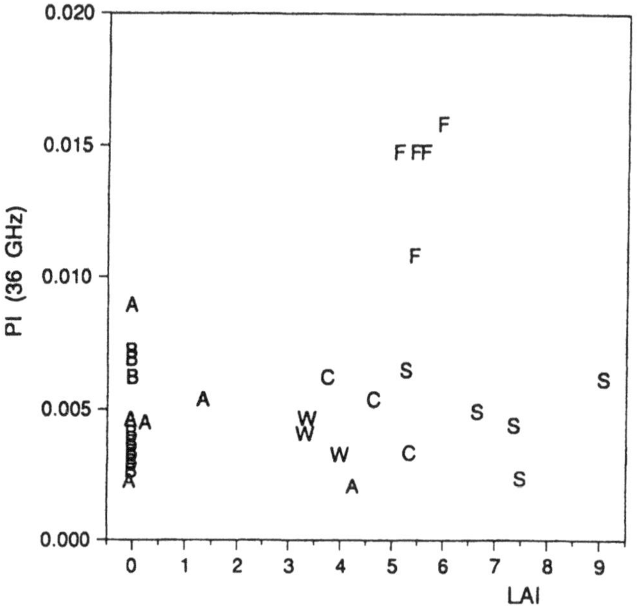

Figure 3. Polarization index (PI) measured by aircraft at 36 GHz ($\theta = 40°$) on different agricultural fields as a function of LAI. Labels are the same as in Fig. 2. (B = bare soil, W = wheat).

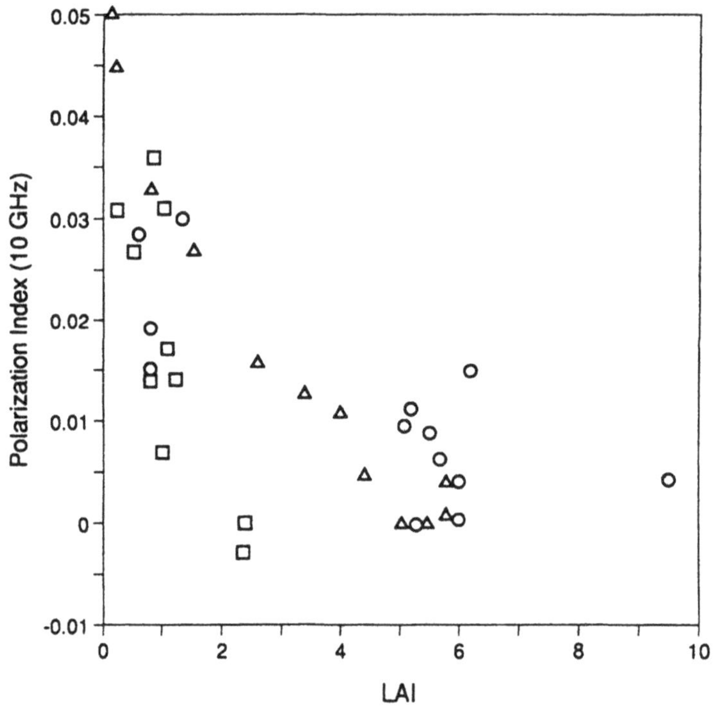

Figure 4. PI measured at 10 GHz ($\theta = 40° - 50°$) on corn (△), alfalfa (□) and sugarbeet/sunflower (○) as a function of LAI.

The PI at 36 GHz seems to be, in contrast, rather insensitive to LAI as it is shown in Fig. 3. However, it should be noted that the conditions indicated with LAI = 0 do not actually correspond to completely bare soil, in that wheat stubble (B) (over very dry soil) and alfalfa sprouts with an unquantifiable amount of green matter, and therefore of water, were present. Moreover, the soil surface under vegetation had the usual range of roughness of a medium texture agricultural soil ($H_{\rm std} > 1.5$ cm). These considerations can explain the low values of PI even on bare soil, in that at this frequency soil emission is easily affected by a soil surface roughness which exceeds a few millimetres and by a small amount of biomass over the ground. A slightly higher value of PI of about 0.02 ($\delta T_{\rm b} \approx 5-6$ K) has in fact been observed on smoother moist natural bare soils [6]. Furthermore, Fig. 3 shows that sunflower had a substantially higher PI when compared to other crops, proving that the plant morphology plays some significant role and cannot be neglected. A significantly different behaviour has been found at 10 GHz [4, 6, 8], where, as expected, a higher value of PI on bare soil, connected with a greater transparency of vegetation layer, causes a more appreciable decreasing trend during the whole crop growth, as plants grow. A good correlation has been pointed out between PI at 10 GHz and both LAI and plant water content, although crop morphology modifies the shape of the curves, as shown in Fig. 4, where the PI at 10 GHz is represented as a function of LAI for four different crops: alfalfa, corn, sugarbeet, and sunflower.

The importance of plant geometry also on polarimetric features of microwave emission is confirmed by observations carried out over several hours following the harvest of an alfalfa field. Indeed the PI, both at 10 GHz and 36 GHz, dramatically changes from negative or near zero to rather high positive values as soon as plants were cut and left on the soil (Fig. 5) [14]. Since biomass and water content of soil and vegetation

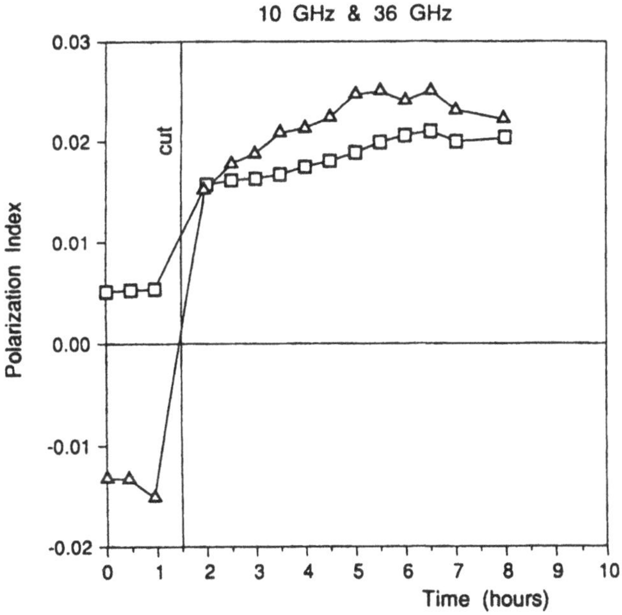

Figure 5. PI measured both at 10 GHz (△) and 36 GHz (□) ($\theta = 50°$) as a function of time (hours) during the harvesting of an alfalfa crop. After being cut, the plants are left on ground.

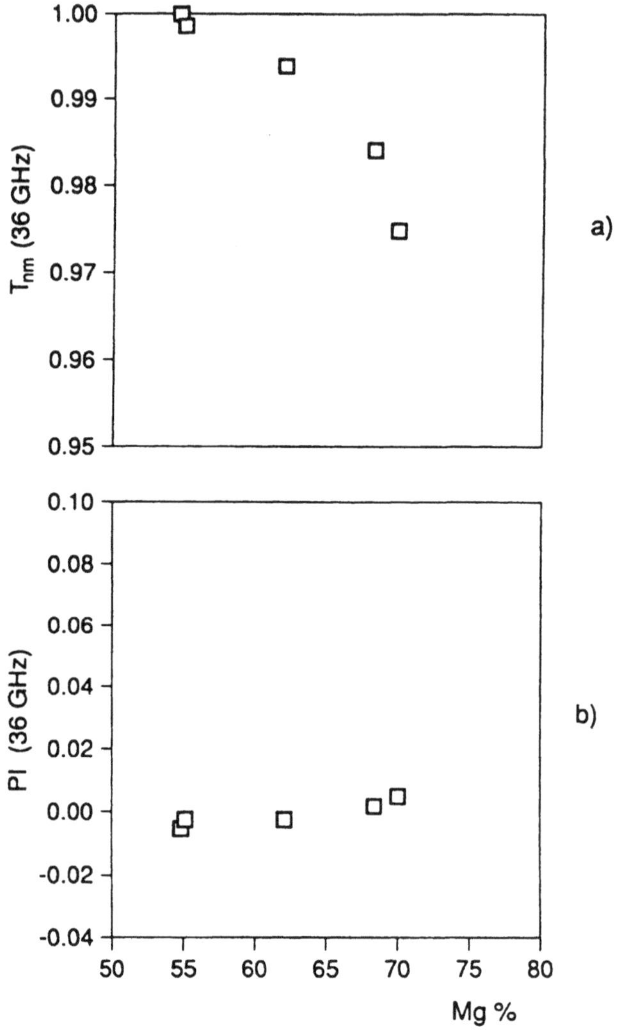

Figure 6. T_{nm} (a) and PI (b) measured at 36 GHz on ripe wheat crops as a function of the plant moisture content (Mg%).

remained almost the same during the short observational time lapse, the sharp variation of PI can be attributed to the change of orientation of stems from vertical to random horizontal position.

The dielectric properties of a vegetation layer can be fruitfully investigated during the senescence which is quite an interesting stage in the life cycle of the plant. During this period, dry matter no longer increases due to the exhaustion of photosynthetic function of leaves, water content decreases and the geometry of the canopy remains almost unchanged; the ripening processis due only to the loss of water. The variations of T_{nm} (a) and PI (b) with plant moisture Mg%, computed as the ratio between the total water content and the fresh weight of wheat crops during their senescent phase, are shown in Fig. 6 (36 GHz) and Fig. 7 (10 GHz). These diagrams pointed out that, when

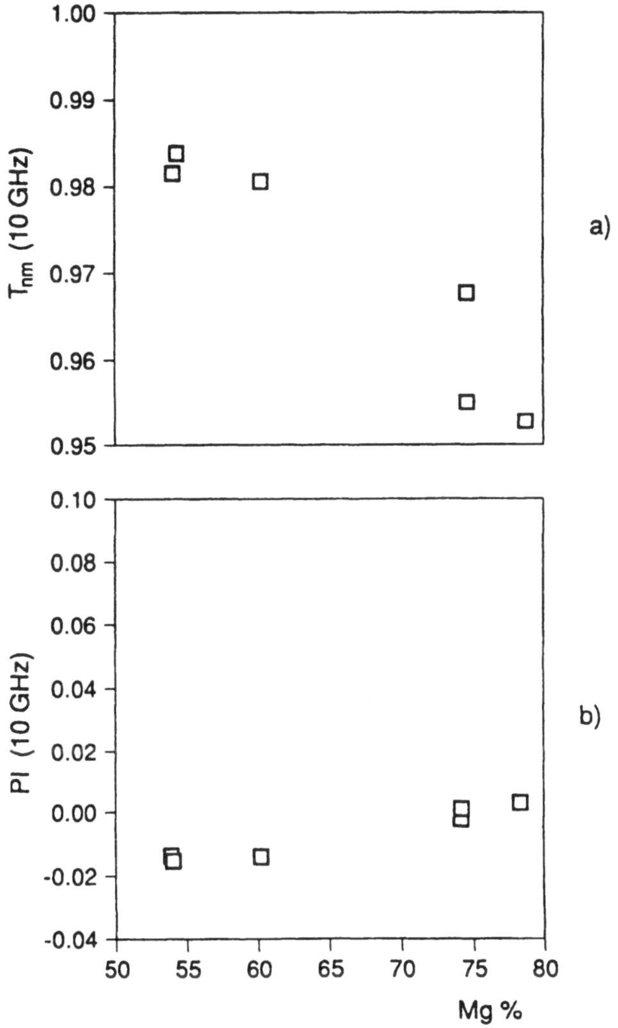

Figure 7. T_{nm} (a) and PI (b) measured at 10 GHz on ripe wheat crops as a function of the plant moisture content (Mg%).

LAI and plant water content (PWC) are no longer correlated to each other, T_{nm} at both frequencies appears to increase according to the progressive dehydration of vegetation, while PI remains basically constant. However, no significant variation of T_{nm} and PI is recognizable in this case as LAI decreases due to the ripening process: the effect of soil is always masked by vegetation, although plants lose water.

2.2. Soil emission

Since vegetation grows on a soil surface and the influence of soil cannot be completely neglected, even at off nadir incidence angles and at frequencies higher than 10–15 GHz, it seems necessary to show how the soil parameters, i.e. moisture and roughness, affect emission and how their effect is similar or different to that of vegetation. From the results shown until now, bare soil appeared both colder or warmer than vegetation according to the observation wavelength and the characteristics of roughness and moisture of the soil itself. In fact, if vegetation grows over a moist wet soil, emission, mainly in H polarization, tends to increase both at 10 and 1.4 GHz, whereas at 36 GHz it generally tends to decrease or at least remain constant. If soil is dry or/and rough, in terms of wavelength, emission of fully-developed vegetation is usually lower.

Emission from soil is mainly influenced by surface roughness and moisture content, the major role played by one or the other of these parameters, depending on the observation frequency. At 1.4 GHz a fairly good sensitivity to the soil moisture, which is about 3 K per percent of soil moisture content SMC on bare and smooth soils and about 1–1.5 K/%SMC on rough and vegetated agricultural soils, has been observed by various scientists who have worked on this topic for several years [1, 2, 13]. At 10 GHz, the sensitivity of T_n to the variations of soil moisture strongly decreases when the soil is covered by well developed vegetation and $H_{std} > 1-1.5$ cm, whereas at 36 GHz a surface roughness with a $H_{std} \approx 1$ cm is enough to mask the effect due to the soil moisture on bare soil as well [13, 15].

The effect of roughness on the emission at different wavelengths is well summarized by the diagram of Fig. 8, where the T_n at L- (1.4 GHz), X- (10 GHz) and Ka- (36 GHz)

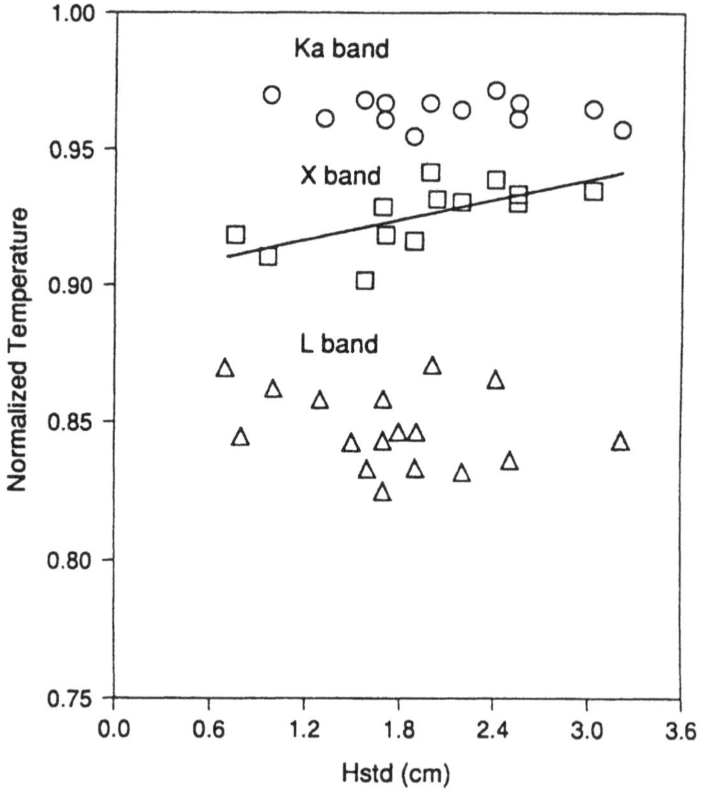

Figure 8. The normalized temperature T_n measured at H polarization ($\theta = 40°$) on bare soil at Ka-band (36 GHz), X-band (10 GHz), and L-band (1.4 GHz) represented as a function of the standard deviation of the surface height (H_{std}). The line represents the best fit for X-band data.

bands are represented as a function of H_{std}. We observe that the sensitivity to roughness depends on the frequency, and emission at 10 GHz seems to be the most affected by the soil surface roughness, in the observed range. At 36 GHz the T_n saturates as soon as the roughness reaches a value close to the corresponding wavelength (0.8 cm), whereas at 1.4 GHz emission does not show noticeable variations, in that height variations are smaller with respect to a wavelength of 20 cm.

In addition, the difference between the vertical and the horizontal polarization is affected both by moisture and roughness. In diagram of Fig. 9, data of PI measured at 36 GHz (a) and 10 GHz (b) on bare soil are represented as a function of the volumetric moisture content (SMCv) of a soil layer of a few centimeters for different classes of

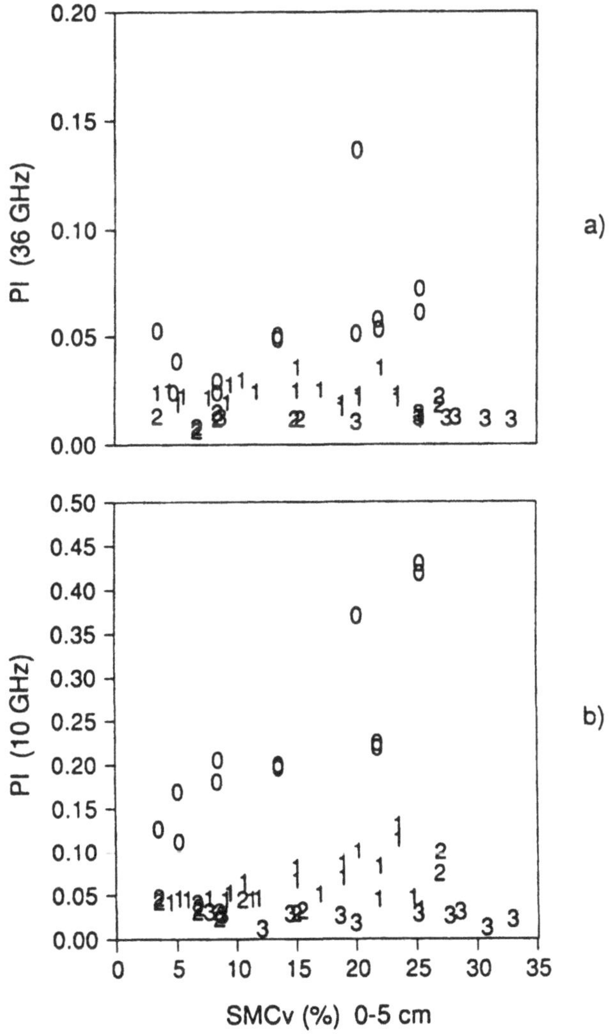

Figure 9. PI at 36 GHz (a) and 10 GHz (b) ($\theta = 50°$) as a function of the volumetric soil moisture content (SMCv%) of the first layer of bare soils (2.5 cm and 5 cm, respectively). 0 = smooth soil ($H_{std} < 0.1$ cm), 1 and 2 = moderately rough soils, corresponding respectively to $H_{std} \leqslant 1$ cm and 1–1.5 cm, and 3 = rough soil with $H_{std} > 1.5$ cm.

roughness represented in the diagram by the labels: 0 = smooth soil ($H_{std} < 0.1$ cm), 1 and 2 = moderately rough soils, corresponding respectively to $H_{std} \leqslant 1$ cm and 1 to 1.5 cm, and 3 = rough soil with $H_{std} > 1.5$ cm. As expected, the PI shows a clear increasing trend as moisture increases only for very smooth soil at both frequencies. The sensitivity of PI, fairly low at 36 GHz (about 0.002/%SMCv, roughly corresponding to δT_b lower than 1 K), is instead rather high at 10 GHz (0.01/%SMCv, i.e. $\delta T_b \approx$ 2.5K/%SMCv) and generally decreases as the roughness increases. At 10 GHz, three different trends can be appreciated corresponding to the different classes of roughness, a certain sensitivity still existing for H_{std} of about 1 cm. For rougher soils no sensitivity at all is shown at this frequency. At 36 GHz, sensitivity becomes pratically zero as H_{std} reaches 0.8 cm.

At the lowest frequency (L-band) the average sensitivity of polarization index to the soil moisture corresponds to δT_b in the order of about 1.5 K/%SMCv, and the dependence on soil roughness (in the considered range) is practically negligible, as we can see in Fig. 10, where the PI at 1.4 GHz is represented as a function of the SMC of the 0–5 cm layer of a bare soil with three different levels of roughness $H_{std} \approx 0$ cm, 0.6 cm and 1.2 cm. Although the measurements appear somewhat dispersed, the increasing trend is clear. The regression line computed for this data set is: PI = 0.24+0.005 SMCv.

A direct comparison of PI with the surface roughness shows a sharp decrease with the H_{std} at both 10 GHz and 36 GHz, when only points with a SMC higher than 15%

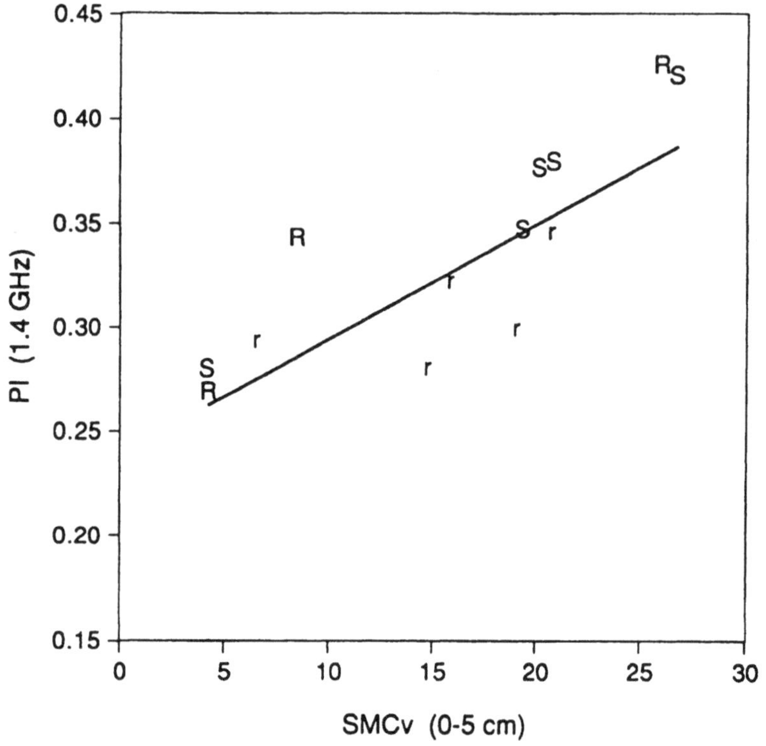

Figure 10. PI at 1.4 GHz ($\theta = 50°$) as a function of the SMCv% of the first 5 cm layer. S = smooth soil ($H_{std} \approx 0$ cm), r = medium rough soil ($H_{std} = 0.6$ cm), R = rough soil ($H_{std} = 1.2$ cm). Line represents the best fit: PI = 0.24 + 0.005 SMCv%.

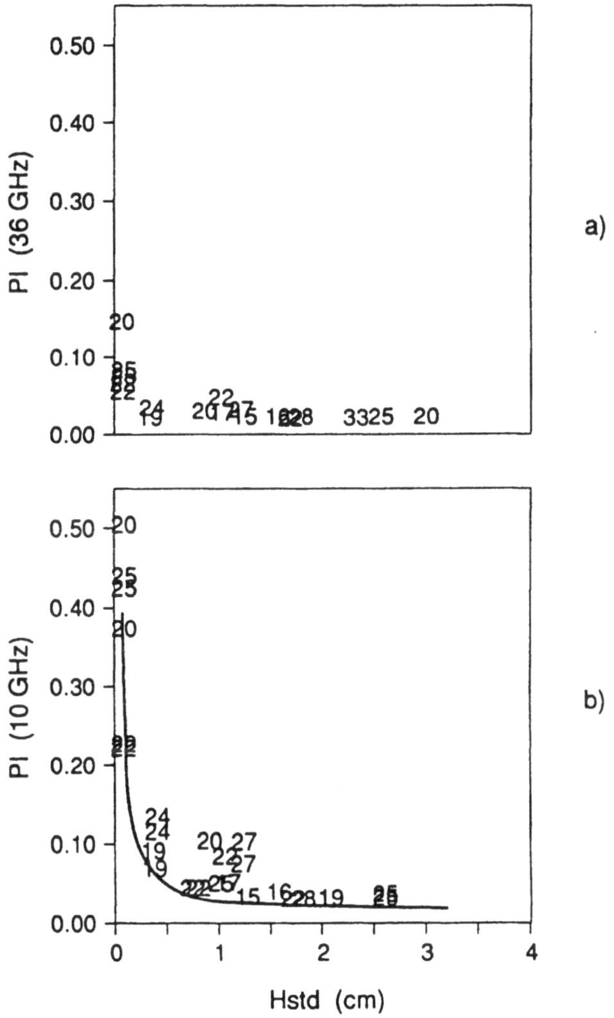

Figure 11. PI at 36 GHz (a) and 10 GHz (b) ($\theta = 50°$) as a function of the H_{std}. Labels represent the soil moisture value of each sample. The line is the best fit of experimental points at 10 GHz obtained by a power function.

have been considered (Fig. 11). At 10 GHz (b) experimental values have been fitted by the function: $PI = (H_{std}^{-0.72}) \times 0.028$. At 36 GHz (a) the decrease of PI values is sharper and the range of variation is smaller than at 10 GHz, showing a high sensitivity to the soil roughness and a saturation as soon as the height standard deviation reaches a value close to the wavelength.

3. DATA ANALYSIS AND MODELING

Experimental results are discussed using a 'slab' model [5, 16] which considers vegetation to be a continuous absorbing and scattering medium upon the soil surface. According to this model and the radiative transfer theory, the emissivity ε_c of a vegetation covered soil in thermal equilibrium (i.e. assuming that the canopy temperature is equal to the soil temperature) can be expressed by the following relation, which takes into account the components of emission from vegetation, from soil attenuated by vegetation and from vegetation reflected by soil:

$$\varepsilon_c = 1 - (1 - \varepsilon_{sp})e^{-2\tau/\mu} - w\left(1 - e^{-\tau/\mu}\right)\left[1 + (1 - \varepsilon_{sp})e^{-\tau/\mu}\right] \tag{1}$$

where:

$\tau = k_e L$ = optical depth of vegetation,

w = single scattering albedo,

ε_{sp} = soil emissivity at p polarization (H or V), which is a function of soil moisture,

$\mu = \cos\theta$ = radiation direction.

Equation (1) has been parametrized (at $\theta = 45°$) as a function of the optical depth of vegetation τ, for different values of single scattering albedo w [17]. It has been generally observed that for pure absorption ($w = 0$), the emissivity monotonically increases with τ up to saturation. As w increases, the shape of the emissivity curves changes: it may have a maximum or a monotonic decrease as τ increases.

Since the optical depth of a vegetation layer has been linearly correlated to the leaf area index LAI of some crop types [17, 18], we can relate the emissivity of the canopy ε_{sp} directly to LAI and furthermore compare the emission, measured for different values of LAI, with this model. In doing this comparison, the measured normalized temperature T_n has been used to represent the emissivity of vegetation.

Figure 12(a) represents the emissivity computed for $w = 0$ and $w = 0.05$, as a function of LAI, compared with experimental values of normalized temperatures obtained

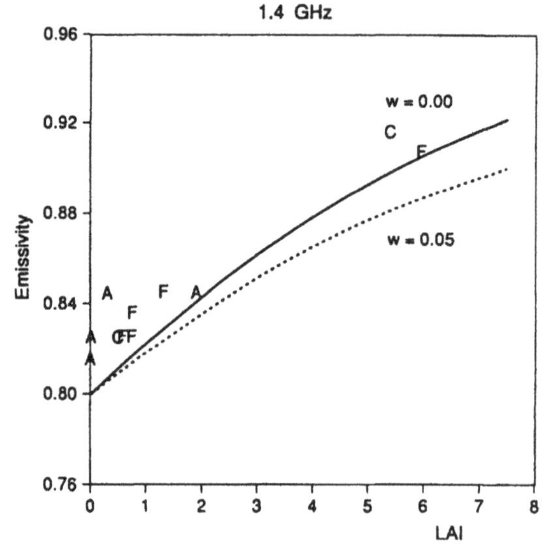

Figure 12a. Emissivity, computed at 1.4 GHz in H polarization using Equation (1) in the text is here represented vs. LAI and compared with experimental values of T_n at the same frequency (labels), already shown in Fig. 2. Emissivity has been computed for different values of single scattering albedo (w) and soil emissivity.

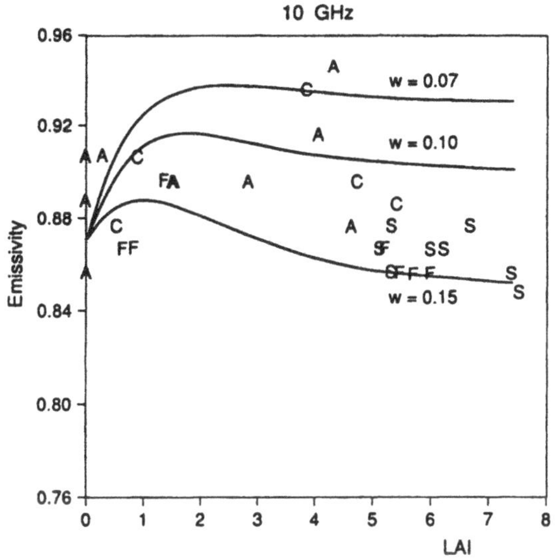

Figure 12b. As in Fig. 12(a), but at 10 GHz.

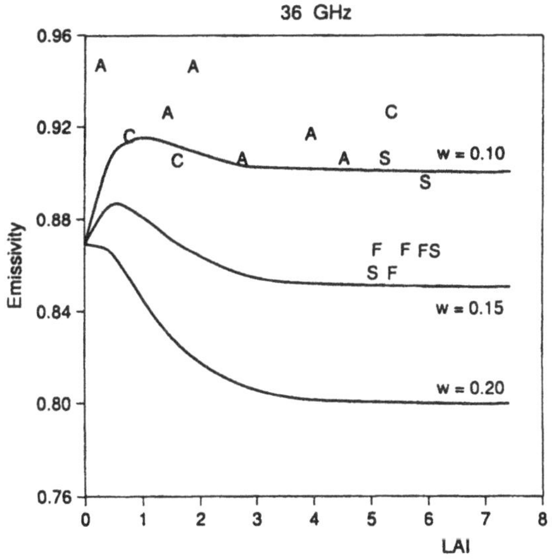

Figure 12c. As in Fig. 12(a), but at 36 GHz.

at 1.4 GHz. The comparison with measurements at 10 GHz and 36 GHz is represented in Figs 12(b) and 12(c) respectively. From these diagrams we see that, at 1.4 GHz, where the observation wavelength is generally larger than the average leaf dimension, experimental data are reproduced fairly well by the model without scattering ($w = 0$) and using a soil emissivity quite low (≈ 0.8). Both model and experimental values show an increase of the emissivity as vegetation grows, pointing out the absorption role of vegetation layer at this frequency. As the frequency increases, the measurements can be approximated only by taking scattering into consideration and choosing a higher value of soil emissivity, due to the decrease of soil permittivity and the the major influence of soil roughness. At 36 GHz a scattering albedo between 0.10 and 0.15 explains the basic decreasing trend of emission as LAI increases due to the relatively high reflectivity of vegetation constituents. At 10 GHz lower values of w (0.07–0.10) model the increasing trend of emissivity for small leaf crops (alfalfa), similar to that observed at 1.4 GHz, while a higher scattering coefficient (0.15) is necessary to follow the decreasing trend for wide leaf crops (sugarbeet and sunflower) with LAI higher than 4. In the latter case, leaf dimensions, in terms of wavelength, are generally large enough to show quite a strong reflectivity.

Although the model is a very simple one, it seems able to reproduce, with some approximations, experimental data, provided the value of soil emissivity is correctly selected. The same model, which can be easily inverted, was found able to relate the polarization index PI, both at 10 GHz and 36 GHz, to LAI (and to plant water content as well), according to a crop factor and provided the PI of bare soil is correctly selected, and to predict the leaf area index from the measured polarization index at 10 GHz [6]. A more sophisticated model, based on the discrete approach, which considers vegetation to be an ensemble of thin layers of single elements (disks and needles) has been recently developed [14] and found capable of successfully reproducing several experimental data sets. However, the complexity of this approach makes an inversion very challenging.

4. FINAL REMARKS

Experiments carried out with passive microwave sensors at 1.4, 10, and 36 GHz demonstrate the sensitivity of microwave emission to the moisture conditions of soil and plants. Two main factors seem to affect microwave emission from vegetation: biomass and crop type. Normalized temperatures at 1.4 GHz and 36 GHz mostly depend on the total amount of biomass, whilst the 10 GHz emission is influenced by crop type too. Generally, when leaves are small in comparison to the observation wavelength, the normalized temperature increases with leaf area index LAI, and vice versa, the emission decreases with LAI when leaves are large enough to show a notable scattering.

These results, interpreted according to a model based on the radiative transfer theory, have been able to reproduce fairly well the experimental data as long as the soil emissivity value is correctly selected. The same model should allow the estimation the leaf area index of a known crop type, from the measurement of polarization index at 10 GHz. Since the leaf area index is a key factor for the estimate of net productivity and biomass of plants at a global scale, the possibility of estimating such a parameter, at least for some crop types, by means of remote sensing technique can be very useful in several disciplines, in particular in the study of the global carbon cycle.

In summary, the experimental and theoretical investigations carried out up to now have demonstrated that the use of multifrequency microwave radiometry for vegetation monitoring over large areas appears to be a promising approach. However, much further research is needed in order to estimate the accuracy and the reliability of the measurement and before the method becomes operational. Airborne radiometers appear to be the most suitable choice for a parametrization of land-surface properties and quantitative estimates of parameters involved in the hydrological cycle. The use of satellite data, although limited by the coarse ground resolution, is advantageous for monitoring of global processes, especially considering that the future instruments will improve the present sensor characteristics considerably. For example the ESA/MIMR (Multifrequency Imaging Microwave Radiometer), a six channel radiometer (at 6.8, 10.6, 18.7, 23.8, 36.5, 90 GHz) [19] will have a ground resolution ranging from about 4 km at the highest frequency to about 50 km at the lowest one. Also, the use of appropriate deconvolution algorithms should offer a significant improvement in the processing of microwave satellite data over regions characterized by a wide spatial heterogeneity.

Acknowledgements
This work was partially supported by the Italian Space Agency (ASI).

REFERENCES

1. T. Schmugge. Remote sensing of soil moisture. In: *Hydrological Forecasting*, M. G. Anderson (Ed.), John Wiley & Sons, New York (1985).
2. A. Shutko. Microwave radiometry of lands under natural and artificial moistening. *IEEE Trans. Geosci. Remote Sensing*, **GE-20** (1), 18 (1982).
3. G. Ijjas, I. Juhasz and Zilahy. Airborne microwave equipment for soil moisture investigation. In: *Microwave Radiometry and Remote Sensing Applications*, P. Pampaloni (Ed.), pp. 321–328, VSP Press, Utrecht, The Netherlands (1989).
4. P. Pampaloni and S. Paloscia. Experimental relationships between microwave emission and experimental results. *Int. J. Remote Sensing*, **6** (2), 315–323 (1985).
5. P. Pampaloni and S. Paloscia. Microwave emission and plant water content: a comparison between field measurements and theory. *IEEE Trans. Geosci. Remote Sensing*, **GE-24** (6), 900–905 (1986).
6. S. Paloscia and P. Pampaloni. Microwave polarization index for monitoring vegetation growth. *IEEE Trans. Geosci. Remote Sensing*, **GE-26** (5), 617–621 (1988).
7. C. Maetzler. Seasonal evolution of microwave radiation from an oat field. *Remote Sens. Environ.*, **31**, 161–173 (1990).
8. S. Paloscia and P. Pampaloni. Microwave vegetation indexes for detecting biomass and water conditions of agricultural crops. *Remote Sens. Environ.*, **40**, 15–26 (1992).
9. P. Ferrazzoli, S. Paloscia, P. Pampaloni, G. Schiavon, S. Solimini and P. Coppo. Sensitivity of microwave measurements to vegetation biomass and soil moisture content: a case study. *IEEE Trans. Geosci. Remote Sensing*, **GE-30** (4), 750–756 (1992).
10. F. Becker and B. J. Choudhury. Relative sensitivity of normalized difference vegetation index (NDVI) and microwave polarization difference index (MPDI) for vegetation and desertification monitoring. *Remote Sens. Environ.*, **24**, 297–311 (1988).
11. B. J. Choudhury. Monitoring global land surface using Nimbus-7 37 GHz data. Theory and examples. *Int. J. Rem. Sensing*, **10** (10), 1579–1605 (1989).
12. Y. Kerr and E. Njoku. A semiempirical model for interpreting microwave emission from semiarid land surfaces as seen from space. *IEEE Trans. Geosci. Remote Sensing*, **GE-28** (3), 384–393 (1990).
13. S. Paloscia, P. Pampaloni, L. Chiarantini, P. Coppo, G. Gagliani and G. Luzi. Multifrequency passive microwave remote sensing of soil moisture and roughness. *Int. J. Remote Sensing*, **14** (3), 467–483 (1993).
14. P. Ferrazzoli, L. Guerriero, S. Paloscia, P. Pampaloni and D. Solimini. Modelling polarization properties of emission from soil covered with vegetation. *IEEE Trans. Geosci. Remote Sensing*, **GE-30**, 157–165 (1992).
15. P. Pampaloni, S. Paloscia and L. Chiarantini. Contribution of passive microwave remote sensing in soil moisture and evapotranspiration measurements. In: *Proc. of Conference on Parameterization of Land Surface Characteristics*, pp. 327–331, Rome, December 2–6, 1985, ESA-SP 248 (1985).

16. T. Mo, B. Choudhury, T. Schmugge, J. Wang and T. Jackson. A model for microwave emission from vegetation covered fields. *J. Geophys. Res.*, **87** (C13), 11229–11237 (1981).
17. P. Pampaloni. Microwave radiometry of soil and vegetation. In: *Proc. of Specialist Meeting on Microwave Radiometry and Remote Sensing Applications, μrad92*, E. Westwater (Ed.), pp. 42–49, Wave Propagation Lab., Boulder (CO), June 1992. US Dept. of Commerce, NOAA, Env. Res. Lab. (1992).
18. F. Ulaby, R. Moore and A. Fung. *Microwave Remote Sensing: Active and Passive*, Vol. III. Artech House Inc., Norwood MA (1986).
19. R. Bernard, M. Hallikainen, Y. Kerr, K. Kuenzi, C. Maetzler, P. Pampaloni, G. Duchossois, Y. Menard and M. Rast. *The Multifrequency Imaging Microwave Radiometer Instrument Panel Report*. ESA SP-1138, August 1990 (1990).

ESA/NASA International Workshop, pp. 375–387
B. J. Choudhury, Y. H. Kerr, E. G. Njoku and P. Pampaloni (Eds)

Canopy opacity models

URS WEGMÜLLER,[1] CHRISTIAN MÄTZLER[1] and ENI NJOKU[2]

[1] *Institute of Applied Physics, University of Berne, Sidlerstr. 5 CH-3012 Bern, Switzerland*
[2] *Jet Propulsion Laboratory, 4800 Oak Grove Drive, Pasadena, CA 91109, USA*

Abstract—Radiative transfer models have been developed to understand and interpret passive microwave remote sensing data. An example is the model described by Kerr and Njoku [1] for spaceborne observations of semi-arid land surfaces. Such models usually incorporate models of vegetation attenuation and emission that are valid at low frequencies only; nevertheless, remotely sensed data are often collected at higher frequencies which contain much information about soil and vegetation. The often-used canopy opacity model of Kirdyashev *et al.* [2] is restricted to frequencies less than about 7.5 GHz. It is shown in this paper that the range of validity can be extended to higher frequencies using geometrical optics. The main improvement is that the electric field within the vegetation components is no longer assumed to be constant. A comparison with measured opacities shows that geometric optics provides an improvement over [2], and should therefore be used preferentially in the radiative transfer models. Geometrical optics may also be used to estimate the importance of the single scattering albedo.The semi-empirical formula of Ulaby and El-Rayes [3], often used to calculate the dielectric constant of vegetation, is restricted in validity to relatively low water contents and frequencies between 1 and 20 GHz. Mätzler and Sume [4] have suggested a more widely applicable empirical formula derived from measurements at 1 to 100 GHz on leaves of various plants including low and high water contents. Finally, the importance of the canopy fractional coverage in estimating vegetation parameters is illustrated.

1. INTRODUCTION

Recent discussions on global climatic change have raised the interest in mapping and monitoring vegetation. Vegetation plays a major role in the carbon cycle, oxygen production combined with carbon dioxide reduction, the water cycle, and the energy balance of the earth. It provides us with many raw materials and it is our primary food resource. In addition, it protects soil against erosion and is responsible for the production of humus. Remote sensing has a high potential for collecting global multi-temporal data and is therefore an ideal tool for mapping and monitoring vegetation. For the optical/near infrared region and the microwave region of the electromagnetic spectrum, many authors have found strong relationships between remote sensing data and vegetation parameters. In this study we concentrate on passive microwave data.

Relations between passive microwave satellite data and parameters such as vegetation water content, wet biomass and leaf area index have been found [1, 5]. Ground-based and airborne measurements have been carried out to improve the understanding of the interaction between microwaves and vegetation [6–10], and radiative transfer models have been developed for the interpretation of the data [1, 11, 12]. A summary of such data was recently published by Jackson and Schmugge [26]. Here we discuss some aspects of the remote sensing model described by Kerr and Njoku [1]. The results are also relevant to other models [11, 12], because of the similarities in the basic assumptions. Our study is focused on the vegetation components of these models. One of the goals is to show the current limitations, and to discuss where the modelling could be improved. Specifically, suggestions improving the canopy opacity model are presented.

2. RADIATIVE TRANSFER MODEL

This section discusses the radiative transfer model described by Kerr and Njoku [1], as a reference point for studying the vegetation attenuation and emission. The model assumes that second and higher order scattering in the vegetation can be ignored and there is no reflection of radiation at the vegetation-air interface. The microwave brightness temperature at the receiving antenna, at p polarization, T_{Bp}, is found by

$$T_{Bp} = T_1 + T_2 + T_3 + T_4 + T_5 \quad (1)$$

where the upward atmospheric emission is T_1;
the soil-surface emission attenuated through the canopy and atmosphere:

$$T_2 = (1 - r_{sp})T_s \exp(-\tau_p - \tau_a) \quad (2)$$

the downward atmospheric T_{ad} and cosmic background radiation T_{sky} attenuated through the canopy and atmosphere:

$$T_3 = r_{sp}\big(T_{ad} + T_{sky} \exp(-\tau_a)\big) \exp(-2\tau_p - \tau_a) \quad (3)$$

the upward canopy emission attenuated through the atmosphere:

$$T_4 = T_c(1 - \omega_p)\big(1 - \exp(-\tau_p)\big) \exp(-\tau_a) \quad (4)$$

and the downward canopy emission reflected at the surface and attenuated through the canopy and atmosphere:

$$T_5 = T_c(1 - \omega_p)\big(1 - \exp(-\tau_p)\big) r_{sp} \exp(-\tau_p - \tau_a) \quad (5)$$

with the temperatures of the canopy (T_c) and the soil (T_s), the brightness temperature of the downward atmospheric radiation (T_{ad}), the opacities of the atmosphere (τ_a) and the canopy (τ_p) the reflectivity of the soil surface (r_{sp}) and the single scattering albedo of the canopy (ω_p). The index p stands for the linear polarization, h or v.

This model formulation was described earlier by Mo *et al.* [11] and others. For a more detailed discussion of the atmospheric part of the model see [13, 14].

The reflectivity of the soil surface is given by Wang and Choudhury [24]

$$r_{sp} = \big(Q r_{oq} + (1 - Q) r_{op}\big) e^{-(2k_0\sigma \cos\theta)^2} \quad (6)$$

θ: incidence angle relative to nadir,
k_0: wave number in free space,
r_{sp}: surface reflectivity (p: polarization v or h),

r_{op}, r_{oq}: Fresnel reflectivities (p, q: polarizations h or v, with p $\neq$ q),
Q: polarization coupling factor,
σ: standard deviation of surface height in cm,
where an empirical expression for Q is given by [1]:

$$Q = 0.35\left(1 - e^{-0.6\sigma^2 f}\right) \tag{7}$$

f: frequency in GHz.

It has been found (e.g. Wang *et al.* [15]), that the values of σ determined by fitting Equation (6) to experimental data over rough surfaces do not agree in general with values of σ derived from direct measurements of the surface roughness (e.g. by a laser profilometer). Thus, Equations (6) and (7) may describe the frequency and polarization dependence of experimental data over rough surfaces in the 1–37 GHz range, but only if artificial values for surface rms height are assumed. This points out a deficiency in the roughness model which should be corrected if the effects of roughness are to be correctly incorporated and estimated. More realistic models can be found by integrating the bistatic scattering coefficient over the upper hemisphere [27, 28] followed by a parametrization using measured data [29].

The vegetation layer is characterized by its optical thickness, also called opacity, τ_p, and its single scattering albedo, ω_p. For small ω_p values τ_p becomes the prime quantity to describe the interaction of microwaves with vegetation. Sometimes ω_p is assumed to be small even at 37 GHz [6, 7]. All the above mentioned authors use a linear relationship between the vegetation water content W [kg/m^2] and the canopy opacity τ_p. In [1] we find

$$\tau_p = A_p k_0 \frac{W}{\rho_{water}} \varepsilon''_{sw} \frac{1}{\cos\theta} \tag{8}$$

with ε''_{sw} the imaginary part of the dielectric constant of saline water having a concentration corresponding to vegetation, k_0 the wave number in air and θ the observation angle relative to nadir. The coefficient A_p depends on the canopy geometry. In [1] the factor $1/\rho_{water}$ was included in the parameter A_p.

3. VEGETATION OPACITY

Originally, as introduced by Kirdyashev *et al.* [2], A_p appearing in Equation (8) was a theoretically derived geometrical parameter. However, there is no simple way of deriving this parameter for actual vegetation such as grasses, trees, or crops and the assumptions used in deriving Equation (8) become invalid at higher frequencies. Hence, for comparing with satellite data, Kerr and Njoku [1] used Equation (8) as an empirical formula, and determined the parameter A_p individually for each frequency and canopy type. In the following sections we shall examine the theoretical origin of Equation (8), which is based on the 'Effective Medium theory', and show that a more accurate frequency dependence can be obtained by considering the 'Geometrical Optics theory'.

3.1. Effective Medium theory for low frequencies

In the Effective Medium theory a heterogeneous medium, in our case plant elements and air, is treated as a homogeneous medium with an effective dielectric constant, ε_{eff}. We will use the low frequency assumptions that the electric fields within the inclusions can be assumed homogeneous and that scattering can be neglected as long as the extinction is concerned. For ellipsoidal inclusions the polarizability tensor, α, can be calculated (see textbooks on electrodynamics, [17]). The low density of plant elements in the canopy in combination with $\varepsilon''_{\text{veg}} \gg \varepsilon''_{\text{air}}$ allows the application of the linear mixing formula. In this case the dipole moments of the individual inclusions are added to the total dipole moment leading to the effective dielectric constant

$$\varepsilon''_{\text{eff,p}} = \nu A_{\text{p}} \varepsilon''_{\text{veg}} \tag{9}$$

with the vegetation volume fraction, ν, defined as the ratio between the vegetation volume and the total canopy volume; a geometrical parameter (at polarization p), A_{p}; and the dielectric constant of the plant elements $\varepsilon''_{\text{veg}}$. For the case of leaves or needles, where the inclusions are very small compared to the wavelength in at least one direction, the geometrical parameter A_{p} has been derived [17]. In Table 1 A_{p} is given for some useful simple situations.

A classic paper on the derivation of effective dielectric constants for heterogeneous media has been written by Polder and van Santen [18]. A good overview is given by Sihvola [19]. For low-density media with $\nu \ll 1$ (typical values in tree crowns are $\nu \leqslant 0.0001$) $\varepsilon'_{\text{eff}}$ is approximately 1. In this case we can use a mixing formula linear in ν. The absorption coefficient, κ_{a}, can be calculated with

$$\kappa_{\text{ap}} = k_0 \varepsilon''_{\text{eff,p}}. \tag{10}$$

Neglecting the contribution of scattering the opacity τ_{p} is

$$\tau_{\text{p}} = \kappa_{\text{ap}} \frac{H}{\cos\theta} \tag{11}$$

with the canopy height H and the observation angle θ. Using the relation

$$\nu H = \frac{B}{\rho_{\text{veg}}} \tag{12}$$

with B the wet biomass per area and ρ_{veg} the vegetation density ($\cong 950\ \text{kg/m}^3$) and from Equations (9) to (12) we find

$$\tau_{\text{p}} = A_{\text{p}} k_0 \frac{B}{\rho_{\text{veg}}} \varepsilon''_{\text{veg}} \frac{1}{\cos\theta}. \tag{13}$$

Table 1.
Geometrical parameters for ellipsoidal inclusions

	A_{h}	A_{ν}
Horizontal leaves	1	$\cos^2\theta$
Isotropic leaves	2/3	2/3
Vertical stalks	0	$\sin^2\theta$
Isotropic stalks	1/3	1/3

The wet biomass is related to the vegetation water content per unit area W

$$B = \frac{W}{1 - m_{\mathrm{d}}} \tag{14}$$

with m_{d} the dry matter fraction of the vegetation, defined as ratio between the mass of the dried vegetation to that of the natural vegetation.

A comparison of Equations (8) and (13) shows that $(W/\rho_{\mathrm{water}})\varepsilon''_{\mathrm{sw}}$ is replaced by $(B/\rho_{\mathrm{veg}})\varepsilon''_{\mathrm{veg}}$. The different formulations result from considering the absorption by only the water in the vegetation (Equation (8)), and by the vegetation as a whole (Equation (13)).

In the derivation the assumptions that the particles are in homogeneous electric fields and that the fields within the particles are homogeneous were used. As long as scattering is negligible the second condition is the more restrictive. Therefore we can use it to determine the range of validity of the Effective Medium theory. The exponential attenuation (assuming there is no scattering) of an electromagnetic wave in a dielectric medium is given by its penetration depth, d_{p}. For green vegetation having a high moisture content and therefore $\varepsilon''_{\mathrm{veg}}$ values on the order of 10 for frequencies between 1 and 36 GHz [3, 4] we find $d_{\mathrm{p}} \cong 0.1\lambda_0$, where λ_0 is the wavelength in free space. At a depth d of $0.05d_{\mathrm{p}}$ the penetrating field is attenuated to 95% of its initial value. For leaves not thicker than d and needles, stems and twigs with diameters smaller than d the above used assumptions of constant fields within the particle are fulfilled. This criterion can be summarised as $\lambda_0 > 200x$, with x being the smallest dimension of the particle. For leaves with a thickness of 0.2 mm this leads to $\lambda_0 > 4$ cm or $f < 7.5$ GHz. The thickness of natural leaves often is between 0.1 and 0.3 mm. We conclude that, for frequencies above 7.5 GHz, it is necessary to correct the model for the inhomogeneity of the electric field within the vegetation.

The dielectric constant of vegetation, and in particular its imaginary part, is dominated by the plant water. Therefore it is often approximated by the dielectric constant of saline water, as in [2]. Dielectric measurements by Ulaby and El-Rayes [3] indicated that the dielectric properties of vegetation can be modelled by representing vegetation as a mixture of saline water, bound water and dry vegetation. They derived a semi-empirical formula [3] from measurements at frequencies between 1 and 20 GHz on corn leaves with relatively high dry matter contents. The extrapolation of the formula to higher frequencies and lower dry matter contents leads to incorrect values. This was shown by Mätzler and Sume [4]. From the data used in [3], and their own data at frequencies up to 94 GHz, they developed a different semi-empirical formula to calculate the dielectric constant of leaves. High and low dry matter contents were included. They found [30]

$$\varepsilon_{\mathrm{veg}} = (0.51 - 0.64m_{\mathrm{d}})\varepsilon_{\mathrm{sw}} + 3.2m_{\mathrm{d}} + 0.49, \qquad (0.1 \leqslant m_{\mathrm{d}} \leqslant 0.5). \tag{15}$$

At frequencies below 20 GHz and for $m_{\mathrm{d}} \geqslant 0.30$ the two models are in fair accordance. Above 20 GHz and for smaller m_{d} the real parts differ quite a bit. To calculate $\varepsilon''_{\mathrm{veg}}$ Equation (15), with the values $m_{\mathrm{d}} = 0.2$, $T = 20$ Celsius, and salinity $= 0.008$, was used in this study.

3.2. Geometrical Optics approach for high frequencies

As discussed above, the validity of the Effective Medium theory is restricted to low microwave frequencies because of the assumption of homogeneous electromagnetic fields within the particles. This limitation can be avoided by using geometrical optics, and the range of validity can be extended to higher frequencies.

In the Geometrical Optics approach a leaf is represented by a flat, homogeneous layer with a uniform dielectric constant ε_{veg} and thickness d. The reflectivity, transmissivity and absorptivity, r_p, t_p, and a_p, where p stands for the polarization, are calculated using the coherent model for layered media discussed in many textbooks [20], [21, pp. 78–82].

$$r_p = \left| \frac{R_p(1 - e^{-j2k_{z1}d})}{(1 - R_p^2 e^{-j2k_{z1}d})} \right|^2 \tag{16}$$

$$t_h = \left| \frac{4k_{z0}k_{z1}e^{j(k_{z0}-k_{z1})d}}{(k_{z0} + k_{z1})^2(1 - R_h^2 e^{-j2k_{z1}d})} \right|^2 \tag{17}$$

$$t_\nu = \left| \frac{4\varepsilon_{veg}k_{z0}k_{z1}e^{j(k_{z0}-k_{z1})d}}{(\varepsilon_{veg}k_{z0} + k_{z1})^2(1 - R_\nu^2 e^{-j2k_{z1}d})} \right|^2 \tag{18}$$

$$a_p = 1 - r_p - t_p \tag{19}$$

where

$$\begin{aligned} k_0 &= 2\pi f/c \\ k_{z0} &= k_0 \cos\beta \\ k_{z1} &= k_0(\varepsilon_{veg} - \sin^2\beta)^{1/2} \\ R_h &= (k_{z0} - k_{z1})/(k_{z0} + k_{z1}) \\ R_\nu &= (\varepsilon_{veg}k_{z0} - k_{z1})/(\varepsilon_{veg}k_{z0} + k_{z1}) \end{aligned}$$

where β is the incidence angle relative to the leaf normal. Using the resistive sheet approximation [22], or from numerical calculations we find that at frequencies below 40 GHz, for $\beta \leqslant 60°$, and for leaves with $d \leqslant 0.4$ mm Equations (16)–(19) can be simplified. Then we can write

$$a_h = k_0 d\varepsilon''_{veg} t_h \frac{1}{\cos\beta} \tag{20}$$

$$a_\nu = k_0 d\varepsilon''_{veg} t_\nu \cos\beta. \tag{21}$$

Diffraction is neglected in the Geometrical Optics approach. The absorption cross section of a single leaf, σ_{ap}, is its absorptivity multiplied by the leaf area normal to the wave propagation direction

$$\sigma_{ap,i} = a_p S_i \cos\beta_i \tag{22}$$

where S_i is the area of the leaf number i, and β_i is the angle between k_0 and the leaf normal. The absorption cross section of the total canopy, $\sigma_{ap,canopy}$, is the sum over the σ_{ap} of the individual components.

We find the absorption coefficient κ_{ap} by dividing $\sigma_{ap,canopy}$ by the canopy volume V_{canopy}:

$$\kappa_{ap} = \frac{\sum_i \sigma_{ap,i}}{V_{canopy}}. \tag{23}$$

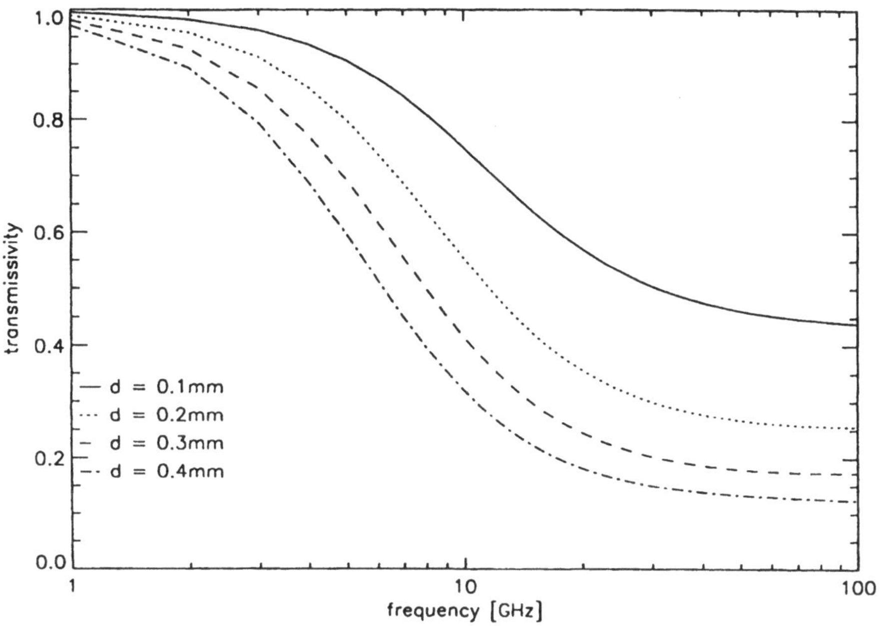

Figure 1. Spectral dependence of the transmissivity through a single leaf (leaf thickness d from 0.1 to 0.4 mm, incidence angle relative to leaf normal $\beta = 0°$).

For horizontal leaves the angles β_i are equal to the incidence angle relative to the canopy layer, θ. Then Equation (23) becomes

$$\kappa_{ap} = \frac{a_p S_{tot} \cos\theta}{V_{canopy}} = \frac{\nu}{d} a_p \cos\theta \tag{24}$$

with the total leaf area, S_{tot}, and the volume fraction of the vegetation, ν, and the leaf thickness, d, as defined earlier. Application of Equations (20)–(24) and Equations (11), (12) leads to

$$\tau_p = A_p k_0 \frac{B}{\rho_{veg}} \varepsilon''_{veg} \frac{1}{\cos\theta} t_p \tag{25}$$

where the same geometrical parameter A_p was used as before for a more direct comparison with the result of the Effective Medium theory, Equation (13). This result is very interesting; it allows a direct connection of the low-frequency with the high-frequency approximation. The only difference with Equation (13) is the additional factor t_p, the transmissivity of a single leaf at polarization p, as given in Equations (17), (18), whose spectral dependence is shown in Fig. 1. This approach can be extended by taking other parts of the vegetation such as twigs and branches into consideration and by adding their absorption coefficients. The Geometrical Optics approach can also be extended to different leaf orientations. Scattering has been neglected in the derivation of Equation (25). A possibility to derive κ_{sp} will be presented in Section 4.

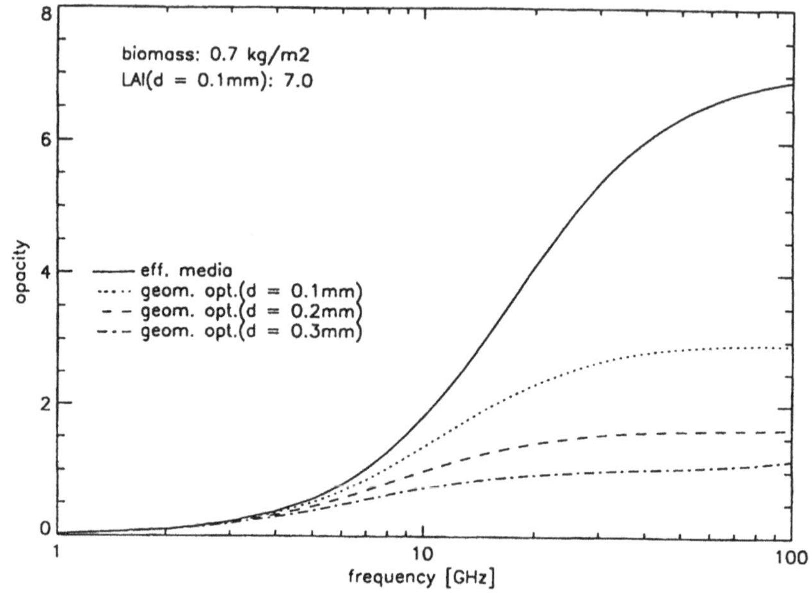

Figure 2. Calculated spectral dependence of the opacity of horizontal leaves at an observation angle $\theta = 0°$, using the Effective Medium theory (solid line) and Geometrical Optics approach. For the latter the leaf thickness d is varied between 0.1 mm and 0.3 mm, keeping vegetation wet biomass B and volume fraction ν constant.

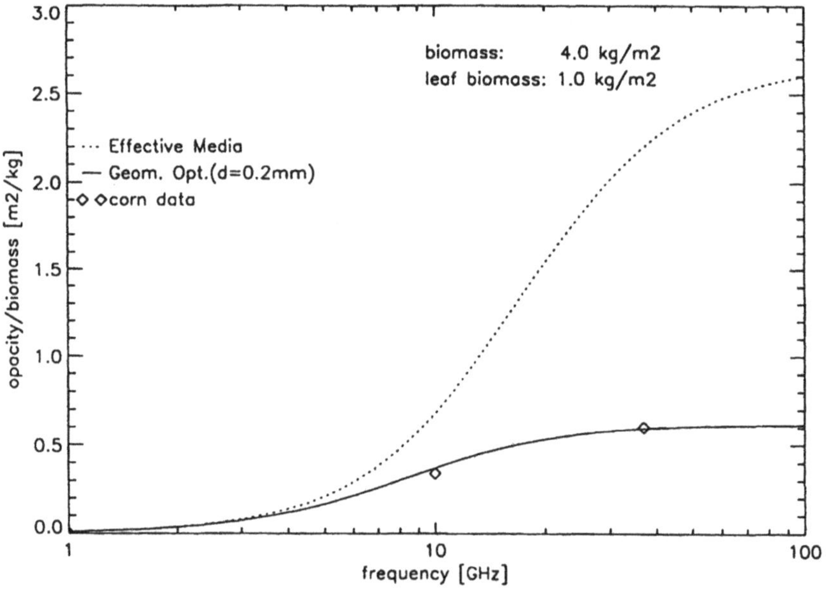

Figure 3. The spectral dependance of the ratio opacity/wet biomass is shown for the Effective Medium theory and the Geometrical Optics approach. The leaves are assumed to be horizontal. Data measured at 1.4 and 5.0 GHz and extrapolated to higher frequencies using the Effective Medium theory lead to too high values. By using Geometrical Optics a good accordance is found with the data measured by [7] at 10 and 37 GHz.

3.3. Comparison of the models with measured opacities

First we compare the Effective Medium theory and the Geometrical Optics approach for a canopy consisting of horizontal leaves. Figure 2 shows that for low frequencies both theories lead to the same opacity because t_p is approximately 1. At frequencies above 5 GHz the values are higher for the Effective Medium theory. With increasing frequency, t_p decreases resulting in a reduction of the opacity calculated with Geometrical Optics. The difference is stronger for thicker leaves. We notice that Geometrical Optics leads to fairly constant opacities above 30 GHz. For all curves in Fig. 2 the wet biomass is constant. This means that the leaf area index decreases with increasing leaf thickness.

With data from ground-based measurements we verified that the Geometrical Optics approach is an improvement over the Effective Medium theory in the description of the spectral behavior of the opacity of vegetation [23]. Choudhury *et al.* [12] discussed a discrepancy between the spectral dependence of measured data [7] and the Effective Medium theory. Figure 3 shows that Geometrical Optics explains the discrepancy. For the thickness of the corn leaves a typical value, $d = 0.2$ mm, was used.

4. SINGLE SCATTERING ALBEDO

The Geometrical Optics approach can also be used to estimate the importance of scattering. Analogous to the derivation of κ_{ap} the scattering coefficient κ_{sp} can be derived. The reflectivities of the leaves r_p Equation (16) are used to determine their scattering cross sections

$$\sigma_{sp,i} = r_p S_i \cos \beta_i \tag{26}$$

where S_i is the area of leaf number i, and β_i is the angle between k_0 and the leaf normal. In this approach a two flux model is used. For this reason scattering into the lower hemisphere is not included; only scattering into the upper hemisphere is considered. The scattering cross section of the total canopy, $\sigma_{sp,canopy}$, is the sum over the σ_{sp} of the individual components. The scattering coefficient κ_{sp} is then

$$\kappa_{sp} = \frac{\sum_i \sigma_{sp,i}}{V_{canopy}}. \tag{27}$$

And for horizontal leaves β_i is equal to θ leading to

$$\kappa_{sp} = \frac{r_p S_{tot} \cos \theta}{V_{canopy}} = \frac{\nu}{d} r_p \cos \theta. \tag{28}$$

The single scattering albedo is ω_p defined as

$$\omega_p = \frac{\kappa_{sp}}{\kappa_{sp} + \kappa_{ap}}. \tag{29}$$

For the case of horizontal leaves we introduce Equations (24) and (28) in Equation (29) and find

$$\omega_p = \frac{r_p}{r_p + a_p} \quad \text{(horizontal leaves).} \tag{30}$$

The single scattering albedo depends exclusively on the single leaf parameters r_p and a_p and therefore also on the leaf orientation as the latter has a significant influence on r_p

Table 2.
Single scattering albedo, ω_p for $\theta = 0°$. Calculated with Geometrical Optics, for frequencies between 1 and 40 GHz.

d [mm]	ω_p	d [mm]	ω_p
0.00	0.00	0.16	0.34
0.02	0.12	0.18	0.36
0.04	0.17	0.20	0.38
0.06	0.21	0.22	0.39
0.08	0.24	0.24	0.41
0.10	0.26	0.26	0.43
0.12	0.29	0.28	0.45
0.14	0.31	0.30	0.46

and a_p. In Table 2, ω_p is calculated for d between 0.0 and 0.3 mm. For $d \leqslant 0.3$ mm and at frequencies between 1 and 40 GHz the spectral dependence of ω_p is very weak. As Geometrical Optics was used for this derivation, diffraction has been neglected. This can be done as long as the area of single leaves is large compared to λ^2. If this condition is not met the more appropriate Physical Optics approach should be used (Senior *et al.* [22]).

5. INCOMPLETE SURFACE COVERAGE

Until now we considered exclusively the case of a homogeneous canopy. Here we consider a canopy where the wet biomass $B(x, y)$ is a function of the location. For simplicity we assume that the geometrical and dielectrical properties of the canopy are constant. With $B(x, y)$ the opacity

$$\tau_p(x, y) = \alpha B(x, y) \tag{31}$$

and the transmissivity

$$t_p(x, y) = \exp\big(-\tau_p(x, y)\big) \tag{32}$$

become functions of the location. The coefficient α depends on the geometrical and dielectrical properties of the canopy (see Section 3).

The wet biomass within an area A, for instance the resolution cell of a radiometer system, is

$$B(A) = \frac{1}{A} \iint B(x, y)\, dx dy \tag{33}$$

and the transmissivity

$$t_p(A) = \frac{1}{A} \iint t_p(x, y)\, dx dy. \tag{34}$$

Introduction of Equations (32) and (33) leads to

$$t_p(A) = \frac{1}{A} \iint \exp\big(-\alpha B(x, y)\big)\, dx dy. \tag{35}$$

Under the assumption of a homogeneous canopy the estimated wet biomass B_{est} for the area A

$$B_{est}(A) = \frac{1}{\alpha} \ln\big(t_p(A)\big) \tag{36}$$

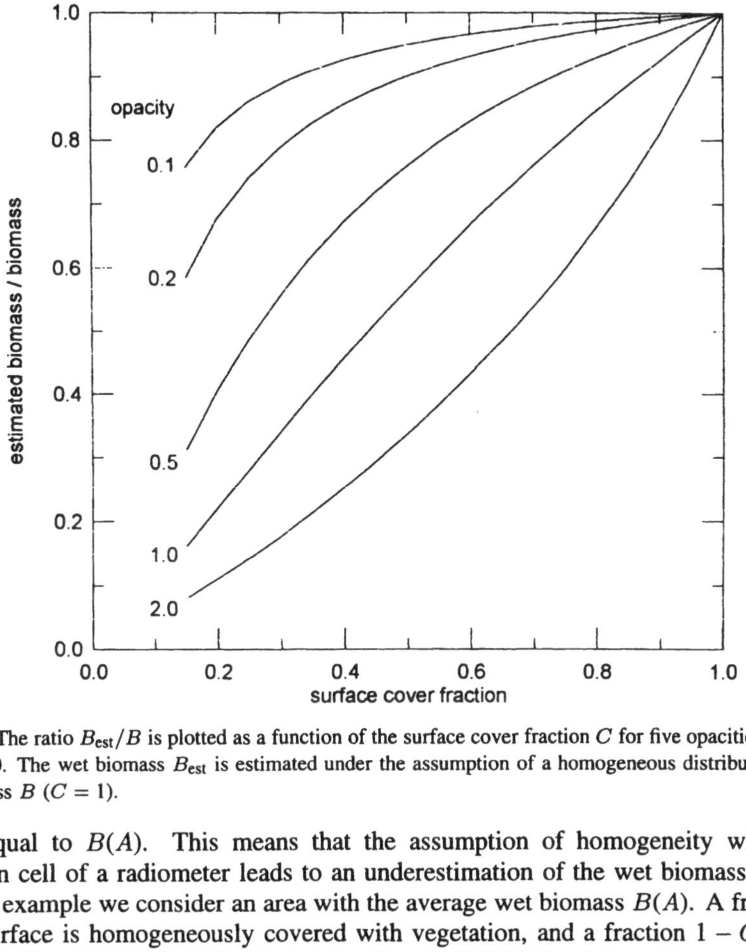

Figure 4. The ratio B_{est}/B is plotted as a function of the surface cover fraction C for five opacities between 0.1 and 2.0. The wet biomass B_{est} is estimated under the assumption of a homogeneous distribution of the wet biomass B ($C = 1$).

is not equal to $B(A)$. This means that the assumption of homogeneity within the resolution cell of a radiometer leads to an underestimation of the wet biomass.

As an example we consider an area with the average wet biomass $B(A)$. A fraction C of the surface is homogeneously covered with vegetation, and a fraction $1 - C$ is bare soil. Therefore $B(x, y)$ is 0 for the bare soil and $B(A)/C$ for the canopy covered part of the area. Using Equation (36) we can calculate the ratio B_{est}/B:

$$\frac{B_{\text{est}}(A)}{B(A)} = \frac{-\ln\left[Ce^{-\frac{\alpha B(A)}{C}} + 1 - C\right]}{\alpha B(A)}, \tag{37}$$

The product $\alpha B(A)$ is the opacity for $C = 1$. Figure 4 shows B_{est}/B as a function of C for various values for $\alpha B(A)$. We see that for high opacities even small errors in the estimation of C lead to a severe error in the estimation of B. For low opacities on the other hand we find, that an accurate knowledge of C is not as important for the estimation of B.

6. CONCLUSIONS

We derived the canopy opacity used in radiative transfer models with the Effective Medium theory. We found that it is restricted to $f \leqslant 7.5$ GHz. We suggested an extension of the model to higher frequencies using Geometrical Optics. The only additional parameter introduced is the leaf thickness. At low microwave frequencies ($f \leqslant 5$ GHz) the results of the two approaches are identical. For $f \geqslant 5$ GHz the Geometrical Optics approximation corrects for the non-constant electric fields in the vegetation. The change in the canopy opacity is, in good approximation, a factor t_p, the transmissivity of a single leaf. Discrepancies between the spectral dependence of measured opacities and values calculated with the Effective Medium theory could be explained with the Geometrical Optics approach.

Geometrical Optics was used to estimate the single scattering albedo ω_p. A strong increase of ω_p with leaf thickness from about 0.1 for thin leaves to about 0.5 for $d = 0.3$ mm was found.

The presence of gaps in a crop layer is very important for both the spectral dependence and the absolute values of the opacity. The estimation of canopy wet biomass is affected strongly by the surface cover fraction as soon as there are saturation effects.

Acknowledgement

Parts of the research presented here were supported by the Schweizerischer Nationalfonds, Bern, Switzerland.

REFERENCES

1. Y. H. Kerr and E. G. Njoku. A semi-empirical model for interpreting microwave emission from semiarid land surfaces as seen from space. *IEEE Trans. Geosci. Remote Sensing*, **GE-28** (3), 384–393 (1990).
2. K. P. Kirdyashev, A. A. Chuckhlantsev and A. M. Shutko. Microwave radiation of the Earth's surface in the presence of vegetation cover. *Radio Science and Electronics*, **24**, 256–264 (1979) (or: *Radio Eng. Electron. Phys.*, **24**, 37–44, (1979)).
3. F. Ulaby and M. El-Rayes. Microwave dielectric spectrum of vegetation. Part II: Dual-dispertion model. *IEEE Trans. Geosci. Remote Sens.*, **GE-25** (5), 550–556 (1987).
4. C. Mätzler and A. Sume. Microwave radiometry of leaves. In: *Microwave Radiometry and Remote Sensing Applications*, P. Pampaloni (Ed.), pp. 133–148, VSP, Utrecht, The Netherlands (1989).
5. B. J. Choudhury. Monitoring global land surface using Nimbus-7 37 GHz data theory and examples. *Int. J. Remote Sensing*, **10** (10), 1579–1605 (1989).
6. F. Ulaby, R. Moore and A. Fung. *Microwave Remote Sensing, Active and Passive.* Vol III. Artech House, Dedham, MA (1986).
7. P. Pampaloni and S. Paloscia. Microwave emission and plant water content: a comparison between field measurements and theory. *IEEE Trans. Geosci. Remote Sensing*, **GE-24**, 900–905 (1986).
8. C. Mätzler. Seasonal evolution of microwave radiation from an oat field. *Remote Sens. Environ.*, **31**, 161–173 (1990).
9. U. Wegmüller. Remote sensing signature studies on agricultural fields with ground-based radiometry and scatterometry. PhD Thesis, Applied Physics, University of Berne (1990).
10. T. J. Jackson, T. J. Schmugge and J. R. Wang. Passive microwave remote sensing of soil moisture under vegetation canopies. *Water Resources Res.*, **18**, 1137–1142 (1982).
11. T. Mo, B. J. Choudhury, T. J. Schmugge, J. R. Wang and T. J. Jackson. A model for microwave emission from vegetation covered fields. *J. Geophys. Res.*, **87**, 11229–11237 (1982).
12. B. J. Choudhury, J. R. Wang, A. Y. Hsu and Y. L. Chen. Simulated and observed 37 GHz emission over Africa. *Int. J. Rem. Sensing*, **11** (10), 1837–1868 (1990).
13. J. W. Waters. Absorption and emission by atmospheric gases. In: *Methods of Experimental Physics*, Vol. 12, Part B, M. L. Meeks (Ed.), Academic Press, New York (1976).
14. J. Hofer and E. Njoku. Regression techniques for oceanographic parameters retrieval using spaceborn microwave radiometry. *IEEE Trans. Geosci. Remote Sensing*, **GE-19**, 178–189 (1981).

15. J. R. Wang, P. E. O'Neill, T. J. Jackson and E. T. Engman. Multifrequency measurements of the effects of soil moisture, soil texture, and surface roughness. *IEEE Trans. Geosci. Remote Sensing*, **GE-21**, 44–51 (1983).
16. B. Reber. Distinct microwave signatures of snow and soil related to model theory. PhD Thesis, Institute of Applied Physics, University of Berne, Sidlerstr. 5, CH-3012 Bern, Switzerland (1990).
17. Van de H. C. Hulst. *Light Scattering by Small Particles*. Dover Publications (1957).
18. D. Polder and J. A. van Santon. The effective permeability of mixtures of solids. *Physica*, **XII** (5), 257–271 (1946).
19. A. Sihvola. Self-consistency aspects of dielectric mixing theories. *IEEE Trans. Geosci. Remote Sensing*, **27** (4), 403–415 (1989).
20. J. A. Kong. *The Theory of Electromagnetic Waves*. J. Wiley & Sons, New York (1975).
21. F. Ulaby, R. Moore and A. Fung. *Microwave Remote Sensing, Active and Passive*. Vol I. Addison-Wesley, Reading MA (1981).
22. T. B. A. Senior, K. Sarabandi and F. T. Ulaby. Measuring and modelling the backscattering cross section of a leaf. *Radio Science*, **22**, 1109–1116 (1987).
23. U. Wegmüller and C. Mätzler. Models to relate canopy opacity to vegetation parameters. In: *Microwave Radiometry and Remote Sensing Applications*, R. Westwater (Ed.), pp. 79–86, US Department of Commerce, NOAA, Boulder, CO (1992).
24. J. R. Wang and B. J. Choudhury. Remote sensing of soil moisture content over bare fields at 1.4 GHz frequency. *J. Geophys. Res.*, **86**, 5277–5282 (1981).
25. J. R. Wang, P. E. O'Neill, T. J. Jackson and E. T. Engman. Multifrequency measurements of the effects of soil moisture, soil texture, and surface roughness. *IEEE Trans. Geosci. Remote Sensing*, **GE-21**, 44–51 (1983).
26. T. J. Jackson and T. J. Schmugge. Vegetation effects on the microwave emission of soils. *Remote Sens. Environ.*, **36**, 203–212 (1991).
27. T. Mo, T. J. Schmugge and J. R. Wang. Calculations of the microwave brightness temperature of rough soil surfaces: bare field. *IEEE Trans. Geosci. Remote Sensing*, **GE-25** (1), 47–54 (1987).
28. S. Saatchi, E. Njoku and U. Wegmüller. Synergism of active and passive microwave data for estimating bare soil surface moisture. In: *ESA/NASA International Workshop*, B. J. Choudhury, Y. H. Kerr, E. G. Njoku and P. Pampaloni (Eds), pp. 205–224, VSP, Utrecht (1994).
29. T. Mo and T. J. Schmugge. A parametrization of the effects of surface roughness on microwave emission. *IEEE Trans. Geosci. Remote Sensing*, **GE-25** (4), 481–486 (1987).
30. C. Mätzler. Experimental and theoretical study of dielectric properties of vegetation. (To be published soon).

ESA/NASA International Workshop, pp. 389–402
B. J. Choudhury, Y. H. Kerr, E. G. Njoku and P. Pampaloni (Eds)

Modeling microwave emission from vegetation-covered surfaces: a parametric analysis

P. FERRAZZOLI and L. GUERRIERO

Dipartimento di Ingegneria Elettronica, Università Tor Vergata, Via della Ricerca Scientifica, 00133 Roma, Italy

Abstract— A model, representing vegetation as an ensemble of lossy dielectric discs and cylinders overlying a homogeneous half-space, is presented. The model is based on the Radiative Transfer Theory and includes computation of multiple scattering effects by means of a matrix algorithm.

The model is used to estimate the emissivity of some agricultural and forest species at both polarizations and in the frequency range extending from 1.5 to 35 GHz. Also emission contributions due to single components within the vegetation medium are computed and the sensitivity to variations of some ground parameters are estimated. The results show that a multifrequency radiometric system is able to discriminate among different vegetation species and is sensitive to ground parameters.

Experimental results available in the literature are in general agreement with the model predictions.

1. INTRODUCTION

Vegetation plays a dominant role in influencing the land–atmosphere interactions. Monitoring the evolutions suffered by physical properties of crops, forests and underlying soils is of crucial interest to understand some relevant phenomena, like water and energy fluxes at the earth surface. To this end, spaceborne microwave passive sensors are a powerful tool, due to their sensitivity to important parameters like soil moisture, plant moisture, vegetation growth, etc. Although these systems are presently limited by ground resolution problems, they may be helpful in understanding large-scale effects.

Vegetation-covered soils are complex media, whose microwave emission characteristics are affected by many factors: soil moisture, soil roughness, total biomass, plant moisture, dimensions and orientation of branches, twigs, stems, leaves. Several experiments have been carried out by ground-based and airborne sensors to investigate the sensitivity of emissivity (at different frequencies and polarizations) to the different medium parameters. However, the parameters suffer simultaneous variations, making it difficult to separate the different effects. The use of electromagnetic models, which can analyze separately the sensitivity to different parameters, is therefore important not only for scientific purposes, but also for applications.

Several models have been proposed, representing vegetation as a continuum [1] or as an ensemble of discrete scattering elements (scatterers) overlying the soil. The first approach is mainly valid at low frequencies, while the latter is in principle valid up to the high frequencies of the microwave spectrum, provided absorption and scattering properties of the elements are correctly represented. In general, when this discrete approach is used, the contributions of the different elements are combined by means of the Radiative Transfer Theory. Simple methods, based on a single-scattering model, have been proposed [2–5]. In that model the scatterers are assumed to be equal in dimensions, permittivity and shape, so that the canopy is characterized by two global parameters: the single-scattering albedo ω and the optical thickness τ. The albedo depends on dimensions, permittivity and shape of the single scatterer while the optical thickness is associated with the total amount of dielectric matter (representing vegetation biomass) overlying the soil. In general ω and τ are related to vegetation parameters by empirical relationships or by fitting experimental data.

To account for multiple scattering effects, a different approach has been proposed in [6], where the Rayleigh approximation is used to describe the single scattering properties, while contributions of the different scatterers are combined by means of the Matrix Doubling method. Also the model developed at Tor Vergata University, and used in this paper, is based on the Radiative Transfer Theory and the Matrix Doubling. Scatterers are here assumed to be dielectric discs and cylinders, with the objective to give, as far as possible, a realistic description of vegetation elements like leaves and branches. For a given wavelength and for given dimensions, shapes, orientations and permittivities of the scatterers, suitable electromagnetic approximations are used to compute the absorption and bistatic scattering cross-sections. The different contributions are then combined by means of the Matrix Doubling Algorithm. The following advantages are offered by this model:

- The range of validity covers the whole microwave spectrum.
- Multiple scattering effects, which at the higher frequencies are not negligible, are taken into account.
- Different kinds of scatterers may be included within the same canopy.
- Physical parameters of soils and plants, which can be measured and are interesting for applications, may be directly given as inputs.

In this paper the model is used to estimate the emissivity of agricultural and forest species typical of the Mediterranean area. Computations are carried out in the frequency range from 1.5 to 35 GHz. For every kind of crop or forest the overall emissivity and the contributions of single emission sources are computed. Ground parameters values measured on average during experimental campaigns are initially assumed as inputs. The sensitivity of the emissivity to variations, within realistic ranges, of the ground parameters is then investigated. Finally, model results are compared with some available experimental results.

2. THE MODEL

2.1. General aspects

The model assumes the vegetation medium to be a homogeneous half-space with rough interface, representing the soil, overlaid by an ensemble of discrete lossy scatterers, representing the plant constituents. Details are given elsewhere [7, 8]; only the main aspects are summarized here.

The electromagnetic properties of the soil are described by its bistatic scattering coefficient. The latter is computed by the Small Perturbation approximation at 1.5 GHz, while the Geometrical Optics approximation, with the shadowing function derived by

Smith, is used at the higher frequencies [2]. This choice seems to us to optimize the use of the available surface models; indeed, the Physical Optics approximation [2] could be more suitable at 5 GHz, but this approximation is valid only under the condition of 'small slope' which, according to our experience, is not satisfied in most of the agricultural fields. The semi-empirical model proposed in [2] is used to compute the soil permittivity.

The electromagnetic properties of the scatterers, which represent the plant constituents, are described by the absorption cross sections $\sigma_a(\theta, \phi)$ and by the bistatic scattering cross sections $\sigma(\theta, \phi; \theta_s, \phi_s)$. θ and θ_s are the incidence and scattering off-normal angles, while ϕ and ϕ_s are the incidence and scattering azimuth angles. Dielectric elements of simple shape, such as discs and cylinders, are assumed. Discs represent small leaves or parts of large leaves. At low frequencies (1.5 GHz) the discs cross-sections are computed by means of the Rayleigh–Gans approximation [9], which assumes the disc to be an oblate ellipsoid with one axis small with respect to the wavelength, so that the inner field may be considered homogeneous along the small dimension. At high frequencies the discs cross-sections are computed by using the Physical Optics approximation [10], which assumes the internal field to be the same as that of an infinitely extended slab. Cylinders may represent trunks, branches, twigs, stems or ears of wheat and barley. For these kinds of scatterers the theory proposed in [11] is used, which assumes the internal field to be the same as that of an infinite length cylinder.

The dimensions of the scatterers (radius and thickness of discs, radius and length of cylinders) are chosen in such a way as to represent, with reasonable accuracy, the real plant constituents. The dielectric constant is expressed as a function of the plant moisture, by empirically fitting the experimental data published by El-Rayes and Ulaby [12]. Once the absorption and bistatic scattering cross sections of the scatterers have been computed for a discrete set of incidence and scattering directions, the electromagnetic behaviour of the ensemble of scatterers is obtained. To this end, the vegetation layer, filled with scatterers of different kinds and proper orientation, is subdivided into thin elementary sublayers, each filled with equal scatterers; the structure is assumed to be symmetrical in azimuth. For each sublayer, both the upper and the lower half-spaces are subdivided into discrete intervals of incidence and scattering off-normal angles θ and θ_s. Due to the azimuthal symmetry assumption, the scattering does not depend on ϕ_s and ϕ separately, but it depends on the difference $\phi_s - \phi$. The sublayer behaviour is characterized by the scatter matrices **S** and the transmission matrices **T**. Each element of an **S**(**T**) matrix represents the ratio I_s/I (I_t/I) between the specific intensity I_s (I_t) scattered into an upper (lower) angular interval of θ_s and the specific intensity I incoming from an upper angular interval of θ. The dependence on $\phi_s - \phi$ is expressed in Fourier series, so that the number of **S** and **T** matrices is equal to the number of the series terms. Only the first series term is needed when the emissivity has to be computed, as shown in [6]. The matrix elements may be obtained as functions of $\sigma_a(\theta, \phi)$ and $\sigma(\theta, \phi; \theta_s, \phi_s)$, as shown in [7].

Once each single sublayer has been characterized by its **S** and **T** matrices, the Matrix Doubling Algorithm is applied [6, 7]. The $\mathbf{S}_t$ matrix of the whole medium is so obtained and the emissivity $e(\theta)$ may be computed through the energy conservation law.

The inclusion of multiple scattering effects, which are particularly important at high frequencies, is a peculiarity of this model.

2.2. Specific aspects

Four canopy structures are considered in the present work, representing different kinds of agricultural and forest species. Input data for agricultural crops are derived from ground parameters measured, on average, over Italian sites in the summers of 1987 and 1988 during passive and active microwave experimental airborne campaigns [13, 14]. A maple forest is also considered, for which ground parameters measured during the Maestro-1 Campaign at the Dutch Flevoland site are used [15]. Details are given below:

- Corn and sunflower crops are described by upper randomly oriented discs, representing the leaves, lower vertical cylinders, representing the stalks, and a half-space with rough interface, representing the soil (Fig. 1a). The cylinders height is related to the total plant water content Q (kg/m^2) by an empirical linear relationship. The total discs area per m^2 is assumed to be equal to the Leaf Area Index. For corn and sunflower crops, characterized by large leaves, the problem of correctly choosing the discs' diameter must be considered. At low frequencies the discs' diameter is assumed to be equal to the average leaf width, while at high frequencies an upper limit of 1 cm is set. In this way the effect of bending and corrugation of leaves, which increases with the frequency, is taken into account, although at a rough level. Finally, the following inputs are derived by average measured crop parameters: plant moisture content = 85%, cylinders density = 7 m^{-2}, cylinders radius = 1.2–1.5 cm, maximum cylinder height for full growth ($Q = 8$ kg/m^2) = 128 cm, discs' thickness = 0.25 mm.
- Wheat and barley crops are described by upper vertical cylinders, representing the ears, lower randomly oriented discs, representing the leaves, and a half-space (Fig. 1b). Stem contribution is assumed to be negligible with respect to ear contribution. Cylinders parameters are: density = 700 m^{-2}, diameter = 1 cm, length = 8 cm. Leaves are represented by discs with 1 cm diameter and 0.15 mm thickness. The moisture content is 70%.
- Alfalfa crops (Fig. 1c) are modelled as a layer of vertical cylinders surmounted by alternate layers of inclined discs and cylinders and assuming that the latter elements may be oriented according to a $\cos^2 \beta$ distribution law (β is the cylinder axis orientation angle with respect to the normal). The following dimensions are assumed: discs' diameter = 1 cm, discs' thickness = 0.15 mm, inclined cylinders' diameter = 1–2 mm, inclined cylinders' length = 5–10 cm, vertical cylinders'

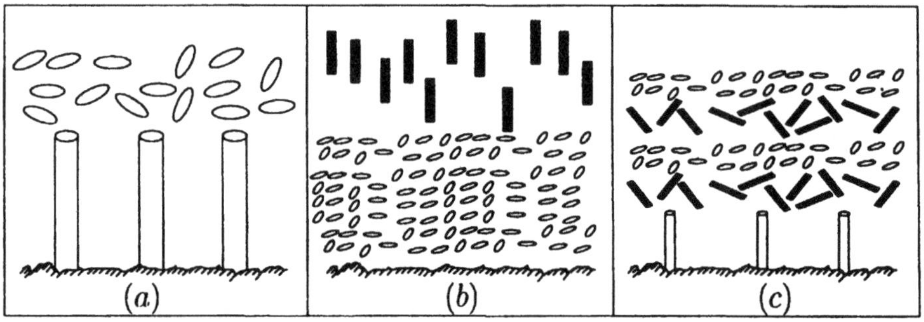

Figure 1. Sketch of model representation for corn/sunflower (a), wheat/barley (b), and alfalfa (c) crops.

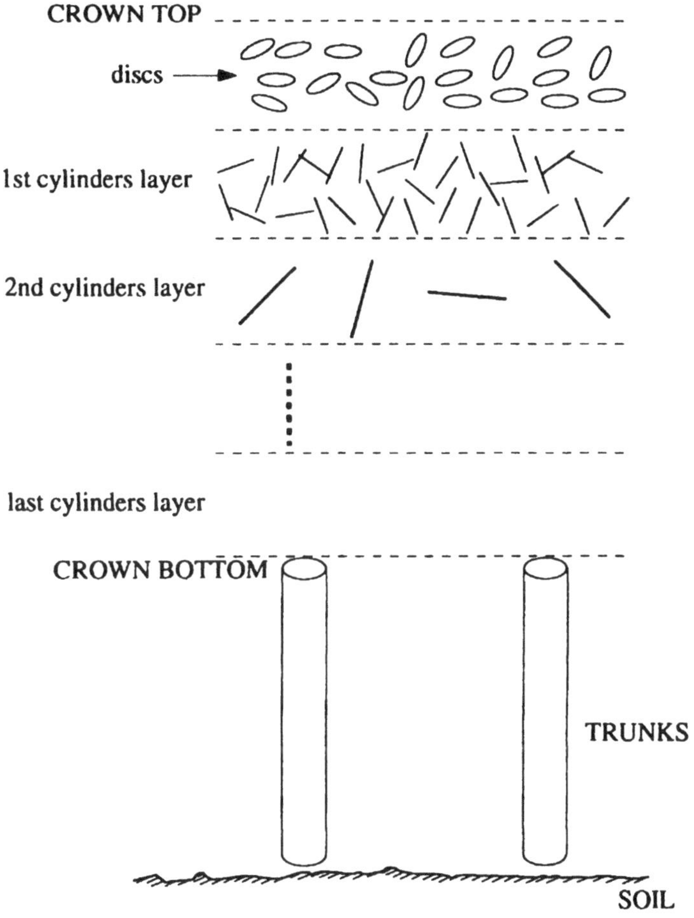

Figure 2. Sketch of model representation for a maple forest.

diameter = 0.2 cm, vertical cylinders' height = 20 cm. The moisture content is 80%.

- A maple forest configuration is also considered. Most of the input data are deduced by [15], as follows. Tree height = 10.2 m, trunks volume = 86×10^{-4} m^3/m^2, tree density = 2770×10^{-4} m^{-2}, wooden matter moisture content = 49%, crown height = 5.6 m, coverage fraction = 97%. Some assumptions are also made to completely characterize the canopy. A sketch of the representation used is given in Fig. 2. Trunks are represented by vertical cylinders, branches and twigs by randomly oriented cylinders subdivided into 5 layers, characterized by different cylinder dimensions. The top layer is filled with 0.4 cm diameter cylinders, the second one with 0.8 cm diameter cylinders,... and so on until the lowest layer, filled with 2 cm diameter cylinders, is reached. For all the cylinders, the length is assumed to be 50 times the diameter. The number of cylinders per unit area in each layer is selected to be inversely proportional to the volume of the single cylinder, so that the total filled volume is the same for all the layers. Discs, representing leaves, are also introduced on top; the total disc area is 5 m^2/m^2 (Leaf Area Index). Finally, it is assumed that the crown volume per unit area, given by the crown height reduced by the coverage fraction, is filled by branches and leaves with a 0.1% filling factor.

Although the model could include scatterers of different kinds within the same thin sublayer, we have chosen, in all the above described structures, to fill each sublayer with equal scatterers and to combine them afterward, by means of the 'doubling'. For the case of alfalfa the two approaches produce the same result, since the sublayers of discs and cylinders are optically thin and are alternated. For the case of wheat and barley the assumption of locating cylinders, representing ears, on top seems to us to be realistic. For the case of corn/sunflower and forest we think our assumptions to be reasonable, since leaves are fundamentally located above stalks and tree twigs are fundamentally located above large branches and trunks. However, more realistic representations, based on a higher flexibility in locating the scatterers, are under study.

3. MODEL RESULTS

3.1. Emission sources

Figures 3–7 show, for the different canopy structures described in 2.2, the model predicted total emissivities and contributions due to single components; results are given as functions of the frequency, in the range 1.5–35 GHz, at both polarizations and for an observation angle of 35°. In the figures, single contributions are shown for the main sources only; however, contributions of the minor sources (e.g. radiation emitted downward and reflected by the soil) are included in the computation of the total emission.

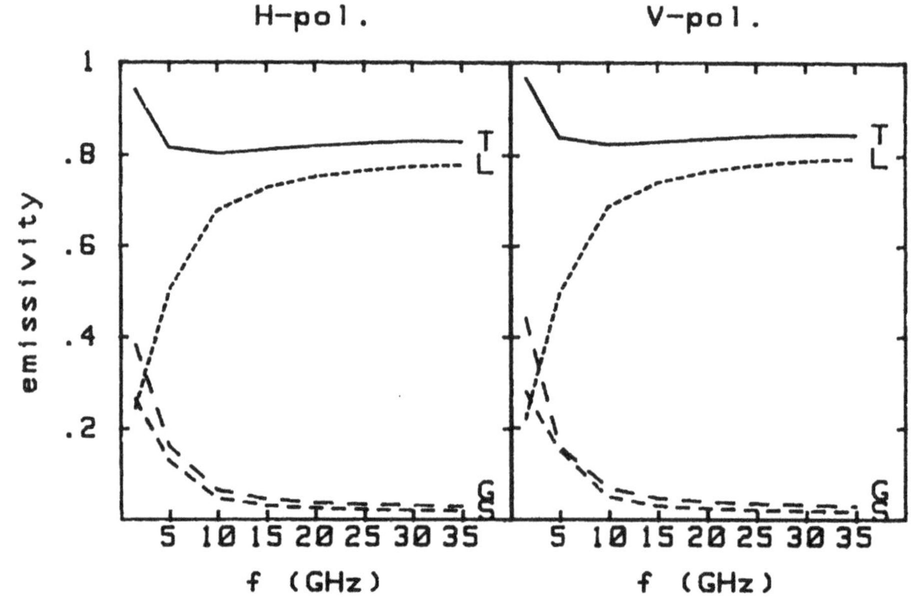

Figure 3. Model predicted emissivity versus frequency for corn/sunflower crops. T: total, L: leaves, S: stalks, G: soil.

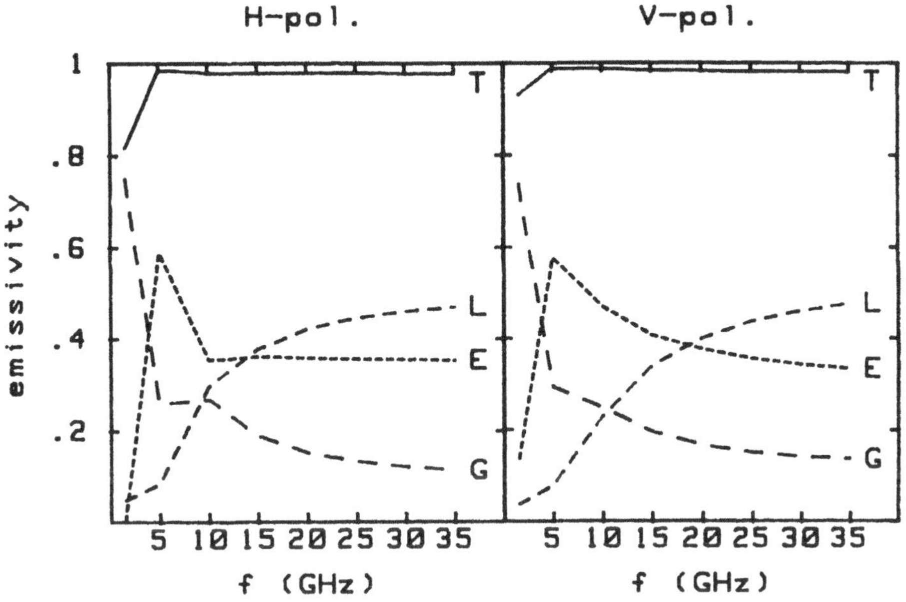

Figure 4. Model predicted emissivity versus frequency for wheat/barley crops. T: total, E: ears, L: leaves, G: soil.

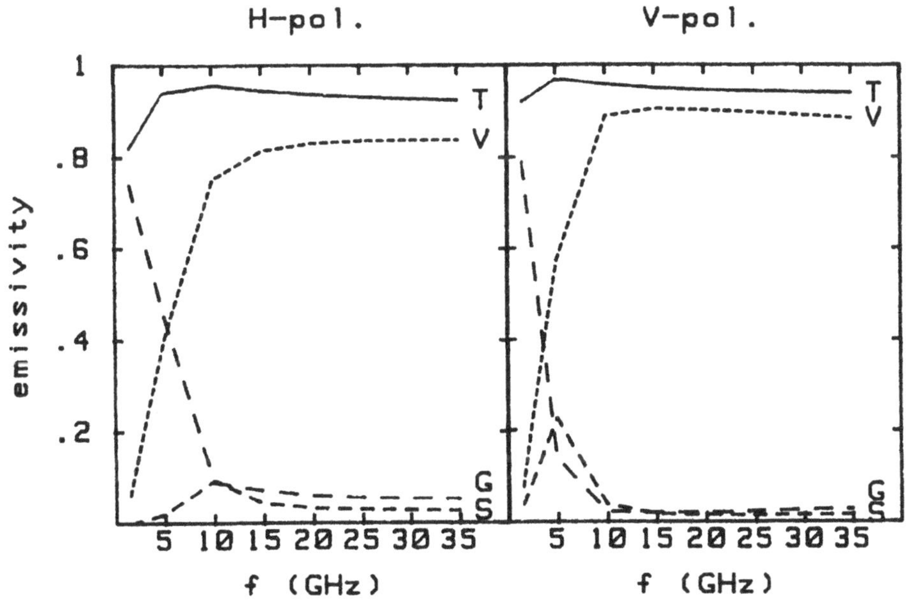

Figure 5. Model predicted emissivity versus frequency for alfalfa crops. T: total, V: leaves+stems, S: vertical stalks, G: soil.

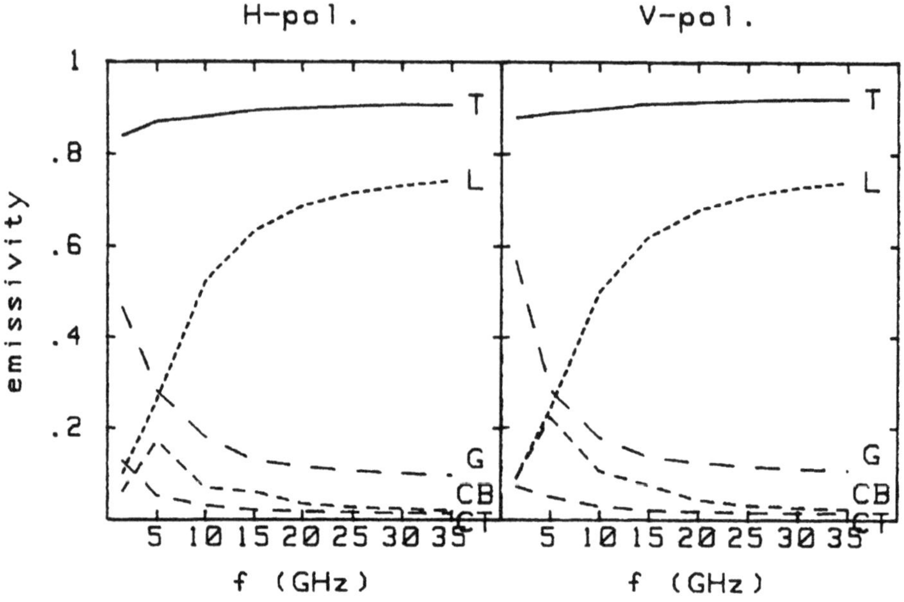

Figure 6. Model predicted emissivity versus frequency for maple forests. T: total, L: leaves, CB: branches, CT: trunks, G: soil.

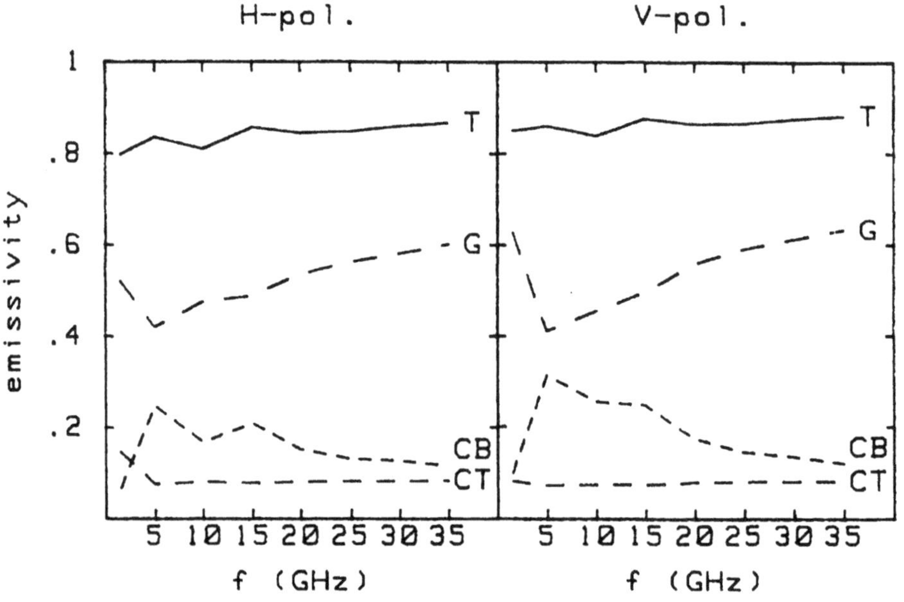

Figure 7. Model predicted emissivity versus frequency for maple forests without leaves. T: total, CB: branches, CT: trunks, G: soil.

Of course, emission contributions from the soil and the lower elements, represented in the figures, include attenuation of the upper elements.

Most of the assumptions concerning vegetation parameters have been described in 2.2. Furthermore, soil moisture content is assumed to be 10% (by volume) for the agricultural crops and 20% (by volume) for the maple forest. These values were measured, on average, in the Italian agricultural fields [13] and in the Flevoland site, respectively. The following soil roughness parameters are chosen: standard deviation = 1 cm, average slope = 0.4.

Figure 3 shows the model computed emissivities for the corn/sunflower configuration. A plant water content $Q = 8$ kg/m^2 is assumed. At 1.5 GHz the highest contribution is due to the soil, as expected; however, contributions from leaves and stalks are also appreciable, in the order of 0.2. At high frequencies leaf emission is dominant while cylinders, representing stalks, contribute weakly due both to their particular geometry (they are few and large) and to leaf attenuation.

Model predicted emission from wheat or barley fields is shown in Fig. 4. Plant water content is 5 kg/m^2. Soil dominates at 1.5 GHz, while leaves are the main emission source at high frequencies, as in Fig. 3. However, the presence of an upper ears layer generates remarkable effects. According to the model, ears seem to be the main source in an intermediate frequency range between $\sim$ 5 and $\sim$ 15 GHz. At high frequencies the total emissivity is relatively high ($>\sim$ 0.97). This result, which is confirmed by some experimental results (see later), may be attributed to the peculiar electromagnetic behaviour of the upper cylindrical elements: their bistatic scattering cross-section is such that emission coming from the lower layers is either absorbed or upward scattered, but very weakly downward scattered. The high emissivity is also due to the relatively low plant moisture content (70%) measured on wheat and barley fields.

Alfalfa emissivity trends vs. frequency are shown in Fig. 5 for a plant water content $Q = 2.5$ kg/m^2. Soil dominates at 1.5 GHz; upper elements (leaves and small stems) dominate at high frequencies, while the contribution of vertical lower stems is relatively low. At high frequencies the total emissivity is higher than that of Fig. 3, because the small upper elements of alfalfa crops are more absorbing than large corn and sunflower leaves, which are strong scatterers and tend to screen the emission coming from the soil and the lower elements.

Model predicted emission from a maple forest is shown in Fig. 6. Trunks and branches are strong scatterers and weak emission sources, so that they give low contributions to the total emissivity. The trends are similar to those of Fig. 3: a low frequency range, where soil dominates, and a high frequency range, where leaves dominate. The total emission at high frequencies is higher than in Fig. 3, because tree leaves are less moist than crop leaves. Figure 7 shows emission contributions for the same maple forest in absence of leaves. Emission from branches is low at 1.5 GHz, it presents multiple peaks between 5 and 15 GHz due to the assumed multilayer cylinders configuration (see Fig. 2), and finally it decreases for $f >\sim 15$ GHz. This result finds explanation in the peculiar electromagnetic properties of cylinders, whose extinction and absorption cross-sectional trends with respect to frequency show a maximum [11]. Soil emission shows an opposite behaviour, characterized by a minimum at $f = 5$ GHz; the positive trend for $f > 5$ GHz may be attributed both to the reduced attenuation by branches and to the reduction of permittivity with frequency, for a given soil moisture content.

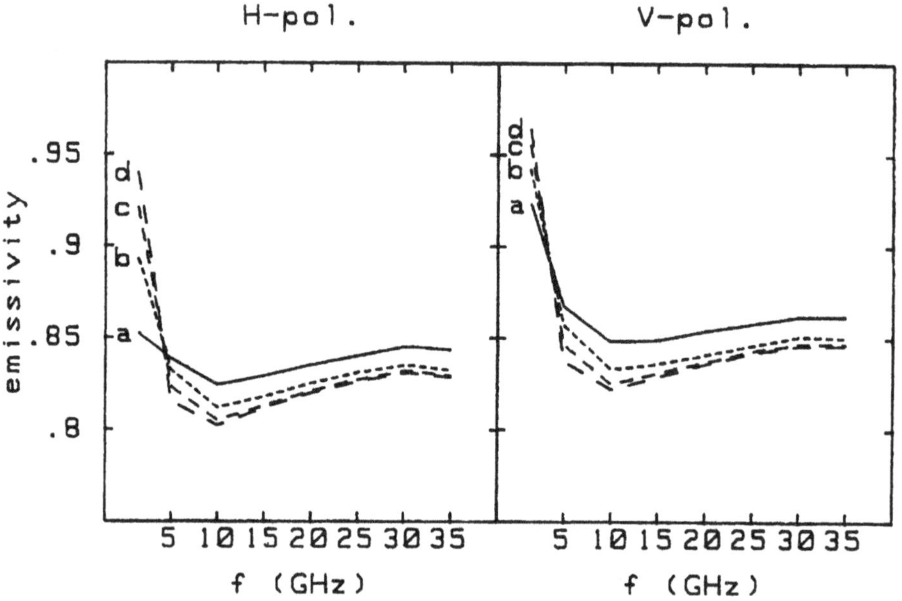

Figure 8. Model predicted emissivity versus frequency for corn/sunflower crops. Plant water content Q (kg/m^2) = 2 (a), 4 (b), 6 (c), 8 (d).

3.2. Sensitivity to ground parameters

For the corn/sunflower configuration, some parameters are now varied with respect to the average values indicated in 2.2, and the corresponding effects are shown.

In Fig. 8 the total emissivity is plotted vs. frequency for different stages of crop growth, corresponding to four values of the total plant water content Q (kg/m^2). At 1.5 GHz the emissivity increases with crop growth because crop elements are more absorbing than the underlying soil; the opposite may be noted at the high frequencies, where the upper leaves are strong scatterers. For this crop configuration and for the assumed soil moisture and roughness, the soil emission is not so far from that of the developed crop, so that the sensitivity to growth parameters is limited.

In Fig. 9 three realistic values of plant gravimetric moisture content M_g are considered, in the situation of full growth ($Q = 8$ kg/m^2). The model predicts a remarkable influence of the M_g parameter on the total emissivity: when the M_g is reduced, the corresponding reduction in the permittivity of scatterers increases the vegetation emissivity while, on the other hand, the soil contribution is still attenuated (at least at the high frequencies).

The effect of soil moisture variations is considered in Fig. 10 for the case of full growth. The high attenuation introduced by the vegetation reduces the sensitivity to the soil moisture, which is appreciable at 1.5 GHz only.

3.3. Comparison with experimental results

Emissivities computed by the model for corn/sunflower, alfalfa, wheat/barley crops are here compared with the experimental results obtained by airborne radiometers operating at L (1.4 GHz), X (9.6 GHz) and Ka (36 GHz) bands [13]. The experiments were carried

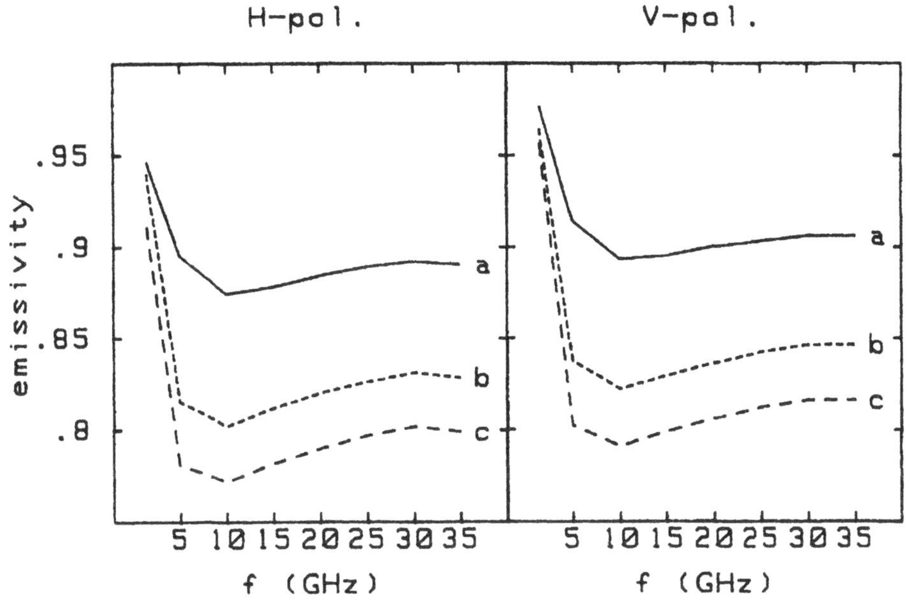

Figure 9. Model predicted emissivity versus frequency for corn/sunflower crops. Plant moisture content $M_g = 70\%$ (a), 85% (b), 90% (c).

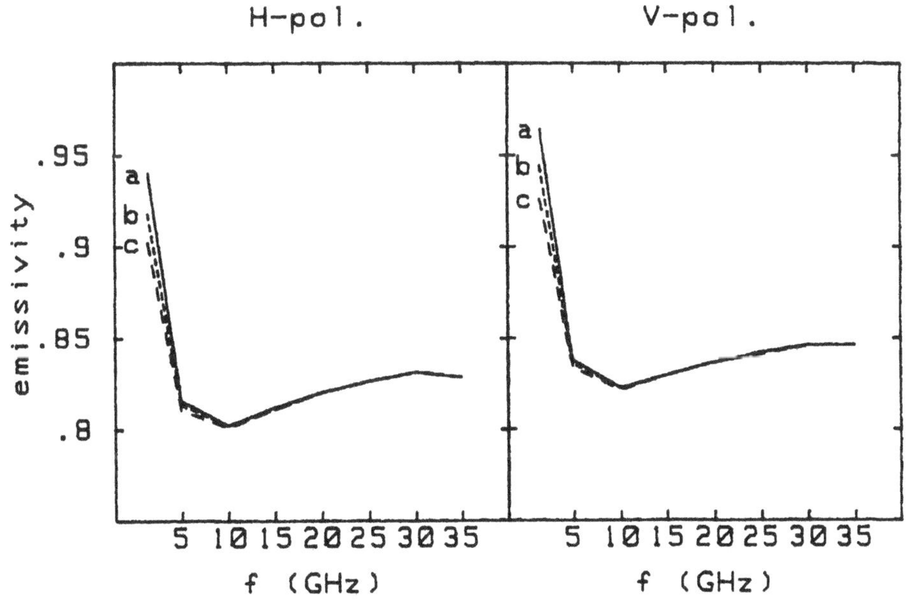

Figure 10. Model predicted emissivity versus frequency for corn/sunflower crops. Soil moisture content (by volume) = 10% (a), 20% (b), 30% (c).

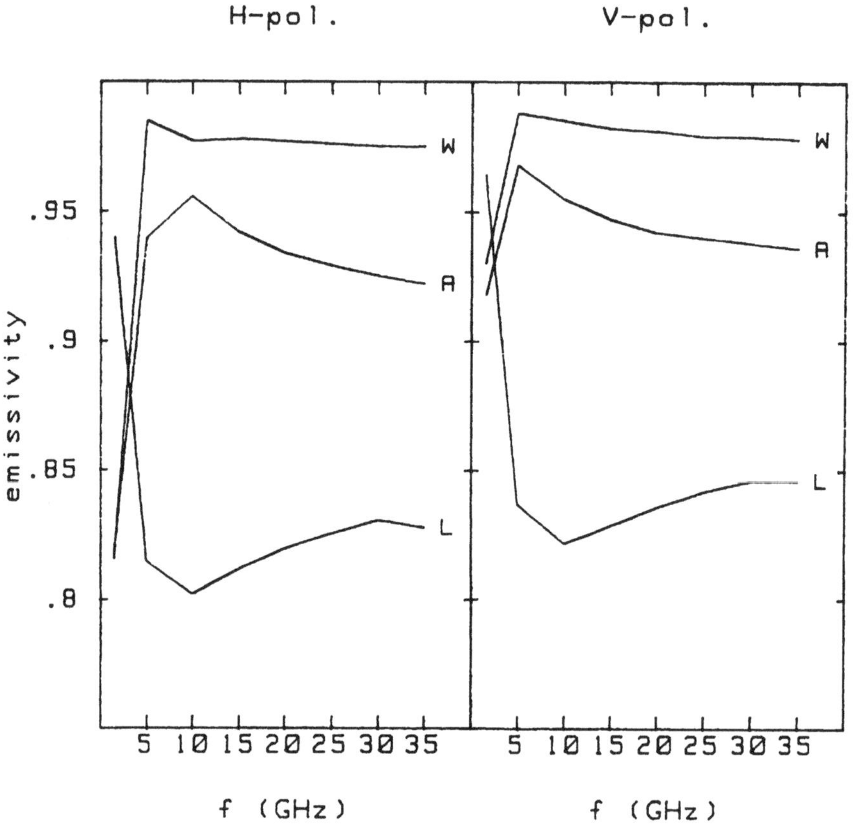

Figure 11. Model predicted emissivity versus frequency. L: corn, sunflower; W: wheat, barley; A: alfalfa.

out over Italian agricultural areas in the summers of 1987 and 1988. Experimental results are expressed in terms of normalized temperature T_n, i.e. ratios between microwave and infrared brightness temperatures. These ratios give an estimate of the emissivities, as discussed in [13]. Theoretical emissivities are shown in Fig. 11, while experimental T_n, measured at $\sim 40°$ are shown in Fig. 12. The label 'L' is associated to crops with large leaves on top like corn, sunflower and sugarbeet. The labels 'W' and 'A' represent wheat/barley and alfalfa, respectively.

In the theoretical data, the emissivity of crops with large leaves on top is the highest at low frequencies, it markedly decreases between 1.5 and 10 GHz, and assumes relatively low values at high frequencies. On the contrary, emissivity of wheat/barley and alfalfa crops show an evident increase between 1.5 and 5 GHz. At the highest frequencies, the emissivity of wheat and barley is close to 1, that of alfalfa is slightly lower, that of crops labeled with 'L' is strongly lower. Experimental results confirm all the main characteristics of theoretical data. The experimental T_n of 'L' crops is higher than the theoretical one; however, it is still appreciably lower than that of alfalfa and wheat.

Other microwave passive measurements performed over agricultural fields [16, 17] are in general agreement with the results of our model. In particular, they confirm that the emissivities of well developed 'grain' crops, like wheat, increase with frequency,

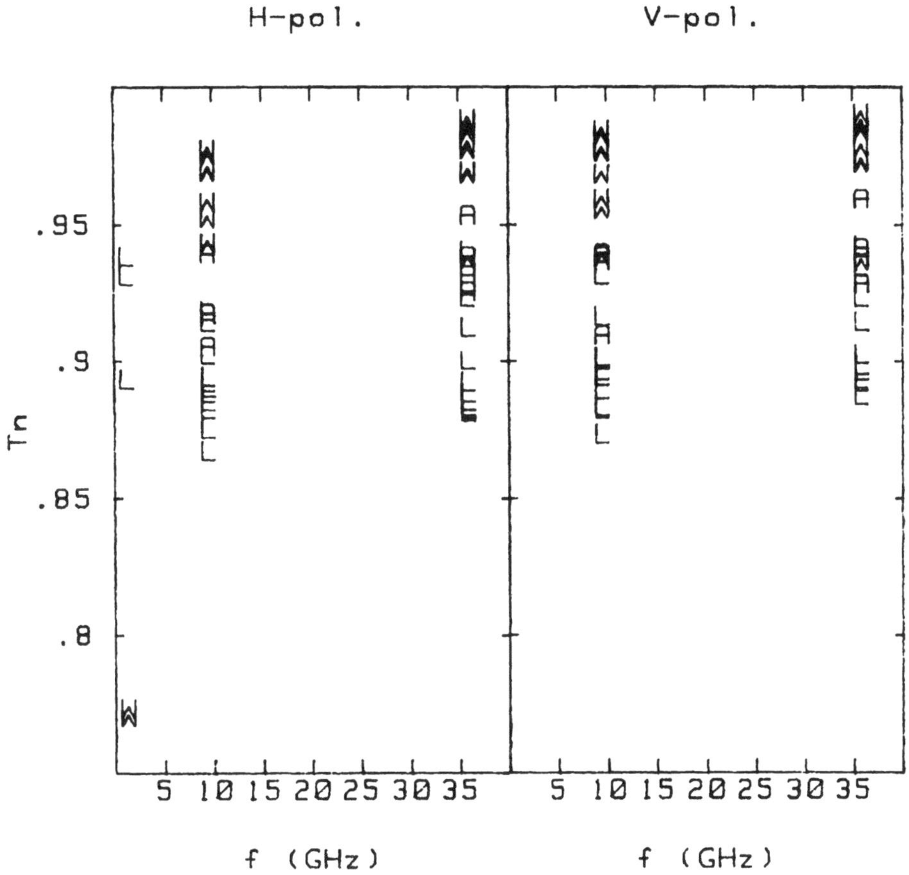

Figure 12. Experimental normalized temperatures versus frequency. L: corn, sunflower, sugarbeet; W: wheat, barley; A: alfalfa.

reaching values close to 1 for $f >\sim 10$ GHz; on the contrary, the 'wide leaf' crops show a general decreasing frequency trend up to 10 GHz, where emissivity values are relatively low.

4. CONCLUSIONS

A model, based on the Radiative Transfer Theory and a matrix algorithm, has been used to compute both the total emissivity and the contributions of single sources for some agricultural and forest species.

As expected, the soil is the dominant emission source at low frequencies, while the upper elements dominate at high frequencies. However, marked differences are noted among the different species, showing the good performance of multifrequency radiometric systems for classification applications.

The sensitivity to plant moisture content is quite good at high frequencies; sensitivities to soil moisture content and crop growth are influenced by variations of other parameters.

Most of the trends predicted by the model are in agreement with experimental results measured over agricultural fields in three different countries.

REFERENCES

1. L. Tsang, J. A. Kong and R. T. Shin. *Theory of Microwave Remote Sensing*. Wiley-Interscience, New York (1985).
2. F. T. Ulaby, R. K. Moore and A. K. Fung. *Microwave Remote Sensing: Active and Passive, Vol. III – From Theory to Applications*. Artech House, Dedham, MA (1986).
3. T. Mo, B. Choudury, T. Schmugge, J. Wang and T. Jackson. A model for microwave emission from vegetation-covered fields. *J. Geophys. Res.*, **87**, 11229–11237 (1982).
4. P. Pampaloni and S. Paloscia. Microwave emission and plant water content: a comparison between field measurements and theory. *IEEE Trans. Geosci. Remote Sensing*, **GE-24**, 900–905 (1986).
5. Y. H. Kerr and E. G. Njoku. A semiempirical model for interpreting microwave emission from semiarid land surfaces as seen from space. *IEEE Trans. Geosci. Remote Sensing*, **GE-28**, 384–393 (1990).
6. A. K. Fung and H. J. Eom. Emission from a Rayleigh layer with irregular boundaries. *J. Quant. Spectrosc. Radiat. Transfer*, **26**, 397–409 (1981).
7. P. Ferrazzoli, L. Guerriero and D. Solimini. Numerical model of microwave backscattering and emission from terrain covered with vegetation. *J. Appl. Comp. Electromagn. Soc.*, **6**, 175–191 (1991).
8. P. Ferrazzoli, L. Guerriero, S. Paloscia, P. Pampaloni and D. Solimini. Modeling polarization properties of emission from soil covered with vegetation. *IEEE Trans. Geosci. Remote Sensing*, **GE-30**, 157–165 (1992).
9. R. Schiffer and K. O. Thielheim. Light scattering by dielectric needles and disks. *J. Appl. Phys.*, **50**, 2476–2483 (1979).
10. D. M. Le Vine, R. Meneghini, R. H. Lang and S. S. Seker. Scattering from arbitrarily oriented dielectric disks in the physical optics regime. *J. Opt. Soc. Am.*, **73**, 1255–1262 (1983).
11. M. A. Karam and A. K. Fung. Electromagnetic scattering from a layer of finite length, randomly oriented, dielectric, circular cylinders over a rough interface with application to vegetation. *Int. J. Rem. Sensing*, **9**, 1109–1134 (1988).
12. M. A. El-Rayes and F. T. Ulaby. Microwave dielectric spectrum of vegetation – Part I: Experimental observations. *IEEE Trans. Geosci. Remote Sensing*, **GE-25**, 541–549 (1987).
13. P. Ferrazzoli, S. Paloscia, P. Pampaloni, G. Schiavon, D. Solimini and P. Coppo. Sensitivity of microwave measurements to vegetation biomass and soil moisture content: a case study. *IEEE Trans. Geosci. Remote Sensing*, **GE-30**, 750–756 (1992).
14. S. Paloscia, P. Pampaloni, L. Chiarantini, P. Coppo, S. Gagliani and G. Luzi. Multifrequency passive microwave remote sensing of soil moisture and roughness. *Int. J. Rem. Sensing*, **14**, 467–483 (1993).
15. W. J. Droesen, D. H. Hoekman, H. J. C. Van Leeuwen, J. J. Van Der Sanden, S. A. M. Bouman, D. Uenk, M. A. M. Vissers and G. G. Lemoine. *Maestro 89 Ground Data Collection Hosterwold/Speulderbos/Flevoland (NL)*. Wageningen Agricultural University Technical Report (1990).
16. U. Wegmüller. *Remote Sensing Signature Studies on Agricultural Fields with Ground-Based Radiometry and Scatterometry*. PhD Thesis, Universität Bern (1990).
17. A. M. Shutko and A. A. Chukhlantsev. Microwave radiative peculiarities of vegetation covers. *IEEE Trans. Geosci. Remote Sensing*, **GE-20**, 27–29 (1982).

ESA/NASA International Workshop, pp. 403–420
B. J. Choudhury, Y. H. Kerr, E. G. Njoku and P. Pampaloni (Eds)

Retrieval of geophysical parameters from multifrequency passive microwave measurements over a soybean canopy

JEAN-PIERRE WIGNERON

INRA, Bioclimatologie, BP91, 84143 MONTFAVET cedex, France

Abstract—The temporal variation of a soybean field microwave emission is investigated using a continuous model. The microwave data (at 1.4, 5.05, and 36.5 GHz) are acquired with a multifrequency dual polarized radiometer. In the continuous approach the vegetation structure is very simply modeled. Two correlation lengths l_z and l_ρ are used to describe volume scattering inside the canopy. A two-step method of inversion is described. It consists in calibrating first these two lengths and then in retrieving the surface parameters: the soil moisture and the vegetation volume fraction. Results of the retrievals and of the direct comparison between simulated and measured data are discussed. A good agreement between measured and inversed surface parameters can be observed. The results of this analysis could be improved by a more accurate modeling of absorption effects inside the vegetated canopy and by extended knowledge about the signification of the correlation lengths.

1. INTRODUCTION

Passive microwave measurements in the remote sensing field are very attractive as the microwave signal is weakly sensitive to the atmospheric and solar illumination effects and highly sensitive to many parameters of the land surface: soil and vegetation water content, green vegetation density and structure, surface temperature. Thus retrievals of microwave remotely sensed data could be very useful to monitor the biomass or the water balance at a regional or global scale. To cope with the numerous parameters which influence the microwave signal, a combination of measurements made with different configurations of the sensor (in terms of frequency, polarization, angle of observation) can be used.

Yet, very few studies describe retrieval attempts. A large set of data is required to perform inversion. Numerous ground truth observations are also required to validate the results. So we still have very little information about the problem of inversion of microwave data concerning the method of use, the accuracy of the retrievals, or the frequency dependence of the retrieved data. The purpose of this study is to investigate this problem using a theoretical modeling approach.

Two different modeling approaches can be used to perform the retrievals: the discrete and the continuous approach which are commonly used to simulate microwave emission of vegetated canopies. The discrete approach takes into account scattering by all the discrete objects of the vegetated canopy (leaves, trunks, stems, branches). It requires a complete description of the vegetation structure in terms of size, shape and probability distribution of orientation of the different discrete scatterers [1–3]. Whereas the continuous approach takes into account a statistical description of the dielectric fluctuations inside the vegetated medium [4–6], the continuous approach suits better the inversion problems since no precise information on the vegetation structure is required. Only four parameters are used to describe the whole vegetation structure: the height and the volume fraction of the vegetation canopy, and the correlation lengths l_z and l_ρ. These two lengths describe the 'dielectric structure' of vegetation and drive volume scattering inside the canopy. No direct measurement of the lengths has yet been attempted. They are most often inverted parameters and their geometric and dielectric significance is not well known. The model which is used to perform the simulations is based on a numerical solution of the radiative transfer equations in a one-layer vegetation medium and on a continuous description of volume scattering [7]. The phase matrix of the continuous random medium with strong permittivity fluctuations is derived from calculations of Jin [8]. The 'strong fluctuation' theory takes into account the large dielectric contrast between the scatterers (leaves, branches, and trunks) and the air inside the vegetated layer [4]. The microwave data which are inverted have been acquired with the multifrequency radiometer PORTOS over a soybean field during the summer 1991 in Avignon [9]. During the experiment, the measurements observed the temporal evolution of the canopy microwave emission for a wide range of configurations of the sensor, six frequencies (1.4, 5.05, 10.65, 23.8, 36.5, and 90 GHz), seven angles of incidence (from nadir to 60°) and the two polarizations H and V (horizontal and vertical).

The soil moisture and the vegetation volume fraction have large effects upon the measured microwave signal evolution, as in a similar experiment over on oat field [10]. These two parameters are retrieved in this study. In a previous paper [7], the continuous model was fully described and the results of the inversion were presented. In the present paper, an analysis of the different steps of the inversion is detailed. We analyse the time variations of the correlation lengths during the crop development and also perform a direct comparison between measured and simulated data. The results of this study are a direct contribution to the use of the continuous modeling approach and to the implementation of microwave measurements inversion. This first step in continuous modeling inversion should be also most useful to interpret multifrequency remotely-sensed microwave data.

2. EXPERIMENT AND DATA DESCRIPTION

The microwave measurements were conducted during the summer of 1991 over a soybean field. They monitored the whole crop development from Day of Year 205 (sowing day) to DoY 273. Microwave radiometric data were acquired with the six frequency dual-polarized radiometer PORTOS (1.4, 5.05, 10.65, 23.8, 36.5, and 90 GHz). The radiometer 3 dB beamwidth is between 13° (at 10.65 GHz) and 9.6° (at 36.5 GHz). The beam efficiency is 90% at 5.05 GHz, 84% at 1.4 GHz, and exceeds 95% at 10.65, 23.8, 36.5, and 90 GHz. The radiometer was installed on a 20 m crane boom, and measurements were obtained over a range of incidence angles from 0° to 60° with a 10° increment [9]. Regular external calibrations have been performed, over liquid water at 45° incidence angle, over eccosorb slabs either at ambient temperature or immersed into liquid nitrogen. These calibrations showed that because of problems in the thermal control of the radiometer, the absolute accuracy of the microwave data was about ±3 K.

The effect of the external temperature variation on the microwave sensor has not been corrected, since it was difficult to take into account steep temperature gradients inside the radiometer. At the present time, these problems are solved and the thermal stability of PORTOS is about ±0.5 K.

The data shown in this study were acquired at midday hour (11:00–13:00 true solar time). Microwave measurements are described at 1.4, 5.05 and 36.5 GHz where sufficient radiometric data where available to perform the retrievals. At 1.4 GHz, after DoY 254 the antenna was changed and measurements after this date are not taken into account.

Measurements of soil moisture content and temperature profile, surface roughness and surface temperature were made concurrently with the microwave measurements. Bi-weekly measurements of the soybean canopy were made to track its temporal evolution. The data include: dry and wet biomass, gravimetric water content, and a description of the vegetation structure. Rainfall and irrigation lead to a large range of soil moisture conditions. The time variations of the volumetric soil moisture content m_v (m^3/m^3) of the top 0.5 cm and 3 cm layer are shown in Fig. 1. The soil dielectric properties are inferred from the moisture content m_v using formulas of Dobson *et al.* [11].

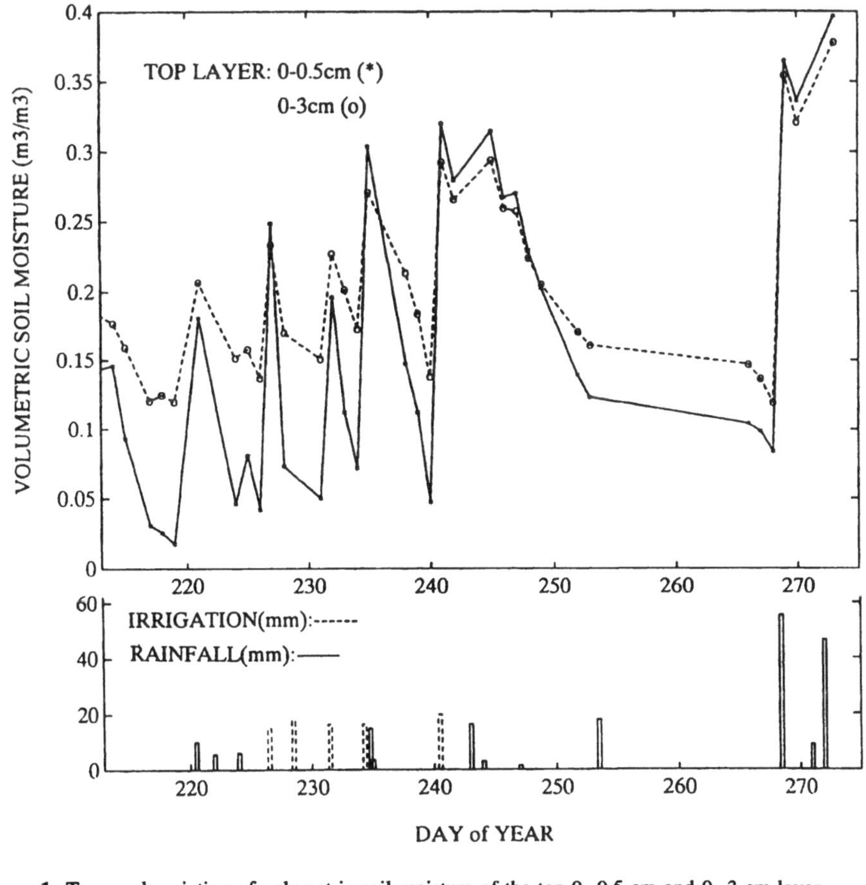

Figure 1. Temporal variation of volumetric soil moisture of the top 0–0.5 cm and 0–3 cm layer.

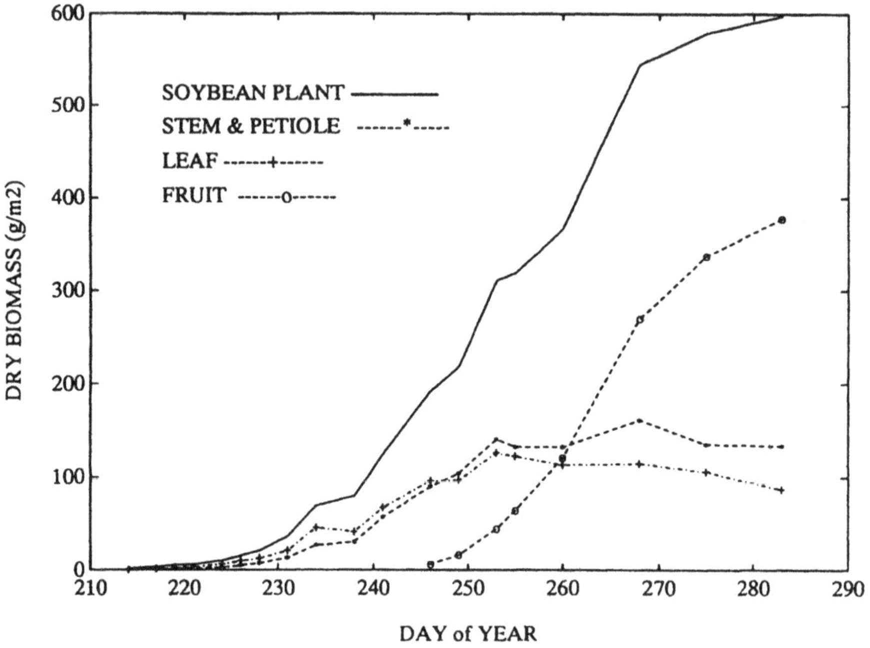

Figure 2a. Temporal variation of total soybean plant, stem & petiole, leaf, and fruit dry biomass (g/m^2).

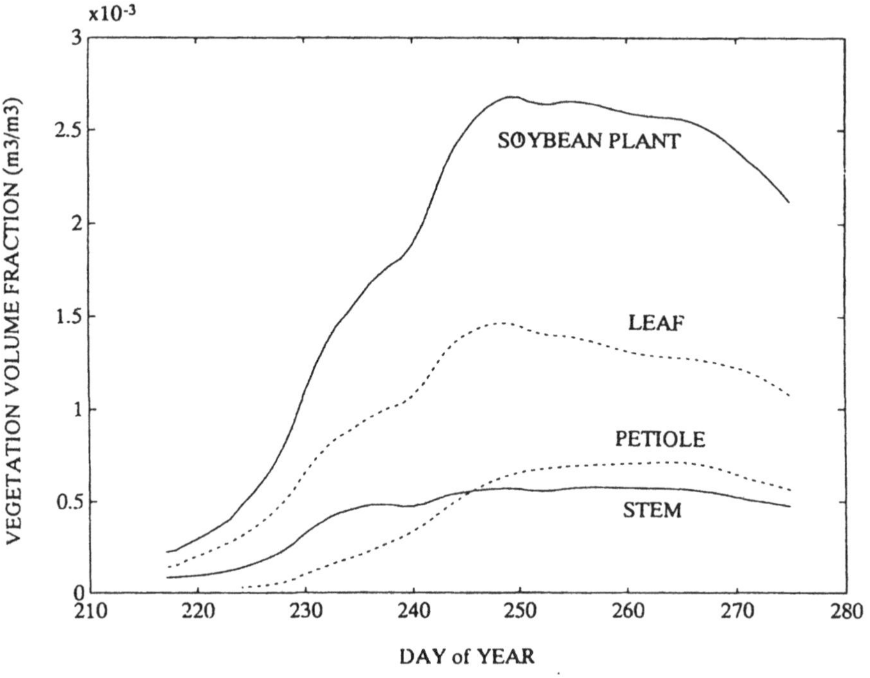

Figure 2b. Temporal variation of soybean plant, petiole, stem, and leaf volume fraction (m^3/m^3).

Table 1.
Retrieved values of soil roughness parameters at 1.4, 5.05, and 36.5 GHz

	h_r	Q
1.4 GHz	0.275	0.13
5.05 GHz	0.014	0.205
36.5 GHz	0.0	0.27

The soil surface was smoothed by a violent rainfall on DoY 211. The smooth soil roughness conditions did not change after this date. A two-parameter h_r and Q modeling of the soil reflectivity is used to describe the soil roughness [12]. The modified p-polarized reflectivity Γ^p is given by:

$$\Gamma^p = \left[(1-Q)\Gamma^s(\theta,\mathrm{p}) + Q\Gamma^s(\theta,\mathrm{q})\right]\exp\left(-h_r\cos^2\theta\right) \tag{1}$$

where $\Gamma^s(\theta,\mathrm{p})$ is the p-polarized specular reflectivity, h_r is a roughness parameter, and Q is a polarization-mixing parameter. Preliminary retrievals of these roughness parameters were carried out when the vegetation biomass was very small (DoY 215 to DoY 226). Retrieved values of h_r and Q are listed in Table 1. At the end of the experiment (DoY 269), wind and a 55 mm rainfall flattened a large part of the soybean field and strongly changed the vegetation structure. The structure became very heterogeneous and these conditions could not be taken into account by the simulations. Vegetation volume fraction f_v and dry biomass B_d are plotted versus time in Figs 2(a)–(b). The leaf volume fraction of a well-developed canopy is approximatively equal to the volume fraction of stems and petioles. The fruit volume fraction was not measured. So total volume fraction is certainly strongly underestimated after DoY 250, as indicated by the steep increase in the fruit dry biomass (Fig. 2a).

Other vegetation parameters used in the simulations are as follows:

$M_g = 0.8\%$	Mean leaf gravimetric water content.
$\rho_s = 0.27\ \mathrm{g\,cm^{-3}}$	Bulk density of dry vegetation.

The calculation of the vegetation dielectric constant using both M_g and ρ_s is based on a dual-dispersion model [13]. The height of canopy d (meters) is deduced from f_v, using a linear relationship:

$$d = 2.5688\times 10^2 f_v + 1.2795\times 10^{-2} \qquad \left(r^2 = 0.97\right). \tag{2}$$

Permittivity modeling of soil and vegetation is valid in a low frequency range (1–20 GHz). This modeling is also used at 36.5 GHz in this study.

3. CONTINUOUS MODEL

The vegetation layer is modeled as a random continuous medium. It is limited by a top vegetation–air interface and a bottom vegetation–soil interface (Fig. 3). Inside this

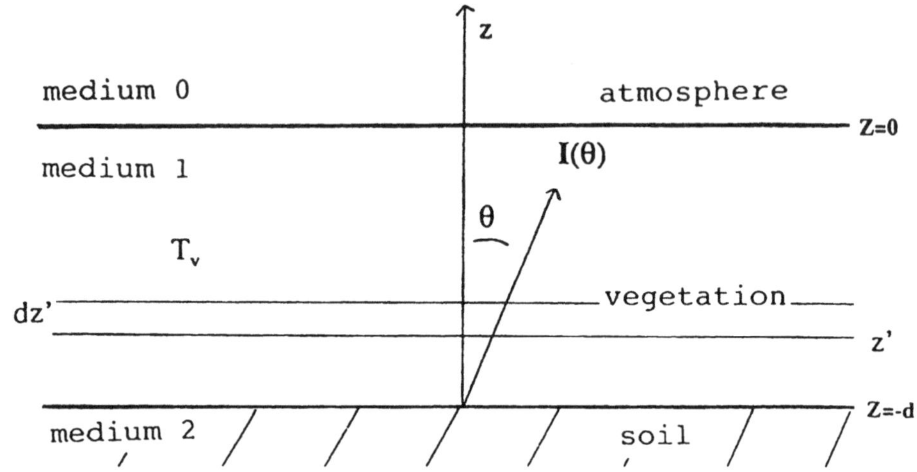

Figure 3. Geometrical configuration of a three-layer medium.

layer the solution of the radiative transfer equations is based on a numerical Gaussian quadrature method [14]. These radiative transfer equations for T_p, the p-polarized brightness temperature (p = v, h), travelling in the direction θ, can be written:

$$\begin{aligned} \cos\theta \frac{dT_v(\theta,z)}{dz} &= k_a T_1(z) - k_{ev}(\theta) T_v(\theta,z) \\ &+ \int_0^{\pi} \big((v,v')T_v(\theta',z) + (v,h')T_h(\theta',z)\big) \sin\theta' d\theta' \\ -\cos\theta \frac{dT_v(\pi-\theta)}{dz} &= k_a T_1(z) - k_{ev}(\pi-\theta) T_v(\pi-\theta,z) \\ &+ \int_0^{\pi} \Big(\big(v(\pi-\theta),v'\big)T_v(\theta',z) + \big(v(\pi-\theta),h'\big)T_h(\theta',z)\Big) \sin\theta' d\theta' \end{aligned} \tag{3}$$

where the subscripts v and h denote the vertical and horizontal polarizations (equations are written for the vertical polarization and are identical for the horizontal polarization); the integration over the azimuth angle is taken into account in the formulation of the phase function elements (p, q'); k_{ep} (p = v, h) is the extinction coefficient of the scatterers:

$$k_{ep} = k_a + k_{sp} \tag{4}$$

where k_a and k_{sp} (p = v, h) are respectively the absorption and scattering coefficients of the vegetated medium.

The random continuous medium is characterized by the dielectric permittivity $\varepsilon_1(r)$, r is the position vector [15]:

$$\varepsilon_1(r) = \varepsilon_{1m} + \varepsilon_{1f}(r) \tag{5}$$

$$\varepsilon_{1m} = \varepsilon'_{1m} + i\varepsilon''_{1m} \qquad \varepsilon''_{1m} \ll \varepsilon'_{1m} \tag{6}$$

where $\varepsilon_{\mathrm{lm}}$ is the mean of $\varepsilon_{\mathrm{l}}(r)$; $\varepsilon_{\mathrm{lf}}(r)$ is the fluctuating part of $\varepsilon_{\mathrm{l}}(r)$. It is characterized by its correlation function $\langle \varepsilon_{\mathrm{f}}\varepsilon_{\mathrm{f}}^{*}\rangle$. To simplify phase matrix calculations, parametric formulations of $\langle \varepsilon_{\mathrm{f}}\varepsilon_{\mathrm{f}}^{*}\rangle$ are commonly used. Following [15], a classical formulation is considered in the present continuous model:

$$\langle \varepsilon_{\mathrm{f}}(r_1)\varepsilon_{\mathrm{f}}^{*}(r_2)\rangle = \delta\varepsilon_{\mathrm{lm}}' \exp\left(-\frac{(x_1-x_2)^2+(y_1-y_2)^2}{l_\rho^2}-\frac{|z_1-z_2|}{l_z}\right) \tag{7}$$

where δ is the variance of the correlation, l_ρ and l_z are the correlation lengths, respectively in the horizontal plane and in the vertical direction.

Tsang and Kong [4] calculated the backscattering coefficients of a half space random medium with strong permittivity fluctuations. The 'strong fluctuation' theory in [4], does not require a weak fluctuation criterion ($\delta \ll 1$) as in previous formulations for random continuous media. So this theory is theoretically more efficient to describe volume scattering inside a vegetated canopy which is characterized by strong dielectric contrasts between the vegetated scatterers and the air. Phase matrix scattering coefficient formulations for strongly-fluctuating media which are used in this study, are derived from calculations of Jin [8].

The scattering coefficients k_{sv} and k_{sh} are deduced from the phase matrix components $(v,v'),(h,v'),(v,h'),(h,h')$ as:

$$\begin{aligned} k_{\mathrm{sv}} &= \int_0^\pi \mathrm{d}\theta' \sin\theta' \left[(v,v')+(h,v')\right] \\ k_{\mathrm{sh}} &= \int_0^\pi \mathrm{d}\theta' \sin\theta' \left[(v,h')+(h,h')\right] \end{aligned} \tag{8}$$

where h and v denote the horizontal and vertical polarization of an incident radiation travelling in the direction θ.

The absorption coefficient k_{a} is expressed as:

$$k_{\mathrm{a}} = 2\mathrm{Im}(k_{\mathrm{l}}) = 2k_0\mathrm{Im}\left[\left(\frac{\varepsilon_{\mathrm{lm}}}{\varepsilon_0}\right)^{1/2}\right] \tag{9}$$

where $\mathrm{Im}(\cdot)$ denotes the imaginary part, k_0 is the free space wave number. This description of absorption is very simplified. In particular, the calculation of k_{a} is not dependent on the polarization and on the incidence angle θ of the radiation. Formulations of scattering coefficients, absorption coefficients and phase matrix components are detailed by Wigneron *et al.* [7]. They are introduced in Equations (3) which are solved using the Gauss quadrature method.

As a conclusion of this section we list in Table 2 the input parameters which are required to calculate the microwave emission of vegetation using the continuous modeling approach.

Table 2.
Input parameters of the continuous modeling

Frequency (GHz): $f = 1.4,\ 5.05,\ 36.5$ GHz	
VEGETATION:	
	*temperature (K): top of canopy, $T_{\sup 1}$; bottom, $T_{\inf 1}$
– *mean dielectric constant of vegetation*: ε_v	
	*moisture content: water mass/fresh mass (%): M_g
	*bulk density of dry vegetation (g/cm^3): $\rho_s = 0.27$
– *mean dielectric constant of the vegetated medium*: ε_1	
– *vegetation structure*:	
	*volume fraction of vegetation: f_v
	*height of canopy (m): d
	*correlation lengths (m): l_z and l_ρ
SOIL:	
	*roughness parameters: h_r and Q
	*temperature (K): T_2
– *dielectric constant of soil*: ε_2	
	*volumetric moisture content (m^3/m^3): m_v
	*fraction of sand (s) and clay (c) (%): $s = 0.11$; $c = 0.27$
	*bulk density (g/cm^3): $\rho_b = 1.35$
ATMOSPHERE:	
	*downwelling atmospheric emission (K): T_{Bsky}

4. RESULTS AND DISCUSSION

4.1. *Time variations of emissivity*

Time variations of the measured soybean field microwave brightness temperature e_p (p = v, h) are shown in Figs 4(a)–(b)–(c) at 1.4, 5.05, and 36.5 GHz. An estimate of the emissivity e_p is calculated by dividing the brightness temperature T_{Bp} (p = v, h) by a surface temperature inferred from thermal infrared measurements. Two main parameters seem to influence the time variations of emissivity. First, steep increase and decrease in e_p are correlated to soil moisture variations. Secondly, during the crop development, the plant growth drives the trend of e_p.

From sowing (DoY 205) to DoY 240–242 when f_v approximates 0.2% (Fig. 2b), signal variations are mainly correlated to soil moisture changes. High soil moisture values after rainfall or irrigation can be related to a steep increase in the polarization ratio PR ($PR = (T_{Bv} - T_{Bh})/(T_{Bv} + T_{Bh})$) (Fig. 5) and to a steep decrease in the emissivity e_v and e_h (DoY 221, 227, 232, 235, 241). The amplitude of these variations is generally higher at 5.05 GHz than at 1.4 GHz. Partly, the explanation is that the upper soil layer contributing to the soil emission becomes thinner as the frequency increases [16, 17], while the amplitude of moisture variations are higher at the soil surface (as shown in Fig. 1). Also, the amplitude of soil moisture variations decreases, following the

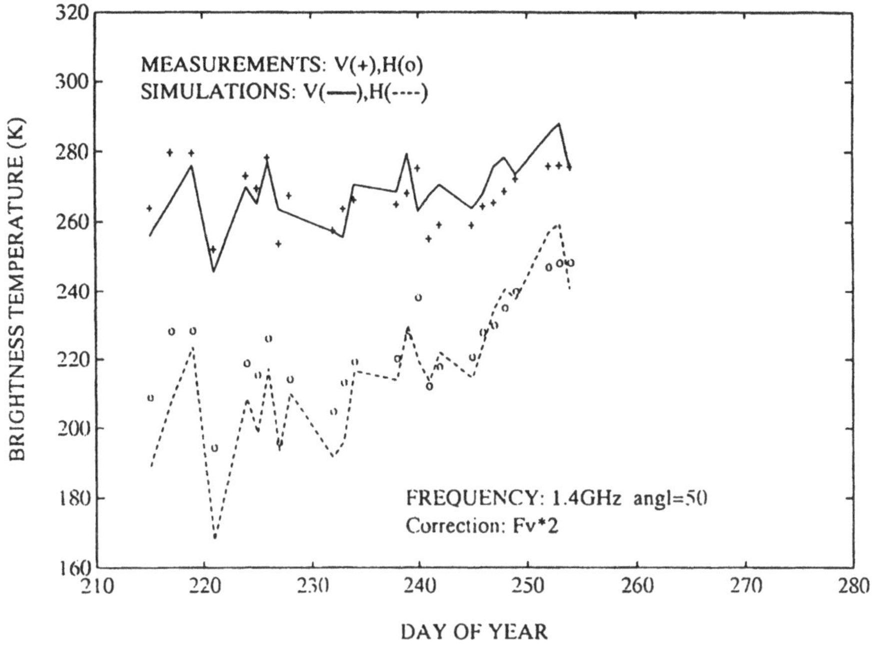

Figure 4a. Measured and simulated microwave soybean canopy brightness temperature as a function of time at 1.4 GHz (doubled values of f_V are used to simulate emissivity).

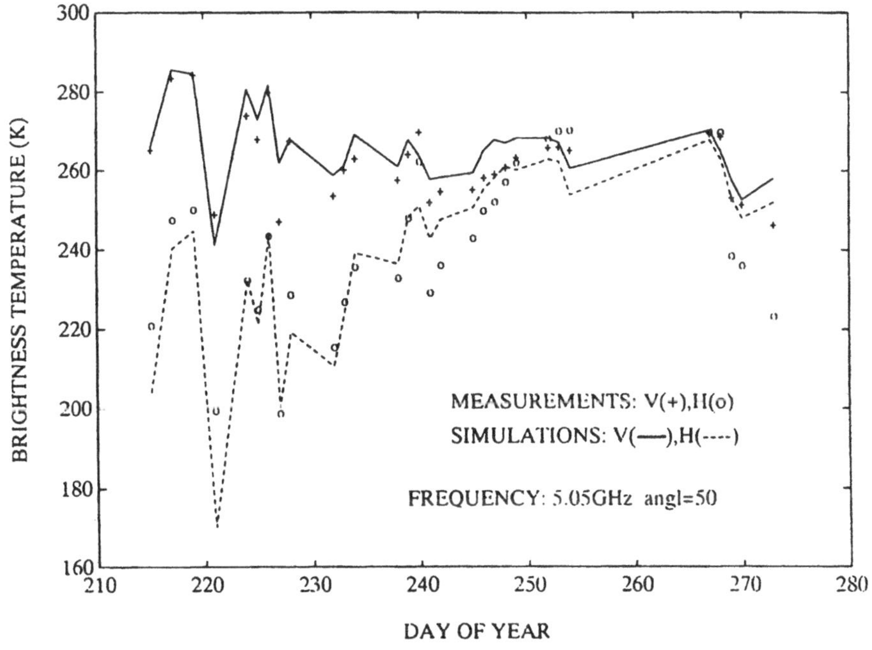

Figure 4b. Measured and simulated microwave soybean canopy brightness temperature as a function of time at 5.05 GHz.

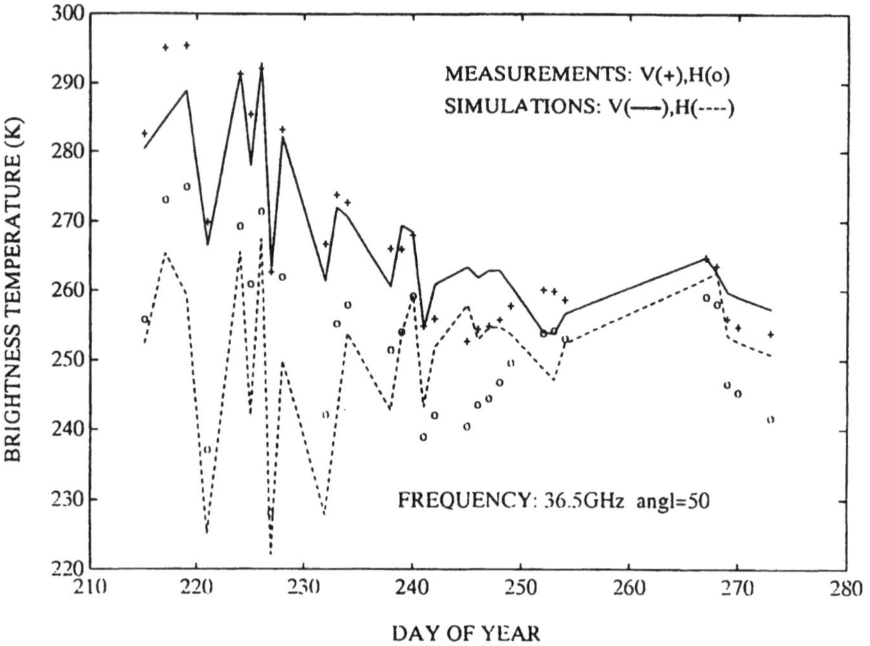

Figure 4c. Measured and simulated microwave soybean canopy brightness temperature as a function of time at 36.5 GHz.

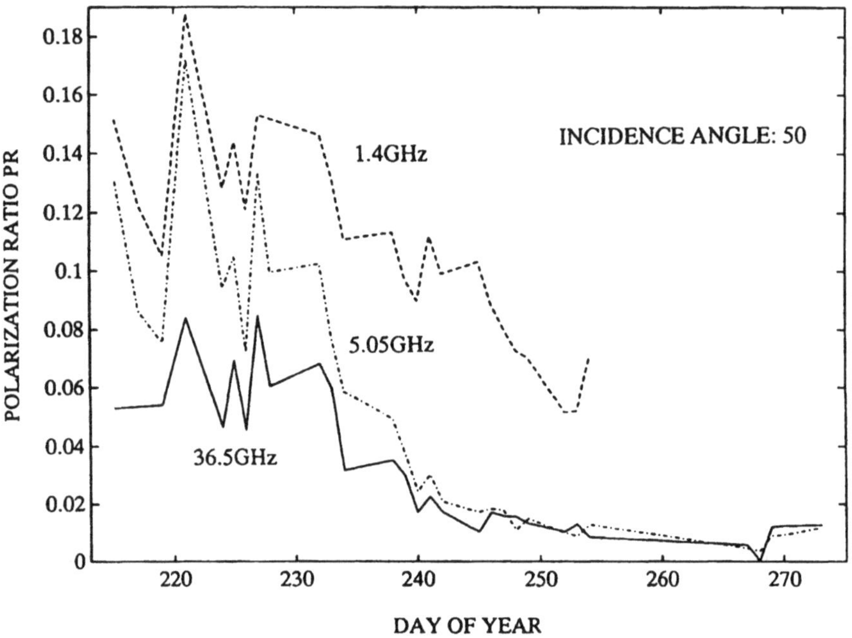

Figure 5. Time variations of the polarization ratio PR inferred from measurements at 1.4, 5.05, and 36.5 GHz.

soybean development. As the vegetation biomass increases, soil emission is more and more attenuated inside the vegetated layer. Moreover, scattering and absorption effects increase with frequency, and consequently attenuation of soil emission by the vegetated layer also increases with frequency. So when f_v goes beyond 0.2% (DoY 241) the influence of m_v upon the microwave surface emission remains perceptible at 1.4 GHz, but is almost zero at 5.05 and 36.5 GHz.

The decreasing trend of the polarization ratio PR is well related to the plant growth. This trend is first perceptible at 36.5 GHz, when f_v goes beyond 0.1% (Figs 2b and 5). This evolution is in relation with the trend of e_p: mainly the increase in e_v and e_h at 1.4 GHz, the increase in e_h at 5.05 GHz, the decrease in e_v at 36.5 GHz. As described by Paloscia and Pampaloni [18] the difference between multifrequency emissivity data improves discrimination of soybean biomass. So, the difference (e_v(36.5 GHz)$-$ e_h(5.05 GHz)) combines two significant effects: the increasing trend of e_h at 5.05 GHz and the decreasing trend in e_v at 36.5 GHz. When f_v goes beyond 0.2% (DoY 241) only the 1.4 GHz microwave signal seems to be sensitive to the biomass increase. Both 5.05 and 36.5 GHz microwave signals seem to saturate (Figs 4b–c).

So both the soil moisture m_v and the vegetation density (parameterized in the model by the vegetation volume fraction f_v) have most perceptible effects upon the temporal evolution of the microwave canopy emissivity. Hence they are the most pertinent parameters to retrieve in this study, using a theoretical modeling approach.

4.2. *Use of the continuous modeling approach*

In this section, an attempt to retrieve the two geophysical parameters m_v and f_v, respectively the volumetric soil moisture content (m^3/m^3) and the vegetation volume fraction (m^3/m^3), is carried out. In order to perform the retrievals, simplifications of the inversion problem have to be made. The height of canopy d is deduced from retrieved values of f_v using the empirical relation [2]. All other parameters are considered as constant and are fixed: h_r and Q (soil roughness parameters), M_g (vegetation parameter), l_z and l_ρ (mixed dielectric and structure parameters of vegetation). Using the continuous modeling approach, scattering inside the vegetation layer is modeled using l_z and l_ρ. The calculation of l_z and l_ρ is uneasy, since it would require an accurate description of the spatial distribution of the vegetation scatterers. So the lengths must be retrieved from microwave measurement data. Attempts to retrieve both the correlation lengths and the geophysical parameters f_v and m_v at the same time were unsuccessful. So the idea we develop in this study is to calibrate first the correlation lengths using ground truth data (Section 4.2.1) and afterwards to perform retrievals of f_v and m_v (Section 4.2.3). In Section 4.2.2 we try to estimate the error related to the fact that the correlation lengths are considered as constant during the crop development.

The method of inversion is based on a minimization routine of the RMSE (Root Mean Square Error) between measured and calculated emissivity e_p. Calculations of RMSE are carried out over polarized data, from the incident angles 0° to 60°, with a 10° increment.

4.2.1. Temporal variation of the correlation lengths l_z and l_ρ. The correlation lengths l_z and l_ρ are retrieved using ground truth data. All soil and vegetation characteristics we used are detailed in Section 2 and Table 2; f_v is shown in Fig. 2b. The soil moisture m_v

which is used in the simulations, must take into account the dependence on frequency of the effective soil depth which contributes to the emission. So top layer 0–3 cm m_v is used at 1.4 GHz, 0–1 cm m_v at 5.05 GHz, and 0–0.5 cm m_v at 36.5 GHz. Time variations of retrieved l_z and l_ρ at 5.05 and 36.5 GHz are shown in Figs 6(a)–(b). At 1.4 GHz, retrievals of l_z and l_ρ did not converge: scattering which is governed by the correlation lengths inside the vegetated layer, is very low. For both frequencies 5.05 and 36.5 GHz, variations of l_ρ are similar. First l_ρ decreases following the vegetation development; then after f_v goes beyond 0.1%, the evolution of l_ρ is rather stable. At 5.05 GHz, variations of l_z are a bit erratic, while at 36.5 GHz l_z slowly increases. The steep increase in l_z at 5.05 GHz at the end of the experiment (DoY 269 to DoY 273) can be related to the storm which partly flattened the soybean field. It is not possible to give a physical meaning to this increase, since at this time, the actual vegetation structure falls outside the range of validity of the model, which takes into account a continuous horizontal vegetation layer.

The correlation lengths are related to the spatial distribution of the vegetation scatterers, to their dimension and dielectric permittivity ε_s. As for a given frequency, ε_s can be considered as constant inside the canopy, in first approximation, the length values are functions of the vegetation structure and are related in particular to the size and distribution of orientation of the vegetation elements. When the vegetation biomass is small, scattering inside the vegetated layer is not significant and the influence of l_z and l_ρ upon the microwave signal is low. Consequently, interpretation of values of the correlation lengths must be careful. Nevertheless, it seems that the vegetation ‘dielectric structure’ becomes more and more vertical (l_ρ decreases and l_z increases), following the soybean growth. When f_v is greater than 0.1%, no clear trend in l_z or l_ρ is perceptible. At 5.05 GHz the canopy microwave emission is that of a vertically structured layer, with a dominant structuring of the vegetation elements along the vertical direction ($l_z \gg l_\rho$): the emission is related to the whole vertical crop layer structure. Whereas at 36.5 GHz, radiations seem to be mostly sensitive to the leafy top surface of the vegetated layer ($l_z \approx l_\rho$).

4.2.2. Comparison between measured and simulated data. To simulate the soybean field microwave emission using the continuous model, we assume that both l_z and l_ρ are constant. At the beginning of the plant growth, this assumption is not valid but the influence of the correlation lengths on the microwave signal is low. We set l_z and l_ρ equal to their mean retrieved values from DoY 245 to DoY 253, when the vegetation is well-developed. These values are listed in Table 3. We assume that the influence of l_z and l_ρ on the microwave signal is low at 1.4 GHz since they could not be retrieved. Values of the correlation lengths at 1.4 GHz are set equal to their values at 5.05 GHz. Comparison

Table 3.
Retrieved values of correlation lengths (l_z and l_ρ) at 1.4, 5.05, and 36.5 GHz

	l_z(mm)	l_ρ(mm)
1.4 GHz	(29.7)	(2.39)
5.05 GHz	29.7	2.39
36.5 GHz	0.75	0.82

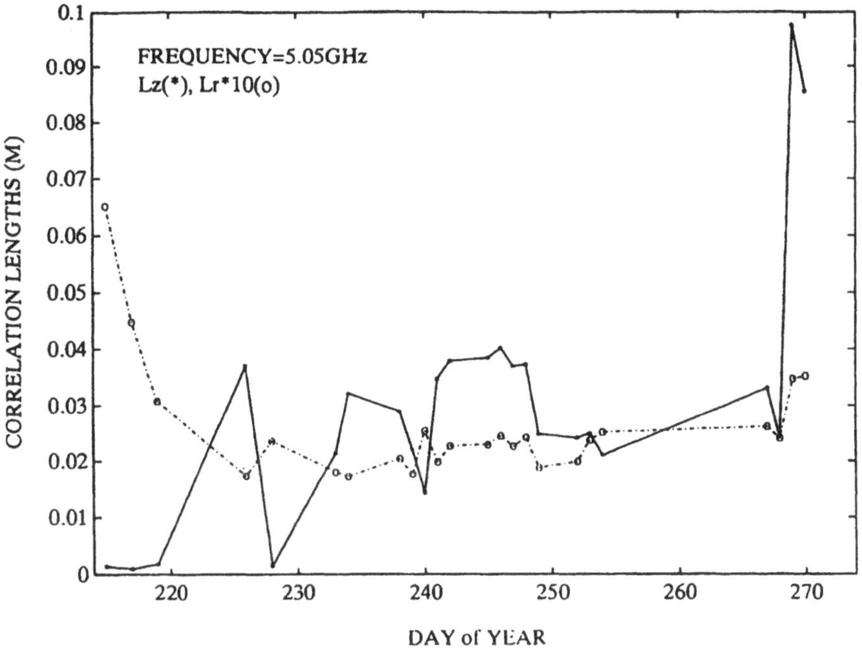

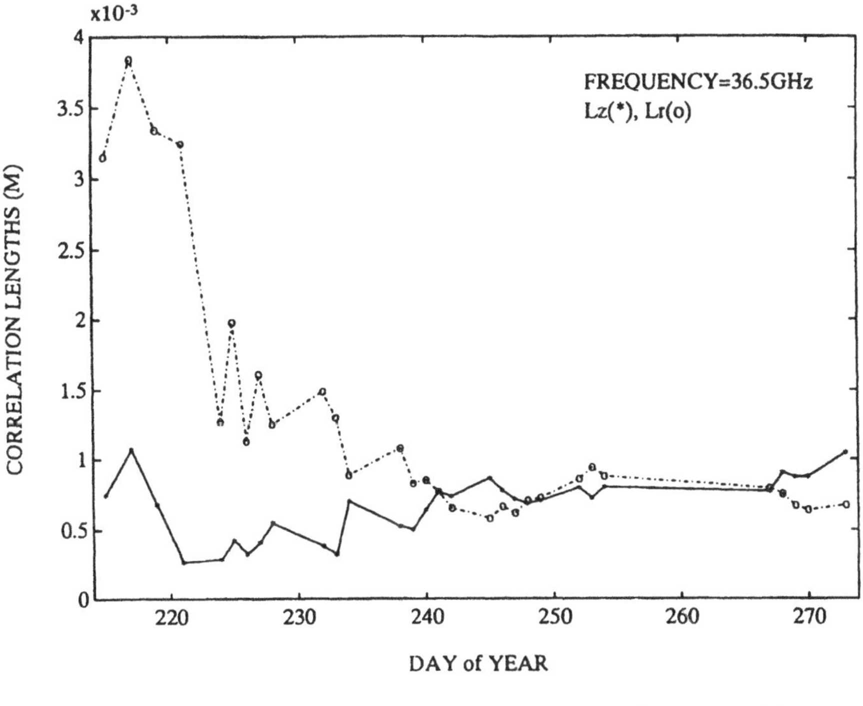

Figure 6. Time variations of retrieved correlation lengths l_z and l_ρ, at 5.05 GHz (a) and 36.5 GHz (b).

Table 4.
Root Mean Square Error (RMSE) between simulated and measured brightness temperatures at 1.4, 5.05, and 36.5 GHz

frequency (GHz):	1.4	1.4 ($2f_v$)	5.05	36.5
RMSE (K):	16.5	11.4	7.6	7.4

between simulated and measured data at the incidence angle $\theta = 50°$ is shown in Figs 4(a)–(c). Also, values of RMSE (Root Mean Square Error) between simulated and measured data are listed in Table 4. For every frequency there are local discrepancies at the end of the experiment: simulations do not take into account the vegetation structural changes on DoY 269–273. Discrepancies are particularly high at 1.4 GHz (RMSE = 16.5 K). This value decreases to 11.4 K if values of f_v are doubled (Fig. 4a). This may indicate either that f_v is under-estimated or that absorption modeling is not accurate. At 5.05 and 36.5 GHz scattering effects are strongly higher than absorption effects, and discrepancies are lower. At 5.05 and 36.5 GHz the accuracy of the continuous model, assuming that both l_z and l_ρ are constant, approximates 7.5 K. Part of this value can be related to the radiometer sensitivity to external climatic conditions which changed from one day to the other. Improvements of the thermal control efficiency of the radiometer are performed at the present time. Future investigations might show improved results.

4.2.3. Retrievals of soil moisture and vegetation volume fraction. Using correlation lengths l_z and l_ρ, calibrated over the well-developed canopy, both the vegetation volume fraction f_v and the soil moisture m_v are retrieved in this section.

The retrieved f_v values, denoted f_{vR}, are shown in Fig. 7, at 1.4, 5.05, and 36.5 GHz. At 5.05 and 36.5 GHz, f_{vR} values are very similar and are in good agreement with the measured data. f_{vR} increases first at 36.5 GHz which appears to be more sensitive to the early soybean development. f_{vR} of the well-developed canopy, is approximately two times higher at 1.4 GHz than at 5.05 and 36.5 GHz. This observation can be related to discrepancies between simulated and measured data, observed at 1.4 GHz in the previous section. Several explanations are plausible:

* The measured total volume fraction of vegetation is perhaps under-estimated. In particular the fruit development is not taken into account to estimate f_v (Figs 2a–b). Yet fruit biomass is only significant after DoY 250 (Fig. 2a).
* The 1.4 GHz microwave radiation easily goes through the whole canopy layer and is still sensitive to the biomass increase after DoY 240, while the information content of the 5.05 and 36.5 GHz radiations may saturate after this date.
* The absorption coefficient calculation is perhaps not accurate and underestimates absorption inside the canopy layer. Effects of this under-estimation would be mainly sensitive at 1.4 GHz where the relative influence of absorption in comparison with scattering is higher than at 5.05 or 36.5 GHz.

It must be noted also that the 1.4 GHz correlation lengths have not been calibrated as for 5.05 and 36.5 GHz radiations. This may also increase discrepancies between retrieved and measured data of f_v.

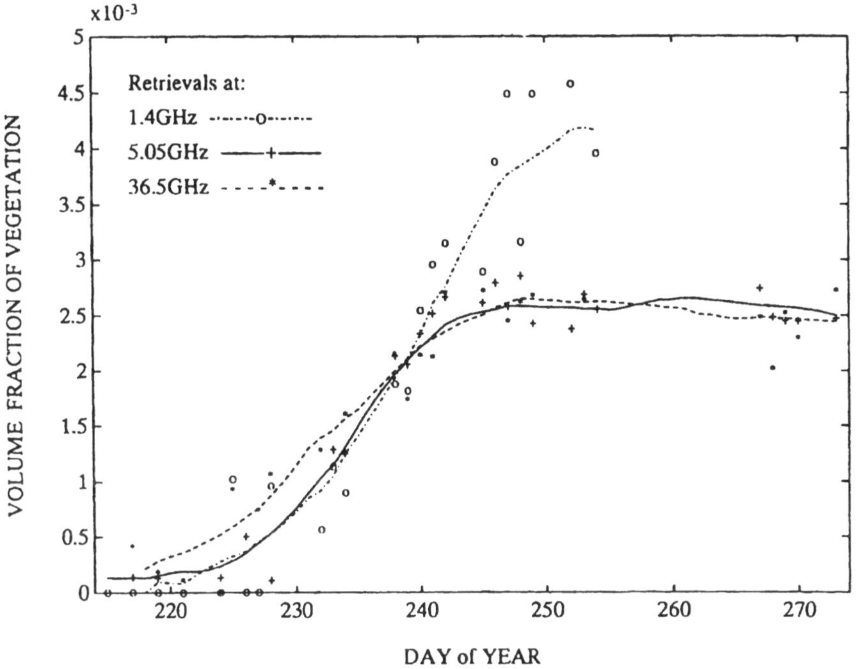

Figure 7. Retrieved values of vegetation volume fraction f_v at 1.4, 5.05, and 36.5 GHz.

Retrieved values of the volumetric soil moisture, denoted m_{vR}, at 1.4 GHz and 5.05 GHz are shown in Figs 8(a)–(b). These retrieved values are compared with measurements of the mean volumetric soil moisture over two different layer thickness: $\langle m_v \rangle_{3cm}$ at 1.4 GHz, $\langle m_v \rangle_{1cm}$ at 5.05 GHz. At 36.5 GHz, absorption and scattering effects inside the well-developed soybean canopy, which are very important at high frequencies, completely attenuate the microwave soil emission. So the 36.5 GHz microwave emission of the well-developed does not contain information about soil moisture.

In Figs 8(a)–(b) the agreement between m_{vR} and $\langle m_v \rangle_{3cm}$ and $\langle m_v \rangle_{1cm}$, respectively, is rather good except a few days. m_{vR} is over-estimated on DoY 221 at 1.4 GHz and on DoY 240 at 5.05 GHz. Radiometric measurements were performed earlier these two days (9:00–10:00 instead of 11:00–13:00 true solar time). The radiometer sensitivity to external temperature possibly explains part of the observed discrepancies. The storm which flattened the soybean field can also be related to discrepancies at the end of the experiment (DoY 269 to DoY 273). In particular the wet soil was partially uncovered, and this fact is not taken into account by the model.

5. CONCLUSIONS

In this paper, the use of a continuous modeling approach to retrieve geophysical parameters from passive microwave measurements is investigated. To perform the retrievals we have to cope with several difficulties. The simultaneous retrieval of both the correlation

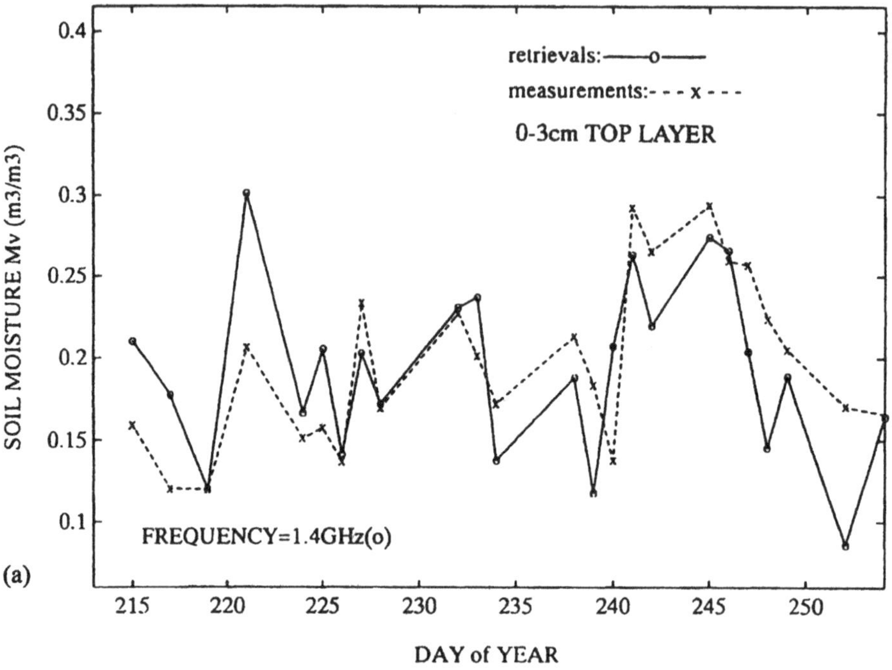

(a)

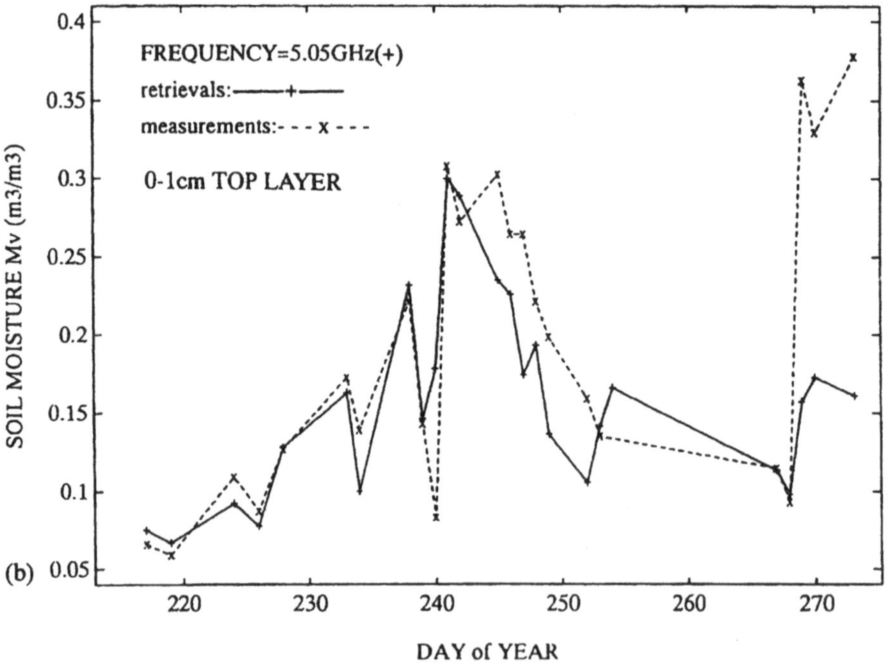

(b)

Figure 8. Comparison between time variations of retrieved and measured soil moisture at 1.4 GHz (a) and 5.05 GHz (b).

lengths and the land surface parameters (the soil moisture m_v and the vegetation volume fraction f_v) was unsuccessful. Retrievals of the correlation lengths l_z and l_ρ when the scattering effects inside the vegetated layer are low (at 1.4 GHz and at the beginning of the crop development) are not reliable. Results of preliminary retrievals indicate that when the vegetation biomass is sufficient ($f_\nu \geqslant 0.1\%$) no trend is perceptible in the correlation length time variations. Hence the method of inversion which is used in this study, consists in calibrating first the correlation lengths over a well-developed canopy using ground truth data. Retrievals of the geophysical parameters are then performed assuming that both l_z and l_ρ are constant during the whole crop development. Results of the study are satisfactory. Comparison between measured and simulated data at 1.4 GHz might indicate that the absorption coefficient value is under-estimated inside the canopy layer. At 5.05 GHz and 36.5 GHz the Root Mean Square Error (RMSE) between measured and simulated data approximates 7.5 K. Retrievals of both f_v and m_v are consistent with measurements.

Future investigations are necessary to increase the continuous model accuracy, when simulating microwave emission of vegetation. First the absorption coefficient calculation inside the canopy must be related more efficiently to vegetation characteristics. Also, analysis of time variations of the correlation lengths must be carried out during the vegetation cycle of different vegetated canopies (forest, sparse vegetation, various crop fields, ...). These aspects appear to be important preliminary steps to better implement continuous microwave modeling.

REFERENCES

1. B. J. Choudhury and J. R. Wang. Simulated and observed 37 GHz emission over Africa. *Int. J. Rem. Sens.*, **11**, 1837–1868 (1990).
2. P. Ferrazzoli, L. Guerriero, S. Paloscia, P. Pampaloni and D. Solimini. Modeling polarization properties of emission from soil covered with vegetation. *IEEE Trans. Geosci. Rem. Sens.*, **30**, 157–165 (1992).
3. J.-P. Wigneron, J.-C. Calvet, Y. H. Kerr, A. Chanzy and A. Lopes. Microwave emission of vegetation: sensitivity to leaf characteristics. *IEEE Trans. Geosci. Rem. Sens.*, **31**, 716–726 (1993).
4. L. Tsang and J. A. Kong. Scattering of electromagnetic waves from random media with strong permittivity fluctuations. *Radio Science*, **16**, 303–320 (1981).
5. R. G. Isaacs, Y. Q. Jin, R. D. Worsham, G. Deblonde and V. J. Falcone. The RADTRAN microwave surface emission models. *IEEE Trans. Geosci. Rem. Sens.*, **27**, 433–440 (1989).
6. J.-C. Calvet, J.-P. Wigneron, E. Mougin, Y. H. Kerr and J. L. Brito. Plant water content and temperature of the amazon from satellite microwave radiometry. (Accepted). *IEEE Trans. Geosci. Rem. Sens.* (1994).
7. J.-P. Wigneron, Y. H. Kerr, A. Chanzy and Y. Q. Jin. Inversion of surface parameters from passive microwave measurements over a soybean field. *Remote Sens. Environ.*, **46**, 61–72 (1993).
8. Y. Q. Jin. The radiative transfer equation for strongly-fluctuating continuous random media. *J. Quant. Spectrosc. Radiat. Transfer*, **42**, 529–537 (1989).
9. A. Chanzy, J.-P. Wigneron and T. J. Schmugge. Multifrequency passive microwave measurements with Portos: experimental design and first results. In: *Meeting on Microwave Radiometry and Remote Sensing Applications*, Boulder, CO (1992).
10. C. Mätzler. Seasonal evolution of microwave radiation from an oat field. *Remote Sens. Environ.*, **31**, 161–173 (1990).
11. M. C. Dobson, F. T. Ulaby, M. T. Hallikainen and M. A. El-Rayes. Microwave dielectric behavior of wet soil. Part II: Dielectric mixing models. *IEEE Trans. Geosci. Rem. Sens.*, **23**, 35–46 (1985).
12. J. R. Wang and B. J. Choudhury. Remote sensing of soil moisture content over bare field at 1.4 GHz frequency. *J. Geophys. Res.*, **86**, 5277–5282 (1981).

13. F. T. Ulaby and M. A. El-Rayes. Microwave dielectric spectrum of vegetation. Part II: Dual-dispersion model. *IEEE Trans. Geosci. Rem. Sens.*, **25**, 550–557 (1987).

14. L. Tsang and J. A. Kong. Thermal microwave emission from a three-layer random medium with three dimensional variations. *IEEE Trans. Geosci. Rem. Sens.*, **18**, 212–216 (1980).

15. L. Tsang and J. A. Kong. Thermal microwave emission from half space random media. *Radio Science*, **11**, 599–609 (1976).

16. J. R. Wang. Microwave emission from smooth bare fields and soil moisture sampling depth. *IEEE Trans. Geosci. Rem. Sens.*, **25**, 616–622 (1987).

17. T. J. Jackson and T. J. Schmugge. Passive microwave remote sensing system for soil moisture: some supporting research. *IEEE Trans. Geosci. Rem. Sens.*, **27**, 225–235 (1989).

18. S. Paloscia and P. Pampaloni. Microwave vegetation indexes for detecting biomass and water conditions of agricultural crops. *Remote Sens. Environ.*, **40**, 15–26 (1992).

Soil moisture

ESA/NASA International Workshop, pp. 423–460
B. J. Choudhury, Y. H. Kerr, E. G. Njoku and P. Pampaloni (Eds)

Passive microwave radiation from soil: examples of emission models and observations

JAMES R. WANG and BHASKAR J. CHOUDHURY

NASA/Goddard Space Flight Center, Greenbelt, MD 20771, USA

Abstract—This paper reviews microwave emission models and observations of surface layer soil moisture. Two general types of emission models have been developed to interpret microwave radiometric observations, which could be grouped as coherent and non-coherent. In the coherent models one follows the propagation of electric and magnetic fields, while in the non-coherent models one analyses the radiation intensities directly. Non-coherent models are mathematically and computationally simpler but less accurate than coherent models, particularly at low microwave frequencies. Results of comparing coherent and non-coherent models for a variety of soil moisture and temperature profiles are given. The basic radiative characteristic determining the emissivity is the dielectric constant of soils, and models relating dielectric constant to soil moisture are discussed. The effective soil layer thickness whose moisture determines the emissivity has been assessed from both theoretical considerations and actual field observations. Estimation of soil moisture from ground-level, aircraft and satellite observations are summarized. Important ancillary data needed to estimate soil moisture from radiometric observations over homogeneous surfaces are soil temperature, vegetation parameters, and surface roughness. Models of varied complexities have been developed to account for the effects of surface roughness and vegetation on the observed brightness temperature at the spatial scale of the ground-level measurements. A better understanding needs to be developed to account for the effects of surface roughness and spatial heterogeneity of vegetation for estimating soil moisture from satellite observations.

1. INTRODUCTION

Remote sensing of soil moisture by microwave radiometry has been a subject of intensive studies in the past two decades [1–30]. Experimentally, measurements have been conducted with microwave radiometers at the ground level, as well as at aircraft and satellite altitudes [1–25]. These measurements cover a wide frequency range of 0.6–90 GHz and, in addition to soil moisture, the effects analyzed include soil texture, surface roughness, vegetation and microwave sampling depth. The spatial scale of the measurements varies from a few square meters to hundreds of square kilometers. Laboratory measurements of the dielectric permittivity of soil-water mixtures were also made at several frequencies [31–34] that were frequently used in conjunction with theoretical models to explain experimental results. A number of theoretical models, with a varying degree of sophistication, have been developed to deal with the radiative transfer in the soil media, surface roughness, and vegetation [26–30]. Some of these models dealing with the radiative transfer in soil media have been compared by Schmugge and Choudhury [30]. A substantial fraction of the experimental efforts has previously been reviewed [35–38].

In this paper some emission models and observations on microwave emission from soils pertaining to soil moisture sensing will be summarized. The emphasis will be placed on the recent results or on results not covered by the previous reviews. The dielectric permittivity of soil water mixtures will first be described as it pertains to discussions on emission models and radiometric measurements in later sections. Discussion on emission models will be limited to those which have previously been compared with measurements. Highlights of results from microwave radiometric measurements at the ground level as well as at aircraft and satellite altitudes will be included. Effects of soil texture, surface roughness, and row structure on microwave emission are discussed. A few radiometric measurements that have implications on the microwave sampling depth of soils are elaborated in some detail. The effect of vegetation cover, which is critical to passive microwave sensing of soil moisture, will be discussed only briefly; the subject is treated in a great length by another paper in this book.

2. DIELECTRIC PROPERTY OF SOILS

Microwave emission from a soil depends on its moisture content because of the large contrast between the real part of dielectric constant of water ($\sim$ 80 at frequencies $\leqslant$ 1 GHz) and that of dry soil ($\sim$ 3). Because of this large dielectric contrast, microwave emissivities at $\sim$ 1 GHz in the direction perpendicular to smooth water and dry-soil surfaces are $\sim$ 0.33 and $\sim$ 0.93, respectively, which translate into a difference of about 180 K in brightness temperature if thermometric temperatures of both water and soil are $\sim$ 300 K. Figure 1 gives a plot of the real (ε') and imaginary (ε'') parts of the dielectric constant for water in the microwave portion of the spectrum. It clearly shows that ε' decreases with increasing frequency (f) at $f > 5$ GHz and assumes a nearly constant value of $\sim$ 6 at $f \geqslant 100$ GHz. The large ε' at $f \leqslant 1$ GHz is caused by the alignment of the dipoles of water molecules along the direction of the oscillating electric field; at high f region, the dipoles cannot change in steps with the rapidly oscillating field and ε' decreases. Consequently, microwave measurements of soil moisture should preferably be made at $f \leqslant 5$ GHz. Both salinity and temperature affect the values of ε' slightly, but are not critical in the estimation of soil moisture from

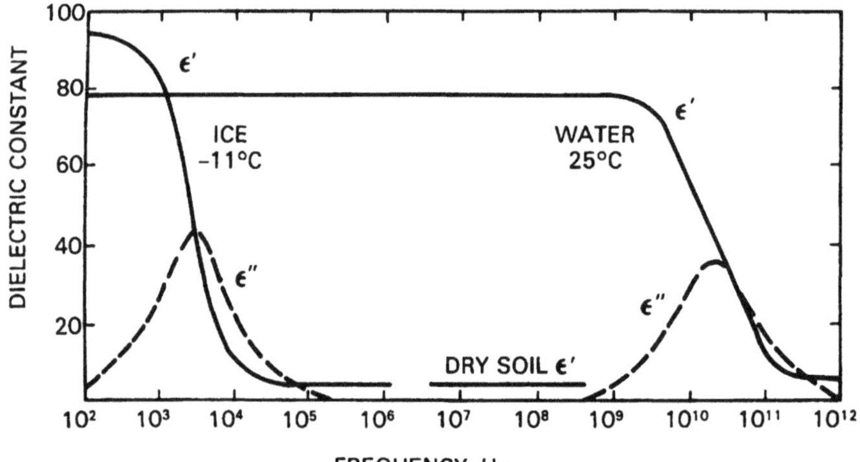

Figure 1. The dielectric relaxation of water and ice at microwave frequencies. Notice that the real parts of dielectric constant for ice and dry soils are quite comparable at frequencies > 100 KHz.

microwave radiometric measurements. Salinity modifies the values of ε'' significantly at $f \leqslant 5$ GHz, but this modification has a little effect on microwave emission from soil–water mixtures.

When water and soil are mixed together, ε' of the mixture increases dramatically from ~ 3 when dry to ~ 30 when wet. This is shown in Fig. 2 [31], where the laboratory measurements of both ε' and ε'' at $f = 1.4$ GHz are plotted against the volumetric moisture content (m_v) for three soils ranging from sand to heavy clay. In general, ε' for all three soils varies from ~ 3 to greater than 20, but there are differences in the variations. Initially, all three soils have regions of slow increase in ε' with m_v and then, after a transition value m_t, a region of sharp increase. This region of slow increase depends on soil texture, being wider for the heavier soils. It has been argued [31, 33] that initially added water to a soil is tightly bound to the surface of soil particles. These tightly bound water molecules are difficult to polarize and their dielectric behavior is similar to that of ice [31]. It is only after a few layers of water on the particle surface that the water molecules begin to behave like freely rotating dipoles and to produce a large change in the dielectric properties of wet soils. A strong correlation has been observed between the surface area per unit weight and the wilting point of soils [34].

Additional measurements ε' and ε'' over a frequency range of 1.4–18 GHz have been reported by Hallikainen *et al.* [32]. Figures 3 and 4 summarize some results from these measurements. The real and imaginary parts of dielectric constants measured at four different frequencies are shown in Fig. 3 for five soils of different textures ranging from sandy loam to silty clay. As expected, the sandy soils have the highest ε' at all frequencies. Of the soils measured, the soils highest in sand content have the least specific surface area and hence the lowest bound-water volume fraction. This corresponds to a small m_t defined above and the measurements are consistent with those of Wang and Schmugge [31].

Figure 4 shows the frequency behavior of the dielectric constant of moist soils at 4, 10, and 18 GHz. For all three soils of different textures, the results indicate that ε' decreases and ε'' increases with increasing frequency from 4 to 18 GHz. This suggests that, in this frequency range, the dielectric constant of a soil–water mixture varies with frequency at a rate similar to that of pure water shown in Fig. 1. At frequencies below 4 GHz the ionic conductivity becomes increasingly important, and ε'' may attain a minimum and increase with decreasing frequency. Measurements by Hallikainen *et al.* [32] suggest a minimum in ε'' in the vicinity of 3 GHz.

Two practical dielectric permittivity models, based on laboratory measurements of ε' and ε'', for soil–water mixtures have been developed [31, 33] that were frequently used for emission model calculations. The earlier model of Wang and Schmugge [31] expressed the complex dielectric constant (ε) of soil–water mixture as

$$\varepsilon = m_v\varepsilon_x + (P - m_v)\varepsilon_a + (1 - P)\varepsilon_r, \qquad m_v \leqslant m_t \tag{1}$$

with

$$\varepsilon_x = \varepsilon_i + (\varepsilon_w - \varepsilon_i)\frac{m_v}{m_t}\eta \tag{2}$$

and

$$\varepsilon = m_t\varepsilon_x + (m_v - m_t)\varepsilon_w + (P - m_v)\varepsilon_a + (1 - P)\varepsilon_r, \qquad m_v > m_t \tag{3}$$

with

$$\varepsilon_x = \varepsilon_i + (\varepsilon_w - \varepsilon_i)\eta. \tag{4}$$

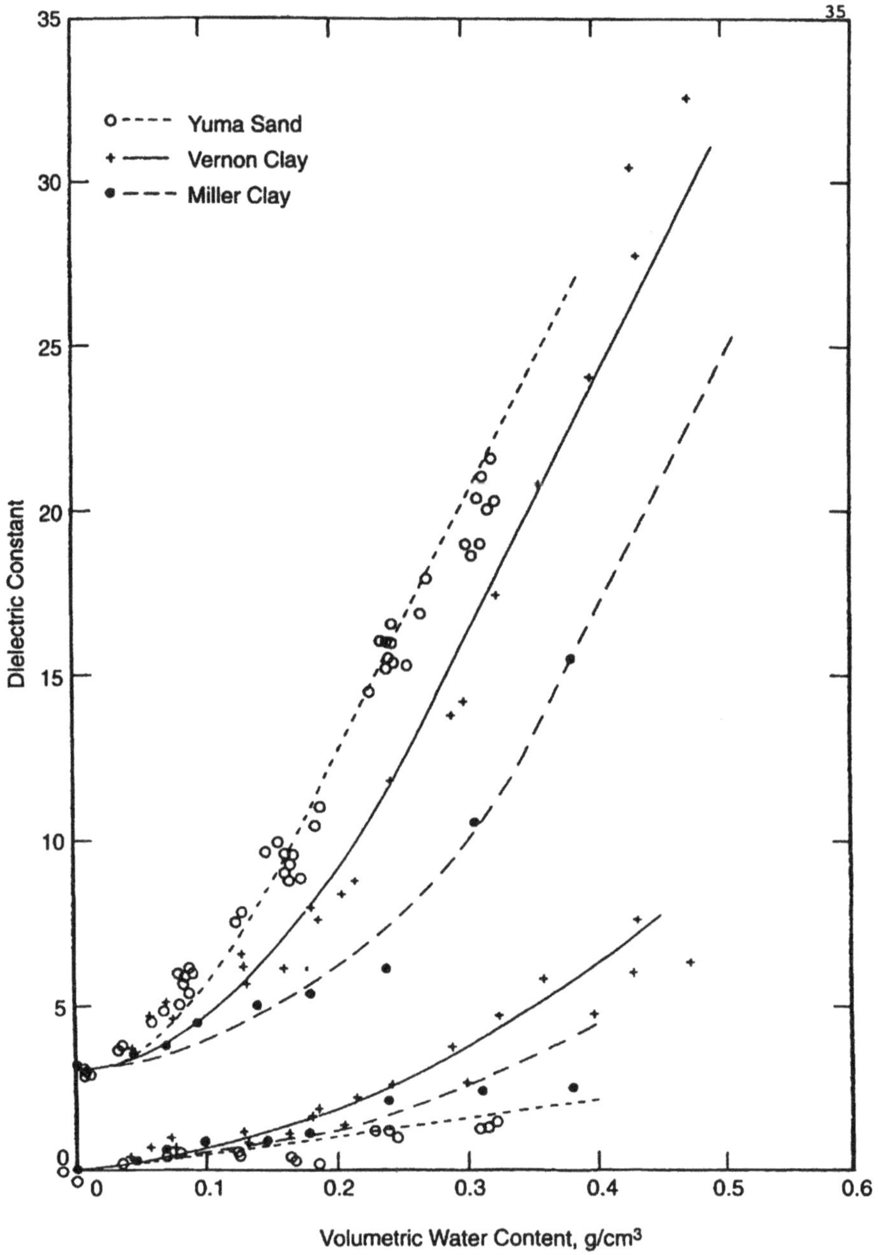

Figure 2. The dielectric constants measured at 1.4 GHz for three soils of different textures. The curves are derived from the empirical models of Wang and Schmugge [31].

Here P is the porosity of the dry soil. ε_a, ε_w, ε_r, and ε_i, in sequential order, are the dielectric constants of air, water, rock, and ice. ε_x stands for the dielectric constant of the initially adsorbed water. η is a parameter chosen to best fit Equations (1) and (3) to the experimental data. Both m_t and η were shown to depend on the wilting point

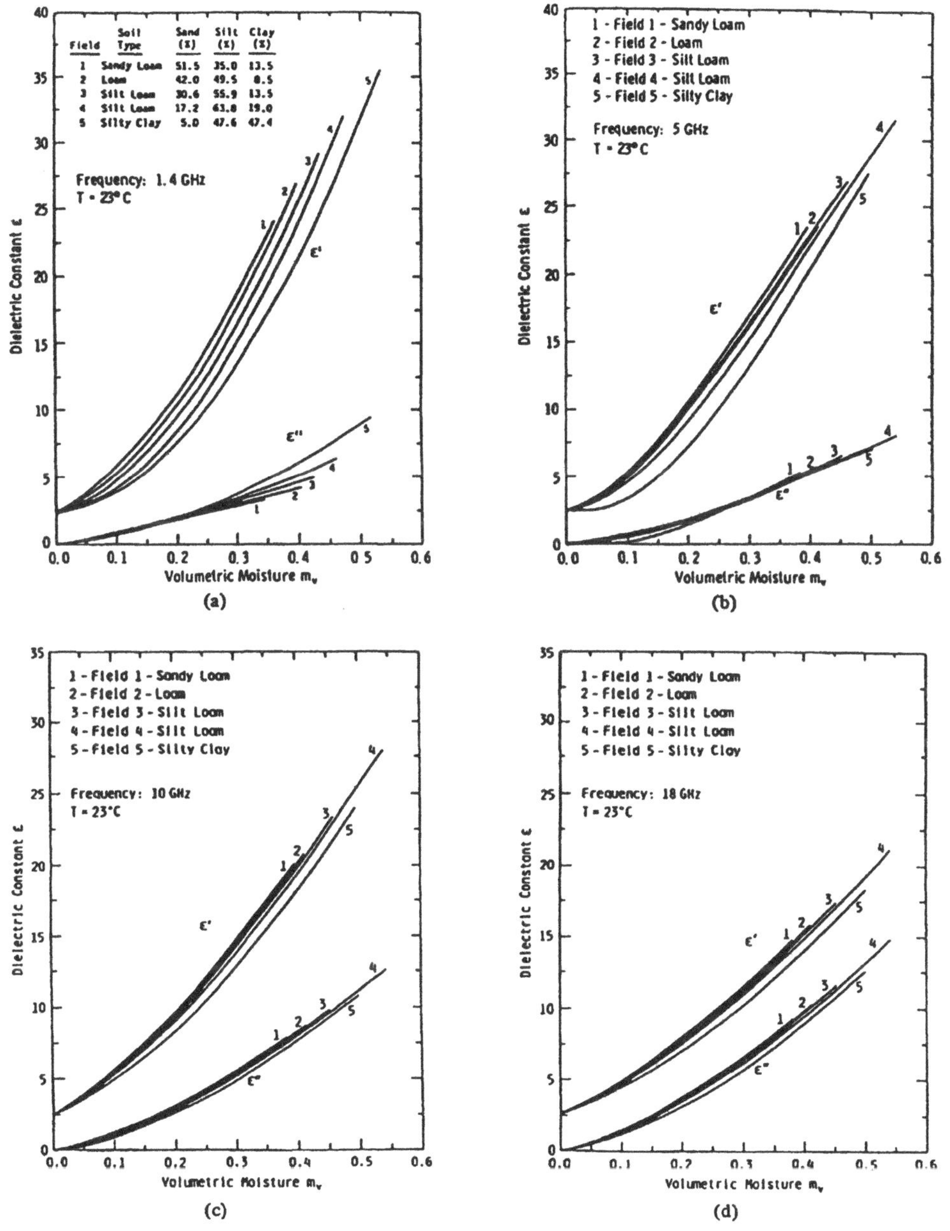

Figure 3. The dielectric constants measured for five soils of different textures at four frequencies by Hallikainen *et al.* [32]: (a) 1.4 GHz, (b) 5.0 GHz, (c) 10.0 GHz, and (d) 18.0 GHz.

of soils and are therefore soil–type dependent. For the imaginary part of the dielectric constant at low frequencies it is also necessary to add a term to account for conductivity loss [33, 34].

The dielectric model of Dobson *et al.* [33] was developed after additional laboratory measurements of ε' and ε'' were made for five different soils at frequencies between

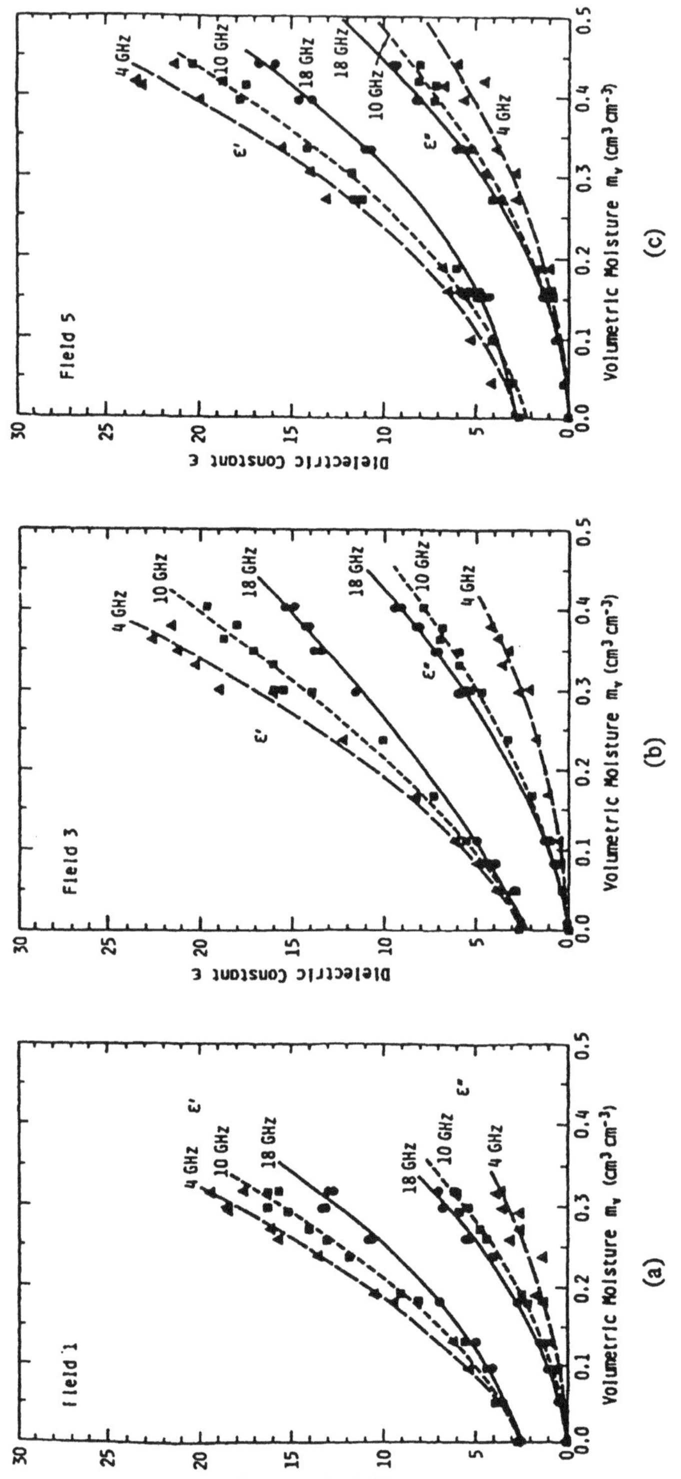

Figure 4. A comparison of dielectric constant measured at three frequencies of 4, 10, and 18 GHz for three fields of different soil textures.

1.4 and 18 GHz [32]. It is a four-component model based on an approach described by de Loor [39] for heterogeneous aqueous mixtures. In this model the complex dielectric constant is written as

$$\varepsilon = \frac{3\varepsilon_s + 2V_{fw}(\varepsilon_{fw} - \varepsilon_s) + 2V_{bw}(\varepsilon_{bw} - \varepsilon_s) + 2V_a(\varepsilon_a - \varepsilon_s)}{3 + V_{fw}\left(\frac{\varepsilon_s}{\varepsilon_{fw}} - 1\right)\left(\frac{\varepsilon_s}{\varepsilon_{bw}} - 1\right)\left(\frac{\varepsilon_s}{\varepsilon_a} - 1\right)} \tag{5}$$

where ε_s, ε_a, ε_{fw} and ε_{bw} are the dielectric constants of soil solids, air, free water, and bound water, respectively. V_a, V_{fw} and V_{bw} are volume fractions of air, free water and bound water respectively. ε_s is determined by an experimental fitting of data acquired from the new measurements [32] and is given by

$$\varepsilon_s = (1.01 + 0.44\rho_s) - 0.062 \tag{6}$$

which yields $\varepsilon_s \cong 4.7$ in a narrow range of specific density ρ_s of 2.56–2.70 g/cm^3 measured for the five soils [32]. $\varepsilon_a = 1.0$ and ε_{fw} is derived at a given frequency assuming a Debye-type relaxation as modified by Lane and Saxton [40] to account for ionic conductivity losses. The values of ε_{bw} have been analyzed in detail by Dobson *et al.* [33]. They concluded that $\varepsilon_{bw} = 35 - j15$ appeared to yield prediction errors within the measurement accuracy of Hallikainen *et al.* [32].

The input parameters to determine ε from Equation (5) are not readily available for specific soils. Therefore, Dobson *et al.* [33] also presented a more convenient empirical model as

$$\varepsilon_m^\alpha = 1 + \frac{\rho_b}{\rho_s}(\varepsilon_s^\alpha - 1) + m_v^\beta \varepsilon_{fw}^\alpha - m_v \tag{7}$$

and

$$m_v^\beta \varepsilon_{fw}^\alpha = V_{fw}\varepsilon_{fw}^\alpha + V_{bw}\varepsilon_{bw}^\alpha \tag{8}$$

where ρ_b is the soil bulk density. The shape factor $\alpha = 0.65$ is determined from the measured data set [32] and β is an empirical constant depending on the textural composition of the soil. ε_{fw} is calculated at a given frequency and temperature using a modified Debye equation

$$\varepsilon_{fw} = \varepsilon_{w\infty} + \frac{\varepsilon_{w0} - \varepsilon_{w\infty}}{1 + j2\pi f\tau_w} - j\frac{\rho_{eff}}{2\pi f\varepsilon_0}\frac{\rho_s - \rho_b}{\rho_s m_v}. \tag{9}$$

Here $\varepsilon_{w\infty} = 4.9$ is the high frequency limit of the dielectric constant of water, ε_{w0} the static dielectric constant of water, ε_0 the permittivity of free space and τ_w is the relaxation time of water. Using Equations (7) and (9), the values of soil–texture dependent coefficient β and effective conductivity σ_{eff} were further optimized for each soil. The resultant values are related to soil texture of the five soils by

$$\beta_{\varepsilon'} = (127.48 - 0.519\,S - 0.152\,C)/100 \tag{10}$$

$$\beta_{\varepsilon''} = (133.797 - 0.603\,S - 0.166\,C)/100 \tag{11}$$

$$\sigma_{eff} = -1.645 + 1.939\,\rho_b - 0.02013\,S + 0.01584\,C \tag{12}$$

where S and C are the percentage of sand and clay respectively.

Both dielectric models of Wang and Schmugge [31] and Dobson *et al.* [33] have been used in the past in the calculations of microwave emission from soils [6, 11, 14, 41–43]. The models also allow the study of the sensitivity of a soil's dielectric properties on parameters like density, texture, and salinity. For example, Schmugge [41] applied the model of Dobson *et al.* [33] to study the dependence of the dielectric constant on density for a sandy loam soil. He found that change in a range of bulk density from 1.0 to 1.5 g/cm^3 could produce a 15% to 20% spread of values in ε' for wet soils. Jackson and O'Neill [11] used both dielectric models in their study of the salinity effect on microwave emission from soils.

3. EMISSION MODELS

Two general types of emission models, one coherent and the other non-coherent, have been developed to interpret microwave radiometric observations. In the coherent models radiation emitted from soils is treated as propagation of electromagnetic waves, with proper accounting of the phases of electric and magnetic fields, and boundary conditions are imposed on these fields to calculate the intensity of emitted radiation. In the non-coherent models, the intensity of the radiation is treated directly and boundary conditions are imposed on the intensity itself. The results of calculations from these two groups of models sometimes can differ by 20 K or more at low frequencies < 5 GHz [30]. This is caused by the interference of the phases of the electromagnetic waves in the coherent models, which can increase or decrease the emitted intensity relative to that calculated from the non-coherent models. The difference between these two groups of models is generally less than 5 K at high frequencies (> 10 GHz) [30]. The non-coherent models are mathematically and computationally simpler, but they are generally less accurate than the coherent models. Scattering can modify the emissivity for soils with pebbles of dimension comparable to the wavelength [44]; the effect of scattering is to decrease the emissivity. Although most soils contain only a small fraction of pebbles, one needs to be careful in the interpretation of surface microwave emission for possible effect of scattering in some specific regions.

3.1. Coherent models

Two coherent models which have been compared and tested against observations were developed by Njoku and Kong [26] and Wilheit [27]. These models considered the propagation of electromagnetic waves in the soil media, given the profiles of temperature and dielectric constant (soil moisture). The Njoku and Kong model is based on the theory of electromagnetic fluctuation developed by Stogryn [45] and extended later by Tsang *et al.* [28] for microwave emission from a stratified medium with subsurface temperature variation. The brightness temperatures at incidence angle θ for horizontal ($T_{\rm bh}$) and vertical ($T_{\rm bv}$) polarizations are given by the following equations:

$$T_{\rm bh} = \frac{k}{\cos\theta} \int_{-\infty}^{0} {\rm d}z\, T(z)\varepsilon''(z)\, |\psi(z)|^2 \tag{13}$$

$$T_{\rm bv} = \frac{1}{k\cos\theta} \int_{-\infty}^{0} {\rm d}z\, T(z)\varepsilon''(z) \left\{ \left| \frac{1}{\varepsilon(z)} \frac{{\rm d}\phi(z)}{{\rm d}z} \right|^2 + \left| \frac{k_x \phi(z)}{\varepsilon(z)} \right|^2 \right\} \tag{14}$$

where $\varepsilon(z) = \varepsilon'(z) + j\varepsilon''(z)$ is the dielectric constant, k is free space wave number $(= 2\pi/\lambda)$, k_x is $k \sin\theta$, $T(z)$ is the vertical temperature profile. The functions $\psi(z)$ and $\phi(z)$ are the solutions of the following differential equations

$$\frac{d^2\psi}{d^2z} + \left[\varepsilon(z)k^2 - k_x^2\right]\psi = 0 \tag{15}$$

$$\varepsilon(z)\frac{d}{dz}\left[\frac{1}{\varepsilon(z)}\frac{d\phi}{dz}\right] + \left[\varepsilon(z)k^2 - k_x^2\right]\phi = 0 \tag{16}$$

which are to be solved with the boundary conditions:

$$\left\{\frac{d\psi}{dz} + ik\cos\theta(2-\psi)\right\}_{z=0} = 0 \tag{17}$$

$$\left\{\frac{d\phi}{dz} + i\varepsilon(z)k\cos\theta(2-\phi)\right\}_{z=0} = 0. \tag{18}$$

The above differential equations were solved by the WKB approximation, and then the brightness temperatures were obtained by numerical integration. Explicit equations for horizontally and vertically polarized brightness temperature as summation of algebraic terms are given in Njoku and Kong [26]. The calculated (T_b at nadir) values at 10.69 and 31.4 GHz were compared with observations in Fig. 5 for a range of soil moisture and temperature conditions; the differences were generally less than 10 K, which was suggested to be due to small-scale surface roughness. Njoku and Kong showed that the above equations for T_b reduce to those of a radiative transfer model when dielectric constant (or soil moisture) varies slowly with depth or when the observations are at high microwave frequencies (> 5 GHz) as shown in Fig. 6.

In Wilheit's model [27], a soil is treated as a layered dielectric medium with known temperature distribution. Solutions of Maxwell's equations, with boundary conditions matching the tangential components of electric and magnetic fields at the interfaces of each layer, were used to calculate the field vectors for each layer. The electromagnetic energy absorbed by each layer f_i ($i = 1, 2, 3, \ldots$ are the layer index) was then calculated from these field vectors. This absorbed energy is a function of angle, polarization and dielectric constant. If the temperature of layer i is T_i then thermal equilibrium of the layer will require that the energy emitted from the layer i be $(f_i T_i)$. The brightness temperature T_b is calculated as the sum of emitted energy from all soil layers,

$$T_b = \sum_{i=1}^{N} f_i T_i. \tag{19}$$

The basic equations for calculating f_i and the related FORTRAN codes are given in Wilheit [27]. The effective reflectivity of soil can be calculated as

$$R_p = 1 - \sum_{i=1}^{N} f_i \tag{20}$$

where subscript p stands for either v or h polarization.

While Njoku and Kong's model is formulated in terms of dielectric constant varying continuously within the soil, Wilheit's model is formulated in terms of discrete changes in dielectric constant (i.e. a layered dielectric medium). To the extent that a continuous dielectric medium can be approximated by a layered dielectric with appropriate choice of layer thickness, it is possible to make a direct comparison of the results obtained from these two models.

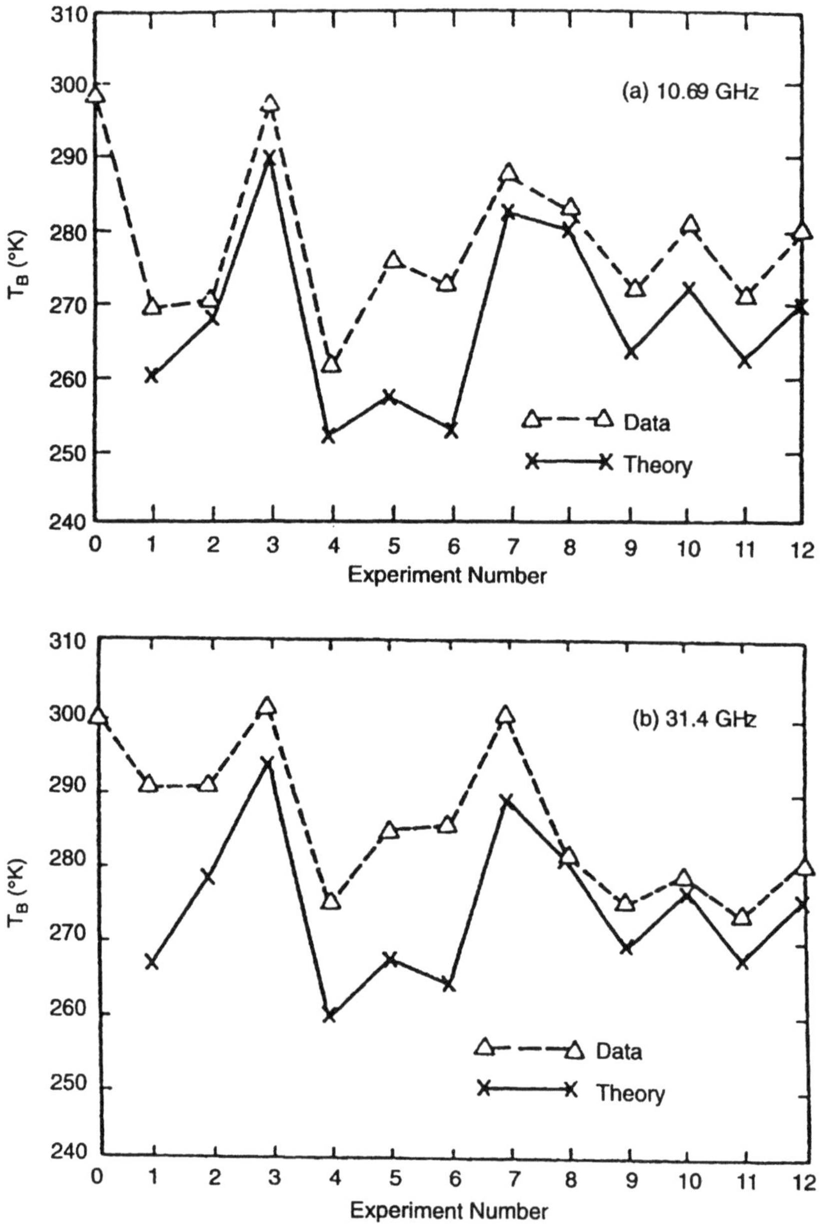

Figure 5. A comparison of the measured and calculated brightness temperatures at the frequencies of (a) 10.7 GHz and (b) 31.4 GHz. The figure was copied from Njoku and Kong [26].

3.2. Non-coherent models

Non-coherent models are based on the solution of the radiative transfer equation. Burke *et al.* [29] proposed a radiative transfer method for calculating T_b of a layer of dielectric medium (i.e. soil) with known temperature distribution. This model was simplified by Schmugge and Choudhury [30] by ignoring all reflections at the interfaces of dielectric

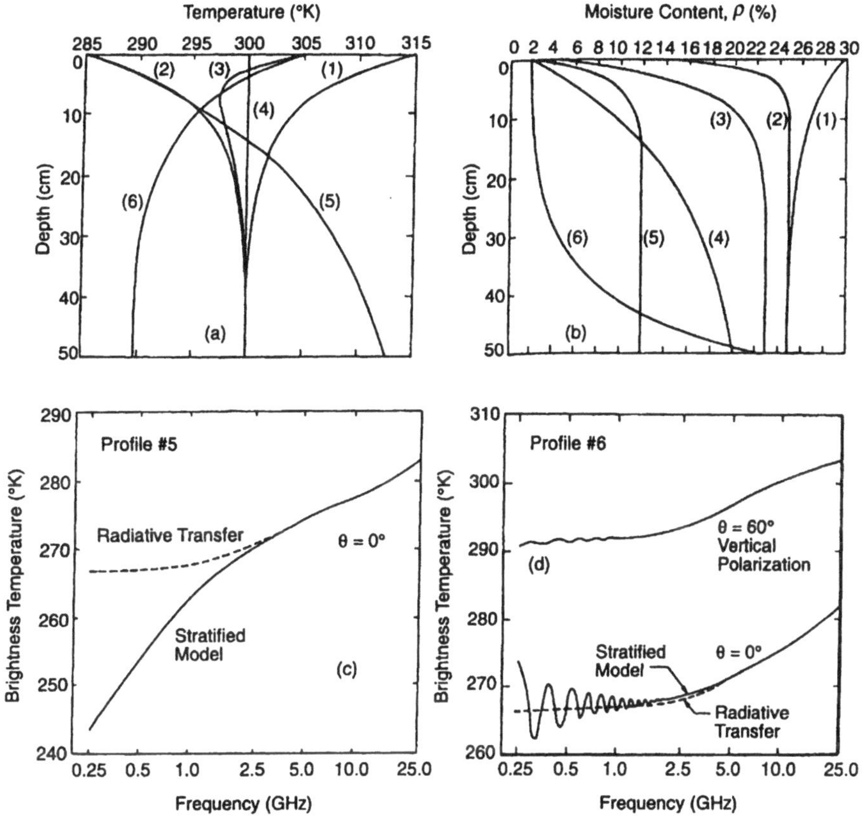

Figure 6. A comparison of T_b values calculated from coherent and non-coherent models at various frequencies [26].

layers within the soil; reflection occurs only at the soil–air interface. The model of Burke *et al.* [29] gives the following expression for T_b at nadir:

$$T_b = \sum_{i=1}^{N} \left\{ T_i \left[1 - \exp(-\gamma_i \Delta z_i)\right] \left[1 + R_{i+1} \exp(-\gamma_i \Delta z_i)\right] \times \prod_{j=1}^{i} \left[1 - R_j\right] \exp\left[- \sum_{j=2}^{i} \gamma_{j-1} \Delta z_{j-1} \right] \right\} \tag{21}$$

where T_i is the temperature of layer i of thickness Δz_i, γ_i is absorption coefficient and R_i is reflectivity at the interface of layers i and $i-1$ as may be calculated from the Fresnel equation. The absorption coefficient γ_i is obtained from the following equation:

$$\gamma_i = 4\pi A_i / \lambda \tag{22}$$

$$A_i = \varepsilon_i'' / 2B_i \tag{23}$$

$$B_i = \left[\frac{1}{2} \{\varepsilon_i' - \sin^2\theta\} \left\{ 1 + \left(1 + \frac{(\varepsilon_i'')^2}{(\varepsilon_i' - \sin^2\theta)^2} \right)^{1/2} \right\} \right]^{1/2}. \tag{24}$$

The simplified model of Schmugge and Choudhury [30] is obtained when $R_i = 0$ for all i except $i = 1$.

The thickness of soil layers could be chosen to approximate variations of dielectric constant as closely as possible. Burke *et al.* [29] had chosen 5 layers of thicknesses 10, 20, 30, 40 and 60 mm, while Schmugge and Choudhury [30] had used layer thicknesses which increased exponentially from the surface.

3.3. Inter-comparison of the emission models

Schmugge and Choudhury [30] gave results of an inter-comparison of the above emission models for a range of soil moisture and temperature profiles. These results were summarized in Fig. 7 and Table 1. The T_b values calculated by coherent or non-coherent models in general did not differ by more than 5 K, with the exception of cases where soil profiles had fairly steep gradients near the surface. The 1.4 GHz T_b values calculated from the non-coherent models could be $\sim$ 20 K higher compared to those calculated from the coherent models for profiles with fairly steep gradients. However, a similar comparison at 19.4 GHz with the same steep profiles gave differences in T_b values of less than 5 K. These differences in the calculated T_b values appear because the reflectivity at the soil–air interface is the controlling factor in the non-coherent models; and in the coherent models, the reflectivity of a soil layer of thickness about one-tenth of the wavelength determines the surface microwave emission. Thus, the differences in T_b values from the two groups of models are rather small at high frequencies, but could be highly significant at low frequencies.

3.4. Surface roughness effect

The above models show that the emissivity e can be calculated to a good approximation from Fresnel equations using dielectric constant corresponding to soil moisture within an effective surface layer (microwave soil moisture sampling depth). Alternatively, soil moisture can be estimated when e is determined from the observed microwave emission. Applicability of Fresnel equations, however, requires that the air–soil interface be smooth within the area viewed by a radiometer. If the surface is not smooth, e calculated for a given soil moisture by Fresnel equations would be modified in two ways: (1) the surface area contributing to the energy received by a radiometer increases with surface roughness, and (2) surface roughness modifies the energy emitted in orthogonal

Table 1.
Comparison of emission models for brightness temperature at nadir at 1.4 GHz calculated for profiles shown in Fig. 7

Profile no.	Radiative transfer	Burke *et al.* [29]	Wilheit [27]
3	177.3	177.5	173.2
4	203.9	204.1	202.7
5	250.7	249.2	229.4
6	262.0	260.4	242.1
8	270.3	269.0	256.4
12	271.6	270.5	268.3
16	273.6	272.6	273.0
23	273.9	272.9	276.1
37	278.9	278.0	279.1

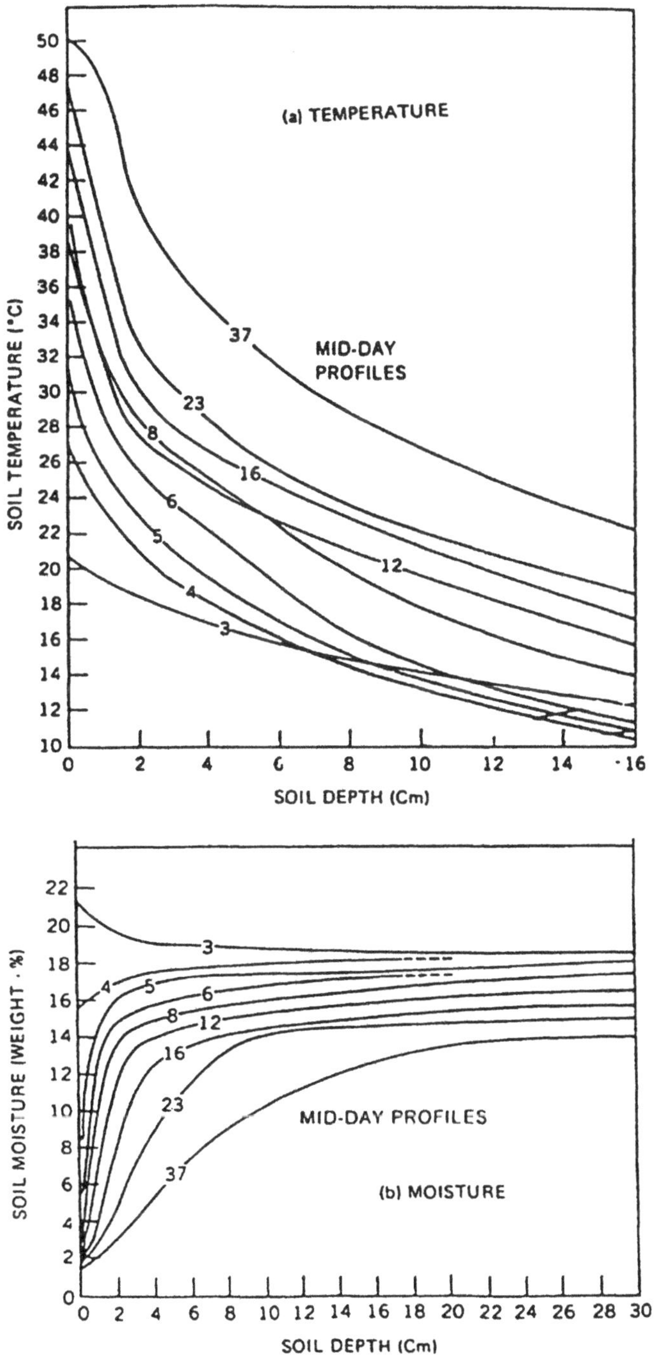

Figure 7. The soil temperature and moisture profiles that were used for brightness temperature calculations from several emission models [30].

polarization states. The net effect from these modifications is reflected in higher T_b values in both polarizations and a smaller difference between T_{bv} and T_{bh}.

Rigorous models to account for the effect of surface roughness on microwave emission have been developed by Tsang and Newton [46] and Fung and Eom [47] and others [48 and references therein]. The basic surface parameters needed to quantify the roughness effect are spatial autocorrelation length and standard deviation of roughness height [49]. The following semi-empirical models proposed by Choudhury *et al.* [15] and Wang and Choudhury [50] have been found to provide a reasonable description of field observations:

$$R'_h(\theta) = \left[(1-Q)R_h(\theta) + QR_v(\theta)\right]\exp(-\delta\cos^2\theta) \tag{25}$$

$$R'_v(\theta) = \left[(1-Q)R_v(\theta) + QR_h(\theta)\right]\exp(-\delta\cos^2\theta) \tag{26}$$

where $R_h(\theta)$ and $R_v(\theta)$ are smooth surface Fresnel reflectivities at incidence angle θ, for horizontal and vertical polarizations respectively. Q and δ are empirical roughness parameters that assume the value 0 for a smooth surface. No quantitative relation has been established between these parameters and measurable surface characteristics. The parameter δ essentially describes the effect of surface roughness to increase e, while the parameter Q, the mixing of energies in orthogonal polarization states. In an analysis of experimental data acquired over rough fields, it was found that the results agreed better with model calculations when $\cos^2\theta$ was set to 1 in the exponential term in the above equations [14]. The horizontal and vertical reflectivities are equal at nadir, and the emissivity in this case is modified by the parameter δ only; i.e. e is increased by an amount

$$\Delta e = R(\theta = 0)\left[1 - \exp(-\delta)\right]. \tag{27}$$

Since $R(\theta = 0)$ increases with increasing soil moisture, the relative increase in e due to roughness is higher for wetter soils.

3.5. Vegetation effect

Vegetation attenuates microwave emission from soils and itself emits essentially depolarized microwave radiation. Estimation of soil moisture requires information about emission and attenuation characteristics of vegetation. Since interaction of microwave radiation with vegetation is being reviewed separately (see Kerr in this volume), only a brief discussion is given below. For soils covered with vegetation, the emissivity e_{veg} becomes [20]

$$e_{veg} = 1 + [e - 1]\exp(-2\tau) \tag{28}$$

where e is again the bare soil emissivity and τ is the opacity of vegetated medium. From microwave observations at low frequencies (e.g. 1–5 GHz), τ is further related to the vegetation water content W by [51]

$$\tau = bW. \tag{29}$$

The parameter b depends on plant characteristics (primarily leaf size) and is almost proportional to the frequency of observations [51]. Measurements at 1.4 GHz over a variety of crops gave $b \approx 0.1$ [51].

4. OBSERVATIONS

Microwave radiometric observations of land surface have been conducted with radiometers mounted on the mobile towers at the ground level as well as the aircraft and satellite platforms [1–25]. Measurements from spaceborne microwave radiometers, owing to the limitation on spatial resolution, could only offer a qualitative interpretation of the large-scale surface emission. They in general could not provide quantitative analyses of the effects of soil texture, roughness and vegetation on surface microwave emission. Recent effort by Choudhury *et al.* [52] suggested the potential use of the polarization difference at 37 GHz measured by the Scanning Multichannel Microwave Radiometer (SMMR) aboard the Nimbus-7 satellite as the Microwave Vegetation Index (MVI). It is not clear, however, which vegetation parameter (e.g. leaf area index, percentage cover or wet biomass) MVI mainly responded to. Studies on the details of the radiometric response to surface parameters could only be conducted rigorously with radiometers at the ground level, as asserted by the results reported in the literature [5, 7, 53, 54]. Because of the difference in the emphasis of these measurement approaches, the discussion in this section is more conveniently divided into three categories below:

4.1. Ground-level measurements.

The measurements at the ground level were mostly made by fixed-beam, multiple-frequency microwave radiometers mounted on a mobile tower [4–9, 11, 13, 14]. These measurements were generally well controlled and the surface treatments well defined so that the effects of important surface parameters like soil texture, roughness, row structure, and vegetation could be studied in great details [8, 9, 14, 15, 17, 54]. The major emphasis was placed on the exploration of efficient passive microwave techniques for remote sensing of soil moisture. Since the dielectric contrast between water and dry soils is most conspicuous at low frequencies, the measurements are mostly confined to the frequency range of 0.6–10 GHz. Typical results from these efforts are displayed in Figs 8 through 14.

Figure 8 shows the variation of normalized brightness temperature with incidence angle (θ) measured from a 21.4 cm ($\sim$ 1.4 GHz) radiometer measurement over a smooth bare field for four different soil moisture contents sampled in the top 5 cm layer [9]. The normalized brightness temperature (T_n) is simply the measured brightness temperature divided by the physical temperature of the surface soil. This was done to minimize the effect of the soil temperature on the microwave radiometric measurements; the strong soil temperature effect on the measured brightness has been demonstrated by the diurnal measurements of Njoku and O'Neill [4]. The family of curves in the figure give both vertical and horizontal polarizations of T_n values in the θ range of 0°–50° degrees. These curves are well behaved over the soil moisture (m_v) range of 0.08–0.35 g/cm^3 and are in excellent agreement with the expected results based on established smooth surface emission models according to Newton and Rouse [9]. The excellent agreement between the radiometric measurements over smooth bare fields and emission models is also reported by Wang *et al.* [6].

Figure 9 shows the effects of soil moisture, soil texture and surface roughness on microwave emission. The figure was taken from Wang *et al.* [14] based on the results of multiple-frequency measurements at $\theta = 20°$ over three bare fields of different surface roughnesses and soil textures. The estimated field capacity for sandy loam is

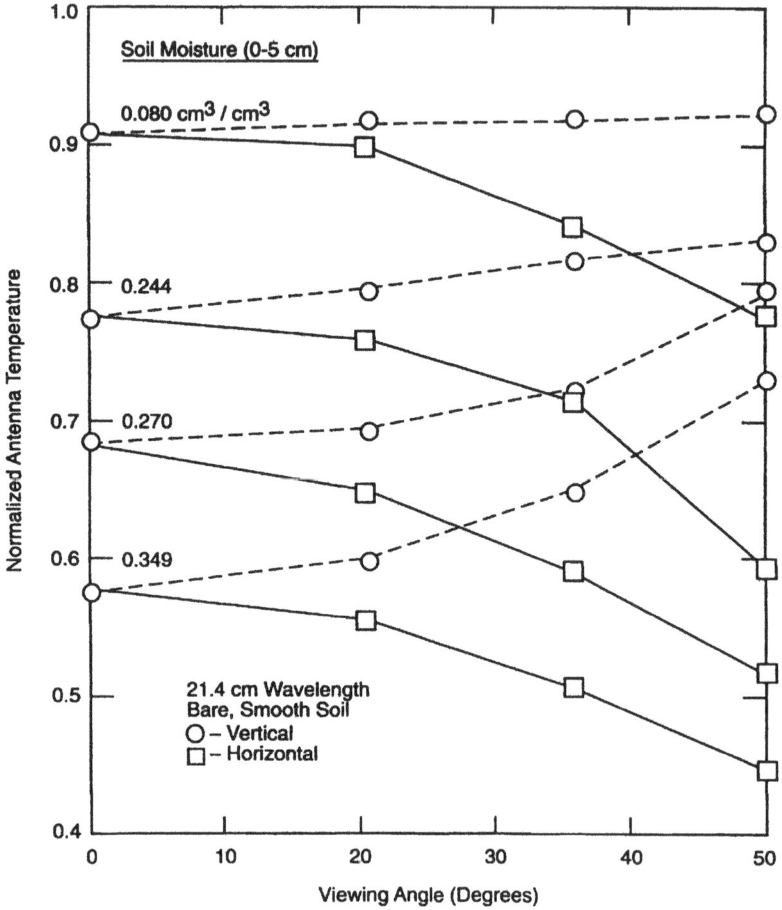

Figure 8. Measured 1.4 GHz emission of bare smooth soil as a function of incidence angle for various soil moisture conditions [9].

$\sim 0.24\ \mathrm{g/cm^3}$ while that for the silty loam is $\sim 0.35\ \mathrm{g/cm^3}$; and the measured RMS roughness heights (d) for the three different fields are approximately 0.21, 0.73 and 2.45 cm [14]. In the figure the measured T_n values are plotted as a function of m_v for three different frequencies of 1.4, 5.0 and 10.7 GHz. The m_v values for the 1.4 GHz measurements are averages over the top 2.5 cm layer, while those for the 5 and 10.7 GHz in plots b and c are averages over the top 0.5 cm layer. The choice of layer thickness is based on both the calculations [27] and the observations [16, 53] that the radiometric moisture sampling depth is approximately $0.06-0.2\,\lambda$, where λ is the wavelength of observation. It is quite clear from the figure that for each frequency of measurements the data taken over three different fields are separable into three different groups depending on soil type and surface roughness. The straight lines and the associated slopes and correlation coefficients are the results of linear regression applied to each group of the measured data. The two nearly parallel but well separated lines in each of the three plots display the effect of soil texture. This observed small difference in the rate of T_n decrease with m_v for the two different soils is consistent with the results discussed in

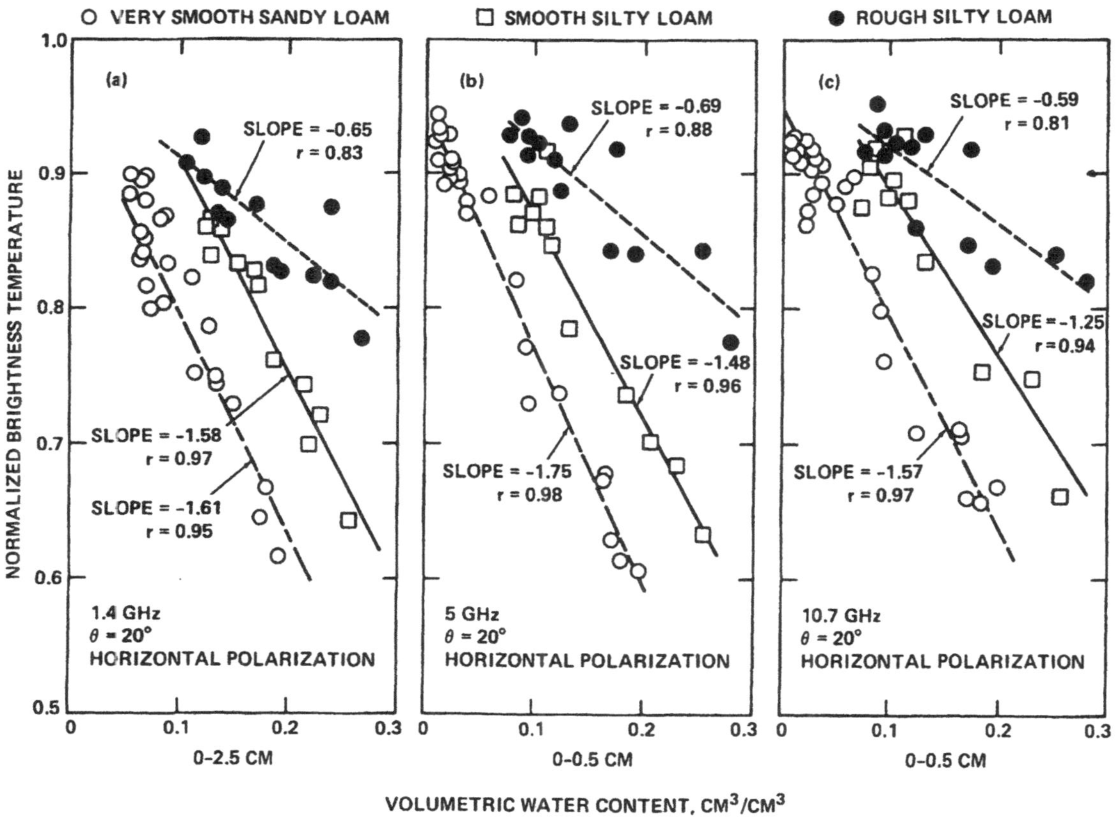

Figure 9. Measured microwave emissions at 1.4 GHz (a), 5.0 GHz (b), and 10.7 GHz (c) as a function of volumetric soil moisture for fields with different soil textures and surface roughnesses [14]. The straight lines are results of linear regression for different groups of data.

Section 2 that the rate of increase of dielectric permittivity with m_v is nearly soil type independent beyond the transition moisture content.

The presence of roughness generally increases surface microwave emission as clearly demonstrated in Fig. 9 by the two groups of data acquired from the two silty loam fields of different roughness heights. The slopes of the regression lines for the rough field are appreciably shallower than those for the smooth fields. Similar results, as shown in Fig. 10, were obtained by Choudhury *et al.* [15] from truck mounted radiometer measurements over three fields of different surface roughnesses. Here the measured brightness temperature T_b at 1.4 GHz is plotted against the gravimetric soil moisture sampled in the top 2 cm layer. Both measured and calculated results are displayed in plots a, b and c of the figure for smooth, medium rough, and rough fields with measured d's of 0.9, 2.6, and 4.3 cm respectively. In general, the measured and calculated T_b values are in good agreement and the roughness parameter δ [15] increases with increasing roughness as expected from the measured d's. The emission model of Choudhury *et al.* [15] relates δ to d through the equation,

$$\delta = 4d^2(2\pi/\lambda)^2. \tag{30}$$

The d's estimated from this equation at 1.4 GHz based on the δ values in Fig. 10 are 0, 0.9, and 1.2 cm for the smooth, medium rough, and rough fields respectively. These estimated d values are about or more than three times smaller than the measured d's reported by Choudhury *et al.* [15]. It is highly conceivable that the measured d's quoted by these authors are roughness heights of these fields. An independent calculation of the same roughness profiles by Wang [55] gave the measured d values of 0.38, 1.10, and 2.15 cm for the smooth, medium rough, and rough fields respectively.

Another effect closely related to surface roughness is the row structure frequently observed in agricultural fields. This effect modifies surface microwave emission in both vertical and horizontal polarizations. Figure 11, taken from Wang *et al.* [8], shows an example of the effect from both field measurements and numerical simulations at 1.4 GHz frequency. Here the brightness temperature T_b is plotted against the incidence angle for two different cases defined by the plane of the radiometer incidence angle scan; (a) the plane was parallel to the row direction and (b) the plane was perpendicular to the row direction. The T_b values measured at vertical and horizontal polarizations are denoted by solid and open circles respectively; the results of numerical simulations are given by the smooth curves [8]. It is clear that the presence of row structure changes the microwave emission in both polarizations and these changes depend on the azimuthal angle. This effect was experimentally verified by Promes *et al.* [17]. A rigorous theoretical model to account for this effect was developed by Kong *et al.* [56].

Another important parameter that merits discussion is the microwave sampling depth of soils. Most research results appeared in the literature [16, 53] show that microwave emission from soils, even at 21 cm wavelength (1.4 GHz), predominantly originates from the top 0–5 cm layer. Figures 12 and 13 give some experimental results leading to this conclusion. Figure 12 , fom Newton *et al.* [16], shows a comparison of dry-down curves measured at 0–2, 0–5, and 0–9 cm layers and those estimated from radiometric measurements at 1.4, 5.0 and 10.7 GHz for the 0–2 cm layer. Clearly, the estimated and measured gravimetric soil moisture values are best matched at the 0–2 cm layer. Figure 13 shows a comparison of m_v's estimated from other independent measurements [53] at 1.4 and 5.0 GHz and those measured at the top 2.5 cm and 0.5 cm layers. The estimated

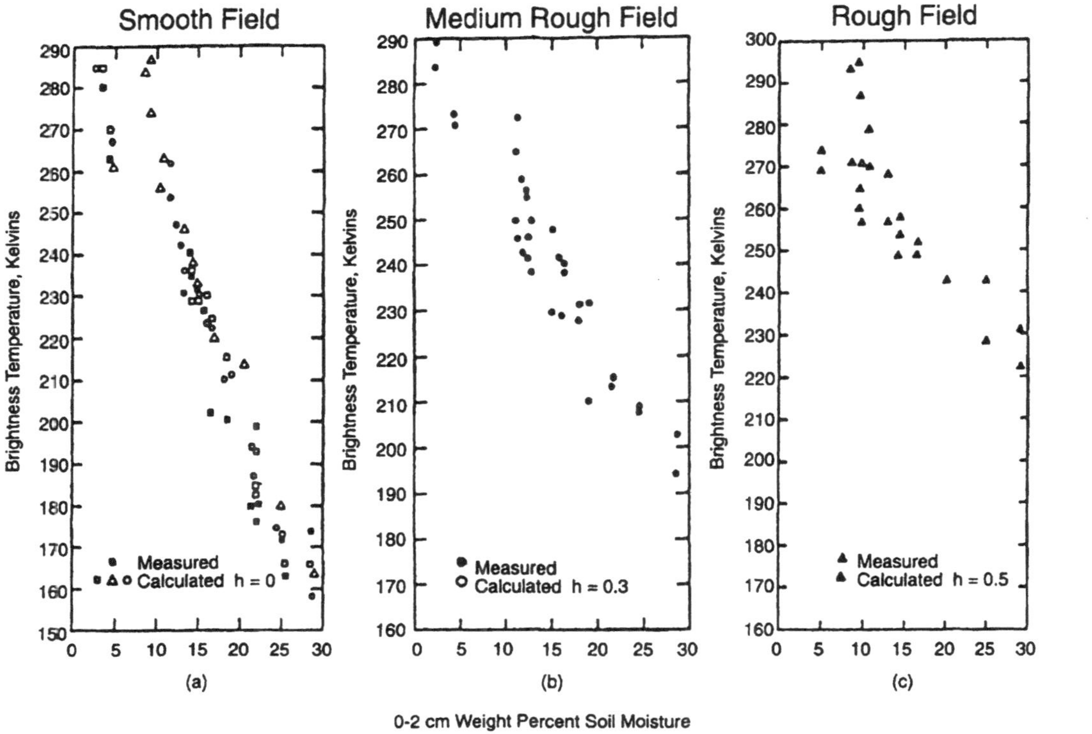

Figure 10. A comparison of calculated and measured values of 1.4 GHz brightness temperatures for fields with three levels of surface roughness [15]. Different values for roughness parameter δ are needed to match calculations with measurements.

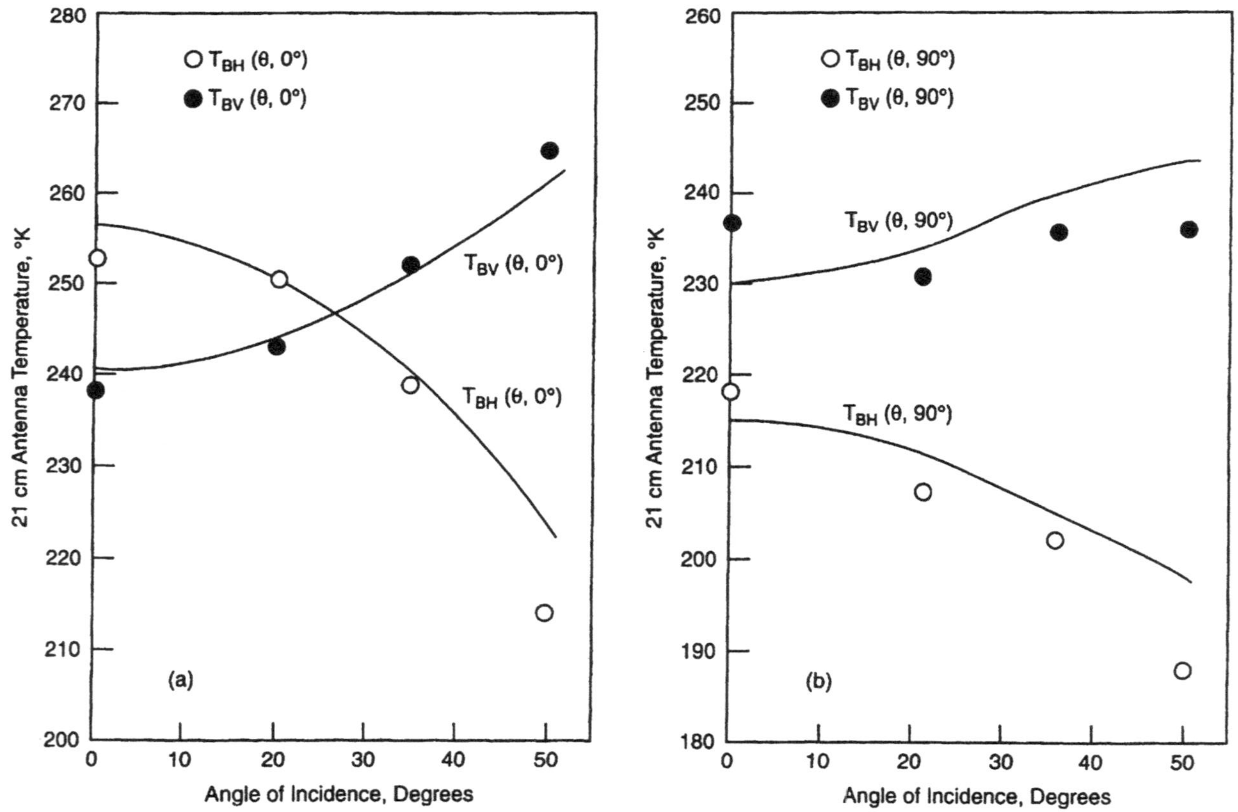

Figure 11. A comparison of the measured and calculated brightness temperatures as a function of incidence angle for (a) plane of measurements parallel to row direction and (b) plane of measurements perpendicular to row direction. The figure was obtained by Wang *et al.* [8].

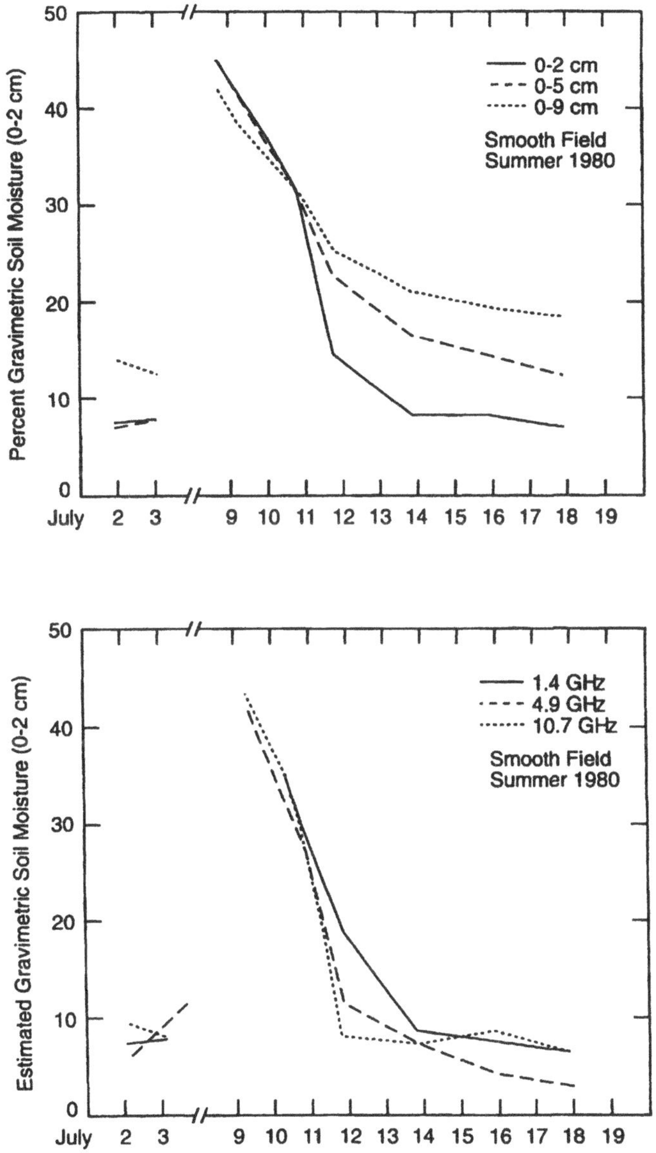

Figure 12. A comparison of measured (a) and estimated (b) gravimetric soil moisture values versus time for a smooth field. Date were obtained from Newton *et al.* [16].

values from the 1.4 and 5.0 GHz measurements correspond well with the measured ones in the top 2.5 and 0.5 cm layers respectively. A numerical analysis by Wilheit [27] also suggests a sampling depth on the order of $\sim 0.1\lambda$.

Jackson and O'Neill [11] and O'Neill and Jackson [13] have also examined the effects of saline and organic matter contents on microwave emission from soils from measurements by truck-mounted microwave radiometers. They found that both of these effects made a negligible contribution to soils' microwave emission.

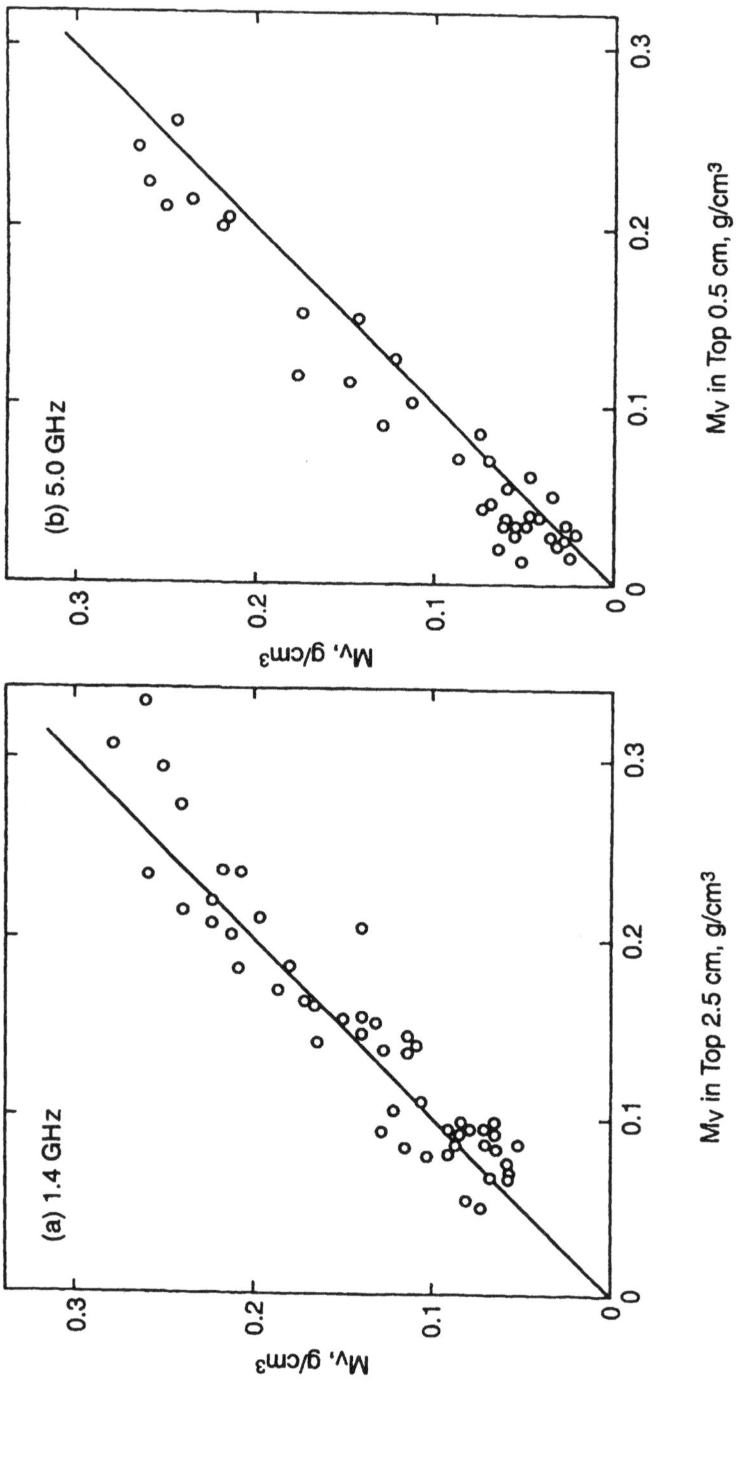

Figure 13. A comparison of volumetric soil moisture contents estimated from the measured emissivities at 1.4 GHz (a) and 5.0 GHz (b) and those measured from the top 2.5 and 0.5 cm layers [53].

4.2. Aircraft measurements

Results from airborne microwave radiometric measurements [1–3, 18–20] generally show a similar dependence of T_b on m_v established from ground-level measurements discussed above. The effects of soil texture and surface roughness on T_b were also observed by airborne radiometric measurements, as discussed respectively by Schmugge [2] and Schmugge [41]. Figures 14 and 15 are derived from the results of these measurements. Figure 14 shows the regression between T_b's measured from an airborne 1.4 GHz radiometer and m_v sampled in the top 1 cm layer of soils along the aircraft flight path. The soils were classified into light, medium and heavy types according to their field capacity and a linear regression was applied to T_b and m_v for each type of soil in the figure [2]. It is clear that the effect of texture on microwave emission from soils is present from these airborne measurements, but was not as conspicuous as that observed from the ground-level measurements.

Figure 15 summarizes the effect of surface roughness on microwave emission observed from an airborne 1.4 GHz radiometer over a test site in Hand County, South Dakota [41]. The data here were segregated into three roughness categories based on the ground photographs of the individual fields. A linear regression was again applied to each of the data groups of different roughness categories. It was found that the regression slopes for T_b versus m_v were not significantly altered but that the roughest fields were about 8 K higher than the smoothest. These results point out that the actual magnitude of roughness effect under normal agricultural conditions may not be as great as the extremes observed in prepared fields for truck-mounted radiometer measurements.

Most of the aircraft radiometric observations over natural terrain have to deal with fields covered by vegetation [3, 20, 57, 58]. For example, a series of measurements were conducted between 1978–80 over native grass pastures in Oklahoma [3]. The surface soil moisture mapping during the First ISLSCP (International Satellite Land Surface Climatology Project) Field Experiment (FIFE) was made with the 1.4 GHz push-broom microwave radiometer (PBMR) over the natural tall grass pastures in and around the Konza Prairie Natural Research Area (KPNRA) [19, 20]. Figure 16 shows the results of both 1.4 and 5.0 GHz measurements from aircraft flights over the Oklahoma test site [3]. A comparison of the two plots in the figure displaying the variation of normalized T_b with m_v indicates a nearly 50% decrease in sensitivity at 5.0 GHz with respect to that at 1.4 GHz. Most of the data were obtained at a flight altitude of $\sim$ 300 m. The two data points acquired at a higher altitude ($\sim$ 3000 m) at 1.4 GHz suggests a radiometer sensitivity to soil moisture variation independent of spatial resolution.

The PBMR measurements over and around KPNRA during FIFE provide an estimation of m_v in the top 5 cm layer of soils. Figure 17 shows the distribution of soil moisture contours for two different days as derived from these measurements over two small adjacent watersheds in KPNRA [19]. A few heavy rainfalls occurred on the test site prior to May 28, 1987, and none occurred within ten days thereafter. The contour plot on May 28, therefore, gives the moisture distibution of saturated soils for the two watersheds, while the one on June 4 gives the moisture distribution after a week of dry-down. The two watersheds differed in surface treatments; the 1D watershed was burned every year and the 2D watershed, every other year (not burned in 1987). It was shown [18, 19] that the radiometric response to soil moisture variation was different for the two watersheds. Furthermore, the contours of May 28 do not show a distinct boundary between the two watersheds. After a week of dry-down, the transition between the two watersheds becomes distinct; a steep gradient in soil moisture develops

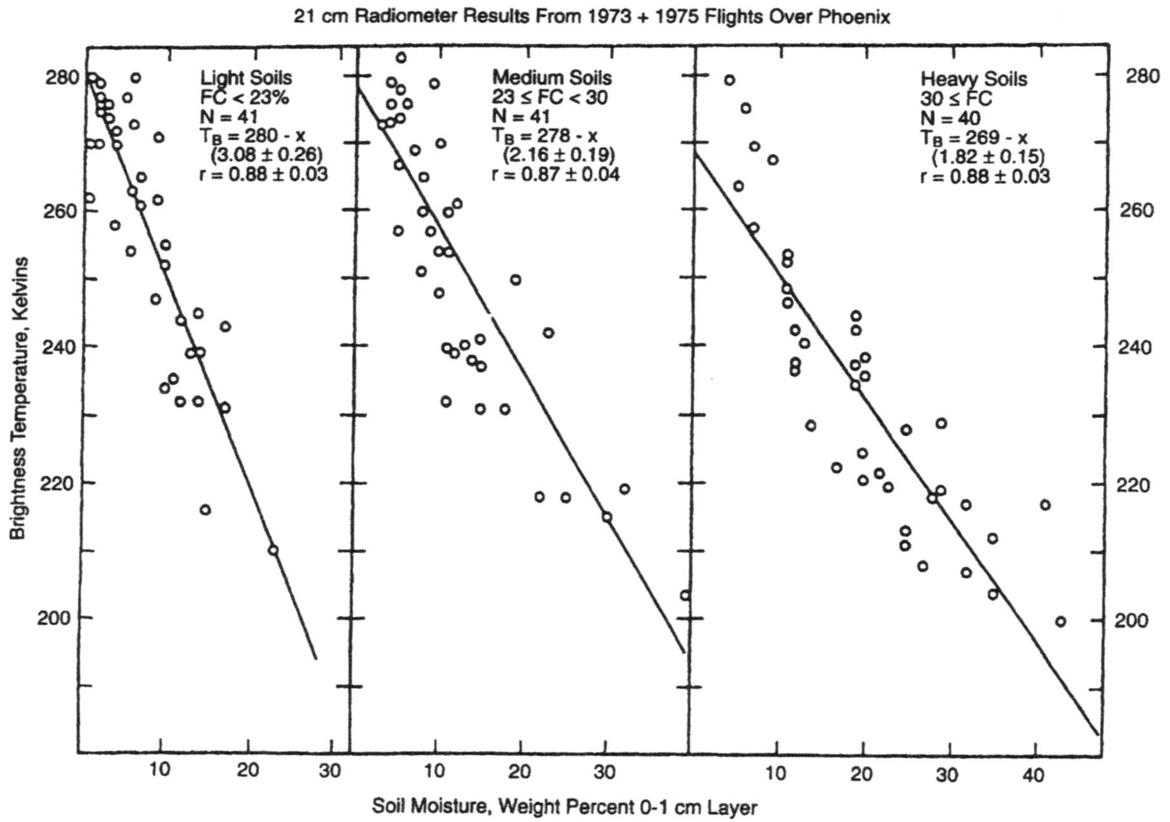

Figure 14. The plot of measured brightness temperatures at 1.4 GHz versus gravimetric soil moisture contents for soils divided into three classes of soil textures defined by the calculated values of field capacity (FC). The figure was obtained from Schmugge [2].

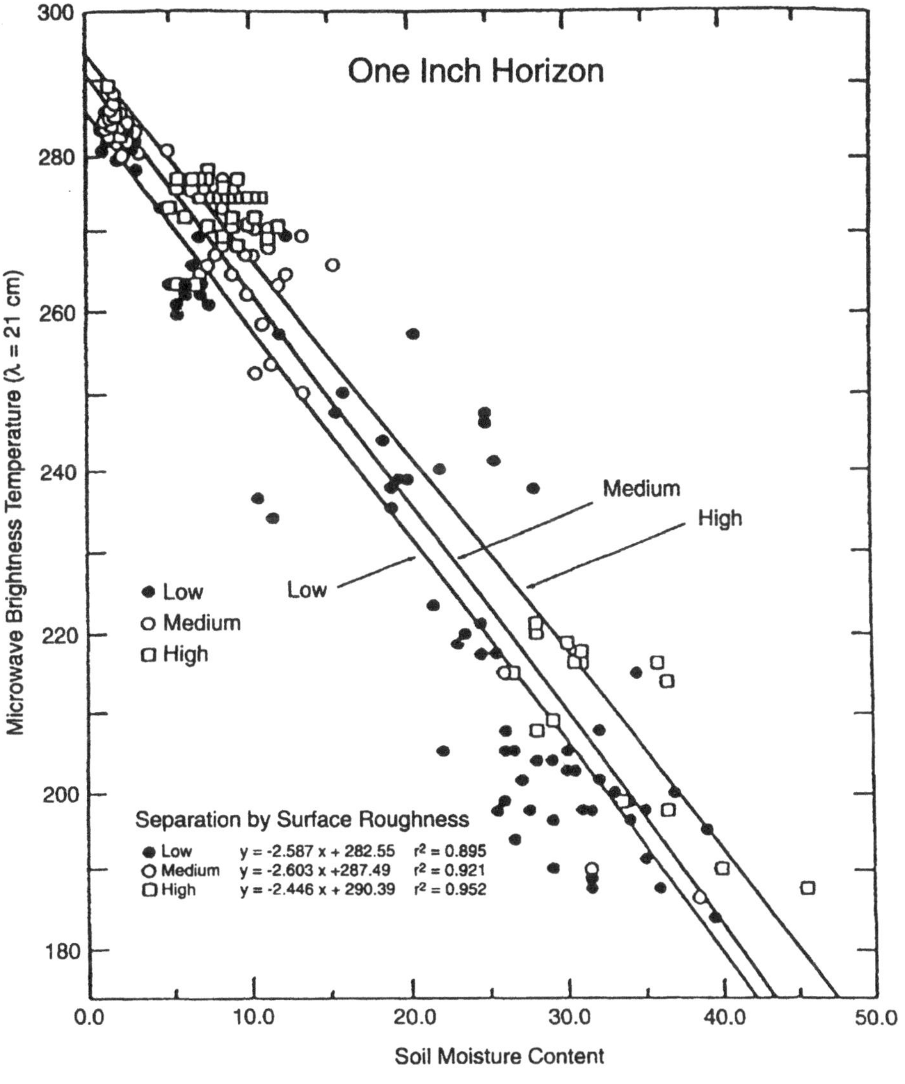

Figure 15. Scatter plot of brightness temperature versus soil moisture for aircraft observations at 1.4 GHz over a test site in South Dakota [42]. The data were segregated into three roughness categories based on ground photographs.

across the middle portion of the boundary. The m_v values in 1D are generally lower than those in 2D. This suggests that the burnt watershed 1D loses soil moisture more readily than the unburnt watershed 2D through direct evaporation from the soil and the higher transpiration losses from the more rapidly growing grass in 1D.

The estimation of m_v in the top 5 cm layer from the PBMR measurements was also extended to an area of 7 km by 15 km, about half of the entire FIFE test site [20]. During the four intensive campaigns conducted between May–October of 1987, sampling of m_v in the 0–5 and 5–10 cm layers was made once every two days at a number of stations widely distributed over the test site [20, 59]. The estimated and measured m_v

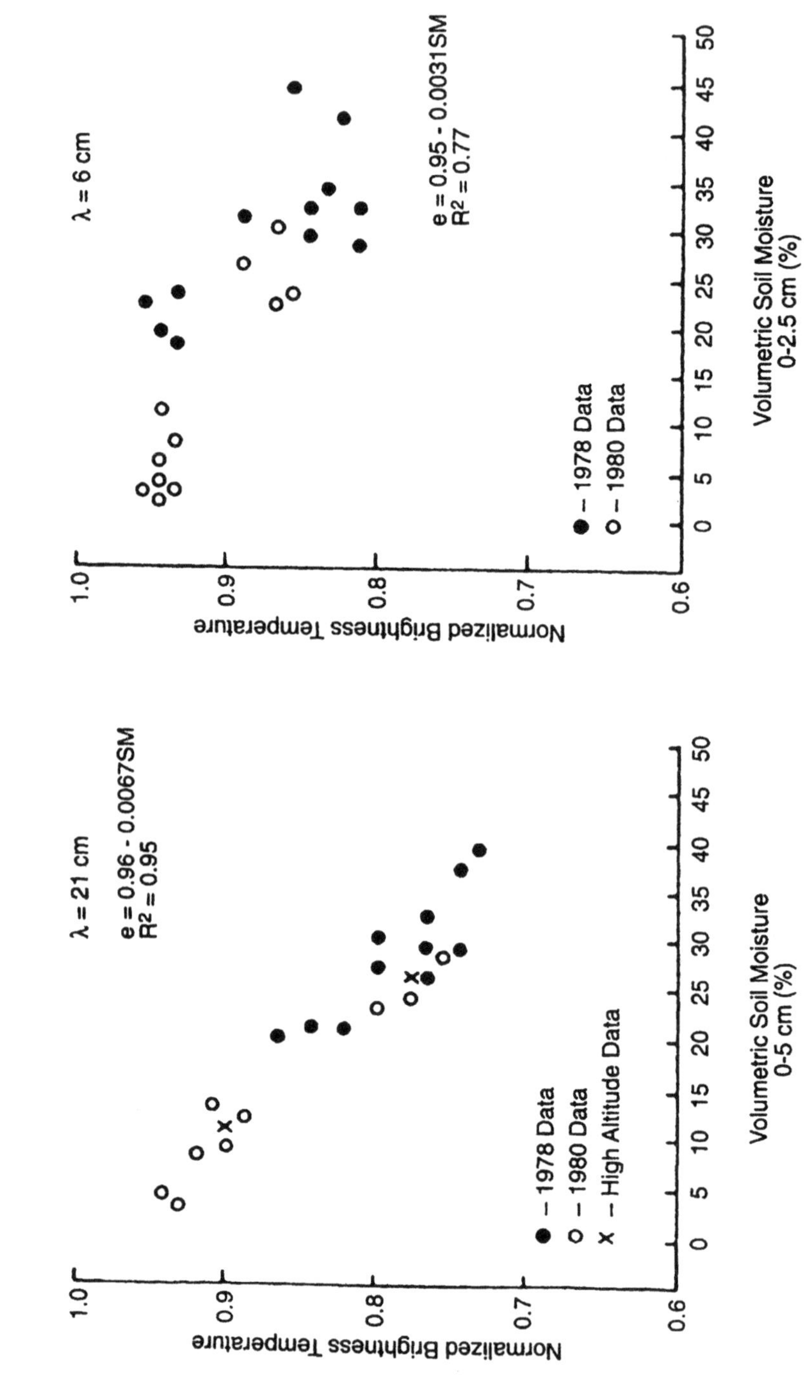

Figure 16. Comparison of 1.4 and 5.0 GHz emissivities measured from an aircraft platform with ground measurements of soil moisture for native grass pastures in Oklahoma [3]. Note the ~ 50% decrease in sensitivity at 5.0 GHz.

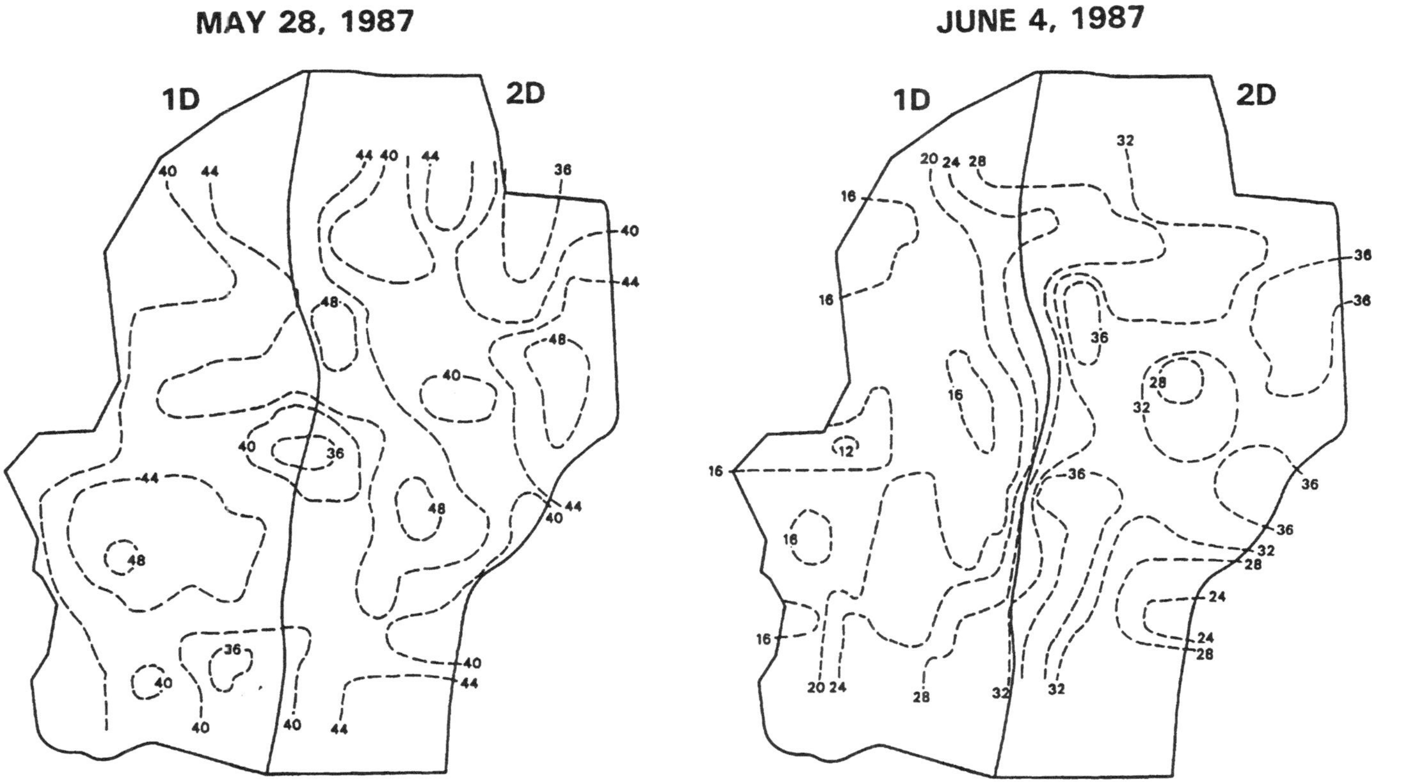

Figure 17. Spatial distribution of soil moisture in the 0–5 cm layer for two watersheds derived from combined measurements of ground sampling and airborne 1.4 GHz push-broom microwave radiometer [19].

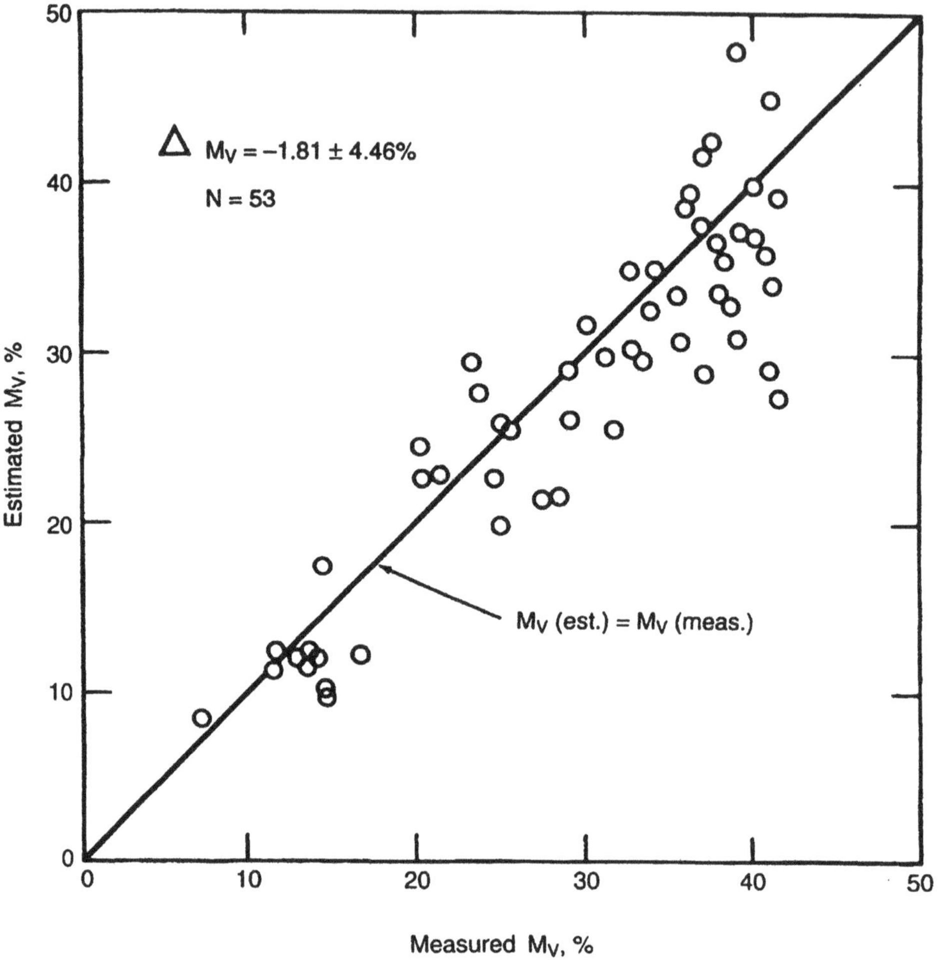

Figure 18. A scatter plot of the estimated and measured values of volumetric soil moisture in the top 5 cm layer obtained by Wang *et al.* [20].

values in the top 5 cm layer for these stations are compared in Fig. 18 [20]. In general, there is a good agreement between the estimated and measured m_v's. Most of the 1.8% bias indicated in the figure could be attributed to the time difference between the ground sampling and the PBMR measurements [20]. The estimation of m_v's from the PBMR measurements, however, was based on a certain assumption of vegetation effect on a day when the soils were saturated [20]. A more subtle approach to take care of the vegetation effect is needed before the radiometric measurements could be used for a large-scale soil moisture mapping.

Recently, a series of aircraft measurements have been made with a new type of L-band imaging radiometer, ESTAR (Electronically Scanning Thinned Array Radiometer), which was developed to demonstrate the capability of measuring soil moisture from space [60, 61]. The measurements at the Walnut Gulch Watershed, Arizona showed a good agreement between the predicted and measured m_v's in the top 5 cm layer; the measurements at Washita River Watershed in Oklahoma showed the effect of soil texture on brightness temperature [62]. A conventional space-borne microwave radiometer at L-band can be heavy and bulky, because it requires a large antenna to achieve a reasonable spatial resolution for soil moisture sensing. The new ESTAR approach may turn out to be more cost effective for space applications.

4.3. Satellite measurements

A number of studies on the relation between T_b and m_v based on satellite radiometric measurements have also been reported in the literature [10, 12, 21–25, 63–65]. Eagleman and Lin [21] correlated the T_b values from a single-beam 1.4 GHz radiometer aboard the Skylab with surface soil moisture. Typical results from these authors are shown in Fig. 19 for two Skylab passes over Kansas and Texas. Here the measured T_b values are plotted against the m_v's sampled in the top 2.5 cm layer and weighted according to the radiometer antenna pattern [21]. Excellent correlation between T_b and m_v for each group of data is evident from the figure. Combining many groups of data similar to the ones in Fig. 19 inevitably introduces more scatter in the correlation, presumably caused by the effects of vegetation cover and surface temperature. Wang [24] has analyzed the vegetation effect from the data acquired by the Skylab at 1.4 GHz and the Nimbus-7 Scanning Multichannel Microwave Radiometer (SMMR) at 6.6 and 10.7 GHz. Figure 20 shows some results of the analysis from the Skylab passes over two regions in Texas. The figure gives a scatter plot between the measured 1.4 GHz T_b and the calculated antecedent precipitation index (API), which is a good indicator of soil moisture [23]. Region A in western Texas contains some arid rangeland that has minimal vegetation. The region B in the eastern Texas is heavily forested throughout. The correlation between T_b and API in each region is reasonably good, but the slopes of the regression lines are quite different showing the effect of vegetation cover. The effect of vegetation on the measured T_b is even more pronounced at 6.6 and 10.7 GHz channels measured by the SMMR [24].

Satellite radiometric measurements at frequencies near 19 GHz or above for soil wetness have also been reported by Blanchard *et al.* [23], Allison *et al.* [10] and Heymsfield and Fulton [63]. Figure 21, from the work of Heymsfield and Fulton, shows a comparison of rainfalls and calculated API's with T_b values and polarization differences (ΔT) measured by SSM/I (Special Sensor Microwave/Imager) at 19, 37 and 86 GHz over bare soil region. A substantial change in both T_b and ΔT values occurred on day 20 following a major rainfall. A 20 day dry-down period followed, during which the calculated API values tracked well with the variations of both T_b and ΔT at all three frequencies. The amplitude of T_b and ΔT variations decreases with increasing frequency, which is consistent with the dielectric models of soil–water mixture discussed in Section 2. These authors also examined the measurements over vegetated region in the same period. They found that there was very little change in both T_b and ΔT following the major rainfall event and the correlation between API and these radiometric parameters was practically non-existent.

Blanchard *et al.* [23] also found a good correlation between API and the 19 GHz T_b's measured by the Nimbus-5 Electrically Scanning Microwave Radiometer (ESMR) over a relatively flat agricultural terrain in Oklahoma. They developed algorithms relating API to soil moisture in the top 23 cm layer for a few well managed watersheds. They concluded that the algorithms could be used as a transfer function to convert data from a

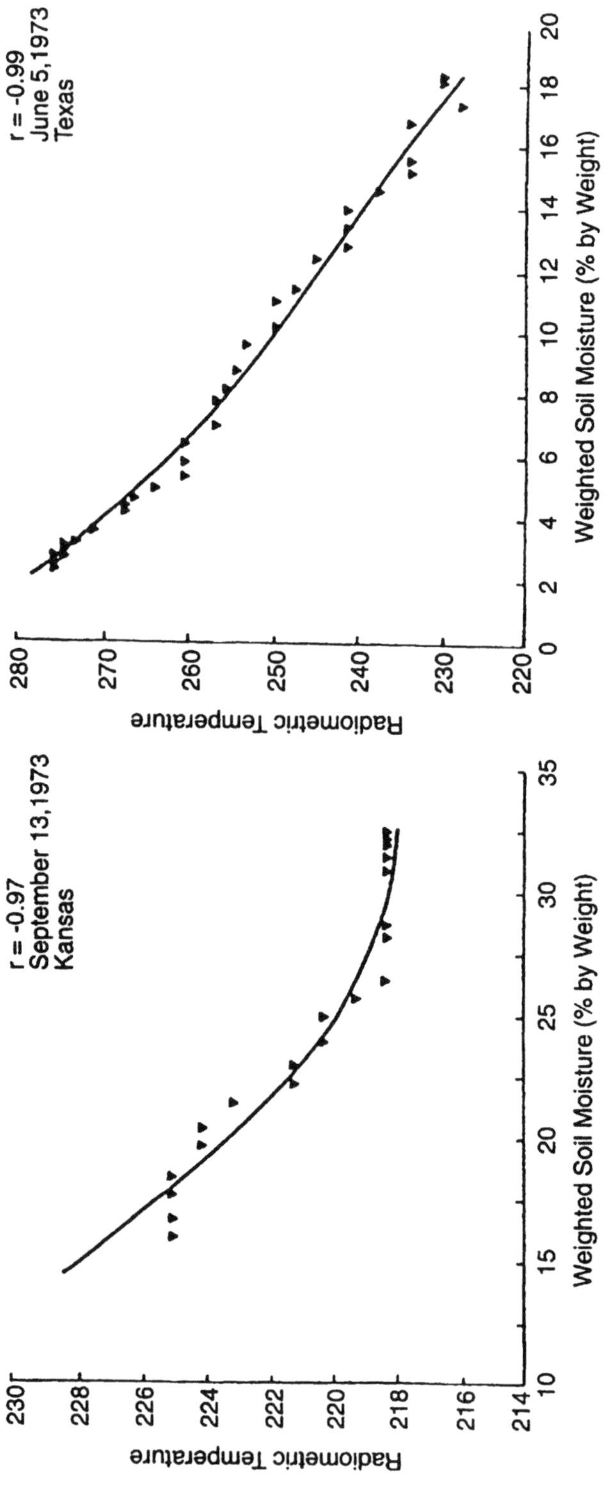

Figure 19. Correlation of Skylab 1.4 GHz brightness temperature with measured soil moisture obtained by Eagleman and Lin [21]. The soil moisture values were weighted by antenna pattern.

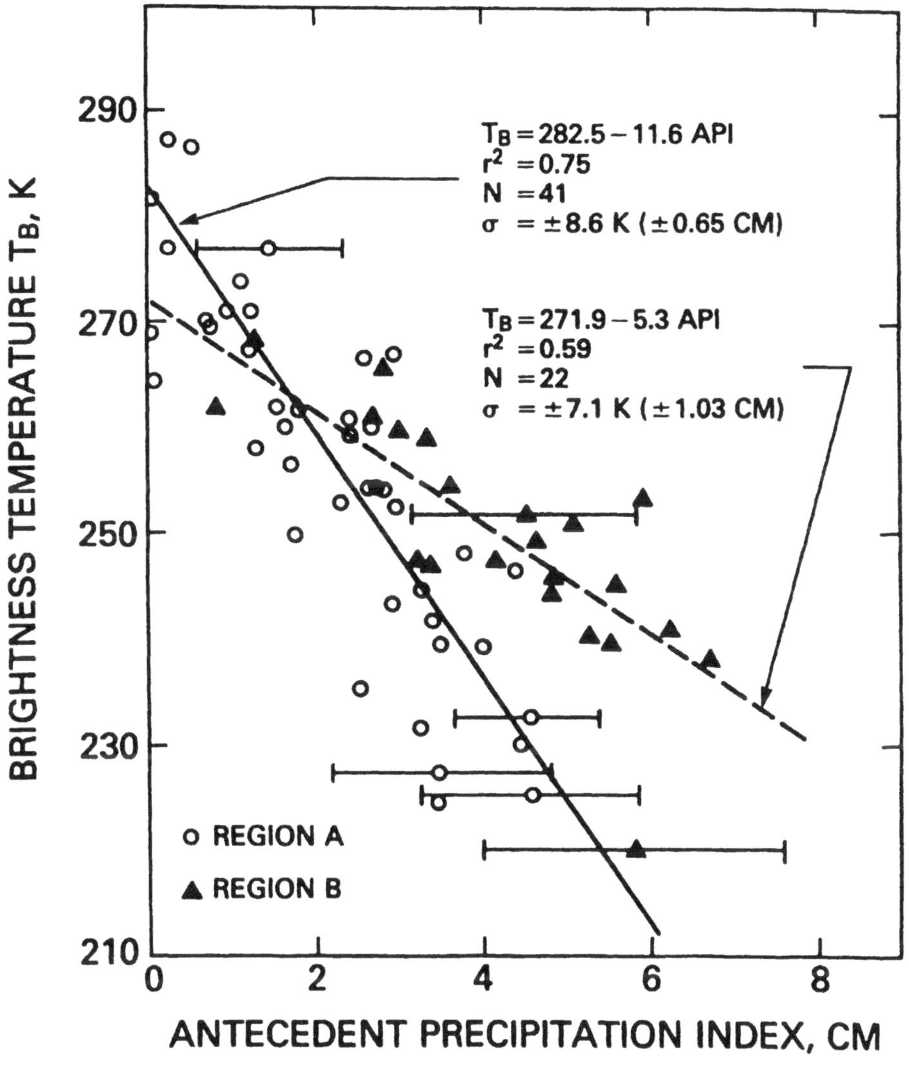

Figure 20. A scatter plot of Skylab 1.4 GHz brightness temperature versus antecedent precipitation index for two regions of different vegetation covers in Texas [24].

passive microwave imager to a reliable estimate of soil moisture at a depth greater than a microwave radiometer can sense, provided that the observations were made repeatedly every three days or less.

A passive microwave imager could also be used as an efficient means for large-scale mapping of flooded areas, as suggested by the work of Allison *et al.* [10]. A series of heavy rains (> 200 mm) fell in Queensland and New South Wales during January 9–26, 1974; these were followed by additional rainfall (35–62 mm) over the area on 19–20 February. Allison *et al.* [10] applied the Nimbus-5 ESMR data to chronicle the eastern Australian floods during January–March of 1974. Figures 22 and 23, supplied by Allison *et al.*, show a sharp contrast in the ESMR images for the near flood area

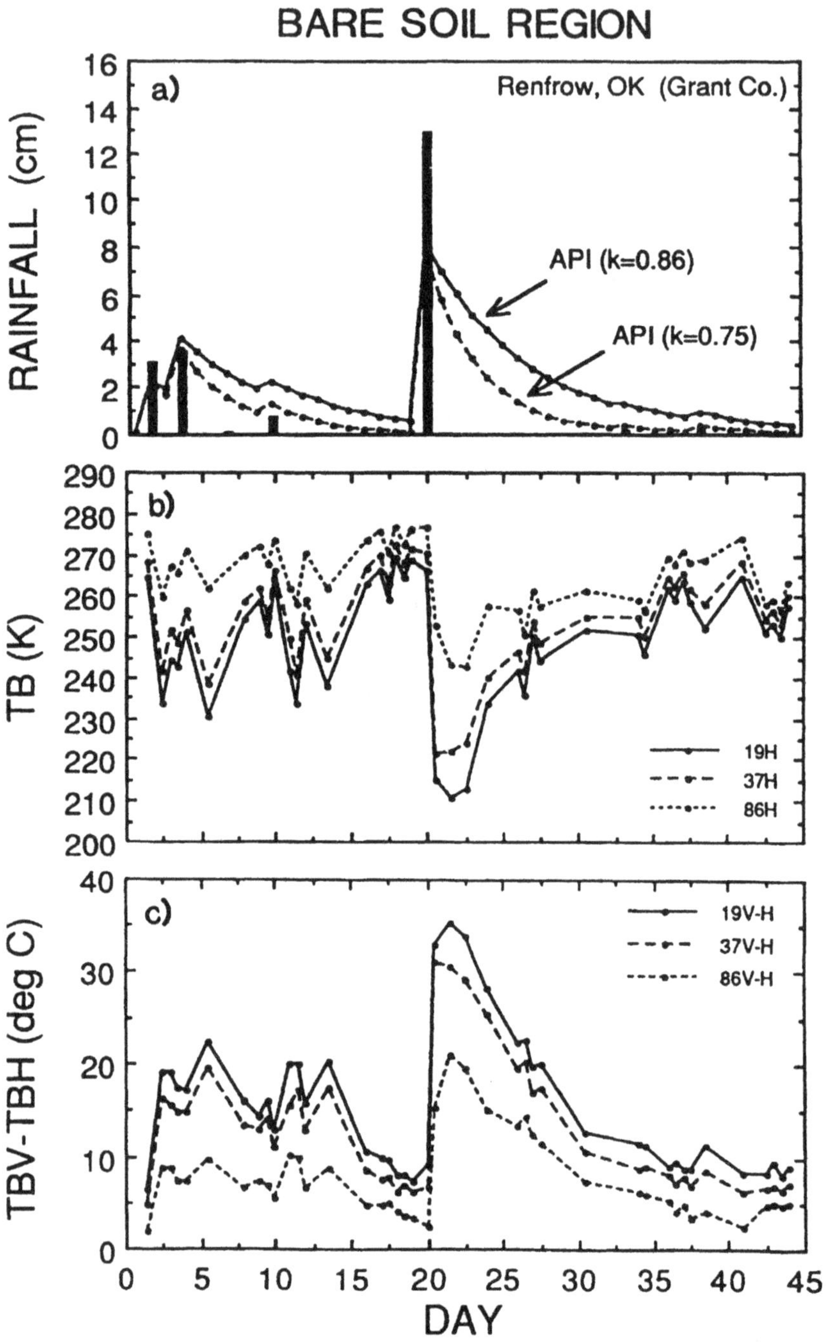

Figure 21. Time series plots of (a) daily raingage measurements (vertical bars) at Renfrow, Oklahoma and computed antecedent precipitation indices for two different recession coefficients, and (b) brightness temperatures and (c) brightness temperature differences derived from SSM/I measurements. The measurements were over a region of sparse vegetation cover [59].

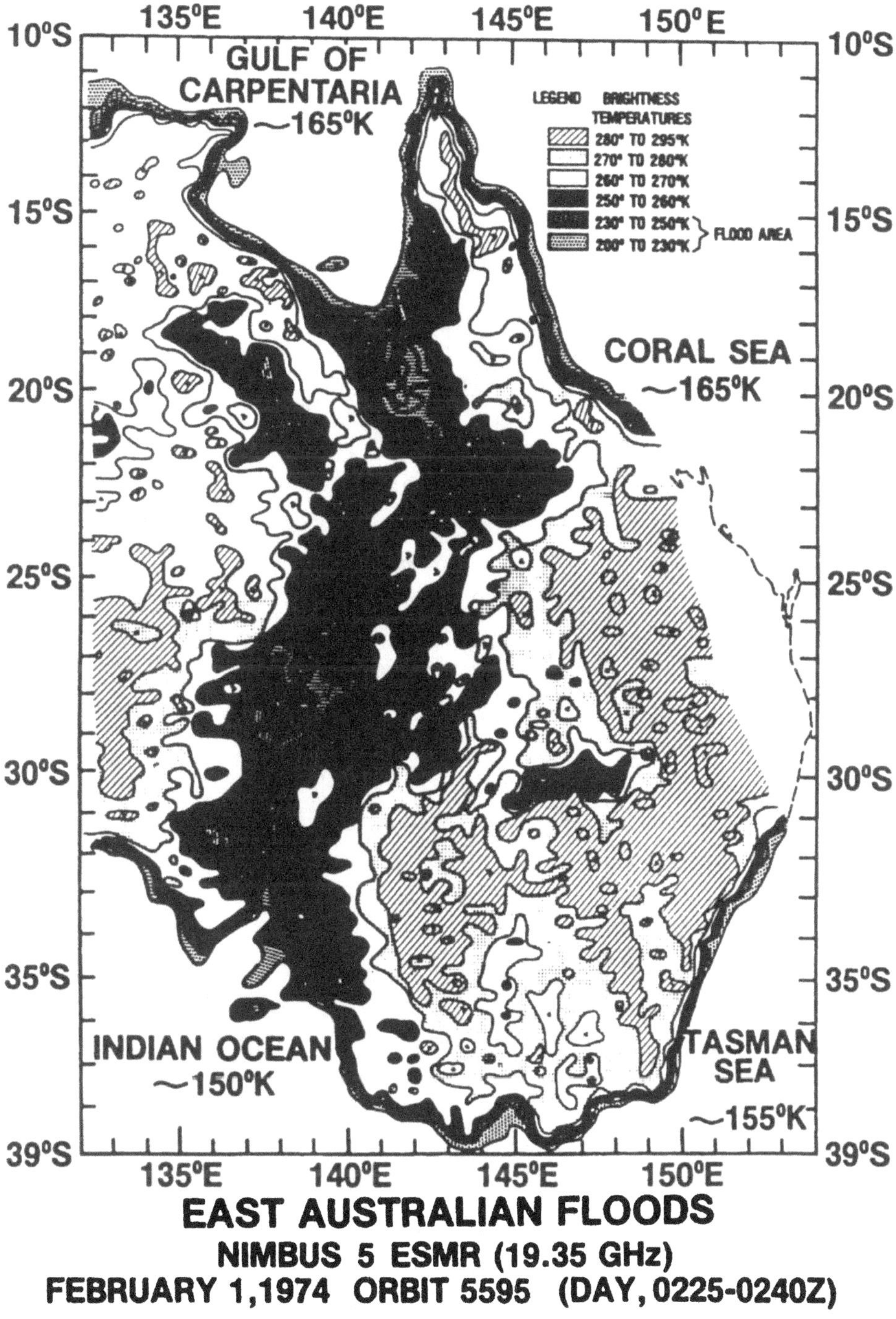

Figure 22. The Nimbus-5 ESMR image of Australia showing the correspondence between the low brightness at 19.35 GHz and the flooded regions [10].

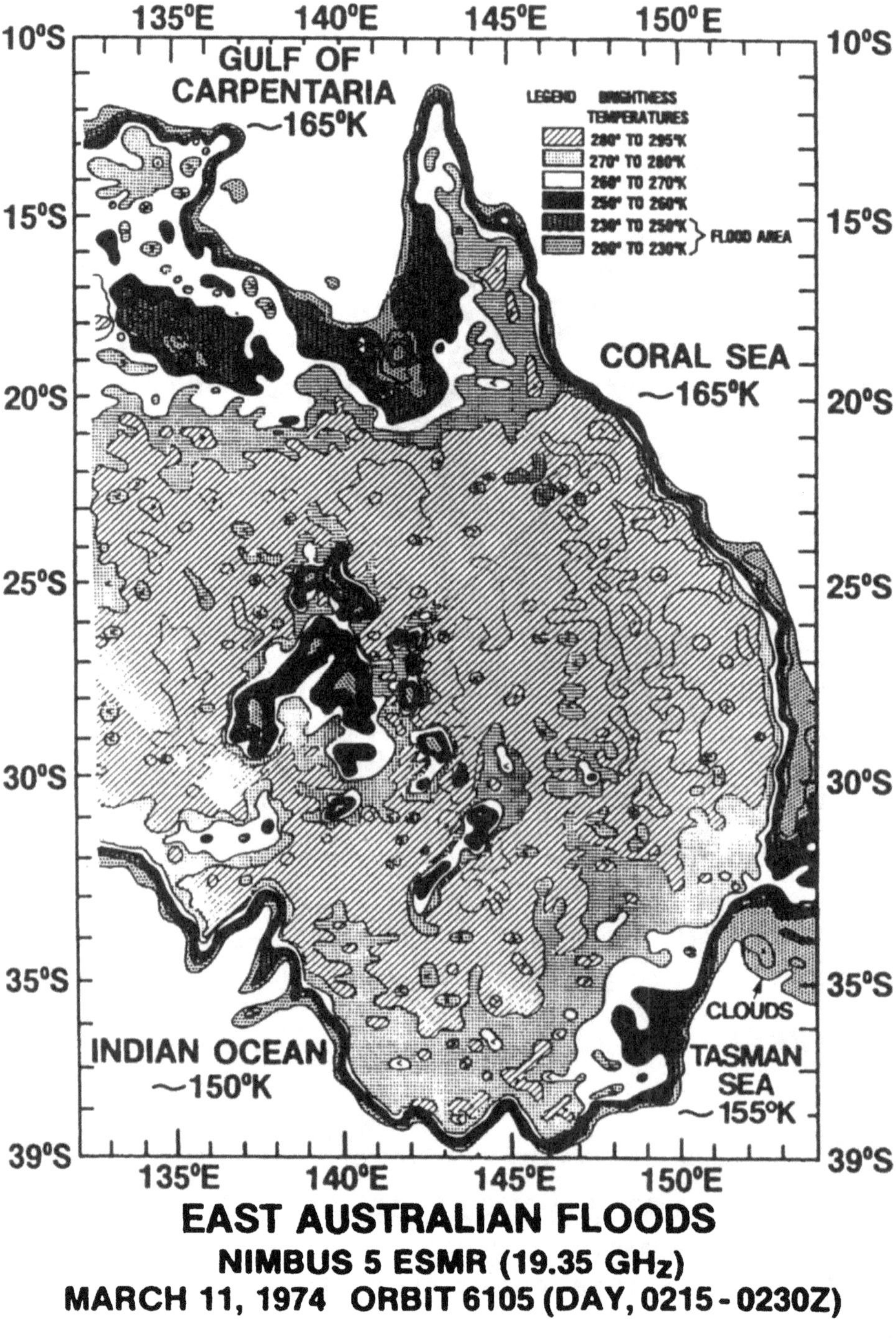

Figure 23. The Nimbus-5 ESMR image of Australia after most of the floods have subsided [10].

maximum on 1 February and near minimum on 11 March respectively. Near the flood area maximum on 1 February a large stripe of land surface extending from Gulf of Carpentaria southward to Indian Ocean was characterized by the low ESMR T_b values (< 250 K). By 11 March the total area of the regions with low T_b values was reduced by a factor of ~ 3.5. Allison *et al.* [10] also found a good correspondence in the temporal variations of T_b for a 50 km by 50 km region and the overbank flow of a nearby river. They concluded that a satellite scanning microwave radiometer would be an efficient tool for monitoring surface water, especially if the sensors with longer wavelengths and better spatial resolution become available.

5. SUMMARY

A good understanding has been developed about various factors affecting microwave emission from soils, which would have to be considered in estimating surface layer soil moisture from microwave radiometric measurements. The effect of soil moisture on microwave emission is reflected through the dielectric constant of soils, which at a given frequency depends primarily on soil texture and to a lesser extent on soil density and salinity. Both theoretical calculations and experimental observations suggest that the microwave soil moisture sampling depth is on the order of 0.1–0.2 times the wavelength. For a given soil moisture microwave emissivity increases with increasing surface roughness. As a first approximation, the ratio of roughness height and wavelength may be considered as a parameter for assessing the surface roughness effect. It is clear from the past efforts on both modeling and field observations that radiometric measurements at low frequencies ($\sim$ 1.4 GHz) are most ideally suited for estimating soil moisture.

Much remains to be learned for interpreting the coarse resolution satellite data at moderate to high frequencies, which are available currently (e.g. 19, 22, 37 and 85 GHz channels from SSM/I) and expected in the future (6.8, 10.7, 18.7, 23.8, 36.5 and 90 GHz from MIMR). Data set acquired from either ground-level or aircraft altitudes at these frequencies is quite limited compared to that near 1.4 GHz. Also, a key aspect of any coarse resolution data is the spatial heterogeneity of surface characteristics (soil moisture, vegetation and surface roughness), and an important question associated with spatial heterogeneity is the effect of aggregation (or spatial scaling) on the observed microwave emission. The relation between microwave emissivity and soil moisture, could be considered as linear for practical purpose; but it is unlikely that the entire scene within a satellite sensor's footprint would consist of bare soils only. The relation between emissivity and surface roughness or vegetation water content is nonlinear. It is non-trivial to devise an effective spatial averaging process to account for the effects of vegetation and surface roughness on microwave emission. Both model simulations and large-scale field observations are needed to better understand the scaling process and improve the accuracy of soil moisture estimation from satellite microwave radiometric measurements.

REFERENCES

1. T. J. Schmugge, P. Gloersen, T. Wilheit and F. Geiger. Remote sensing of soil moisture with microwave radiometers. *J. Geophys. Res.*, **79**, 317–323 (1974).

2. T. J. Schmugge. Effect of texture on microwave emission from soils. *IEEE. Trans. Geosci. Remote Sensing*, **GE-18**, 353–361 (1980).

3. T. J. Jackson, T. J. Schmugge and P. E. O'Neill. Passive microwave remote sensing of soil moisture from an aircraft platform. *Remote Sens. Environ.*, **14**, 135–151 (1984).

4. E. G. Njoku and P. E. O'Neill. Multifrequency microwave radiometer measurements of soil moisture. *IEEE Trans. Geosci. Remote Sensing*, **GE-20** (4), 468–475 (1982).

5. J. R. Wang, T. J. Schmugge, J. E. McMurtrey, W. I. Gould, W. S. Glazar and J. E. Fuchs. A multi-frequency radiometric measurement of soil moisture content over bare and vegetated fields. *Geophys. Res. Lett.*, **9** (4), 416–419 (1982).

6. J. R. Wang, J. C. Shiue and J. E. McMurtrey. Microwave remote sensing of soil moisture over bare and vegetated fields. *Geophys. Res. Lett.*, **7** (10), 801–804 (1980).

7. P. Pampaloni and S. Paloscia. Experimental relationships between microwave emission and vegetation features. *Int. J. Rem. Sensing*, **6** (2), 315–323 (1985).

8. J. R. Wang, R. W. Newton and J. W. Rouse. Passive microwave remote sensing of soil moisture: the effect of tilled row structure. *IEEE Trans. Geosci. Remote Sensing*, **GE-18** (4), 296–302 (1980).

9. R. W. Newton and J. W. Rouse. Microwave radiometer measurements of soil moisture content. *IEEE Trans. Ant. Prop.*, **AP-28** (5), 680–686 (1980).

10. L. J. Allison, T. J. Schmugge and G. Byrne. A hydrological analysis of east Australian floods using Nimbus-5 Electrically Scanning Radiometer data. *Bull. Amer. Meteor. Soc.*, **60** (12), 1414–1427 (1979).

11. T. J. Jackson and P. E. O'Neill. Salinity effects on the microwave emission of soils. *IEEE. Trans. Geosci. Remote Sensing*, **GE-25** (2), 214–220 (1987).

12. G. D. Wilke and M. J. McFarland. Correlations between Nimbus-7 scanning multichannel microwave radiometer (SMMR) data and an antecedent precipitation index. *J. Climate Appl. Meteor.*, **25**, 227–238 (1986).

13. P. E. O'Neill and T. J. Jackson. Observed effects of soil organic matter content on the microwave emissivity of soils. *Remote Sens. Environ.*, **31**, 175–182 (1990).

14. J. R. Wang, P. E. O'Neill, T. J. Jackson and E. T. Engman. Multifrequency measurements of the effects of soil moisture, soil texture, and surface roughness. *IEEE Trans. Geosci. Remote Sensing*, **GE-12** (1), 44–51 (1983).

15. B. J. Choudhury, T. J. Schmugge, A. Chang and R. W. Newton. Effect of surface roughness on the microwave emission from soils. *J. Geophys. Res.*, **84**, 5699–5706 (1979).

16. R. W. Newton, Q. R. Black, S. Makanvand, A. J. Blanchard and B. R. Jean. Soil moisture information from thermal microwave emission. *IEEE Trans. Geosci. Remote Sensing*, **GE-20**, 275–280 (1982).

17. P. M. Promes, T. J. Jackson and P. E. O'Neill. Significance of agricultural row structure on the microwave emissivity of soils. *IEEE Trans. Geosci. Remote Sensing*, **GE-26** (5), 580–589 (1988).

18. T. J. Schmugge, J. R. Wang and G. Asrar. Results from the push broom microwave radiometer flights over Konza Prairie in 1985. *IEEE Trans. Geosci. Remote Sensing*, **GE-26** (5), 590–596 (1988).

19. J. R. Wang, J. C. Shiue, T. J. Schmugge and E. T. Engman. Mapping surface soil moisture with L-band radiometric measurements. *Remote Sens. Environ.*, **27**, 304–312 (1989).

20. J. R. Wang, J. C. Shiue, T. J. Schmugge and E. T. Engman. The L-band PBMR measurements of soil moisture in FIFE. *IEEE Trans. Geosci. Remote Sensing*, **GE-28** (5), 906–914 (1990).

21. J. R. Eagleman and W. C. Lin. Remote sensing of soil moisture by a 21 cm passive radiometer. *J. Geophys. Res.*, **81**, 3660–3666 (1976).

22. T. J. Schmugge, J. M. Meneely, A. Rango and R. Neff. Satellite microwave observations of soil moisture variations. *Water Res. Bull.*, **13**, 265 (1977).

23. B. J. Blanchard, M. J. McFarland, T. J. Schmugge and E. Rhoades. Estimation of soil moisture with API algorithm and microwave emission. *Water Res. Bull.*, **17** (5), 767–774 (1981).

24. J. R. Wang. Effect of vegetation on soil moisture sensing observed from orbiting microwave radiometers. *Remote Sens. Environ.*, **17**, 141–151 (1985).

25. B. J. Choudhury and R. E. Golus. Estimating soil wetness using satellite data. *Int. J. Rem. Sensing*, **9**, 1251–1257 (1988).

26. E. G. Njoku and J. A. Kong. Theory for passive microwave remote sensing of near surface soil moisture. *J. Geophys. Res.*, **82** (20), 3108–3118 (1977).

27. T. T. Wilheit. Radiative transfer in a plane stratified dielectric. *IEEE Trans. Geosci. Electron.*, **GE-16**, 138–143 (1978).

28. L. Tsang, E. Njoku and J. A. Kong. Microwave thermal emission from a stratified medium with non-uniform temperature distribution. *J. Appl. Phys.*, **46**, 5127–5133 (1975).

29. W. J. Burke, T. Schmugge and J. F. Paris. Comparison of 2.8 and 21 cm microwave observations over soils with emission model calculations. *J. Geophys. Res.*, **84**, 287–294 (1979).

30. T. J. Schmugge and B. J. Choudhury. A comparison of radiative transfer for predicting emission from soils. *Radio Sci.*, **16** (5), 927–938 (1981).

31. J. R. Wang and T. J. Schmugge. An empirical model for the complex dielectric permittivity of soil as a function of water content. *IEEE Trans. Geosci. Remote Sensing*, **GE-18**, 288–295 (1980).

32. M. T. Hallikainen, F. T. Ulaby, M. C. Dobson. Microwave dielectric behavior of wet soil – Part I: Empirical models and experimental observations. *IEEE Trans. Geosci. Remote Sensing*, **GE-23** (1), 25–34 (1985).

33. M. C. Dobson, F. T. Ulaby, M. T. Hallikainen and M. A. El-Rayes. Microwave dielectric behavior of wet soil – Part II: Dielectric mixing models. *IEEE Trans. Geosci. Remote Sensing*, **GE-23**, 35–46 (1985).

34. J. R. Wang. The dielectric properties of soil–water mixtures at microwave frequencies. *Radio Sci.*, **15** (5), 977–985 (1980).

35. T. J. Schmugge. Remote sensing of soil moisture: Recent advances. *IEEE Trans. Geosci. Remote Sensing*, **GE-21** (3), 336–344 (1983).

36. T. J. Schmugge, P. E. O'Neill and J. R. Wang. Passive microwave soil moisture research. *IEEE Trans. Geosci. Remote Sensing*, **GE-24** (1), 12–22 (1986).

37. T. J. Jackson and T. J. Schmugge. Passive microwave remote sensing of soil moisture. *Adv. Hydroscience*, **14**, 123–159 (1986).

38. T. J. Schmugge, T. J. Jackson and H. L. McKim. Survey of methods for soil moisture determination. *Water Resources Res.*, **16** (6), 961–979 (1980).

39. G. P. de Loor. Dielectric properties of heterogeneous mixtures containing water. *J. Microwave Power*, **3**, 67–73 (1968).

40. J. A. Lane and J. A. Saxtin. Dielectric dispersion in polar liquids at very high radar frequencies, III, The effects of electrolytes in solutions. *Proc. Roy. Soc. London*, **214**, 531–545 (1952).

41. T. J. Schmugge. Measurements of surface soil moisture and temperature. In: *Remote Sensing of Biosphere Functioning*, R. J. Hobbs and H. A. Mooney (Eds), Springer-Verlag, New York (1990).

42. J. R. Wang, J. C. Shiue, S. L. Chuang, R. T. Shin and M. Dombrowski. Thermal microwave emission from vegetated fields: A comparison between theory and experiment. *IEEE Trans. Geosci. Remote Sensing*, **GE-22** (2), 143–150 (1984).

43. T. J. Jackson and P. E. O'Neill. Microwave dielectric model for aggregated soils. *IEEE Trans. Geosci. Remote Sensing*, **GE-24** (6), 920–929 (1986).

44. A. W. England. Thermal microwave emission from a scattering layer. *J. Geophys. Res.*, **80**, 4484–4496 (1975).

45. A Stogryn. The brightness temperature of a vertically structured medium. *Radio Sci.*, **5**, 1397–1406 (1970).

46. L. Tsang and R. W. Newton. Microwave emission from soils with rough surfaces. *J. Geophys. Res.*, **87**, 9017–9024 (1982).

47. A. K. Fung and H. J. Eom. A study of backscattering and emission from closely packed inhomogeneous media. *IEEE Trans. Geosci Remote Sensing*, **GE-23**, 761–767 (1985).

48. L. Tsang, J. A. Kong and R. T. Shin. *Theory of Microwave Remote Sensing*. John Wiley & Sons, New York (1985).

49. F. T. Ulaby, R. K. Moore and A. K. Fung. *Microwave Remote Sensing: Active and Passive*. Vol. II, Addison-Wesley, Reading, MA (1982).

50. J. R. Wang and B. J. Choudhury. Remote sensing of soil moisture content over bare field at 1.4 GHz frequency. *J. Geophys. Res.*, **86** (C6), 5277–5282 (1981).

51. T. J. Schmugge and T. J. Jackson. A dielectric model of the vegetation effects on the microwave emission from soils. *IEEE Trans. Geosci. Remote Sensing*, **GE-30** (2), 757–760 (1992).

52. B. J. Choudhury, C. J. Tucker, R. E. Golus and W. W. Newcomb. Monitoring vegetation using Nimbus-7 scanning multichannel microwave radiometer's data. *Int. J. Rem. Sensing*, **8**, 533–538 (1987).

53. J. R. Wang. Microwave emission from smooth bare fields and soil moisture sampling depth. *IEEE Trans. Geosci. Remote Sensing*, **GE-25** (5), 616–622 (1987).

54. T. J. Jackson and T. J. Schmugge. Vegetation effects on the microwave emission of soils. *Remote Sens. Environ.*, **36**, 203–212 (1991).

55. J. R. Wang. Passive microwave sensing of soil moisture content: the effects of soil bulk density and surface roughness. *Remote Sens. Environ.*, **13**, 329–344 (1983).

56. J. A. Kong, S. L. Lin and S. L. Chuang. Microwave thermal emission from periodic surfaces. *IEEE Trans. Geosci. Remote Sensing*, **GE-22** (4), 377–382 (1984).

57. T. J. Jackson and P. E. O'Neill. Temporal observations of surface soil moisture using a passive microwave sensor. *Remote Sens. Environ.*, **21**, 281–296 (1987).

58. T. J. Schmugge, T. J. Jackson, W. Kustas and J. R. Wang. Passive microwave remote sensing of soil moisture: results from HAPEX, FIFE and MONSOON 90. *ISPRS J. Photo. Remote Sens.*, **47**, 127–143 (1992).

59. F. G. Hall, P. J. Sellers, I. MacPherson, R. D. Kelly, S. Verma, B. Markham, B. Blad, J. Wang and D. E. Strebel. FIFE: Analysis and results – A review. *Adv. Space Res.*, **9** (7), 275–293 (1989).

60. D. M. Le Vine, T. Wilheit, R. Murphy and C. Swift. A multifrequency microwave radiometer of the future. *IEEE Trans. Geosci. Remote Sensing*, **GE-27** (2), 193–199 (1989).

61. D. M. Le Vine, M. Kao, A. Tanner, C. Swift and A. Griffis. Initial results in the development of a synthetic aperture microwave radiometer. *IEEE Trans. Geosci. Remote Sensing*, **GE-28** (4), 614–619 (1990).

62. T. J. Jackson, D. M. Le Vine, A. Griffis, D. Goodrich, T. Schmugge, C. T. Swift and P. O'Neill. Soil moisture and rainfall estimation over a semiarid environment with the ESTAR microwave radiometer. *IEEE Trans. Geosci. Remote Sens.*, **31** (4) (1993).

63. G. M. Heymsfield and R. Fulton. Modulation of SSM/I microwave soil radiances by rainfall. *Remote Sens. Environ.*, **39**, 187–202 (1992).

64. Y. H. Kerr and E. G. Njoku. A semiempirical model for interpreting microwave emission from semiarid land surface as seen from space. *IEEE Trans. Geosci. Remote Sensing*, **GE-28**, 384–393 (1990).

65. M. Owe, A. A. van de Griend and A. T. C. Chang. Surface moisture and satellite microwave observations in semiarid southern Africa. *Water Resources Res.*, **28**, 829–839 (1992).

ESA/NASA International Workshop, pp. 461–474
B. J. Choudhury, Y. H. Kerr, E. G. Njoku and P. Pampaloni (Eds)

Estimation of soil moisture profiles and root zone moisture content by means of microwave radiometry and *a priori* information

A. M. SHUTKO, E. A. REUTOV and S. P. GOLOVACHEV

Institute of Radioengineering and Electronics, Russian Academy of Sciences, Moscow, Russia

Abstract—Temporal and spatial changes of volumetric water content in soil are the major source of intensity variations of microwave radiation. Thickness of the effectively emitting layer is a function of the volumetric water content as well as wavelength. The longer the wavelength, the thicker the skindepth, with other conditions being equal.

The transition layer between the upper dry and lower moist soil, the moisture content under the transition layer, and the gradient of soil moisture profile are the main parameters which influence the spectrum of microwave radiation.

Shallow water table manifests itself as passive microwaves due to reflections from the stratified media and the properties of the capillary fringe.

Thus, the radiation spectrum characteristics, their changes in time during the drying process as well as *a priori* knowledge of soil properties are the starting points in developing software for estimation of soil moisture profiles and total water content in the root zone.

The paper discusses the physical background and mathematical approaches for developing the software for estimation of some parameters of soil moisture vertical profile and net water content in the root zone using both multichannel microwave radiometry and *a priori* information. Theoretical models, laboratory data, and data of field and aircraft experiments are presented.

1. INTRODUCTION

Characteristics of non-uniformly moistened soils at passive microwaves have been studied at the Institute of Radioengineering and Electronics, USSR Academy of Sciences [6–8, 12–16], General Geophysical Observatory, USSR Department of Hydrometeorology [5], Institute of Civil Aviation Engineers in Moscow [4], Beltsville Agricultural Research Center and NASA Goddard Space Flight Center [2, 9–11], Institute of Electronics, Bulgarian Academy of Sciences [14], and a number of other institutions [1, 3, 4, 17].

These studies have been conducted with due regard for the theory of electromagnetic wave interaction with plane stratified media [1–4] with different dielectric constant profiles as well as for specific characteristics of soils including microwave dielectric properties of water in soil, dependent on its temperature, salinity and bulk density; peculiarities of soil moisture profiles in a layer of aeration in different soil climatic zones include those with a shallow water table [6, 7, 12, 13, 15]. Representative laboratory experiments on the influence of the moisture transition layer on microwave response of the medium [12], as well as field and aircraft experiments with extensive *in situ* measurements have been conducted [6–8, 12–15]. As the results of these investigations, the principal peculiarities of passive microwave spectroradiometry of nonuniformly moistened soils, the feasibilities and limitations of software for estimating the soil moisture in top 30 to 100 cm layer using passive microwave remotely sensed information and *a priori* knowledge about soils have been studied [13–15].

2. RADIATION-SOIL PARAMETER DEPENDENCIES FOR A HOMOGENEOUS MEDIUM

In accordance with [12] an approximation of the complex dielectric permittivity by means of a 'refraction' model leads to an analytical expression for emissivity in the form

$$\kappa = \frac{4(B_w \rho_w + C - 1)}{(A_w - 2B_w)\rho_w^2 + 2B_w C \rho_w + C^2} \tag{1}$$

where

$$A_w = |e_w| - 1$$
$$B_w = \sqrt{|e_w|}\cos \beta_w/2 - 1$$
$$\beta_w = \mathrm{arctg}(e''_w/e'_w)$$
$$C = 0.5\rho_d + 2$$
$$e_w = e'_w - ie''_w$$

and e_w is the complex dielectric permittivity (dielectric constant) of water. For soil moisture $0 < \rho_w = \rho_{wa} < \rho_{wp}$, which characterizes the volumetric content of absorbed (bound) water, $e_w = e_{wa}$ (wp – wilting point – maximum absorbtive soil moisture capacity); the range $\rho_{wp} < \rho_w = \rho_{wg} < \rho_{sat}$ characterizes the volumetric content of free (gravimetric) water in the soil, $e_w = e_{wg}$ (sat – saturation – full soil moisture capacity), where the e_{wg} is dependent on the water temperature T, mineralization (salt concentration) of the soil solutions S, and the wavelength λ.

The thickness of the effective radiating layer at a wavelength λ may be expressed in the form:

$$L_e(\rho_w, \rho_d, T, S) = L_{ew}(T, S)/\rho_w \tag{2}$$

where

$$L_e = 1/\gamma_e$$
$$L_{ew} = 1/\gamma_{ew}$$

and γ_e and L_{ew} are the absorption coefficients and the thickness of the water skin-layer, respectively.

As it follows from (2), absorbing properties of the soil are determined wholly by the absorbing properties of water (soil solution) at this wavelength with given values of temperature and salinity, and also with the value of the relative volumetric water content which do not practically depend on soil density. This result corresponds to the physical representations of the special features of the electromagnetic field interaction with the continuous media, whose dielectric heterogeneities allow one to disregard the effect of scattering. That is, the particles of soil and the pores are much smaller than the wavelength. Taking into account spectral features of the dielectric permittivity of water, in particular, in the case of fresh water (in the absence of ionic conductivity), with $T = 18\,°C$, in the soil moisture range $0.1 < \rho_w < 0.35\,g/cm^3$, it follows from (2)

$$L_e(\rho_w) = 0.1\lambda^2\rho_w^{-1} \tag{3}$$

with λ in cm and ρ_w in units of relative volume.

Dry soil is characterized by an emissivity coefficient of 0.9, which corresponds to a brightness temperature $T_b = 250–270$ K, which relates to a change in soil temperature from 0 to 30 °C. The emissivity coefficient of a considerably moistened ground, which is characterized by the moisture content value $\rho_w = 0.35\,g/cm^3$ (the free moisture), is close to 0.6, corresponding to a brightness temperature of 165–180 K.

Evidently, an increase in the moisture content of 0.35 g/cm^3 is accompanied by a decrease in the ground brightness temperature of approximately 110 K. Taking into account the sensitivity of microwave radiometric receivers (i.e. 0.1–0.3 K for 1 sec integration time) we will obtain for the soil moisture range from the value of maximum to the 'signal/noise' ratio in the range of 300–1000.

Sensitivity of the radiation field to variations of ground density is 15–20 times less than the sensitivity to moisture content variations (Fig. 1a). Thus, for instance, with an increase in the density by the value $\rho_d = +0.35\,g/cm^3$ the reduction in brightness temperature is less than 5 K. In nearly saturated soil, the effect of variations in the

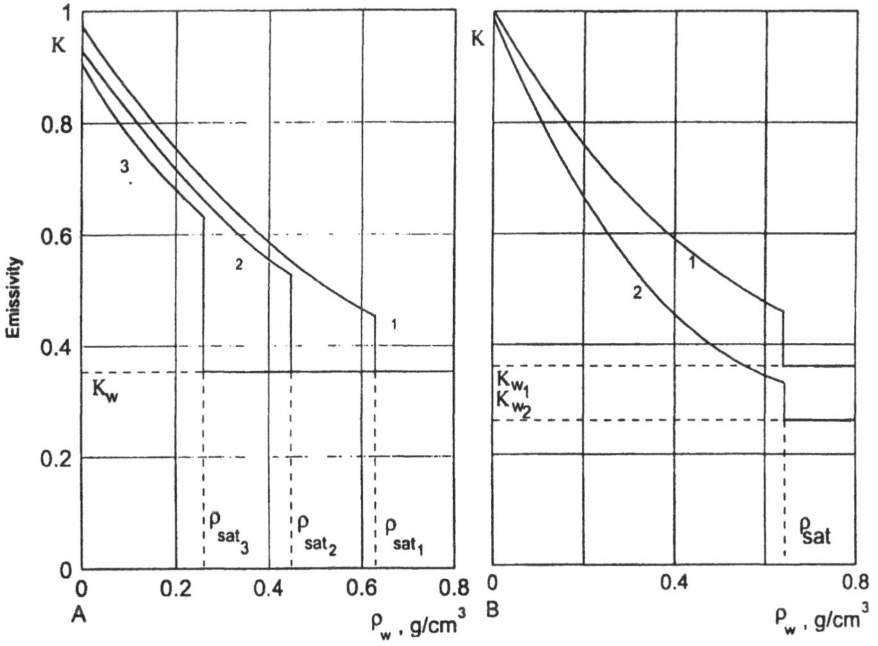

Figure 1. Dependence of emissivity at $\lambda = 30$ cm, $\Theta = 0°$ (nadir viewing) for temperature $T = 20\,°C$ on volumetric free moisture content in soil under variations of dry soil density (a) and salinity (NaCl) of water in soil (b). (a): salinity $S = 0$ ppt; 1, 2, and 3 are the dry soil density data corresponding to values of 1.0, 1.5, and 2.0 g/cm^3, respectively; (b): dry soil density $\rho_d = 1.0\,g/cm^3$; 1 and 2 are salinity data corresponding to values of 0 and 60 ppt, respectively. κ_w is the emissivity of open water, ρ_{sat} is the soil moisture under condition of saturation when all pores in the soil are filled with water.

salinity of soil solution on the intensity of the radiation field in the range of decimeter wavelengths is noted in Fig. 1(b).

Analyses show that daily changes in ground surface temperature of 10–30 °C cause variations in T_b within a range of 2–7 K, depending on wavelength and soil moisture. These changes of T_b commensurate with the variations of T_b with a change of the moisture content approximately 0.01–0.02 g/cm^3.

3. TYPICAL VERTICAL SOIL MOISTURE PROFILES IN ZONE OF AERATION AND THEIR APPROXIMATIONS

The vertical distribution of soil moisture in the zone of aeration is determined by ground type, by ground water (level of ground water — LGW), by meteorological conditions, evaporation, and (in integrated areas) by mode of irrigation. The quantity of applied water in irrigated areas, the duration of irrigation, the temperature of air and soil, all have an influence.

The object that was studied is a nonuniformly moistened soil with typical vertical profiles of moisture content. The purpose of this study was to determine the microwave radiometry response of different types of the above soils modelled by respective profiles of the dielectric constant $|e|$.

Four major types of vertical soil moisture profiles have been identified by an analysis of the moistening characteristics of the various climatic zones. These types of profiles are presented in Fig. 2(a) and relate to the following climatic zones and moistening conditions:

Type 1: unirrigated farm areas of any climatic zone in the absence or with low value of precipitation;

Type 2: soils with a mechanical composition strongly non-uniform vertically;

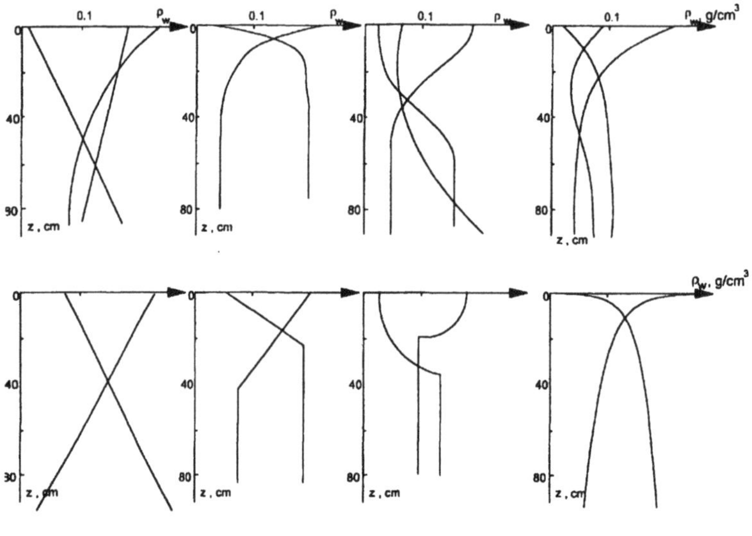

Figure 2. Basic types of soil moisture profiles (a) and corresponding moisture profile approximations (b) (see explanations in the text).

Type 3: 1) unirrigated farm lands with subsurface water levels below 2 or 4 m; 2) the soil condition in any climatic zone in the first few hours following rainfall or irrigation;

Type 4: soils in any climatic zone during a post-irrigation or post-rainfall drying-up period.

Typical profiles of moisture content above ground water level are shown in Fig. 3. Typical moisture profile for a soil with surface transition layer is shown in Fig. 5.

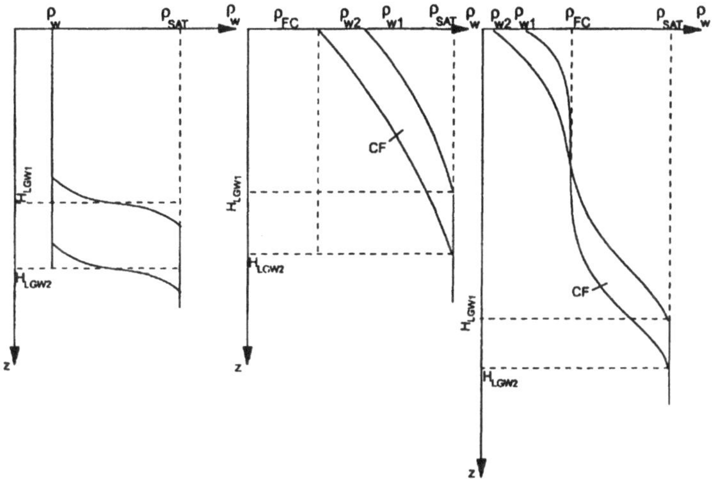

Figure 3. Characteristic profiles of moisture content distribution in zone of aeration, which justify models of interconnection of intensity of emission with a level of ground water (LGW): (a) arid areas, coarse grained soils, LGW is within a few decimeters; (b) the humid, overmoistened areas, wetland; (c) arid areas under conditions, when evaporation is not compensated by field current of moisture (schematic representation), LGW is within a range 1 to 4 m. CF is a capillary fringe; ρ_{fc} is the soil moisture under condition of field capacity.

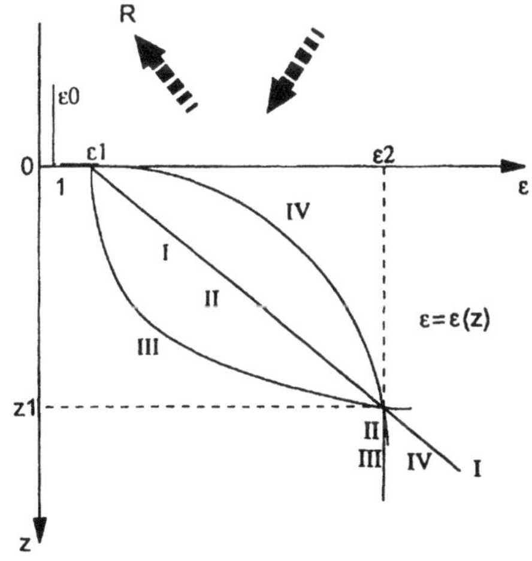

Figure 4. The models of dielectric constant profiles used in this study, corresponding to the moisture profiles approximations shown in Fig. 2(b) and described by the formulas (4) to (7). I – linear, II – piecewise, III – exponential, IV – parabolic.

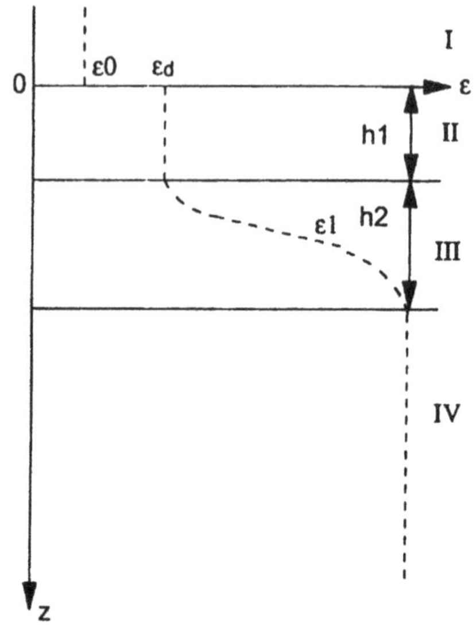

Figure 5. Electrodynamic model of nonuniformly moistened soil with surface transition layer: Medium I is air with $e_0 = 1$; Medium II is a dry soil with correspondent e_d; Medium III is the soil layer with e varying exponentially from e_d to e_1 corresponding to some moisture content; Medium IV is the soil layer with moisture close to constant.

4. MICROWAVE RADIATION MODELS

The quantity $\kappa = 1 - |R|^2$ varies with moisture content since the latter affects the complex dielectric constant of soil (e), that enters in the relations for $|R|^2$ (R is the reflection coefficient). Our analysis of these expressions for various mixtures of dielectrics compared with laboratory measurements of the dielectric constants for moistened soils has shown that e of a moistened soil as a function of moisture content can be best represented at the microwave range by the 'refractive' model. Hence this model will be used in the discussion to follow.

Figure 2(b) shows electrodynamic models representing each of the four types of moisture profile:

Model 1: Linear variation of the dielectric constant with depth

$$e(z) = e_1 + az \qquad 0 \leqslant z < \infty \tag{4}$$

where e_1 is the value of e at the air/soil interface and z is the vertical distance from the surface.

Model 2: Piecewise linear variations

$$e(z) = \begin{cases} e_1 \pm az & 0 \leqslant z < z_1 \\ e_2 = e_1 \pm az_1 & z_1 \leqslant z < \infty. \end{cases} \tag{5}$$

Model 3: Exponential variation

$$(z) = \begin{cases} e_1 \exp(\pm az) & 0 \leqslant z < z_1 \\ e_2 = e_1 \exp(\pm az_1) & z_1 \leqslant z < \infty. \end{cases} \tag{6}$$

Model 4: Parabolic variation

$$e(z) = e_1 \pm az \mp bz^2 \quad 0 \leqslant z < \infty. \tag{7}$$

To evaluate the thermal radiation characteristics of an inhomogeneous medium, we took advantage of the basic premise that the thermal radiation of the medium is determined by the heat losses experienced in this medium by a plane electromagnetic (EM) wave. It means that we had to deal with the electromagnetic problem of plane wave reflection from an inhomogeneous medium with the emissivity κ determined as $\kappa = 1 - |R|^2$. In the models discussed here the vertical variations of the dielectric constant were represented by (4), (5), (6), or (7) and a plane EM wave was incident at the air/medium interface, as shown in Fig. 4. This figure shows the models used in this study where I is linear, II is piecewise linear, III is exponential, and IV is parabolic.

These models have been applied to derive, via simulations, reflectivities in the 3 to 60 cm wavelength range for various parameters of the profile $e(z)$.

We will characterize the effect of the profile parameters on the reflectivity of a plane EM wave either by $\Delta R = R - R_f$, where R is the reflectivity of an inhomogenous medium and R_f is the Fresnel reflection coefficient for the same e_1, or by the radiation brightness temperature contrast $\Delta T_b = \Delta R T_0$, where T_0 is the thermodynamic temperature of the medium.

4.1. Model I

It is clear that ΔR is higher for longer wavelength, higher gradients a or lower values of e_1. From an analysis of simulated results, one can derive

$$\Delta R \approx c_1 |e_1|^{-3/2} |a| \tag{8}$$

with c_1 being a constant factor.

4.2. Model II

Analysis of the data demonstrates that a change in the transition layer thickness z_1 influences the emissivity κ, provided z_1 does not exceed 0.4 wavelength. The intensity of fluctuations of κ due to the oscillatory terms is higher for greater wavelengths and smaller e_1's. A study of type-2 real-world profiles has shown that the moisture profile 'inflection' typically occurs at a depth of 15 cm or more which conforms to the 0.4 λ requirement at wavelength up to 40 cm. For this reason expression (8) for ΔR obtained for model I also holds for model II.

4.3. Model III

It is clear that at a fixed e_1 the emissivity oscillates around the respective value of the Fresnel coefficient in the expression for e_2. The magnitude of these oscillations is determined by z_1 as well as by e_1 and e_2. The layer reflectivity is equal to the Fresnel reflection coefficient. This conclusion agrees with the previous results indicating that a nonreflecting layer can be found for any frequency. The oscillatory behaviour of the radiative characteristics of the medium represented by (6) leads to a rather uncertain relationship between ΔR and the profile parameters when $z/\lambda > 0.3$. Then the expression for ΔR is:

$$\Delta R \approx F_1 \sin(F_2) \tag{9}$$

where F_1 is a function of z_1/λ, e_1, and e_2. Also, $F_2 = ((e_1 - e_2)/c,\ z_1/\lambda)$, and c is a constant. With $z_1/\lambda < 0.2$, the behavior of $\kappa(e_2)$ closely resembles that of the same function for a double-layer medium. In this case z_1 can be determined in addition to e_1 and e_2 from data measured at three wavelengths. The relationship of ΔR and z_1 and e_2 can be represented in matrix form while e_1 can be determined from the microwave measurements.

4.4. Model IV

An analysis of simulated reflectivities of the medium whose dielectric properties are represented by (7) has shown ΔR to be determined by

$$\Delta R \approx c_1\lambda|e_1|^{-3/2}|a_1| - c_2\lambda|e_1|^{-3/2}|a_2| \tag{10}$$

c_1 and c_2 being some positive factors, a_1 and a_2 being coefficients in (7).

A comparison of (8) and (10) clearly demonstrates that (8) corresponds to the first term in (10) and obviously represents the linear effect of parabolic layer. The second term describes the deviation of a parabolic layer. The second term describes the deviation of a parabolic profile from a linear one and is negative since the effect of deeper layers on ΔR is weaker than in the case of a layer with $e = e_1 \pm az$. The second term is just 10 to 20 percent of the value of the first term in (10) which is never greater than 0.01.

An examination of expressions (8)–(10) shows that for all of the above models, save Model III, spectral measurements furnish information on the dielectric constant e (i.e. moisture content) at the air/soil interface and on the gradient of e (i.e. the moisture gradient) in the upper soil layer. This kind of dependence is more complex for Model III.

When the transition layer thickness is greater than 0.3 wavelength this layer behaves essentially as a matching layer so that the reflectivity is determined by moisture at the air/soil interface; mean while, a thickness of 0.2 wavelength or less, measured spectral data permit estimation of the above parameters of the profile of e (or moisture content). Laboratory measurements have been made to verify these conclusions.

4.5. Model of soil medium with transition layer

The models III in Fig. 2 and in Fig. 3(a) describe this type of soil.

There is a thin surface or near surface transition layer. Studies show that overdried layer is formed under the soil surface in the course of a drying up process within 1.5 to 6 days (depending on the particular climatic zone) after watering or rainfall. Capillary bonds break at the lower boundary of this layer due to evaporation and a dry soil layer up to several centimeters thick and more (layer II in Fig. 5) may develop above the transition layer. The thickness of transition layer varies from a few millimeters to 3–5 cm depending on the climatic region and mechanical composition of soil.

Figure 5 shows the electrodynamic model of nonuniformly moistened soil surface transition layer, where

medium I is an air with $e = 1$;

medium II is dry soil with known e_d;

medium III is a soil layer with e, varying exponentially from e_d to e_l corresponding to some moisture content;

medium IV is a soil layer with e_l and moisture close to constant.

A jump in the moisture content (and thus in the dielectric constant) takes place between the upper and the lower boundaries of the transition layer due to the fact that the layer is rather thin and some capillary bonds are broken. Thus, for dielectric constant of medium

$$e(z) = \begin{cases} e_d & \text{if } 0 < z < h_l \\ e_d \exp(az) & \text{if } h_l < z < h_l + h_2 \\ e(z) & \text{if } z > h_l + h_2. \end{cases} \tag{11}$$

Calculations based on the model II showed that the thickness h_l of upper layer II (in Fig. 5) (transparent for microwaves) and the form of $e(z)$ in this layer have practically no effect on R.

It was found that when the normalized thickness of the transition layer III h_2/λ is less than 0.15, the form of $e(z)$ has practically no influence on emissivity at fixed values of e_d and e_l while the effect is more significant at higher thicknesses.

Transition layer is essentially a matching medium when h_2/λ is greater than 0.15 to 0.20. With h_2/λ around 0.07 to 0.10, the fractional error in an estimated e (and thus ρ_w) is more than 30% when measurements are made at a single wavelength.

5. EXPERIMENTS

5.1. *Laboratory measurements*

Laboratory measurements included use of microwave radiometers to measure κ at 2.25, 18, 21, 27, and 30 cm wavelengths for soil samples whose structures matched that in the model.

The instruments were calibrated with respect to two reference samples (one absorbing, another reflecting) whose emissivities were known. Soil structures with different top layer moisture contents and different moisture gradients were studied under laboratory conditions. A dry transition layer h_2 thick was modelled at the surface using pretempered sand and dry lumps of soil having a diameter from a few mm to 2 or 3 cm. The transition layer depth h_2 varied between 1 and 10 cm.

Comparison between measured and simulated data shows good agreement.

5.2. *Field experiments*

Field experiments have been carried out in different climatic zones mainly in the Uzbekistan, Turkmenia, Moldavia, Estonia, Ukraine (Odessa and Dnepropetrovsk regions), Russia (Saratov and Krasnodar region), East Germany, Bulgaria, Poland, and Hungary.

Detailed *in situ* measurements of soil moisture content at 50 and 90 cm below the soil surface were made along test routes using a thermogravimetric technique. Calibration ground measurements were separated by 300 to 600 m, and 3 to 5 repeated measurements were made at each point. The total number of routes was about 40 with total of 1500 *in situ* measurement points. The flight measurements of the microwave emissive performance of soils were complemented by concurrent *in situ* measurements of the moisture content under the test routes.

The field data compared with model predictions for moisture content ρ_w between 0.2 and 0.3 g/cm^3 for Estonia, Bulgaria, Saratov region, Uzbekistan, and Turkmenia are given in Fig. 6.

Predicted and measured data are compared for a typical case in this figure. It is seen that h_2 strongly depends on the climatic zones, decreasing from 2 and 4 cm for arid regions to 0 or 1 cm for humid regions. Model calculations agree fairly well with the measured data.

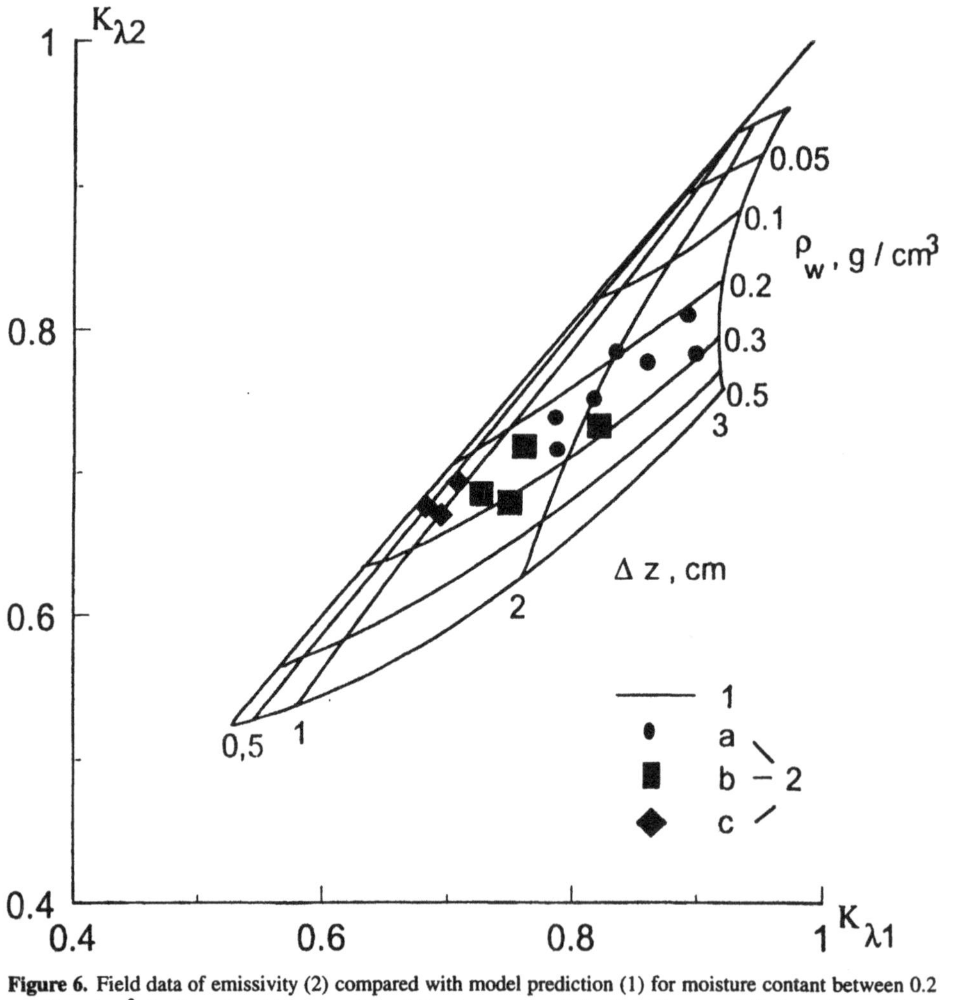

Figure 6. Field data of emissivity (2) compared with model prediction (1) for moisture contant between 0.2 and 0.3 g/cm^3 for Estonia (2c), Bulgaria and Saratov region (2b), Uzbekistan and Turkmenistan (2a) at X-band (λ_2) and L-band (λ_1).

6. DETERMINATION OF NET WATER CONTENT, MOISTURE GRADIENT, AND PARAMETERS OF VERTICAL MOISTURE PROFILE

6.1. Determination of soil moisture content

A single band measurement can provide only the net water content of a soil layer. By recalculation, the reflectivities of inhomogeneous structures of the type shown in (4) through (7) into effective moisture parameters and by comparing the resulting values with the net water content of different layers of an inhomogeneous soil structure, one can determine the average thickness of a layer within which the moisture level can be determined to a specified accuracy. The results of calculation show that the thickness of a layer within which the moisture content can be derived from single band measurements is determined primarily by the top layer moisture content, moisture gradient, and the wavelength, while being essentially independent (within the framework of the above models) of the vertical moisture profile.

The moisture content of nonuniformly moistened type II soils have been measured in the field to compare the theoretical and experimental data obtained. Remote sensing estimates of moisture content, obtained by spectral processing of microwave radiometric data at 18 cm and 30 cm, are compared with single band (18 or 30 cm) measurements of moisture content at depths between 2 and 40 cm from the surface.

Field experiments in the different regions lead to the conclusion that the proposed technique enables the moisture content ρ_w at the lower boundary of a transition layer with a thickness h_2 to be determined within 0.04 g/cm^3, and the thickness h_2 itself within 0.8 cm.

Moisture content values obtained for different types of soils through spectral processing of radiometric data are by a factor of 2 or 3 more accurate than those derived from single-band measurements.

6.2. Moisture gradient estimates

It follows from (8) and (10) that in the above models the moisture gradient a can be estimated, in addition to the net water content of a soil layer, using spectral data. In particular, the sensitivity of the microwave brightness temperature contrast ΔT_b to a variation and the uncertainty in the top-layer moisture content ρ_1 are the main sources of error in the moisture gradient a determination from microwave radiometric measurements.

The magnitudes of the error in a are higher for higher top-layer moisture content, higher gradient magnitudes and shorter wavelengths, while fractional errors decrease with gradient.

6.3. Moisture profile estimates

The authors have obtained qualitative relationships between variations of plane-wave reflectivity and the vertical profile of the soil dielectric constant. An analysis of these relationships shows that spectral microwave radiometric measurements permit estimation of ρ_w (equivalent to some value of the dielectric constant) and of the moisture profile in a linear approximation (i.e. for $\partial\rho_w(z)/\partial z\,|_{z=0}$). However, many applications require near real time data on soil moisture profiles down to depths as low as 70 to 100 cm or at least on the net water content of the top 70 to 100 cm of the soil.

A technique for determining the net water content of the top 100 cm layer of soil and for recovering the moisture profile of the same layer is worked out. This technique relies on microwave radiometric data and *a priori* knowledge of the hydrophysical properties of the soil. The main *a priori* knowledge includes data of wilting point and field capacity at the depths of 0, 50, and 100 cm, precipitation (watering), diurnal mean soil temperature. It applies to virtually any real world moisture profile as well as to any real type of soil.

For details, please see [13].

7. RADIATION-SOIL MOISTURE DEPENDENCIES UNDER THE DYNAMICS OF EVAPORATION

As it has been pointed out in paragraph 2, for a constant with depth soil moisture profile, the drying up of soil results in emissivity changes in agreement with (1) as seen in Fig. 1.

For a model of soil moisture profile with dry surface layer and constant value of moisture below it, emissivity, in accordance with the previously given explanation, is a function of soil moisture below the dry layer. It is obvious that a slope of these curves is a function of the thickness of upper layer and the moisture content below.

Some examples of soil moisture profiles are shown in Fig. 7(a′). Their approximations are given in Fig. 7(a). For each profile, variation of a soil moisture below the upper layer under fixed data of thickness and moisture content result in changes of emissivity, as seen in Fig. 7(b) (solid lines).

Thus we have a family of 'radiation vs. soil moisture below upper dryer layer' curves fixed for each curve data of soil moisture and thickness of upper layer. There are static $\kappa(\rho_w)$ dependencies. Under the dynamics of soil drying the 'working point' changes

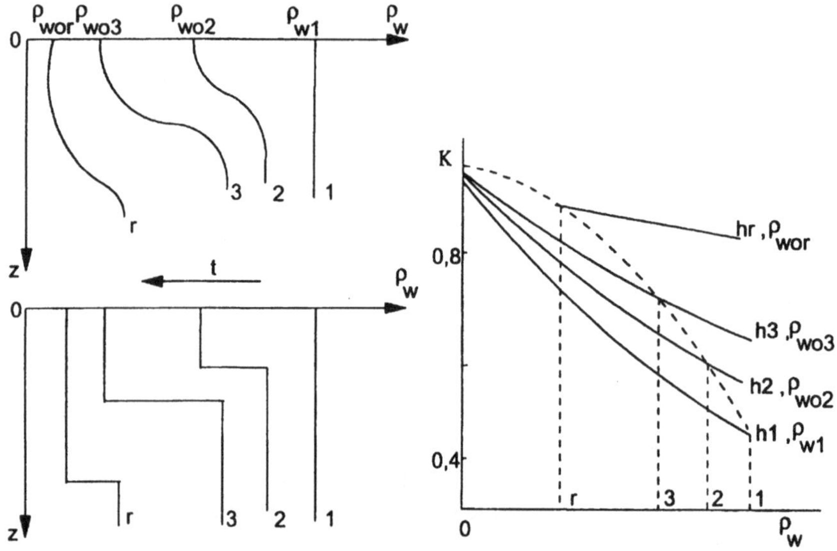

Figure 7. The static (solid lines) and dynamic (dashed lines) dependencies of the emissivity (b) on soil moisture below upper dry transition layer (a, a′), (schematic representation).

its position from curve to curve thus forming dynamic $\kappa(\rho_w)$ dependence shown with a dashed line in Fig. 7(b).

The dynamic curves described above may be used for determining soil moisture below upper layers during the drying up process.

8. CONCLUSION

The results of the described theoretical and experimental investigations show:

- emissivity at microwaves is sensitive to specific peculiarities of soil moisture profile, first of all to soil moisture at the top of medium and subsurface gradient of soil moisture;
- sensitivity of radiation to the parameters of soil moisture profile differs along spectrum;
- quantitative data of sensitivity are dependent on the type of soil-climatic zone, the presence of shallow water table, the conditions of moistening and drying; this shows a necessity of some *a priori* knowledge about an area to get the correct interpretation of obtained results;
- the dynamic 'radiation vs. soil moisture content' dependence differs from one for static conditions;
- approaches for solving the inverse problem of the moisture profile determination in top 100 cm soil layer have been formulated and the feasibilities and limitations of their determination studied;
- the method for determination of the characteristics of moisture profile and net water content in root zone based on spectral microwave radiometric data and *a priori* knowledge about specific hydrophysical properties and type of soils have been worked out;
- it has been shown that dynamic 'radiation-moisture content' calibration curve may be used to obtain the soil moisture estimates below upper transition zone;
- representative experiments conducted in different areas of the world have proven the principal theoretical results and have revealed that the dual-channel measurements along with prior-knowledge-based information about the soil properties can be practically used for near-real-time measurements of 3 to 5 gradations of moisture content in top 100 cm layer.

REFERENCES

1. A. Stogrin. The brightness temperature of vertically structured medium. *Radio Science*, **5**, (12), 1397 (1970).
2. T. Wilheit. *Radiative Transfer in a Plane Stratified Dielectric*. NASA Rep. X-911-75-66, Greenbelt, USA, GSFC (1975).
3. L. E. Tsang, E. G. Njoku and J. A. Kong. Microwave thermal emission from a stratified medium with nonuniform temperature distribution. *J. Appl. Phys.*, **46**, 5127 (1975).
4. V. V. Bogorodskij, A. I. Kozlov and L. T. Tuchkov. *Thermal Radiation of the Earth's Surface*. Leningrad, Gidrometeoizdat (in Russian) (1977).
5. K. Ya. Kondratiev, A. A. Grigoriev, Yu. I. Rabinovich and E. M. Shulgina. *Space-Borne Meteorological Sensing of the Earth Surface*. Leningrad, Gidrometeoizdat (in Russian) (1979).

6. E. A. Reutov and A. M. Shutko. The influence of subsoil moistening and the depth to a shallow water table on microwave emission characteristics of terrain covers. In: *Proc. 12th All Union Conference on Radio Waves Propagation, Abstracts*, Vol. 2, p. 249, Moscow, Nauka Publishing House, (in Russian) (1978).
7. A. M. Shutko. Microwave radiometry of land under natural and artificial moistening. *IEEE Trans. Geosci. Remote Sensing*, **GE-20**, 18–26 (1982).
8. A. M. Shutko and E. A. Reutov. Statistical processing of microwave radiometric data for sites with shallow water table. In: *Proc. of Workshop on 'Statistical methods for processing of remotely sensed data'*, pp. 34–38, Moscow, IRE AS USSR (in Russian) (1983).
9. T. Schmugge. Remote sensing of soil moisture: recent advances. *IEEE Trans. Geosci. Remote Sensing*, **GE-21** (3), 336–344 (1983).
10. T. Schmugge, P. O'Neill and J. Wang. Passive microwave soil moisture research. *IEEE Trans. Geosci. Remote Sensing*, **GE-24** (1), 12–22 (1986).
11. T. Jackson. Soil water modelling and remote sensing. *IEEE Trans. Remote Sensing*, **GE-24** (1), 37–46 (1986).
12. A. M. Shutko. *Microwave (SHF) Radiometry of Water Surface and Ground*. Moscow, Nauka Publishing House (in Russian) and NASA TT, Doc. 89058401 (in English) (1986).
13. E. A. Reutov and A. M. Shutko. Prior-knowledge-based soil-moisture determination by microwave radiometry. *Soviet J. Remote Sensing*, **5** (1), 100–125 (1986).
14. F. A. Mkrtchjan, E. A. Reutov, A. M. Shutko, K. G. Kostov, M. A. Mikhalev, N. M. Nedelchev, A. Ya. Spasov and B. I. Vichev. Experiment in Bulgaria for determination of soil moisture in the top one-meter layer using microwave radiometry and a priori information. In: *Proc. IGARSS '88 Symposium, ESA Publ. Division*, pp. 665–666 (1988).
15. E. A. Reutov and A. M. Shutko. Microwave spectroradiometry of water content of nonuniformly moistened soil with a surface transition layer. *Soviet J. Remote Sensing*, **6** (1), 72–79 (1990).
16. A. M. Shutko. *Soil/Vegetation Characteristics at Microwave Wavelengths, TERRA 1: Understanding the Terrestrial Environment*. Paul M. Mather (Ed.). London, Washington, DC, Taylor & Francis (1992).
17. H. Weichelt, K. Schmidt, A. Shutko, E. Reutov, G. Ijash, T. Jackson, T. Schmugge, C. Swift, P. Pampaloni, S. Paloscia and K. Rao. Progress and problems in microwave radiometric remote sensing of soilmoisture (Instruments and application). In: *Proc. ISPRS '92 Congress*, Washington (1992).

ESA/NASA International Workshop, pp. 475–484
B. J. Choudhury, Y. H. Kerr, E. G. Njoku and P. Pampaloni (Eds)

Soil moisture estimation from multifrequency microwave radiometer data based on statistical inversion technique

K. S. RAO,[1] Y. S. RAO[1] and G. IJJAS[2]

[1] *Centre of Studies in Resources Engineering, Indian Institute of Technology, Powai, Bombay 400 076, India*
[2] *Department of Microwave Telecommunication, Technical University of Budapest, Goldman Gy ter 3, H1111, Budapest, Hungary*

Abstract—An attempt is made to retrieve the vertical profile of soil moisture based on the three-frequency statistical inversion technique. A three-parameter exponential function was used to represent the soil moisture profile. Initially, the model was tested by retrieving simulated soil moisture profiles using theoretically computed three-frequency brightness temperature T_B data. Good correlation was found between simulated and estimated soil moisture profiles. The model was further verified by retrieving measured soil moisture profiles using experimentally measured multifrequency T_B data. It is observed from the results that the estimation of the vertical profile of soil moisture is possible for bare soils within a standard deviation of ±5% by weight.

1. INTRODUCTION

Several investigators demonstrated the potential of passive microwave remote sensing for the estimation of surface soil moisture (0–5 cm) through field, aircraft and satellite data analysis, which was summarized by Schmugge *et al.* [1] and subsequently by Jackson and Schmugge [2]. For some of the hydrological and agricultural applications, the soil moisture information at greater depths is preferable. Several attempts [3–6] were made to develop models to retrieve soil moisture at greater depths, based on correlation techniques and a priori information. Biswas and Dasgupta [3] and Arya *et al.* [4] made attempts to find a correlation between surface and subsurface soil moisture. Since the surface soil moisture can be measured using remote sensing techniques, the above correlation will help to predict the subsurface soil moisture. Reutov and Shutko [6] used the multifrequency approach coupled with *a priori* information of the test site to map subsurface soil moisture.

Our attempt for the estimation of subsurface soil moisture is based on multifrequency statistical inversion technique which was formerly reported by Njoku and Kong [7]. They retrieved soil moisture and temperature profile parameters based on the four-frequency statistical inversion technique. In this paper, we used the inversion technique for the retrieval of the vertical profile of soil moisture. The subsequent sections deal with the details of the parameterization of soil moisture profile, statistical inversion technique and results and discussion.

2. PARAMETERIZATION OF SOIL MOISTURE PROFILE

The vertical profile of soil moisture can be represented by a model which takes into account various soil and climatic conditions, agricultural practices, etc. [5]. An approximate approach to represent the soil moisture profile is as follows [6, 7].

$$\rho(z) = \rho_s + \Delta\rho \frac{\left(e^{\beta z} - 1\right)}{\left(e^{-\beta d} - 1\right)} \tag{1}$$

$$\rho(z) = a_0 + a_1 z + a_2 z^2. \tag{2}$$

Equation (1) is an exponential function with parameters ρ_s, $\Delta\rho$ and β [7]. $\rho(z)$ is the soil moisture at depth z, ρ_s is the surface soil moisture, $\Delta\rho$ is the variation of soil moisture from surface to the depth d, and β is the profile trend parameter. Equation (2) is a second degree polynomial with parameters a_0, a_1, and a_2.

The above two functions were examined to represent the experimentally observed soil moisture profiles. Two such profiles are shown in Fig. 1. One profile represents a steep gradient near the surface, characterized by high soil moisture content. Another profile has a very low gradient near the surface, characterized by low soil moisture content. The fitted lines using Equations (1) and (2) are superimposed on the experimental data points. It can be seen from Fig. 1 that the exponential function fits well with the experimental

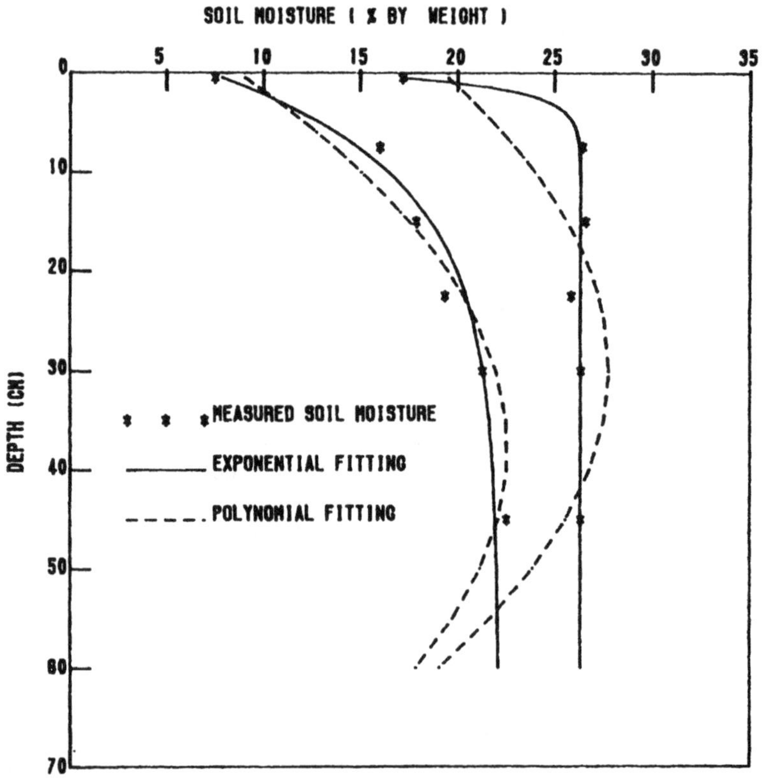

Figure 1. Measured soil moisture values are fitted to both exponential and second degree polynomial functions.

data for both the profiles compared to the polynomial expression. Therefore, we prefered an exponential function to represent soil moisture profiles, as recommended by Njoku and Kong [7].

3. INVERSION TECHNIQUE

The soil moisture profile parameters can be formed into parameter vector **p** and can be written as:

$$\mathbf{p} = \begin{bmatrix} \rho_s \\ \Delta\rho \\ \beta \end{bmatrix}.$$

The multifrequency brightness temperature data are transformed into emissivity (ε) values using effective soil physical temperature [8]. The data vector can be written as:

$$\mathbf{d} = \begin{bmatrix} T_{B1} \\ T_{B2} \\ T_{B3} \end{bmatrix} \longrightarrow \begin{bmatrix} \varepsilon_1 \\ \varepsilon_2 \\ \varepsilon_3 \end{bmatrix}.$$

The statistical inversion technique reported by Njoku and Kong [7] is as follows:

$$\mathbf{p}^* = \mathbf{D} \cdot \mathbf{d} \tag{3}$$

where $\mathbf{p}^*$ is the estimated profile parameter vector. **D** is a matrix given by

$$\mathbf{D} = \mathbf{C}(\mathbf{p}, \mathbf{d}) \cdot \mathbf{C}^{-1}(\mathbf{d}, \mathbf{d}) \tag{4}$$

where $\mathbf{C}(\mathbf{X}, \mathbf{Y}) = \langle \mathbf{X} \cdot \mathbf{Y}^{\mathrm{T}} \rangle$.

The attenuation of microwaves in the soil medium and the soil moisture profiles are nonlinear in nature, whereas the inversion technique discussed above is a linear type. Therefore, the data vector **d** is augmented to include the nonlinear terms [7]. The augmented data vector $\mathbf{d}'$ is given by

$$\mathbf{d}' = \left[1,\ \varepsilon_1,\ \varepsilon_2,\ \varepsilon_3,\ \varepsilon_1^2,\ \varepsilon_2^2,\ \varepsilon_3^2,\ \varepsilon_1\varepsilon_2,\ \varepsilon_1\varepsilon_3,\ \varepsilon_2\varepsilon_3\right]^{\mathrm{T}}.$$

To estimate the **D** matrix, a set of known soil moisture profiles (simulated or measured) and corresponding multifrequency brightness temperature data (computed theoretically or experimentally measured) will be required. This **D** matrix along with T_B data will be used subsequently for the retrieval of the soil moisture profile.

4. RESULTS AND DISCUSSION

A set of 15 soil moisture profiles were simulated up to a depth of 45 cm, based on the knowledge of observation of weekly soil moisture profiles for a period of 2 years [9] over an Indian test site, and published representative soil moisture profiles [7]. These profiles are a cumulative effect of all soil characteristics and environmental conditions. Therefore, this acts as a training set for retrieving soil moisture profiles for this test site.

A radiative transfer model developed by Wilheit [10] coupled with four-component dielectric model [11] were used to estimate emissivities at the frequencies 1.4, 5.0, and 10.7 GHz. The multifrequency emissivity data along with profile data were used in the inversion technique to estimate the **D** matrix. This **D** matrix in turn is used to retrieve the simulated 15 soil moisture profiles. The retrieved soil moisture profile parameters were compared with the simulated profile parameters. Figures 2(a)–(c) show the actual versus the estimated profile parameters ρ_s, $\Delta\rho$ and β, respectively. The solid line represents the 1:1 line. It can be seen from Fig. 2(a) that the surface soil moisture ρ_s could be estimated very accurately as all the points are close to 1:1 line. The correlation coefficient is 0.99. The difference in soil moisture between surface and greater depth $\Delta\rho$ also could be estimated with acceptable accuracy as can be seen from Fig. 2(b). However, the errors in the estimation of profile shape parameter β are very high as can be observed from Fig. 2(c). The correlation coefficient between actual and estimated β values is 0.87.

It is well known that the penetration ability of microwaves depends on frequency, in addition to several other parameters. The lower frequencies penetrate deeper than the higher frequencies. To study the effect of errors in brightness temperature on the estimated profile parameters, random noise was generated in T_B and the profile parameters were re-estimated. It has been observed that an error in T_B at 1.4 GHz affects the estimation of both $\Delta\rho$ and β parameters. An error in T_B at 5.0 GHz affects

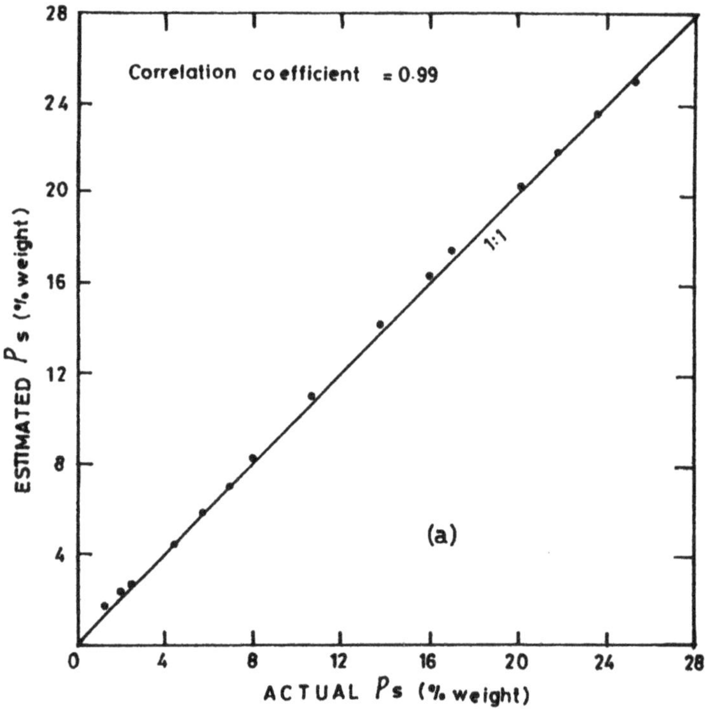

Figure 2a. Estimated versus actual soil moisture profile parameters: surface soil moisture (ρ_s). The estimated values were obtained using three frequency inversion technique.

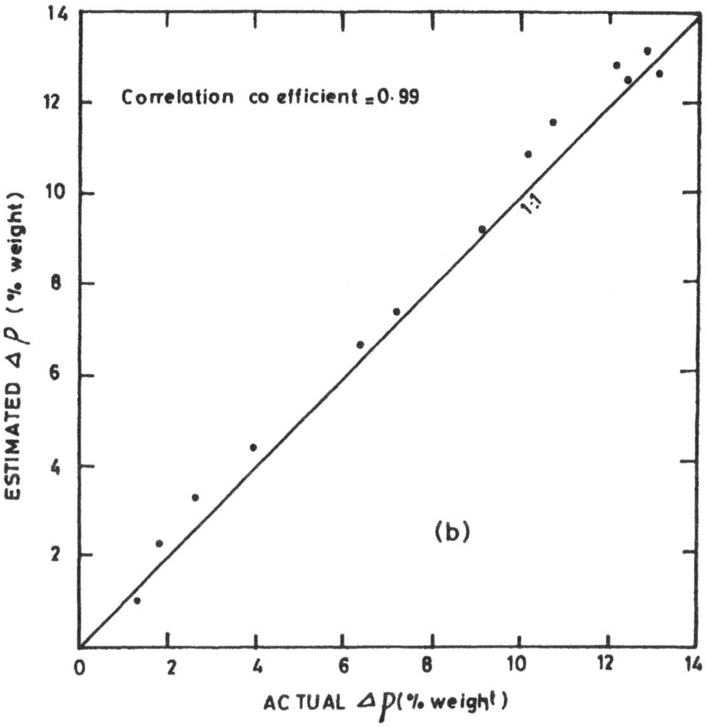

Figure 2b. Estimated versus actual soil moisture profile parameters: variation in soil moisture between surface and deeper depth ($\Delta\rho$). The estimated values were obtained using three frequency inversion technique.

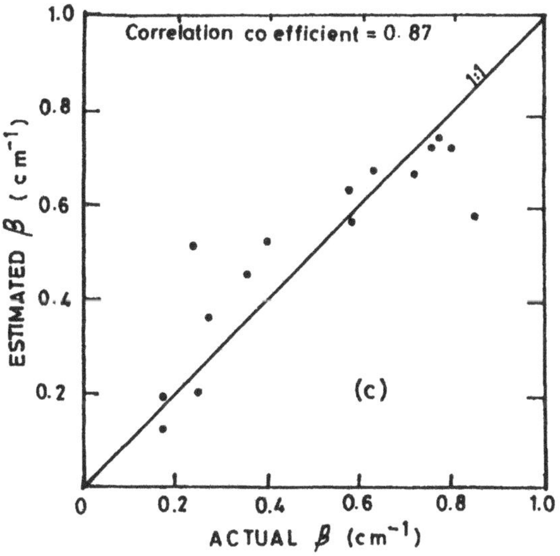

Figure 2c. Estimated versus actual soil moisture profile parameters: profile trend parameter (β). The estimated values were obtained using three frequency inversion technique.

the estimation of ρ_s. An error in T_B at 10.7 GHz has least affect on the estimation of profile parameters.

The above discussion implies that it will be possible to estimate the surface soil moisture and the greater depth soil moisture very accurately. However, its distribution with depth, which depends on β could not be estimated with good accuracy. The vertical profile is computed using the estimated profile parameters and compared with the simulated profiles and good agreement between the two was found. The standard deviation between estimated and actual soil moisture profile is $\pm 2\%$ by weight. The applicability of **D** matrix for a totally different set of data is examined for the retrieval of soil moisture profile with standard deviation of $\pm 5\%$ [12].

Inversion technique was further verified using available experimental data. Systematic field experiments were conducted during August 1990 using truck mounted 1.4 and 2.7 GHz microwave radiometers and a thermal infrared radiometer over three test plots near Budapest. The vertical profile of soil moisture upto 40 cm depth was measured synchronous with the T_B measurements. Two-frequency inversion technique was applied to retrieve soil moisture profiles upto 40 cm, and compared with the measured profiles. We have used the **D** matrix derived from USDA drying down profiles [7] as the soil types are close to each other. It was observed that the soil moisture could be estimated upto 20 cm depth accurately. For further deeper depths, there was a divergence between the actual and estimated profiles. The measured profiles showed an increase in soil moisture upto 20 cm depth and then started decreasing sharply. Therefore, the estimated profiles agreed up to that depth and showed constant soil moisture beyond 20 cm depths [13].

The results discussed so far use the **D** matrix based on simulated data, though it was verified with experimental data. Due to the limitations of the theoretical models used for the simulation of emissivities, there is a possibility of getting larger errors. Here, the **D** matrix is completely based on experimental data and also verified with experimental profiles. Wang *et al.* [14] carried out field experiments at Beltsville Agricultural Research Centre (BARC), USA, test site during summer 1981, using truck mounted multifrequency radiometers operating at frequencies 0.5, 1.4, 5.0, and 10.7 GHz. The measurements were made at multiple incidence angles from 10° to 70°. Ground truth data in terms of vertical profile of soil moisture and temperature upto 15 cm depth, and other related soil and vegetation parameters were measured. Some of these fields were bare and smooth with sandy loam soils. The **D** matrix is determined for the three-frequency inversion model, based on the experimental data of Wang *et al.* [14]. To achieve better statistics, it is required to consider a large number of samples. However, due to lack of proper data, we considered only 11 profiles and their corresponding experimental T_B data. These profiles consist of surface soil moisture which varies from 2% to 17% and a slope ranging from very low to high with both increasing and decreasing profile trends. The T_B data at frequencies 1.4, 5, and 10.7 GHz and at observation angle 10° were considered for the three-frequency inversion technique, and the **D** matrix is calculated. The **D** matrix is given below:

$$\begin{bmatrix} 39.4 & 139.3 & -83.6 & -38.6 & 35.8 & -105.4 & -10.7 & 28.8 & -72.0 & 64.2 \\ 24.6 & -330.7 & 219.9 & 58.8 & -67.8 & 62.3 & 13.9 & -47.2 & 170.4 & -86.2 \\ -1219 & -4179 & 2027 & 1459 & -81.3 & 2536 & -1367 & -1943 & 4854 & -2016 \end{bmatrix}$$

The above 11 soil moisture profiles were retrieved based on the three-frequency model and shown in Figs 3(a)–(d) along with the experimentally measured profiles. It can

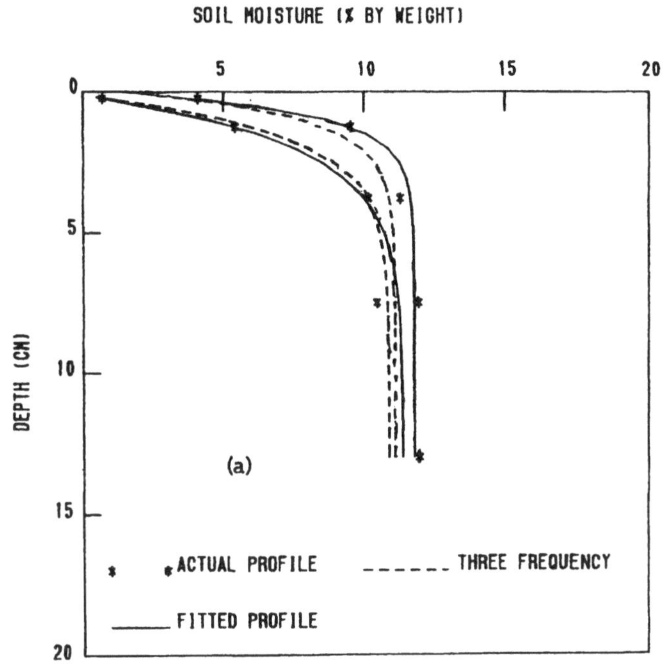

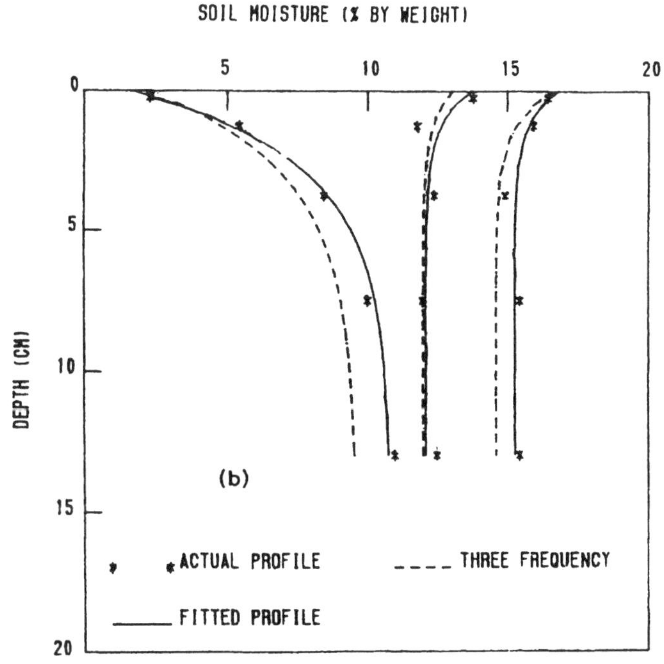

Figure 3. Comparison of retrieved soil moisture profiles with actual profiles reported by Wang *et al.* [14].

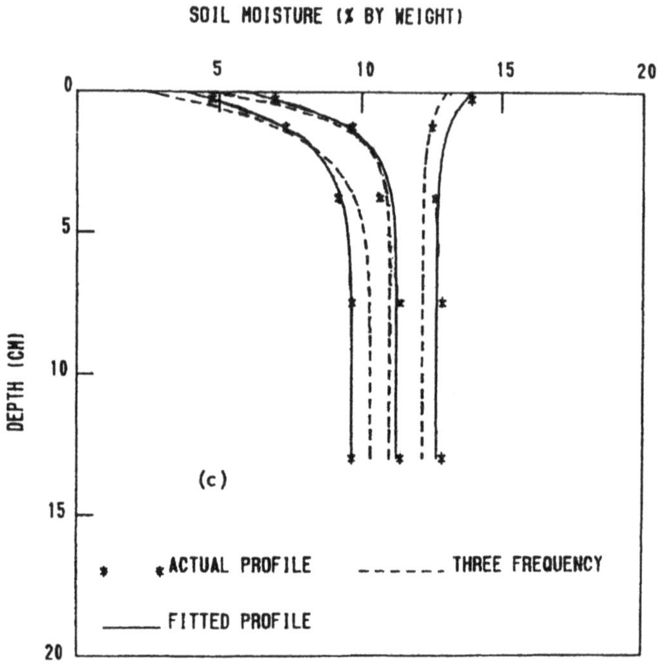

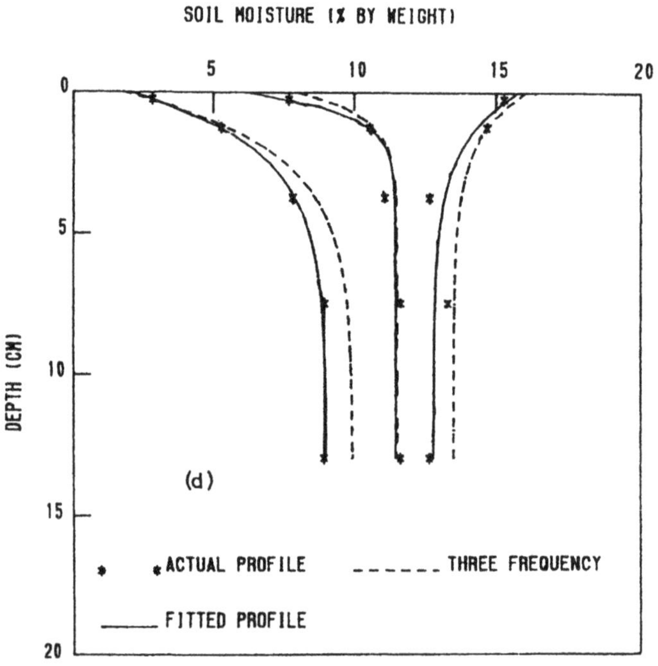

Figure 3. (Continued).

be seen from Fig. 3 that the surface soil moisture and the slope near the surface are estimated with good accuracy. The standard deviation in retrieving soil moisture at greater depth is ±2% by weight. The high accuracy in the retrieval of soil moisture profile can be understood since the same profile data was used to compute the **D** matrix. When this **D** matrix was applied on another set of T_B data to retrieve soil moisture profiles, a decrease in the retrieval accuracy of soil moisture (±5%) at greater depths was noticed. It can be seen from Fig. 3 that there are both increasing and decreasing soil moisture profiles with depth. Since multifrequency T_B data contains the profile information, both types of profiles could be retrieved using inversion technique.

5. CONCLUSIONS

The use of the three-frequency inversion technique for the retrieval of vertical profile of soil moisture is demonstrated. This technique is limited to bare and smooth fields. Under high surface roughness and vegetation cover, the model needs substantial modification. Our further investigations are directed at this problem.

Acknowledgements
The authors are grateful to Prof. C. Natarajan, Head, CSRE for providing necessary facilities and Department of Science and Technology (DST) for sponsoring the project under Science and Engineering Research Council (SERC) program (No. D.O. No. ES/23045/88).

REFERENCES

1. T. Schmugge, P. E. O'Neill and J. R. Wang. Passive microwave soil moisture research. *IEEE Trans. Geosci. Remote Sensing*, **GE-24**, 12–22 (1986).
2. T. J. Jackson and T. J. Schmugge. Passive microwave remote sensing system for soil moisture: Some supporting research. *IEEE Trans. Geosci. Remote Sensing*, **GE-27**, 225–235 (1989).
3. B. C. Biswas and S. K. Dasgupta. Estimation of soil moisture at deeper depth from surface layer data. *Mausam*, **30**, 511–516 (1979).
4. L. M. Arya, J. C. Richter and J. F. Paris. Estimating profile water storage from surface zone soil moisture measurements under bare field conditions. *Water Resources Res.*, **19**, 403–412 (1983).
5. T. J. Jackson. Soil water modelling and remote sensing. *IEEE Trans. Geosci. Remote Sensing*, **GE-24**, 37–46 (1986).
6. E. M. Reutov and A. M. Shutko. Prior-knowledge based soil moisture determination by microwave radiometry. *Sov. J. Remote Sensing*, **5** (1), 100–125 (1986).
7. E. G. Njoku and J. A. Kong. Theory of passive microwave remote sensing of near-surface soil moisture. *J. Geophys. Res.*, **82**, 3108–3118 (1977).
8. B. J. Choudhury, T. J. Schmugge and T. Mo. A parameterization of effective soil temperature for microwave emission. *J. Geophys. Res.*, **87**, 1301–1304 (1982).
9. Reports on soil moisture and temperature profiles, 1986 to 1988. Indian Meteorological Department, Agromet Division, Pune, India.
10. T. T. Wilheit. Radiative transfer in a plane stratified dielectic. *IEEE Trans. Geosci. Electron.*, **GE-16**, 138–143 (1978).
11. M. C. Dobson, F. T. Ulaby, M. T. Hallikainen and M. A. El-Rayes. Microwave dielectric behavior of wet soil. Part II: Dielectric mixing models. *IEEE Trans. Geosci. Remote Sensing*, **GE-23**, 35–46 (1985).

12. K. S. Rao and Y. S. Rao. Three frequency model for the retrieval of vertical soil moisture profile. *Internal Report,* Ictp, Trieste, Italy, No. IC/90/113 (1990).

13. G. Ijjas, K. S. Rao and Y. S. Rao. Efficiency of 1.4 and 2.7 GHz radiometers data for the retrieval of soil moisture profile. In: *Proc. of IGARSS Symp.*, pp. 765–768, Espoo, Finland (1991).

14. J. R. Wang, T. J. Jackson, E. Engman, W. Gould, J. Pauches, W. Glazer, P. E. O'Neill and T. Schmugge. Microwave radiometers experiment of soil moisture at BARC test site during summer 1981. *NASA Tech. Memo 86056* (1984).

ESA/NASA International Workshop, pp. 485–498
B. J. Choudhury, Y. H. Kerr, E. G. Njoku and P. Pampaloni (Eds)

Retrieval of soil moisture profile by combined remote sensing and modeling

DARA ENTEKHABI,[1] HAJIME NAKAMURA[1] and ENI G. NJOKU[2]

[1]*Department of Civil and Environmental Engineering, Massachusetts Institute of Technology, Cambridge, Massachusetts 02139, USA*
[2]*Jet Propulsion Laboratory, California Institute of Technology, Pasadena, California 91106, USA*

Abstract—The objective of this paper is to develop a statistically optimal assimilation algorithm which estimates the entire system state, i.e. vertical soil moisture and temperature profiles, from information which is only available for a portion of the system (namely, the near-surface soil moisture and temperature). A physical model of coupled heat and water flux is used to propagate the information to the unobserved portion of the system. This problem will be regarded as the inverse problem, combining the radiative transfer equation as the observation equation and the water and heat transport equation as the system equation. The Kalman filter statistical scheme is then used to propagate (forecast) the best estimate of the system (conditional mean) based on the initial conditions and the system equation. Occasionally, as remote sensing observations become available (in the form of brightness temperature), the observation equation is used to update the states with the information extracted from the measurements.

1. INTRODUCTION

The space-based measurement of emitted radiation in various spectral intervals contains relevant information about the state of the surface and the intervening atmosphere. Operational and research satellite and aircraft programs have expanded to become the major source of observational data for earth and atmospheric science applications.

The measured emitted radiation in itself is of little value; there is a need for geophysical calibration whereby the radiation intensity in a specific spectral range, often in combination with others or transformed functionally, is related to an environmental variable of interest. The mappings between the radiation readings and the environmental variable are often established on empirical bases.

The purpose of this paper is to introduce a theoretical assimilation scheme that serves three major purposes: 1) it provides the geophysical calibration for the remote sensing of an important state in land–atmosphere interaction, namely the soil moisture profile, 2) it produces the mapping between passive microwave emission measurements and the volumetric soil moisture content profile based on first principles in radiative transfer and without reliance on empirical relations, and 3) it provides the framework for multi-spectral and multi-sensor data assimilation in a simultaneous and consistent manner.

Soil moisture at the surface is instantaneously important in partitioning the net radiation into latent and sensible heat fluxes; the soil moisture at depth is critically important in determining the response of the land system to cycles of intermittent storm and interstorm forcing. During a storm, the infiltration rate (that partitions the incoming precipitation into runoff and infiltration) is dependent on the soil moisture content ahead of the wetting front at depth. During interstorm periods, the evapotranspiration rate is dependent on the moisture state below the shallow depth that controls the instantaneous response. Determination of water balance over land regions is critically dependent on the availability of accurate information on the soil moisture down to a depth of one meter or more.

Selective spectral ranges in the microwave (less that 10 GHz) has been identified as capable of providing some information on the moisture content of the soil. This microwave radiation is mostly unaffected by the intervening atmosphere and clouds. It is a rather weak emission; the dielectric constant properties of water stored in the soil affect its emission and attenuation significantly and it thus contains some information on soil moisture.

The emitted radiation and its brightness temperature is a weighted integral of the temperature of the soil column. The weighting is according to the abundance of water and other factors. The major problem is, however, that the radiation is significantly attenuated in the soil matrix; the e-folding depth scale of the weighting function is a function of the microwave frequency and the volumetric moisture content. This depth scale varies between a few centimeters to ten centimeters for the microwave range indicated above over a variety of hydrologic conditions. Clearly the soil moisture profile down to a depth of a meter or more cannot be determined directly by the use of microwave data alone.

In the next section, a review is made of some of the published techniques and schemes introduced in the literature that are designed to infer the soil moisture profile at depth from surface microwave (passive and active) observations. In the continuing sections, an algorithm is introduced that provides the geophysical calibration for soil moisture profile based on multi-spectral observations. It is a combined heat and moisture transport modeling and radiative transfer modeling algorithm. In this paper the focus is on algorithm development and testing; a synthetic example is used to illustrate a few characteristics of the retrieval system. The system is posed as an inverse-problem where only a few observations (brightness temperatures) are available as some integrated measures of a system with many degrees of freedom (profiles of soil moisture and temperature). Due to the excess of the degrees of freedom, the problem has no unique solution. An algorithm is developed to track the evolution of the soil heat and moisture transport in order to find constraining factors that will reduce the uncertainty on the solution to the inverse-problem. In the same process, observations are sequentially assimilated and used to update the knowledge of the states of the system. The final result is that by observing portions of the system (microwave observations of near-surface soil moisture), the information is transferred to lower depths by the transport model and the soil moisture profile is estimated down to a meter depth.

In the final section, concluding remarks are made regarding the algorithm. Several additional issues are discussed in this last section; the field-validation of the algorithm, the inclusion of vegetation and other factors are discussed and outlined as the focus of future investigations based on this preliminary study.

2. BACKGROUND

There have been a number of significant research efforts directed towards developing the capability of monitoring soil moisture by remote sensing techniques [1–5]. Schmugge *et al.* [6] find that the multi-sensor approach that includes a low frequency microwave component (1.4 GHz 21 cm L-band) is best suited for land-atmosphere interaction studies.

The critical need for estimating the soil moisture and temperature profile down to depths beyond the penetration reach of remote sensing radiometric observations has resulted in numerous investigations into this topic. These investigations mostly rely on estimates of the near-surface volumetric soil moisture that are made by applying empirical relations between the measured brightness temperature and surface soil moisture. Each study then uses a different scheme to infer the soil moisture at depth.

Jackson [7] developed an algorithm based on the assumption that the soil moisture profile in the prolonged interstorm period is in hydrostatic equilibrium. In this case the matrix head profile $\psi(z)$ has a slope of -1 with depth. When the surface soil moisture is determined by passive microwave remote sensing, a first-order estimate of the rest of the soil moisture profile may be made by invoking the hydrostatic assumption. Jackson [7] shows that this model is mostly applicable when the surface flux is small. When the interstorm period is characterized by strong drying due to evaporation, significant departures from hydrostatic equilibrium may occur. In response to this issue, Camillo and Schmugge [8] extended this model to include a root sink term. In a later investigation Camillo and Schmugge [9] relate cumulative rainfall (surrogate variable for water content in a column of soil) to the brightness temperature in the microwave range. A number of studies (e.g. in Russia) use multifrequency observation in conjuction with parameterized profiles of soil moisture state to infer moisture in the root zone (referenced in [3, 5]).

A number of studies use the surface soil moisture observation, as determined by empirical relations between soil moisture and microwave brightness temperature and reflectivities, as the concentration boundary condition on the vertical soil moisture diffusion equation. In this manner, sequential observations are used to infer the soil moisture profile. Bruckler and Witono [10] and Prevot *et al.* [11] use reflectivities from active microwave systems as the upper boundary condition for the soil water diffusion equation. Stroosnijder *et al.* [12] and Newton *et al.* [13] use passive microwave observations in a similar manner.

The purpose of this paper is to introduce an assimilation scheme that similarly combines transport modeling and remote sensing. The chief focus is nevertheless to reduce the reliance on empirical geophysical calibrations and instead solve the inverse-problem directly. The inverse-problem is essentially contained in the radiative transfer equation. The moisture heat and moisture transport equations are used to place constraints on the system. The two are combined in the context of a filter that merges them into one model. In this respect the report of Milly and Kabala [14] is of fundamental importance. In that study, the Kalman filter including a soil heat and moisture transport model is used to estimate the surface moisture content by tracking the diurnal surface temperature range changes. The thermal inertia of the surface is a function of the moisture content and thus the diurnal range measures degree of water saturation.

In the next section, the radiative transfer equations that define the inverse-problem and the heat and moisture transport and storage equations that provide the constraints are introduced. Finally the filtering algorithm that combines them is outlined and its features are demonstrated through an example.

This paper builds on results reported in [15] and provides a test of the algorithm under various forcing conditions. Since the focus of this paper is sensitivity studies on the algorithm, the full equations that constitute the algorithm are only summarized. The details may be found in [15] where the algorithm and its components are outlined.

3. THE RADIATIVE TRANSFER PROBLEM

The microwave brightness temperature is a complicated function of the soil dielectric constant and temperature profiles. Njoku and Kong [16] outline the principles behind the determination of the brightness temperature based on theory of electromagnetic fluctuation. By using these fluctuations as the source, Maxwell's equations, with appropriate boundary conditions, can be used to solve for radiation intensity emitted from a surface bounding a medium.

The analytical solutions exist only for a few simple functional forms of the soil temperature and dielectric constant profiles. For more complicated or arbitrary shapes for the profiles, the stratified medium approach and the WKB approximation are considered by Njoku and Kong [16]. The stratified model is a discrete and an approximate solution for the integrals in the coherent wave model. In general, the sensing depth and relative contributions of the various depths to overall brightness temperature varies with different microwave frequencies. The definition of a temperature weighting function allows one to quantify the relative contribution of the subsurface regions to the overall brightness temperature. The p-polarized brightness temperature observed at angle α is:

$$T_{\mathrm{B_p}}(\alpha) = \int_{-\infty}^{0} T(z) F_{\mathrm{p}}\{\varepsilon_{\mathrm{r}}(z), \alpha\}\, \mathrm{d}z \tag{1}$$

where $\varepsilon_{\mathrm{r}}(z)$ is the dielectric constant profile and $T(z)$ is the temperature profile. The functional $F_{\mathrm{p}}\{\varepsilon_{\mathrm{r}}(z), \alpha\}$ is a weighting function and it indicates the relative contributions of the subsurface regions to the overall brightness temperature. These weighting functions can be computed from the stratified model and are presented in [16]. The dielectric constant for a partially saturated volume of soil is related to the volumetric soil moisture content; here we use a linear model similar to that found in [17].

Equation (1) is essentially the inverse-problem. The brightness temperature observation is related to an integral of the states that need to be estimated. For a discretized profile, this integral translates into one equation and many unknowns. The problem has too many degrees of freedom to yield a unique solution. Entekhabi *et al.* [15] further discuss the properties of the inverse-problem posed in Equation (1).

Since the temperature and moisture profiles, especially their evolution, are constrained to conform to equations governing heat and moisture storage and transport, these latter relationships may be used to further reduce the excess degrees of freedom in the inverse-problem. In the following sub-section, the constraining relations are outlined.

4. MOISTURE AND HEAT TRANSPORT IN UNSATURATED POROUS MEDIA

The coupled flow of heat and moisture in the soil matrix occurs in both vapor and liquid phases but the vapor term is orders of magnitude smaller than the liquid flux and may thus be neglected. The coupling between the heat and liquid moisture equations are through the heat capacity of the soil and through the thermal conductivity dependence on soil moisture. Using the well-known Darcy equation and mass-continuity, we may write the Richards equation for moisture transport in partially saturated porous media as

$$\frac{\partial \theta}{\partial t} = \nabla \cdot [K \nabla \psi + K\mathbf{k}] \tag{2}$$

where θ is the volumetric moisture content, K is the unsaturated hydraulic conductivity (a function of θ), ψ is the matrix head (also a function of θ), and $\mathbf{k}$ is the unit vector in the vertical.

The heat transport equation may be similarly written in terms of a heat capacity (composed of volumetric contributions from soil particles, air and water content θ) and a heat flux equation containing the thermal conductivity term λ. The thermal conductivity is a function of the soil moisture content.

Storage and transport coefficients in the equations are functions of the state variables. They are the major source of nonlinearity and coupling between the heat and moisture transport components. Entekhabi *et al.* [15] outline the detailed forms of the parameterized storage and transport coefficients (e.g. hydraulic conductivity, heat capacity, soil-water retention, thermal conductivity, etc.).

The heat and moisture transport and storage equations form a set of nonlinear partial differential equations (PDEs). Integrating over finite-element space, those nonlinear PDEs are reduced to the following set of nonlinear ordinary differential equations (ODEs) whose unknowns are the values of the state variables at a finite number of points (after [18]),

$$\mathbf{A}\dot{\Psi} + \mathbf{B}\dot{\mathbf{T}} + \mathbf{C}\Psi + \mathbf{D}\mathbf{T} + \mathbf{E} = 0 \tag{3}$$

where Ψ and $\mathbf{T}$ are $n \times 1$ vectors defined by $\Psi = [\psi_1 \cdots \psi_n]^{\mathrm{T}}$ and $\mathbf{T} = [T_1 \cdots T_n]^{\mathrm{T}}$. The subscripts of Ψ and $\mathbf{T}$ indicate the nodal locations in the vertical soil column. The dot operation implies time derivative. $\mathbf{A}, \mathbf{B}, \mathbf{C}, \mathbf{D}$ and $\mathbf{E}$ are coefficient matrices that are functions of the state variables. The ODE state space representations shown above are still nonlinear with respect to the state vector.

Equation (3) together with the discrete stratified model for Equation (1) (in [16]) form a set of equations that govern the augmented state vector,

$$\mathbf{X} = [\Psi \vdots \mathbf{T}]^{\mathrm{T}}. \tag{4}$$

Now we wish to extend the system to include multi-spectral and multi-sensor observations and then outline the algorithm that yields a unique solution for the state $\mathbf{X}$ given the observations. We define the nonlinear observation vector function $\mathbf{h}$ for one polarized passive microwave channel and an infrared channel,

$$\mathbf{h}(\mathbf{X}) = \begin{bmatrix} T_{\mathrm{B_h}}(\Psi, \mathbf{T}) \\ T_{\mathrm{B_v}}(\Psi, \mathbf{T}) \\ IR(T_{\mathrm{s}}) \end{bmatrix} \tag{5}$$

as an example. Furthermore, we write Equation (3) as

$$\frac{d\mathbf{X}(t)}{dt} = \mathbf{F}(\mathbf{X}). \tag{6}$$

Equations (5) and (6) now need to be combined in a consistent manner.

There is uncertainty associated with both the observation Equation (5) and the propagation Equation (6). The uncertainty stems from a number of sources ranging from instrument error to heterogeneity in soil properties. We assume that the errors result in zero-mean additive noise to both (5) and (6). The noise variables are independent (both serially and cross-serially) and have a Gaussian marginal distribution. This last attribute allows us to define their characteristic and strength through their covariance matrix alone. Let $\mathbf{R}$ be covariance of the observation errors and $\mathbf{Q}$ the covariance of the error in the propagation system. In the context of the forthcoming example, the role of adding noise to the system and its practical consequences will be revisited.

Equations (5) and (6) with noise added no longer define a unique solution for the augmented state $\mathbf{X}$; rather, they define the probability distribution of the state. The best-estimate of the state may be obtained by establishing the conditional-mean of the distribution. The Kalman filter may be used to estimate and track the conditional-mean of the noisy system [19, 20]. This algorithm is optimal in the statistical sense in that it conforms to the least-squares of the error residuals.

The observation and propagation equations in (5) and (6) are nonlinear functions of the state $\mathbf{X}$. A straight-forward application of the Kalman filtering algorithm is not possible. It is necessary to first linearize the equations so that they may be combined in a Bayesian framework as required by the Kalman filter. Entekhabi *et al.* [15] discuss the various issues associated with the differing approaches to linearization; here we simply report that the observation and propagation equations were linearized using a first-order Taylor-series expansion around the latest available estimate of the conditional mean,

$$\mathbf{F}(\mathbf{X}(t)) = \mathbf{F}(\widehat{\mathbf{X}}_k) + \mathbf{A}(\widehat{\mathbf{X}}_k)(\mathbf{X}(t_k) - \widehat{\mathbf{X}}_k) + \dots \tag{7}$$

and

$$\mathbf{h}(\mathbf{X}_k) = \mathbf{h}(\widehat{\mathbf{X}}_k) + \mathbf{H}(\widehat{\mathbf{X}}_k)(\mathbf{X}(t_k) - \widehat{\mathbf{X}}_k) + \dots \tag{8}$$

where the conditional mean is

$$\widehat{\mathbf{X}}_k = E[\mathbf{X}(t_k)] \tag{9}$$

at discrete time t_k. The matrix $\mathbf{A}(\cdot)$ is the Jacobian of the coupled heat and moisture transport and storage model [18]. The matrix $\mathbf{H}(\cdot)$ is the Jacobian of the Njoku and Kong [16] stratified model of radiative transfer. These Jacobians are determined by the use of *Mathematica*, a symbolic algebra application for the digital computer [21].

Given the initial state $\mathbf{X}_0$ with covariance matrix $\mathbf{P}_0$, the soil moisture and temperature profiles are propagated (denoted by minus symbol) in time by

$$\widehat{\mathbf{X}}_k(-) = \Xi_{k-1}\widehat{\mathbf{X}}_{k-1} \tag{10}$$

where Ξ_{k-1} is the transition matrix resulting from the linearized and discretized coupled moisture and heat flow in the vertical column. The associated error covariance is propagated according to

$$\mathbf{P}_k(-) = \Xi_{k-1}\mathbf{P}_{k-1}\Xi_{k-1}^{T} + \mathbf{Q}_{k-1} \tag{11}$$

where the system covariance $\mathbf{Q}$ is first used. If the system is 'controllable', the above equations should provide the necessary conditions to estimate the conditional mean of the system based on the inputs (such as boundary conditions). Formally, controllability refers to when $\Xi\mathbf{Q}\Xi^{\mathrm{T}}$ is bounded and positive-definite.

Initial estimates for the state vector variance will decrease with occasional remote sensing observations if the system is 'observable'. The state vector is updated (update using observations $\mathbf{z}_k$; updates denoted by positive sign) whenever the observation is available according to,

$$\widehat{\mathbf{X}}_k(+) = \widehat{\mathbf{X}}_k(-) + \mathbf{K}_k\left[\mathbf{z}_k - \mathbf{H}_k\left(\widehat{\mathbf{X}}_k(-)\right)\right]. \tag{12}$$

The Kalman Gain matrix $\mathbf{K}_k$ weights the observation against the model forecast. Its weighting is determined by the relative magnitudes of model uncertainty embodied in $\mathbf{P}_k(-)$ with respect to observation error $\mathbf{R}_k$. The Kalman Gain is given by,

$$\mathbf{K}_k = \mathbf{P}_k(-)\mathbf{H}_k^{\mathrm{T}}\left(\widehat{\mathbf{X}}_k(-)\right)\left[\mathbf{H}_k\left(\widehat{\mathbf{X}}_k(-)\right)\mathbf{P}_k(-)\mathbf{H}_k^{\mathrm{T}}\left(\widehat{\mathbf{X}}_k(-)\right) + \mathbf{R}_k\right]^{-1}. \tag{13}$$

Associated error covariance update is given by,

$$\mathbf{P}_k(+) = \left[\mathbf{I} - \mathbf{K}_k\mathbf{H}_k\left(\widehat{\mathbf{X}}_k(-)\right)\right]\mathbf{P}_k(-) \tag{14}$$

where $\mathbf{I}$ is the identity matrix. Derivations for these algorithms are outlined in [19, 20].

5. APPLICATION OF THE RETRIEVAL ALGORITHM

Synthetic environmental conditions are generated and the algorithm is tested with this example for which the true situation is numerically known. The finite-element model of Milly and Eagleson [18] is used to generate one week of soil moisture and temperature profiles beginning with an initial condition of uniform −50 cm matrix suction and 20 degrees Celsius temperature throughout the meter-deep soil. The boundary conditions are assigned as 0.5 cm/day evaporation and a diurnal soil heat flux at the surface. No heat and moisture flux are allowed at depth for this clay soil column.

Figure 1 shows the drying profiles under these conditions sampled every half-day. Figure 2 illustrates the diurnal temperature fluctuations for the same synthetic example. The stratified model Njoku and Kong [16] is next used to generate a time-series of brightness temperature for 1.4 and 9.2 GHz passive radiation. Figure 3 is the time-series of the brightness temperatures and the surface temperature time-series. The latter time-series are used for infra-red observations. The diurnal cycle is apparent in all the time-series; the upward trend in the brightness temperatures is due to the drying of the soil and its changing dielectric constant properties.

The retrieval algorithm is provided with an intentionally poor initial guess of the states. For example, the initial soil moisture is guessed to be −300 cm while in reality it is considerably lower at −50 cm. In order to demonstrate the capability of the algorithm in using observations in correcting for the poor initial guess and eventually retrieving the true soil moisture profile down to a depth of at least one meter, two comparisons are made. The first is the comparison between the algorithm and the true values (unknown to the algorithm) when at every hour an update is made using observations. The second is a comparison between the true situation and an 'open-loop'

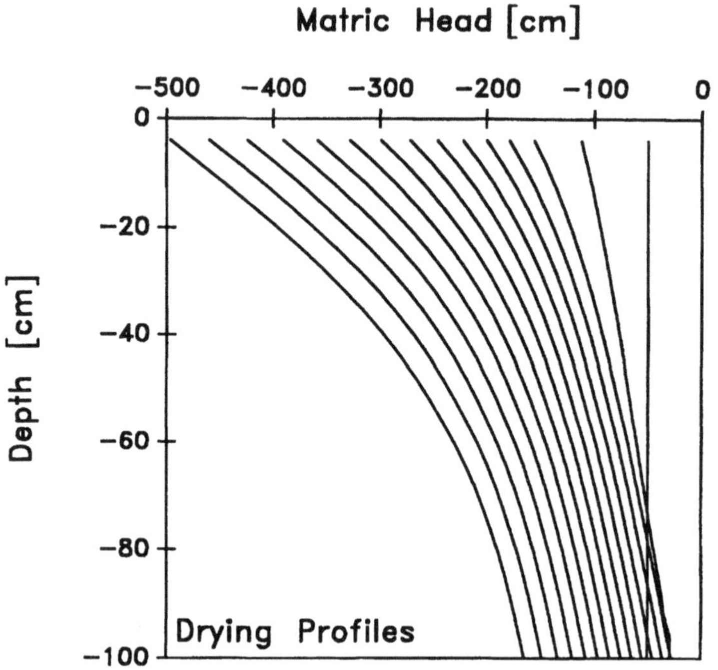

Figure 1. Twice-daily profiles of soil matrix head under conditions of 5 mm/day evaporation and initially $\psi = -50$ cm uniform conditions. The coupled remote sensing-modeling retrieval algorithm is used to estimate these synthetic profiles starting from a bad initial and occasional passive microwave and infra-red observations.

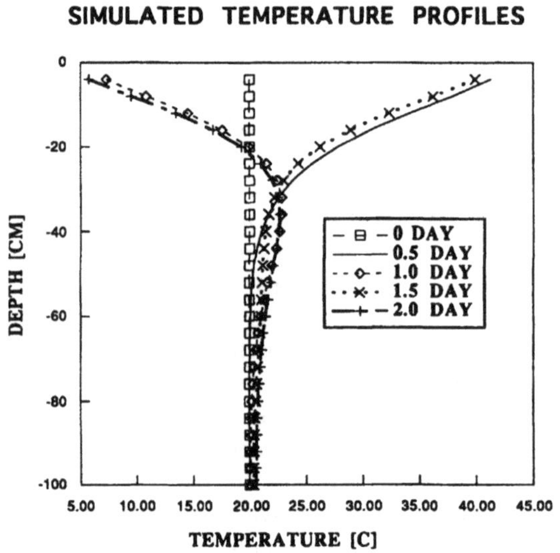

Figure 2. Same as Fig. 1 but for the soil temperature profile. The soil column is subject to periodic (diurnal) radiative forcing.

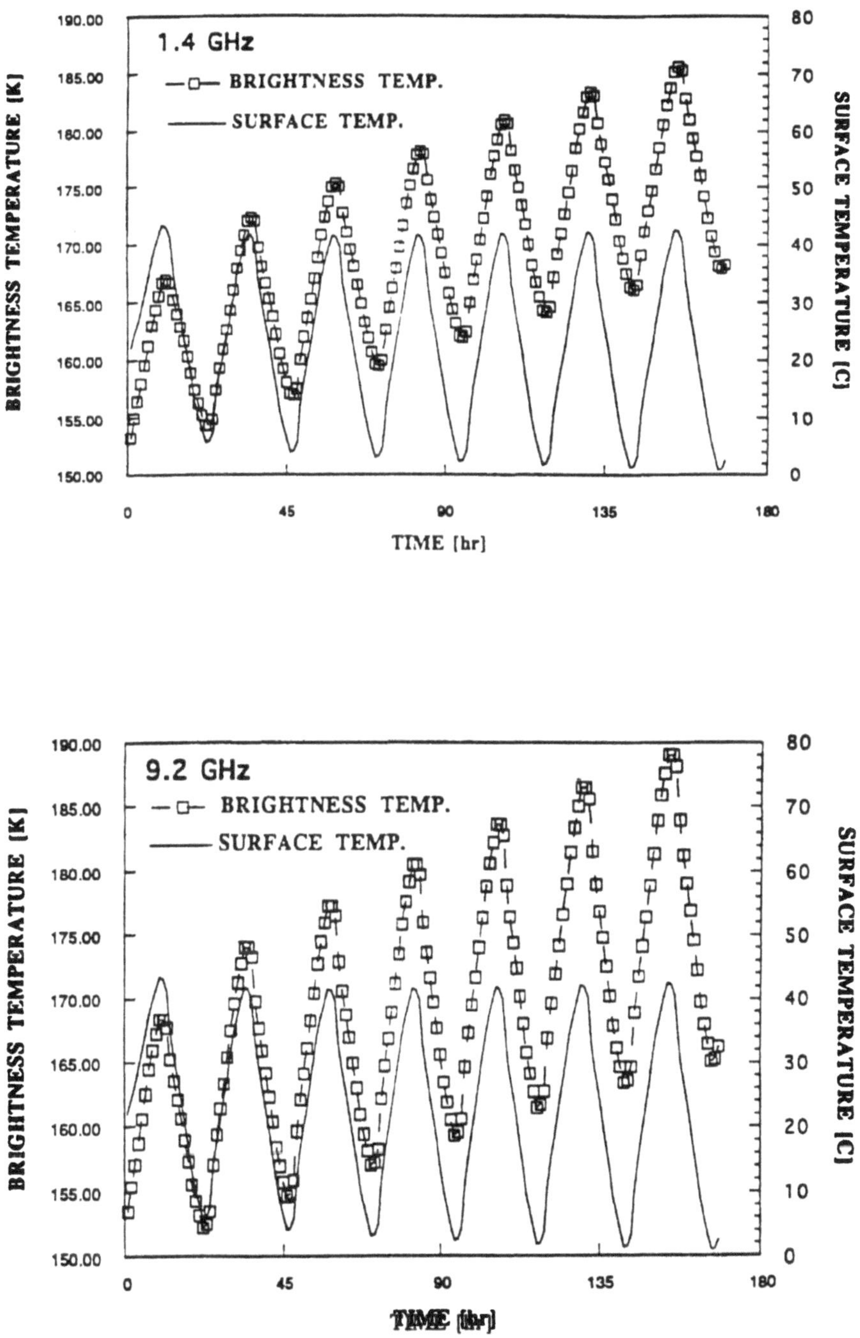

Figure 3. The surface temperature and brightness temperature for the synthetic example. The brightness temperature follows the same periodic pattern as the diurnally-forced surface temperature. The rising trend in the brightness temperature is due to the drying of the soil column. Brightness temperatures are generated by the Njoku and Kong [16] model.

run of the algorithm where no observations are supplied to the system. If the algorithm is truly using the observations and assimilating them in order to improve its estimates of the soil moisture at depth, then the improvements seen in the updated algorithm must not be present in the open-loop results.

The sensitivity study that is the main objective of this paper relates to the added noise. In the previous section, it was illustrated that the noise levels (as measured by the covariance matrices **R** and **Q** for the observation and the propagation systems) determine the level of controllability and observability in the retrieval algorithm. Here we will return to these issues in the context of the example. For illustration purposes, we assign a white noise level equivalent to 5% of the state as the error in the propagation system. The observation noise is taken to be 2% of the 9.2 GHz brightness temperatures and a very high value of 10% for the 1.4 GHz brightness temperatures. The latter value is rather extreme; it is a value used to illustrate the sensitivity clearly. Entekhabi *et al.* [15] discuss in more detail the criteria for the choice of noise and the role of noise in insuring the stability of the retrieval algorithm.

Figure 4 summarizes the results by the time-series representation of the soil matrix head (surrogate for volumetric soil moisture content and related to it uniquely by the soil-water retention curve) for five discrete depths. The whole profile is modeled by twenty-five nodes down to a depth of one-meter. This example refers to the multi-spectral retrieval when a 9.2 GHz and an infra-red observation are used. The true soil moisture states are denoted by open circles; they begin at −50 cm at the time-origin. Drying is apparent at all vertical levels due to the evaporation. The algorithm with the updates and the open-loop integration both begin in time at the intentionally poor estimate of −300 cm matrix head. As evident in Fig. 4, beginning at the surface where the microwave and infrared information are provided directly, the updated algorithm (filled circles) begins to correct the initial guess towards the true profile (the algorithm is unaware of the existence of the true profile data). The open-loop results (open triangles) continue the drying trend since they do not have the benefit of observations. As a result, there are no corrections to the initial poor estimate of the profile apparent. After the third day, the algorithm has found the true profile (the true and unique solution of the inverse-problem) and after that time, the algorithm tracks the true situation exactly. The 2% noise level in this example is considered to be the upper limit of what is expected of instruments and environmental conditions. The next example uses a 10% noise level which is unrealistic but it helps us to understand an important sensitivity in the algorithm. This high noise level illustrates the intrinsic behavior of the system and the role of noise in the retrieval.

Figure 5 shows the near-surface soil moisture estimates; the remainder of the profile reflects these conditions. The true values (open circles) are the same as before. The updated algorithm (filled circles) begin to improve on the poor initial guess (as evident in the divergence from the open-loop results) but they do not capture the true profile as much as the previous case. This is due to the high amount of variance associated with the observations; the algorithm simply does not trust the observations and prefers to make its best estimate of the soil moisture state by relying more on propagating the initial state.

Mo and Schmugge [22] determined the effect of noise on microwave emissivity and soil moisture using simulations. Their results, information on heterogeneity in the natural environment, and sensor specifications may be used to further define the range of

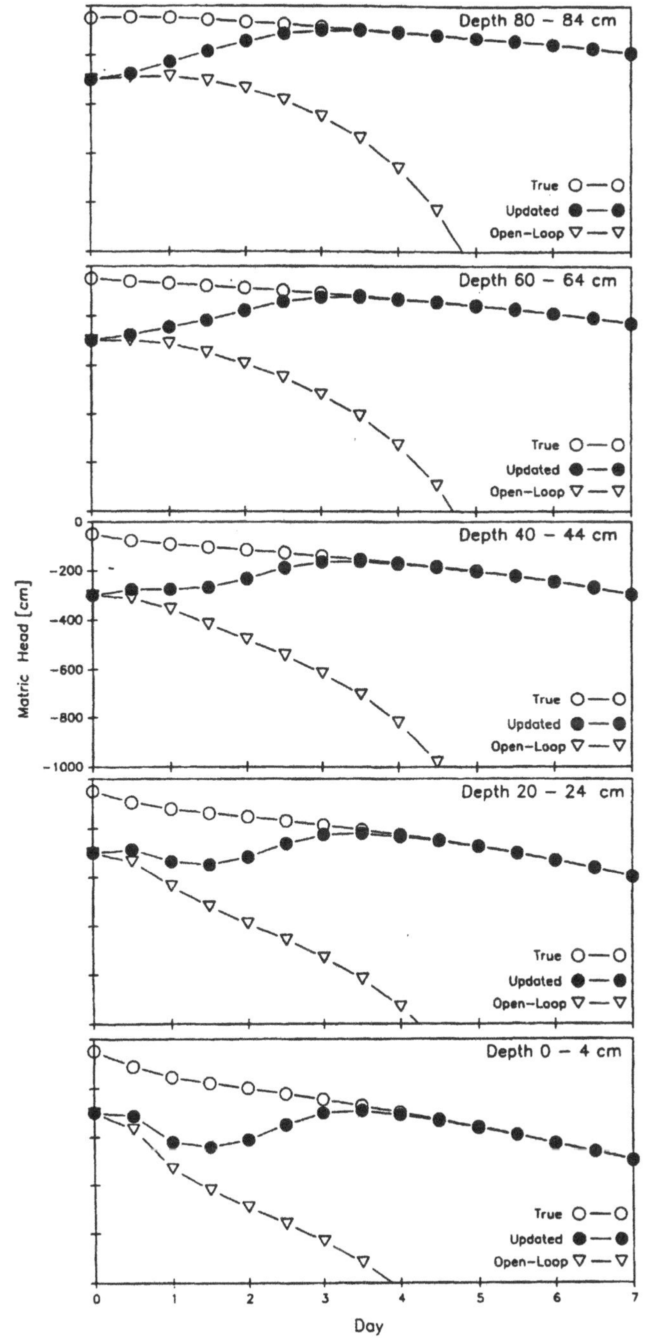

Figure 4. The matrix head at different depths using the 9.2 GHz microwave observations. The retrieval algorithm is making a weighted average of the 'open-loop' (trusting soil heat and moisture model forecasts based on initial guesses) and the observations. Since in this case the observations are given high confidence (low variance), the algorithm succeeds in tracking the true profile.

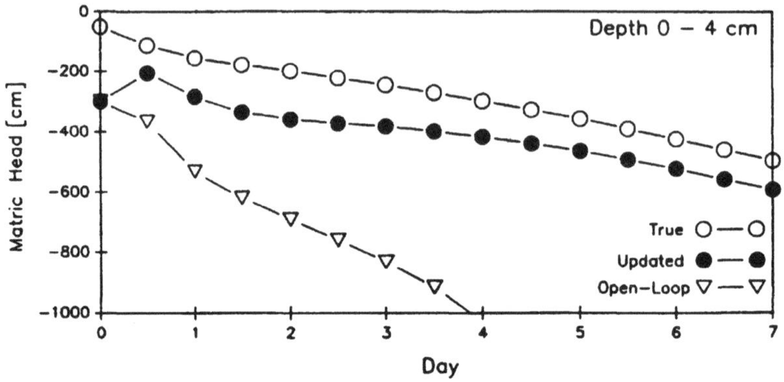

Figure 5. Same as Fig. 4 but for 1.4 GHz observations with more uncertainty (larger variance). Now the retrieval algorithm places more confidence in the forecast using the soil heat and moisture diffusion equations.

values for the noise intensities in the algorithm. The noise level in the observation and the propagation equations may be used to insure the stability and efficiency of the retrieval algorithm. The controllability and observability of the system may be engineered using the noise levels. One important output of this algorithm is the probability distribution of the estimates for soil moisture and temperature at all depths. In this brief report we have focused only on the mean values. The variance and distribution of the estimates may be used to very effectively perform important error analyses on remote sensing applications.

6. CONCLUSIONS

An algorithm is developed that solves the inverse-problem associated with the upwelling brightness temperature from a column of soil. The algorithm combines radiative transfer equations with the equations governing the coupled transport and storage of heat and moisture in the vertical soil column.

The examples of the retrieval in this paper show that the algorithm is capable of 1) extracting the soil moisture information from brightness temperatures without empirical geophysical calibration formulae, 2) effectively assimilate multi-spectral and multi-sensor observations, and 3) estimate the soil moisture at depth far beyond the direct sensitivity range of low-frequency passive microwave observations.

In future applications, the role of vegetation in altering the emission signature needs to be included [23–25]. The soil texture also has strong influence on microwave emission and its influence must be considered [26–28]. Testing with field observations is an ongoing research activity.

Acknowledgements
This research has been supported by National Aeronautics and Space Administration grant NAGW 2942. The authors thank Drs Tom Jackson, P. Christopher D. Milly and Dennis B. McLaughlin for their comments and advice.

REFERENCES

1. T. J. Schmugge. Remote sensing of soil moisture: Recent advances. *IEEE Trans. Geosci. Remote Sensing*, **GE-21** (3), 336–344 (1983).
2. T. J. Jackson. Research toward an operational, passive microwave remote sensing system for soil moisture. *J. Hydrol.*, **102**, 95–112, Elsevier Science Publ., Amsterdam (1988).
3. T. J. Jackson and T. J. Schmugge. Passive microwave remote sensing system for soil moisture: Some supporting research. *IEEE Trans. Geosci. Remote Sensing*, **27** (2), 225–235 (1989).
4. B. J. Choudhury. Passive microwave remote sensing contribution to hydrological variables. In: *Land Surface Atmosphere Interactions for Climate Modeling: Observations, Models and Analysis*, E. Wood (Ed.), pp. 63–84, Kluwer Academic Publ. (1991).
5. T. J. Jackson. Measuring surface soil moisture using passive microwave remote sensing. *Hydrologic Processes*, **7**, 139–152 (1993).
6. T. Schmugge, P. E. O'Neill and J. R. Wang. Passive microwave soil moisture research. *IEEE Trans. Geosci. Remote Sensing*, **GE-24** (1), 12–22 (1986).
7. T. J. Jackson. Profile soil moisture from surface measurements. *J. Irrigation Drainage Division, Proc. ASCE*, **106**, (IR2), 81–92 (1980).
8. P. Camillo and T. J. Schmugge. Estimating soil moisture storage in the root zone from surface measurements. *Soil Science*, **135** (4), 245–264 (1983).
9. P. Camillo and T. J. Schmugge. Correlating rainfall with remotely sensed microwave radiation using physically based models. *IEEE Trans. Geosci. Remote Sensing*, **22** (4), 415–423 (1984).
10. L. Bruckler and H. Witono. Use of remotely sensed soil moisture content as boundary conditions in soil–atmosphere water transport modeling. 2: Estimating soil water balance. *Water Resources Res.*, **25** (12), 2437–2447 (1989).
11. L. Prevot, R. Bernard, O. Taconet, D. Vidal-Madjar and J. L. Thony. Evaporation from a bare soil evaluated using a soil water transfer model and remotely sensed surface moisture data. *Water Resources Res.*, **20** (2), 311–316 (1984).
12. L. Stroosnijder, R. L. Lascano, R. W. Newton and C. H. M. van Bavel. *Estimating Net Rainfall, Evaporation and Water Storage of a Bare Soil from Sequential L-band Emissivities*. Proceedings of IGRASS '84 Symposium, European Space Agency SP-215 Publications (1984).
13. R. W. Newton, J. L. Heilman and C. H. M. van Bavel. Integrating passive microwave measurements with a soil moisture/heat flow model. *Agricul. Water Manag.*, **7**, 379–389 (1983).
14. P. C. Milly and J. Kabala. Integrated modeling and remote sensing of soil moisture. In: *Hydrol. Appl. Space Technology*, pp. 331–339, (Proceedings of the Cocoa Beach Workshop, Florida, August 1985), IAHS Publ. No. 160 (1986).
15. D. Entekhabi, H. Nakamura and E. G. Njoku. Solving the inverse-problem for soil moisture and temperature profiles by sequential assimilation of multifrequency remotely sensed observations. *IEEE Trans. Geosci. Remote Sensing*, **32** (2) (1994).
16. E. G. Njoku and J.-A. Kong. Theory for passive microwave remote sensing of near-surface soil moisture. *J. Geophys. Res.*, **82** (20), 3108–3118 (1977).
17. J. R. Wang and T. J. Schmugge. An empirical model for the complex dielectric permittivity of soils as a function of water content. *IEEE Trans. Geosci. Remote Sensing*, **GE-18** (4), 288–295, October (1980).
18. P. C. D. Milly and P. S. Eagleson. A coupled transport of water and heat in a vertical soil column under atmospheric excitation. (1980).
19. R. L. Bras and I. Rodriguez-Iturbe. *Random Functions and Hydrology*. Addison-Wesley (1985).
20. A. Gelb (Ed.). *Applied Optimal Estimation*. MIT Press (1974).
21. S. Wolfram. *Mathematica*. Addison-Wesley (1988).
22. T. Mo and T. J. Schmugge. Monte Carlo simulation of the effect of soil moisture variation on the microwave emission from soils. *IEEE Trans. Geosci. Remote Sensing*, **GE-21** (4), 473–474 (1983).
23. T. Mo, B. J. Choudhury, T. J. Schmugge, J. R. Wang and T. J. Jackson. A model for microwave emission from vegetation-covered fields. *J. Geophys. Res.*, **87** (C13), 11 229–11 237 (1982).

24. T. J. Jackson, T. J. Schmugge and J. R. Wang. Passive microwave sensing of soil moisture under vegetation canopies. *Water Resources Res.*, **18** (4), 1137–1142 (1982).

25. F. T. Ulaby, M. Razani and M. C. Dobson. Effects of vegetation cover on the microwave radiometric sensitivity to soil moisture. *IEEE Trans. Geosci. Remote Sensing*, **GE-21** (1), 51–61 (1983).

26. T. Mo, T. J. Schmugge and J. R. Wang. Calculations of the microwave brightness temperature of rough soil surfaces: Bare field. *IEEE Trans. Geosci. Remote Sensing*, **GE-25** (1), 47–54 (1987).

27. J. R. Wang, P. E. O'Neill, T. J. Jackson and E. T. Engman. Multifrequency measurements of the effects of soil moisture, soil texture, and surface roughness. *IEEE Trans. Geosci. Remote Sensing*, **GE-21** (1), 44–51 (1983).

28. L. Tsang and R. W. Newton. Microwave emissions from soils with rough surfaces. *J. Geophys. Res.*, **87** (11), 9017–9024 (1982).

29. B. J. Choudhury, T. J. Schmugge and T. Mo. A parameterization of effective soil temperature for microwave emission. *J. Geophys. Res.*, **87** (C2), 1301–1304 (1982).

30. T. J. Schmugge and B. J. Choudhury. A comparison of radiative transfer models for predicting the microwave emission from soils. *Radio Science*, **16** (5), 927–938 (1981).

31. H. Witono and L. Bruckler. Use of remotely sensed soil moisture content as boundary conditions in soil–atmosphere water transport modeling. 1: Field validation of a water flow model. *Water Resources Res.*, **25** (12), 2423–2435 (1989).

ESA/NASA International Workshop, pp. 499–505
B. J. Choudhury, Y. H. Kerr, E. G. Njoku and P. Pampaloni (Eds)

Identification of snowcover using SSM/I measurements

NORMAN C. GRODY[1] and ALLAN BASIST[2]

[1]*National Environmental Satellite Data and Information Service (NESDIS)*
[2]*Climate Analysis Center (CAC), National Oceanic and Atmospheric Administration (NOAA), Washington, DC 20233, USA*

Abstract—Satellite observations of the earth's surface provide information for use in meteorology, hydrology, agriculture, and oceanography. The primary advantage of passive microwave measurements over those in the visible and infrared is their ability to probe through non-precipitating clouds, enabling observations under nearly allweather conditions. This paper discusses the use of the Special Sensor Microwave Imager (SSM/I) aboard the DMSP satellites to identify snowcover. A procedure is developed to identify snowcover and discriminate between a number of other surface and atmospheric parameters based on the SSM/I measurements.

1. INTRODUCTION

Snowmelt is a major component of the total annual water supply in many parts of the world. The amount of water stored as snow is important in forecasting its effect on the water supply, flooding, irrigation and management of agricultural commodities. Snowcover information is also required to specify the initial state of regional and mesoscale models for numerical weather prediction. In general, the areal extent of snow and its water equivalent or depth is required on a daily basis.

Snowcover measurements from space began in 1966, with the launch of the first polar orbiting satellite capable of obtaining visible images of the earth. Beginning about that time, an operational snowcover product was produced weekly by subjectively analyzing the visible measurements from the NOAA polar orbiting satellites for the latest clear conditions during the week [1]. As a major limitation, the data acquisition is hampered by cloud cover, sometimes at critical times when new snowfall occurs or old snow melts. Furthermore, information on water equivalent, free water content, and other properties pertinent to accurate runoff predictions can not be obtained using visible or infrared measurements.

The primary advantage of passive microwave measurements over those in the visible and infrared is their ability to probe through non-precipitating clouds, providing daily observations. Besides penetrating clouds, the subsurface microwave emission can propagate through various snow depths depending on the frequency, thereby enabling a measure of snow depth. Furthermore, as shown in Fig. 1, the spectral characteristics of emitted radiation changes dramatically depending on the snow condition [2]. Microwave measurements can therefore provide an analysis of the internal snowpack properties under all weather conditions, excluding precipitation.

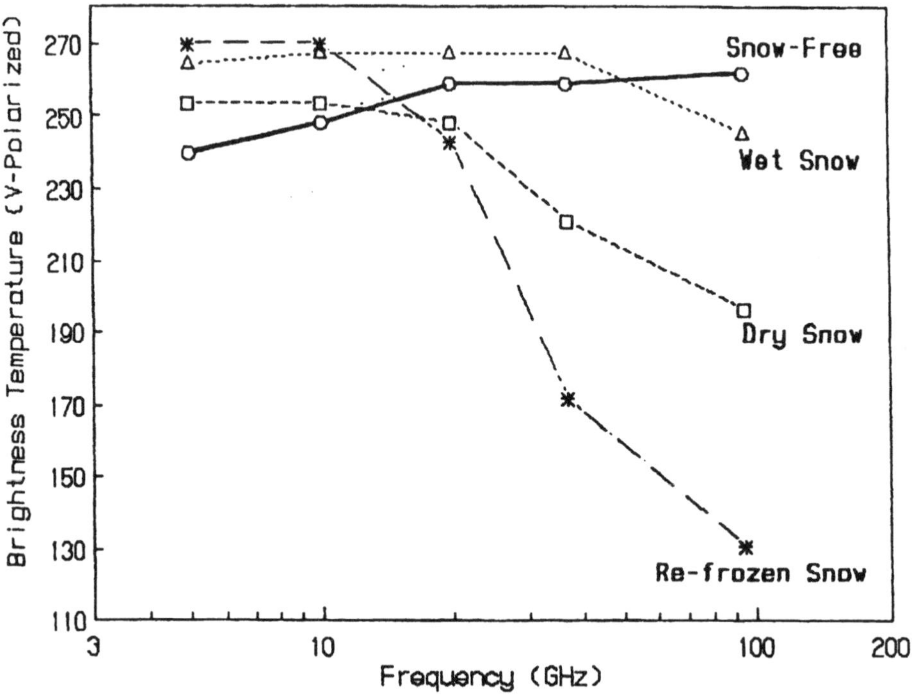

Figure 1. Vertically polarized brightness temperature (50 degree nadir angle) as a function of frequency for different snow conditions. Based on Fig. 1 of paper by Mätzler and Hüppi [2] using a snow temperature of 270 K.

Snowcover measurements from space using microwave radiometery were first obtained using the Nimbus-E Microwave Spectrometer (NEMS), launched aboard Nimbus-5 on December, 1972. This experimental, nadir viewing instrument had a spatial resolution of 200 km and contained window channels at 22 and 31 GHz to view the surface; it also included three channels within the 60 GHz oxygen band for obtaining temperature soundings. Snow and ice coverage was identified by their scattering signatures, i.e. when the 22 GHz measurements exceeded the 31 GHz measurements [3]. Although more advanced instruments have been developed over the past twenty years (containing more channels with higher spatial resolution and greater area coverage), the same dual frequency approach is used today for snow identification [4–6].

The latest, and most advanced instrument is the Special Sensor Microwave Imager (SSM/I), which was developed as part of the Defense Meteorological Satellite program (DMSP) and flown aboard the Air Force Block 5D satellites in 1987, 1989 and again in 1991. The SSM/I is a conically scanning radiometer having channels at 19, 22, 37 and 85 GHz [7]. All measurements are obtained with dual polarization except for the 22 GHz channel. The addition of an 85 GHz channel makes the SSM/I unique to previous instruments. This channel has the highest frequency and highest resolution (15 km) ever flown for a microwave radiometer. Due to its high frequency, the 85 GHz channel provides the highest sensitivity for measuring scattering materials such as snowcover and the ice particles formed aloft in the case of precipitation. While snowcover was previously observed at frequencies around 37 GHz [3–6], the 85 GHz measurements are particularly important for identifying the weaker scattering signature due to shallow snow and melting snow (see Fig. 1). Both snowcover and precipitation have been identified using the 85 GHz channel in conjunction with the lower frequency SSM/I measurements [8].

2. CLASSIFICATION TECHNIQUE

As mentioned above, the scattering due to snowcover produces a positive difference between low and high frequency channels. Although this scattering signature is the primary identifier of snow, it must be combined with other information to account for precipitation, deserts and frozen ground, which can also scatter microwave radiation. Furthermore, melting snow removes the ice-scattering signature at low frequencies. These atmospheric and surface features must be accounted for when developing a global snowcover product.

Snowcover and precipitation were identified using a decision tree approach that incorporated the vertically polarized SSM/I measurements at 22 and 85 GHz as primary classifiers [8]. Revisions to the decision tree are needed to account for frozen ground and surfaces containing glacial ice. Also, in that study [8] the 85 GHz measurements were convolved to the 22 GHz resolution to reduce mixed pixel effects. However, to be consistent with the data used by most investigators, the full resolution SSM/I measurements are used here.

Scattering materials are often identified by the simple condition $T_{22V} > T_{85V}$, where T_{22V} and T_{85V} denotes the vertically polarized antenna temperatures at the respective frequencies. Figure 2 shows the regions where T_{22V} exceeds T_{85V} by more than 1 K for June and February 1992; based on SSM/I antenna temperature measurements from the F-11 satellite. Note that over land, snow and ice cover, precipitation, deserts and frozen ground display the scattering signature. Over oceans the condition $T_{22V} > T_{85V}$ identifies multi-year sea ice and precipitation. To better identify precipitation over oceans a more refined parameter called a scattering index must be used [8]; this empirically derived index discriminates between rain and rain-free areas by combining low frequency (19V, 22V) and high frequency measurements (85V). While the scattering index approach can also be used over land to discriminate between snow and snow-free regions, we have chosen to only consider a relationship between the 22 and 85 GHz channels with appropriate threshold conditions. As discussed next, and illustrated in Fig. 3, various geophysical parameters can be identified using the 22 and 85 GHz measurements.

2.1. Precipitation signature

Precipitation and snowcover contain ice particles which scatter the upwelling radiation. Atmospheric and surface scatterers are best delineated by comparing the combined scattering signature obtained using window channels with the atmospheric scattering contribution obtained using tropospheric sounding channels. Since the SSM/I does not contain sounding channels (i.e. within the oxygen band or near the 183 GHz water vapor line), a different approach, based only on window channel measurements is used to separate snowcover from precipitation.

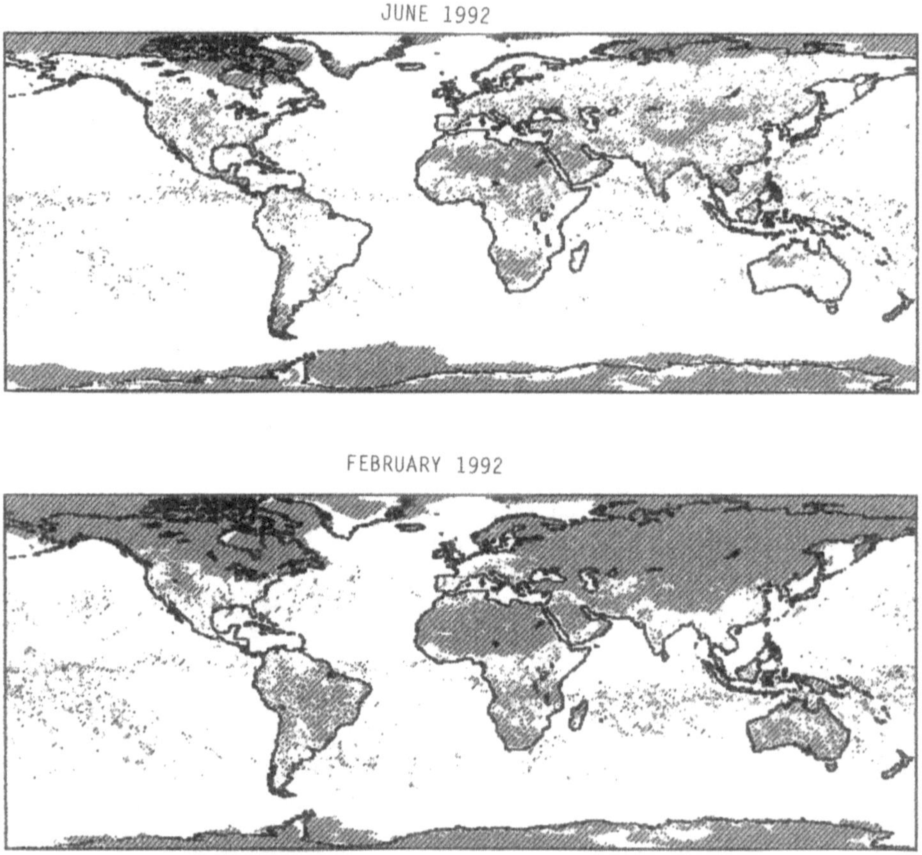

Figure 2. Monthly average scattering measurements for June and February, 1992. The hatched areas show the regions where the 22 GHz SSM/I antenna temperature is greater than the 85 GHz vertically polarized measurements. Daily observations of the channel difference are below 10 K for frozen ground, less than 25 K for deserts, and generally more than 25 K for precipitation, snow and ice covered surfaces.

The density and size of ice particles accompanying precipitating clouds are often smaller than that comprising snowcover. Precipitation therefore produces smaller scattering effects (i.e. higher low-frequency measurements) than snowcover. Even in the absence of scattering, the low frequency radiation is generally higher for precipitating atmospheres than for the frozen surface below the snow. Based on these considerations, precipitation (also deserts) can be separated from snowcover by requiring the 22 GHz measurements to be less than a given threshold (i.e. $T_{22V} < 257$ K). It should be noted that 22 GHz is used, rather than the 19 GHz channel, since it is a better snow-rain discriminator for light rain over wet ground [8]. This is due to the larger opacity at 22 GHz due to water vapor which increases the measurements beyond the 257 K threshold required to properly classify rain from snow.

The 257 K threshold represents average conditions and can deviate by ± 3 K depending on the surface and atmospheric conditions, i.e. cold rain can reduce T_{22V} to 254 K while melting snow increases T_{22V} to 260 K. An additional relationship given by $T_{22V} < 168\,\mathrm{K} + 0.49\,T_{85V}$ is used to account for intense rain events, which reduce

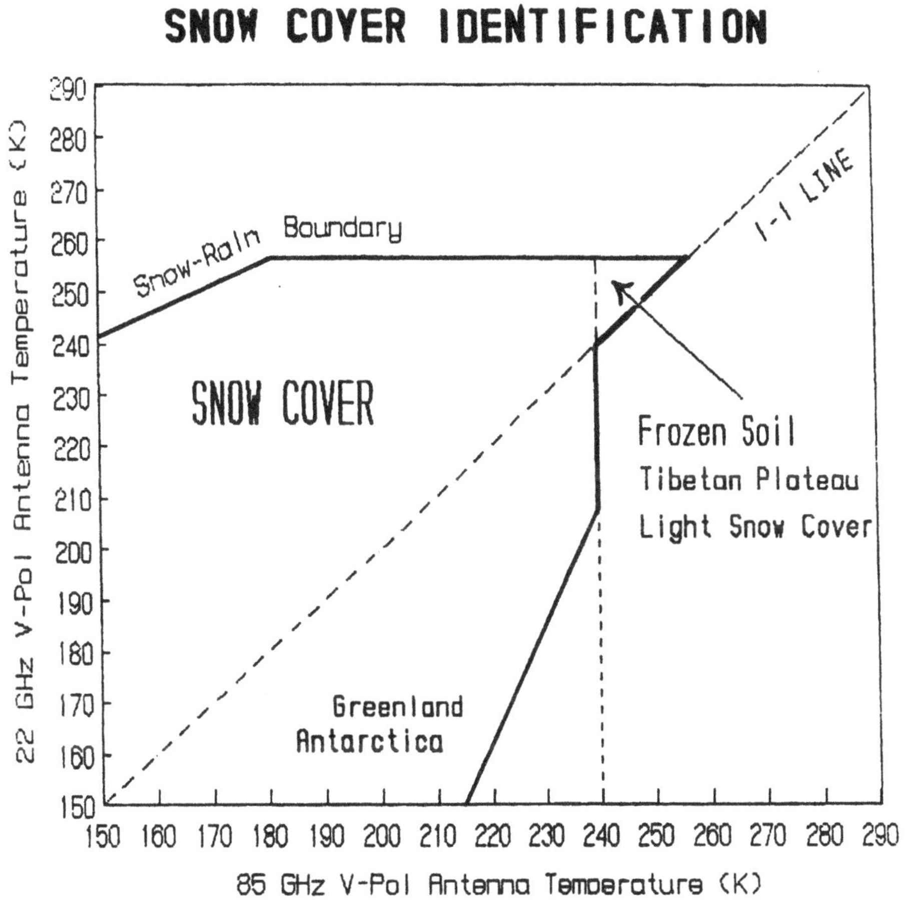

Figure 3. Schematic diagram showing the different parameters occupying the space defined by the 22 and 85 GHz, vertically polarized antenna temperatures. The region within the solid boundary is classified as snowcover and has been used to generate the snowcover maps shown in Fig. 4.

the 22 GHz measurements below 257 K due to scattering by large ice particles. These results are summarized in Fig. 3 and based on qualitative comparisons of the SSM/I measurements with the global maps of weekly composite snowcover derived from satellite imagery [1], in addition to surface observations of daily snowcover and rainfall over the United States.

2.2. Frozen ground signature

The largest geographical regions containing snowcover are bound by the three line segments discussed above; $T_{22V} = T_{85V}$, $T_{22V} = 257$ K and $T_{22V} = 168\,\mathrm{K} + 0.49\,T_{85V}$. However, as indicated in Fig. 3, the small triangular area formed by the lines $T_{22V} = T_{85V}$, $T_{22V} = 257$ K and $T_{85V} = 240$ K results from weak scattering due to shallow snowcover in addition to frozen ground [9]. Frozen ground occurs in high elevated areas during their corresponding cold season as well as regions adjacent to snowcover; this can result in the misclassification of frozen ground as snowcover, resulting in an extension of snow covered regions. Also, some measurements within the Tibetan Plateau reside in this triangular space, suggesting the possibility of frozen ground for this anomalous region. These ambiguities due to frozen ground are removed by considering all measurements within the triangle as snow-free when 19V – 19H > 7. Unfortunately, this test needs further improvement since it has the adverse effect of also removing some regions of shallow snow.

2.3. Greenland and Antarctica

The third and final signature to be discussed is that observed over large parts of Greenland and Antarctica which appear scatter-free in Fig. 2. Although these surfaces contain deep snow above a glacial subsurface, large portions of these surfaces have measurements lying outside the snow-designated area in Fig. 3. In fact, the polarization and frequency spectrum of the SSM/I measurements over Greenland (particularly the southern region) resemble a water surface. Excluding the possibility of melting ice (in February), and for lack of a better understanding, we have designated this area in Fig. 3 by the two geographical regions. The area is bound by the three lines $T_{22V} = T_{85V}$, $T_{85V} = 240$ K and $T_{22V} = -344\,\mathrm{K} + 2.3\,T_{85V}$. Note that the measurement space occupied by these geographical regions is quite large, although the measurements do not overlap any other features such as wet ground.

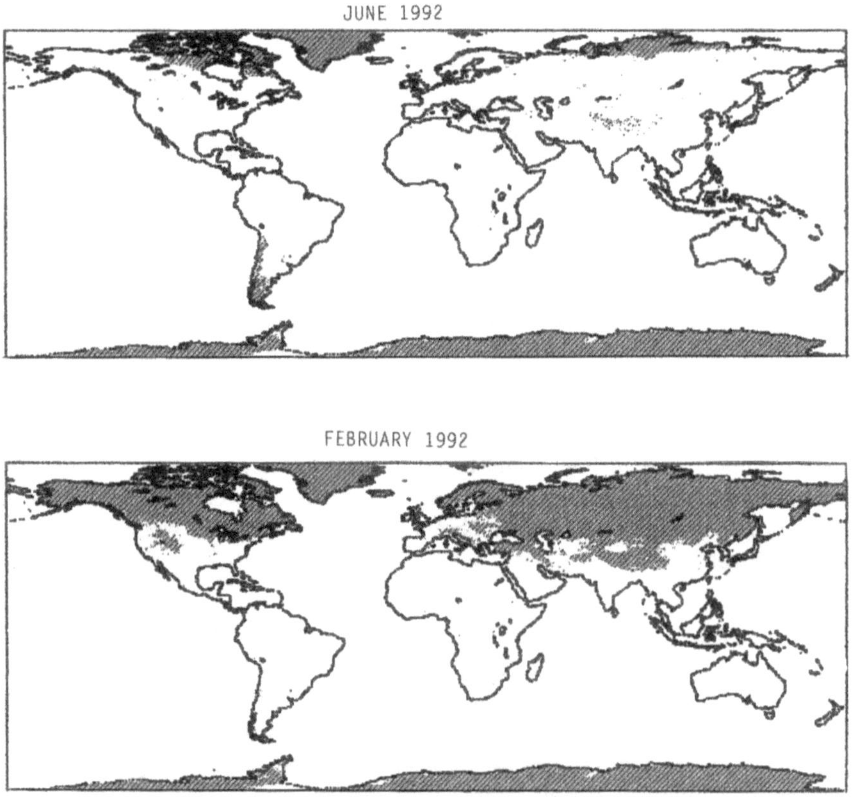

Figure 4. The hatched areas show the regions of snow cover for June and February 1992 obtained from the SSM/I measurements. The classification procedure is described in this paper and shown in Fig. 2.

3. CONCLUSIONS

This paper applies the SSM/I measurements in the form of a decision tree to identify snowcover. The decision tree is summarized in Fig. 3 and delineates between the microwave scattering signatures of snowcover, precipitation, desert and frozen soil using thresholds based on empirical observations. Snowcover is currently being produced daily as an experimental product using the decision tree method. As an example, Fig. 4 shows the monthly averaged product for June and February 1992. The classification procedure identifies much of the scattering found over deserts, precipitation and frozen ground surfaces (refer to Fig. 2) as snow-free. Also, the non-scattering areas over Greenland and Antarctica are properly classified as snowcover. However, the Tibetan Plateau appears to contain too much snowcover in February. In addition to this problem, further investigations are needed to consider melting snow and snow within forested areas which are sometimes undetected or misclassified by our analysis.

REFERENCES

1. M. Matson, C. F. Ropelewski and M. S. Varnadore. *An Atlas of Satellite-Derived Northern Hemispheric Snow Cover Frequency*. Washington, DC, US Department of Commerce (1986).
2. C. Mätzler and R. Hüppi. Review of signature studies for microwave remote sensing of snowpacks. *Adv. Space Res.*, **9** (1), 253–265 (1989).
3. K. F. Künzi, A. D. Fisher, D. H. Staelin and J. W. Waters. Snow and ice surfaces measured by the Nimbus-5 microwave spectrometer. *J. Geophys. Res.*, **881**, 4965–4980 (1976).
4. K. F. Künzi, S. Patil and H. Rott. Snow-cover parameters retrieved from Nimbus-7 Scanning Multichannel Microwave Radiometer (SMMR) data. *IEEE Trans. Geosci. Remote Sensing*, **20**, 452–467 (1982).
5. J. L. Foster, D. K. Hall and A. T. C. Chang. An overview of passive microwave snow research and results. *Rev. Geophys. Space Phys.*, **22**, 195–207 (1984).
6. R. R. Ferraro, N. C. Grody and J. A. Kogut. Classification of geophysical parameters using passive microwave satellite measurements. *IEEE Trans. Geosci. Remote Sensing*, **GE-24**, 1008–1013 (1986).
7. J. R. Hollinger, B. Lo, G. Poe, R. Savage and J. Pierce. *Special Sensor Microwave/Imager User's Guide*. Naval Research Laboratory, Washington, DC (1987).
8. N. C. Grody. Classification of snowcover and precipitation using the special sensor microwave imager. *J. Geophys. Res.*, **96**, 7423–7435 (1991).
9. A. W. England. Radiobrightness of diurnally heated, frozen soil. *IEEE Trans. Geosci. Remote Sensing*, **28**, 464–476 (1990).

Surface temperature

ESA/NASA International Workshop, pp. 509–530
B. J. Choudhury, Y. H. Kerr, E. G. Njoku and P. Pampaloni (Eds)

Surface temperature estimation over land using satellite microwave radiometry

ENI G. NJOKU

Jet Propulsion Laboratory, California Institute of Technology, Pasadena, CA 91109, USA

Abstract—Surface temperature is an important parameter of the physical processes that govern the fluxes of energy and water at the biosphere-atmosphere interface. The energy and water fluxes have major influences on the Earth's weather and climate. Models ranging from local to global scales are required to understand and predict effects of the energy and water fluxes. Satellite-derived land surface temperature measurements can be used to develop and calibrate these models, and to monitor long-term environmental changes. Infrared and microwave techniques have been applied successfully to the measurement of surface temperature over the oceans. The measurement of surface temperature over land using similar techniques has received less attention. This is due partly to the complexity of the electromagnetic interactions with the land surface and to the variability in land surface types. Microwave measurements are feasible under nearly all weather conditions (except for precipitation). This is an advantage in the tropics and desert regions where clouds, dust, and aerosols frequently obscure the surface at infrared wavelengths. This advantage is offset, for some applications, by the relatively low spatial resolution of microwave radiometers. The purpose of this paper is to review the status of microwave sensing of land surface temperature for bare and vegetation-covered soils, and to indicate current limitations in applications to global climate and environmental monitoring.

1. INTRODUCTION

The land surface temperature plays an important role in the Earth's surface energy balance and hydrologic cycle. Energy is provided to the Earth in the form of incident solar radiation, a portion of which is reflected and absorbed by the atmosphere, while the remainder reaches the surface. At the surface there is a balance between the net radiation flux (solar and longwave) entering the surface, the sensible and latent heat fluxes from the surface to the atmosphere, and the heat flux into the surface. These fluxes are influenced by surface characteristics such as soil type, soil moisture, and vegetation cover, and by atmospheric variables such as air temperature, humidity, wind speed, cloudiness, and precipitation.

Understanding the dynamics of the surface and atmospheric variables, their feedback effects, and human-induced changes, is critical for modeling and predicting climate and environmental change. Early simulations with climate models showed that reductions in soil moisture and vegetation cover could alter the regional balances of energy and water fluxes, leading to global patterns of decreased precipitation and increased surface temperature over land (Walker and Rowntree [1]; Shukla and Mintz [2]). Recent simulations using more sophisticated models have shown similar results, e.g. Dickinson and Henderson-Sellers [3]. Chahine *et al.* [4] have used data from the High-resolution Infra-Red Sounder (HIRS) on the NOAA polar-orbiting satellites to produce monthly mean global maps of surface skin temperature (defined as the top infrared radiating layer of the surface) illustrating the diurnal and seasonal global temperature changes. Such data may be used routinely to validate climate models. Studies by Carlson *et al.* [5], and Price [6], using data from the Heat Capacity Mapping Mission (HCMM), and by Klaassen and van den Berg [7], Taconet *et al.* [8], and others, using data from the Advanced Very High Resolution Radiometer (AVHRR) on the NOAA satellites, have shown that infrared-derived surface temperatures may be used with one-dimensional boundary layer models to derive surface fluxes of sensible and latent heat. Similarly, Wetzel and Woodward [9], have used data from the Visible-Infrared Spin-Scan Radiometer (VISSR) on the NOAA geostationary satellites to infer surface soil moisture. These studies have demonstrated the usefulness of satellite-derived surface temperature measurements for global monitoring and flux studies.

Satellites provide the only viable means for measuring land surface temperatures globally and with adequate temporal sampling. Conventional (*in situ*) measurements of near-surface air temperature, such as collected by the worldwide network of surface meteorological stations, are inadequate for these purposes since the spatial sampling and coverage are limited in many regions of the world. Furthermore, the near-surface air temperatures typically differ significantly (i.e. by more than a few degrees) from the ground and canopy surface temperatures that parameterize the surface energy balance models. Satellite measurements are better suited to provide data for these models since they measure the temperature of the ground surface, and provide area-averaged rather than point values. The main limitations in using satellite data are the need to correct for variable surface emissivity and atmospheric water vapor attenuation, and the difficulty in quantifying the temperature measurement accuracy due to the lack of comparable *in situ* data.

For the above reasons, increasing attention is being paid to the problem of retrieving accurate satellite measurements of land surface temperature (LST). Aspects of the infrared and microwave techniques have been studied by several investigators, e.g. Price [10], Wan and Dozier [11], Becker and Li [12], Kerr *et al.* [13], and McFarland *et al.* [14]. Most studies have addressed infrared techniques, primarily due to the higher spatial resolutions available from infrared sensors ($\sim$ 1 km versus $\sim$ 10 to 100 km for microwave sensors), and also to the historical availability of data from satellite infrared sensors such as the HCMM, AVHRR, HIRS, and VISSR. Microwave measurements, however, are feasible under nearly all weather conditions (except for precipitation). This is an advantage in the tropics and deserts where clouds, dust, and aerosols frequently obscure the surface at infrared wavelengths. This advantage is offset, for some applications, by the relatively low spatial resolution of microwave radiometers. Long-term microwave data sets are available from the Scanning Multichannel Microwave Radiometer (SMMR) on the Nimbus-7 satellite and the Special Sensor Microwave Imager (SSM/I) on the DMSP satellites. The availability of these data sets, and the potential of continued data from the SSM/I series and future sensors such as the Multifrequency Imaging Microwave Radiometer (MIMR) on the Earth Observing System (EOS), has renewed interest in the potential of microwave measurements of LST for climate monitoring and model development and validation.

In this paper the status of satellite microwave measurements and retrieval techniques for LST are reviewed. In the first part of the paper the surface energy balance is discussed briefly as a means for understanding the role of surface temperature in the fluxes of energy and water at the land surface. This is followed by a discussion of the theoretical basis for microwave measurement of LST. In subsequent sections some results of satellite microwave observations are discussed, and simple model simulations and regression techniques are used to examine the potential for multichannel microwave surface temperature estimation.

2. BACKGROUND

Surface fluxes (in particular evapotranspiration) have major influences on weather and climate. Thus, it is essential to understand the physical processes that govern these fluxes, and, in particular, to represent the processes of evaporation and transpiration realistically in the climate models. Evaporative cooling is a major determinant of the land surface temperature, and depends not only on the precipitation input to the surface, and how this is apportioned into evaporation, soil storage, and runoff, but also on how the net radiative energy is divided between latent and sensible heat fluxes. The surface energy balance determines the relationship between these fluxes.

The surface energy balance equation may be expressed in the form

$$R_n = H + LE + G \tag{1}$$

where R_n is the net radiation into the surface, H is the sensible heat flux from the surface to the atmosphere, LE is the latent heat flux into the atmosphere, and G is the flux of heat into the soil. The latent heat flux LE is the product of the latent heat of evaporation L and the evaporation rate E. For conceptually simple surfaces, straightforward expressions can be derived for the energy balance components. These are

$$R_n = R_{S\downarrow}(1 - A) + R_{L\downarrow} - \varepsilon\sigma T_s^4 \tag{2}$$

$$H = \rho C_p(T_s - T_a)/r_a \tag{3}$$

$$E = \rho[q_s(T_s) - q_a]/(r_c + r_a) \tag{4}$$

$$G = -\lambda\frac{\partial T}{\partial z}. \tag{5}$$

In Equation (2), $R_{S\downarrow}$ is the incident solar energy at the surface, A is the surface albedo, $R_{L\downarrow}$ is the incident thermal infrared energy (minus any reflected), ε is the surface emissivity, σ is the Stefan–Boltzmann constant, and T_s is the surface temperature. For the sensible heat flux, T_s and T_a are the surface temperature and the near-surface air temperature, respectively, C_p and ρ are the specific heat and density of air, respectively, and r_a is a 'resistance' coefficient. For the evaporation, q_a is the water vapor mixing ratio of the near-surface air and q_s is the saturation water vapor mixing ratio of the surface. The canopy resistance, r_c, is zero for bare soil. Note that q_s is a function of the surface temperature T_s. For the soil heat flux, λ is the thermal conductivity and T is the soil temperature as a function of depth z. In climate models the soil heat flux is of minor significance and is usually ignored. The above expressions are discussed in more detail elsewhere (e.g. Dickinson *et al.* [15] and Kustas [16], who also discusses appropriate expressions for r_a that account for the difference between 'aerodynamic' surface temperature and 'skin' surface temperature as measured by an infrared sensor; these distinctions will not concern us further here).

Equations (1) through (5) illustrate the role of LST in affecting both the energy and the moisture fluxes at the surface. For natural surfaces, which may include mixtures of bare soil, vegetation canopy of different types, and undergrowth, Equations (2)–(5) can be modified to take into account the radiative transfer and fluxes within the canopy (e.g. Sellers *et al.* [17]; Taconet *et al.* [8]). A unique definition of surface temperature is difficult in this case since different temperatures are involved in the equations. These include the soil surface temperature, the soil near-surface temperature, the temperature of the air within the canopy, the temperature of the leaf surfaces of the canopy, and the internal leaf tissue temperature. In remote sensing, the surface temperature is normally defined as a weighted mean temperature over the radiating ('skin' or 'penetration') depth in the medium. The penetration depth depends on the electromagnetic attenuation in the medium, and, for microwaves, is a function primarily of the moisture content of the medium and the radiation wavelength. In microwave sensing, this depth may vary from a millimeter or less in moist soils at short wavelengths, to over a meter in sparse vegetation or dry soil at long wavelengths. Furthermore, spaceborne sensor antenna footprints, of spatial dimensions 10 to 100 km, encompass a variety of canopy types and soils. Thus, the satellite brightness temperatures often represent a rather complicated weighted mean temperature over the surface radiating volume within the footprint. These considerations show that care must be taken in relating the surface temperature measured by a satellite radiometer to the appropriate surface temperature parameters of surface flux and climate models.

3. THEORETICAL BASIS

3.1. *Surface 'effective' temperature*

The microwave radiation received by a spaceborne radiometer viewing land may by expressed approximately as

$$T_{\mathrm{Bp}} = T_{\mathrm{u}} + \exp(-\tau_{\mathrm{a}})[T_{\mathrm{bp}} + r_{\mathrm{p}} T_{\mathrm{d}}] \tag{6}$$

where T_{bp} is the surface brightness temperature, T_{u} and T_{d} are the upwelling and downwelling atmospheric radiation components, τ_{a} is the atmospheric opacity, r_{p} is the surface reflectivity, and T_{Bp} is the brightness temperature received at the radiometer. The subscript p refers to polarization which may be $\mathrm{p} = \mathrm{h}$ (horizontal) or $\mathrm{p} = \mathrm{v}$ (vertical). The dependence on incidence angle θ is understood. If we assume that the subsurface dielectric properties are horizontally homogeneous and vary slowly with depth the surface brightness temperature can be written in approximate radiative transfer form

$$T_{\mathrm{bp}} = e_{\mathrm{p}} T_{\mathrm{e}} \tag{7}$$

where $e_{\mathrm{p}} = (1 - r_{\mathrm{p}})$ is the surface emissivity, and T_{e} is the 'effective' (or 'skin') temperature of the surface given by

$$T_{\mathrm{e}} = \int_{-\infty}^{0} T(z)\alpha(z) \exp\left[-\int_{z}^{0} \alpha(z')\mathrm{d}z'\right] \mathrm{d}z. \tag{8}$$

The integration is performed for radiation emanating from all depths up to the surface. The attenuation coefficient in the medium, $\alpha(z)$, depends on the dielectric constant, and, for nadir viewing, can be expressed simply as $\alpha(z) = 4\pi n''(z)/\lambda$, where $n''(z)$ is the imaginary part of the refractive index of the medium, and λ is the wavelength. (For a medium whose dielectric constant profile varies rapidly over the distance of a wavelength in the medium, the more accurate coherent model formulation must be used in place of the radiative transfer equation (Njoku and Kong [18]).)

The effective temperature is a weighted mean of the vertical temperature profile in the medium, and may differ substantially from the actual surface temperature. Examples of weighting functions for a bare soil are shown in Fig. 1 (Njoku and Kong [18]), where the weighting function is defined as the factor multiplying $T(z)$ in the integrand of Equation (8). The weighting functions shown, F_N, have been normalized to a maximum value of unity, for ease of comparison, since their absolute magnitudes include the effects of surface emissivity. Weighting functions at shorter wavelengths (higher frequencies) sample the temperature close to the surface, whereas at longer wavelengths the weighting functions show that the radiation originates from deeper in the medium, and the effective temperature is therefore characteristic of the weighted mean temperature over a greater depth. Note that for dry surface conditions with a positive moisture gradient the weighting functions may not peak at the surface. For a medium of uniform dielectric constant (i.e. uniform soil moisture profile) the weighting functions are exponentials, and are characterized by the penetration depth, $d = \lambda/4\pi n''$, at which the weighting functions decrease to 1/e of their values at the surface. Using this expression, Fig. 2 shows a plot of the penetration depth versus wavelength in sandy soil, for various (uniform) moisture contents. At 1.5 GHz the penetration depth varies from approximately 10 cm to 1 cm for soil conditions ranging from saturated to dry, while at 30 GHz the penetration depth varies from less than 1 mm to a little over 1 cm for similar conditions.

It should be noted that the penetration depth, as defined above, is a characteristic of the exponential weighting function for a uniform dielectric. For weighting functions of more general, nonuniform moisture profiles, such as shown in Fig. 1, a generalized penetration depth (or temperature sensing depth) may be defined as the depth from above which a fraction $(1 - e^{-1})$, or approximately 63%, of the emitted radiation originates. This depth, d_g, can be found as the solution of the expression:

$$\int_{-d_g}^{0} F_N(z)\,dz = \left(1 - e^{-1}\right) \int_{-\infty}^{0} F_N(z)\,dz. \tag{9}$$

The moisture sensing depth on the other hand is defined as the depth within which moisture variability affects the emission, and has been observed experimentally, for some naturally occurring soil moisture profiles, to range from approximately 0.1 to 1.0 wavelengths in the medium (e.g. $\sim$ 2–20 cm at 1.4 GHz), depending on the moisture content of the near-surface layer (see also Njoku and O'Neill [19]).

Choudhury *et al.* [20] have computed the differences between the surface temperature and the effective temperature for a range of naturally occurring soil moisture and temperature profiles. Figure 3 shows scatterplots of effective temperature versus surface temperature for these profiles. At 10.7 GHz, ($\lambda = 2.8$ cm), the effective temperature

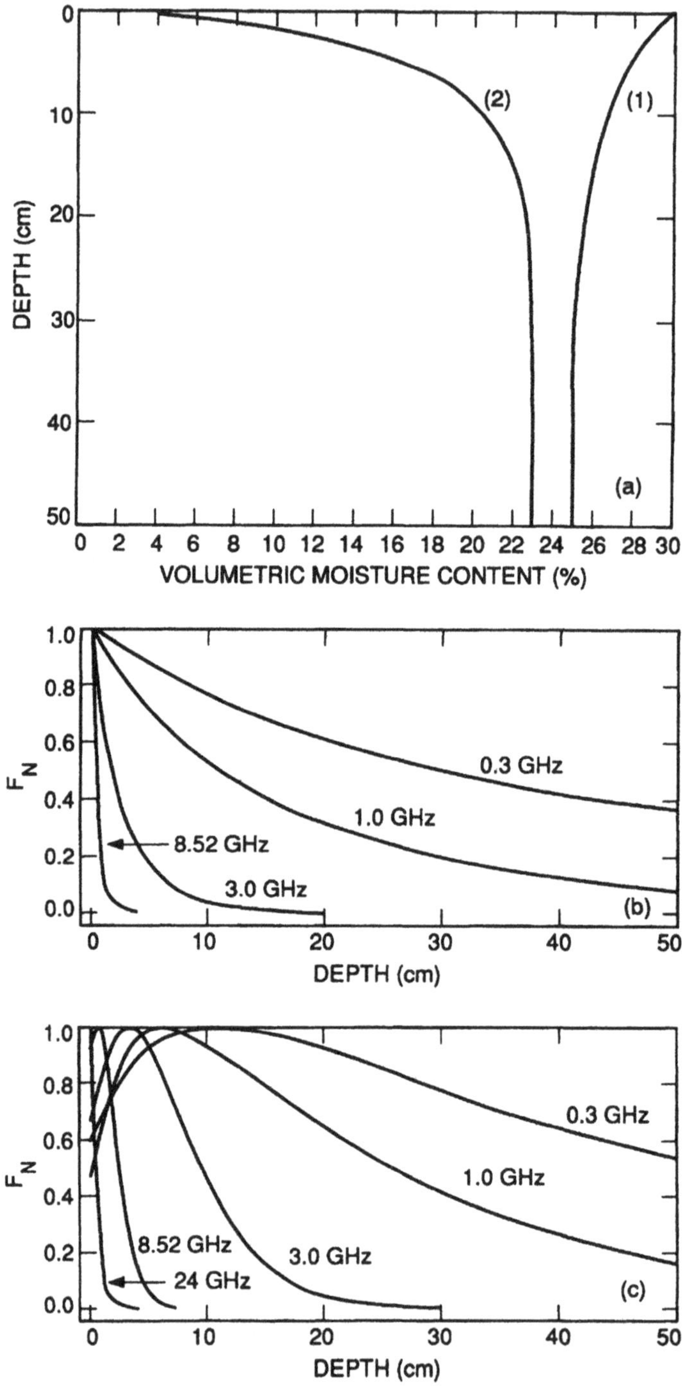

Figure 1. (a) Soil moisture profiles corresponding to (1) wet, and (2) dry conditions. (b) Normalized temperature weighting functions F_N for moisture profile (1). (c) Normalized temperature weighting functions F_N for moisture profile (2). (Njoku and Kong [18].)

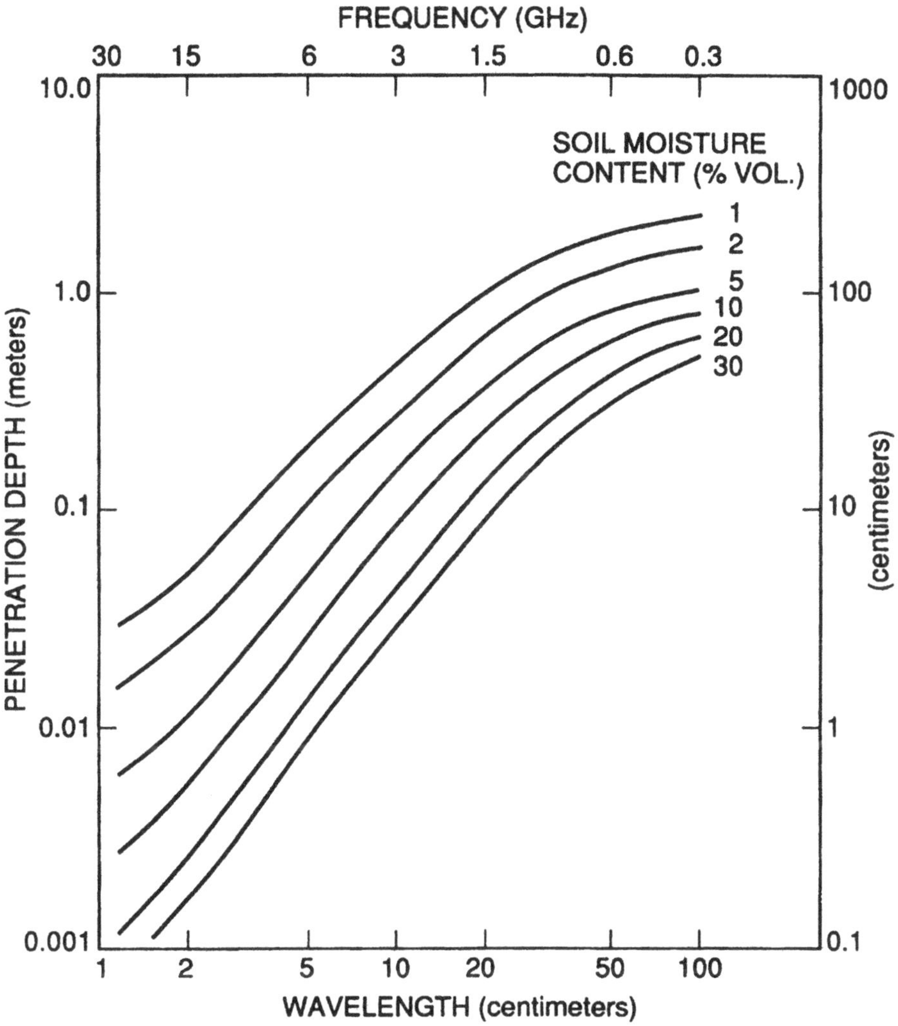

Figure 2. Penetration depth as a function of wavelength for sandy soil with uniform moisture content between 1 and 30% by volume.

agrees with the surface temperature to within a few degrees centigrade over the diurnal temperature cycle. However, at 1.4 GHz, ($\lambda = 21$ cm), the effective temperature differs from the surface temperature by as much as 15 to 20°C near mid-day for these semi-arid zone profiles. Choudhury *et al.* [20] suggest that the effective temperature can be related empirically to the surface temperature T_s and the temperature at large (infinite) depth T_∞ by the expression

$$T_e = T_\infty + (T_s - T_\infty)C \tag{10}$$

where C is a least-squares-fitted parameter that depends on wavelength and to a lesser degree on soil type. For a given location and soil type, if T_∞ and C can be computed or estimated, then T_s may be determined from satellite-derived estimates of T_e.

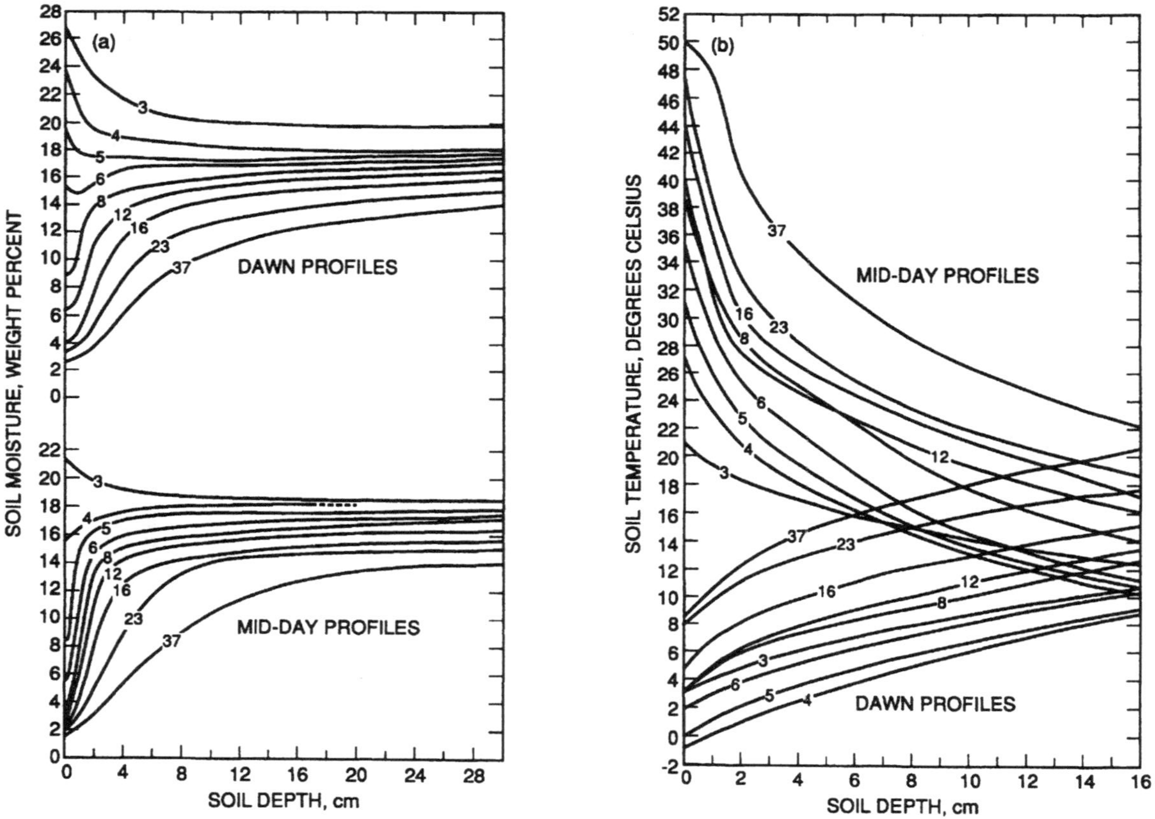

Figure 3. Representative soil profiles of (a) moisture and (b) temperature, for Phoenix, Arizona soil. Curves are labeled by number of days after irrigation. (Choudhury *et al.* [20].)

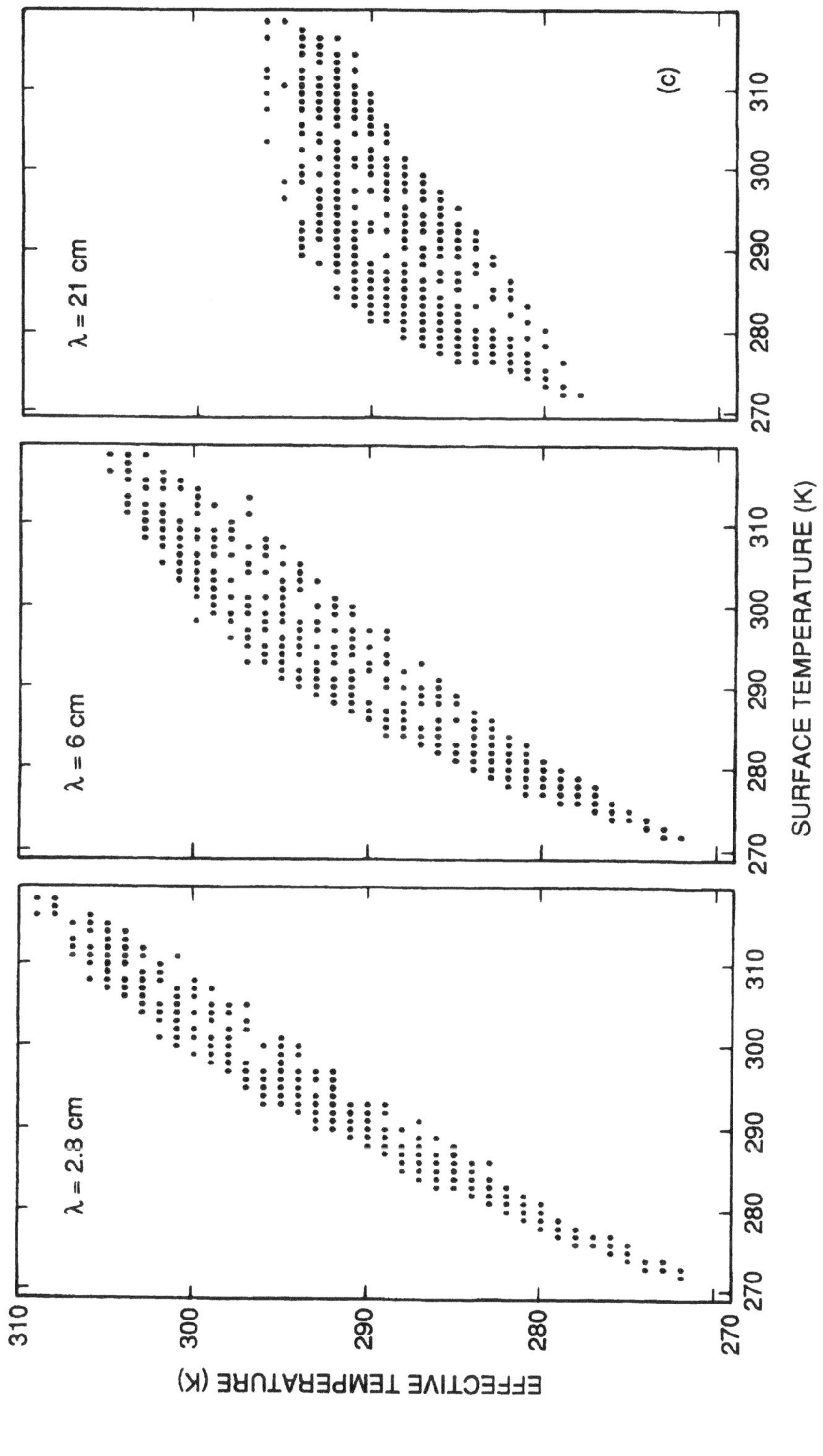

Figure 3. (Continued). (c) Scatter plots of effective temperature versus surface temperature computed for profiles shown in (a) and (b) at various wavelengths. (Choudhury *et al*. [20].)

When the surface is covered by vegetation, similar considerations apply, except that the weighting functions and penetration depth depend on the attenuation coefficient of the vegetation volume, which in turn depends on the vegetation biomass (density and water content). In the following discussions, we shall refer to T_e as the surface temperature, for simplicity, with the understanding that the effective temperature is implied.

3.2. Atmosphere and emissivity effects

The atmospheric emission contributions to the received radiation may be expressed approximately as $T_u \cong T_d \cong T_a[1 - \exp(-\tau_a)]$, where T_a is a weighted mean temperature over the absorbing region in the atmosphere (lower troposphere) that varies with wavelength. Values for T_a are typically between 10 and 20°C lower than the surface air temperature at atmospheric window frequencies. Using these approximations, Equations (6) and (7) lead to the expression

$$T_{Bp} = T_a\left[1 - \exp(-\tau_a)\right]\left[1 + r_p \exp(-\tau_a)\right] + e_p T_e \exp(-\tau_a). \tag{11}$$

At frequencies below approximately 10 GHz, the opacity, τ_a, is usually less than 0.03 for non-precipitating conditions, and the first term in Equation (11) is small. The observed brightness temperature T_{Bp} is then approximately equal to the surface brightness temperature T_{bp}, given by Equation (7). At higher frequencies, and especially for the 37 and 85 GHz channels of the SMMR and SSM/I instruments, the atmospheric contributions cannot be safely ignored and Equation (11) is a more accurate representation. Figure 4

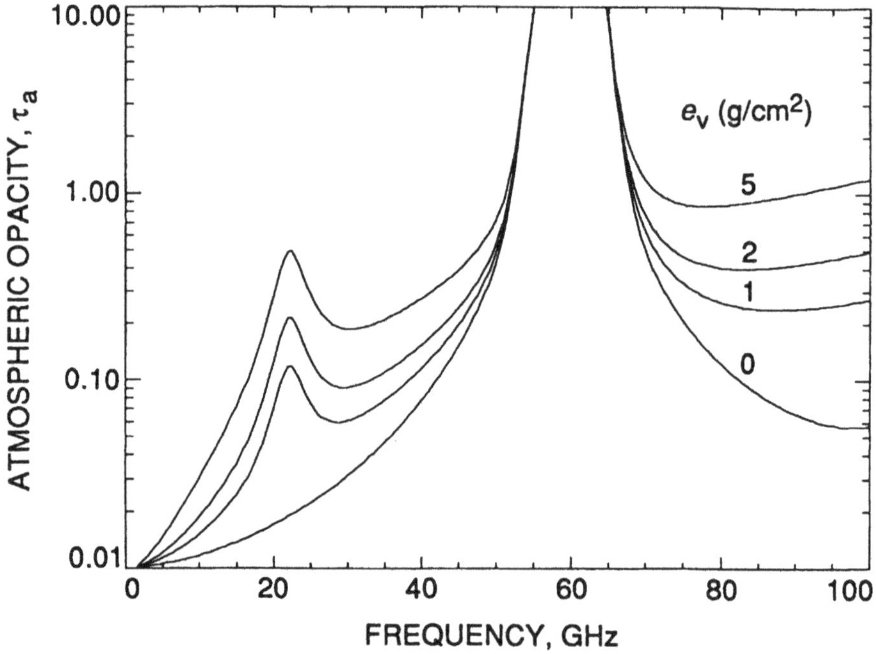

Figure 4. Atmospheric opacity as a function of frequency in the range 1–100 GHz, for a cloud-free annual tropical atmosphere viewed at an incidence angle of 50°. Curves are for integrated water contents e_v of 0, 1, 2, and 5 g/cm^2.

shows the opacity as a function of frequency at $\theta = 50°$ incidence angle (typical of the SMMR and SMM/I sensors), for an annual tropical, cloud-free atmosphere with 0 to 5 g/cm^2 integrated water vapor.

The effects of surface temperature, emissivity, and atmospheric opacity on brightness temperature, computed by Equation (11), are shown in Fig. 5. The expected increases of brightness temperature with increasing emissivity and surface temperature are observed. From Fig. 5(b) we note that increasing atmospheric opacity increases the brightness temperature at low emissivities, since the atmospheric temperature (T_a) is higher than the surface brightness temperature ($T_{bp} = e_p T_e$). Conversely, the brightness temperature decreases with opacity at high emissivities ($e_p \cong 1$). The crossover point depends on the difference between the surface and atmospheric temperatures ($T_e - T_a$).

3.3. Surface temperature error sensitivities

Equation (11) may be rearranged to express the surface temperature in terms of the emissivity, atmospheric variables, and the observed brightness temperature

$$T_e = \frac{T_{Bp} - T_a[1 - \exp(-\tau_a)][1 + (1 - e_p)\exp(-\tau_a)]}{e_p \exp(-\tau_a)}. \tag{12}$$

If the atmospheric variables T_a and τ_a and the surface emissivity e_p are known or can be estimated, then the surface temperature T_e may be obtained from single-channel measurements of the observed brightness T_{Bp}. In practice, the variability in τ_a may be small (at low frequencies) or may be estimated from coincident radiosonde measurements. In some cases the variability in e_p may also be estimated from knowledge of the surface type and moisture content. To investigate the accuracies with which e_p, τ_a, T_a, and T_{Bp} must be known to obtain a reliable estimate of T_e, one may study the partial derivatives of Equation (12), i.e. the sensitivities of T_e to the observed or estimated quantities

$$\frac{\partial T_e}{\partial e_p} = \frac{-\exp(\tau_a)[T_{Bp} - T_a(1 - \exp(-2\tau_a))]}{e_p^2} \tag{13}$$

$$\frac{\partial T_e}{\partial \tau_a} = \frac{\exp(\tau_a)\{T_{Bp} - T_a[1 + (1 - e_p)\exp(-2\tau_a)]\}}{e_p} \tag{14}$$

$$\frac{\partial T_e}{\partial T_a} = \frac{-\exp(\tau_a)[1 - e_p \exp(-\tau_a) - (1 - e_p)\exp(-2\tau_a)]}{e_p} \tag{15}$$

$$\frac{\partial T_e}{\partial T_{Bp}} = \frac{\exp(\tau_a)}{e_p}. \tag{16}$$

The functional forms of these derivatives are plotted in Fig. 6 for representative ranges of values. Figures 6(a) and 6(b) show that to obtain an accuracy of 3 °C in T_e the emissivity must be known to an accuracy of between 0.006 and 0.012, depending on the values of T_e and e_p. This corresponds to an accuracy of approximately 1%, since one may assume approximate values of $T_e = 300$ K and $e_p = 0.8$, and, for small τ_a,

$$\frac{\Delta T_e}{T_e} \cong \frac{\Delta e_p}{e_p}.$$

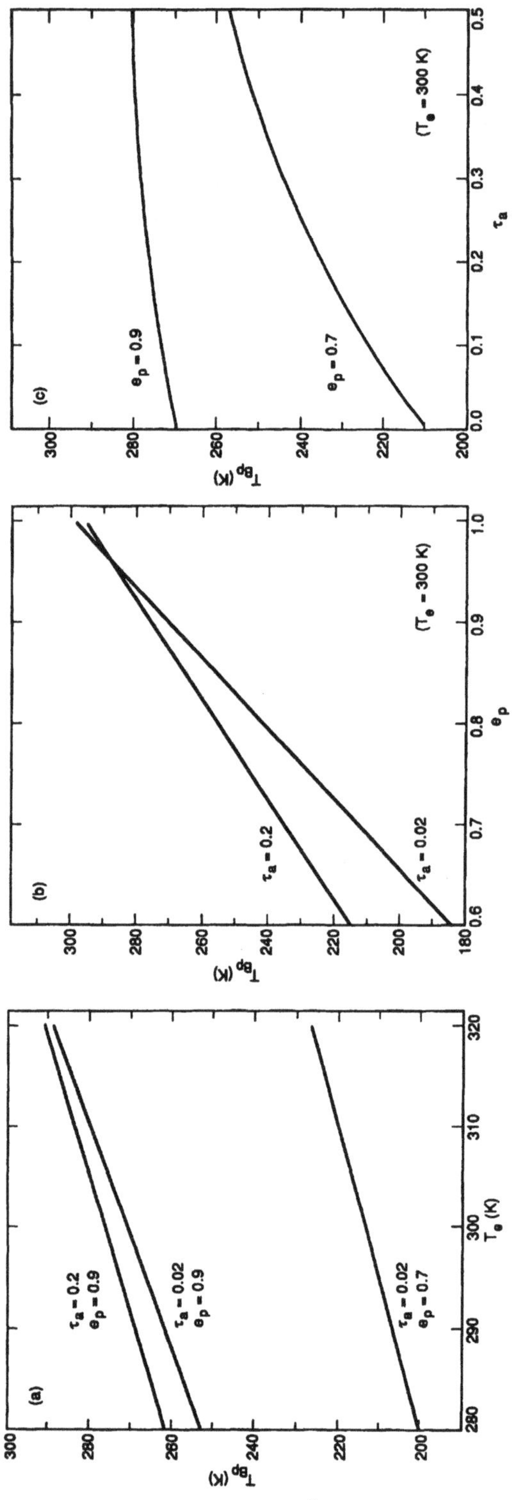

Figure 5. Brightness temperature dependence on: (a) surface temperature T_e; (b) surface emissivity e_p; and (c) atmospheric opacity τ_a; computed from Equation (10) with $T_a = 280$ K.

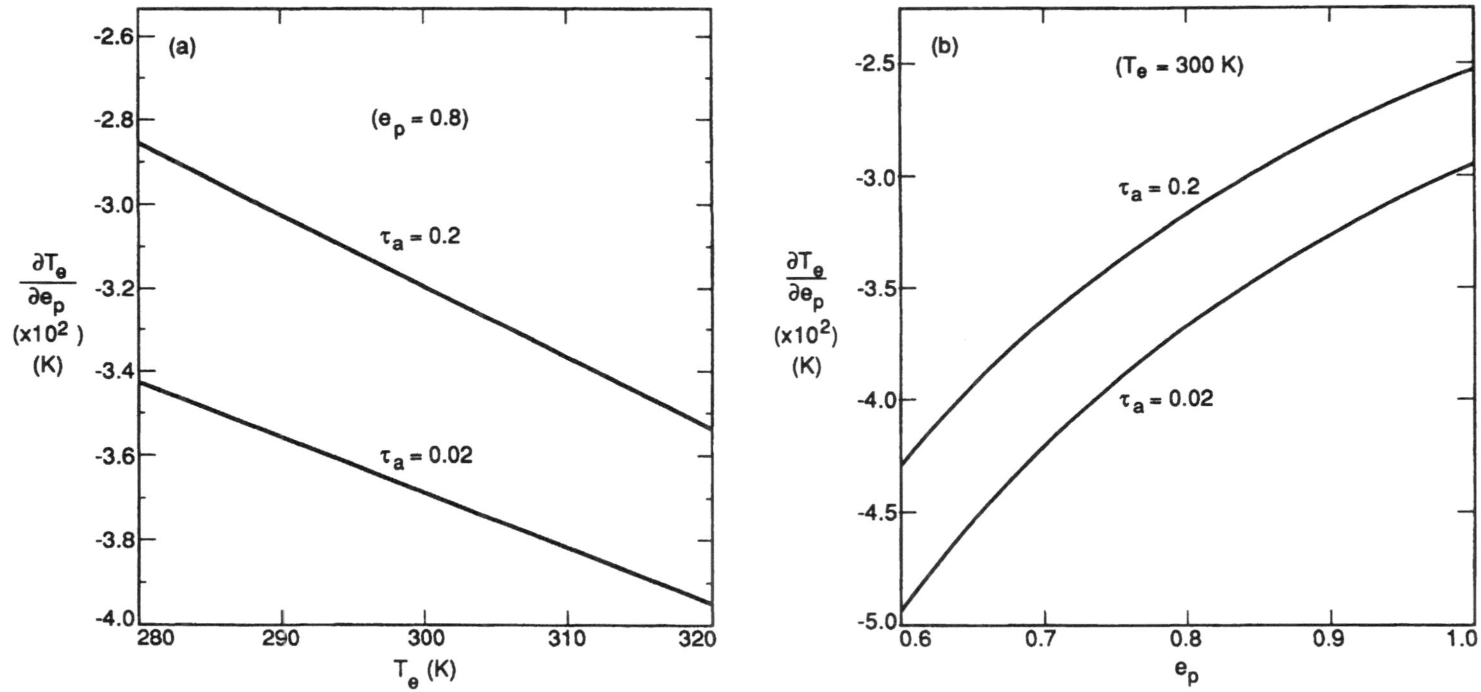

Figure 6. (a) and (b): Sensitivity of estimated surface temperature T_e to surface emissivity e_p as a function of surface temperature and emissivity, respectively.

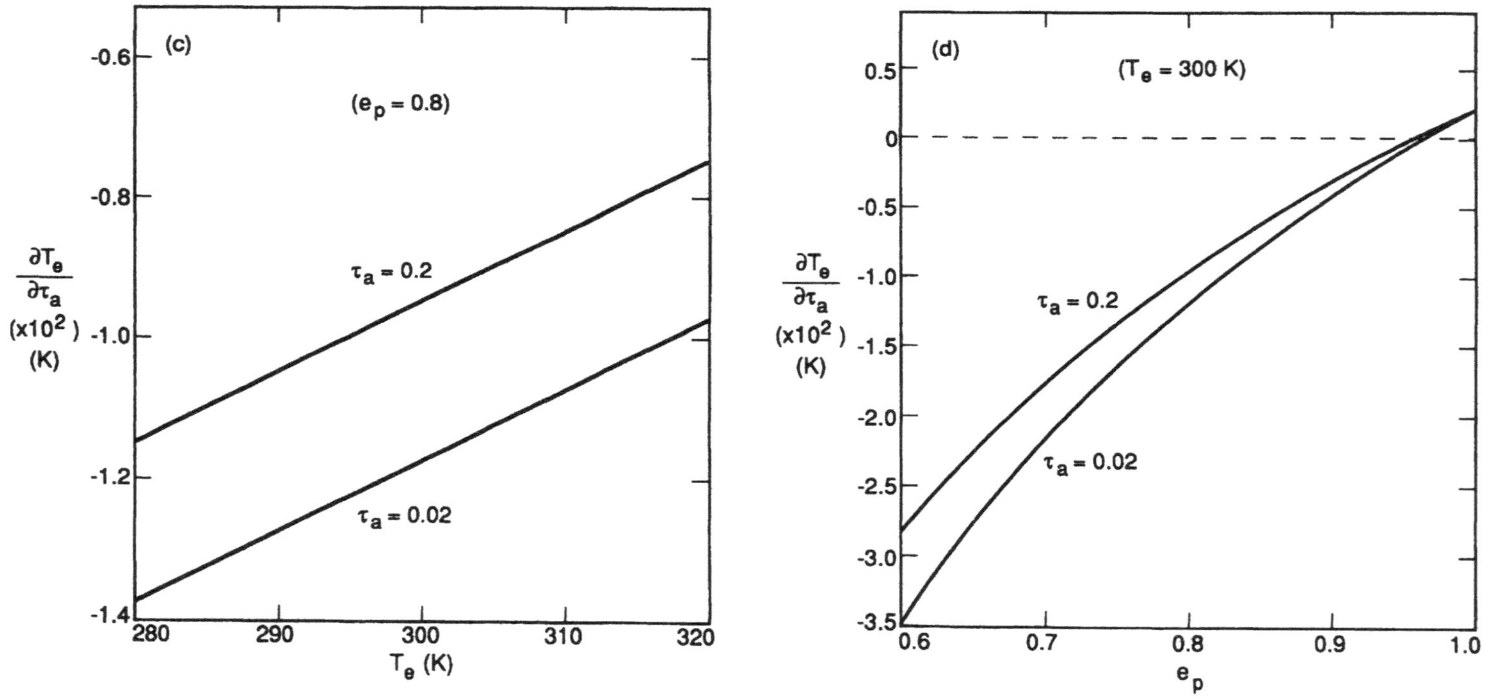

Figure 6. (c) and (d): Sensitivity of estimated surface temperature T_e to atmospheric opacity τ_a as a function of surface temperature and emissivity, respectively.

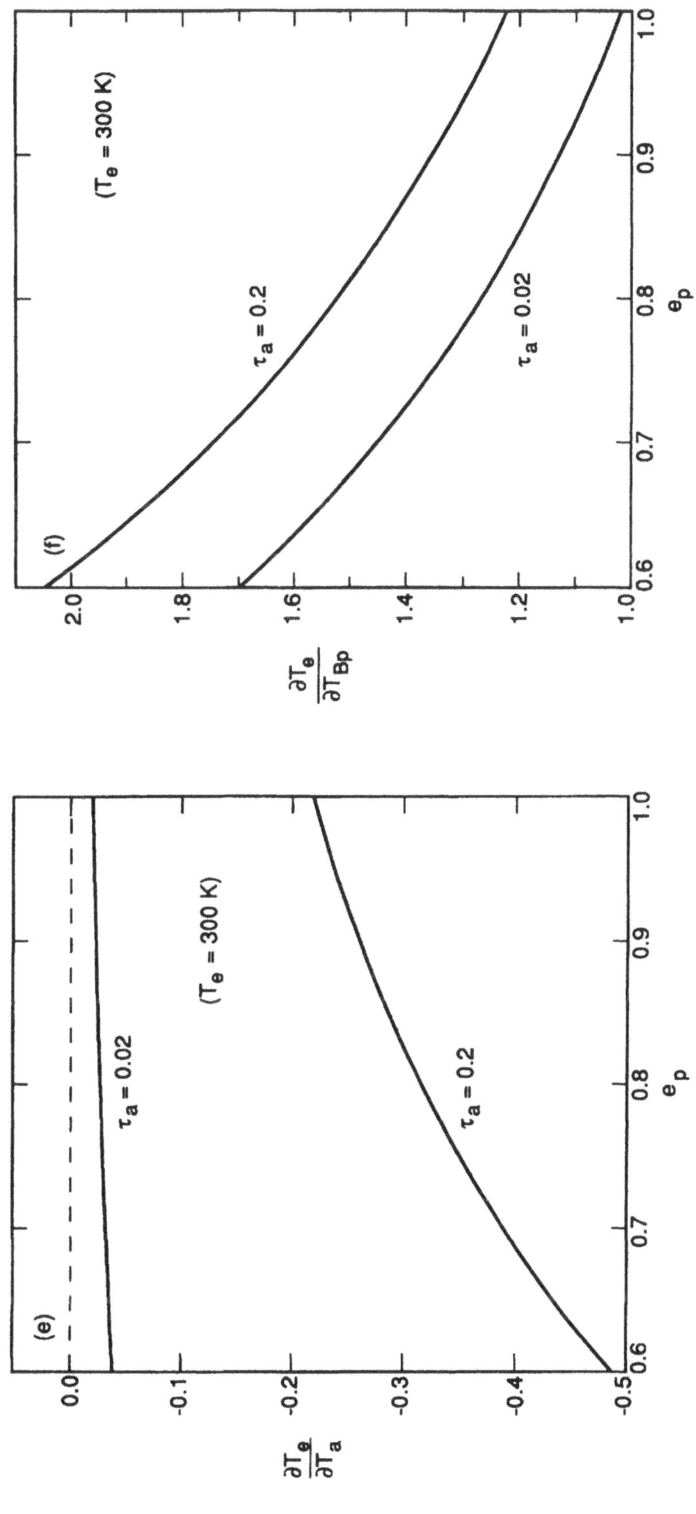

Figure 6. (e) and (f): Sensitivity of estimated surface temperature T_e to atmospheric temperature T_a and brightness temperature T_{Bp}, respectively, as functions of emissivity.

Such accuracy in *a priori* knowledge of emissivity appears unrealistic given the natural variability of 10 to 25% or greater found for e_p in many regions. On the other hand, surface classification schemes (e.g. Townshend *et al.* [21] and Neale *et al.* [22]) can be used to identify the gross surface type. Within each surface type the emissivity is influenced primarily by moisture. The use of multiple channels may permit the moisture, and thereby the emissivity, to be estimated sufficiently accurately to retrieve the surface temperature. This procedure may be carried out using multiple regression or iterative retrieval methods as discussed in the next section.

At frequencies less than 10 GHz where the variability in atmospheric opacity is small ($\Delta\tau_a < 0.01$), Figs 6(c) and (d) show that the effect of uncertainties in opacity on T_e will also be small. At higher frequencies this may not be the case. For example, with reference to Fig. 4, a radiosonde-derived water vapor measurement uncertainty of 0.1 g/cm^2 leads to uncertainties of approximately 0.003 and 0.02 in opacity at 37 and 85 GHz, respectively, with corresponding errors in T_e of 0.3 and 2 °C for an emissivity of 0.8. The presence of clouds will further increase the errors. These errors decrease as the emissivity increases. For low opacities, Fig. 6(e) shows that the effect on T_e of uncertainties in T_a is small. Finally, Fig. 6(f) shows that radiometer noise will be amplified in the estimation of T_e, typically by a factor of between 1 and 2, with higher values occurring in regions of low emissivity and high opacity.

4. MULTICHANNEL RETRIEVALS

The magnitudes of the corrections necessary to account for effects of variable surface emissivity and atmospheric opacity when estimating surface temperature were discussed in the previous section. These corrections may be attempted using independent water vapor (radiosonde) and cloud measurements, and *a priori* emissivity estimates, or alternatively they may be made by using multichannel data such as available from the SMMR and SSM/I, or a combination of both approaches may be used. There has been limited research in studying these approaches however.

4.1. Satellite retrievals

Few studies of surface temperature using satellite microwave data have been reported in the literature. One study is that of McFarland *et al.* [14] who investigated the rms errors resulting from multiple linear regressions of SSM/I data against air temperatures measured at surface meteorological stations across the US Great Plains. The air temperatures were obtained from the NOAA operational climatological database and represent measurements at screen height (1.2 m). By using SSM/I data from ascending (morning) overpasses occurring between 0600 and 0700 local solar time, and comparing these with the minimum air temperatures reported for the same morning from the meteorological stations, an attempt was made to minimize the differences between the ground surface temperature and the screen height air temperature. For the range of temperatures encountered (1.1 to 26.7 °C), the authors report possible rms errors as high as 2.85 °C in using point-located minimum air temperatures in this manner as comparisons for the footprint-average SSM/I data at the exact time of overpass. The regressions between surface air temperatures T_{as} and SSM/I brightness temperatures T_{Bi} can be expressed as

$$T_{as} = a_0 + \sum_{i=1}^{N} a_i T_{Bi} \tag{17}$$

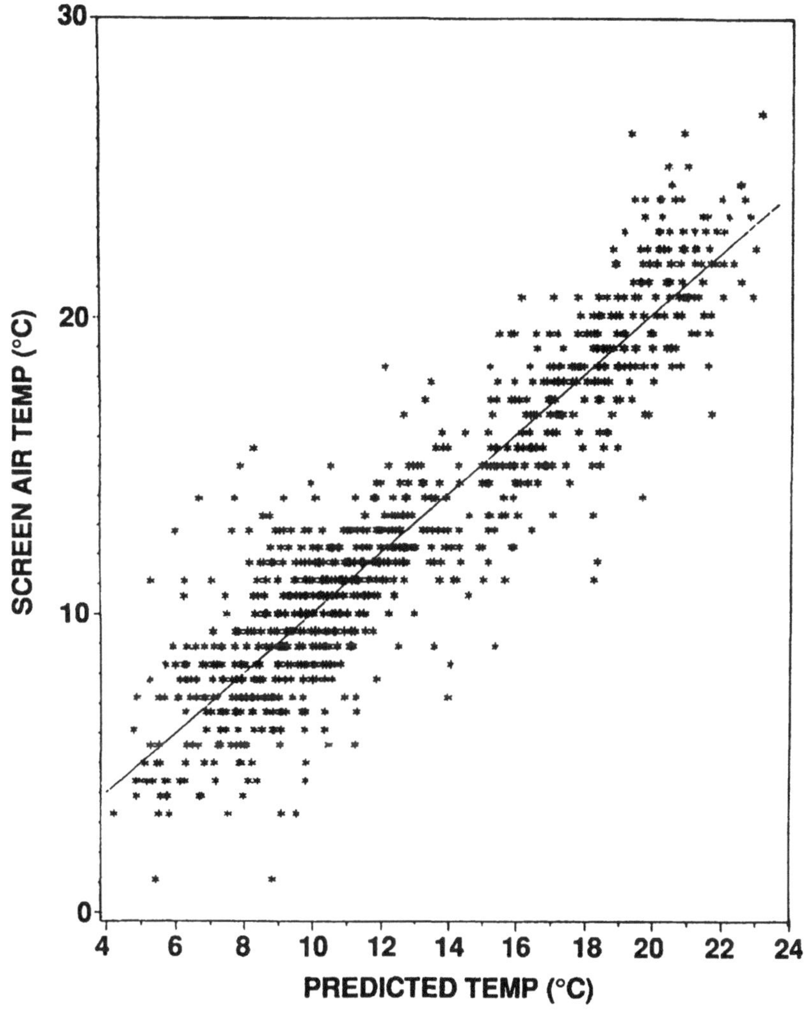

Regression Coefficients and Statistics for 4-Channel Model:

Land Surface Type	R^2	RMSE	Intercept	85 V	37 V	22 V	19 H
Crops/range	0.81	2.58	22.40	1.239	-0.396	0.275	-0.174
Moist soils	0.85	1.99	41.69	1.245	-0.724	0.415	-0.063
Dry soils	0.62	2.60	76.28	-0.367	-0.318	1.408	0.025
All	0.79	2.37	58.07	0.811	-0.555	0.730	-0.170

Figure 7. Regression fit of 'predicted temperature' (linear combination of brightness temperature channels) to screen air temperature, for moist soils, covering central US study region described by McFarland *et al.* [14]. Statistics for other surface types are also shown. Brightness temperature data are from the DMSP SSM/I satellite instrument.

where a_i are regression coefficients, and N is the number of channels used in the regressions. Figure 7 shows the results of these regressions for the surface type classified as 'moist soil', displayed as the predicted temperature from Equation (17) versus the actual values of screen air temperature. The channels selected in the regression, 85 V, 37 V, 22 V, and 19 H, have the coefficients tabulated in the figure. Also shown is the rms error of the regression, and similar statistics for the other land surface types within the study region. For moist soil, the magnitude of the coefficients indicates that the surface temperature is most highly correlated with the 85 V channel, while the remaining channels provide information to correct for emissivity and atmospheric variability. This is reasonable since the 85 GHz radiation originates close to the surface, and for vertical polarization the 53° incidence angle of SSM/I is close to the Brewster angle, reducing the variability due to surface emissivity for this channel. For dry soils, however, 22 V appears to be the most significant channel, indicating perhaps that what is being sensed is the lower atmosphere rather than the surface.

From these results a number of observations can be made. First, the rms errors of the regression fits appear promising since they are in the range 2–2.6°C over a dynamic range of 4–23°C, and the correlation coefficients in some cases are above 0.8. However, the rms errors mask a variety of error sources as mentioned earlier. The regression coefficients have been tuned to the conditions of the specific study region and hence the equations cannot be applied reliably to other regions. The variability of the coefficients from one surface type to another also indicates that surface classification errors may lead to corresponding surface temperature errors.

4.2. *Model simulations*

The problem may be analyzed from another perspective by performing simulations using a simple radiative transfer model. In this model, vegetation is represented as an absorbing layer above the soil, characterized by an opacity τ_c, with soil reflectivity r_{sp}. If the soil surface and vegetation are assumed to be at the same temperature, Kerr and Njoku [23] have shown that Equation (11) describes the brightness temperature viewing this two-layer medium if r_p is replaced by $r_{sp}\exp(-2\tau_c)$, i.e. Equation (11) is replaced by

$$T_{Bp} = T_a\left[1-\exp(-\tau_a)\right]\left[1+r_{sp}\exp(-\tau_a-2\tau_c)\right] + T_e\exp(-\tau_a)\left[1-r_{sp}\exp(-2\tau_c)\right]. \tag{18}$$

The canopy opacity can be expressed as

$$\tau_c = A\frac{\varepsilon''_{sw}}{\lambda}W/\cos\theta \tag{19}$$

where W is the vegetation water content, ε''_{sw} is the imaginary part of the dielectric constant of the vegetation water, λ is the wavelength, θ is the incidence angle, and A is a coefficient that depends on vegetation type. This expression is strictly valid at low frequencies only ($\lambda \geqslant 3\,\text{cm}$). The soil reflectivity depends on surface roughness and on soil dielectric constant which can be related to the soil moisture, m_v, using data given by Hallikainen *et al.* [24] and Njoku and Kong [18]. The atmospheric opacity can be related directly to the atmospheric integrated water vapor, e_v (Waters [25]). These relationships are discussed more fully in Kerr and Njoku [23].

Equation (18) allows the satellite-observed brightness temperature to be computed as a function of a limited number of atmospheric and surface parameters, i.e. $T_{Bp} = f(e_v, T_e, m_v, W)$. A simulated data set can then be generated by computing T_{Bp} using Equation (18) for various combinations of parameter values chosen to cover a suitable dynamic range. This was done for values covering the ranges: $e_v = 0.5$ to $5.0\,g/cm^2$; $T_e = 4$ to $40°C$; $m_v = 0.03$ to $0.3\,g/cm^3$; and $W = 0$ to $0.2\,kg/m^2$. Ten values were selected to cover the ranges in equal increments for each parameter, and brightness temperatures were computed for all possible combination of each parameter value, for a total of 10^4 combinations. For each combination, ten brightness temperatures were computed at frequencies of 6.6, 10.7, 18, 21, and 37 GHz, vertical and horizontal polarizations, and 50° incidence angle, to simulate measurements of the Nimbus-7 SMMR satellite instrument. The SMMR frequencies are of interest also because they correspond closely to the frequencies that will be used for the future EOS Multifrequency Imaging Microwave Radiometer (MIMR) instrument (ESA [26]); frequencies higher than 37 GHz were not computed since the assumption in Equations (18) and (19) of vegetation as an absorbing (non-scattering) layer is not valid at the higher frequencies. (Strictly speaking the validity is in question at frequencies greater than about 10 GHz, but the error is not expected to be too great up to 37 GHz.)

The simulated data set was used to develop a regression algorithm for surface temperature of a form similar to Equation (17), i.e.

$$T_e = a_0 + \sum_{i=1}^{N} a_i F_i(\overline{T}_B) \tag{20}$$

where $\overline{T}_B$ is a vector of the ten brightness temperatures. The F_i are functions of the ten brightness temperatures, and allow for regressions on functions of the brightness temperatures as well as on the brightness temperatures themselves. This helps mitigate some of the nonlinearities in the model Equation (18). Examples of such functions are $F_i = \ln(315 - T_{Bj})$ and $F_i = (T_{Bj} - T_{Bk})/(T_{Bj} + T_{Bk})$, chosen to account for exponential nonlinearities, in the first case, and to investigate the use of functions that are independent of surface temperature, in the second. Regression techniques for microwave radiometry have been discussed elsewhere (Hofer and Njoku [27]). Caution must be used in interpreting rms errors of regressions on nonlinear functions since the errors will in general not be normally distributed.

Application of the regression formula of Equation (20) to the simulated data set gave the results shown in Fig. 8. A stepwise regression procedure was used in which the single brightness temperature channel T_{Bi} (or function of channels F_i) was first selected that was most highly correlated to the surface temperature. Additional channels (or functions) were then selected for the regression in successive steps, such that the residual error was minimized at each step for that selection. The stepwise procedure was terminated when the addition of further channels to the regression did not reduce the residual error significantly from the previous step. This led to the following equation

$$\begin{aligned} T_e^* = {} & a_0 + a_1 \ln(315 - T_{B37V}) + a_2\left(\frac{T_{B10V} - T_{B10H}}{T_{B10V} + T_{B10H}}\right) \\ & + a_3 T_{B37V} + a_4 \ln(315 - T_{B18H}). \end{aligned} \tag{21}$$

The rms residual error between the predicted temperature T_e^* from Equation (21) and the actual temperature T_e was 2.07 °C. This simulation did not take into account the effects of radiometer noise in the brightness temperatures which would increase the rms residual error slightly. If only brightness temperatures, and not nonlinear functions,

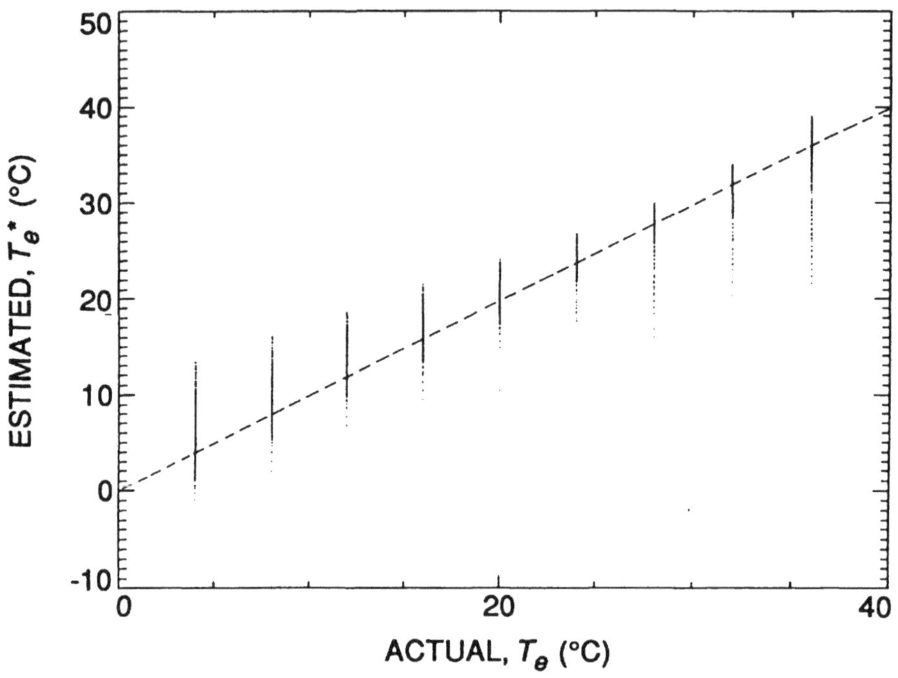

Figure 8. Comparison between actual T_e and estimated T_e^* surface temperature using multiple regression on a simulated brightness temperature data set. The dashed line is the perfect agreement line. An rms residual error of 2.07 °C was obtained. However, due to the nonlinearity of the model, the errors are not normally distributed.

were used in the regressions, then the residual rms error increases to 2.9 °C. This value seems to be significantly larger than observed in Fig. 7. However, the dynamic range of surface temperature used for the simulations was much larger, and no attempt was made to classify the simulations into groupings of dry, wet, and low vegetated soils.

The intent of the regression simulations is to show that the results obtained using satellite data are consistent with model simulation results. Limitations of the nonscattering vegetation model preclude analysis of the suitability of the 85 GHz channel for surface temperature measurement, but the selection of the higher frequency channels in the simulation regressions results directly from the fact that the higher frequencies are less sensitive to surface emissivity variability (in the simulations the 37 V channel showed the highest correlation to surface temperature). Due to the nonlinearity of the problem, it is likely that iterative or neural network methods would be more appropriate for developing a useful surface temperature retrieval agorithm.

5. CONCLUSIONS

Surface temperature has a major effect on land surface microwave emission and can be estimated from satellite measurements of brightness temperature if the natural variability in surface emissivity and atmospheric opacity can be accounted for. Multichannel measurements are one method of estimating and correcting for the emissivity and atmospheric effects. Limited satellite data analyses and theoretical simulations have shown that surface temperature estimation accuracies of 2 to 2.5 °C should be feasible using a nonlinear retrieval algorithm. The accuracy may be improved if *a priori* information on the surface emissivity is available. Additional research and data analysis are required to investigate these approaches. It is difficult to assess the accuracy of satellite-derived temperature estimates since comparable *in situ* data are generally unavailable over spatial scales typical of the satellite footprints. If satellite temperature measurements are used in energy balance models care should be taken in defining the effective temperature consistently in the models and radiative transfer formulations. Since several years of global satellite data are now available from the SMMR and SSM/I instruments, a concerted effort should be made to develop surface temperature products from these data to compare with GCM model output and to examine regional seasonal and interannual temperature variability. Techniques developed using these data will be directly applicable to forthcoming data from the EOS MIMR instrument.

Acknowledgment
This work represents one phase of research carried out at the Jet Propulsion Laboratory, California Institute of Technology, under contract with the National Aeronautics and Space Administration.

REFERENCES

1. J. Walker and P. R. Rowntree. The effect of soil moisture on circulation and rainfall in a tropical model. *Quart. J. Roy. Met. Soc.*, **103**, 29–46 (1977).
2. J. Shukla and Y. Mintz. Influence of land-surface evapotranspiration on the Earth's climate. *Science*, **215**, 1498–1500 (1982).
3. R. E. Dickinson and A. Henderson-Sellers. Modelling tropical deforestation: A study of GCM land surface parameterizations. *Quart. J. Roy. Met. Soc.*, **114**, 439–462 (1988).
4. M. T. Chahine, R. Haskins, J. Susskind and D. Reuter. Remote sensing of land-surface temperature from the HIRS/MSU data. In: *Proc. ISLSCP Conference*, Rome, Italy, ESA SP-248, European Space Agency, Paris (1986).
5. T. N. Carlson, J. K. Dodd, S. G. Benjamin and J. N. Cooper. Satellite estimation of the surface energy balance, moisture availability, and thermal inertia. *J. Appl. Meteor.*, **20**, 67–87 (1981).
6. J. C. Price. On the use of satellite data to infer surface fluxes at meteorological scales. *J. Appl. Meteor.*, **21**, 1111–1122 (1982).
7. W. Klaassen and W. van den Berg. Evapotranspiration derived from satellite observed surface temperatures. *J. Climate Appl. Meteor.*, **24**, 412–424 (1985).
8. O. Taconet, R. Bernard and D. Vidal-Madjar. Evapotranspiration over an agricultural region using a surface flux/temperature model based on NOAA-AVHRR data. *J. Climate Appl. Meteor.*, **25**, 284–307 (1986).
9. P. J. Wetzel and R. H. Woodward. Soil moisture estimation using GOES-VISSR infrared data: A case study with a simple statistical method. *J. Climate Appl. Meteor.*, **26**, 107–117 (1987).
10. J. C. Price. Land surface temperature measurements from the split window channels of the NOAA-7 Advanced Very High Resolution Radiometer. *J. Geophys. Res.*, **89**, 7231–7237 (1984).
11. Z. Wan and J. Dozier. Land-surface temperature measurement from space: Physical principles and inverse modeling. *IEEE Trans. Geosci. Rem. Sens.*, **27**, 268–278 (1989).
12. F. Becker and Z.-L. Li. Towards a local split window method over land surfaces. *Int. J. Rem. Sens.*, **11**, 369–393 (1990).
13. Y. Kerr, J. P. Lagouarde and J. Imbernon. Accurate land surface temperature retrieval from AVHRR data with use of an improved split window algorithm. *Remote Sens. Environ.*, **41**, 197–209 (1992).

14. M. J. McFarland, R. L. Miller and C. M. U. Neale. Land surface temperature derived from the SSM/I passive microwave brightness temperatures. *IEEE Trans. Geosci. Rem. Sens.*, **28**, 839–845 (1990).
15. R. E. Dickinson, A. Henderson-Sellers, C. Rosenzweig and P. J. Sellers. Evapotranspiration models with canopy resistance for use in climate model: A review. *Agric. Forest Meteorol.*, **54**, 373–388 (1991).
16. W. P. Kustas. Estimates of evapotranspiration with a one- and two-layer model of heat transfer over partial canopy cover. *J. Appl. Meteor.*, **29**, 704–715 (1990).
17. P. J. Sellers, Y. Mintz, Y. C. Sud and A. Dalcher. A simple biosphere model (SiB) for use within general circulation models. *J. Atmos. Sci.*, **43**, 505–531 (1986).
18. E. G. Njoku and J. A. Kong. Theory for passive microwave remote sensing of near-surface soil moisture. *J. Geophys. Res.*, **82**, 3108–3118 (1977).
19. E. G. Njoku and P. E. O'Neill. Multifrequency microwave radiometer measurements of soil moisture. *IEEE Trans. Geosci. Rem. Sens.*, **GE-20**, 468–475 (1982).
20. B. J. Choudhury, T. J. Schmugge and T. Mo. A parameterization of effective soil temperature for microwave emission. *J. Geophys. Res.*, **87**, 1301–1304 (1982).
21. J. R. G. Townshend, C. O. Justice, B. J. Choudhury, C. J. Tucker, V. T. Kalb and T. E. Goff. A comparison of SMMR and AVHRR data for continental land cover characterization. *Int. J. Rem. Sensing*, **10**, 1633–1642 (1989).
22. C. M. U. Neale, M. J. McFarland and K. Chang. Land-surface-type classification using microwave brightness temperatures from the Special Sensor Microwave/Imager. *IEEE Trans. Geosci. Rem. Sens.*, **28**, 829–838 (1990).
23. Y. H. Kerr and E. G. Njoku. A semiempirical model for interpreting microwave emission from semiarid land surfaces as seen from space. *IEEE Trans. Geosci. Rem. Sens.*, **28**, 384–393 (1990).
24. M. T. Hallikainen, F. T. Ulaby, M. C. Dobson, M. A. El-Rayes and L. K. Wu. Microwave dielectric behaviour of wet soil. Part I: Empirical models and experimental observations. *IEEE Trans. Geosci. Rem. Sens.*, **GE-23**, 25–34 (1985).
25. J. W. Waters. Absorption and emission by atmospheric gases. In: *Methods of Experimental Physics*, Vol. 12, Part B, M. L. Meeks (Ed.), New York, Academic Press (1976).
26. ESA. *The Multifrequency Imaging Microwave Radiometer*. Instrument Panel Report, Publication No. ESA SP-1138, European Space Agency, Paris (1990).
27. R. Hofer and E. G. Njoku. Regression techniques for oceanographic parameter retrieval using spaceborne microwave radiometry. *IEEE Trans. Geosci. Rem. Sens.*, **GE-19**, 178–189 (1981).

ESA/NASA International Workshop, pp. 531–550
B. J. Choudhury, Y. H. Kerr, E. G. Njoku and P. Pampaloni (Eds)

Evaporation monitoring over land surface using microwave radiometry

A. CHANZY[1] and W. P. KUSTAS[2]

[1]*INRA Station de Science du Sol, BP91, 84143 Montfavet CEDEX, FRANCE*
[2]*USDA-ARS Hydrology Lab, Beltsville, Maryland, USA*

Abstract—Among the surface parameters that affect the microwave emission from land surfaces, the surface soil moisture θ_s and the surface temperature are useful for the estimation of the evapotranspiration. The physical processes involved in the soil evaporation and the plant transpiration are different. Thus, bare soil and vegetated surfaces are considered separately. For bare soils, evaporation models are based on θ_s. The depth of the surface soil layer is important with respect to estimating evaporation. It was found that at least the upper 2 cm should be considered, which requires the use of low frequency (L-band) microwave radiometers. It is shown that the soil dependent parameters of an evaporation model can be determined by inversion when both microwave and thermal infrared observations are available. The capability of the microwave radiometry for predicting the evapotranspiration for vegetated surfaces is less evident. If soil moisture is a relevant quantity for inferring the evapotranspiration on sparse vegetation, the success of the microwave radiometry relies on its capability to estimate accurately the temperature of land surfaces.

1. INTRODUCTION

Actual evapotranspiration (ET) is a key parameter in many fields of application such as hydrology, agronomy and meteorology. At the regional scale, evapotranspiration is a major component of the water balance and may represent 80% of the water fluxes. The evapotranspiration flux is affected by the availability of the water within the soil and could be considered as an indicator of the vegetation stress which can reduce biomass production or crop yield. Evapotranspiration is also involved in the energy budget of the surface which allows the coupling between the surface and the atmosphere in General Circulation Models. These applications require a regular monitoring of ET at regional scales. For agronomic or hydrological purposes, ET should be determined at a daily time scale, while the diurnal variations of the ET should be accounted for in meteorological applications. ET is composed of direct evaporation of the soil and the transpiration of the vegetation. Both components are governed by climatic conditions and regulated by water availability. But the mechanisms that govern water availability are different between the evaporation and the transpiration. In the case of direct evaporation from the soil, the water vaporisation and the water transport from deeper layers to the evaporating surface are directly related to the moisture of the top layer. Hence, when the soil dries up, the top dry layer acts as a barrier to moisture flows which limits the evaporation. In the case of transpiration, the water is extracted by the roots and all the layers occupied by the roots contribute to the water supply of the evaporating surface. Consequently, the factors that affect water availability for evaporation and transpiration are different. Therefore most models of ET using remotely sensed quantities generally deal with the evaporation and the transpiration separately.

Because remote sensing provides spatial information of the earth's surface, it is well suited to map and monitor ET at regional scale. Among the wavelengths used in remote sensing, regular acquisition of surface parameters can only be achieved by microwave sensors, owing to the cloud transparency to the microwave radiation. Such a regularity in the observation of surface parameters is of great importance to monitor ET, which presents strong temporal variations. Until now, the use of the microwave radiometry for predicting ET has mainly focused on the soil evaporation E [1–8]. These studies are based in the capability of microwave radiometry to infer the surface soil moisture. However, most land surfaces are covered with vegetation. Only a few studies have dealt with the utility of microwave radiometry for estimating the transpiration component of the ET [9]. In this paper, it will be shown how the surface parameters derived from the microwave radiometry can be introduced in models of ET.

1. ESTIMATION OF BARE SOIL EVAPORATION USING SURFACE SOIL MOISTURE

1.1. Material and methods

As a first step in developing an evaporation model using microwave data, the relationship between the brightness temperature and the moisture of the top soil layer was investigated. The results presented in this study were obtained experimentally with the PORTOS radiometer mounted on a crane [10]. PORTOS is a six frequency radiometer (1.4, 5.05, 10.65, 23.8, 36.5, and 90 GHz). For each frequency, the measurements in both horizontal and vertical polarizations can be performed concurrently. The experiment was carried out during the summer of 1991 over silty clay loam fields (0.2 ha) near Avignon (France) [10]. The use of the surface moisture in evaporation models was analysed with a separate set of data which combined ground-based measurements and numerical simulations [8, 11]. The latter was performed with a mechanistic model of heat and mass flows according to Philip and de Vries theory [12]. The mechanistic model was calibrated and validated on three experiments of about 20 days which were carried out on three soils (loam, silty clay loam and clay). The calibrated model was then used to simulate the soil hydrodynamic behaviour for each of the three soils under various climatic and moisture conditions.

1.2. Evaporation models involving surface soil moisture (θ_s)

There are generally two approaches to use θ_s for estimating evaporation. In one approach, θ_s is an output of the evaporation model. The surface soil moisture derived from remotely sensed measurement is used to check the results given by the model. This allows either a calibration of the model parameters or a correction of the outputs in order to minimize the drifts in the evaporation estimate. This approach is the only alternative for monitoring evaporation with infrequent remote sensing acquisitions. Nevertheless, the model should simulate a realistic evolution of the remotely sensed parameters. This is possible with physically based comprehensive models which require a rather detailed description of the soil. The non-applicability of such models at regional scales is the main drawback to this approach. Because of the potential regularity of microwave observation, this paper will focus on the other approach, when remotely sensed parameters are used as input variables in the evaporation model.

1.2.1. The water balance approach. The surface moisture, θ_s, is introduced as the upper boundary condition of a mechanistic model of water flow in soils [1, 3]. This approach is time consuming and highly sensitive to the hydraulic conductivity which is difficult to estimate over large areas.

1.2.2. The climatic approach. In this approach the limitation to evaporation compared to the climatic demand is expressed as a function of the surface soil moisture. In the physically based models, θ_s is related either to relative humidity of the air near the soil surface H_r (1) or to the soil resistance r_s of the water diffusion from the evaporation front to the surface (2) and (4):

$$\lambda E = \frac{\rho C_p}{\gamma r_a}\left[H_r(\theta_s)e^*(T_s) - e_a\right] \tag{1}$$

where e_a is the partial vapor pressure of the air at a reference level, $e^*(T_s)$ is the saturated vapor pressure at temperature T_s and r_a is the aerodynamical resistance of the vapor flows, ρC_p is the volumetric specific heat of the air, γ the psychrometric constant and λ the latent heat of vaporisation. In the soil resistance approach, the water vapor fluxes from the evaporation front, which stands below the surface in dry condition, to the soil surface is given by:

$$\lambda E = \frac{\rho C_p}{\gamma r_s}\left[e^*(T_e) - e_s\right] \tag{2}$$

where e_s is the partial water vapor pressure at the soil surface and T_e is the temperature at the evaporation front. The water vapor fluxes could be also expressed in the lower part of the atmosphere as:

$$\lambda E = \frac{\rho C_p}{\gamma r_a}[e_s - e_a]. \tag{3}$$

By combining Equations (2) and (3) and assuming that $T_e = T_s$, we obtain the Penman Monteith expression [13]:

$$\frac{E}{E_p} = \left[1 + \frac{\gamma}{\gamma + \Delta}\frac{r_s}{r_a}\right]^{-1} \tag{4}$$

where E_p is the potential evaporation and Δ is the rate of change of saturation vapor pressure with temperature. In Equations (1), (2) and (4), r_s and H_r are related to θ_s through empirical equations [4, 5, 6, 14]. These relationships depend strongly on the soil type and are also affected by climatic conditions [5, 14].

In a more simplified approach, $E = \alpha E_p$, where α is determined using the concept of limiting evaporation [2] or related to θ_s by an empirical relationship [7]. With the concept of limiting evaporation α is given by

$$\alpha = \min(1,\ E_{lim}/E_p) \tag{5}$$

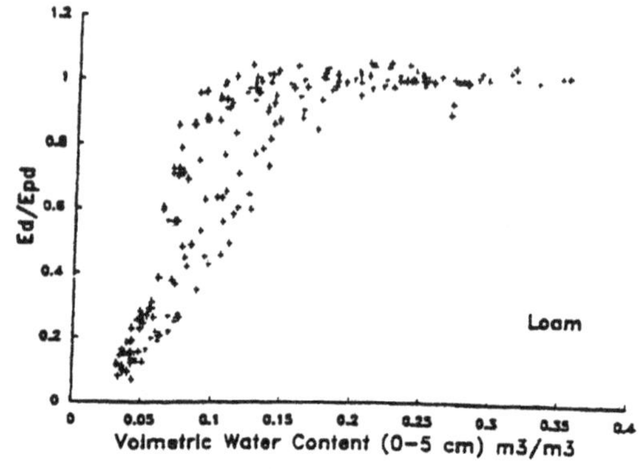

(a)

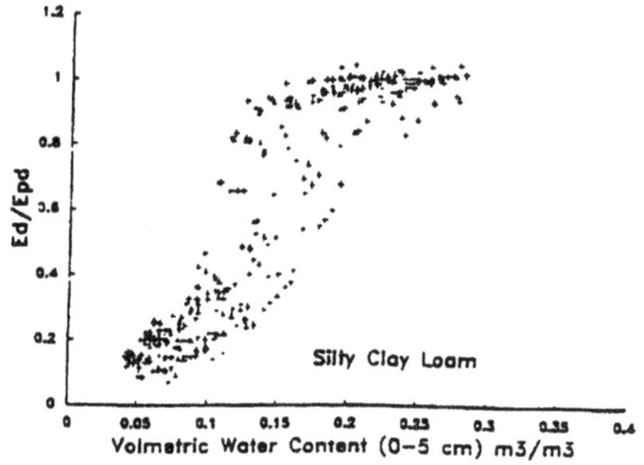

(b)

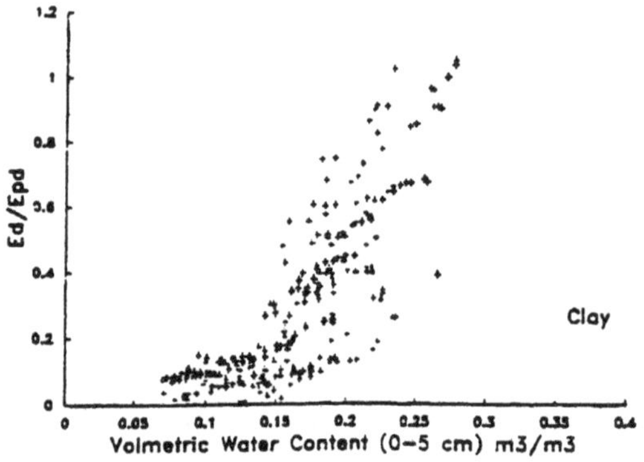

(c)

Figure 1. $E_d/E_{pd} = f(\theta_{0-5})$ relationships for three different soils. Data were simulated by the mechanistic model under various climatic conditions.

where E_{lim} is the limiting evaporation. It corresponds to the maximum water flow through the upper layers of the soil that allows the water supply of the evaporating surface. Although E_{lim} can be calculated from the soil hydrodynamics properties, E_{lim} is mostly given by an empirical form involving θ_s [2].

The above models describe the instantaneous evaporation. As mentioned in the introduction, a daily time period is sufficient for many applications. Thus, in some models the daily ratio E_d/E_{pd} is related to θ_s [7, 8, 11]. Such a relationship is displayed in Fig. 1(a–c) with the set of data obtained from loam, silty clay loam and clay soils. The variability of E_d/E_{pd} for a given moisture value clearly shows the need of additional information in order to provide an accurate estimate of E_d/E_{pd}. Therefore a relationship based on θ and classical climatic data has been proposed [11]:

$$E_d/E_{pd} = f(\theta_s,\ U_{ad},\ E_{pd}) \tag{6}$$

where U_{ad} is the daily wind velocity. Equation (6) includes three parameters which govern the soil dependence of the relationship between E_d/E_{pd} and θ_s as shown in Fig. 1(a–c).

Most of the evaporation models presented in this section were evaluated with different data sets, therefore it is difficult to compare their performances. Nevertheless, this review has emphasised three crucial points which have to be addressed in order for the implementation of a method to estimate soil evaporation by using surface soil moisture observations:

1) Since θ_s is introduced in the evaporation model through empirical relationships which require soil dependent parameters, how can they be determined?
2) To provide an accurate estimation of the evaporation, climatic quantities such as E_{pd} and U_{ad} are necessary. How accurately do we need to know these quantities and what is their availability?
3) The evaporation models are established on local data sets and/or are based on local physical laws. It is shown in Fig. 1(a–c) that $E_d/E_{pd} = f(\theta_s)$ are non-linear and soil dependent. Therefore, how can such models be applied to heterogeneous surfaces containing different moisture conditions and soil types?

1.3. Influence of the microwave frequency for estimating the evaporation

The soil microwave thermal emission of the soil is known to be sensitive to the soil surface moisture within a large range of frequency. Wang *et al.* [15] have shown good relationships between the microwave brightness temperature T_B and θ_s for L-, C-, and X-bands. The main difference observed between the frequencies is the skin depth. It is defined as the depth of the surface layer which contributes to the microwave emission. Newton *et al.* [16] results indicate that the skin depth in L-band is between the 2 and the 5 cm level, while in C- and X-bands the skin depth is some where between 0 and 2 cm. The L-band is oftenly presented as the best band to monitor the soil surface moisture [17]. This is mainly due to the lover sensitivity of the L-band to the presence of vegetation and the possibility of detecting the moisture on a deeper layer. With the aim of monitoring the evaporation, it is also necessary to consider the choice of the microwave frequency with respect to the evaporation estimate. Therefore, we have displayed in Fig. 2 the relationships between the ratio of E_d to E_{pd} and the volumetric

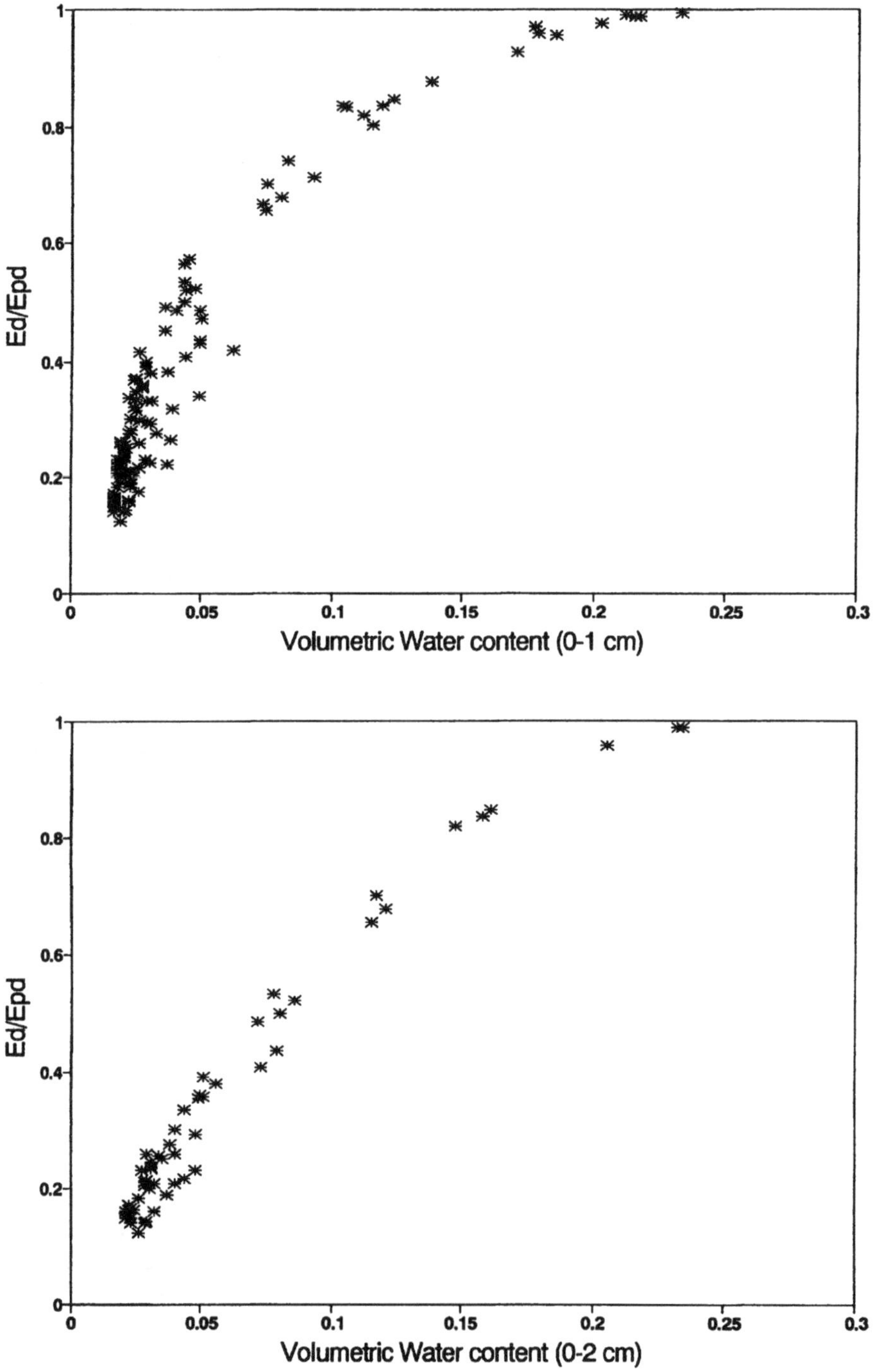

Figure 2. Influence on $E_d/E_{pd} = f(\theta_s)$ relationships of the surface layer depth in which the volumetric water content is characterized. The data were simulated by the mechanistic model on the silty clay loam with a high climatic demand.

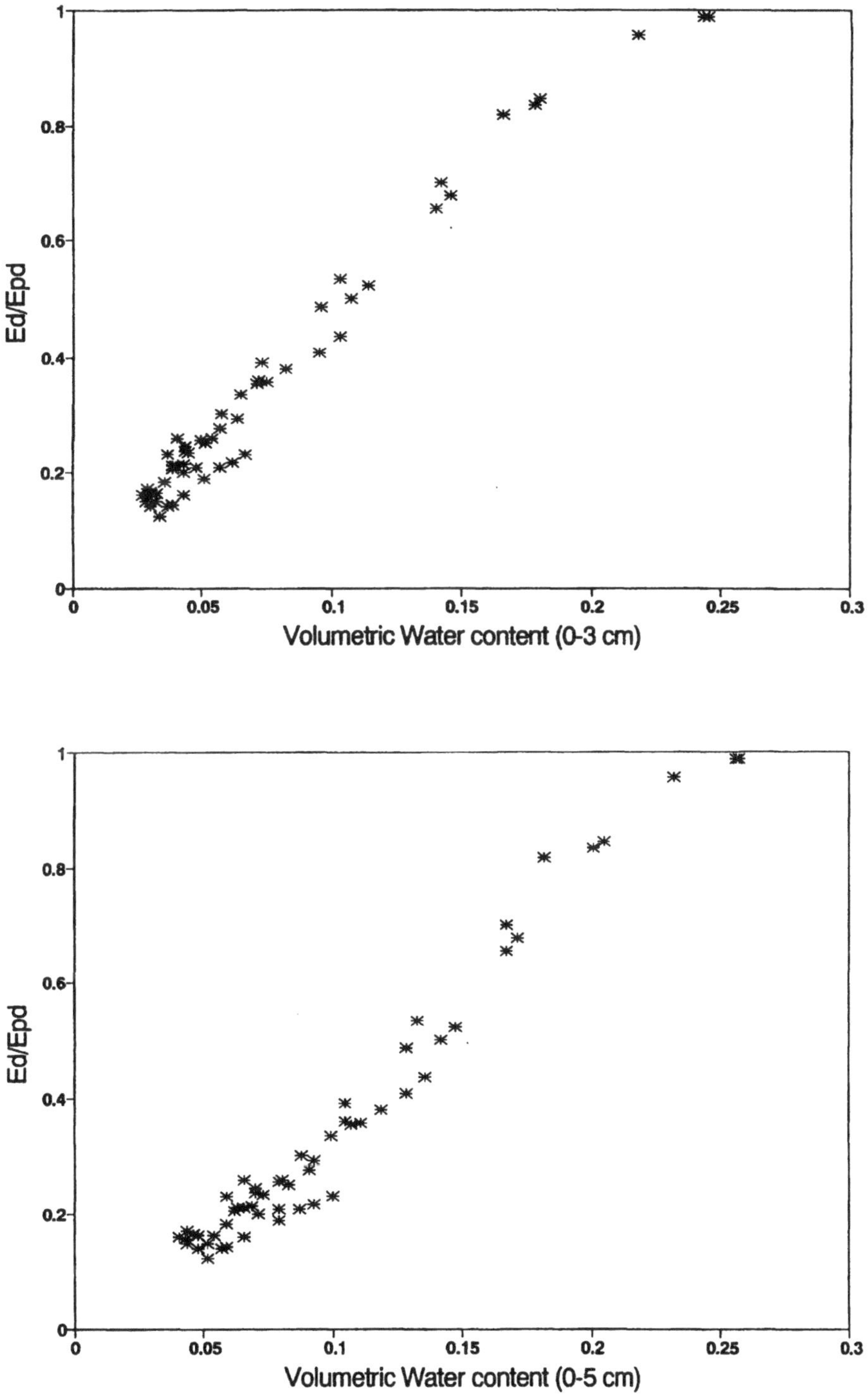

Figure 2. (Continued).

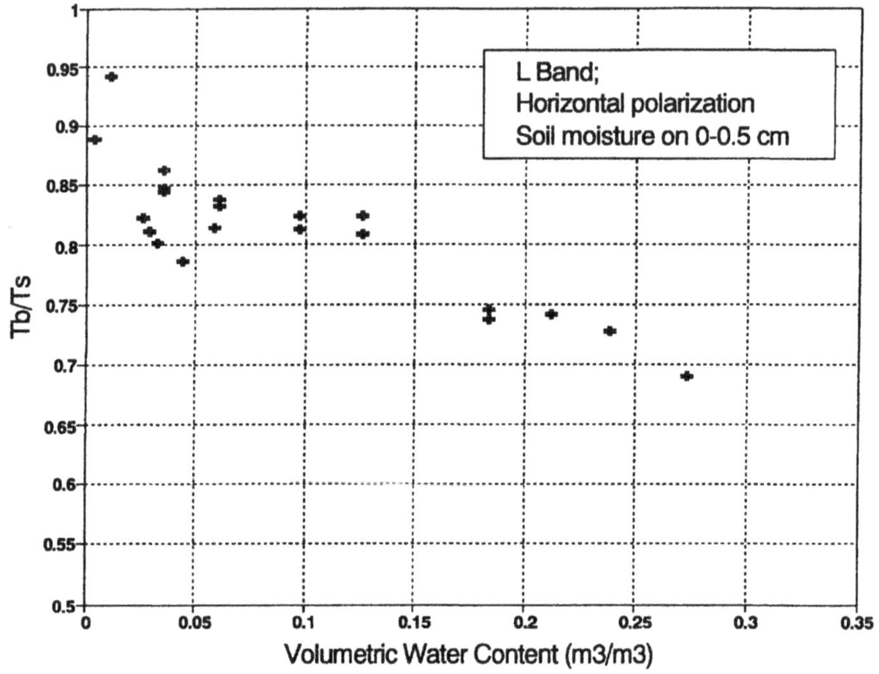

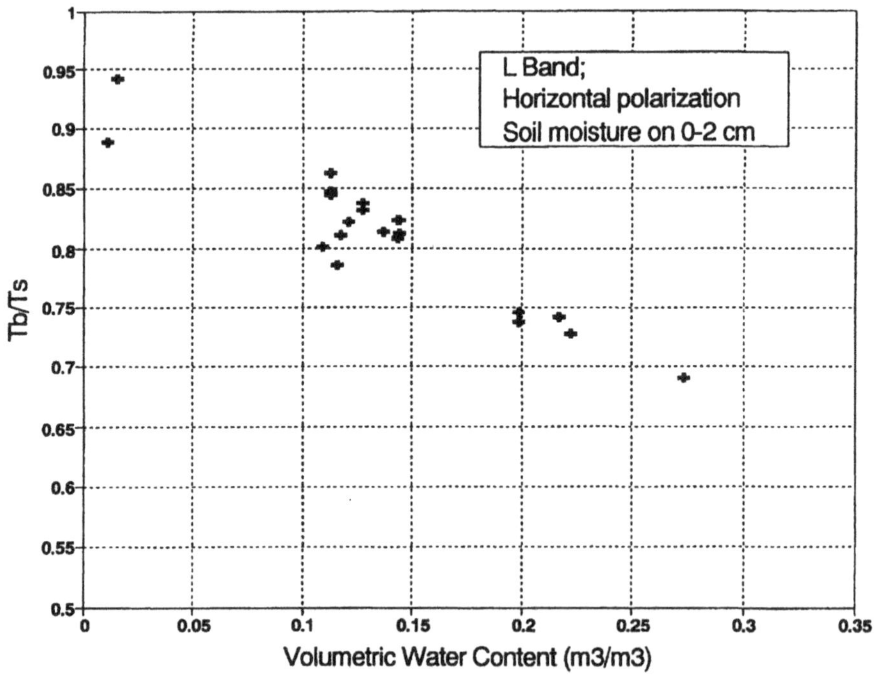

Figure 3. Experimental values of T_B/T_s versus the surface moisture at different frequencies (1.4, 5.05 and 10.65 GHz) with horizontal polarization and an angle of incidence of 20°.

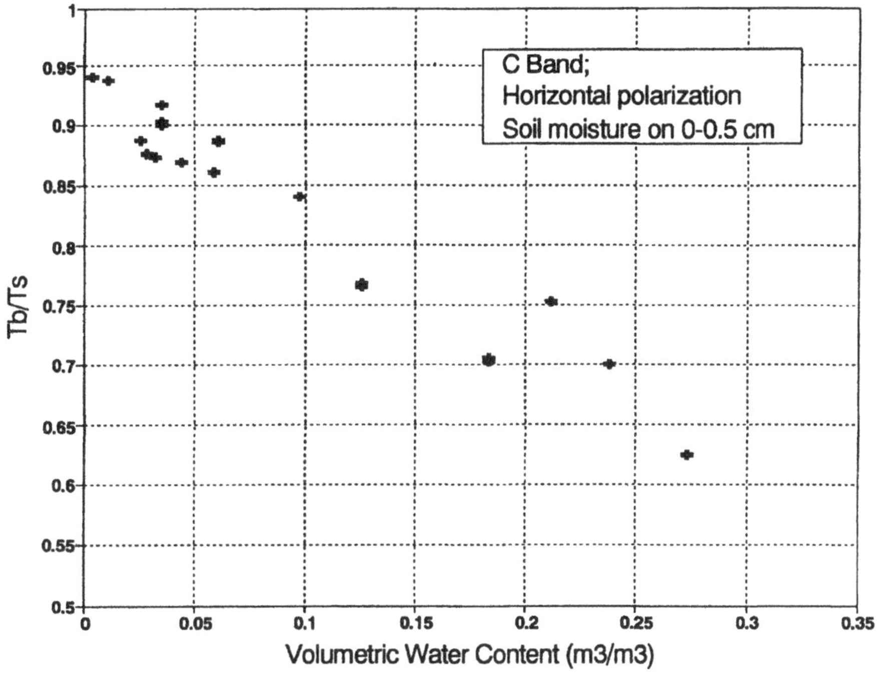

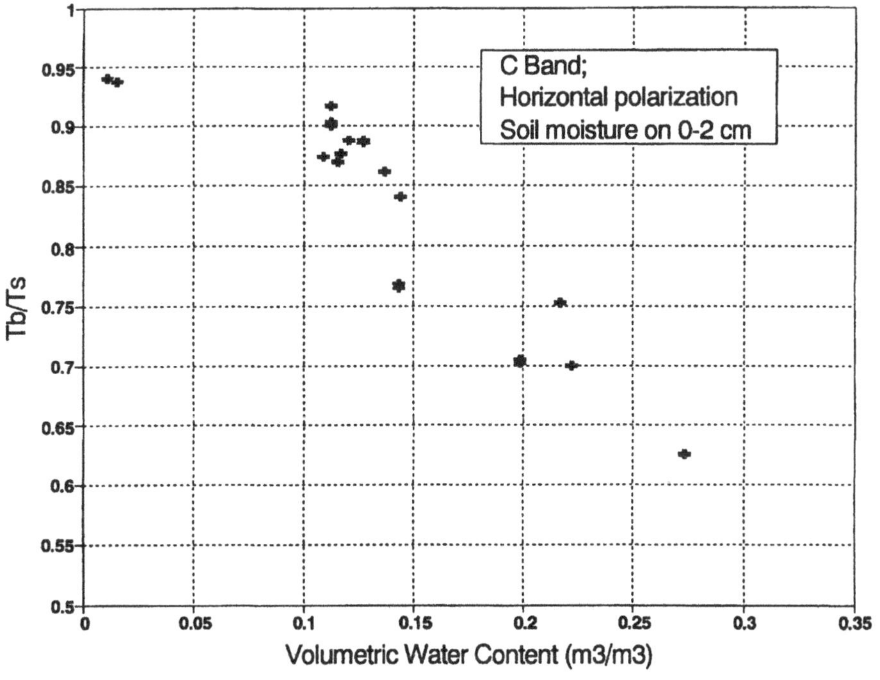

Figure 3. (Continued).

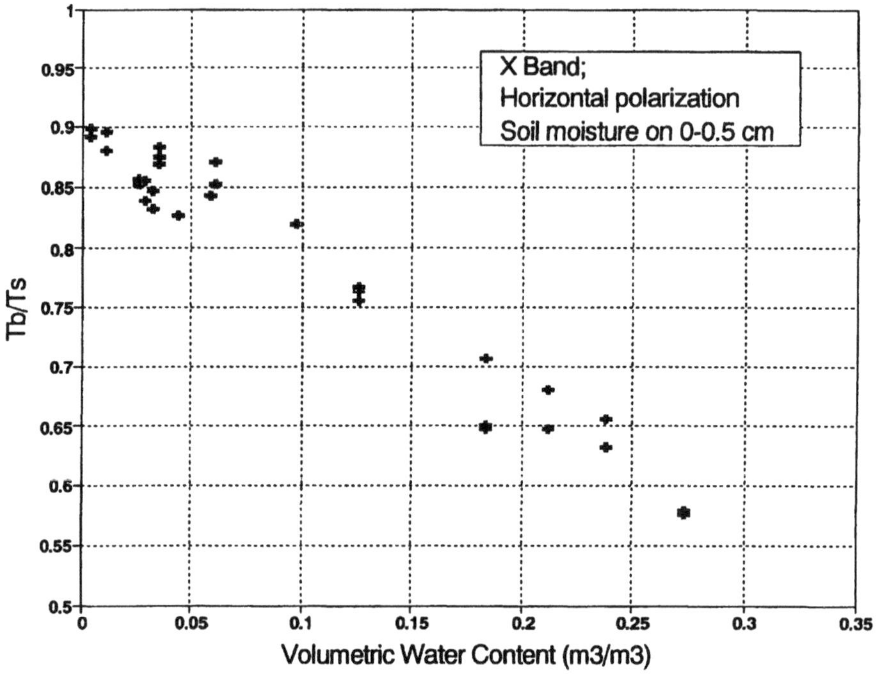

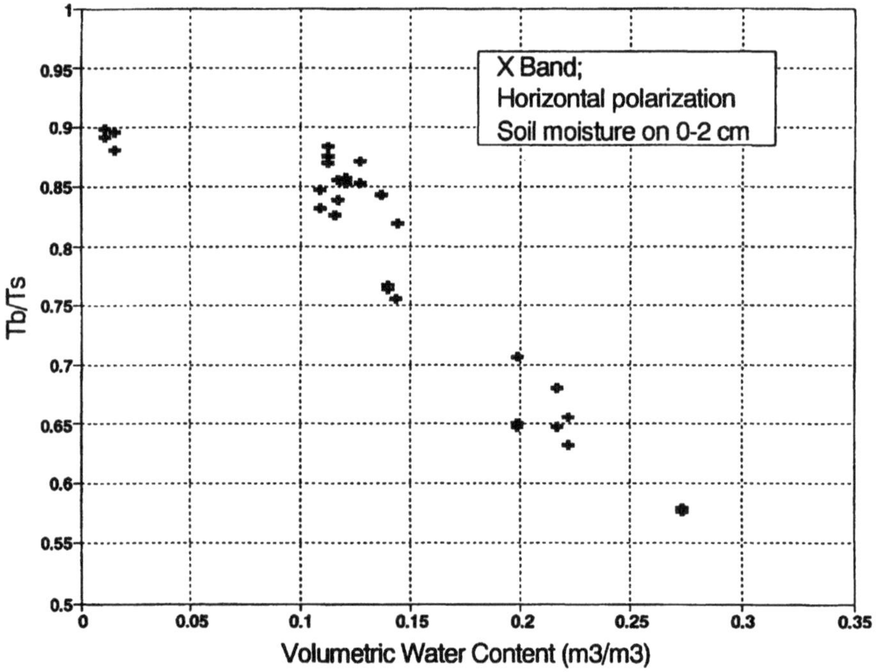

Figure 3. (Continued).

water content $\theta_{0-\delta}$ (δ is the skin depth which is assumed to be equal to 1, 2, 3 and 5 cm) in the case of the silty clay loam. When we consider a thin top layer, i.e. 0–1 cm, there is a high sensitivity of E_d/E_{pd} to θ_{0-1} at the dry end. The E_d/E_{pd} ratio decreases from 0.5 to 0.1 while θ_{0-1} only varies from 0.05 to 0.02 m^3/m^3. With thicker top layers, the steepness of the $E_d/E_{pd} = f(\theta_{0-\delta})$ relationship is reduced. Thus the sensitivity of E_d/E_{pd} to $\theta_{0-\delta}$ is lower which improves the capability of estimating evaporation. The results in Fig. 2 suggest that the microwave observations must detect variations of the water content over the depth of the least 2 cm in order to reliably compute evaporation. Relationships between the ratio of T_B to be surface temperature T_s and the volumetric water content of either the 0–0.5 cm or the 0–2 cm soil layers are displayed in Fig. 3. They were obtained on a smooth (standard deviation of the height equal to 3 mm) silty clay loam field. The three lowest PORTOS frequencies are presented (1.4, 5.05 and 10.65 GHz). The ranges of variation of T_B/T_s are consistent with results presented by Schmugge *et al.* [17] on similar wetness conditions. Figure 3 exhibits the interest of the low frequencies, particularly the L-band. As a matter of fact, while the sensitivity of T_B/T_s to $\theta_{0-0.5}$ is constant over the range of soil moisture conditions for the X- and C-bands, T_B/T_s in these bands is no longer sensitive to θ_{0-2} at drier conditions. Consequently, measurements of T_B/T_s in X- and C-bands do not allow an accurate estimation of θ_{0-2} when $\theta_{0-2} < 0.1$ m^3/m^3. Thus, the L-band appears as the most adequate frequency to estimate the evaporation. With C- and X-bands, the skin depth is too thin to discriminate E_d/E_{pd} variations when E_d/E_{pd} is below 0.5.

1.4. Implementation of the evaporation models

Since T_B is often related to the moisture of the 0–5 cm soil layer [18, 19], θ_{0-5} is used as surface soil moisture input in the evaporation models. According to the results displayed in Fig. 2, the analysis presented with a 5 cm layer depth is available for layer depth ranging from 2 to 5 cm. Two models will be used in this section. One was the Penman–Monteith physical model (4) in which $r_s = f(\theta_{0-5})$. (Hereafter referred to as PM.) The second one was an empirical model developed by Chanzy and Bruckler [8] (CB) which is designed to provide daily evaporation from one measurement of θ_s per day performed at midday and using standard climatic data.

$$\frac{E_d}{E_{pd}} = \frac{\exp(A\,\theta_{0-5} + B)}{1 + \exp(A\,\theta_{0-5} + B)}\,C + (1 - C) \tag{7}$$

$$A = a + 5\max(3 - E_{pd},\ 0) \tag{8}$$

$$B = b - 5(-0.025\,b - 0.05)\max(3 - E_{pd},\ 0) + (U_{ad} - 3)\,\alpha \tag{9}$$

$$C = 0.90 - 0.05\,\alpha\,(U_{ad} - 3) \tag{10}$$

where a, b and α are three soil dependent parameters. This model relies on a large data set in order to account for the yearly climatic variations and the hydrodynamic behaviour of very different soils.

1.4.1. Range of validity of the models of bare soil evaporation. Both PM and CB models require several empirical parameters. Therefore their validity is constrained by data used to fit the parameters. How $r_s = f(\theta_{0-5})$ (PM) and $E_d/E_{pd} = f(\theta_{0-5})$ relationships are

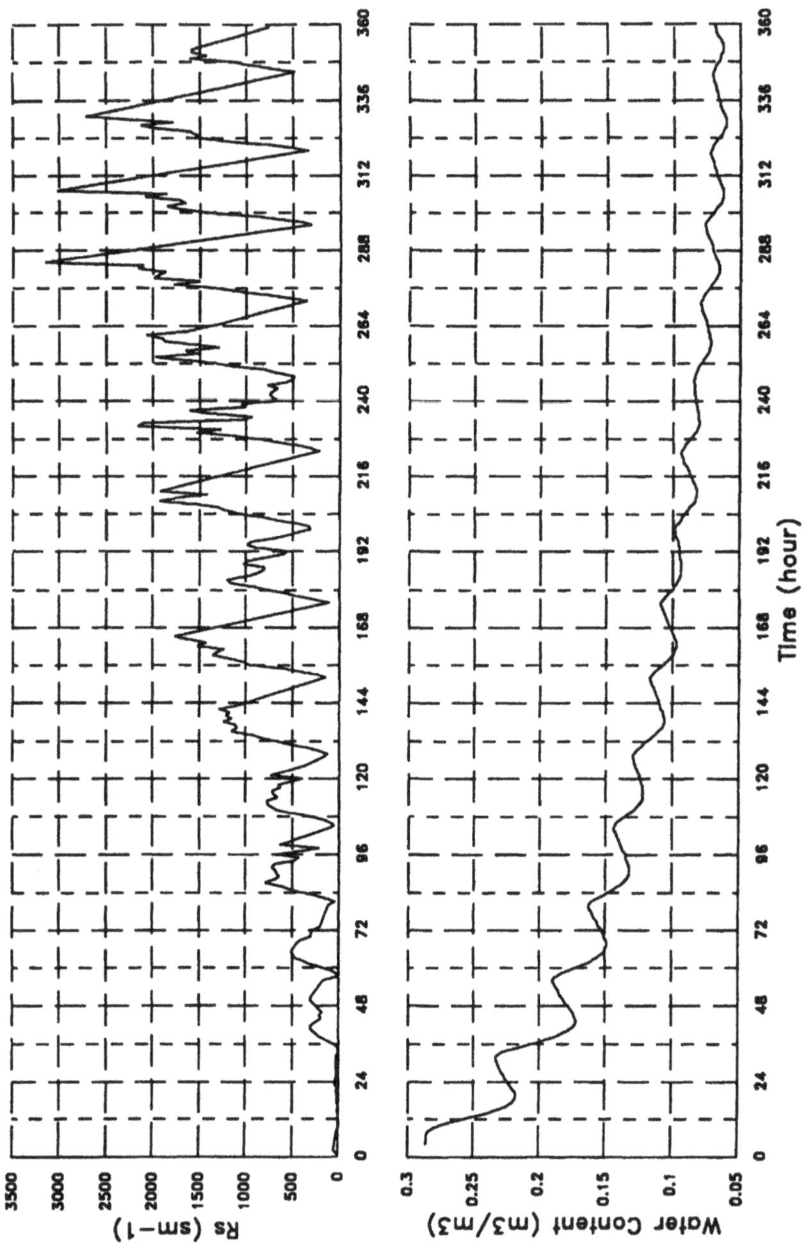

Figure 4. Evolution of the surface resistance in comparison with the evolution of the surface moisture. Data were simulated by the mechanistic model on the silty clay loam under a high climatic demand (E_{pd} varied from 3.7 to 7.5 mm/d).

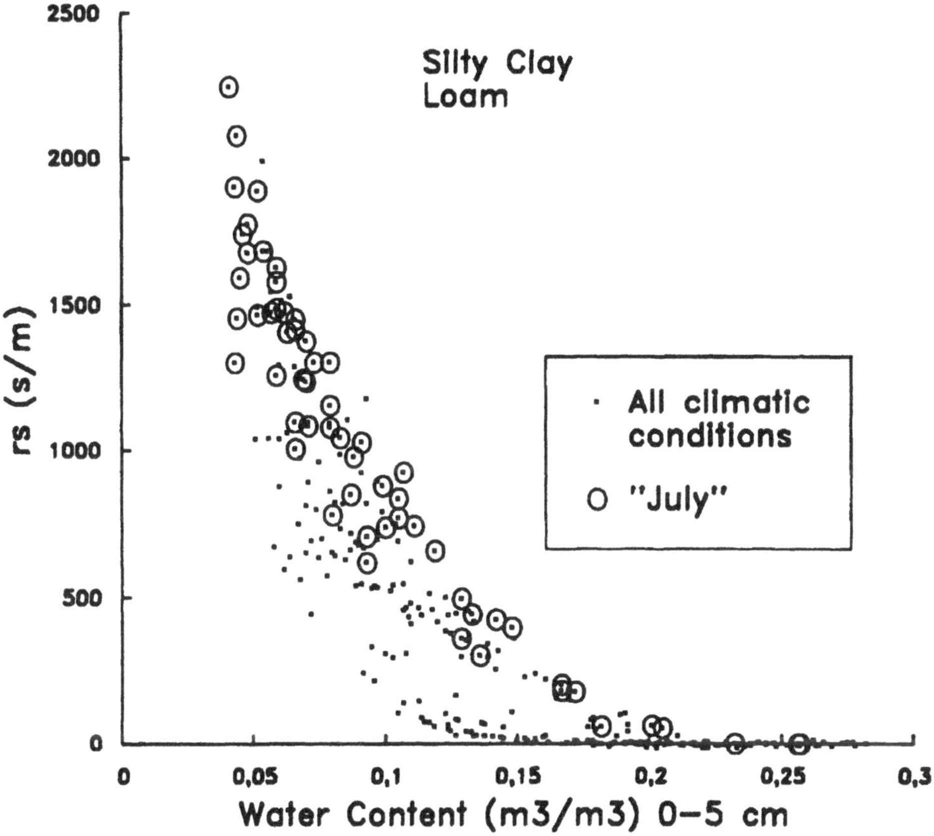

Figure 5. The surface resistance obtained at midday versus θ_{0-5} relationship. Data were simulated by the mechanistic model on the silty clay loam under various climatic conditions.

influenced by external factors such as climatic conditions? Let us investigate in Fig. 4 the hourly evolution of r_s compared to the evolution of θ_{0-5}. It can be seen that the soil resistance is strongly affected by the diurnal cycle. The range of daily variations of r_s at drier soil moisture conditions covers the full range in r_s values, while the daily variation of the surface moisture is significantly damped. This means that the $r_s = f(\theta_{0-5})$ relationships will be strongly dependent on the time of day. Moreover, when a given time is selected (midday in Fig. 5), the variability of r_s at a given value of θ_{0-5} is significant. This variability is partly explained by the climatic conditions. As a result, the highest values of r_s obtained for a given value of θ_{0-5} correspond to a day of high climatic demand ('July' days in Fig. 5). Thus, to implement PM, it is necessary to establish $r_s = f(\theta_{0-5})$ relationship on sets of data that cover the range of climatic variability expected for that region and period under investigation (mostly the time of year). In the design of CB model [8], this was accounted for by using two climatic quantities (E_{pd} and U_{ad}). Both quantities are required to provide an accurate estimate of E_d/E_{pd} through a wide range of climatic conditions. The effect of the wind velocity on E_d/E_{pd} is displayed in Fig. 6. It shows, in particular, that E_d/E_{pd} is significantly lower for a given value of θ_{0-5} when the wind velocity is high.

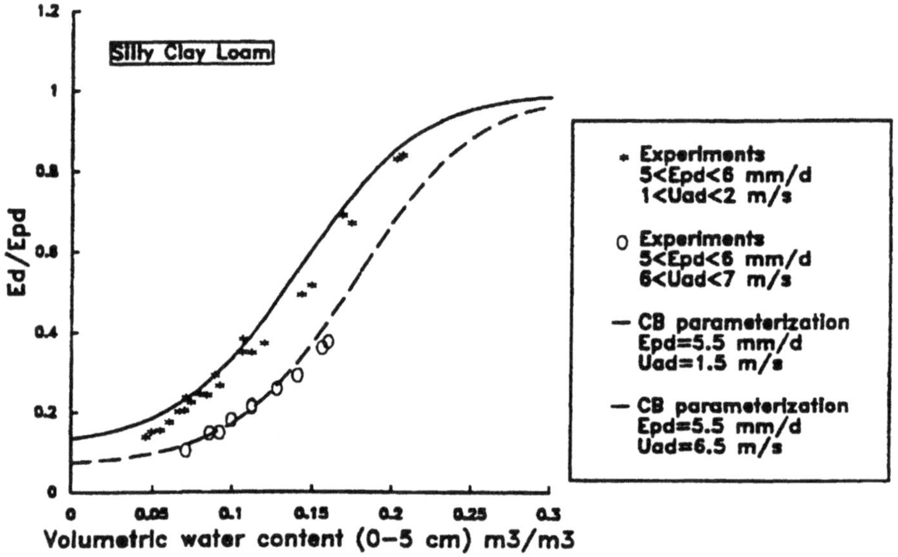

Figure 6. Wind effect on $E_d/E_{pd} = f(\theta_{0-5})$ relationship. Symbols correspond to data simulated by the mechanistic model.

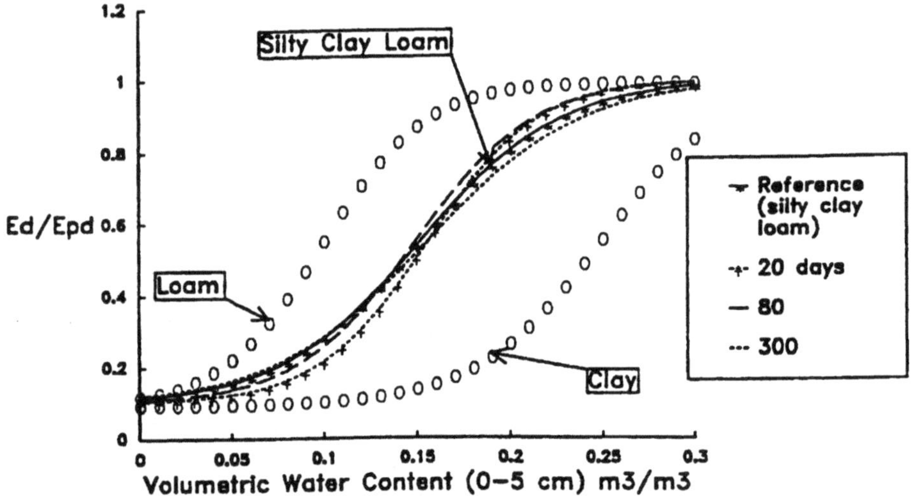

Figure 7. E_d/E_{pd} relationship predicted by CB parameterization (Equations (7) to (10), with $E_{pd} = 3$ mm/d and $U_{ad} = 3$m/s) when CB parameters are obtained by an inversion procedure involving a variable number (20 to 300) of thermal infrared observations. Reference (silty clay loam), loam and clay relationships correspond to CB results when its parameters are fitted with data simulated by the mechanistic model.

1.4.2. Determination of the soil dependent empirical parameters. The variability of $E_d/E_{pd} = f(\theta_{0-5})$ relationships as a function of soil type is illustrated in Fig. 1. This variability is mostly treated in evaporation models via empirical parameters. However, the way to determine these parameters is rarely analysed. Chanzy [11] has proposed a way to calculate E_d when thermal infrared measurements are available by inverting the CB model to yield the corresponding parameters. Specifically, E_d was estimated by the simplified model of Jackson *et al.* [20] based on the temperature difference between the surface and the air ($T_s - T_a$). The following equation was used:

$$E_d = -(R_n + G)_d - 0.98 + 0.275\,(T_s - T_a)_{14H} \qquad (11)$$

where R_n and G are the net radiation and the soil heat flux, all expressed in mm of water. Subscript 14H refers to the hour of ($T_s - T_a$) measurement (2 pm). To test the inversion procedure, from 20 to 300 days were randomly selected in the data base which gathers results of numerical simulations by the mechanistic model and the corresponding climatic conditions. For each soil, the data base includes a wide range of climatic and moisture conditions and includes the required inputs to run the CB and Jackson *et al.* [20] models. For each sampled day, E_d is estimated by using the model of Jackson *et al.* [20] with $T_s - T_a$, the daily net radiation R_{nd} and the daily soil heat flux G_d. This estimation of E_d was then combined with θ_{0-5}, E_{pd} and U_{ad} to fit a, b, and α from CB parameterization (Equations (7)–(10)) by a non linear fitting procedure [27]. To account for errors in the inputs, an unbiased random error term was added to each variable ($T_s - T_a$, R_{nd}, G_d, E_{pd}, U_{ad} and θ_{0-5}). The errors were determined according to Gaussian laws with a standard deviation for $T_s - T_a$ and θ_{0-5} equal to 2 K and 0.04 m^3/m^3, respectively. Such errors may represent the expected accuracy on surface parameters when observed from satellite platforms. Figure 7 shows the results of the inversion procedure and suggests that even with rough (but unbiased) estimates of the input variables, we can obtained reasonable values of the CB model parameters that differentiate the three soils. Figure 7 also shows that a relatively small number of available days (20 days) of measurements seems sufficient to discriminate the silty clay loam soil from the other soils. On the contrary, biased estimates of the inputs would lead to a biased set of CB parameters and results in significant deviations in long term monitoring of the evaporation. This inversion procedure is a synergism example of thermal infrared and microwave as it has been developed by Schmugge and Becker [22].

1.4.3. Expected accuracy of the evaporation estimate. By assuming that the parameters of Equations (7)–(10) are known, the variations in E_d estimate affected by errors in the

Table 1.
Expected accuracy on E_d estimate for 4 error scenarios. Results obtained when E_d are averaged over a five day period are given in italic

Scenario	Inputs quantities are known exactly (*)	Carefully controlled experimental conditions	θ_{0-5} is observed from satellite	θ_{0-5} is observed from satellite 'pessimistic case'
θ_{0-5}(**)	0	0.02 m^3/m^3	0.04 m^3/m^3	0.06 m^3/m^3
E_{pd}(**)	0	0.2 mm/d	0.5 mm/d	1 mm/d
U_{ad}(**)	0	0.1 m/s	0.5 m/s	1.0 m/s
E_d(***)	±0.4 mm/d	±0.8 mm/d	±1.1–1.7 mm/d *(0.5 mm/d)*	±2 mm/d

(*) the error on E_d corresponds to the simplification of CB model in comparison with the mechanistic model.
(**) standard deviation.
(***) 95% confidence interval.

inputs (θ_{0-5}, E_{pd} and U_{ad}) were computed. As for the inversion procedure described in the previous section, error terms were added to θ_{0-5}, E_{pd} and U_{ad}, according to a Gaussian law. Four different error scenarios were considered (see Table 1).

For the two more favourable scenarios, a quantitative estimate of daily evaporation is possible while results obtained for the worst cases are rather qualitative. Nevertheless, a quantitative estimate of E_d can be obtained when E_d is averaged over a five day period. The error is then in the same order of magnitude than the one obtained with approaches using thermal infrared observations [26].

2. ESTIMATION OF THE EVAPOTRANSPIRATION OF VEGETATION COVERED-SURFACE

As was mentioned in the Introduction, the evaporation and the transpiration components of the ET are regulated by different mechanisms. With regard to the evaporation of the soil, the main difference between the bare soil case and a soil covered with vegetation is the magnitude of the climatic demand at the soil surface. As a matter of fact the amount of radiative energy that reaches the soil surface is affected by the interception by the vegetation and the convective exchanges are modified by the plant canopy. Nevertheless, the soil surface moisture remains as an important factor to describe the ratio of the evaporation to the climatic demand at the ground level. The rate of the transpiration involves a greater number of factors, in that transpiration is not only governed by the water available in the root zone, but also by the incoming solar radiation, the development of the canopy (characterised by the height of the canopy, Leaf Area Index, biomass, etc. . . .), plant physiology (phenologic states) and atmospheric demand (vapour pressure, air temperature and wind speed).

Therefore, the contribution of the microwave radiometry to the ET monitoring can be achieved through the detection of:

* the surface soil moisture for estimating the soil evaporation (L-band);
* the surface temperature (T_s); it results from energy balance of the surface which involves the ET. T_s, derived from thermal infrared observations, is already widely used to infer the ET. Nevertheless, thermal infrared observations are only available by clear sky conditions whish prevents a regular monitoring of T_s.
* the vegetation parameters [29]; they are necessary to implement the ET model. For instance, they are useful to quantify the transpiration (or the evaporation) contribution to the ET or to characterise the climatic demand at the ground level;

The advantage of microwave radiometry in comparison with the other wavelength regions is its all weather capability to observe the temperature and moisture of the surface. We will then focus on these parameters.

2.1. Use of the surface soil moisture

Camillo and Schmugge [9] showed a linear relationship between the volumetric water content near the soil surface and water storage within the root zone. This study was performed under specific conditions. The vegetation was well developed and the depth of the root zone was constant. Although the implementation of such a method was not investigated in detail, the results from Camillo and Schmugge [9] provide a possible way for predicting the ET over vegetated land using the microwave radiometry. More

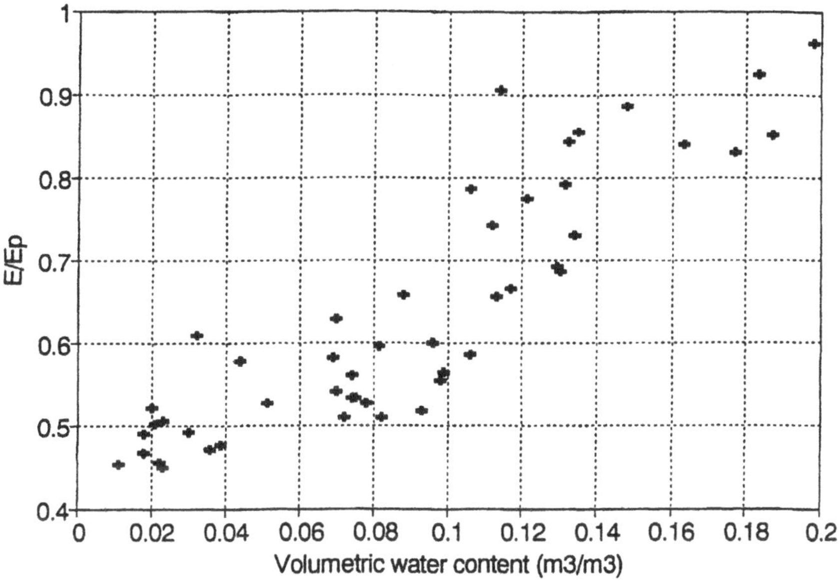

Figure 8. $E_d/E_{pd} = f(\theta_{0-5})$ relationship in case of sparse vegetation. Data were collected during MONSOON 90 experiment [17].

recent studies which combine modelization and microwave radiometry are presented in this issue [30].

In the case of land covered with a sparse vegetation, the contribution of the evaporation to the ET is important. To illustrate this, a comparison of daily ET (ET_d) and soil moisture measurements are shown in Fig. 8. The MONSOON 90 experiment was conducted over the Walnut Gulch Watershed in southern Arizona [22]. The vegetation is mixed grass-brush rangeland typical of southeastern Arizona and southwestern New Mexico. Microwave, soil moisture, climatic and ET measurements were made for six dates during a ten day period at eight sites. Figure 8 represents the relationship between the daily ratio of the ET to the potential ET (ET_{pd}) and the surface soil moisture. It can be seen from Figure 8 that there is a strong dependence of the ET_d/ET_{pd} ratio on the surface moisture when a short period (10 days) is under investigation. This means that rapid changes in ET_d are mainly related to the variations in the soil evaporation while the water availability for the transpiration varies more slowly. Thus, the measurement of the surface moisture can be useful to track the variations in ET_d of sparse vegetation over a small time period.

2.2. *Use of the surface temperature*

There are only a few studies dealing with the extraction of the surface temperature from the microwave brightness temperature [24, 25, 28]. For the cases of homogeneous pixels in terms of their microwave emissivity (US central plains [24], Amazon forest [25]), the authors have shown a significant relationship between surface temperature and microwave observations. However, there is no attempt, to our knowledge, to use microwave observations to estimate the ET through the detection of the surface temperature. We have to refer to studies dealing with the use of thermal infrared radiometry to infer ET. Among these studies, an approach based on the simplification of the energy balance at the surface is widely investigated [20, 26]. This approach leads to equations similar to the Equation (11). It involves two parameters (equal to −0.98 and 0.275 in Equation (11)) which depend on the wind velocity and the surface roughness [11, 22]. Surface roughness could be investigated by other remote sensing sensors (radar, optical radiometer), while the wind velocity should be obtained from ground observations or meteorological simulations. The main limitation of using microwave radiometry to measure the surface temperature is the expected low accuracy [28]. Further researches are then necessary to establish: (1) the sensitivity of the calculated ET to rough estimates of the surface temperature and (2) the benefit we can take from multitemporal observations.

3. CONCLUSIONS

The use of surface soil moisture derived from microwave observations for monitoring the evaporation of bare soils or soils covered with a sparse vegetation shows great potential. Nevertheless, the evaporation models discussed in this paper need to be tested in different environments. Two questions remain particularly critical. How can the soil-dependent parameters of the evaporation models be determined over large areas (the presented synergism with the thermal infrared observations is promising)? At what spatial scale are the models applicable? In particular, how can we implement a model established at a local scale to large pixels as provided by space-borne radiometers? We think that microwave radiometry to infer the evaporation through the detection of the soil surface moisture is more adapted to experiments based on airborne sensors. Therefore, microwave radiometry may be a useful tool for upscaling studies as there are investigated in HAPEX, FIFE, MONSOON 90 or HAPEX SAHEL experiments.

The extent of bare soils and soil covered with sparse vegetation is limited in space and in time. In this context, the surface temperature appears as the most useful parameter owing to its capability of being applicable in both bare soil and vegetation cases. Furthermore, the linearity between the surface temperature and the ET [26] is an interesting property to provide estimate of the ET over large areas. Consequently, we feel that the success of the microwave radiometry to monitor the ET will ultimately rely on the accuracy with which the surface temperature can be obtained by space-borne sensors.

Acknowledgement

The field experiments in Avignon, France, were supported by the 'Programme National de Télédétection Spatiale'. The co-operation and assistance of the USDA–ARS Southwest Watershed Research Center in Tucson, Arizona and of site personnel who maintain the Walnut Gulch Experimental Watershed are gratefully acknowledged. Also funding from NASA Interdisciplinary Research Program in Earth Science (NASA Reference Number IDP-88-086) and funds from USDA-ARS Beltsville Area Office provided the necessary financial support to conduct this study.

REFERENCES

1. L. Prévot, R. Bernard, O. Taconet, D. Vidal Madjar and J. L. Thony. Evaporation from a bare soil evaluated using a soil water transfer model and remotely sensed surface soil moisture data. *Water Resources Res.*, **20** (2), 311–321 (1984).

2. J. V. Soarès, R. Bernard, O. Taconet, D. Vidal Madjar and A. Weill. Estimation of bare soil evaporation from airborne measurements. *J. Hydrology*, **99**, 281–296 (1988).
3. L. Bruckler and H. Witono. Use of remotely sensed soil moisture content as boundary conditions in soil-atmosphere water transport modeling. 2. Estimating soil water balance. *Water Resources Res.*, **25** (12), 2437–2447 (1989).
4. I. J. Barton. A parameterization of the evaporation from non saturated surfaces. *J. Appl. Meteor.*, **18**, 43–47 (1979).
5. J. Kondo, N. Saigusa and T. Sato. A parameterization of evaporation from bare soil surfaces. *J. Appl. Meteor.*, **29**, 385–389 (1990).
6. J. F. Mahfouf and J. Noilhan. Comparative study of various formulations of evaporation from bare soil using *in-situ* data. *J. Appl. Meteor.*, **30** (9), 1354–1365 (1991).
7. M. Owe and A. A. Van de Griend. Daily surface moisture model for large area semiarid land application with limited climate data. *J. Hydrology*, **121**, 119–132 (1989).
8. A. Chanzy and L. Bruckler. Significance of soil surface moisture with respect to bare soil evaporation. *Water Resources Res.*, **29** (4), 1113–1125 (1993).
9. P. J. Camillo and T. J. Schmugge. Estimating soil moisture storage from surface measurements. *Soil Science*, **135** (4), 245-265 (1983).
10. A. Chanzy, J. P. Wigneron and T. Schmugge. Multifrequency passive microwave measurements with PORTOS, experimental design and first results. In: *Proceedings of Specialist Meeting on Microwave Radiometry and Remote Sensing Applications*, Boulder, 14–16 January 1992, pp. 67–71 Ed, NOAA (1992).
11. A. Chanzy. *Modélisation Simplifiée de l'Évaporation d'un Sol nu Utilisant l'Himidité et la Température de Surface Accessibles par Télédétection.* Thèse de docteur de l'Institut National Agronomique de Paris Grignon, Paris (1991).
12. J. R. Philip and D. A. de Vries. Moisture movements in porous material under temperature gradients. *Trans. Ann. Geophys. Union*, **38**, 222–232 (1957).
13. J. L. Monteith. Evaporation and environment. In: *The State and Movement of Water in Living Organisms*, 19th Symp. Soc. Exp. Biol. (1964).
14. P. J. Camillo and Gurney. A resistance parameter for bare soil evaporation models. *Soil Science*, **104** (2), 95–105 (1986).
15. J. R. Wang, P. E. O'Neill, T. J. Jackson and E. T. Engman. Multifrequency measurements of the effect of soil moisture, soil texture and surface roughness. *IEEE Trans. Geosci. Remote Sensing*, **GE-21** (1), 44–51 (1983).
16. R. W. Newton, Q. R. Back, S. Makanvand, A. J. Blanchard and B. R. Jean. Soil moisture information and thermal microwave emission. *IEEE Trans. Geosci. Remote Sensing*, **GE-20**, 275–281 (1982).
17. T. J. Schmugge, P. E. O'Neill and J. R. Wang. Passive microwave soil moisture research. *IEEE Trans. Geosci. Remote Sensing*, **GE-24** (1), 12–22 (1986).
18. J. R. Wang, T. J. Schmugge, J. C. Shiue, E. T. Engman. Mapping surface soil moisture with L-band radiometric measurements. *Remote Sens. Environ.*, **27**, 305–312 (1989).
19. T. J. Schmugge, T. J. Jackson, W. P. Kustas and J. R. Wang. Passive microwave remote sensing of soil moisture: results from HAPEX, FIFE and MONSOON 90. *ISPRS J. Photogramm. Remote Sensing*, **47**, 127–143 (1992).
20. R. J. Jackson, R. J. Reginato and S. B. Idso. Wheat canopy temperature: a practical tool of evaluating water requirements. *Water Resources Res.*, **13**, 651–656 (1977).
21. A. Bard. *Non-Linear Parameter Estimation.* Academic Press, New York (1974).
22. T. J. Schmugge. Synergism at infrared and microwave frequencies. In: *ESA/NASA International Workshop*, B. J. Choudhury, Y. H. Kerr, E. G. Njoku and P. Pampaloni (Eds), pp. 193–204, VSP, Utrecht (1994).
23. W. P. Kustas and 35 co-authors. An interdisciplinary study of the energy and water fluxes in atmosphere-biosphere system over semi-arid rang elands: Description and some preliminary results. *Bull. Amer. Meteor. Soc.*, **72**, 1683–1705 (1991).

24. M. J. Mac Farland, R. I. Miller and C. M. U. Neale. Land surface temperature derived from SSM/I passive microwave brightness temperatures. *IEEE Trans. Geosci. Remote Sensing*, **28**, 839–845 (1990).

25. J. C. Calvet, J. P. Wigneron, E. Mougin, Y. H. Kerr and J. L. Brito. Plant water content and temperature of the Amazon forest from satellite microwave radiometry. Accepted in *IEEE Trans. Geosci. Remote Sensing* (1993).

26. B. Seguin, S. Baelz, J. M. Monget and V. Petit. Utilisation de la thermographie IR pour l'estimation de l'évaporation régionale. II: Résultats obtenus à partir de données satellite. *Agronomie*, **2** (2), 113–118 (1982).

27. E. Njoku. Surface temperature estimation over land using satellite microwave radiometry. In: *ESA/NASA International Workshop*, B. J. Choudhury, Y. H. Kerr, E. G. Njoku and P. Pampaloni (Eds), pp. 509–530, VSP, Utrecht (1994).

28. A. Vidal and A. Perrier. Analysis of a simplified relation for estimating daily evapotranspiration from satellite thermal IR data. *Int. J. Rem. Sensing*, **10**, 1327–1337 (1989).

29. S. Paloscia. Microwave emission from vegetation. In: *ESA/NASA International Workshop*, B. J. Choudhury, Y. H. Kerr, E. G. Njoku and P. Pampaloni (Eds), pp. 357–374, VSP, Utrecht (1994).

30. D. Entekhabi, H. Nakamura and E. Njoku. Retrieval of soil moisture profile by combined remote sensing and modeling. In: *ESA/NASA International Workshop*, B. J. Choudhury, Y. H. Kerr, E. G. Njoku and P. Pampaloni (Eds), pp. 485–498, VSP, Utrecht (1994).

ESA/NASA International Workshop, pp. 551–567
B. J. Choudhury, Y. H. Kerr, E. G. Njoku and P. Pampaloni (Eds)

A model for the radiobrightness of Northern Prairie

A. W. ENGLAND, J. F. GALANTOWICZ, Y. A. LIOU, E. J. KIM and P. A. DAHL

Department of Electrical Engineering and Computer Science, The University of Michigan, Ann Arbor, MI 48109-2122, USA

Abstract—This paper is a review of our progress in developing the Michigan Cold Region Radiobrightness (MCRR) model. Our objective is that a mature MCRR model will link satellite radiobrightness to the land-atmosphere hydrologic parameters of soil moisture, quantity, and state which influence atmospheric circulation models. The current MCRR model includes diurnal and annual insolation, long-wave radiance, and sensible heat transfer, but omits the effects of snow and vegetation. Because of these omissions, the current MCRR model is too primitive to capture the full complexity of the land–atmosphere processes, but it does illustrate dominant temporal characteristics of the diurnal and annual radiobrightness signature. Predictions of the model suggest that Radiobrightness Thermal Inertia might be used to estimate soil moisture, that a combination of atmospherically corrected, 37 GHz radiobrightness and the 19.35–85.5 GHz H-polarized spectral gradient can be used to classify prairie soils as frozen or thawed, and that seasonal radiobrightness anomalies might be used to map regions of discontinuous permafrost in arctic wetlands.

1. INTRODUCTION

Continental atmospheric circulation models are driven by radiant energy, sensible heat, latent energy, momentum, and moisture flux boundary forcing at the land–atmosphere interface [1–7]. There is ample evidence that mesoscale and global atmospheric circulation are particularly sensitive to the hydrologic boundary forcing of moisture and latent energy flux [8–16]. Hydrologic parameters that are interactive with circulation models have evolved from the bucket model of Manabe [17] and Budyko [18] to complex parameterizations like the Biosphere–Atmosphere Transfer Scheme (BATS) [19, 20] and the Simple Biosphere Model (SiB) [21]. While BATS involves more than 80 parameters and SiB involves 44, many are predictable for large, relatively homogeneous regions like the northern prairie or the arctic tundra. For example, soil thermal conductivity, density, and specific heat are readily estimated for either environment.

However predictable many of the hydrologic parameters are, the dynamic hydrological variables must be obtained through observation. Of these variables, soil moisture is critical because it represents the integrated effects of rainfall, runoff, and evapotranspiration. Because soil moisture tends to be relatively slowly varying with time at mid- and high-latitudes [13, 15], its estimation is particularly suited to daily, or even to weekly, satellite observation. Given that some satellite observation might yield an estimate of soil moisture, producing such estimates on something like the 25 km spatial grid of a mesoscale atmospheric circulation model is relatively undemanding of sensor spatial resolution or of satellite data management systems.

Estimates of soil moisture and state would be based upon a combination of satellite observations and a model that places the observations in a physical or temporal context. For example, we expect L-band (1.4 GHz) radiobrightness to correlate well with soil moisture in situations where a physical model of emission from a moist soil halfspace is plausible (e.g. [69]). Complications caused by rough surfaces, by vegetation, or by non-uniform distributions of moisture in the soil are viewed as perturbations to the basic halfspace model. Without such physical models, we cannot readily extrapolate to untested situations. As models become more sophisticated, they might include the dynamical effects of diurnal and annual insolation, of vegetation, and of snow. The Michigan Cold Region Radiobrightness (MCRR) model is a moist soil radiobrightness model that includes the effects of diurnal and annual insolation and the effects of freezing and thawing soils. This model has provided a basis for classifying soils as frozen or thawed and has suggested that soil moisture might be inferred from the diurnal extremes in radiobrightness. Because the MCRR model does not yet include the effects of vegetation or snow, it is not yet a suitable link between radiobrightness observations and the land-atmosphere fluxes that drive the continental boundary layer of a General Circulation Model (GCM). That is, the current MCRR model is more a qualitative than a quantitative model.

2. THE MICHIGAN COLD REGION RADIOBRIGHTNESS (MCRR) MODEL

Because so little radiant energy is transferred at microwave frequencies, the problems of thermal modeling and of microwave emission modeling are readily separable. There are several models for diurnal, near-surface soil temperature. These include the Watson model [22] for diurnal heating of dry soils, the Kahle model [23] for diurnal heating of moist soils, and the England model [24] (MCRR/Diurnal) for diurnal heating of moist soils that are subject to freezing and thawing. The Watson and Kahle models predict thermal infrared signatures, while the MCRR/Diurnal model predicts radiobrightness signatures. We recently expanded the MCRR/Diurnal model to an annual model for dry soils [25], and, more significantly, to the MCRR/Annual model for wet, freezing, and thawing soils [26]. These models include both diurnal and annual insolation.

The thermal module of the MCRR model solves the one-dimension heat flow equation in terms of soil depth z, and time t,

$$\frac{\partial E(T(z,t))}{\partial t} = \frac{\partial}{\partial z}\left(K(T(z,t))\,\frac{\partial T(z,t)}{\partial z}\right) \tag{1}$$

for a moist soil halfspace. Both $E(T)$, moist soil enthalpy, and $K(T)$, moist soil thermal conductivity, are functions of thermal temperature T. Permitting the moisture in soil to freeze and thaw means that the heat flow equation cannot be linearized. The problem is particularly difficult because phase boundaries propagate in time, and because soils, particularly clay-rich soils, freeze over a range of temperatures rather than at 0 °C: that is, their phase boundaries are diffuse. We employ a modified Chernous'ko method [24] — a finite element method that tracks isotherms within the soil — to achieve an annual solution to Equation (1). This annual model typically converges to within 0.01 K after 5 iterations for all 10 minute intervals of an annual cycle. Model predictions are tested by their reasonableness and, for non-freezing soils, by comparison with numerical solutions to a variable time interval form of the Laplace analytical method [27].

Figures 1 and 2 show examples of solutions for prairie soils near Bismarck, North Dakota. The 7% moist soil is atypically dry while the 25% moist soil is atypically wet.

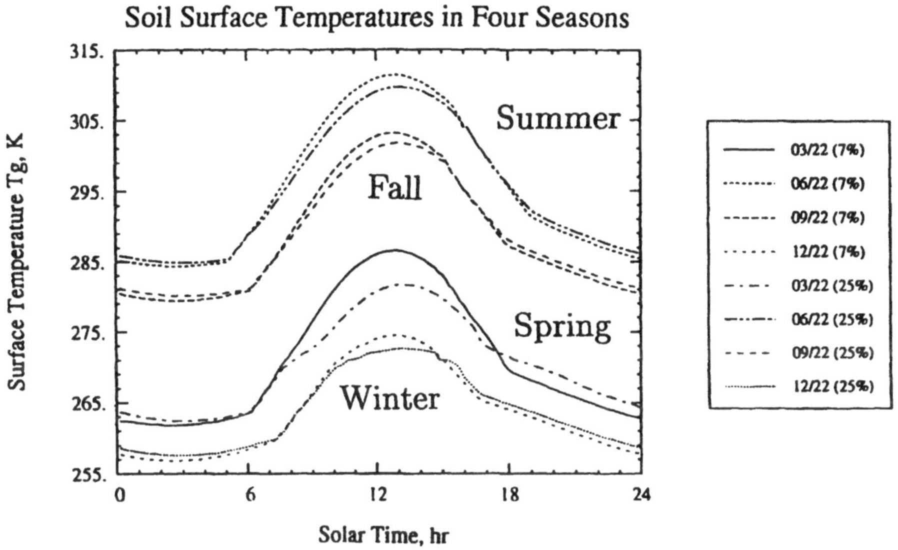

Figure 1. Predicted diurnal surface temperature for prairie near Bismarck, North Dakota.

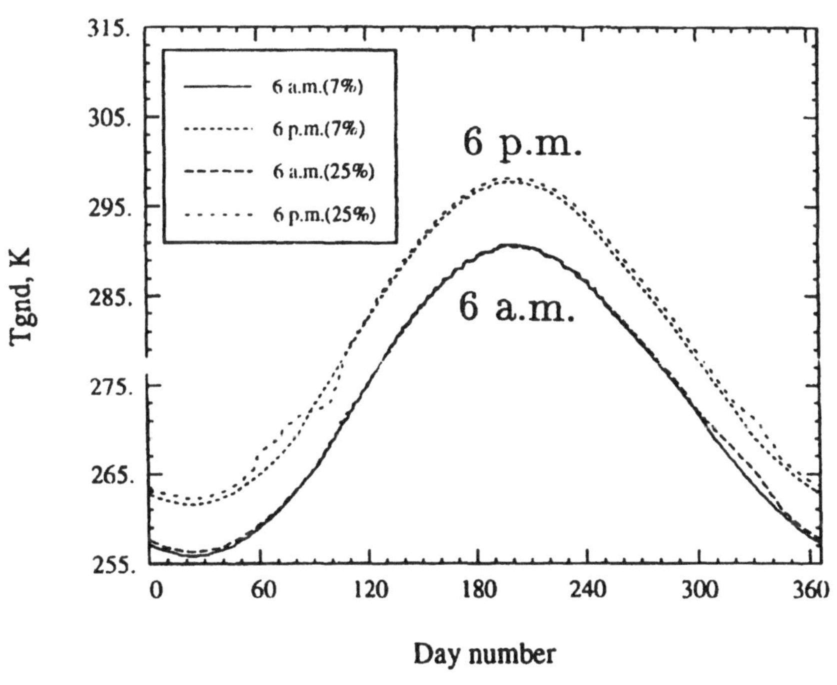

Figure 2. Predicted annual surface temperature for prairie near Bismarck, North Dakota.

We note that diurnal extremes in soil temperature are relatively sensitive to moisture content but annual extremes are not. Moisture dependent anomalies in the annual cycle occur during fall freezing and spring thawing.

The radiobrightness module of the MCRR/Annual model is a first-order approximation to emission from a semi-transparent halfspace whose temperature varies with depth [24]. That is, the radiobrightness as a function of time t, becomes

$$T_b(t) = e(t)\left\{T(0,t) + \mu(t)\,\eta(t)\left(\frac{\partial T}{\partial z}\right)_0\right\} \tag{2}$$

where radiobrightness T_b and emissivity $e(t)$ are implicit functions of polarization (V or H) and incidence angle. Emissivity is based upon the Fresnel coefficients for a plane interface. The assigned permittivity of the soil varies with surface moisture content and temperature [28]. The MCRR model predicts V- and H-polarized radiobrightnesses for a choice of incident angle and frequencies appropriate to either the Nimbus-7 Scanning Multichannel Microwave Radiometer (SMMR), (50° and 10.7, 18.0, and 37.0 GHz [29]), or the Defense Meteorological Satellite's Special Sensor Microwave/Imager (SSM/I), (53.1° and 19.35, 37.0, and 85.5 GHz [30]). $T(0,t)$ is the thermal temperature at the soil surface, $\mu(t)$ is the direction cosine with respect to vertical of the ray in the soil, $\eta(t)$ is optical depth, and $(\partial T/\partial z)_0$ is the thermal gradient at the surface. The complex permittivity of moist soil varies with temperature and, thus, with time so that emissivity, ray direction, and optical depth become time dependent.

Figure 3 shows examples of the 19 GHz, H-polarized radiobrightness during a diurnal cycle. Note that moisture causes an extreme variation in emissivity for winter and spring days when freezing and superficial thawing occur. Figure 4 shows the annual radiobrightness at 6:00 a.m. and 6:00 p.m. — times of SSM/I overflight. The precipitous drop in radiobrightness during the spring thaw or rise during the fall freeze should be evident among the SSM/I data.

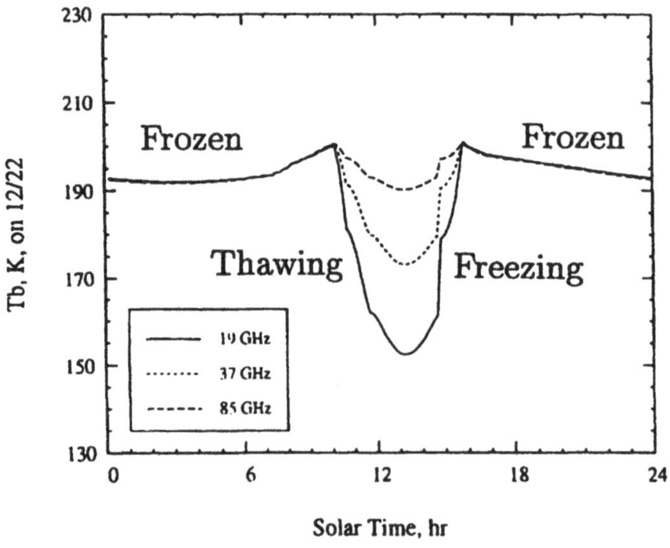

Figure 3. December 19.35, 37.0, and 85.5 GHz, H-polarized diurnal radiobrightness of 25% moist soil.

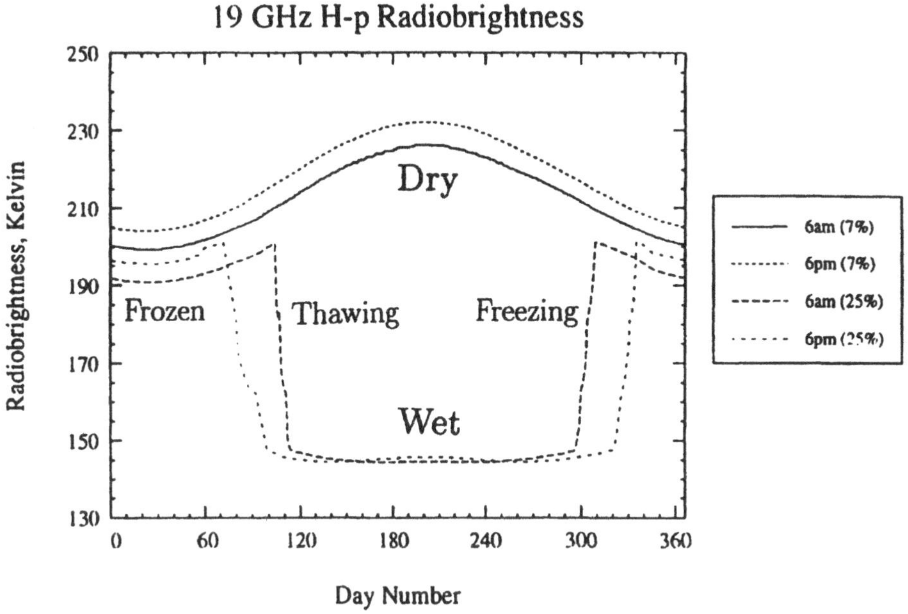

Figure 4. 19.35 GHz, H-polarized annual radiobrightness.

Figure 5(a) shows a plot of SSM/I-derived, 19 GHz, H-polarized, radiobrightness averages during August through December, 1988, for a 100 km square region centered on Fargo, North Dakota. Radiobrightness was estimated by correcting 6:00 a.m. SSM/I scene temperatures for a standard atmosphere using differences between 19.35 and 22.235 GHz data, and assuming a water vapor scale height of 3 km. A 7-day running boxcar filter was used to reduce daily scatter. Precipitation, snowpack thickness, and minimum and maximum air temperatures also appear in Fig. 5. Meteorological data are not temporally filtered. Qualitatively, rainfall in mid-September correlates with reduced radiobrightness during the second half of September; the snowpack during the second half of November correlates with the anomalously low radiobrightness during the same period; low air temperatures beginning in mid-November and extending through December correlate with the rise in radiobrightness during December; and the new snowpack in late December correlates with the precipitous drop in radiobrightness in late December. We note that the period of cold temperatures in December that would freeze snow-free soil does correspond to an increase in radiobrightness as predicted by the model. We also note the scatter darkening in dry snow during November and during late December.

Snow and vegetation must be added to the MCRR model. With vegetation comes evapotranspiration, scattering, and emission, and with snow comes scattering. We are borrowing from the GCM land–atmosphere parameterizations to incorporate evapotranspiration. Many volume scattering models might be applied to microwave emission from snow [31–35], and from grass and sedge vegetation [36–38]. The simplest would be to treat the snow or the canopy as a Rayleigh scattering layer. The most complex for vegetation would be to develop scattering matrices for leaves and stems based upon their dielectric properties, size, orientation, and distribution, and incorporate these scattering matrices in a multi-layer, radiative transfer model, e.g. the Michigan Microwave

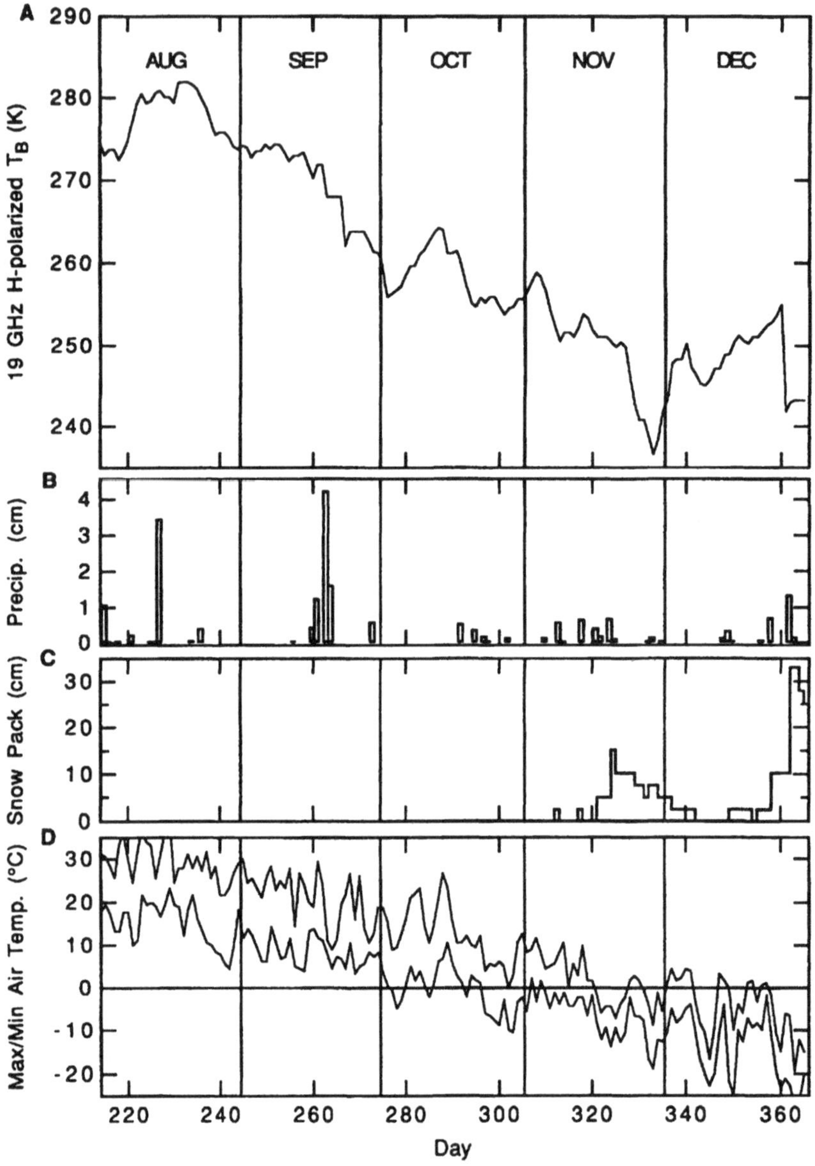

Figure 5. (a) 19 GHz, H-polarized, 6:00 a.m., SSM/I radiobrightness near Fargo, North Dakota, (b) daily precipitation, (c) snow pack depth, and (d) high and low air temperatures, August through December, 1988.

Canopy Scattering Model (MIMICS) [39, 40]. The Rayleigh scattering layer would be unrealistically simple, but the MIMICS models are likely to be unnecessarily complex. MIMICS models are designed to predict radar backscatter where scattering is the dominant process. Emission, absorption, and scattering are equally important to a radiobrightness model of a canopy so that second order scattering might safely be ignored. That is, first order emission and scattering models based upon leaf and stem statistics may predict scatter-induced darkening without the complexity of solving the radiative transfer equation. Dry snow is not absorbing so that one of the many radiative transfer models must be used. There is, as yet, no adequate model for wet snow.

3. THE REMOTE MEASURE OF SOIL MOISTURE

Moisture influences soil reflectivity and emissivity at all wavelengths so that, potentially, there are several remote sensing methods that might be used to infer moisture content or state. For example, soil wetness governs the visible reflectance of bare soil [41]. However, such estimates are limited by a dependence on soil type, by surficial drying, by vegetation cover, and by clouds.

Radiobrightness is sensitive to soil moisture in prairie and arctic tundra — regions of relatively short vegetation — through the dominant influence of liquid water upon microwave emissivity in soil and in the grass canopy. The Debye relaxation of liquid water (which is centered around 12 GHz at 280 K [42]) causes the microwave emissivity of moist soil to increase with frequency so that the radiobrightness spectral gradient, $\partial T_b/\partial f$, where f is frequency, becomes increasingly positive with increasing moisture content. Single frequency estimates of moisture content are possible at microwave frequencies near to or below the Debye relaxation frequency if the thermal temperature of the soil is known or can be independently measured. L-band, single frequency estimates are least prone to errors caused by volume scatter darkening and emission by vegetation. Volume scatter darkening in senescent grass causes a slightly negative bias in the 1–30 GHz spectral gradient [43].

Vegetation is a primary physical link between soil moisture and the atmosphere in northern prairie. Any correlation between soil moisture and atmospheric moisture would involve the integrated moisture content from the soil's surface through the root zone of its vegetation at a spatial scale which is smaller than a drainage basin (significantly less than 100 km). High correlation between radiobrightness and root zone moisture occurs at frequencies below 5 GHz [44, 45], but moderate spatial resolution at frequencies below 5 GHz from satellite altitudes requires large antennas. This combination of frequency and spatial resolution from satellites has not been achieved with passive imaging systems. For example, the lowest frequency and spatial resolution of the SMMR was 6.6 GHz and 150 km, and that for the SSM/I is 19.35 GHz and 43 km. At these frequencies, scattering and emission by plant canopies certainly competes with soil moisture in governing the spectral gradient.

Radar measures of soil moisture have been examined by, for example, Ulaby and Batlivala [46], Blanchard and Chang [47], and McDonald *et al.* [48]. When compared to imaging radiometers, operational satellite radars, like ESA's ERS-1, the Japanese JERS-1, and, eventually, the Canadian Radarsat, offer the advantages of much better spatial resolution (typically a few tens of meters), and measurements that are, to first order, independent of thermal temperature. Their disadvantage includes significantly greater cost for both the spaceborne system and for subsequent data processing. Beyond economics, radar differs from radiometry in that radar backscatter is more strongly influenced by scattering in the plant canopy and by rough soil surfaces. If canopy over soil is pictured as a scattering layer over a rough-surfaced halfspace, then a radar signal has to pass through the scattering layer at least twice in the process of being reflected by the soil. Radiation emitted by the soil passes through the scattering layer only once.

The difference is analogous to viewing an object through frosted glass when the light is on the viewer's side (radar), in contrast to viewing the object when the light is on the object side (radiometry).

Differences caused by surface roughness can be even more striking. For example, if a vegetation-free, moist soil were effectively a homogeneous, quasi-specular halfspace at the microwave wavelength, then there would be no radar backscatter at off-nadir incidence for any moisture content. The equivalent radiobrightness would decrease monotonically with increasing wetness. Radar and microwave radiometry are complementary. Radar more effectively discriminates among plant species where there may be differences in the scattering characteristics of foliage. Microwave radiometry is more usefully sensitive to soil moisture in regions of relatively short vegetation that are homogeneous on the scale of the spatial resolution.

There are alternative methods for obtaining soil moisture from satellite imaging radiometers. Moisture increases the apparent thermal inertia of soil by increasing its thermal conductivity, density, and specific heat, and by daytime cooling through evapotranspiration and nighttime warming through condensation. That is, as the moisture content of soil increases, its day–night difference in thermal temperature tends to decrease, and, consequently, its day–night difference in radiometric brightness also decreases. These effects have been examined in the thermal infrared spectrum [49–53], and were the basis of the Heat Capacity Mapping Mission (HCMM) [54, 55], a thermal infrared experiment that, in part, used differences in the near-surface storage of moisture to discriminate among various rock and soil types.

The thermal microwave day–night signature will exceed the equivalent thermal infrared signature because soil moisture reduces microwave emissivity, but increases (slightly) thermal infrared emissivity. While a thermal infrared technique more easily achieves high spatial resolution, the twice-daily coverage required by either thermal inertia method favors a radiobrightness technique for its tolerance to cloud cover. In essence, the Radiobrightness Thermal Inertia (RTI) method uses the diurnal thermal pulse to probe the soil, and its interpretation uses temperature's influence upon radiobrightness. The effective penetration depth comes from the thermal pulse while the radiometric sensitivity comes from thermal and dielectric changes in the surficial soils. RTI is a product of the MCRR/Diurnal model [25]. Predicted sensitivities to soil moisture are shown in Fig. 6.

The RTI method is particularly well suited to satellite remote sensing by Sun-synchronous satellites because they provide successive day and night observations for properly phased orbits. Overflight times of SMMR were local midnight and noon which are nearly optimum for observing maximum day–night temperature differences. Overflight times of the current SSM/I are 6:00 a.m. and 6:00 p.m. which are particularly poor for RTI because they correspond to thermal crossover.

Vegetation will tend to mask both thermal infrared and microwave soil signatures. Diurnal variations of moisture within extensive plant canopies have complex signatures of their own [56, 57]. Additionally, there are diurnal changes in the vertical distribution of moisture in soil. Njoku and O'Neill [58] have investigated the diurnal biases of single frequency measures of soil moisture caused by diurnal variations in the effective emission depth for frequencies of 0.6–0.9, 1.4, and 10.7 GHz. At the SSM/I frequencies of 19.35, 37.0, and 85.5 GHz, penetration is slight, but such variations may be important. The effects of vegetation — volume scattering, emission, and absorption — are expected

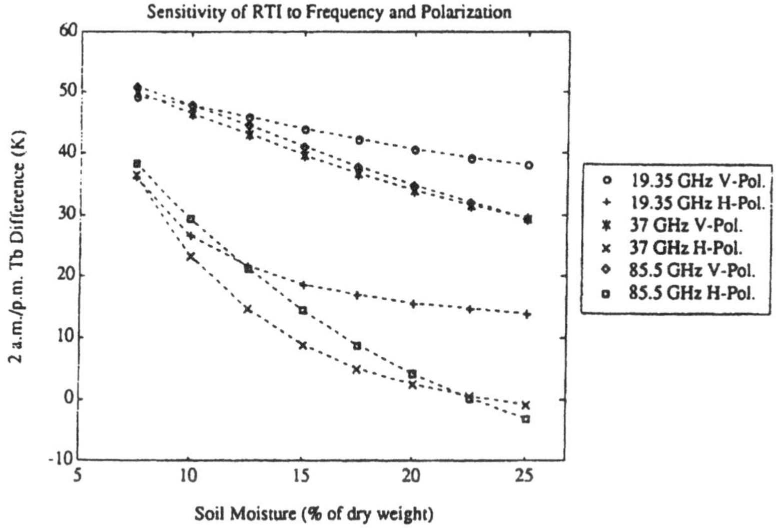

Figure 6. Potential sensitivity of RTI to soil moisture [28]. These summertime day-to-night differences were derived from the MCRR model. Nighttime soil temperatures were always above freezing.

to be less severe in prairie or in arctic regions where vegetation is short or sparse, or when there is little moisture in the canopy. Because RTI uses the difference between day and night radiobrightness, an underlying dependence upon moisture should emerge through a bias caused by the canopy if the bias is relatively constant over time. In fact, day-to-night effects are not constant because canopy moisture varies diurnally.

The RTI measure of soil moisture is, as yet, theory with weak observational corroboration. The theory is being tested in our Prairie Experiment near Sioux Falls, South Dakota, during the fall and winter of 1992–93.

4. REMOTE CLASSIFICATION OF FROZEN SOILS

The radiobrightness signatures of frozen and thawed soils were first studied theoretically and through satellite observations by Zuerndorfer *et al.* [59, 60], and experimentally by Wegmüller *et al.* [61]. The satellite observations refer to data from SMMR for a test area that included North and South Dakota, and parts of surrounding states and southern Canada.

Classification of frozen and thawed soils is possible because moisture state in soil is expressed through unique radiometric characteristics. Diurnal radiobrightness at the SMMR frequencies of 10.7, 18, and 37 GHz were computed for one day of each month. The MCRR/Diurnal model predicted that the 37 GHz radiobrightness would best track soil surface temperature, the 10.7–37 GHz spectral gradient of thawed soil would be strongly positive, the spectral gradient of frozen soil would be slightly negative, and the daytime spectral gradient of frozen soil would be more positive than the nighttime gradient. SMMR data for relatively snow-free days confirmed these predictions except that the observed frozen soil spectral gradients were often strongly negative (Fig. 7). England *et al.* [43] explained the observed negative gradients as being caused by increased scatter darkening in frozen soils and vegetation at higher microwave frequencies.

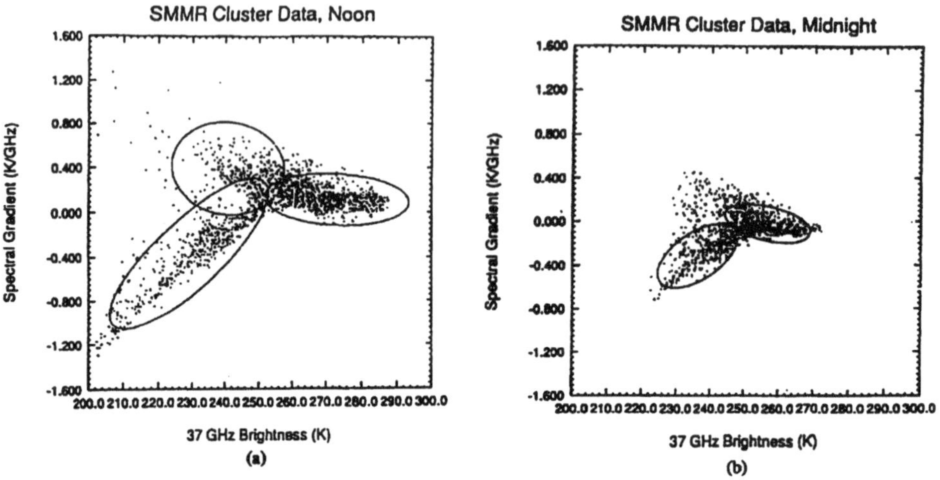

Figure 7. Scatter diagram of $\partial T_b/\partial f$ versus $T_b(37)$ throughout North Dakota and the surrounding region. Data were collected from 8/1/84 to 12/31/84 at (a) noon, and (b) midnight [62].

The decision criteria for classifying frozen soil were empirically derived from migrating means clustering of the noon and midnight SMMR data shown in the scatter diagrams of Figs 7(a) and 7(b). These data are from sixteen SMMR passes over the Dakotas during August through December, 1984. The 18 and 37 GHz radiobrightness data were resolution compensated to the (coarse) resolution of the 10.7 GHz channel before computing spectral gradients. Spectral gradients were obtained from linear regressions of radiobrightness versus frequency. Data at all frequencies were averages of vertical and horizontal polarization. Surfaces were classed as frozen, wet, hot, or mixed. There were few wet surfaces at midnight during the test period so that wet and mixed surface types became inseparable. Frozen surfaces were characterized by relatively low spectral gradients and low 37 GHz radiobrightnesses. Wet surfaces were characterized by high positive spectral gradients and low 37 GHz radiobrightnesses. Hot surfaces exhibited high 37 GHz radiobrightnesses but 'dry', nearly neutral spectral gradients.

Through Mahalanobis minimum distance classification (maximum likelihood classification) based upon assumed bivariant normal distributions, we obtained the cluster centroids shown in Table 1. Constant-deviation, single-class ellipses were drawn in decision space for frozen, wet, and hot surfaces (at noon) and for frozen and hot surfaces

Table 1.
Cluster centroid in decision space

	Noon data		Midnight data	
Surface type	37 GHz (K)	Gradient (K/GHz)	37 GHz (K)	Gradient (K/GHz)
Frozen	227	−0.43	234	−0.35
Hot	277	0.11	258	−0.01
Wet	238	0.37		
Mixed	250	0.14	243	0.015

Table 2.
Freeze/thaw criteria in decision space; σ are standard deviations of the data within the ellipses

	37 GHz (K)	Refined 37 GHz (K)	Gradient (K/GHz)	Deviaton σ at intersection
Noon	252	249	0.0625	3.1
Midnight	247	244	−0.044	2.55

(at midnight). The freeze/thaw criteria were obtained by allowing the deviation of these ellipses to expand equally until all ellipses intersected as shown in Figs 7(a) and 7(b). The corresponding freeze/thaw criteria appear in Table 2. Note that the gradient criterion was more positive at noon than at midnight as predicted by the MCRR/Diurnal model.

If the freeze/thaw criteria from clustering are viewed as initial estimates, then they can be refined for greater self-consistency by requiring minimum scatter of the 37 GHz radiobrightness, $T_b(37)$, along freeze/thaw boundaries. The process involves adjusting the 37 GHz freeze/thaw criterion, $T37$, to minimize the sum square error, SSE,

$$\mathrm{SSE} = \sum_i \left[T_{bi}(37) - T37\right]^2 \tag{3}$$

where subscript i refers to the ith pixel along a freeze/thaw boundary. Refined $T37$ are also shown in Table 2. The process is first-order since we do not iterate SSE minimization with refined criteria. An interpreted history of ground-freezing for the Dakota region during the fall of 1984 appears in a paper by Zuerndorfer and England [62].

The MCRR model, the satellite observations, and the experimental data support use of SMMR's 37 GHz radiobrightness and its 10.7–37 GHz spectral gradient as discriminants in frozen soil classification of high latitude prairie. A similar argument applies to the SSM/I's 37 GHz radiobrightness and its 19.35–85.5 GHz spectral gradient [63]. The thawed soil gradient is less positive because the lowest SSM/I frequency lies well above the Debye relaxation frequency of liquid water. In contrast, the frozen soil spectral gradient is more negative because of increasingly severe scatter darkening at the higher SSM/I frequencies. An example of the sensitivity of the SSM/I data to frozen soils is shown in Figs 8(a) and 8(b). These are atmospherically corrected observations near Fargo, North Dakota, for the fall and early winter of 1988. Radiobrightnesses at 37.0 and 85.5 GHz were spatially filtered to achieve an effective beam pattern equivalent to that of the 19.35 GHz channel. Small dots in Figs 8(a) and 8(b) represent all data points. The squares in Fig. 8(a) correspond to periods that are known from meteorological reports to be snow-free and cold enough for frozen soil. The crosses in Fig. 8(b) correspond to periods known to have greater than 2 cm of snow and air temperatures cold enough to expect the snow to be dry. We include both figures to emphasize that frozen soil and dry snow appear radiometrically similar and, consequently, may be misclassified.

The boundaries in the SSM/I decision space between frozen and thawed soil shown in Fig. 8 are based upon meteorological data. A better estimate of these boundaries will be a product of our Prairie Experiment. Figures 9(a), 9(b), and 9(c) are examples of frozen soil classification maps for the Dakotas based upon the boundaries shown in Fig. 8.

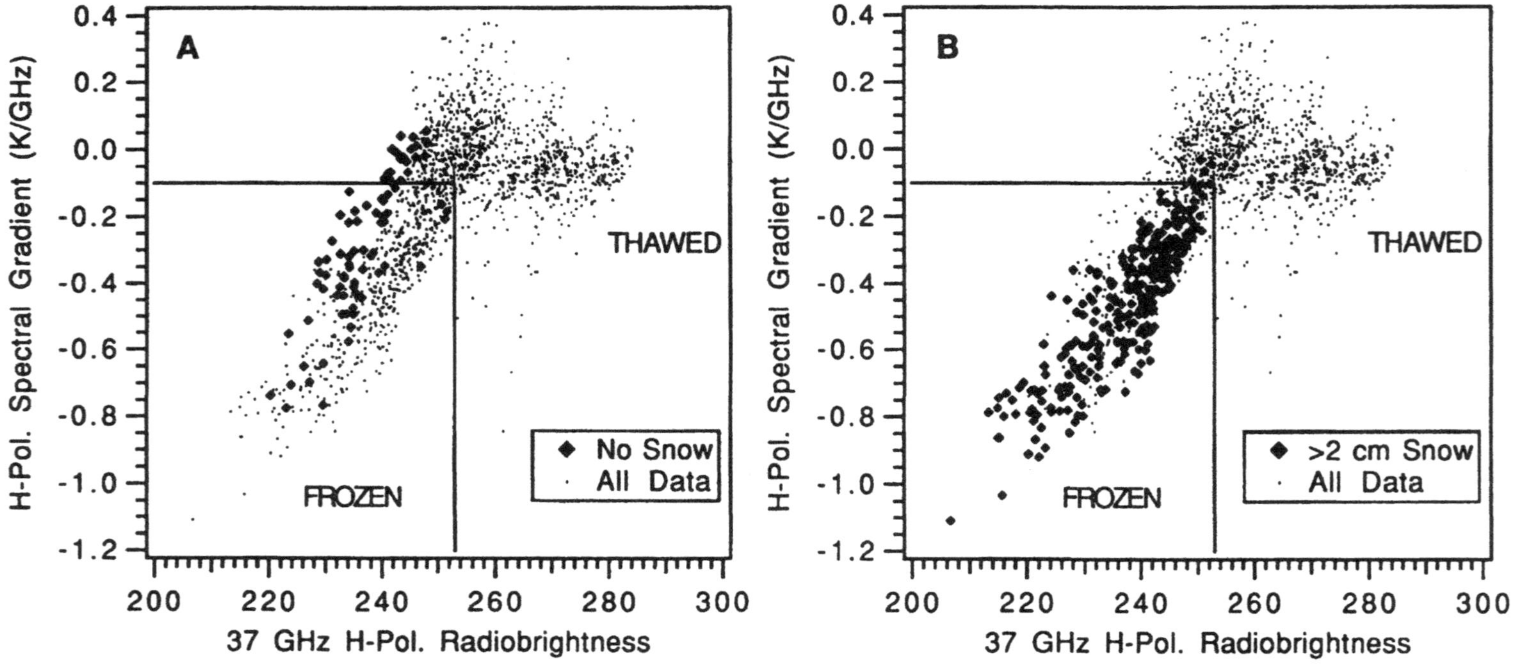

Figure 8. Scatter diagrams of SSM/I, H-polarized data for a 100 km square region centered on Fargo, North Dakota. Data were collected from 8/1/88 to 12/31/88 at 6:00 am local solar time. Data marked are for (a) frozen ground, no snow, and (b) more than 2 cm snow, as reported at Fargo.

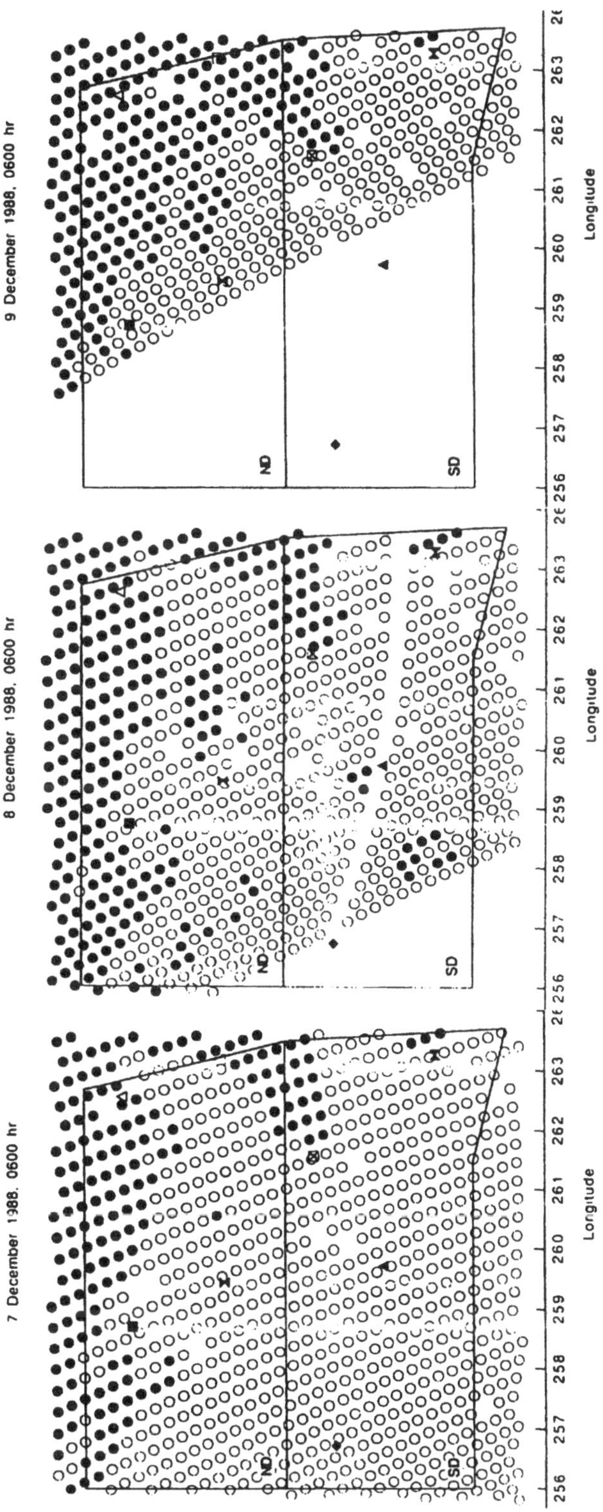

Figure 9. SSM/I-derived freeze-thaw patterns for the Dakotas. Filled circles indicate frozen soil. These patterns are consistent with the meteorological data. Data taken at 06:00 local solar time on (a) 12/7/88, (b) 12/8/88, and (c) 12/9/88.

5. REMOTE DETECTION OF PERMAFROST

Periglacial processes associated with the freezing and thawing of boreal soils are fundamental to understanding high latitude, land environments [64, 65]. Frozen soils appear dry to land-atmosphere processes. Furthermore, boundaries between continuous and discontinuous permafrost represent divisions between climate regions which are sufficiently cold that they support perennial ground ice, and regions which are less cold. Because permafrost is the integrated effect of many cold seasons, migrations of these boundaries are potential indicators of climate change [66].

Regional shifts in permafrost boundaries reflect regional variations in dominant weather patterns. Relating such regional shifts to global change will require monitoring continuous and discontinuous permafrost boundaries on a polar scale over many years. Polar scale phenomena are best monitored by operational satellite sensors like the SSM/I. It is not enough to monitor changes of periglacial morphology or of boreal forest tree lines because such features will not reflect recent gains or losses of permafrost. There are no satellite remote sensing techniques that would directly probe soil depths below an active layer. However, there may be a temporal surface temperature signature associated with underlying permafrost.

Thermal characteristics of permafrost terrains have been modeled by Lachenbruch [67], and Kazemi and Perkins [68] among others, but their primary objective was to understand the gross thermal structure at depth. Their approximations to the land-atmosphere energy budget were too coarse to provide reliable ground surface temperatures. Predictions from the MCRR/Annual model suggest that permafrost will influence the temporal characteristics of the surface temperature and, consequently, of the temporal radiobrightness signature. Qualitatively, permafrost increases the apparent thermal inertia of soil. Spring thaws in permafrost terrains require more thermal energy than they would in seasonally frozen terrains, and the thaw process that creates the active layer actually extends throughout the summer. Similarly, a fall freeze occurs more quickly because the permafrost acts as a thermal sink. The effect is that permafrost terrains should exhibit later thaws, earlier freezes, and persistently lower surface temperatures throughout the growing season. An evolutionary version of the MCRR model will be tested in a field experiment on tussock tundra near the Toolik Field Station, Alaska, during the summer and fall of 1994.

Acknowledgment
This work has been supported by NASA grant NAGW-1983 and by a grant from the US Geological Survey.

REFERENCES

1. C. M. Bhumralkar. Parameterization of the planetary boundary layer in atmospheric General Circulation Models. *Reviews of Geophys.*, **14** (2), 215–226 (1976).
2. M. F. Wilson and A. Henderson-Sellers. Cover and soils data sets for use in general circulation climate models. *J. Climatology*, **5**, 119–143 (1985).
3. F. Abramopoulos, C. Rosenzweig and B. Choudhury. Improved ground hydrology calculations for Global Climate Models (GCMs): Soil water movement and evapotranspiration. *J. Climate*, **1**, 921–941 (1988).

4. M. M. Verstraete. Land surface processes in climate models: Status and prospects. In: *Climate and the Geosciences: A Challenge for Science and Society in the 21st Century*, A. Berger, S. Schneider and J. C. Duplessy (Eds), pp. 321–340, Kluwer, Dordrecht, Netherlands (1989).
5. R. E. Dickinson, R. M. Errico, F. Giorgi and G. T. Bates. A regional climate model for the Western United States. *Climatic Change*, **15**, 383–422 (1989).
6. R. Avissar and M. M. Verstraete. The representation of continental surface processes in atmospheric models. *Reviews of Geophys.*, **28** (1), 35–52 (1990).
7. F. Giorgi and L. O. Mearns. Approaches to the simulation of regional climate change: A review. *Reviews of Geophys.*, **29** (2), 191–216 (1991).
8. J. Namias. Persistence of mid-tropospheric circulations between adjacent months and seasons. In: *The Atmosphere and Sea in Motion*, P. Polin (Ed.), pp. 240–248, Oxford University Press (1958).
9. J. Namias. Surface–atmosphere interactions as fundamental causes of droughts and other climatic fluctuations. In: *Arid Zone Research, 20, Changes of Climate Proc. of Rome Symp.*, pp. 345–359, UNESCO, 75700 Paris (1963).
10. J. Walker and P. R. Rowntree. The effect of soil moisture on circulation and rainfall in a tropical model. *Quart. J. Roy. Meteor. Soc.*, **103**, 29–46 (1977).
11. D. Rind. The influence of ground moisture conditions in North America on summer climate as modeled in the GISS GCM. *Mon. Weat. Rev.*, **110**, 1487–1494 (1982).
12. J. Shukla and Y. Mintz. Influence of land-surface evapotranspiration on the Earth's climate. *Science*, **215**, 1498–1501 (1982).
13. T. C. Yeh, R. T. Wetherald and S. Manabe. The effect of soil moisture on the short-term climate and hydrology change – A numerical experiment. *Mon. Wea. Rev.*, **112**, 474–490 (1984).
14. R. J. Oglesby and D. J. Erickson III. Soil moisture and the persistence of North American drought. *Climate*, **2**, 1362–1380 (1989).
15. T. Delworth and S. Manabe. The influence of potential evaporation on the variability of simulated soil wetness and climate. *Climate*, **1**, 523–547 (1988).
16. T. Delworth and S. Manabe. The influence of soil wetness on near-surface atmospheric variability. *Climate*, **2**, 1447–1462 (1989).
17. S. Manabe. Climate and the ocean circulation: I. The atmospheric circulation and the hydrology of the earth's surface. *Mon. Wea. Rev.*, **97**, 739–744 (1969).
18. M. I. Budyko. *Climate and Life*. Academic Press, NY (1974).
19. R. E. Dickinson. Modeling evapotranspiration for three-dimensional Global Climate Models. *Climate Processes and Climate Sensitivity*, **5**, 58–72 (1984).
20. R. E. Dickinson, A. Henderson-Sellers, P. J. Kennedy and M. F. Wilson. *Biosphere–Atmosphere Transfer Scheme (BATS) for the NCAR Community Climate Model*. NCAR Technical Note NCAR/TN-275+STR, National Center for Atmospheric Research, Boulder, Colorado (1986).
21. P. J. Sellers, Y. Mintz, Y. C. Sud and A. Dalcher. A Simple Biosphere model (SiB) for use within general circulation models. *J. Atmos. Sci.*, **43**, 505–531 (1986).
22. K. Watson. Geologic application of thermal infrared images. *Proc. IEEE*, 128–137 (1975).
23. A. B. Kahle. A simple thermal model of the Earth's surface for geologic mapping by remote sensing. *J. Geophys. Rcs.*, **82**, 1673 1680 (1977).
24. A. W. England. Radiobrightness of diurnally heated, freezing soil. *IEEE Trans. Geosc. Rem. Sens.*, **GE-28** (4), 464–476 (1990).
25. A. W. England and Y. A. Liou. Diurnal and annual radiobrightness thermal inertia of desert soils. In: *Proc. URSI Microwave Signature Conf.*, p. 3B-4, Innsbruck, Austria, July 1–3 (1992).
26. Y. A. Liou and A. W. England. Annual model of the energy flux and the radiobrightness of Northern Prairie soil. In: *Am. Geophysical Union Fall Meeting*, San Francisco, December 7–11 (1992).
27. Y. Liou and A. W. England. An annual model of SSM/I radiobrightnesses for dry soil. In: *Proc. of IGARSS'92*, Houston, Texas, March 16–20 (1992).
28. A. W. England, J. F. Galantowicz and M. S. Schretter. The radiobrightness thermal inertia measure of soil moisture. *IEEE Trans. Geosci. Rem. Sens.*, **30**, 132–139 (1992).

29. P. Gloerson and L. Hardis. The Scanning Multichannel Microwave Radiometer (SMMR) experiment. In: *The Nimbus-7 Users' Guide*, C. R. Madrid (Ed.), pp. 213–246, Goddard Space Flight Center, Greenbelt, MD (1978).

30. J. Hollinger, R. Lo, G. Poe, R. Savage and J. Pierce. *Special Sensor Microwave/Imager User's Guide*. Naval Research Laboratory, Washington, DC (1987).

31. A. S. Gurvich, V. I. Kalinin and D. T. Matveyev. Influence of the internal structure of glaciers on their thermal radio emission. *J. Atmos. Oceanic Phys.*, **9** (12), 1247–1256 (1973).

32. A. W. England. Thermal microwave emission from a scattering layer. *Geophysical Research*, **80**, 4484–4496 (1975).

33. L. Tsang and J. A. Kong. Microwave remote sensing of a two-layer random medium. *IEEE Trans. Ant. Prop.*, **AP-24**, 283–287 (1976).

34. A. K. Fung and M. F. Chen. Emission from an inhomogeneous layer with irregular interfaces. *Radio Science*, **16**, 289–389 (1981).

35. L. Tsang and A. Ishimaru. Radiative wave equations for vector electromagnetic propagation in dense nontenuous media. *J. Electr. Waves Appl.*, **1**, 52–72 (1987).

36. R. H. Lang and J. S. Sidhu. Electromagnetic backscattering from a layer of vegetation: A discrete approach. *IEEE Trans. Geosci. Rem. Sens.*, **GE-21**, 62–71 (1983).

37. F. T. Ulaby, T. H. Haddock and Y. Kuga. Measurement and modeling of millimeter-wave scattering from tree foliage. *Radio Science*, **25**, 193–203 (1990).

38. K. Sarabandi, F. T. Ulaby and T. B. A. Senior. Millimeter wave scattering model for a leaf. *Radio Science*, **25**, 9–18 (1990).

39. K. C. McDonald, M. C. Dobson and F. T. Ulaby. Using MIMICS to model microwave backscatter from tree canopies. In: *Proc. of IGARSS'89*, Vol. 4, pp. 2491–2494 (1989).

40. F. T. Ulaby, K. Sarabandi, K. McDonald, M. Whitt and M. C. Dobson. Michigan Microwave Canopy Scattering Model. *Int. J. Rem. Sensing*, **11**, 1223–1253 (1990).

41. S. B. Idso, R. Jackson, B. Kimball and F. Nakayama. The dependence of bare soil albedo on soil water content. *J. Appl. Meteor.*, **14**, 109–113 (1975).

42. J. B. Hasted. Dielectric properties of water and aqueous solutions. In: *Dielectric and Related Molecular Processes*, pp. 121–162, The Chemical Society, London (1972).

43. A. W. England, J. F. Galantowicz and B. W. Zuerndorfer. A volume scattering explanation for the negative spectral gradient of frozen soil. In: *Proc. of IGARSS'91*, pp. 1175–1177, Espoo, Finland, June 3–6 (1991).

44. W. J. Burke, T. Schmugge and J. F. Paris. Comparison of 2.8- and 21-cm microwave radiometer observations over soils with emission model calculations. *J. Geophys. Res.*, **84**, 287–294 (1979).

45. J. R. Wang, T. J. Schmugge, W. I. Gould, W. S. Glazar and J. E. Fuchs. A multi-frequency radiometric measurement of soil moisture content over bare and vegetated fields. *Geophys. Res. Let.*, **9**, 416–419 (1982).

46. F. T. Ulaby and P. P. Batlivala. Optimum radar parameters for mapping soil moisture. *IEEE Trans. Geosci. Electr.*, **GE-14** (2), 81–93 (1976).

47. B. J. Blanchard, and A. T. C. Chang. Estimation of soil moisture from SEASAT SAR data. *Water Res. Bulletin*, **19** (5), 803–810 (1983).

48. K. C. McDonald, M. C. Dobson and F. T. Ulaby. Determination of soil moisture beneath a stalk or trunk-dominated canopy. In: *Proc. of IGARSS'88*, pp. 1593–1598, Edinburgh, Scotland, September 13–16 (1988).

49. S. B. Idso, T. J. Schmugge, R. D. Jackson and R. J. Reginato. The utility of surface temperature measurements for the remote sensing of surface soil water status. *J. Geophys. Res.*, **80**, 3044–3049 (1975).

50. R. J. Reginato, S. B. Idso, J. F. Vedder, R. D. Jackson, M. B. Blanchard and R. Goettelman. Soil water content and evaporation determined by thermal parameters obtained from ground-based and remote measurements. *J. Geophys. Res.*, **81**, 1617–1620 (1976).

51. J. C. Price. The potential of remotely sensed thermal infrared data to infer surface soil moisture and evaporation. *Water Resources Res.*, **16**, 787–795 (1980).

52. J. L. Heilman and D. G. Moore. Thermography for estimating near-surface soil moisture under developing crop canopies. *J. Appl. Meteor.*, **19**, 324–328 (1980).

53. J. Vleck and D. King. Detection of subsurface soil moisture by thermal sensing: Results of laboratory, close-range and aerial studies. *Photogr. Engin. and Rem. Sens.*, **49**, 1593–1597 (1983).

54. J. L. Heilman and D. G. Moore. HCMM detection of high soil moisture areas. *Rem. Sens. Environment*, **11**, 73–76 (1981).

55. J. L. Heilman and D. G. Moore. Evaluating near-surface soil moisture using Heat Capacity Mapping Mission data. *Remote Sens. Environ.*, **12**, 117–121 (1982).

56. H. K. Burke and T. J. Schmugge. Effects of varying soil moisture contents and vegetation canopies on microwave emission. *IEEE Trans. Geosci. Rem. Sens.*, **GE-20**, 268–274 (1982).

57. J. R. Wang. Effect of vegetation on soil moisture sensing observed from orbiting microwave radiometers. *Remote Sens. Environ.*, **17**, 141–151 (1985).

58. E. G. Njoku and P. E. O'Neill. Multifrequency microwave radiometer measurements of soil moisture. *IEEE Trans. Geosci. Rem. Sens.*, **GE-20**, 468–475 (1982).

59. B. Zuerndorfer, A. W. England and G. H. Wakefield. The radiobrightness of freezing terrain. In: *Proc. of IGARSS'89*, pp. 2748–2751, Vancouver, Canada, July 10–14 (1989).

60. B. W. Zuerndorfer, A. W. England, C. M. Dobson and F. T. Ulaby. Mapping freeze/thaw boundaries with SMMR data. *J. Agriculture Forest Meteorology*, **52**, 199–225 (1990).

61. U. Wegmüller, C. Mätzler and E. Schanda. Microwave signatures of bare soil. *Adv. Space Res.*, **9**, 307–316 (1989).

62. B. W. Zuerndorfer and A. W. England. Radiobrightness decision criteria for freeze/thaw boundaries. *IEEE Trans. Geosci. Rem. Sens.*, **30**, 89–101 (1992).

63. P. A. Dahl, J. F. Galantowicz and A. W. England. Frozen soil classification from SSM/I radiobrightness. In: *Am. Geophysical Union Fall Meeting*, San Francisco, December 7–11 (1992).

64. A. L. Washburn. *Periglacial Processes and Environments*. St. Martin's Press, NY (1973).

65. J. C. F. Tedrow. *Soils of the Polar Landscapes*. Rutgers University Press, NJ (1977).

66. A. L. Washburn. Permafrost features as evidence of climatic change. *Earth-Sci. Rev.*, **15**, 327–402 (1980).

67. A. H. Lachenbruch. Periodic heat flow in a stratified medium with application to permafrost problems. In: *US Geological Survey Bulletin 1083-A*, pp. 1-36, US Government Printing Office, Washington, DC (1959).

68. H. Kazemi and T. K. Perkins. A mathematical model of thaw-freeze cycles beneath drilling rigs and production platforms in cold regions. *J. Petroleum Technology*, 381–390 (1971).

69. T. J. Schmugge, P. E. O'Neil, and J. R. Wang. Passive microwave soil moisture research. *IEEE Trans. Geosci. Remote Sensing*, **24**, 12–22 (1986).

Atmospheric correction and precipitation

ESA/NASA International Workshop, pp. 571–598
B. J. Choudhury, Y. H. Kerr, E. G. Njoku and P. Pampaloni (Eds)

Overland precipitation

ERIC C. BARRETT and DOMINIC KNIVETON

Remote Sensing Unit, Department of Geography, University of Bristol, Bristol BS8 1SS, UK

Abstract—Rainfall is a rare event, spatially and temporally noisy. It is inadequately monitored at the surface by either conventional means (raingauges) or remote sensing (radar). Indeed, evidence strongly suggests that surface measurements of rainfall are globally less numerous today than they have been in the past.

Fortunately, data from Earth-orbiting satellites can be used to provide estimates of rainfall which are useful for many applications in areas relatively poorly served by raingauges and/or radar. Infrared imagery has been used widely for such purposes, but results suffer particularly from the physical indirectness and inconsistency of the relationship between cloud-top temperatures and both rain areas and rain rates.

Passive microwave imagery has opened up new opportunities for improved rainfall monitoring by satellites. In some microwave wavebands emitted radiation from hydrometeors within the clouds can be measured, and in others reflected radiation.

Overwater precipitation, at least instantaneously on the one hand, and climatically on the other, is already better measured by passive microwave methods than by surface data in most ocean and sea areas of the world. Unfortunately, the retrieval of overland precipitation is more difficult because land backgrounds can be radiometrically more extreme and more variable than water backgrounds. The problems are most acute over areas of bare soil (mainly deserts) and snow and ice. Research is proceeding in an increasing number of individual laboratories, and in new co-ordinated projects, to address the difficulties and identify the most promising passive microwave algorithms; the next stage will be to develop 'community algorithms' for the preparation of global climatological data sets.

This paper reviews the need for such activities, the nature of present state-of-the-art algorithms, their levels of performance, and attempts to see the way ahead.

1. THE NEED FOR IMPROVED RAINFALL DATA

Precipitation is a vital, but rare and poorly-monitored atmospheric parameter. Over land it is the ultimate source of the water which is required by all plants and animals for life itself, and is a prerequisite for many kinds of industry, commerce, transportation and recreation. The area of the globe covered by falling precipitation has been estimated from surface observations as below 1% [1] and although recent satellite-based estimates have suggested that this an underestimate [2], there can be no doubt that the true area of instantaneous precipitation is less than 5%. Primary reasons for the precipitation measurement problem are that precipitation is notoriously variable in both the *spatial* and *temporal* domains. At any one instant precipitation intensities around the globe range from zero to > 125 mm h^{-1}. And, since most precipitation is generated by mobile cloud and weather systems, accumulated precipitation totals through time range very widely: for example, mean annual rainfall is near zero in some desert areas but > 12 500 mm y^{-1} in some localities where mountains frequently tap moisture-laden winds.

The above facts alone ensure that precipitation is difficult to monitor satisfactorily. However, several other types of problems exacerbate this problem. These include *physical* and *practical* problems which stem from the nature and behaviour of precipitation particles, and their interactions with both the Earth surfaces (e.g. land or water) on which they impinge, and the means used to collect or measure them. But they include *economic* and *political* problems also, for neither precipitation gauge networks nor meteorological radars are cheap, and different standards and strategies for precipitation monitoring have been favoured by different nations and groups of nations. The World Meteorological Organisation (WMO) has helped establish precipitation measuring networks globally, but it has been estimated that if all the world's currently operational raingauges were placed side by side in one place they would cover only the area of one soccer pitch, and that the world's operational rainfall radars would cover proportionately only about the area of its centre circle. To make matters worse, rainfall totals are no longer reported mandatorily via the Global Telecommunication System (GTS) of the WMO, and for near real-time uses of such data it has to be realised that data may be sparse or absent not only from ocean areas and thinly-populated land areas (e.g. over ice-caps and hot deserts), but increasingly from some countries with long traditions of comparatively good surface precipitation data collection.

Unlike surface-based observations of precipitation, which involve measurements of precipitation falling or fallen from the bases of clouds, satellite-based estimates of precipitation depend more upon characteristics of the clouds themselves. Three types of electromagnetic radiation have proved to be of special value in this respect, namely visible (VIS), infrared (IR) and microwave (MW), the last-named wholly to date through the operation of passive microwave (PMW) radiometers, although plans exist for the use of active microwave (AMW) rainfall radars on future satellites, e.g. the Tropical Rainfall Monitoring Mission (TRMM) [3]. Because the VIS and/or IR techniques depend on radiation from cloud tops, whereas MW techniques depend on radiation from ice and/or hydrometeors within the clouds, it will be convenient to review these two categories of methods separately. Because it is not yet possible for satellite algorithms to distinguish between the different types of precipitation, and because special problems are encountered when attempting to discriminate between water falling to the ground and water on the ground where this is in the form of snow or ice, the remainder of this review will focus on precipitation measurement over unfrozen areas, and will refer to all precipitation as 'rain'. By the early 1980s numerous techniques for rainfall monitoring by satellites had already been developed and tested (see [1]); over the last decade the level of interest in satellite rainfall monitoring has rapidly increased (see, e.g. [4–6]). Recent suggestions that climate changes due to global warming may lead to water shortages in some areas historically well supplied by rainfall are amongst the reasons why there is new urgency in the search for improved methods for monitoring rainfall on all scales from the local and regional to the global.

2. VISIBLE/INFRARED RAINFALL MEASUREMENTS BY SATELLITES

2.1. *Principles and practices*

Because all techniques for rainfall measurement in the VIS and IR regions of the electromagnetic spectrum rely on radiation from cloud tops the results can be misleading so far as rain falling from cloud bases is concerned. The best performance of such techniques is found in *convectional regimes* where there are good relationships between the temperatures of cumulonimbus cloud tops and the instantaneous areas and rates of rainfall. Such techniques encounter greatest difficulties in *frontal zones* and other situations where thick layered clouds (especially cirriform, stratiform and multi-layered stratiform) are present.

In recent years the greatest focus of attention in satellite rainfall monitoring has been on infrared imagery to the exclusion of visible imagery, for it is only the former which are available during night-time as well as day-time hours, and some of the physical implications of infrared imagery are the more consistent, especially the relationship between cloud top temperature and cloud top height. Key components of current infrared-based schemes include the following:

- *a cold cloud identifier* (to distinguish rain from no-rain areas), based on the understanding that colder clouds are more likely to rain than warmer clouds;
- *cold cloud duration* (usually assessed in terms of the numbers of satellite-observed cold cloud occurrences per geographic location per day), based on the understanding that the longer the duration of raincloud locally the more rain which is likely to fall;
- *cold cloud type discrimination* (commonly using some kind of texture discriminator), based on the expectation that, for example, rough textured cumulonimbus clouds are likely to produce more rain than smoother textured cirriform clouds;
- *background fields* (e.g. terrain models and/or climate fields), necessitated because precipitating clouds in humid and/or hilly areas are more likely to rain heavily than those in dry and/or flat areas;
- *meteorological (weather) reports* (e.g. daily rainfall observations from GTS stations); if available these provide opportunities to calibrate the satellite/terrain/climate 'initial' estimates on the basis of actual weather.

Many visible and/or infrared satellite rainfall monitoring techniques have been developed over the last two decades [7]. (See Tables 1a and 1b.) At first these were dominantly *manual*, in which the analyst would draw probable raincloud province boundaries by hand on hard-copy satellite images. These techniques were generally superseded by *interactive* techniques, in which characteristically the analyst would outline different probable rain cloud provinces on imagery displayed on an image processing system screen, the associated computer then permitting collateral data to be involved in the evaluation process in a relatively consistent and detailed fashion. More recently, *objective* or *automatic* techniques have come to dominate, in which the image data from the satellite are processed entirely within the computer, perhaps using a wide range of ancillary data within a detailed Geographic Information System (GIS) context to provide the most detailed results possible in both space and time.

In the tropics, *geostationary* satellite data have become more and more widely used for rainfall monitoring purposes. They are best whenever short period rainfall estimates are required, and in areas dominated by convectional raincloud development. In middle to high latitudes recourse often has to be made to data from *near polar orbiting* satellite data. Unfortunately, these are obtained less frequently, and more ambiguities exist in such regions because of the greater incidence, even dominance, of layered cloud within which rain areas may be embedded only rather locally. Fortunately, polar orbiting

Table 1.
Classifications of VIS/IR satellite rainfall measurement algorithms according to (a) technical complexity (top), and (b) their space/time scales (bottom). (After [7])

(a)

Channels:	ONE		TWO
Predictors:	Visible	Infrared	Visible & infrared
ONE	Kilonsky-Ramage Follansbee-Oliver	GPI/EPI (Arkin) Bristol	Lovejoy-Austin
TWO OR MORE	Linear	DelBeato-Barrell Whitney-Herman Negri *et al.* (1984) Kruger *et al.* Stout *et al.* Fenner Motell and Weare Inoue	Martin-Howland Callis-LeComte FAO
	Non-Linear	TAMSAT PERMIT EARS-bv Griffith-Woodley Weiss-Smith Adjusted Griffith-Woodley B4	BIAS Heitkemper *et al.* Scofield-Oliver Neil Wu *et al.* Tsonis-Isaac ADMIT

(b)

Spatial resolution					
25 km–2.5° squares					Global Precipitation Index FAO Kilonsky-Ramage Garcia
5–25 km	Kruger *et al.* Weiss-Smith	Bristol Fenner Whitney-Herman	Follansbee Stout *et al.* Martin-Howland B4	Creutin *et al.*	LeComte-Callis Motell-Weare
1–5 km	Scofield-Oliver Neil Lovejoy-Austin Robertson Tsonis-Isaac Wu *et al.* Lee *et al.* Griffith-Woodley Adler-Negri	BIAS	Heitkemper *et al.* Doneaud *et al.* Delbeato-Barell Inoue	ADMIT PERMIT TAMSAT EARS-bv	
Temporal resolution	0.5–1 hour	1–6 hours	daily	1–10 days	10 days–1 month

satellite systems such as NOAA provide greater spectral resolution than the geostationary satellites and this has been exploited by some workers to help differentiate between probably precipitating and probably non-precipitating (layered) clouds.

2.2. *Visible and/or infrared techniques: recent advances*

The dominant trend in VIS/IR technique development has been towards greater algorithm *complexity*. This has found a variety of different expressions, including the following:

2.2.1. Objective cloud type recognition. In estimating rainfall from satellite imagery in convectional (tropical) regimes it is important to distinguish between *active cumulonimbus* and associated *stratiform* and *cirriform* cloud, for rain probabilities and intensities are much higher in the first case than the second or third. It has been established that the heaviest rains fall from rough-textured rather than smooth-textured clouds, this distinction forming the basis of the Cumuliform–Stratiform Technique (CST), in which different rain rates are assigned to these two broad categories of clouds [8]. The CST has proved useful both as a stand-alone technique, and as an adjunct to other simpler techniques. For monitoring rainfall from satellite imagery in stratiform/mixed (mid-latitude) regimes multiple-category cloud recognition schemes have been developed using multispectral feature recognition techniques applied to NOAA AVHRR data, with the primary purpose of serving numerical weather prediction models with appropriate initial cloud and precipitation information at the time of the polar-orbiting satellite overpasses (e.g. [9]).

2.2.2. Objective selection of key model parameters and/or model calibration. Key questions in any satellite rainfall monitoring techniques are how best to select the *rain/no-rain boundaries*, and suitable *rain rates* for the areas identified as raining. In a new geostationary infrared method now being tested pre-operationally over West Africa, raingauge data from real-time reporting Global Telecommunication System (GTS) stations are being used to set rain/no-rain boundaries at cloud top temperatures giving the same number of rain days from the satellite image analyses as those reported from the same regions on the ground. Rain rates are then calculated using either mean rain per day statistics from ('background field') climate records, and/or intensity/duration relationships derived from the GTS station reports if these are available and adequate. Therefore, this approach, named 'B4' (Bristol/Barrett, Beaumont, Bellerby) is both objective and internally self-calibrating: the rain/no-rain threshold temperatures and rain intensity/duration factors are automatically adjusted to the observed changes in the weather [10]. In the US AID Upper Nile Project designed to improve forecasting of the flow of the River Nile for more efficient water management in Egypt a somewhat similar approach is being developed using streamflow data for satellite rainfall algorithm calibration (J. Schaake, personal communication).

2.2.3. Interactive recognition of key mesoscale and local rainfall amplifiers. One of the first fully operational uses of satellite rainfall monitoring techniques for hydrological applications has been through the Interactive Flash Flood Analysis (IFFA) system of the US National Weather Service. This is based on interactive analyses of geostationary cloud images when these reveal possible high-intensity rain-producing weather situations, and includes quite detailed schemes for several categories of synoptic, sub-synoptic and mesoscale storms [11]. Present research is focussing on means whereby high intensity rains may be not only recognised and evaluated, but also forecast better so as to further improve the forecasting of associated flash floods.

Notwithstanding the advances summarised above, satellite rainfall estimation from VIS and/or IR imagery alone is still unable to provide the dependability of results and the breadth of coverage which is required *to make good the global deficiencies in surface rainfall measurement*; therefore, in recent years attention has turned increasingly to the microwave region of the spectrum to provide additional, or even possibly better alternative, sources of information.

3. PASSIVE MICROWAVE RAINFALL MEASUREMENT BY SATELLITES

3.1. General conditions

We have seen that rainfall indications derived from visible and/or infrared images can be no better than *physically indirect*. Fortunately, passive microwave data from sensors summarised in Table 2 give *physically more direct* evidences of hydrometeors. Foremost amongst the passive microwave systems which provide these data, have been the Electrically Scanning Microwave Radiometer (ESMR) and Scanning Multichannel Microwave Radiometer (SMMR) instruments of the Nimbus Research and Development ('R&D') family and, since July 1987, the Special Sensor Microwave Imager (SSM/I) instruments on the USAF-DMSP Block 5D family of satellites. The SSM/I in particular has stimulated a big upsurge of interest in satellite rainfall monitoring in the passive microwave, helped by the NASA-led 'WetNet' Project [12], of whose Precipitation Working Group the present first author is the Chairman.

Although rainfall measurement has not been the sole aim of any of the passive microwave sensor missions summarised in Table 2, rainfall is a parameter whose monitoring was expected to be significantly improved thereby (see, e.g. the list of geophysical parameters which the SSM/I sensor was designed to elucidate [13]). However, the spatial resolution of the PMW sensors has been low relative to that of the VIS and IR sensors, having been at best about 15 km by 13 km in the case of the 85 GHz channel in the SSM/I. This must be remembered throughout the following discussion, because it limits the uses to which these data can be put: a major problem is that partial beam-filling by rain-clouds becomes increasingly significant above spatial resolutions of a few square kilometres.

It must also be noted that such sensors have flown only on low-Earth orbiters, and their swath widths have supported less than two imaging overpasses per day below about 55° latitude (SSM/I); in the case of the long-operating SMMR on Nimbus-7 power sharing between this instrument and the Coastal Zone Color Scanner (CZCS) unfortunately ensured that most areas on the globe were imaged only once every three days. Therefore, the past and present uses of the PMW for rainfall measurement *climatological* rather than *meteorological*, although there is now increasing interest in the synergistic use of VIS, IR and PMW data, the latter being specially useful for initialising models and/or calibrating (especially geostationary-based) VIS and/or IR rainfall algorithms.

Table 2.
Satellite microwave imaging instruments useful for rainfall monitoring

	ESMR-5	ESMR-6	SMMR	SSM/I
Spacecraft	Nimbus-5	Nimbus-6	Nimbus-6	DMSP 5D
Launch dates	10/72	06/75	10/78	06/87 & 12/90
Frequencies (GHz)	*Polarization* (V, H, X) *and footprint size* (km)			
6.6	–	–	VH 121 × 79	–
10.7	–	–	VH 74 × 49	–
18.0	–	–	VH 44 × 29	–
19.35	X 25 × 25	–	–	VH 69 × 43
21.0	–	–	VH 38 × 25	–
22.235	–	–	–	V 50 × 40
37.0	–	VH 20 × 45	VH 21 × 14	VH 37 × 28
85.5	–	–	–	VH 15 × 13
Imaging frequency	2 passes per day.	Incomplete global coverage.	1 midday and 1 midnight pass every six days (more at high latitudes).	At least 1 pass per day (more at high latitudes.
Primary aims of mission	Rain-rate, sea ice, and ice concentration.	Liquid water content of clouds, sea ice and open sea coverage. Surface composition and soil type with surface features and surface moisture.	Sea surface temperature, near sea surface winds. Water vapour, liquid water content and cloud droplet size. Rainfall rate.	Ocean wind speed and ice coverage, age and extent. Intensity of precipitation cloud water content, and land surface moisture.

Key: ESMR = Electrically Scanning Microwave Radiometer;
SMMR = Scanning Multispectral Microwave Radiometer;
SSM/I = Special Sensor Microwave Imager;
V = vertical polarization;
H = horizontal polarization;
X = non-polarized.

A further problem which has restricted the use of PMW data has been the lack of real-time access to such data excepting by a few fortunate organizations; ESMR and SMMR were R&D instruments whose data were subject to the restrictions normal to such instruments, and the SSM/I is a military instrument whose data are available in real-time to the US Military (Air Force and Navy) and NOAA, but to few other bodies.

Finally, and of even greater significance in the context of the present Workshop, it is recognized that PMW rainfall algorithms are thought to perform best over *sea* surfaces (where background conditions are optimal for such purposes), and least well over *land* surfaces (particularly where bare soil, ice and snow are present).

3.2. Principles and techniques

Figure 1 shows that a number of frequency-related atmospheric windows are found in the microwave region of the spectrum. Of particular value for rainfall monitoring are those just below 20 GHz, around 30–40 GHz, and around 90 GHz. Based on radiation measurements in these regions a growing range of different types of PMW rainfall measurement techniques have developed, as summarised in Table 3. All are underpinned

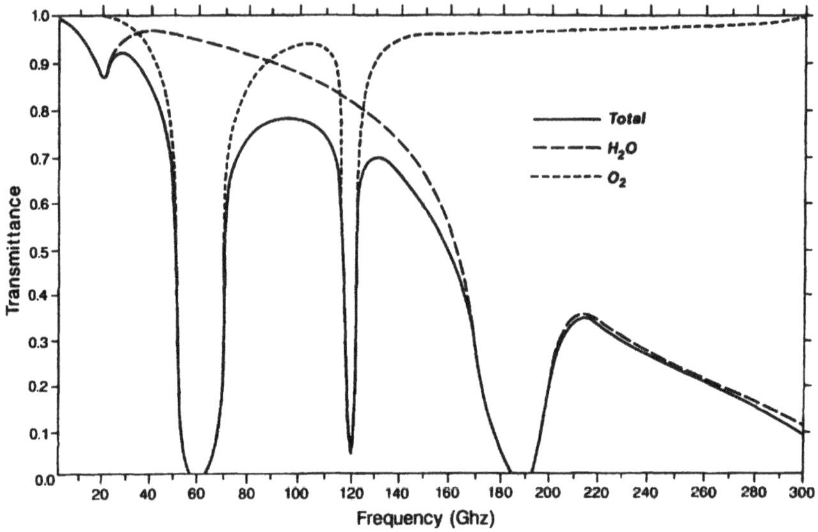

Figure 1. Transmittance of radiation at microwave frequencies indicating the strong absorption lines at around 60 GHz, at 120 GHz and around 185 GHz [14].

by the physical understanding that, below about 20 GHz, evidence of rain is provided by rainfall *absorption/emission* processes, through which areas of active rainfall over water surfaces (e.g. oceans and seas) are warmer than the background radiation; above about 60 GHz evidence of rainfall comes predominantly from *scattering* processes, whereby areas of heavy rainfall are much colder than their backgrounds; and between about 20–60 GHz *some combination of absorption/emission and scattering* processes dominate in different situations.

Two basic approaches have been used to detect rain over land using the passive microwave. The first of these uses cloud models and structures to generate inputs into radiative transfer models that then relate simulated energy brightness temperatures to simulated rain rates. These are then compared to real measurements to ascertain the real rain rates [15, 16]. Such algorithms are called '*inversion type algorithms*' as they generate what is thought to be occurring, then compare the results to the real world situation. They are characterized by high degrees of microphysical sophistication of the cloud models and structures [17], and some by large computational requirements especially where broad-area, real-time, estimates are involved. At the moment they are largely experimental, and unsuitable for applications other than research, but rapid progress is being made in this direction, as other papers in this Workshop confirm.

The second, and currently more widely used approach uses real observations and a theoretical basis to derive relationships (often through regression analysis) between combinations of microwave channels and rain rates. We may refer to these as '*empirical algorithms*'. This second type of approach is subdivisible into two categories. The first uses the emission/absorption properties, and the second the scattering properties, of rain.

Emission-based empirical techniques for rainfall detection have been used successfully over oceanic regions. Relationships have been developed from radiative transfer

Table 3.
A classification of satellite passive microwave (PM) rainfall monitoring techniques

Types	Basic relationships	Requirements	Special comments
1. Brightness temperature	$R = f(T_B)$	Single frequency, non-polarized brightness temperature data.	Low frequency PM technique for use over oceans only for medium/heavy rain rates; can identify very heavy rain over land at high frequencies.
2. Frequency	$R = f(T_B\text{V1} - T_B\text{V2})$ or $= f(T_B\text{H1} - T_B\text{H2})$	Brightness temperature data at two frequencies, and the same single.	Best for over-land use in middle latitude; evidences rain over water also, if rain.
3. Calibrated frequency	$R = a^*T_B\text{V1} - b^*T_B\text{V2} + c$ or $= a^*T_B\text{H1} - b^*T_B\text{H2} + c$	As in (2) above.	Collateral data necessary to evaluate a, b and c; use as in (2) above.
4. Polarization	$R = f\left(T_B\text{V} + \frac{\beta}{1-\beta}{}^*(T_B\text{V} - T_B\text{H})\right)$	Single frequency dual-polarized brightness temperature data.	Can be used over water and over land in tropics; local calibration of β desirable.
5. Brightness temperature/ polarization	$R = a(b^*T_B\text{H} - T_B\text{V} + c)$	Single frequency polarized brightness temperature data.	Useful over water for most rains rates.
6. Frequency/ polarization	Combination of (3) & (4) above	Brightness temperature data at two frequencies, and two polarisations for the higher frequency.	Developed for use over land and over middle latitudes. Applicable to all rain rates, including very light rain?
7. Physical (multi-channel)	$R = f(T_B\text{V1} + T_B\text{V2} + \ldots + T_B\text{Vn})$	Brightness temperature data for several frequencies.	Designed to provide components from each significant layer within the (convective) cloud.

Key: R : Rainfall;
β : Factor to equalize microwave responses from land and sea areas;
T_B : Microwave brightness temperatures;
V, V1, V2 : Vertically-polarized radiances;
H, H1 : Horizontally-polarized radiances;
a, b, c : Empirical weights.

models linking the brightness temperatures of selected channels or combination of channels and rain rates (e.g. [18, 19]). An example of an operationally used scheme of this kind is given in Fig. 2. Although the results are directly related to rain, these techniques are severely limited in their use over land surfaces where emissivities are higher and more variable. Emissivities over land are known to be influenced by many factors, including soil type, moisture content, vegetation, the presence of condensation, snow and ice (see [18, 20]). Therefore emission-based techniques are generally of less value over land areas than *scattering-based techniques*. In order to monitor rainfall over land with confidence the scattering properties of ice particles associated with precipitation are utilized; ice particles scatter energy out of the incident radiation (either upwelling radiation, or cold, downwelling, cosmic radiation) with negligible absorption or emission.

In defining the degree of depression of a brightness temperature due to scattering, the scattering cross-section is the single most important parameter. This is a function of the radiation wavelength. For the SSM/I, the 85 GHz channel is the most sensitive to scattering by ice particles in raining clouds.

One of the simplest and most widely used bases for overland passive microwave rainfall algorithms is that of the *frequency difference*, i.e. the difference between brightness temperatures at different microwave frequencies, for the same polarization (for example, V37–V85). This forms the basis of almost all regression-based (empirical) algorithms. The rationale behind its use was outlined by Grody [21]. Basically the use of one frequency is used to isolate the particular influence of scattering on another frequency. Research using Nimbus-7 SMMR data confirmed that it was possible to detect rainfall from the decreases in brightness temperature of the 37 GHz channel compared with brightness temperatures of lower frequencies [22, 23].

The arrival of the SSM/I sensor and the 85 GHz channel increased the scope for rainfall detection, many of the new approaches using radars to provide calibration and/or validation data. For example Olson *et al.* [24] and Hollinger [25] have used radar measurements and multichannel PMW observations to form the basis of multiple regression relationships to derive rainfall over land. Their published work has not provided details of the coefficients for different channels, but it is suggested the 85 GHz depression provides the bulk of the rain rate detection.

A rather different technique for rainfall detection over land was developed by Spencer *et al.* [26], involving '*polarization corrected temperatures*' (PCT). This has the advantage of use over land and ocean simultaneously. The PCT is based on the assumption that the surface emissivities for vertical and horizontal polarization are highly correlated and that a rain rate relationship for both can be derived from the radiative transfer equation.

This approach has been tested by Spencer *et al.* [26], and Barrett and Kidd [27] using FRONTIERS radar data for calibration/validation, and has given encouraging results. However, work by Kidd [20] suggests that the parameter Beta, which is used to unify responses from land and sea surfaces, is not constant as originally assumed, but varies both regionally and seasonally. Recent work at Bristol has used an automatic Beta calculation procedure to accommodate these variations for the production of instantaneous rain rates for AIP-1 project (see Bellerby and Barrett (in press), and below) but much wider testing of this approach needs to be undertaken before its validity at the global scale can be confirmed. Meanwhile, the SSM/I sensor has also stimulated increased efforts to detect and measure rainfall based on frequency differencing, for example the Bristol frequency difference algorithm, which was based on the (V37–V85) difference [20]. However various problems have become evident with such an approach, particularly the misclassification of some land surface such as sand, snow and ice as rain.

An attempt to correct for such anomalous signals was made by Fiore and Grody [28], who used a 'scattering index' based on the 19, 22 and 85 GHz channels to determine precipitation. Here the 19 and 22 GHz channels perform the same role as the 37 GHz channel in the Bristol (V37–V85) frequency difference algorithm: in this regard, Fiore and Grody's scattering index remains a frequency difference type algorithm (see Fig. 2). In this algorithm various combinations of channels are included in a decision tree form to identify 'false rain' signals from surfaces such as those covered by snow, glacial ice, old sea ice and desert sand. However, not all false rain signals are masked in

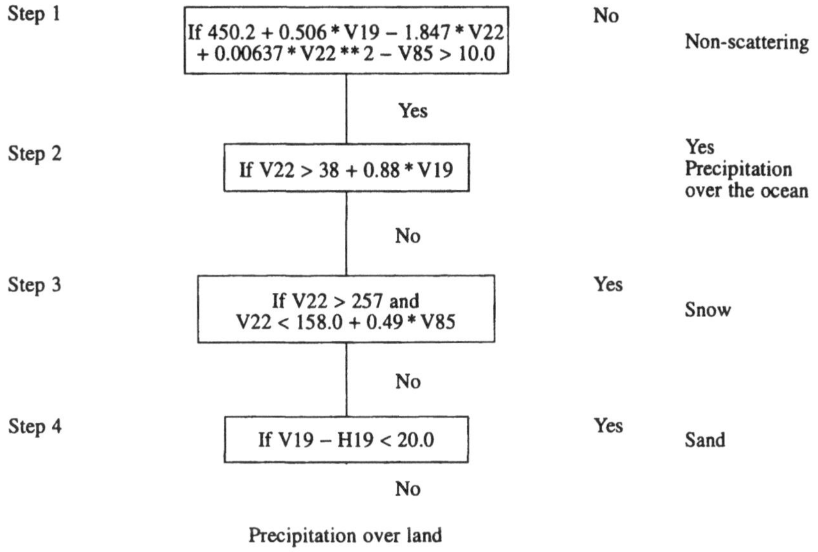

Figure 2. Flow diagram showing the steps of the Fiore/Grody algorithm.

this process as Grody [29] recognizes, e.g. through its registration of false rain over desert regions in winter. Barrett *et al.* [30] used a slightly different approach in tackling this problem. Rather than identifying sand *or* precipitation they choose to identify the problem region as sand, and then identify precipitation within these areas using different algorithms. This approach has the added advantage of extending the area of sand mask to fully cover problematic land types, as explained more fully below.

3.3. False rain signals

3.3.1. Desert sand. As explained earlier, overland regression type algorithms attempt to measure the degree of scattering produced by ice particles in precipitating clouds and thus estimate the associated rain rate. Where rain rates/totals are estimated by techniques like that of Barrett *et al.* [30] over regions where no rain is present this may be termed *'false rain scattering'* as the scattering-like signals are being produced in some way not yet fully understood, by land surfaces rather than precipitation related ice particles.

The occurrence of 'false rain scattering' over deserts at any one time is relatively low compared with that produced by precipitation, but because of its near permanent nature it gives the impression of large rain totals over monthly periods. This low intensity 'false rain scattering' arises from a combination of factors: first, as the soil moisture decreases the brightness temperatures of the lower frequency channels (used to simulate the non-scattering brightness temperatures at V85 GHz in frequency difference type algorithms) increase more than those at V85 GHz. This was confirmed theoretically by Strawbridge [31] using a non-scattering model to assess the effect of soil moisture on brightness temperature at various microwave frequencies (Fig. 3). This relationship

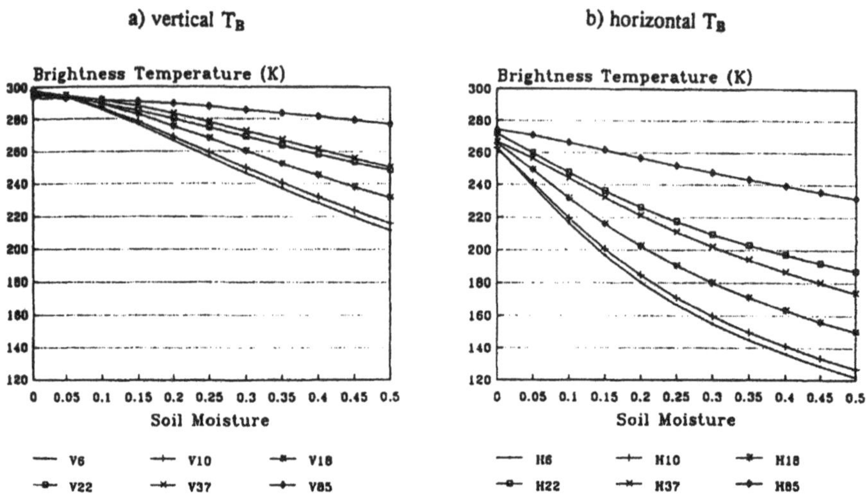

Figure 3. Simulated dependence of: (a) vertical brightness temperature and (b) horizontal brightness temperature at 6.6, 10.7, 18.6, 22, 37 and 85 GHz frequencies upon volumetric soil moisture contents. (From [32].)

between brightness temperatures and soil moisture variations is not observed over vegetated surfaces, for which the frequency difference algorithm was initially developed. Second, in certain cases of extreme dryness and surface roughness the uppermost surface layers scatter radiation at 85 GHz, thereby depressing the brightness temperatures at this frequency.

To identify regions where soil moisture and roughness effects might combine to produce 'false rain scattering' the 19 GHz polarization difference is utilized. Bare soil is characterised by a high polarization difference at this frequency compared with vegetation, the magnitude of this difference increasing with soil moisture and varying with surface roughness. Fiore and Grody [28] first used the 19 GHz polarization difference with a threshold of 20 K to identify 'false rain scattering' areas in dry regions.

The rationale behind such an approach is that, when rain occurs, the polarization difference falls, due to volume scattering properties of rain, leaving the area unmasked. A 'scattering index' or frequency difference can then be applied to ascertain the rain rate. The problems with this approach are as follows. First, although the polarization difference has dropped because of the presence of rain, and the sand mask has been 'deactivated', the surface still gives a 'false rain scattering' signal which combines with precipitation related scattering to give falsely high rates. Second, the frequency difference algorithm threshold of 20 K has been found to be too high in some cases, leaving some areas of 'false rain scattering' unmasked.

The 19 GHz polarization difference was also used by Barrett *et al.* [30] but with a lower threshold of 9 K. This increases the area of potential 'false rain scattering' detection, but necessitates the use of another precipitation retrieval algorithm within the problem area. This is because over some dry regions the polarization difference at 19 GHz can reach as high as 35 K. Such a high polarization reduces the extent of rain detection in such locations as few rain systems are strong enough to reduce the polarization by the 26 K required. The algorithm chosen by Barrett *et al.* [30] for such a supplementary role was the horizontal frequency difference (H37 – H85). An alternative approach would be to alter the threshold of the vertical frequency difference algorithm. However, various cases over the Sahara and Simpson (Australia) Deserts have shown that the range of values from those produced at the surface to those produced by higher rain rates is larger in the horizontal frequency difference than in the vertical. Potentially this allows for clarity in the distinction between rain rates and between rain rates and the surface.

Previously the vertical channels were those most used in frequency differencing because of alignment problems between the different frequency channels resulting from different footprint sizes [22]: the smaller alignment variations in the vertical polarization channels compared to the horizontal channels results in reduced residual differences. The noise from this source is thought to affect low rain rate detection. However, over desert regions rainfall appears to be relatively intense, reducing the significance of the alignment problem, in that low rain rates are comparatively less frequent than over, say, frontal regions.

An example of the use of the horizontal frequency difference is shown in Fig. 4. This shows the positive horizontal and vertical frequency differences over Central and Northern Africa for Julian Day 269, 1987, descending pass. The vertical frequency is depicted in mid-grey and the horizontal in light-grey. Where the differences coincide is coloured white. The extensive mid-grey tone over the Sahara desert illustrates the false scattering problem. This is severely reduced by the use of the horizontal frequency difference. The box in the picture highlights the area around two of the very few

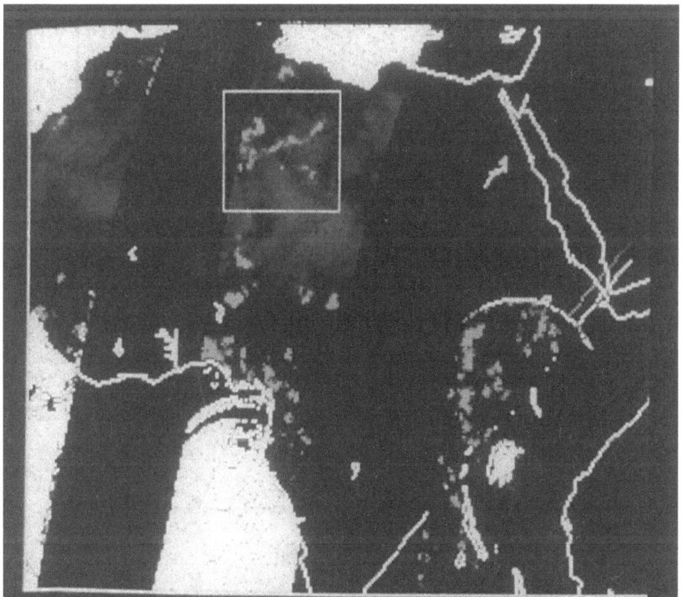

Figure 4. The frequency differences of (37 GHz – 85 GHz) the vertically polarized channels (mid-grey) and the horizontally polarized channels (light-grey, and white when the differences are co-located) over Central and Northern Africa on JD 269, 1987, descending pass. The box highlights the area around the meteorological stations of Tamanrassat and Gat.

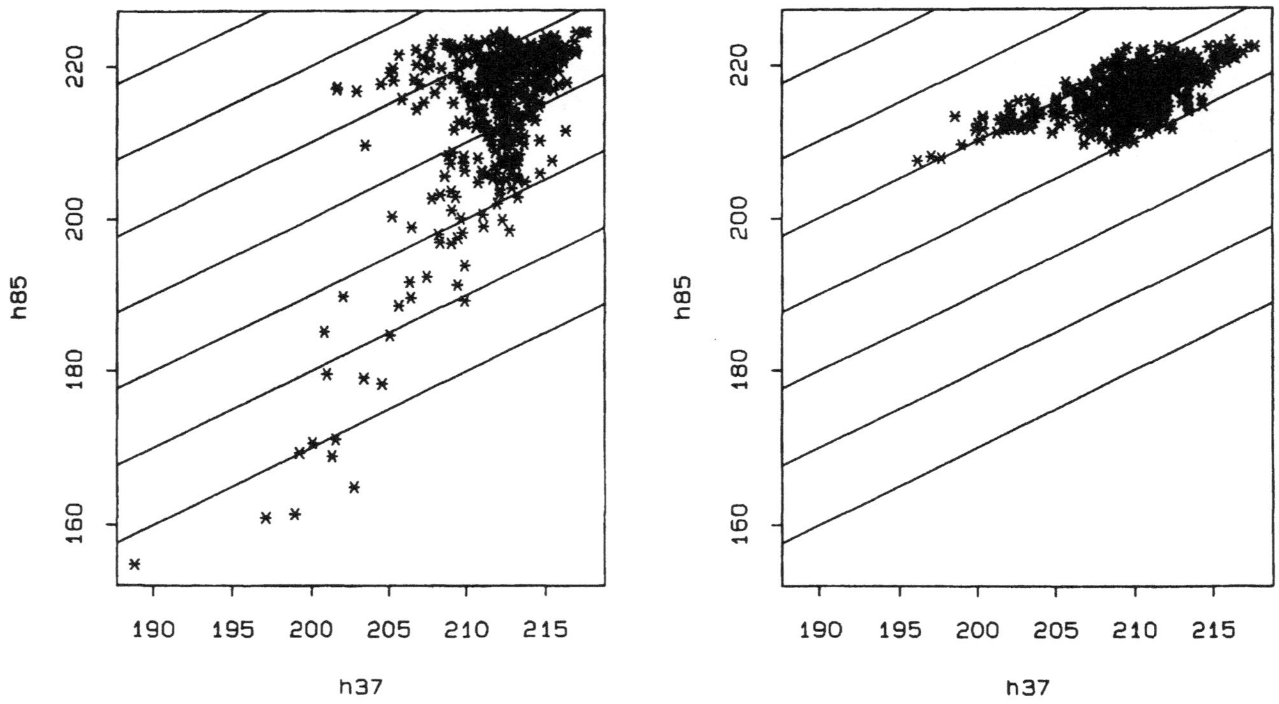

Figure 5. Plots of the brightness temperatures of the horizontally polarized channels at 85 GHz and 37 GHz, of and around the meteorological station of Tamanrassat, on JD 264, descending pass, (left hand side) and JD 260, 1987, descending pass (right hand side). The lines running diagonally across the plots are lines of equal frequency difference.

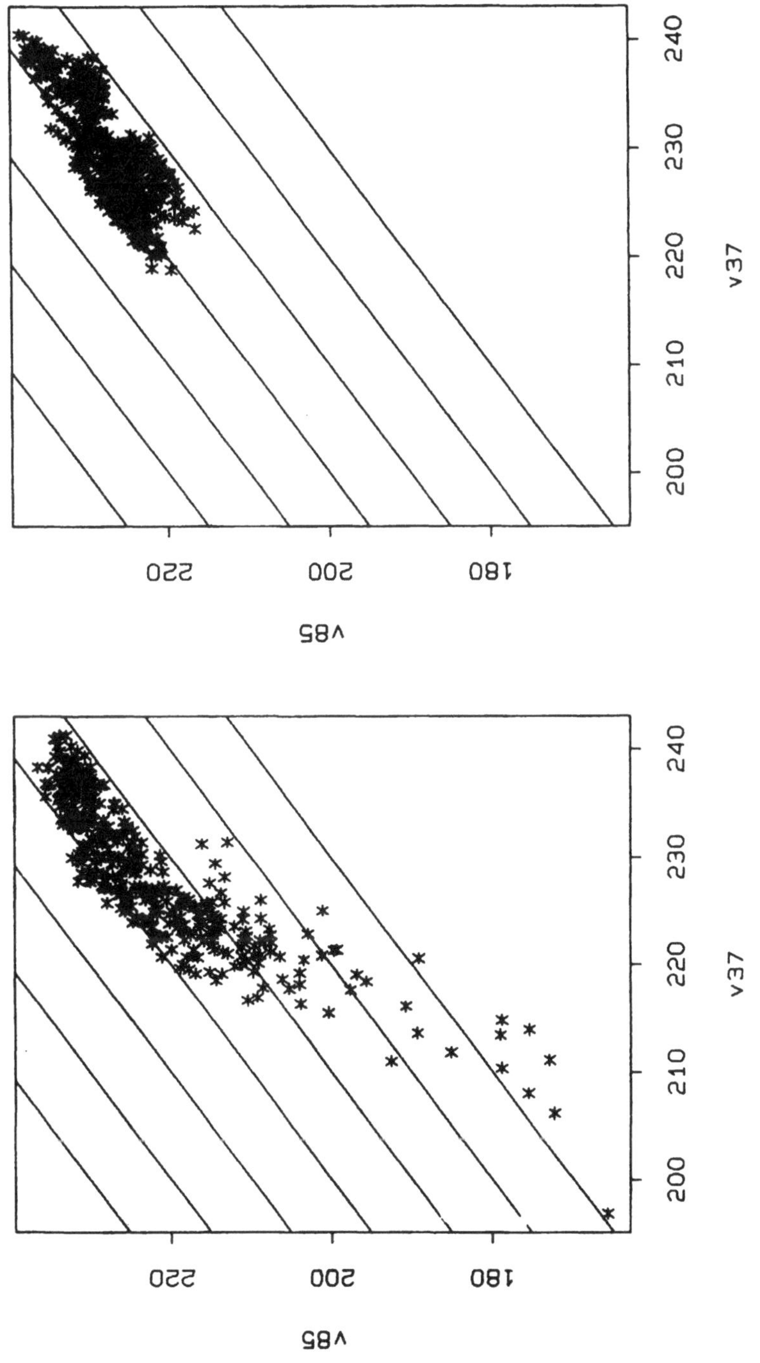

Figure 6. The same as Fig. 5 but for the vertically polarized channels at 85 GHz and 37 GHz.

meteorological stations in the region, namely Tamanrassat and Gat. Surface charts showed these stations to be recording precipitation near the time of the SSM/I overpass. Figures 5 and 6 show plots of the T_B's of the horizontal and vertical channels respectively at 85 GHz and 37 GHz for an area around the meteorological station of Tamanrassat on Day 264 (left hand side) and 260 (right hand side). Taking an area around the station means that the left hand plot shows both non-raining and raining points. On Julian Day 260 no rain was recorded at Tamanrassat and this plot is taken to represent a purely non-raining situation. By subtracting the region of non-raining points on day 260 plot from that of Julian Day 264 it is possible to estimate the raining portion of the plot and the rain/no rain threshold in terms of frequency difference. On the horizontal frequency plots this appears to be approximately zero, while on the vertical ones about 8 K. The highest frequency difference values are approximately the same for both the vertical and horizontal channels, giving a larger range of values for rain rate detection in the horizontal. Elsewhere on the globe a threshold of zero may be used with the vertical frequency difference to detect rain.

Figure 7 shows global monthly rain totals from August 1987 using the vertical frequency difference over land except regions where the 19 GHz polarization difference is above 9, when the horizontal frequency difference is used. The 37 GHz polarization difference is used over the ocean. Figure 8 shows global continental rainfall for August 1987 from raingauge data (GPCC): the degree of similarity with Fig. 7 is encouraging.

3.3.2. Frozen ground. Another misclassification evident from global monthly rainfall studies is that produced by *frozen ground.* For example, frequency difference-derived monthly rainfall totals for October and November 1987 over the northwestern United States have been up to about 200 mm. These compare with equivalent conventional rain gauge estimates of up to 50 mm (GPCC data). The 'false rain scattering' signal responsible for this misinterpretation is thought to be due to ice particles in the frozen ground which depress the 85 GHz channel T_B. As with desert regions, the magnitude of this effect at any given time is small in comparison with precipitation scattering but its near constant nature results in high monthly rainfall total estimates.

Recent work in the RSU, University of Bristol, has addressed this problem through the use of monthly background coefficients derived from the 37 GHz and 85 GHz brightness temperatures. As mentioned above, this group has used the 37 GHz T_B in the frequency difference to simulate the non-scattering brightness temperature at 85 GHz. In the case of frozen ground where the 85 GHz is permanently depressed a coefficient can be applied to fine-tune the accuracy of the 37 GHz estimate. In effect this coefficient attempts to take into account differences in emisivities of the different frequencies.

In order to accommodate the spatial and temporal emissivity changes on a global scale the monthly mode ratio of the 85 to the 37 GHz vertical T_Bs was calculated for every $0.5° \times 0.5°$ pixel over land: the mode values indicate the ratio one would expect for non-precipitation events in individual pixels. These values can then be applied to the frequency algorithm as a multiple of the 37 GHz T_B. The adjusted 37 GHz T_B then becomes a more realistic estimator of the non-scattering 85 GHz T_B. However, the mode ratio is not coincident with the rain/no rain boundary and another 'threshold' coefficient is applied, relating the mode value to the rain/no rain boundary. This can be taken as a global constant. However, evidence from the PIP-1 global study suggests

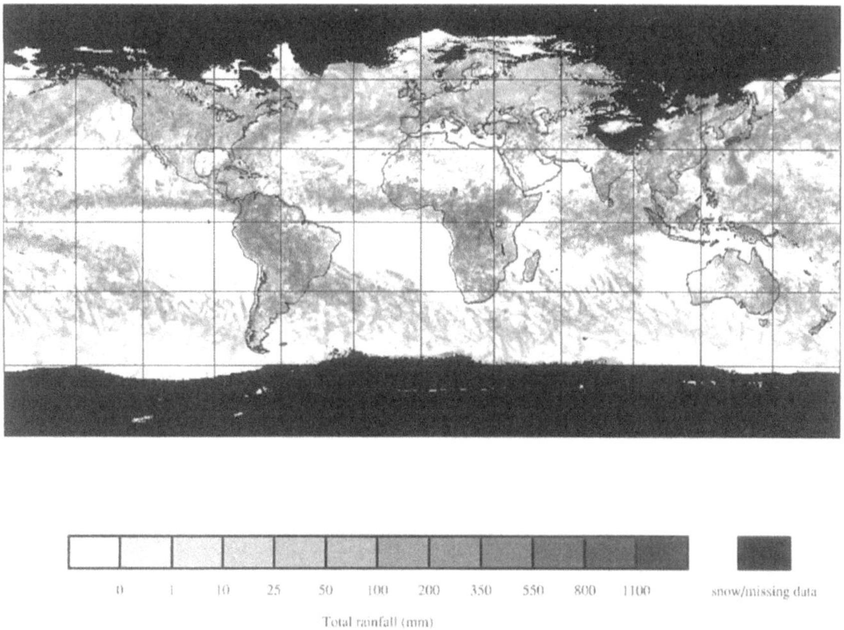

Figure 7. Global rainfall estimates for August, 1987, obtained by a multiple-component PMW technique treating land, sea, desert, and ice and snow areas differently. (Courtesy C. Kidd and D. Kniveton, RSU, University of Bristol.)

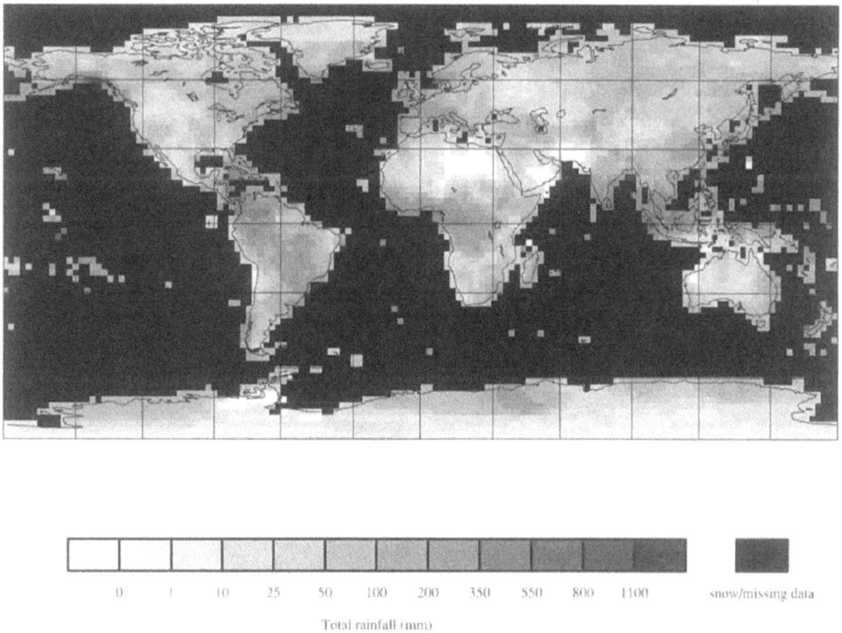

Figure 8. Global continental rainfall mapped for August 1987 from raingauge data. (Courtesty GPCC.)

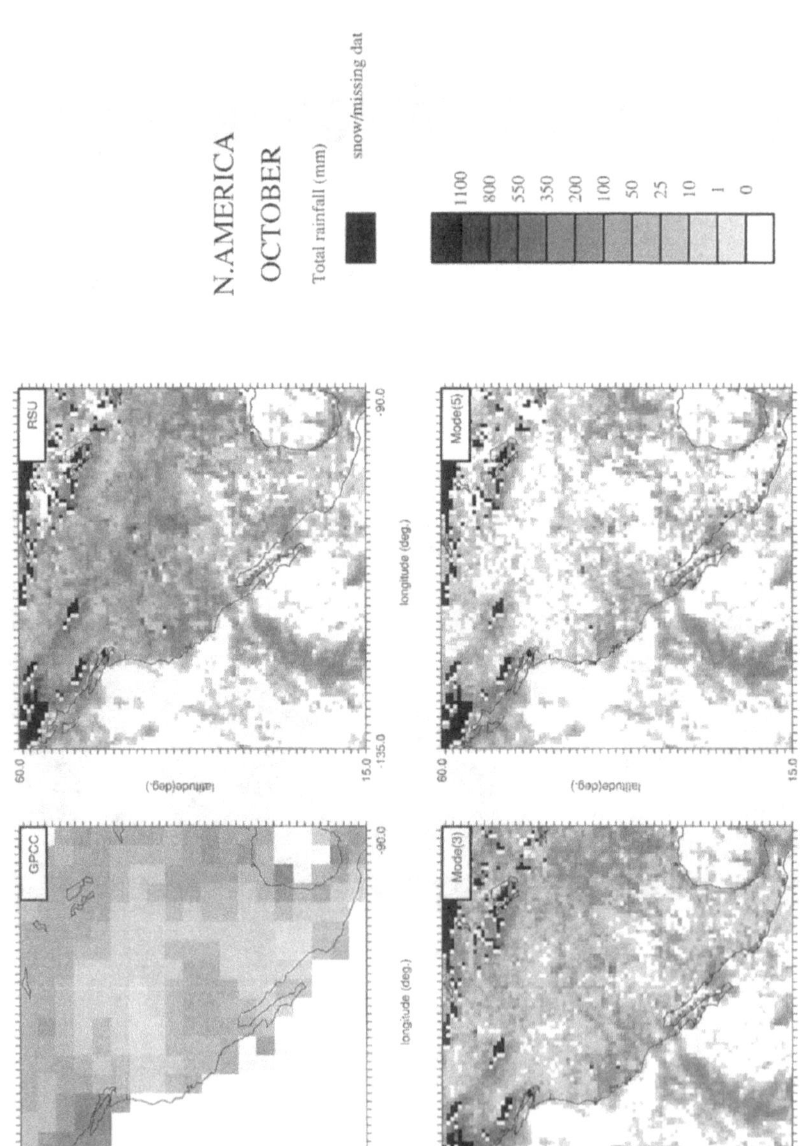

Figure 9. Estimated monthly rainfall totals over the west of the USA for October 1987.

it may well vary a little regionally, although with far less influence than mode derived coefficient.

Figure 9 shows results from the GPCC (rain gauge interpolation) data set over the western USA at 2.5° resolution, the vertical frequency difference rainfall product at 0.5° resolution (named 'RSU'), the mode approach, detailed, above taking the threshold coefficient to be equal to 3 at 0.5° resolution, and the mode approach as above taking the threshold coefficient to be equal to 5, also at 0.5° resolution. A snow mask has been used to eliminate snow and ice areas on the algorithm results. The region of frozen land induced 'false rain scattering' can be seen in the vertical frequency difference algorithm estimates, whereas it is largely absent in the mode approaches, without any loss of rain total magnitude in non-frozen ground pixels to the east and south.

An interesting feature picked out by the above approach is the apparent medium-intensity rainfall area lying across the otherwise relatively dry regions of Texas and New Mexico. This may be due to misclassification by the algorithm, but more probably to inadequacies in the GPCC data set due to the relatively few reporting SYNOP and CLIMAT stations in these areas. The high coastal rainfall totals suggested by the algorithm are due to a recently confirmed mislocation problem between high and low resolution channels on ascending passe in the row SSM/I data (W. Olson, personal communication).

3.3.3. Snow. Figure 9 also illustrates the use of Fiore and Grody's [28] snow mask. Unlike over desert regions, there is at present no algorithm able to retrieve precipitation over snow and ice due to similar scattering signals from ice in the atmosphere and on the ground. Research into this problem has recently begun, and suggests that *time differencing* may afford hope at least in some cases [32].

4. THE WAY AHEAD

4.1. Development of passive microwave techniques

4.1.1. Multiple-component algorithms. A significant consequence of the physical principles of microwave satellite rainfall monitoring outlined earlier is that at present different rainfall estimation procedures are necessary to provide the best results over land and sea. Furthermore, we have seen that frozen and bare soil (especially desert) surfaces also presently require individual treatment. Thus, for global rainfall measurement, multiple-component techniques yield best rain area results (see Figs 7 and 8). It is not yet certain whether these will further benefit from rain rate calibration schemes varied by climatic region and/or by weather system type and structure. The WetNet Precipitation Working Group is exploring these questions.

4.1.2. Physically-based algorithms for individual storm evaluation. Increasing attention is being paid to the development of PMW techniques based more strongly on the physics of the atmosphere than on empirical relationships between PMW brightness temperatures and rain rates. In this work *cloud radiation* models are being used (see Fig. 10) in attempts to improve rain rate estimation from different categories of clouds [16], and

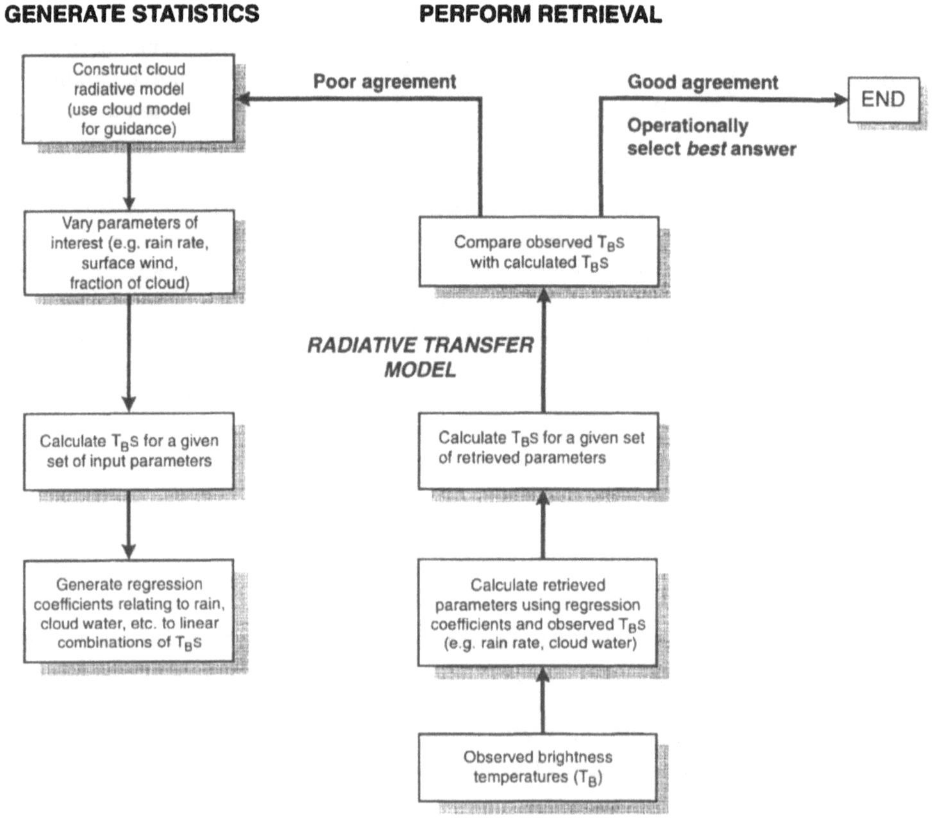

Figure 10. Flow-chart for a 'physical' algorithm for rainfall estimation from PMW SSM/I data. (Courtesy, C. Kummerow.)

to explore relationships between PMW radiation emerging from cloud tops and the vertically-distributed mixtures of liquid and frozen hydrometeors within the clouds (e.g. cloud drop, rain drops, ice crystals, aggregates and graupel) [33]. These multichannel investigations may be expected to underpin a future generation of improved global rainfall measurement operations.

4.2. Merged data techniques

Since VIS, IR and PMW data are all known to provide different types and degrees of information on rainfall systems observed from satellite altitudes, it was suggested in 1988 that better satellite rainfall algorithm performance should be achievable by combining the rain/no-rain and rain intensity evaluation abilities of data from all these different wavebands in single, *unified or merged ('trispectral') techniques* [34]. Significant progress in this direction is now being achieved by combining geostationary IR and polar-orbiting PMW data so as to capitalise simultaneously on the relatively high ability of PMW algorithms to identify instantaneous rain areas and evaluate rain rates, and to capitalise on the utility of the much more frequent geostationary IR data to provide necessary evidence of cloud growth and movement even when these are rapid

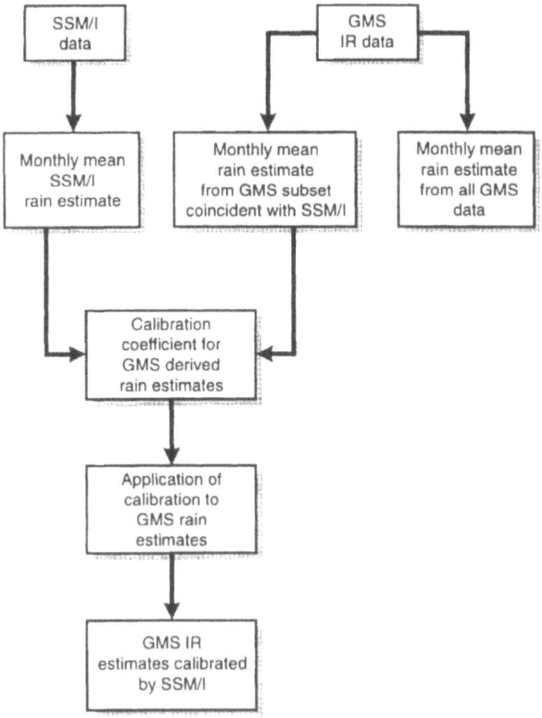

Figure 11. Schematic diagram describing method for combining polar-orbiting geosynchronous IR estimates of monthly total rainfall. (Courtesy, R. F. Adler, A. J. Negri and I. M. Hakkarinen.)

(see Fig. 11). More examples of this will be provided by other papers in this Workshop.

Under the auspices of the Global Precipitation Climatology Project (GPCP) attention is now being paid to strategies for the best possible combination(s) of data not only from different satellite sources but also ground station (initially raingauge) sources. Meanwhile, regional rainfall monitoring schemes are increasingly seeking to combine satellite data, climate records, real-time raingauge reports, radar data, and streamgauge (runoff) data – or selections of these – for best possible measurements of rainfall. Such trends are sure to accelerate in the intensifying search for 'seamless' rainfall products of the highest possible quality over large areas and different types of underlying surfaces.

4.3. Satellite rainfall algorithm intercomparison projects

As the number of research workers in the satellite rainfall monitoring field has grown, and interest in satellite-improved rainfall products has widened amongst the end-user community up to and including global modellers, so the need for objective identification of the best technique(s) to use for different purposes under different circumstances has also grown. An important new development to satisfy this need is the organisation of typically *international intercomparison projects* to compare results of different algorithms generated from the same input data, and wherever possible to 'validate' the results using independent data sets. Following a prematurely-terminated attempt to undertake such an intercomparison under the auspices of the AGRISTARS (Agriculture and Resource Inventory Surveys through Aerospace Remote Sensing) Program in the mid-1980s, effectively the first such project was carried out by the Food and Agriculture Organisation (FAO) of the United Nations in 1987/88, following a proposal by the present first author; more recently, such projects, or plans for them, have become more popular. The more important of these include:

- *FAO's Sahel Study*. This evaluated the products from nine Meteosat IR objective techniques plus FAO's own 'ARTEMIS' Meteosat IR interactive technique against a large set of surface raingauge data. Results were evaluated for daily, 10 daily and monthly estimates. The conclusions were that 'none of the (objective) methods gave overall better results than the others, but there were distinct differences in performance' in respect of different criteria; however, it was also concluded that 'the use of Meteosat data can contribute significantly to the effectiveness of rainfall monitoring over the Western Sahel' [35].
- *The First GPCP Algorithm Intercomparison Programme (AIP-1)*. The WMO/ICSU World Climate Research Programme (WCRP) has established a Global Precipitation Climatology Project (GPCP) initially to provide climate researchers with global precipitation statistics for the period 1986–95. One expression of this has been the selection and implementation of a simple geostationary-based IR technique for quasi-global rainfall monitoring on a monthly basis for 2.5° lat/long squares, namely the GOES Precipitation Index (GPI); another has been the establishment of a Global Precipitation Climatology Centre (GPCC) in Offenbach, Germany to develop the best possible rainfall climate data bases from gauge records. A third important activity of the GPCP has been the inauguration of an Algorithm Intercomparison Programme (AIP) 'to evaluate alternative (satellite rainfall) estimation techniques, and determine if improvements can be made upon the present algorithms and to evaluate new precipitation estimation methods.' The first intercomparison was carried out over Japan and the surrounding oceanic regions for two periods in 1989, viz. 1–30 June, and 15 July–15 August, based on Japanese Geostationary Meteorological Satellite (GMS) VIS and IR imagery, SSM/I imagery, radar and raingauge data, and numerical forecast outputs from the Japan Meteorological Agency. Results have been presented in an '*Atlas of Products*' published in August 1991 [36]. Considerable differences were revealed between the performances of different algorithms (see Fig. 12) and, in the case of SSM/I-based algorithms, the types of areas over which they were applicable.
- *The Second GPCP Algorithm Intercomparison Programme (AIP-2)*. Building on the success of AIP-1, a similar activity is under way over parts of northwest Europe, covering a mid-latitude area during the local cold season (R. Allam, UK Meteorological Office, personal communication). The period selected for AIP-2 is from 1 February 1991 to 9 April 1991, based on a broadly similar data set to that available over Japan in AIP-1. Precipitation estimates from Meteosat imagery are requested from participants on a daily and monthly basis, plus hourly estimates from the same data base for special areas. Instantaneous rain areas and rain rates were requested for selected cases using AVHRR data (34 cases) and SSM/I (27 cases). Estimates are prepared by the participants for delivery at the end of 1992. These results processed and intercompared by

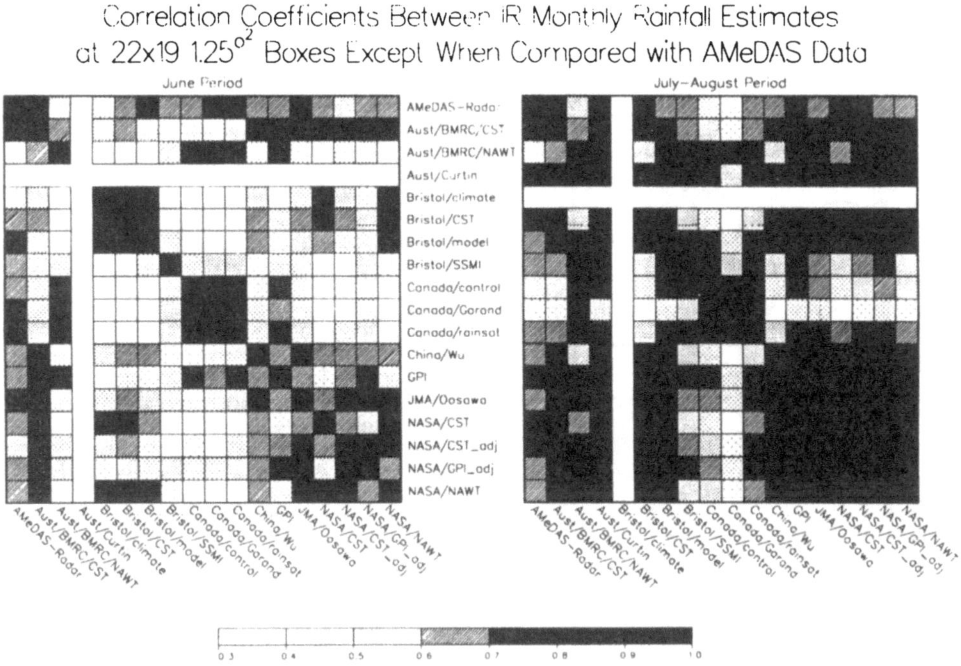

Figure 12. AIP-1 matrix designed to show major similarities and dissimilarities between the performances of different IR algorithms. (Courtesy, P. Arkin and J. Janowiak.)

the UK Meteorological Office and a final Workshop was held in Shinfield Park, Reading, UK from 23–26 August 1993.

- *The Third GPCP Algorithm Intercomparison Programme (AIP-3).* Results are now being generalized for AIP-3, over an area in the western equatorial Pacific, within the general area of the TOGA-COARE study region. Data have been specially obtained from two ship-board radars for three periods each 20 days long to help validate the satellite estimates for an area which is mainly oceanic, and much more poorly provided with raingauges than either of the two areas investigated in the previous two AIP Studies. A Results Discussion Workshop is being planned for Melbourne, Australia in 1995.
- *The First WetNet Precipitation Intercomparison Project (PIP-1).* As part of the programme of the WetNet Precipitation Working Group, the largest of the six WetNet Working Groups, two passive microwave satellite rainfall algorithm intercomparison projects have been planned, the first of which (PIP-1) was completed in 1993. The principal objectives of PIP-1 have been 'to demonstrate, test, intercompare, and where possible, validate techniques for global monthly rainfall monitoring by SSM/I' and 'by validation with climatology, to establish which SSM/I technique(s) function best over land, oceans, and other surfaces'. The test data set included SSM/I digital data on standard WetNet optical discs for August through November 1987. Deliverables requested from participating laboratories included global arrays of estimated total monthly rainfall for 0.5° grid squares, histograms of estimated monthly rainfall totals for land, oceans, and

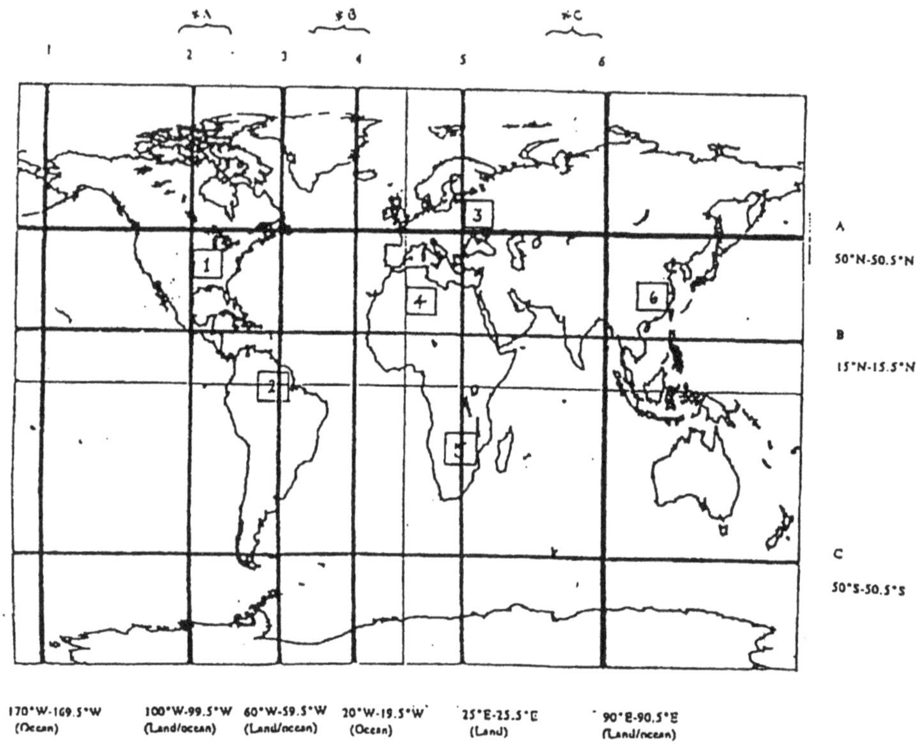

Figure 13. WetNet PIP-1 map showing latitude and longitude lines selected for global cross-sectional studies and algorithm intercomparisons, and boxes within which algorithm validations are being undertaken.

land plus oceans, and statistics for a number of selected latitude and longitude bands. NASA's Marshall Space Flight Center, after receipt of results data sets from participants, extracted sub-sets for special analysis drawn from a number of window regions over the continents (validated by GPCC raingauge data) and the tropical Pacific Ocean (validated by a Pacific Atoll raingauge data set) as shown in Fig. 13. A Results Discussion Workshop was held in Bristol, UK in May 1993.

- *The Second WetNet Precipitation Intercomparison Project (PIP-2).* In order to investigate the design and performance of passive microwave algorithms in different, but complementary, contexts to those of PIP-1, a second WetNet Precipitation Intercomparison Project is being undartake during 1994. A key aim of this Project is 'to identify the common physical and mathematical retrieval concepts embodied in the retrieval algorithms which either strengthen or weaken the ability to measure precipitation from space at high time and space resolutions' and also 'to help shed light on the question of how much uncertainty in estimates of time-accumulated rainfall can be attributed to the performance of any algorithm in retrieving instantaneous rain rates, and how much should be attributed to limitations on satellite sampling of the rain events.' Thus, the emphasis in PIP-2 is on analyses of individual rain events, rather than on global rainfall monitoring. Some 30 to 40 oceanic and continental rainfall case studies are being studied, for which high quality radar (both ground and aircraft) and high density raingauge validation data are being made available for as many cases as possible. The intention is that the cases should embody the varied rainfall types over continents and oceans with enough latitudinal, topographical and air mass diversity so that the final results will have significance at the global scale, and help the WetNet Precipitation Working Group advance to its ultimate Stage-I goal, namely to develop a passive microwave algorithm or algorithms which represent the best possible at this stage in the development of the science.

The importance of this wave of intercomparison projects is underlined by the present-day search for a PMW rainfall algorithm which merits application to the new SSM/I 'Pathfinder' global data/product sets now under development (C. Neale, personal communication). Because the Pathfinder data sets will be widely-distributed research community resources, any algorithm-derived, parameter-related products included in them will assume at least something of the mantle of community standards. Whilst it is possible that a suitable algorithm may be available already for global overland rainfall measurement based on satellite data, it seems more likely that community experience, if properly coordinated and capitalised upon, could lead to a more successful algorithm than any currently available. It is with this kind of need in mind that NASA's WetNet Project was established, and one of the specific objectives of its Precipitation Working Group is to attempt 'to formulate and test a community passive microwave rainfall algorithm' before Stage-I of WetNet ends in mid-1995.

5. A SCENARIO FOR FUTURE GLOBAL OVERLAND RAINFALL MEASUREMENT INVOLVING PASSIVE MICROWAVE SATELLITE ALGORITHMS

In conclusion, four things are clear:

- There is an urgent need for improved rainfall monitoring at all scales from the local to the global especially over land, a need which is growing because some surface data sources are in decline whilst global water supply demands are escalating.
- Satellites represent a still relatively under-explored and generally under-exploited source of data of great value in rainfall measurement, but, despite considerable effort, satellite rainfall monitoring techniques are still insufficiently mature for many quantitative applications.
- Much is being learned about the strengths and limitations of satellite rainfall monitoring techniques from the newly-instituted intercomparison projects, which should lead to improved algorithms in the future.
- Greater emphasis must be placed on the calibration of passive microwave algorithms intended for fully operational use.

Polar-orbiting SSM/I passive microwave sensors, although very well-suited to relatively broad-area, long-term rain area identification and rain-rate evaluation are, as indicated earlier, poorly suited to measurement of rain in areas of rapid rain-cloud development and/or movement, and/or where greater detail is required on the ground. Therefore, a broad scenario for global rainfall measurement for hydrological purposes will require:

- *Raingauge* and *radar* data for temporally and spatially detailed local/regional area studies, and for calibration and verification of satellite-derived rainfall estimates, as well as
- *Various types of satellite data* to provide relatively uniform wide-area coverage of rain events, particularly over areas from which raingauge and radar data are deficient.

The satellite rainfall algorithms which will be required as a consequence of all the above considerations include:

- *Geostationary satellite-based cold cloud duration IR techniques* for the general estimation of rainfall from convectional cloud areas, especially in tropical air masses and air streams, ideally being objective techniques designed to treat different classes of clouds differently, and to undertake 'rolling' internal calibration using surface weather or river flow data (where available), and to provide outputs on a pixel-by-pixel basis.
- *Geostationary satellite-based cloud growth techniques* for the estimation, as and when necessary, of rainfall from rain clouds likely to generate relatively high-intensity rain events, and, perhaps, associated flash floods. These techniques may retain a significant interactive element, although the advance of 'expert systems' may help reduce the present high reliance on trained analysts.
- *Polar-orbiting satellite-based instantaneous rain area/rain rate for estimation techniques* in which passive microwave satellite data will replace VIS/IR satellite data increasingly as the former become more widely accessible. Different algorithms are likely to be required over land and sea surfaces. The measurement of precipitation over areas of ice and snow will continue to be particularly problematic.
- *Merged techniques.* Where fully-integrated global rainfall products are required, results from two or more of the above may be assembled, or rainfall products generated from mergers of the input data sets prior to processing.

This scenario will be capable of supporting outputs in a variety of formats as required by the end-user. Whilst it may not satisfy all end-user requirements, e.g. initially in respect of tight confidence limits, it should provide data bases far superior to those available in the pre-satellite era. Thus, rainfall measurement will benefit increasingly from the use of techniques involving satellite data from a variety of different sources, facilitating much-needed improvements in the management of water from rainfall – this rare yet vital global resource.

REFERENCES

1. E. C. Barrett and D. W. Martin. *The Use of Satellite Data in Rainfall Monitoring*. Academic Press, London (1981).
2. E. C. Barrett. Diagnostic, historic and predictive analyses of rainfall using passive microwave data. *Palaeogeography, Palaeoclimatology and Palaeoecology*, **90**, 99–106 (1991).
3. J. Simpson (Ed.). *TRMM: A Satellite Mission to Measure Tropical Rainfall*. NASA (1988).
4. J. S. Theon and N. Fugono (Eds). *Tropical Rainfall Measurements*. Deepak, Hampton, VA (1988).
5. S. L. Callis and W. L. Decker. *Report on the International Workshop on Satellite Techniques for Estimation of Precipitation*. NOAA Cooperative Institute for Applied Meteorology, Univ. of Columbia-Missouri (1989).

6. E. C. Barrett (Ed.). Satellite remote sensing for operational hydrology. *Remote Sens. Reviews*, **4** (2), 256 pp. (1990).
7. G. D'Souza, E. C. Barrett and C. H. Power. Satellite rainfall estimation techniques based on visible and infrared imagery. *Remote Sens. Reviews*, **4** (2), 379–414 (1990).
8. R. F. Adler and A. J. Negri. A satellite infrared technique to estimate tropical convective and stratiform rainfall. *J. Appl. Meteor.*, **27**, 30–51 (1988).
9. E. Liljas. Multispectral classification of cloud, fog and haze. In: *Remote Sensing Applications in Meteorology and Climatology*, R. A. Vaughan (Ed.), pp. 301–319, Reidel, Dordrecht (1987).
10. E. C. Barrett and M. J. Beaumont. Satellite rainfall monitoring for agrometeorology: operational problems, practices and prospects. *Adv. Remote Sensing*, EARSeL (in press).
11. R. A. Scofield. The NESDIS operational convective precipitation technique. *Mont. Weat. Rev.*, **115**, 1773–1792 (1987).
12. NASA. *WetNet User's Manual*. Marshall Space Flight Center, AL (1990).
13. J. P. Hollinger. Passive microwave measurements of the surface. *J. Geophys. Res.*, **75**, 5200–5213 (1970).
14. Kuo-Nan Liou. *An Introduction to Atmospheric Radiation*. Academic Press, New York (1980).
15. R. F. Adler, W. K. Tao, N. Prasad, J. Simpson, and H. Y. Yeh. Satellite microwave rainfall simulations with a 3-D dynamical cloud model. In: *IIIrd Conference on Satellite Meteorology and Oceanography*, Atlanta, pp. 292–295 (1988).
16. C. D. Kummerow. Rain and rain system structure from SSM/I measurements. In: *Preprint Volume, American Meteorological Society Fifth Conference on Satellite Meteorology and Climatology*, London, pp. 241–246 (1990).
17. E. A. Smith, A. Mugnai, H. J. Cooper, G. J. Torpoli and X. Xiang. Foundations for statistical-physical precipitation retrieval from passive microwave satellite measurements. Part I: Brightness temperature properties of a time-dependent cloud radiation model. *J. Appl. Meteorol.*, **31**, 506–531 (1992).
18. N. C. Grody and R. R. Ferraro. *A Comparison of Passive Microwave Rainfall Retrieval Methods*. Atlanta, AMS, pp. 60–65 (1992).
19. J. T. Wilheit and A. T. C. Chang. Retrieval of monthly rainfall indices from microwave radiometric measurements using probability distribution functions. *J. Atmosph. Oceanic Technology*, 118–136 (1990).
20. C. Kidd. *Passive Microwave Monitoring of Rainfall over Land*. Unpublished PhD Thesis, University of Bristol (1988).
21. N. C. Grody. Precipitation monitoring over land from satellite by microwave radiometry. In: *International Geoscience and Remote Sensing Symposium*, (IGARSS '84), Strasbourg, France, ESA SP-215, pp. 417–423 (1984).
22. R. J. Ferraro, N. C. Grody and J. A. Kogut. Classification of geophysical parameters using passive microwave satellite measurements. *IEEE Trans. Geosci. Remote Sensing*, **GE-24** (6), 1008–1013 (1986).
23. R. W. Spencer. Satellite passive microwave rain rate measurement over croplands during spring, summer and fall. *J. Climate and Appl. Meteor.*, **23**, 1553–1562 (1984).
24. W. S. Olson, F. J. Lafontaine, W. I. Smith, B. A. Roth and C. D. Kummerow. Multispectral retrieval of rainfall rates using the Special Sensor Microwave/Imager (SSM/I). In: *IVth Conference on Satellite Meteorology and Oceanography*, pp. 263–266 (1989).
25. J. P. Hollinger. *DMSP Special Sensor Microwave/Imager Calibration/Validation*. Final Report, Vol. II, Naval Research Laboratory, Washington DC, 117–1121 (1991).
26. R. W. Spencer, H. M. Goodman and R. E. Hood. Precipitation retrieval over land and ocean with the SSM/I: Identification and characteristics of the scattering signal. *J. Atmosph. Oceanic Technology*, **6** (2), 254–273 (1989).
27. E. C. Barrett and C. Kidd. Rainfall monitoring by the SSM/I in middle latitudes. In: *Vth AMS Conference on Satellite Meteorology and Oceanography*, London, England, September 1990, p. 5 (1990).
28. J. V. Fiore and N. C. Grody. A classification algorithm for monitoring snowcover and precipitation using SSM/I measurements. In: *Proc. 5th Conference on Satellite Meteorology and Oceanography*, American Meteorological Society, London, pp. 237–240 (1990).

29. N. C. Grody. Classification of snow cover and rain using the Special Sensor Microwave Radiometer. *J. Geophys. Res.*, **96**, 7423–7435 (1991).

30. E. C. Barrett, C. Kidd and D. Kniveton. *Global Rainfall Monitoring by the SSM/I: Products, Problems and Intercomparison Projects*. Final Report (Stage III), USRA, RSU, University of Bristol (1990).

31. F. Strawbridge. *Satellite Passive Microwave Remote Sensing of Vegetation*. Unpublished Doctoral Thesis, University of Bristol (1991).

32. J. O. Bailey, E. C. Barrett, M. J. Beaumont, R. W. Herschy, R. J. Kelly, C. Kidd and H. Xu. *Remote Sensing of Snow by Satellites*. Project Report to the National Rivers Authority RSU, University of Bristol (1991).

33. E. A. Smith, A. Mugnai, H. J. Cooper, G. T. Tripoli and X. Xiang. Foundations for statistical-physical precipitation retrieval from passive microwave satellite measurements. Part I: brightness temperature properties of a time-dependent cloud radiation model. *J. Appl. Meteor.*, **31** (6), 506–531 (1992).

34. E. C. Barrett, G. D'Souza, C. H. Power and C. Kidd. Towards trispectral rainfall monitoring algorithms. In: *Tropical Rainfall Measurements*, J. S. Theon and N. Fugono (Eds), pp. 285–292, Deepak, Hampton, VA (1988).

35. F. L. Snijders. Rainfall monitoring based on meteosat data: a comparison of techniques applied to the western sahel. *Int. J. Rem. Sensing*, **12** (6), 1331–1347 (1988).

36. T. H. Lee, J. E. Janowiak and P. A. Arkin. *Atlas of Products from the Algorithm Intercomparison Project 1: Japan and Surrounding Oceanic Regions June–August 1989*. University Corporation for Atmosphere Research, Washington, DC (1991).

ESA/NASA International Workshop, pp. 599–621
B. J. Choudhury, Y. H. Kerr, E. G. Njoku and P. Pampaloni (Eds)

Theoretical foundations and verification of a multispectral, inversion-type microwave precipitation profile retrieval algorithm

ERIC A. SMITH,[1] ALBERTO MUGNAI[2] and GREGORY TRIPOLI[3]

[1]*Department of Meteorology and Supercomputer Computations Research Institute, Florida State University, Tallahassee, FL 32306, USA*
[2]*Istituto di Fisica dell'Atmosfera, Consiglio Nazionale delle Ricerche, Via Galileo Galilei, 00044 Frascati, Italy*
[3]*Department of Atmospheric and Ocean Sciences, University of Wisconsin, Madison, WI 53706, USA*

1. INTRODUCTION

In a systematic set of theoretical studies, we have determined how the microphysical properties of convective storms and tropical cyclones determine the nature of multispectral passive microwave brightness temperatures (T_B's) measured from spaceborne satellite radiometers [1–7]. In the more recent studies, the highly resolved thermodynamical and microphysical outputs from a three-dimensional, nonhydrostatic cloud model [8, 9] were used as input to a vertically and angularly detailed radiative transfer model, in order to characterize the impact of cloud and precipitation microphysics on upwelling radiation at selected frequencies in the microwave spectrum. To quantify the vertical sources of radiation, two different types of vertically dependent structure functions were defined which describe the processes by which radiation originates and then reaches a satellite radiometer through absorption–emission–scattering interactions. These are referred to as the emission-source and generalized weighting functions. The combination of these two functions completely describes how the vertical distribution of various liquid and frozen hydrometeors (cloud drops, rain drops, graupel particles, ice crystals, snow flakes, and aggregates), along with other relevant radiative sources (surface, space and gaseous H_2O and O_2), contribute component-wise to the total upwelling brightness temperatures.

In turn, the weighting function framework is utilized to guide the retrieval of liquid and ice hydrometeor profile information from a given set of satellite or aircraft-derived frequency dependent microwave brightness temperatures. Using SSM/I frequencies as an example, Smith and Mugnai [3] and Mugnai *et al.* [6] show how in the presence of precipitation, each frequency senses a discrete and well defined portion of the cloud column. From this perspective, it has been shown by Smith *et al.* [5] and Mugnai *et al.* [6] how a physically-based, inversion-type cloud and precipitation retrieval scheme can be designed to produce radiatively and microphysically consistent results. In the latter paper, a discussion is also provided on how a numerical optimization scheme can be constrained, so as to ensure optimal performance from a forward radiative transfer (RTE) model as first guess microphysical profiles are perturbed to produce RTE model agreement with the measured brightness temperatures.

In the study presented here, a description of the inversion algorithm is provided along with the results from various preliminary verification studies, including two cases over a land background. One of the continental studies incorporates Special Sensor Microwave/Imager (SSM/I) satellite data; see [10] for a description of the polarized 7-channel/4-frequency (19, 22, 37, 85 GHz) conically scanning SSM/I instrument. The other uses high spatial resolution (2.8 km) Advanced Microwave Precipitation Radiometer (AMPR) data obtained from a NASA ER-2 high-flying research aircraft platform; see [11] for a description of the polarized 4-channel/4-frequency (10.7, 19, 37, 85 GHz) cross-track scanning AMPR instrument. In the former case, the Florida State University (FSU) version of the algorithm is used with the 19, 37 and 85 GHz measurements to examine a severe storm/flash-flood event that impacted the mountainous Valtellina region of northern Italy in August, 1987; see [12, 13]. In the latter case, the Italian Istituto di Fisica dell'Atmosfera–Universitá La Sapienza (IFA-SAP) version of the algorithm is used with all 4 AMPR frequencies to examine a severe convective storm which occurred within the experiment array of the Convection and Precipitation/Electrification Experiment (CaPE) held at Cape Canaveral, Florida during July–August, 1991; see [14]. For the satellite study, measurements at 19 and 37 GHz are spatially enhanced to the 85 GHz resolution with an energy-preserving deconvolution scheme prior to their use in the retrievals; see [15, 16].

We also discuss the performance and results from the inversion scheme in conjunction with global scale processing. We point out here that direct application of the inversion–optimization procedure, on a pixel by pixel basis, is computationally prohibitive. For example, in this mode, one year of global processing of SSM/I measurements would require 50 000 hours on a single Cray-YMP processor using optimized vector code. Therefore, we have developed an alternative, lower fidelity but computationally efficient strategy to overcome the computational barrier set up by the brute force approach. In the alternative approach, we employ large pre-calculated, 3-dimensional (3-frequency) precipitation bench mark grids, generated once at 5 °C resolution for each month of the year and for each 10° latitude band, to circumvent the computational overhead associated with footprint by footprint radiative iterations. Using a tri-linear interpolation procedure, instantaneous rainrates are then extracted for processing into grid-box averages of preferred dimensions. The month-latitude band stratification is related to climatologically varied atmospheric temperature–moisture profiles and surface temperature conditions, needed for specifying thermodynamic state and lower boundary conditions in the radiative transfer model.

2. EMISSION-SOURCE AND GENERALIZED WEIGHTING FUNCTIONS

The emission-source and generalized weighting functions represent two distinct but equally legitimate means to characterize the different roles of atmospheric constituents along with surface and space boundaries to emerging brightness temperatures. Both functions are defined in terms of relative contributions by unit thickness layers to brightness temperatures measured at the radiometer level. The application of such functions requires the use of component-based radiative transfer procedures, where the components fall into three classes: (1) hydrometeors (i.e. cloud drops, rain drops, graupel particles, ice crystals, snow flakes, and aggregates); (2) absorbing gases (i.e. H_2O and O_2 treated in a combined fashion); and (3) inward boundary sources (i.e. surface emission–reflection or cosmic illumination). If we assume horizontal homogeneity, component-based radiative transfer for the above $N_s = 9$ sources requires solutions to 9 radiative transfer equations for the azimuthally averaged radiation I_s arising from each emission source s:

$$\mu \frac{\mathrm{d}I_s(z,\mu)}{\mathrm{d}\tau(z)} + I_s(z,\mu) = \varepsilon_s(z)\left[1 - \omega_0(z)\right] B\left[T(z)\right] + \frac{\omega_0(z)}{2} \int_{-1}^{1} \overline{P}(z,\mu,\mu')\, I_s(z,\mu')\, \mathrm{d}\mu' \tag{1}$$

where z, μ, τ, ω_0, and $\overline{P}$ are the altitude, cosine of zenith angle, optical depth, single scatter albedo, and azimuthally averaged scattering phase function, respectively. The Planck function is denoted by $B(T)$ and $\varepsilon_s = \beta_{\mathrm{abs},\,s}(z)/\beta_{\mathrm{abs}}(z)$ is the height-dependent emissivity factor for source s, where $\beta_{\mathrm{abs},\,s}$ is the volume absorption coefficient due to source s ($\beta_{\mathrm{abs},\,s} = 0$ for cosmic background radiation or radiation emitted by the Earth's surface), while β_{abs} is the volume absorption coefficient due to all atmospheric constituents. Boundary conditions are given by:

$$\begin{aligned} I_s(z_{\mathrm{R}}, -\mu) &= B(2.7\ \mathrm{K}) && \text{for cosmic background radiation} \\ &= 0 && \text{otherwise} \end{aligned} \tag{2}$$

$$\begin{aligned} I_s(0,\mu) &= \bar{\varepsilon} B(T_s) + 2(1-\bar{\varepsilon}) \int_0^1 I_s(0,-\mu')\mu'\,\mathrm{d}\mu' && \text{for emitted Earth surface radiation} \\ &= 2(1-\bar{\varepsilon}) \int_0^1 I_s(0,-\mu')\mu'\,\mathrm{d}\mu' && \text{otherwise.} \end{aligned}$$

Here, z_{R}, T_s, and $\bar{\varepsilon}$ are the altitude of the radiometer, the skin temperature, and the average beam emissivity of the Earth's surface, respectively. Thus at the satellite radiometer, the cumulative radiance is given by:

$$I(z_{\mathrm{R}}, \mu_{\mathrm{R}}) = \sum_{s=1}^{N_s} I_s(z_{\mathrm{R}}, \mu_{\mathrm{R}}) \tag{3}$$

where μ_{R} is the cosine of the radiometer zenith angle. By taking the Rayleigh–Jeans representation (T_{RJ}) of the upwelling radiance:

$$T_{\mathrm{RJ}} = \frac{c^2}{2k} \int_{\Delta\nu} \frac{I(z_{\mathrm{R}}, \mu_{\mathrm{R}})}{\nu^2\, \mathrm{d}\nu} \tag{4}$$

the relative contribution to brightness temperature ($W_{\mathrm{BT},\,s}$) can be expressed by:

$$W_{\mathrm{BT},s} = \frac{T_{\mathrm{RJ},\,s}}{T_{\mathrm{B}}} \tag{5}$$

where the following normalization property applies:

$$\sum_{s=1}^{N_s} W_{\mathrm{BT},s} = 1. \tag{6}$$

Note in Equation (4), c, k, ν, and $\Delta\nu$ are velocity of light, Boltzmann's constant, frequency, and channel bandpass, respectively.

An emission-source weighting function (ESWF) can now be defined that describes the relative contribution to T_B due to emission from layers along the z axis, indicating where radiation reaching the radiometer has originated, and in which emission is considered the only true source of radiative energy. For each of the $N_{s,\mathrm{A}} = 7$ atmospheric constituents (absorbing gases and the 6 hydrometeor species), N_L transfer equations for emission within layer L ($L = 1, \ldots, N_L$) must be solved. This is accomplished by solving an RTE similar to that given in Equation (1), provided emission by single layers is taken one at a time instead of all layers taken at the same time. In this framework, the source-dependent emissivity factor ($\varepsilon_{s,L}$) is defined as:

$$\varepsilon_{s,L}(z) = \begin{cases} \beta_{\mathrm{abs},s}(z)/\beta_{\mathrm{abs}}(z) & \text{for } z \text{ within layer } L \\ = 0 & \text{for } z \text{ outside of layer } L \end{cases} \tag{7}$$

and the emission-source weighting function becomes:

$$\mathrm{ESWF}(z) = \frac{1}{T_\mathrm{B}} \frac{\mathrm{d}T_\mathrm{RJ}(z)}{\mathrm{d}z} \tag{8}$$

where $\mathrm{d}T_\mathrm{RJ}(z)$ is the total contribution to the upwelling T_RJ due to emission by all $N_{s,\mathrm{A}}$ atmospheric constituents within a layer $\mathrm{d}z$ at altitude z.

The foremost properties of the ESWF are that it exhibits: (1) group normalization; (2) conservative decomposition; and (3) component-wise normalization. These properties are stated mathematically by:

$$\int_0^{z_\mathrm{R}} \mathrm{ESWF}(z)\,\mathrm{d}z + W_{\mathrm{BT,SPACE}} + W_{\mathrm{BT,SURF}} = 1 \tag{9.1}$$

$$\mathrm{ESWF}(z) = \sum_{s=1}^{N_{s,\mathrm{A}}} \mathrm{ESWF}_s(z) \tag{9.2}$$

$$\int_0^{z_\mathrm{R}} \mathrm{ESWF}_s(z)\,\mathrm{d}z = W_{\mathrm{BT},s} \tag{9.3}$$

Furthermore, a generalized weighting function (GWF) can be defined which describes the relative contribution to T_B due to emission and scattering from layers along the z axis, indicating where radiation reaching the radiometer last encountered an interaction, and in which scattering is considered as a pseudo-source of radiative energy. In this framework, the contribution by each atmospheric constituent s within a thin layer $\mathrm{d}z$ at altitude z is given by:

$$\mathrm{d}C_s(z, \mu_\mathrm{R}) = \frac{\beta_{\mathrm{ext}}(z)}{\mu_\mathrm{R}} J_s(z, \mu_\mathrm{R}) \exp\left[-\frac{\tau(z)}{\mu_R}\right] \mathrm{d}z \tag{10}$$

where β_{ext} is the volume extinction coefficient, and the source function $J_s(z, \mu_R)$ is given by:

$$J_s(z,\mu_R) = \varepsilon_s\left[1-\omega_0(z)\right]B\left[T(z)\right] + \frac{\sigma_s(z)\,\omega_0(z)}{2}\int_{-1}^{1}\overline{P}_s(z,\mu_R,\mu')I(z,\mu')\,\mathrm{d}\mu'. \quad (11)$$

Here, $\overline{P}_s$ is the azimuthally averaged scattering phase function of the selected atmospheric constituent, while $\sigma_s(z) = \beta_{sca,\,s}(z)/\beta_{sca}(z)$ is its height-dependent scattering factor (note $\beta_{sca,\,s}$ and β_{sca} are the volume scattering coefficients due to constituent s only and not to all $N_{s,\,A}$ atmospheric constituents). Then by taking the Rayleigh–Jeans representation T'_{RJ} of the total dC (i.e. due to all $N_{s,\,A}$ atmospheric constituents) according to Equation (4), the generalized weighting function becomes:

$$\mathrm{GWF}(z) = \frac{1}{T_B}\frac{\mathrm{d}T'_{RJ}(z)}{\mathrm{d}z}. \quad (12)$$

Finally, the group normalization, decomposition, and component-wise normalization properties are stated mathematically by the following three equations:

$$\int_0^{z_R}\mathrm{GWF}(z)\,\mathrm{d}z + W'_{BT,\,SURF} = 1 \quad (13.1)$$

$$\mathrm{GWF}(z) = \sum_{s=1}^{N_{s,\,A}}\mathrm{GWF}_s(z) \quad (13.2)$$

$$\int_0^{z_R}\mathrm{GWF}_s(z)\,\mathrm{d}z = W'_{BT,\,s} \quad (13.3)$$

where the W'_{BT}'s are relative T_B contributions due to both emission and scattering.

As an example of the use of the above weighting functions, Fig. 1 shows the ESWF's and GWF's at four frequencies associated with a mature summertime multicell continental thunderstorm over the southeastern US taken from the simulation study of Mugnai *et al.* [4]. This diagram reveals that different microwave wavelengths are sensitive to different depths within the precipitating cloud, and thus to various microphysical processes within the storm. It is this feature of a sequence of microwave window frequencies that provides the basis for use of a profile-based inversion-type retrieval scheme.

3. ALGORITHM DESIGN

The retrieval algorithm utilizes a large number (thousands) of distinct microphysical profile situations, derived from a cloud model and used as input to a forward radiative transfer model, to generate a cloud-radiation database of profile cases and their radiative signatures (i.e. the upwelling brightness temperatures at the top-of-atmosphere at the selected microwave frequencies and polarizations) at a horizontal spatial scale of 10 km — to be generally consistent with the highest ground resolution currently available from

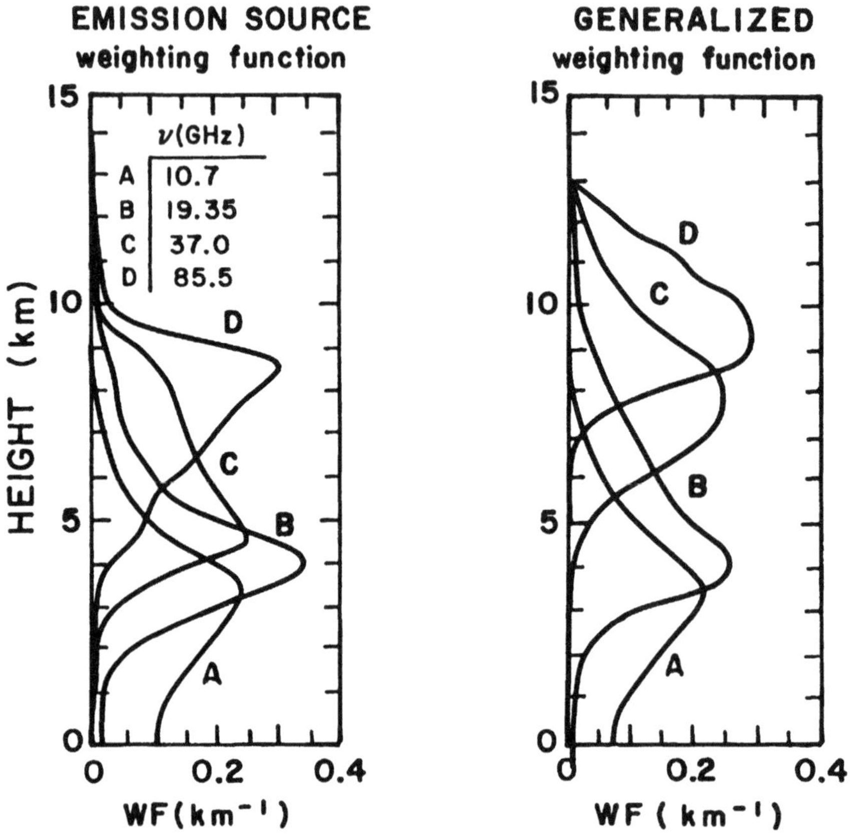

Figure 1. Emission-source and generalized weighting functions at four microwave frequencies (10.7, 19, 37, and 85 GHz) associated with an intense continental convection cell (left and right panels, respectively). The ESWFs depict the relative contributions to upwelling brightness temperatures at the top-of-atmosphere due to purely emission processes. The GWFs depict the contributions due to both emission and scattering assuming no further interactions with the atmosphere or surface.

satellite. By using different cloud model simulations for different types of precipitation systems, the database can be tailored to the different precipitating environments that occur in nature. To obtain the detailed cloud microphysics, the algorithm uses a 3-dimensional, nonhydrostatic mesoscale model run at storm scales, the University of Wisconsin Nonhydrostatic Modeling System (UW-NMS) as described by Tripoli [8, 9]. The algorithm is derived from classical inversion concepts based on explicit radiative transfer, for multispectral inversion of constituent profiles, in which the vertical distributions of liquid and ice water contents of various liquid and frozen hydrometeor species are the sought-after variables. The theory and design of the algorithm have been given by Smith *et al.* [5] and Mugnai *et al.* [6]. Applications have been described by Smith *et al.* [7, 17], Mugnai *et al.* [12], Levizzani *et al.* [13], Marzano *et al.* [14], Farrar *et al.* [16], and Xiang *et al.* [18]. Because of the specific nature of its initialization and inversion–optimization techniques, the algorithm is referred to as a hybrid statistical–physical scheme, somewhat analogous to the terminology used for various contemporary infrared-based temperature retrieval schemes; see e.g. [19]. Kummerow *et al.* [20, 21] have also experimented with an inversion technique based on a different type of optimization approach, and using radar-derived microphysics as a source of initialization data. Our algorithm is designed to accept any combination of polarized or unpolarized passive microwave measurements. The algorithm determines vertical profiles of liquid/ice water contents (LWC/IWC) for suspended cloud drops, precipitating rain drops, precipitating ice particles, and suspended ice hydrometeors.

Once the profiles are calculated, the surface and vertically distributed rain and graupel precipitation rates are diagnosed from gravity fallout equations consistent with those used in the cloud model. The LWC/IWC profiles can also be transformed to condensation heating/deposition cooling rates by evaluating vertical derivatives of the mass fluxes of the precipitating water or ice hydrometeors, thus yielding the latent heating profile. The vertical mass fluxes are derived directly from the gravity fallout equations. The vertical rainrate profile $\mathrm{RR}(z)$ is given by:

$$\mathrm{RR}(z) = -\rho_0(z)\left[\overline{w_{\mathrm{rd}}}(z) + \overline{W}\right]\frac{q_{\mathrm{rd}}(z)}{\rho_{\mathrm{w}}} \tag{14}$$

where ρ_0 is the basic state air density, ρ_{w} the density of water (10^3 kg m^{-3}), q_{rd} the rain water mixing ratio, $\overline{W}$ the externally specified environmental mean vertical velocity, and $\overline{w_{\mathrm{rd}}}$ the vertically dependent mean terminal velocity, where:

$$\overline{w_{\mathrm{rd}}} = -1.94\left[\frac{4\,\rho_{\mathrm{w}} D_{\mathrm{m}} g}{3\,C_{\mathrm{D}}\rho_0(z)}\right]^{1/2} \tag{15}$$

and g, C_{D} and D_{m} are the gravitational acceleration, the water drop drag coefficient (0.588) and the characteristic diameter of the rain drops (540 μm). Thus:

$$\mathrm{RR}(z) = 3.6\times 10^6 \times \rho_0(z)\left[289.2\left(\frac{D_{\mathrm{m}}}{\rho_0(z)}\right)^{1/2} + \overline{W}\right]\frac{q_{\mathrm{rd}}(z)}{\rho_\omega} \tag{16}$$

where the terms on the right hand side are given in SI units and RR is in mm hr^{-1}. A similar expression can be developed for a graupel fall rate using 0.45 for the ice particle drag coefficient, 1000 μm for D_{m} and 600 kg m^{-3} for the graupel density. The surface mass flux of water is given by:

$$R(z_s) = 289.2\left(\frac{D_{\mathrm{m}}}{\rho_0(z_s)}\right)^{1/2} l_{\mathrm{rd}}(z_s) \tag{17}$$

where $l_{\mathrm{rd}}(z_s)$ is the rain LWC at the surface. Finally, the latent heating due to condensation or freezing/deposition is given by:

$$Q_l(z) = L\,\frac{\partial R(z)}{\partial p}\,\frac{g}{c_{\mathrm{p}}} \tag{18}$$

where L is the latent heat of the designated phase change, and c_{p} is the specific heat of moist air at constant pressure.

Prior to the retrieval process, the raw satellite observations are spatially deconvolved such that there is footprint conformity for the different measuring frequencies. The technique has been described by Farrar and Smith [15]; see also [22, 23] for alternative approaches of spatial deconvolution with passive microwave measurements. Spatial deconvolution partially offsets problems due to incomplete beam filling; see [24–26] for a discussion of the heterogeneous beam filling problem. For the applications presented here, the 19 and 37 GHz SSM/I measurements have been spatially enhanced to the $\sim$ 15 km resolution of the 85 GHz channels, whereas the 37 and 85 GHz AMPR measurements have been spatially convolved down to the $\sim$ 2.8 km resolution of the 10.7 and 19 GHz channels. Appropriate sea surface or land surface emissivity boundary conditions and temperature-moisture profiles needed for the radiative transfer calculations are specified for the scene, the appropriate cloud-radiation database generated (from the simulated microphysics for ocean or continental storms pre-processed in the forward RTE model), and the individual pixel packets then processed one at a time.

In what follows, we shall focus on the FSU version of the algorithm which utilizes the cloud-radiation database for initializing a two-stream/polarized radiative transfer model, run in a perturbative forward mode to seek an optimal match between the modeled and measured brightness temperatures. The cloud-radiation model calculations forming the initial guess database, along with spatially deconvolved SSM/I satellite measurements or coincident AMPR measurements, and the RTE model, represent the three main components of the algorithm. The multi-spectral brightness temperature measurements for a given pixel coordinate, in combination with the theoretically derived microphysical-brightness temperature database, allow for the selection of initial guess profiles to be used in the radiative transfer model, which is then run in a forward-perturbative mode until there is optimal agreement between the modeled and measured brightness temperatures. The perturbation process is guided by an explicit optimization scheme which seeks a unique solution profile from the initial guess profile domain. The optimization scheme is based on the method of Golden Section Search described by Gill *et al.* [27]. The radiative transfer model is a polarized Sobolev two-stream solution of the azimuthally independent radiative transfer equation, which allows for arbitrary top and bottom boundary conditions, and decomposition of the zenith-dependent radiance field for match-up with the radiometer look-angle. Mathematical details of the RTE model have been described by Xiang [28] and Xiang *et al.* [29].

An individual retrieval for a measurement packet consists of first scanning the database for a set of one or more initial-guess profile pairs which are in proximity to the brightness temperature measurement vector (in this context, 'profile' actually signifies four consistently derived vertical profiles representing the four precipitating and suspended liquid and frozen hydrometeor categories — i.e. crystals, snow flakes, and aggregates are treated in a combined fashion). This step is referred to as 'bracketing', designating that the resultant first-guess profile set surrounds or brackets the measurement. This procedure is carried out with respect to pairs of profiles — order 2 bracketing implies a single pair is used, order 4 implies two pairs are used, etc. The first profile pair to bracket the solution involves the profile whose RMS measured–modeled brightness temperature difference for the selected frequencies is minimum (over the entire database), and the next smallest RMS case whose measured–modeled T_B differences are of opposite sign to the first profile with respect to each frequency. The second and following profile pairs are also obtained using the 'opposite sign rule'.

Once the initial-guess profile domain is obtained, the optimization procedure is applied which perturbs the first-guess solution profile set toward a final solution. The optimization involves a transform in which the final solution is represented by a linear combination of the bracketing profiles involving separate weights ($w_{i,j}$) for the constituent hydrometeors, where i is a function of the order of bracketing and j is the constituent index. For example, if using order 2 bracketing for 4 hydrometeor constituents, the final solution would be represented by a linear combination of 4 sets of a single initial-guess profile pair in which the 8 weights ($w_{1,1}, 1 - w_{1,1}$; $w_{1,2}, 1 - w_{1,2}$; $w_{1,3}, 1 - w_{1,3}$; $w_{1,4}, 1 - w_{1,4}$) would be used as multipliers on the individual profiles. Thus, for order 2 bracketing, the optimization involves 4 unknowns ($w_{1,1}$, $w_{1,2}$, $w_{1,3}$, $w_{1,4}$). The scheme is based on the Golden Section Search method [27], and guides perturbation of the $w_{i,j}$'s until the RMS brightness temperature difference between the forward model calculations and measurements is minimized. At the conclusion of the optimization process, the final set of linear combinations for the 4 constituent hydrometeors represents the LWC/IWC profile solution. The flow chart in Fig. 2 summarizes the FSU algorithm design as applied to SSM/I measurements.

4. APPLICATIONS WITH SSM/I AND AMPR

The algorithm has been exercised in various preliminary verification studies. On a global scale, we have used the algorithm in the First WetNet Precipitation Intercomparison Project (PIP-1) which involved 14 algorithm investigators and 17 different algorithms [30]. Accumulated monthly rainfall maps for a 4-month period (August to November 1987) were calculated from SSM/I measurements at a half-degree spatial scale and compared to various conventional and climatological data sets. Figure 3 provides the results for August-87 in conjunction with the multi-year ocean climatology of Legates [31] and the August-87 compilation from land stations developed by Eischeid *et al.* [32]; see also [33–36]. Only the oceanic rainrates have been derived with the profile version of the algorithm; the continental rainrates are derived from a simpler multi-variate regression scheme [5]. It can be noted that the difference in global means between the satellite and climatology is less than 10% in the northern hemisphere where the climatologies are more robust, but almost 100% in the southern hemisphere where the climatologies are weaker. Because there should be little hemispheric bias in the satellite result, we speculate that validations based on southern hemisphere climatologies must be given circumspect consideration.

Table 1 provides a direct comparison between the monthly SSM/I estimates and the specialized west Pacific oceanic atoll-station monthly accumulation data set developed by Morrissey [37] and Morrissey and Greene [38]; see also [39]. Note that monthly differences are as small as 1% and as large as 18%. Given that the satellite results are

Table 1.
Comparison of SSM/I derived monthly rainfall accumulations (mm) with raingage accumulations from approximately 60 west Pacific atoll stations

Month	Average accumulation (mm)				
(1987)	SSM/I	Gauge	Mean difference (%)	Slope factor	Correlation
Aug.	227	251	11	0.94	0.74
Sep.	174	180	3	0.88	0.85
Oct.	204	240	18	1.00	0.78
Nov.	196	199	1	0.87	0.78

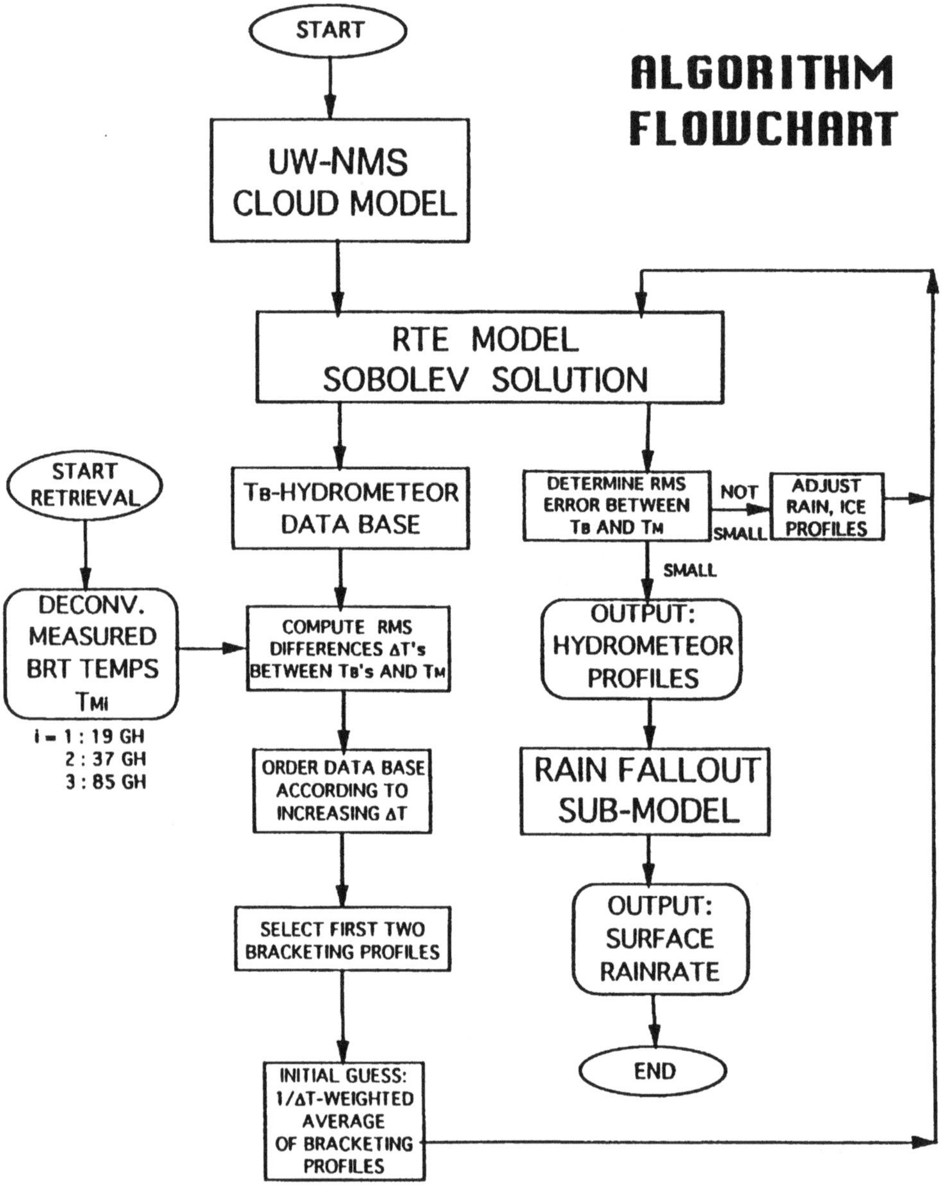

Figure 2. Flow chart of the FSU version of profile-based inversion-type precipitation retrieval algorithm for SSM/I applications.

based on averaging rainrates for the relatively few overpasses that cross at fixed times of the day over the 88 stations in the atoll data set, and that these latter results are based on nearly continuous accumulations in the gauge collectors, the differences must be considered quite small. However, this is only a 4-month sample involving time-space averages, so that the assessment of uncertainty in the satellite estimates must be treated cautiously until an entire annual cycle is examined.

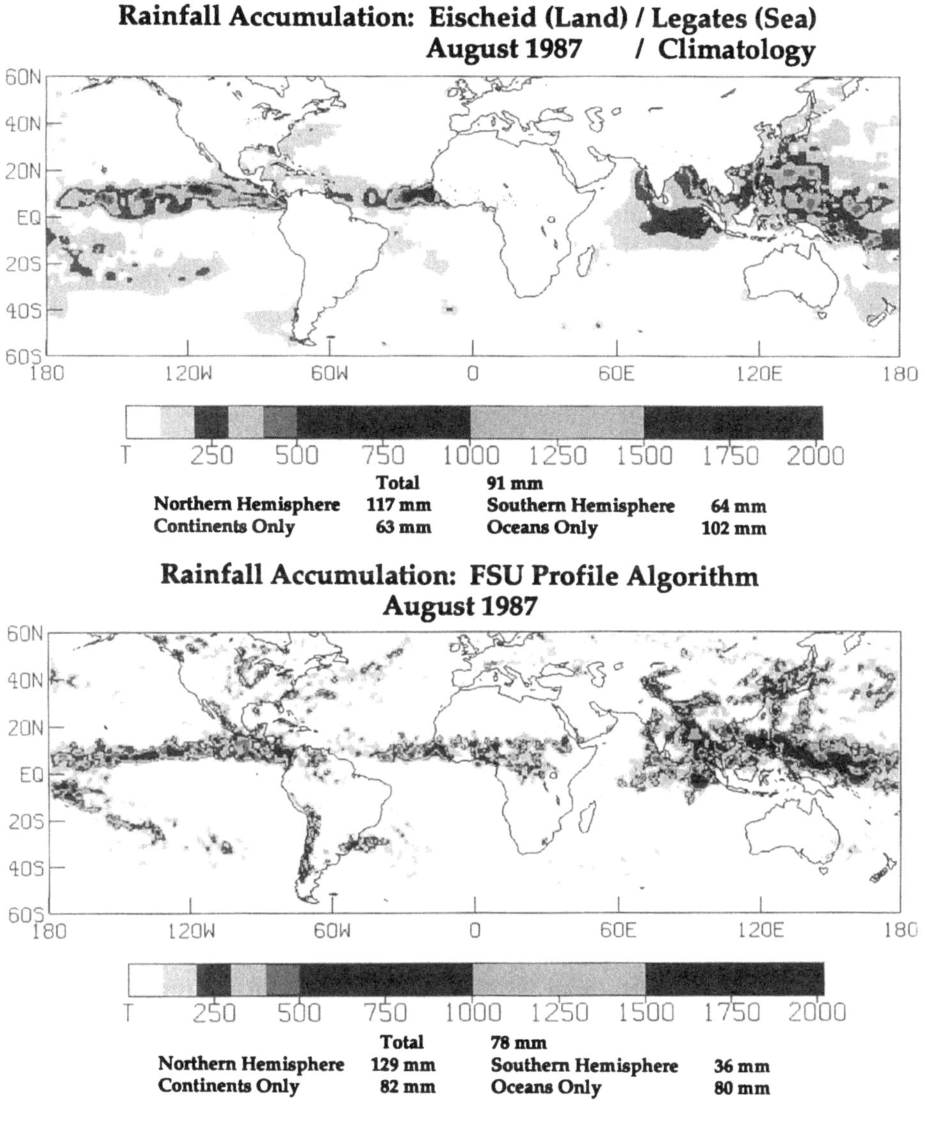

Figure 3:

Figure 3. Comparison of SSM/I derived monthly rainfall accumulation for August-1987 (bottom panel) with multi-year climatology of Legates [31] for ocean and August-87 compilation of Eischeid *et al.* [32] for land (top panel).

Additional intercomparisons have been carried out with the climatological data sets of Jaeger [40, 41], and the IR-satellite global precipitation index (GPI) data of P. Arkin (see [42, 43]). These comparisons show good qualitative agreement. One of the main difficulties with these types of comparisons is that the SSM/I sampling frequency is at most twice a day, whereas the climatologies are based on rainfall accumulating within collecting gauges continuously, or in the case of the Arkin retrievals, on a 3-hourly basis. Thus there is a serious sampling mismatch between what goes into the microwave averages and what goes into the comparison data sets. The implication of the sampling problem on errors in the monthly estimates is an issue which has yet to be resolved and is therefore of major concern in trying to generate accurate global data sets based on microwave measurements alone. Adler *et al.* [44] have developed and tested a technique with qualified success, in which geosynchronous satellite infrared retrievals are used in conjunction with the microwave retrievals, and in which the IR estimates are bias-adjusted to agree with the microwave estimates where there are time-matchups. The bias-adjust factors are then employed for all other IR times where there are no coincident microwave data. These provides a means to diurnally sample the rain signal with the calibration underpinnings of the microwave algorithm. However, problems remain with this technique also related to the concept of statistical stationarity, and so sampling continues to be a thorny problem for the microwave algorithms.

Motivated by the sampling dilemma, we are now engaged in a new WetNet Precipitation Intercomparison–Validation Project called PIP-2 which is examining instantaneous rainrates from a collection of physical and empirical precipitation retrieval algorithms contributed by approximately 15 individuals and teams, to be applied to 28 multi-overpass precipitation case studies extending from July 1987 to February 1993, and drawing from measurements obtained by all three in-flight SSM/I platforms (F8, F10, F11). In the PIP-2 project, validation data will be derived from ground-based and airborne radar systems, as well as from high-density surface raingauge networks.

As an example of the advantage of a profile algorithm in this type of intercomparison study, we have compared SSM/I rainrate retrievals for Hurricane Hugo (a devastating Atlantic tropical cyclone occurring during September, 1989, which severely damaged the coastal city of Charleston, South Carolina) with S-band radar retrievals from the Charleston weather radar system. In this intercomparison, we could not obtain an exact time matchup between the archived radar data and the satellite data; however, two retrievals on September 22, near the time Hugo was making landfall, are within ~ 1.5 hours from one another (the SSM/I overpass was at 00:18 GMT, whereas the radar acquisition was at 01:53 GMT). Figure 4 is shown to demonstrate that the hurricane was exhibiting relatively steady state behavior in terms of both domain mean rainrate and rainrate variance statistics from 02:00 to 03:15 GMT near the nighttime period for which the intercomparison has been carried out. The intercomparison examines two sets of calculations from the satellite profile algorithm. The first set is obtained by taking the retrieved rainrates across a horizontal plane through the storm running along the surface, thus representing the surface rainrates. The second set is obtained by taking the retrieved rainrates along a scan plane equivalent to what the Charleston radar scan plane would have been, had the storm at the time of the SSM/I overpass been positioned at the storm location associated with the time of the Charleston radar image acquisition. Rainrates over the entire storm for these two different retrieval procedures are accumulated in histograms and then compared to the rainrate histogram obtained from the radar reflectivity data using the Z–R relationship of Jorgensen and Willis [45]. These histogram results are presented in Fig. 5. There is a startling improvement in the comparison when going from the surface plane case to the radar scan plane case, both in terms of the shape of the histogram and the mean value of the rainrates within the histogram. It is obvious from this result, which may be partly fortuitous in terms of the closeness of the agreement, that the altitude from which rainrates are obtained can determine how well a radar result and a satellite result compare. That is a main

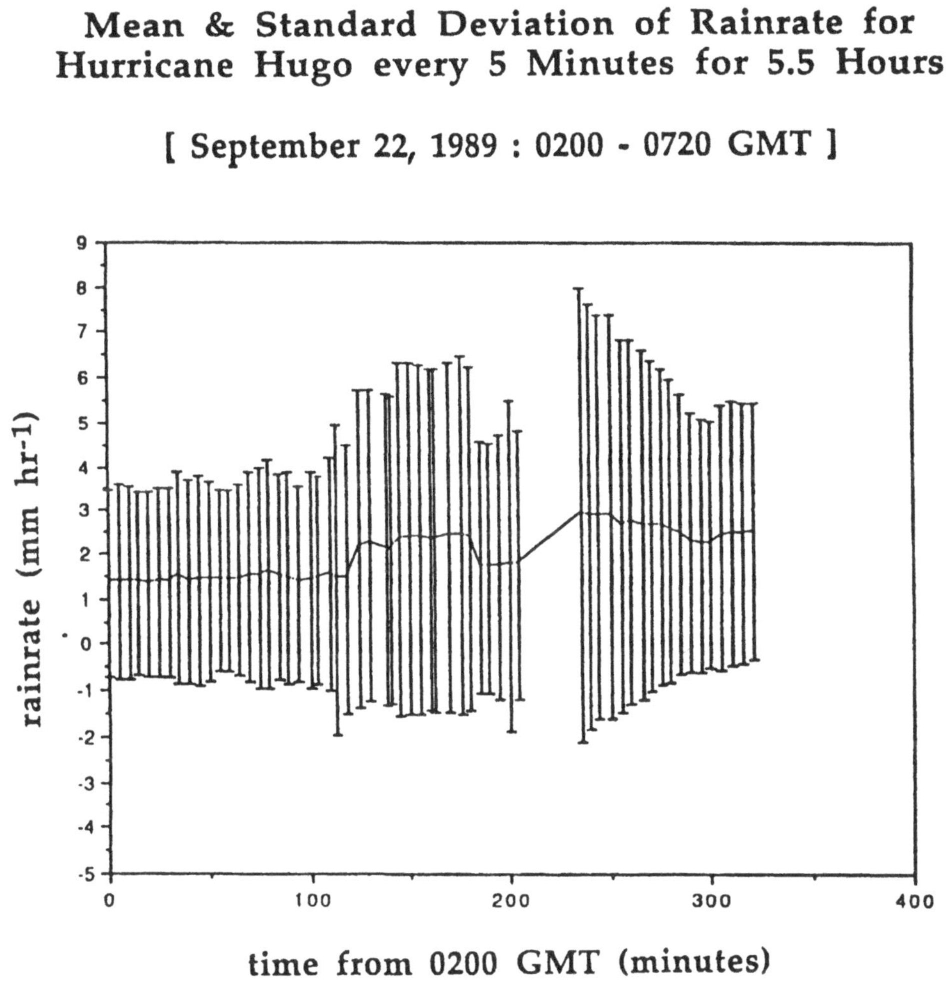

Figure 4. Means and standard deviations of rainrate at 5-minute intervals for Hurricane Hugo based on Charleston weather radar on September 22, 1989 from 02:00 to 07:20 GMT. The Jorgensen and Willis [45] Z–R relationship has been used to convert the S-band reflectivity to rainrate.

advantage of the profile-type algorithms to the rainmap-type algorithms. In fact, it is not clear that single-parameter rainmap-type algorithms can really be compared properly to plan-view radar retrievals given the mismatch between the vertical levels considered by the two retrieval schemes.

In a recent study of Levizzani *et al.* [13], they used the FSU version of the profile algorithm to retrieve rainrate estimates over an alpine area in northern Italy called the Valtellina region, during the passage of an intense mountain storm which produced damaging flash flooding in the valley areas (the storm took place on the afternoon of August 23, 1987). The left-hand panel of Fig. 6 taken from their paper shows the surface rainrates estimated from the SSM/I measurements along with raingauge-derived rainrate estimates from the remaining pluviometer network that survived the August 23 storm and an even more intense system that occurred approximately one month earlier in the

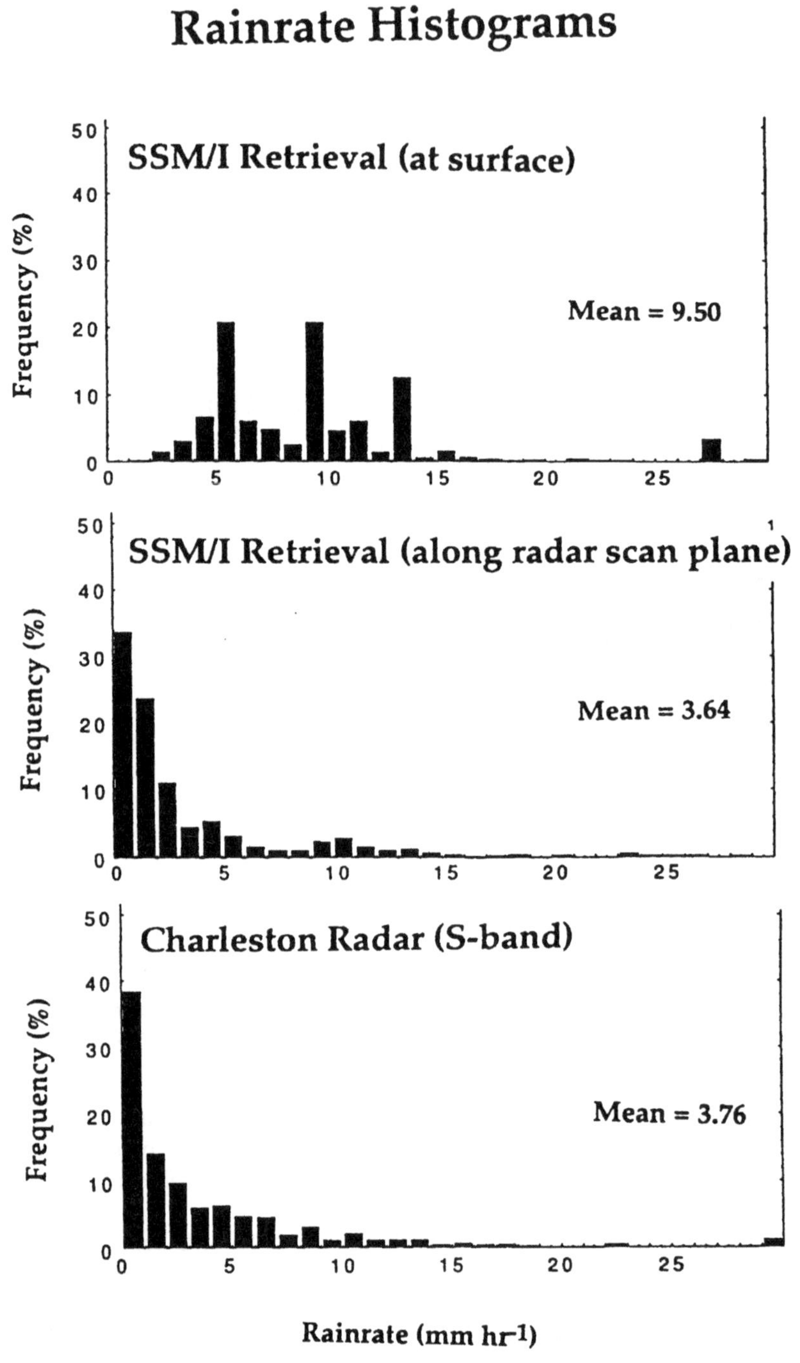

Figure 5. Rainrate histograms for Hurricane Hugo. Top panel corresponds to SSM/I retrieval (Sept. 22, 1989 – 00:18 GMT) of rainrates taken at surface; middle panel corresponds to SSM/I retrieval at same time of rainrates taken along a plane equivalent to elevation scan plane of Charleston weather radar; bottom panel corresponds to radar retrieval (Sept. 22, 1989 – 01:53 GMT).

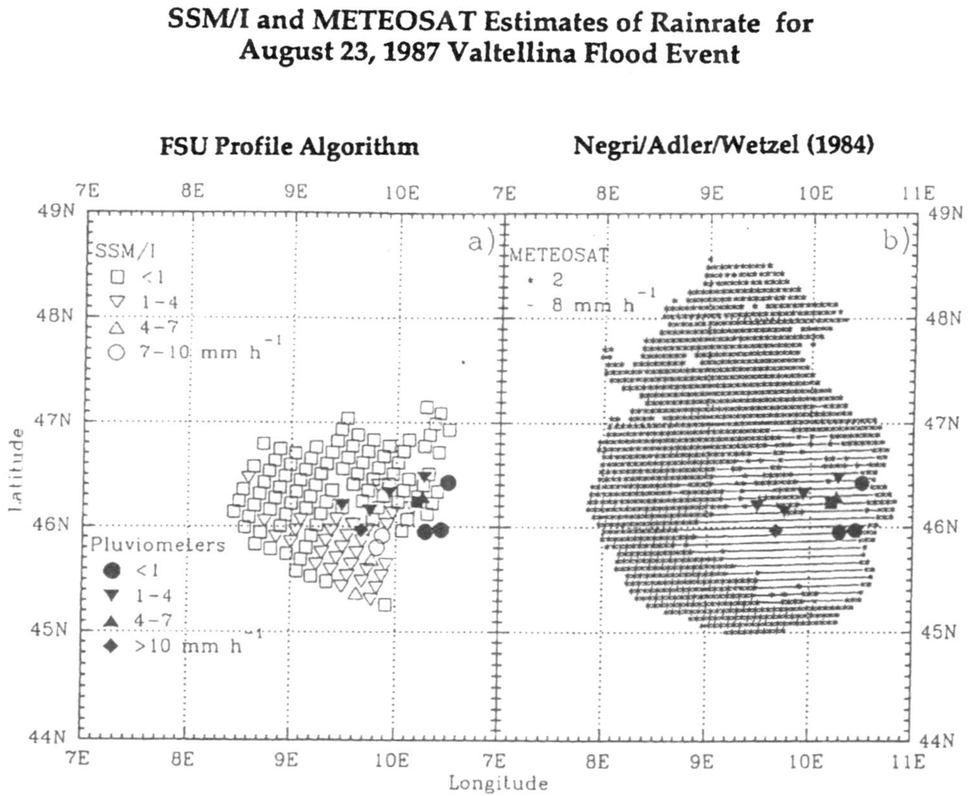

Figure 6. Rainrate estimates for August 23, 1987 flash flood event in Valtellina alpine region of northern Italy obtained from profile algorithm applied to SSM/I scene (left panel) and IR algorithm of Negri *et al.* [47] applied to METEOSAT scene (right panel). Rainrate estimates based on pluviometer data are indicated for 10 locations within storm area. (From [13].)

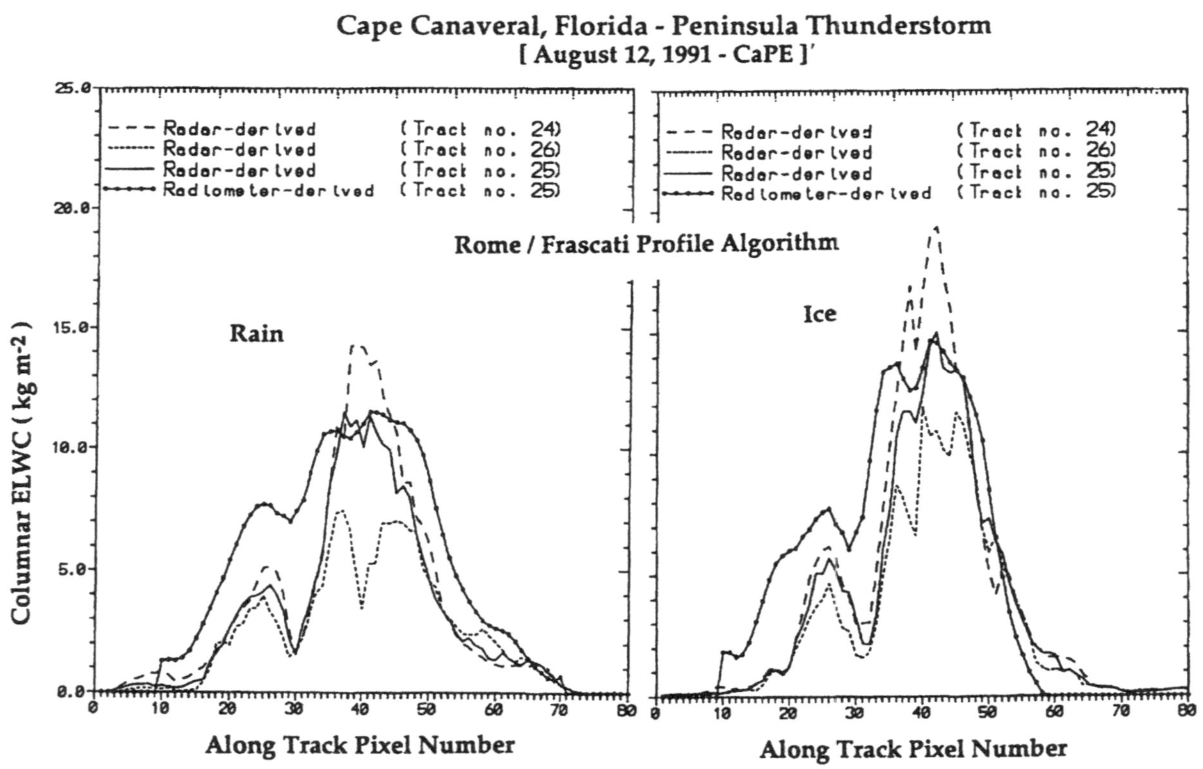

Figure 7. Comparison of integrated liquid water and ice water contents (LWC, IWC) based on the IFA-SAP (Rome/Frascati) version of profile algorithm adapted for the aircraft-mounted 4-frequency Advanced Microwave Profile Radiometer described by Marzano *et al.* [14] and a dual-frequency/dual-polarization NCAR/CP-2 multiparameter radar algorithm described by Turk *et al.* [49]. The algorithms have been applied to AMPR and CP-2 data sets for an intense thunderstorm situated over the Florida mainland in the vicinity of Cape Canaveral which occurred on August 12, 1991 during the Convection and Precipitation/Electrification Experiment. Left panel shows AMPR-derived LWC's for the nadir scan direction along the ER-2 aircraft track (track no. 25) at 80 pixel locations, compared to CP-2 retrievals for the same 80 along-track locations and for adjacent tracks (24 and 26). Right panel shows similar comparisons for the integrated IWC's.

same general area. There is good qualitative agreement between the satellite estimates and the gauge measurements, particularly when considering that this is an over-land retrieval calculation in a complex terrain setting. In fact, a precipitation screening procedure had to be used as a front-end to the retrieval algorithm to avoid pathological retrievals from occurring over rain-free areas because of the poor contrast between the background and rain areas. The screening procedure was based on the use of an 85 GHz polarization corrected temperature (PCT) scheme following the approach of Spencer *et al.* [46]. The right hand panel of this figure shows an analogous intercomparison based on the use of METEOSAT infrared data and the IR retrieval algorithm of Negri *et al.* [47]. It is more difficult to assess this result because the IR algorithm, which is basically sensitive to cloud top temperature behavior, tends to smooth over sharp gradient features picked by the surface gauges and to some extent by the microwave algorithm.

As a final example of the type of validation studies that can be done with this algorithm, we draw from the results of Marzano *et al.* [14], in which the IFA-SAP version of the algorithm has been employed with very high resolution passive microwave measurements obtained from the high-flying ER-2 aircraft. The instrument generating the data is the 4-frequency (10.7, 19, 37, 85 GHz) Advanced Precipitation Microwave Radiometer (AMPR) developed at Georgia Tech. University and the NASA-Marshall Space Flight Center in Huntsville, Alabama; see [11] for a description. Smith *et al.* [17] have described an FSU version of the profile algorithm designed for application with AMPR data, and tested on a data set obtained from the initial AMPR flights over the Gulf of Mexico. The validation products for the Marzano *et al.* study are obtained from multi-parameter radar retrieval techniques applied to the dual-polarization/dual-frequency CP-2 radar system of the National Center for Atmospheric Research (NCAR). The multi-parameter radar retrieval algorithm was developed at Colorado State University and has been described by Vivekanandan *et al.* [48] and Turk *et al.* [49]. The storm case under study occurred on August 12, 1991 in the vicinity of Cape Canaveral, Florida during the Convection and Precipitation/Electrification Experiment (CaPE); see [50] for a description of CaPE. Figures 7 and 8 illustrate how well the columnar rain liquid water contents (LWC) and ice water contents (IWC), and the surface rainrates agree with respect to the CP-2 retrievals and the AMPR retrievals. The retrievals are taken with respect to an along-track path of the aircraft for the nadir position of the AMPR when the degree of polarization in the measurements is exactly zero. The aircraft is passing over a tight but intense thunderstorm at the time the NCAR CP-2 system was viewing it in vertical incidence mode. It is clear from these comparisons through the outer boundaries and into the core of the storm, that although there is not perfect agreement, the qualitative agreement is excellent.

It should be stressed here that none of the so-called validation products, whether they be the climatological data sets, radar data, raingauge data, or multi-parameter radar retrievals, can be considered ground truth, as all of these data sets and measuring systems have intrinsic flaws and shortcomings. Therefore, the extent to which the satellite retrievals and the validation products do agree is extremely encouraging and highly suggestive that profile-type inversion schemes are credible alternatives to the more limited rainmap algorithms.

The algorithm is currently being tested insofar as its ability to derive estimates of the latent heating profile by taking vertical derivatives of the water and ice mass fluxes for the precipitating rain drops and graupel particles. Figure 9 presents a set of condensation

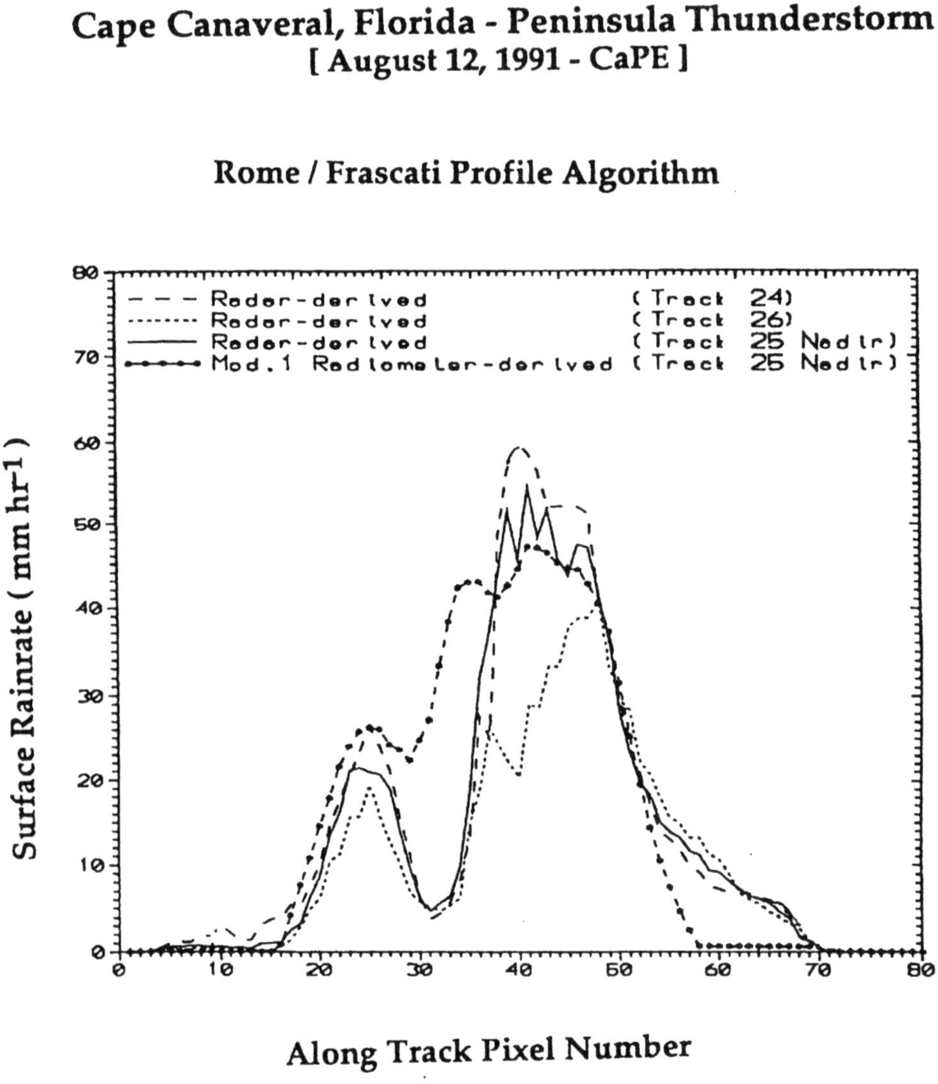

Figure 8. Same as Fig. 7, except for surface rainrate.

heating and deposition cooling profiles for the a west Pacific Ocean region centered at 1.5°S latitude/155.5°E longitude and bounded by the $\sim 5 \times 5$ degree Inner Flux Array (IFA) coordinates used during the 1992–93 TOGA COARE experiment; see [51] for a description. Vertical latent heating structures are given in deg hr^{-1} based on four months of 1987 SSM/I measurements (August through November). The top panel illustrates profiles for 29 suppressed days during the 4-month TOGA COARE intensive observational period, the middle panel illustrates profiles for 41 active convection days, and the bottom panel indicates profiles for 18 very enhanced convection days. The five sub-panels within each main panel consider average profiles made up from the

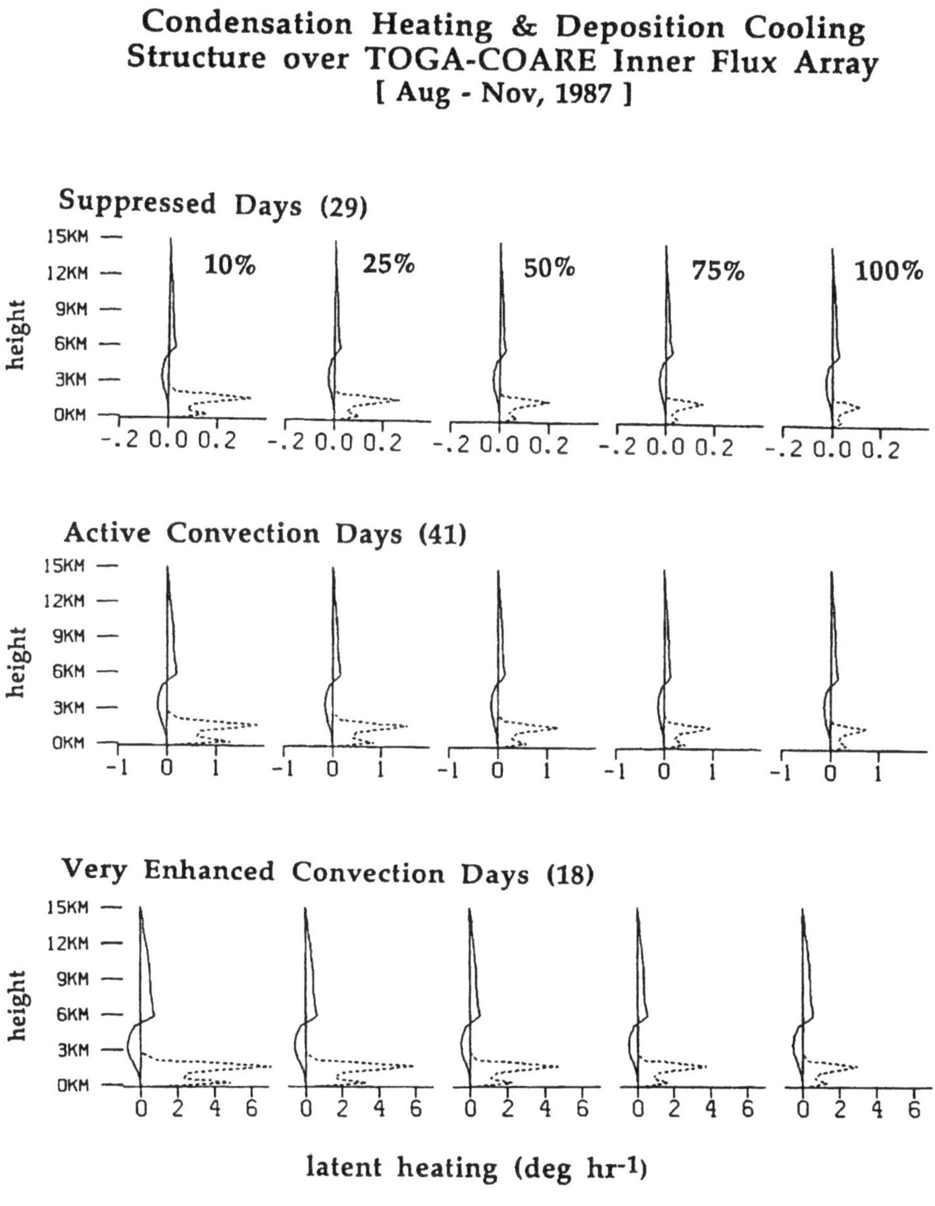

Figure 9. Vertical latent heating structure in deg hr^{-1} over TOGA COARE Inner Flux Array (IFA) derived from profile algorithm applied to four months of SSM/I measurements (Aug., Sept., Oct., Nov., 1987). The IFA is an approximately 5 × 5 degree box centered at 1.5°S latitude/155.5°E longitude. Top panel indicates profiles for 29 suppressed days during the 4-month period; middle panel indicates profiles for 41 active convection days; bottom panel indicates profiles for 18 very enhanced convection days. The five subpanels within each main panel illustrate average profiles made up from the pixel group within the IFA for a given convection category representing the largest 10% surface rainrate cases, then the largest 25%, 50%, 75% and all pixels (100%). Solid lines represent deposition cooling profiles; dashed lines represent condensation heating profiles.

pixel group within the IFA (for a given convection category) representing the largest 10% surface rainrate cases, then the largest 25%, 50%, 75% and all pixels (100%). Profiles given as solid lines represent the deposition cooling structures, whereas dashed lines represent condensation heating structures. It is clear from the abscissa scales used on these diagrams that maximum latent heating rates increase from $\sim 0.3\ \mathrm{deg\,hr^{-1}}$ for the 10% stratification/suppressed convection category up to $\sim 7\ \mathrm{deg\,hr^{-1}}$ for the 10% stratification/very enhanced convection category. The altitude of peak heating is found at approximately 2 km, which is most likely too low resulting from setting the environmental vertical velocity term in Equations (14) and (16) to zero, rather than negative and non-zero, which would be more characteristic of convective storms. One of the interesting features of these profiles is the double maximum structure found in the lower atmosphere.

If this is an accurate account of the heating function structure within the Pacific warm pool, it is somewhat unique relative to heating profiles found in other parts of the tropics; see [52, 53] for summaries. On a last note concerning latent heating, Tao and Simpson [54] and Tao *et al.* [53] have tested LWC/IWC profile output from our algorithm for a west Pacific super typhoon case in conjunction with the Tao *et al.* [55] latent heating model and a more parameterized latent heating model. The main conclusion drawn from those studies vis-a-vis our retrievals, is that they appear to produce at least qualitatively correct latent heating structures, and although these are in partial quantitative disagreement with the hurricane simulation study of Lord *et al.* [56] concerning anvil heating, there are no suggestions that this stems from quantitative problems with the satellite retrievals. However, as indicated above, we must be cautious in interpreting these intercomparisons at this stage, particularly in terms of assessing absolute accuracy, because these are such new techniques and robust ground truth data are virtually impossible to obtain.

5. CONCLUSIONS

In this study, we have attempted to demonstrate that the inversion approach to precipitation retrieval based on cloud-radiation modeling is physically, numerically and computationally viable, capable of directly generating the latent heat profile, and in some sense more physically consistent with multispectral passive microwave measurements than one-level rainmap algorithms. It is also portable to different frequencies, different instruments and different platforms. An important advantage of the algorithm is that, on the basis of an initialization procedure derived from cloud model microphysics, no empirical tuning or empirical adjustments have had to be applied. At this juncture, we have conducted various preliminary validation studies to demonstrate that the technique is sound and likely within the uncertainty of validation measurements. Comparisons with radar data along the scan plane demonstrate that validating a physical retrieval algorithm with plan-view radar data should not be done on an *ad hoc* basis, and that the profile approach has the advantage of being able to match vertical levels. We have successfully applied the profile technique to two continental cases: (1) the mountainous Valtellina region of northern Italy; and (2) the central Florida Peninsula. In the future, in addition to our participation in the WetNet PIP-2 Intercomparison-Validation Project, we also plan to apply our technique to a 2-month West Africa SSM/I data set collected during the 1992 HAPEX-Sahel experiment. In this study, the results will be validated against *in situ* rainfall measurements obtained from the high density EPSAT raingauge network outside of Niamey, Niger (approximately 90 raingauges in a 1 × 1 degree area).

Of final note, it is important to recognize in the course of the current and future validation studies that we and others are conducting and are taking place on an international basis, that there is a strong possibility that the reliability of microwave retrievals of rainfall may soon outpace conventional ground truth measurements as a point of reference for validating the satellite estimates. That is both a tantalizing and disturbing prospect.

Acknowledgments

This research has been supported by National Aeronautics and Space Administration grants NAG-991 and NAG5-1602, and by Gruppo Nazionale per la Difesa dalle Catastrofi Idrogeologiche and Agenzia Spaziale Italiana of Italy. NATO grant CG-890894 has contributed essential travel support. A portion of the computing resources has been provided by the Supercomputer Computations Research Institute at Florida State University under US Department of Energy contract DOE-FC05-85ER250000.

REFERENCES

1. A. Mugnai and E. A. Smith. Radiative transfer to space through a precipitating cloud at multiple microwave frequencies. Part I: Model description. *J. Appl. Meteor.*, **27**, 1055–1073 (1988).
2. E. A. Smith and A. Mugnai. Radiative transfer to space through a precipitating cloud at multiple microwave frequencies. Part II: Results and analysis. *J. Appl. Meteor.*, **27**, 1074–1091 (1988).
3. E. A. Smith and A. Mugnai. Radiative transfer to space through a precipitating cloud at multiple microwave frequencies. Part III: Influence of large ice particles. *J. Meteor. Soc. Japan*, **67**, 739–755 (1989).
4. A. Mugnai, H. J. Cooper, E. A. Smith and G. J. Tripoli. Simulation of microwave brightness temperatures of an evolving hail storm at the SSM/I frequencies. *Bull. Amer. Meteor. Soc.*, **71**, 2–13 (1990).
5. E. A. Smith, A. Mugnai, H. J. Cooper, G. J. Tripoli and X. Xiang. Foundations for statistical–physical precipitation retrieval from passive microwave satellite measurements. Part I: Brightness temperature properties of a time dependent cloud-radiation model. *J. Appl. Meteor.*, **31**, 506–531 (1992).
6. A. Mugnai, E. A. Smith and G. J. Tripoli. Foundations for physical–statistical precipitation retrieval from passive microwave satellite measurements. Part II: Emission source and generalized weighting function properties of a time dependent cloud radiation model. *J. Appl. Meteor.*, **32**, 17–39 (1993).
7. E. A. Smith, X. Xiang, A. Mugnai and G. Tripoli. A cloud radiation model for spaceborne precipitation retrieval. In: *Extended Abstract Vol. of International TRMM Workshop on the Processing and Utilization of the Rainfall Data Measured from Space*, Communications Research Laboratory, Tokyo, Japan, pp. 273–283 (1992).
8. G. J. Tripoli. A nonhydrostatic model designed to simulate scale interaction. *Mon. Wea. Rev.*, **120**, 1342–1359 (1992).
9. G. J. Tripoli. An explicit three-dimensional nonhydrostatic numerical simulation of a tropical cyclone. *Meteorol. Atmos. Phys.*, **49**, 229–254 (1992).
10. J. P. Hollinger, J. L. Peirce and G. A. Poe. SSM/I instrument evaluation. *IEEE Trans. Geosci. Rem. Sens.*, **GE-28**, 781–790 (1990).
11. R. W. Spencer, R. E. Hood, F. J. LaFontaine, E. A. Smith, J. Galliano and E. Lobl. High resolution imaging of rain systems with the Advanced Microwave Precipitation Radiometer. *J. Atm. Oceanic Tech.*, submitted (1993).
12. A. Mugnai, E. A. Smith and X. Xiang. Passive microwave precipitation retrieval from space: A hybrid statistical–physical algorithm. In: *Microwave Radiometry and Remote Sensing Applications*, E. R. Westwater (Ed.), NTIS, Springfield, VA 22061, pp. 237–244 (1992).

13. V. Levizzani, A. Mugnai, F. Porcu, F. Prodi. E. A. Smith and X. Xiang. Comparison of rainfall estimation using SSM/I and Meteosat measurements. In: *IRS–92: Current Problems in Atmospheric Radiation*, Proceedings of the IAMAP International Radiation Symposium, pp. 65–68, A. Deepak Pub., Hampton, VA (1993).

14. F. S. Marzano, A. Mugnai, E. A. Smith, X. Xiang, J. Turk and J. Vivekanandan. Active and passive remote sensing of precipitating storms during CaPE. Part II: Intercomparison of precipitation retrievals from AMPR radiometer and CP-2 radar. *Meteorol. Atmos. Phys.*, in press (1994).

15. M. R. Farrar and E. A. Smith. Spatial resolution enhancement of terrestrial features using deconvolved SSM/I microwave brightness temperatures. *IEEE Trans. Geosci. Rem. Sens.*, **30**, 349–355 (1992).

16. M. R. Farrar, E. A. Smith and X. Xiang. The impact of spatial resolution enhancement of SSM/I microwave brightness temperatures on rainfall retrieval algorithms. *J. Appl. Meteor.*, **33**, 313–333 (1994).

17. E. A. Smith, X. Xiang, A. Mugnai, R. Hood and R. W. Spencer. Behavior of an inversion-based precipitation retrieval algorithm with high resolution AMPR measurements including a low frequency 10.7 GHz channel. *J. Atmos. Oceanic Tech.*, **11**, (in press for October issue) (1994).

18. X. Xiang, E. A. Smith and G. J. Tripoli. A cloud and radiation model-based algorithm for rainfall retrieval from SSM/I multispectral microwave measurements. In: *Preprint Vol., Sixth AMS Conf. on Satellite Meteorology and Oceanography [Jan. 5–10; Atlanta, GA]*, pp. 286–289, Amer. Meteor. Soc., Boston, MA (1992).

19. C. M. Hayden. GOES–VAS simultaneous temperature–moisture retrieval. *J. Appl. Meteor.*, **27**, 705–733 (1988).

20. C. Kummerow, R. A. Mack and I. M. Hakkarinen. A self-consistency approach to improve microwave rainfall estimates from space. *J. Appl. Meteor.*, **28**, 869–884 (1989).

21. C. Kummerow, I. M. Hakkarinen, H. F. Pierce and J. A. Weinman. Determination of precipitation profiles from airborne passive microwave radiometric measurements. *J. Atm. Oceanic Tech.*, **8**, 148–158 (1991).

22. W. S. Olson, C.-L. Yeh, J. A. Weinman and R. T. Chin. Resolution enhancement of multichannel microwave imagery from the Nimbus-7 SMMR for maritime rainfall analysis. *J. Atmos. Oceanic Tech.*, **3**, 422–432 (1986).

23. W. D. Robinson, C. Kummerow and W. S. Olson. A technique for enhancing and matching the resolution of microwave measurements from the SSM/I instrument. *IEEE Trans. Geosci. Rem. Sens.*, **GE-30**, 419–429 (1992).

24. T. T. Wilheit. Some comments on passive microwave measurement of rain. *Bull. Amer. Meteor. Soc.*, **67**, 1226–1232 (1986).

25. L. S. Chiu, G. R. North, D. A. Short and A. McConnell. Rain estimation from satellites: Effect of finite field of view. *J. Geophys. Res.*, **95**, 2177–2185 (1990).

26. D. A. Short and G. R. North. The beam filling error in the Nimbus-5 Electrically Scanning Microwave Radiometer observations of Global Atlantic Tropical Experiment rainfall. *J. Geophys. Res.*, **95**, 2187–2193 (1990).

27. P. E. Gill, W. Murray and M. H. Wright. *Practical Optimization*. Academic Press, New York (1981).

28. X. Xiang. *The Delta-Sobolev Approach for Modeling Solar Spectral Irradiance and Radiance*. PhD Dissertation, Georgia Inst. of Tech., Atlanta, GA (1989).

29. X. Xiang, E. A. Smith and C. G. Justus. A rapid radiative transfer model for reflection of solar radiation. *J. Atmos. Sci.*, **51**, (in press for June issue) (1994).

30. E. C. Barrett, J. Dodge, M. Goodman, J. Janowiak and E. A. Smith. The First WetNet Precipitation Intercomparison Project. *Remote Sensing Reviews*, in press (1994).

31. D. R. Legates. A climatology of global precipitation. *Publications in Climatology*, **40** (1), Newark, DE (1987).

32. J. K. Eischeid, H. F. Diaz, R. S. Bradley and P. D. Jones. *A Comprehensive Precipitation Data Set for Global Land Areas*. DOE Technical Report DOE/ER-69017TH1, NTIS, US Dept. of Commerce, Springfield, VA (1991).

33. C. E. Dorman and R. H. Bourke. Precipitation over the Pacific Ocean, 30°S to 60°N. *Mon. Wea. Rev.*, **107**, 896–910 (1978).

34. C. E. Dorman and R. H. Bourke. Precipitation over the Atlantic Ocean, 30°S to 70°N. *Mon. Wea. Rev.*, **109**, 554–563 (1981).

35. D. R. Legates and C. J. Willmott. Mean seasonal variability in gauge-corrected global precipitation. *Int. J. Climatol.*, **10**, 111–127 (1990).

36. D. R. Legates. An evaluation of procedures to estimate monthly precipitation probabilities. *J. Hydrol.*, **122**, 129–140 (1991).

37. M. L. Morrissey. Using sparse raingauges to test satellite-based rainfall algorithms. *J. Geophys, Res.*, **96**, 18561–18571 (1991).

38. M. L. Morrissey and F. S. Greene. *The Pacific Atoll Raingauge Data Set.* Planetary Geosciences Division Contribution No. 648, Univ. of Hawaii, Honolulu, HI (1991).

39. M. L. Morrissey and J. S. Greene. Comparison of two satellite-based rainfall algorithms using Pacific atoll raingauge data. *J. Appl. Meteor.*, **32**, 411–425 (1993).

40. L. Jaeger. Monatskarten des Niederschlags für die ganze Erde. *Ber. Dt. Wetterd.*, **18**, 139–157 (1976).

41. L. Jaeger. Monthly and areal patterns of mean global precipitation. In: *Variations in the Global Water Budget*, A. Street-Perrot *et al.* (Eds), pp. 129–140, D. Reidel (1983).

42. J. E. Janowiak and P. A. Arkin. The Global Precipitation Climatology Project: Description, goals, and progress. In: *Proc. 15th Annual Climate Diagnostics Workshop*, pp. 343–346, NOAA, Asheville, NC, NTIS PB89-178115AS (1988).

43. J. E. Janowiak. Tropical rainfall: A comparison of satellite-derived rainfall estimates with model precipitation forecasts, climatologies, and observations. *Mon. Wea. Rev.*, **120**, 448–462 (1992).

44. R. F. Adler, A. J. Negri, P. R. Keehn and I. M. Hakkarinen. Estimation of monthly rainfall over Japan and surrounding waters from a combination of low-orbit microwave and geosynchronous IR data. *J. Appl. Meteor.*, **32**, 335–356 (1993).

45. D. P. Jorgensen and P. T. Willis. A Z–R relationship for hurricanes. *J. Appl. Meteor.*, **21**, 356–366 (1992).

46. R. W. Spencer, H. M. Goodman and R. Hood. Precipitation retrieval over land and ocean with the SSM/I: Identification and characteristics of the scattering signal. *J. Atm. Oceanic Tech.*, **6**, 254–273 (1989).

47. A. J. Negri, R. F. Adler and P. J. Wetzel. Satellite rain estimation: An analysis of the Griffith–Woodley technique. *J. Clim. Appl. Meteor.*, **26**, 1565–1576 (1984).

48. J. Vivekanandan, J. Turk and V. N. Bringi. Comparisons of precipitation measurements by the Advanced Microwave Precipitation Radiometer and multiparameter radar. *IEEE Trans. Geosci. Rem. Sens.*, **31**, 860–870 (1993).

49. J. Turk, J. Vivekanandan, F. S. Marzano, R. E. Hood, R. W. Spencer and F. J. LaFontaine. Active and passive microwave remote sensing of precipitating storms during CaPE. Part I: Advanced Microwave Precipitation Radiometer and polarimetric radar measurements and models. *Meteorol. Atmos. Phys.*, in press (1994).

50. H. J. Cooper and E. A. Smith. The importance of short-term forecasting of thunderstorms to launch operations at Cape Canaveral. *Bull. Amer. Meteor. Soc.*, **74**, 81–86 (1993).

51. P. J. Webster and R. Lukas. TOGA COARE: The coupled ocean–atmosphere response experiment. *Bull. Amer. Meteor. Soc.*, **73**, 1377–1416 (1992).

52. W. M. Gray. Hurricanes: Their formation , structure and likely role in the tropical circulation. In: *Supplement to Meteorology Over the Tropical Oceans*, D. B. Shaw (Ed.), pp. 155–218, Royal Meteorological Soc., James Glaisher House, Granville Place, Bracknell, Berkshire, RG12 1BX (1979).

53. W. K. Tao, S. Lang, J. Simpson and R. Adler. Retrieval algorithms for estimating the vertical profiles at latent heat release: Their applications for TRMM. *J Meteor. Soc. Japan*, **71**, 685–700 (1993).

54. W. K. Tao and J. Simpson. Goddard cumulus ensemble model: Its application to TRMM. In: *Extended Abstract Vol. of International TRMM Workshop on the Processing and Utilitzation of the Rainfall Data Measured from Space*, pp. 79–83, Communications Research Laboratory, Tokyo, Japan (1992).

55. W. K. Tao, J. Simpson, S. Lang, M. McCumber, R. Adler and R. Penc. An algorithm to estimate the heating budget from vertical hydrometeor profiles. *J. Appl. Meteor.*, **29**, 1232–1244 (1990).

56. S. J. Lord, H. E. Willoughby and J. M. Piotrowicz. Role of parameterized ice-phase microphysics in an axisymmetric non-hydrostatic tropical cyclone model. *J. Atmos. Sci.*, **41**, 2836–2848 (1984).

ESA/NASA International Workshop, pp. 623–634
B. J. Choudhury, Y. H. Kerr, E. G. Njoku and P. Pampaloni (Eds)

Radiometric correction for atmospheric effects in surface sensing from aircraft and satellites

J. B. SNIDER, E. R. WESTWATER and L. S. FEDOR

NOAA/ERL/Wave Propagation Laboratory, 325 Broadway, Boulder, CO 80303, USA

Abstract—The Wave Propagation Laboratory (WPL) has pioneered the use of surface-based microwave radiometry for measurement of atmospheric water substance and temperature. The performance of WPL groundbased radiometers in measurement of water vapor and liquid water in clouds is reviewed. Recently WPL increased its capabilities to include sensing of the earth's surface and the atmosphere from an aircraft. The airborne instrumentation consists of upward- and downward-viewing infrared radiometers and a dual-channel microwave radiometer, operating at 23 and 31 GHz. Both infrared instruments and the microwave system, which is also equipped with an upward/downward viewing antenna system, can perform near-simultaneous measurement of upwelling emission from the surface and downwelling emission from the atmosphere. An experiment is proposed that uses a combination of airborne and ground-based radiometers to measure the emissivity of the land surface and to provide atmospheric corrections for the surface measurements.

1. INTRODUCTION

Dual-channel microwave radiometers for continuous, unattended measurement of precipitable water vapor and cloud liquid water have been employed operationally in the Wave Propagation Laboratory (WPL) since 1979. Measurement accuracy for water vapor, assessed through comparison with radiosonde observations, has been shown to be approximately 1 mm RMS or of the same order as the radiosonde accuracy. The accuracy of the liquid measurement has been more difficult to evaluate because liquid sensors are not routinely carried on radiosondes. However, an indirect method, described below, indicates that the liquid measurement accuracy is in the 15–20% range. Recently, WPL has developed a new dual-channel radiometer for operation in a NOAA King Air or WP-3D aircraft. The airborne instrument, when combined with ground-based data, provides a powerful new tool for study of the interaction of the atmosphere and the surface. In this paper, we review the capabilities of ground-based and airborne radiometry for sensing of atmospheric water vapor and liquid water. We conclude by proposing an experiment to perform aircraft measurements of surface emissivity over a network of ground-based radiometers. The ground-based data provide information for correction of the emissivity data for atmospheric absorption.

2. PERFORMANCE OF GROUND-BASED RADIOMETERS

Linear statistical methods are used operationally to retrieve precipitable water vapor (V) and liquid water (L) from WPL radiometer data. Details of the technique have

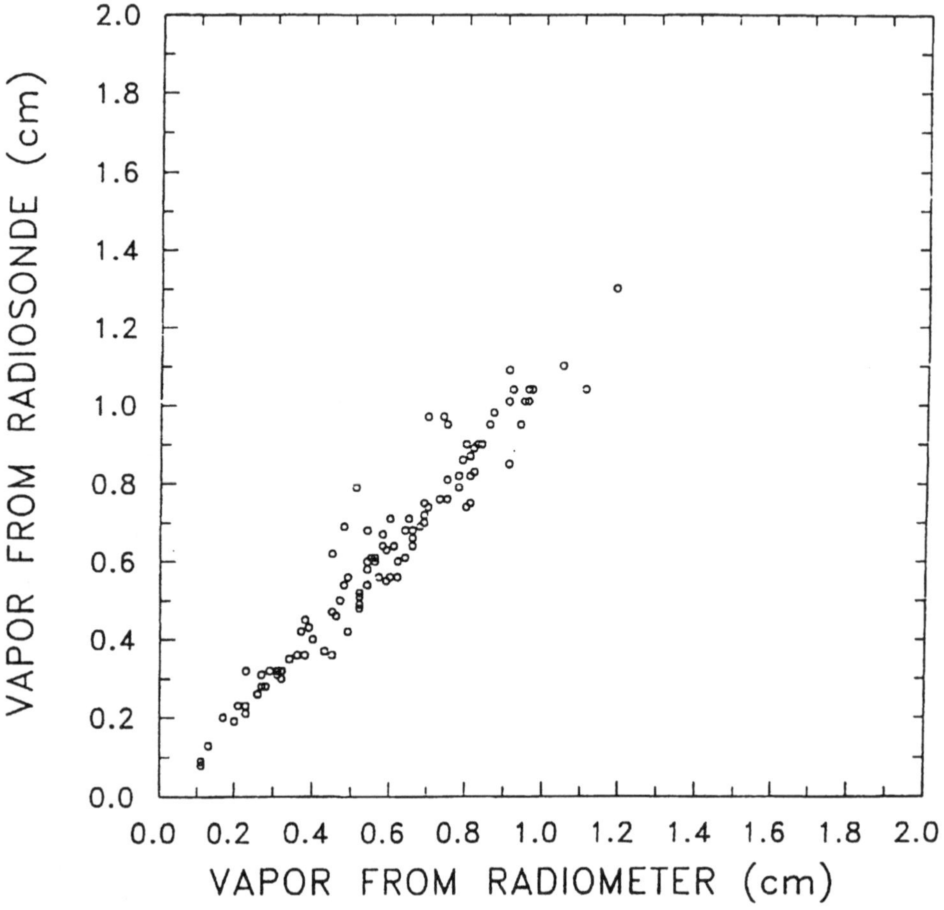

Figure 1. Comparison of precipitable water measured by microwave radiometer and by radiosonde for 121 soundings at Platteville, Colorado.

been described in [1] and [2]. The statistical procedure results in the simple relationships (1) and (2) between atmospheric absorption measured at the radiometer operating frequencies and the retrieved parameter:

$$V = C_0 + C_1\tau_{\nu_1} + C_2\tau_{\nu_2} \tag{1}$$

$$L = C_3 + C_4\tau_{\nu_1} + C_5\tau_{\nu_2}, \tag{2}$$

where C_n are coefficients calculated from a history of radiosondes for the location of interest, and the τ terms are absorptions calculated using (3) from mean radiating temperatures (T_{mr}) and brightness temperatures (T_b) measured by the microwave radiometer at frequency ν_n. T_c is the cosmic background temperature (2.7 K):

$$\tau = -\ln\left[\frac{T_{mr} - T_b}{T_{mr} - T_c}\right]. \tag{3}$$

Microwave frequencies employed in WPL systems are $\nu_1 = 20.6$ or 23.87 GHz for water vapor and $\nu_2 = 31.65$ GHz for liquid water. In calculation of the retrieval coefficients, brightness temperatures and mean radiating temperatures are calculated from historical radiosonde data using an absorption model [3]. During observations, vapor and liquid data may be retrieved using the historical calculation or estimated from the surface temperature. Because unresolved discrepancies exist between observations and models, we adjust the measured brightness temperatures to agree with the model using a linear regression between theoretical brightness temperatures as the independent variable and the observed brightness as the dependent variable [4]. In this manner, we achieve consistency between the observations and the historical radiosonde data through the absorption model. Figure 1 presents the agreement between radiometer and radiosonde water vapor for 121 simultaneous measurements at Platteville, Colorado. For the comparison, frequent (every 1.5–3 h) balloon releases were obtained using a Crosschain Loran Atmospheric Sounding System (CLASS) that employed HUMICAP humidity sensors. Agreement of the order of 1 mm seen at Platteville is typical of that obtained during clear or light cloud cover at all locations.

As discussed above, it is often necessary to employ indirect methods of assessing the accuracy of liquid water measurements. One such method is described in [5]. Values of liquid water in stratocumulus clouds over San Nicolas Island (SNI), California, were calculated by assuming adiabatic ascent of a parcel of moist air from cloud base to cloud top of the relatively thin (~200 m) stratocumulus clouds. The rationale for the adiabatic calculation is given in [5]; a synopsis of the technique follows. The adiabatic liquid water content at any level in a cloud can be estimated if the cloud base height and temperature are known. At SNI, cloud base height was measured by a laser ceilometer; cloud top height was assumed to be located 50 m below the maximum reflectivity observed by an acoustic sounder; cloud temperatures were estimated from the surface temperature and cloud base and top heights. Because the stratocumulus cloud is thin, it may be assumed that the liquid water mixing ratio varies linearly with height in the cloud. The time series of Fig. 2 compares adiabatic (calculated) and liquid water values measured by a dual-channel radiometer for a 24-h period. An offset has been added to the calculated liquid water for clarity. Similar comparisons for 19 days of observations show that the measured liquid water was close to adiabatic values for non-precipitating stratocumulus clouds less than 300 m in depth. These comparisons suggest that the actual measurement accuracy of the radiometric method is better than 20% .

3. APPLICATIONS OF VAPOR AND LIQUID MEASUREMENTS

Because of their demonstrated accuracy in measurement of water vapor and cloud liquid, microwave radiometers have been employed in a variety of atmospheric studies. Among these are studies of cloud modification for precipitation enhancement, lake effect snow storms, aircraft icing, cloud-radiative transfer interactions, and as a source of data for validation of satellite measurements. Typically, the radiometers operate continuously and unattended with the antenna beams directed vertically. The resulting data constitute a time history of the weather systems advecting overhead. Another WPL system, with antenna scan capability, has been employed to measure the spatial distribution of water vapor and liquid water in the above studies.

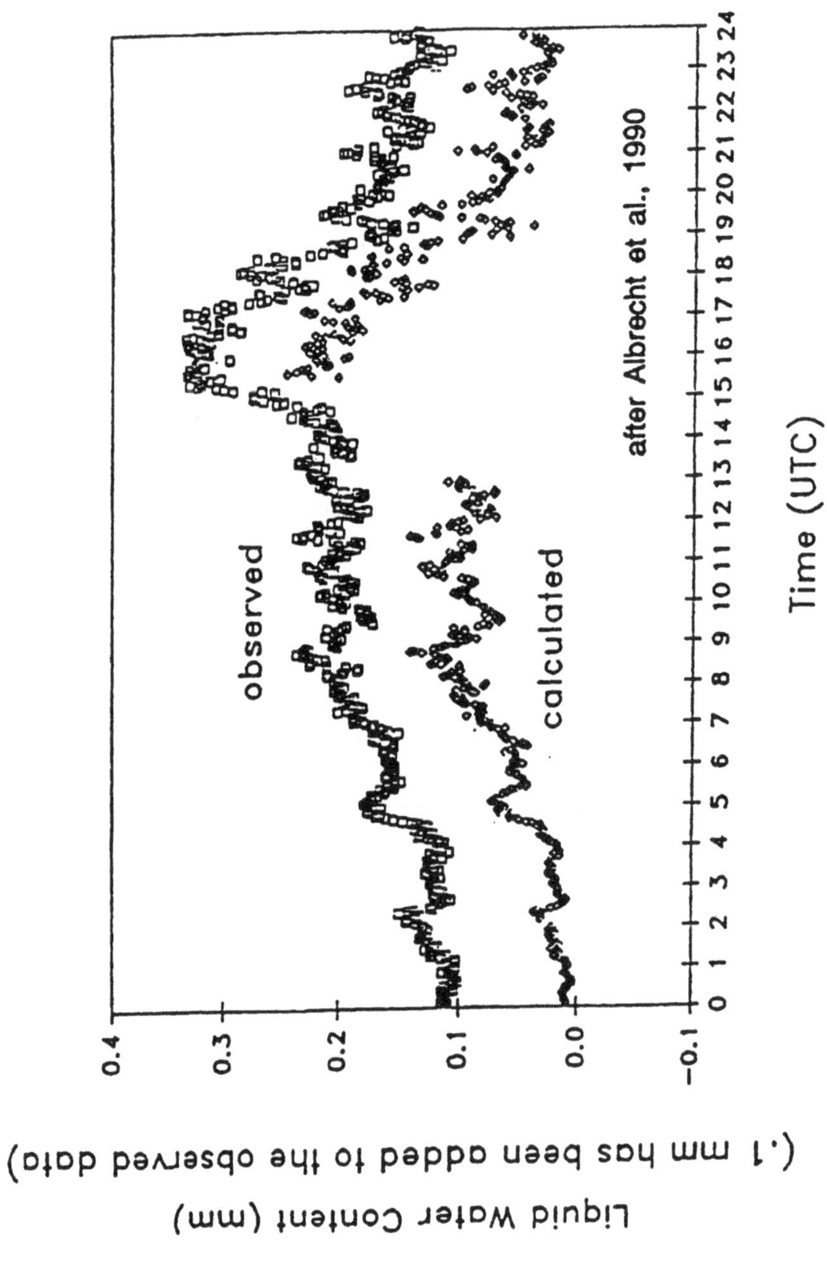

Figure 2. Observed liquid water path measured by microwave radiometer and the adiabatic liquid water amounts calculated for 13 July 1987 at San Nicolas, Island, CA.

4. USE OF VAPOR AND LIQUID DATA FOR ATMOSPHERIC CORRECTION OF SATELLITE OBSERVATIONS

Data from ground-based radiometers are capable of making corrections for atmospheric absorption and emission to satellite or aircraft measurements of surface characteristics at frequencies above 15 GHz that are absorbed substantially by water vapor and cloud liquid. During non-precipitating conditions, measurements of vapor and liquid by a dual-channel, ground-based, upward-viewing radiometer allow the accurate calculation of both brightness temperature and absorption at most frequencies above 15 GHz. With the exception of the emission in the 50–60 GHz region, a high degree of predictability exists. For example, it has been shown [4] that 90 GHz absorption could be predicted to about 5% from brightness temperature measurements near 20 and 31 GHz. Similar conclusions about the predictability of brightness temperatures, absorption, and electrical phase were reached in [6]. Therefore, it appears feasible to make atmospheric corrections for satellite observations at frequencies other than those of the WPL airborne and ground-based radiometers.

Microwave radiometers generally measure moisture in a column of relatively small angular extent directly above the radiometer (see Table 1, which lists radiometers operated by WPL). Because most satellite observations are made over footprints of tens of kilometers, there is the potential for a large inhomogeneity in the atmosphere over the area observed by the satellite. Indeed, observations by a WPL radiometer with an antenna steerable in azimuth have shown inhomogenous water vapor distributions over such areas. Figure 3 indicates the variability of the horizontal distribution of integrated water vapor at a 15° elevation angle over a 3-h period. The change in spatial distribution of the water vapor is associated with the advection of a dry air mass into the Los Angeles basin. Similar inhomogeneity in the horizontal distribution of water vapor has been observed in the Boulder–Denver, Colorado area where the Great Plains meet the Rocky Mountains. Orographic lifting often causes higher amounts of vapor over the mountains than over the plains.

These observations of inhomogeneity of the water vapor distribution suggest that point radiometric measurements, or radiosondes, may at times be of limited value in making atmospheric corrections to satellite measurements. Two possible solutions to the problem exist. The first is to employ a network of ground-based radiometers over a spatial scale comparable to a satellite footprint. For example, an array of four radiometers operated by WPL in northeastern Colorado has provided data to calibrate water vapor measurements by the Visible and Infrared Spin–Scan Radiometer (VISSR) Atmospheric Sounder (VAS) [7]. A second solution is to employ an aircraft instrumented with a water substance radiometer. WPL recently completed construction and initial flight tests of a two-frequency (23.87 and 31.65 GHz) radiometer system with upward-looking antenna. The system consists of two radiometers which are coupled into a common antenna feed through an orthogonal mode transducer. The illumination of the parabolic reflector by the feed horn is tapered so as to produce concentric, equal beamwidths at each frequency [1]. The system has been designed to be installed in both NOAA King Air and WP-3D aircraft. Low altitude flights by these aircraft would permit determination of the spatial distribution of water vapor and cloud liquid over footprint-size areas. Because of the time required to sample the area, the data would be of greatest use for correcting measurements by geostationary satellites. However, in combination with a network of ground-based radiometers, the data would also be capable of generating corrections for measurements by polar orbiters.

Table 1.
Summary of microwave radiometers operated by the wave propagation laboratory as of 3 January 1993

Location	Equipment (ν in GHz)	Viewing	Status
Stapleton International Airport Denver, CO	6-channel (20.6, 31.65, 52.85, 53.85, 55.45, 58.8)	Zenith, equal beamwidths at all channels (2.5°)	Operating (50 GHz channels to be replaced by MARK II)
Platteville, CO	2-channel (20.6, 31.65)	Zenith, equal beamwidths at both channels (5.0°)	Operating
Transportable: Porto Santo Island, Portugal	2-channel (20.6, 31.65)	Zenith, equal beamwidths at both channels (5.0°)	Operating
Transportable: Shipboard	2-channel (23.87, 31.4)	Zenith	Operating on RV Moana Wave
Airborne	2-channel (23.87, 31.65)	Zenith	Operating on NOAA P3 and NOAA King Air
Transportable	3-channel (20.6, 31.65, 90.0)	Steerable Full Sky Coverage, equal beam-widths (2.5°)	Operating in Erle, CO
Transportable	2-channel (23.87, 31.65)	Zenith equal beamwidths (2.5°)	Operating in Kavieng Papua NG

An example of an aircraft observation of an inhomogeneous airmass is given in Figs 4 and 5. Figure 4a shows the humidity profile measured by National Weather Service (NWS) radiosonde at 11:00 UTC, 24 January 1992. Figures 4b and 4c show humidity profiles measured by sensors on board the NOAA King Air aircraft during an ascent at 12:06 UTC and a descent at 12:29 UTC. The increase in upper level moisture with time is clearly evident. Figures 5a and 5b show the corresponding brightness temperature profiles (solid lines) at 23.87 and 31.65 GHz, respectively, measured by the aircraft radiometer. The aircraft profiles were flown in a spiral ascent at a zenith angle of approximately 75°. From left to right, the dotted-dashed-dotted lines are the brightness temperatures at elevation angles of 90°, 70°, and 60°, respectively, calculated from the 11:00 UTC radiosonde data updated by the humidity data from the aircraft sounding and an absorption model. The increase in brightness temperatures above the calculations was later found by *in situ* sensing by the aircraft to be associated with an increase in upper-level moisture (2.5–5 km) located to the southwest of the aircraft

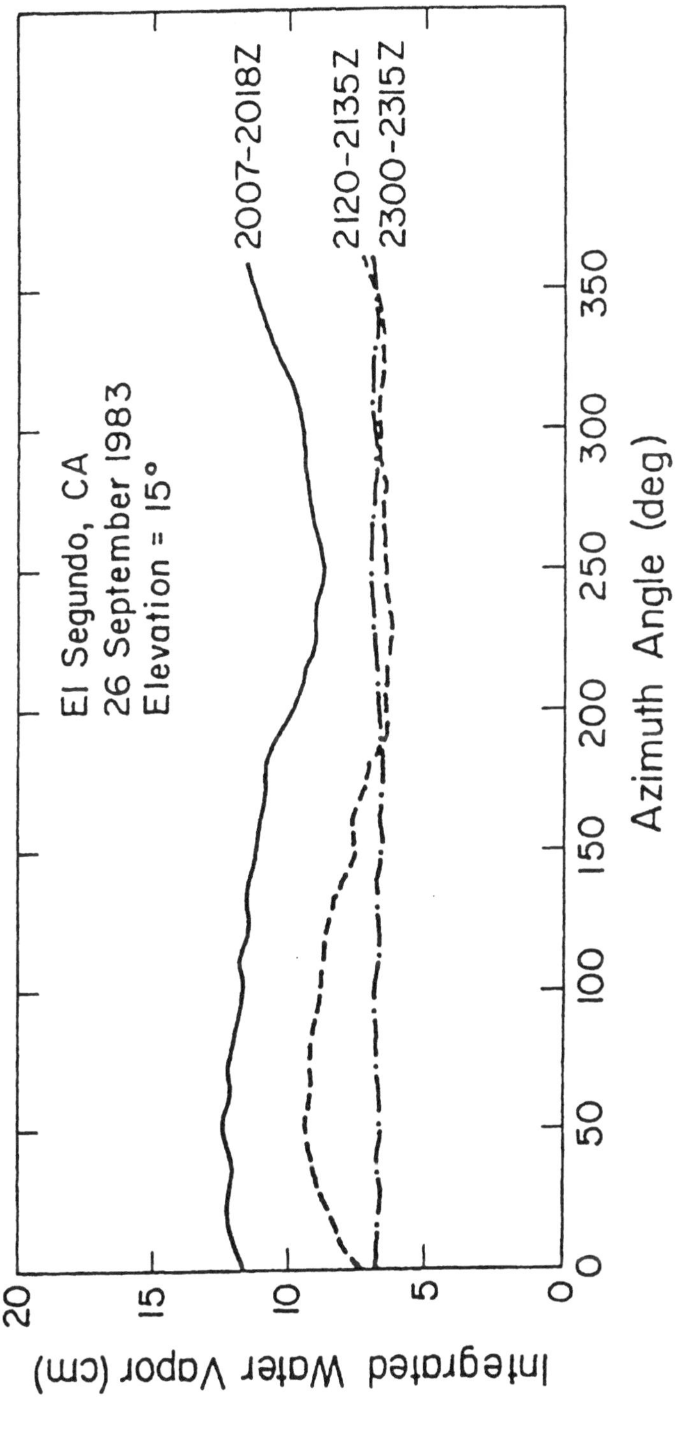

Figure 3. Azimuth scans by microwave radiometer showing movement of a dry air mass into the Los Angeles Basin.

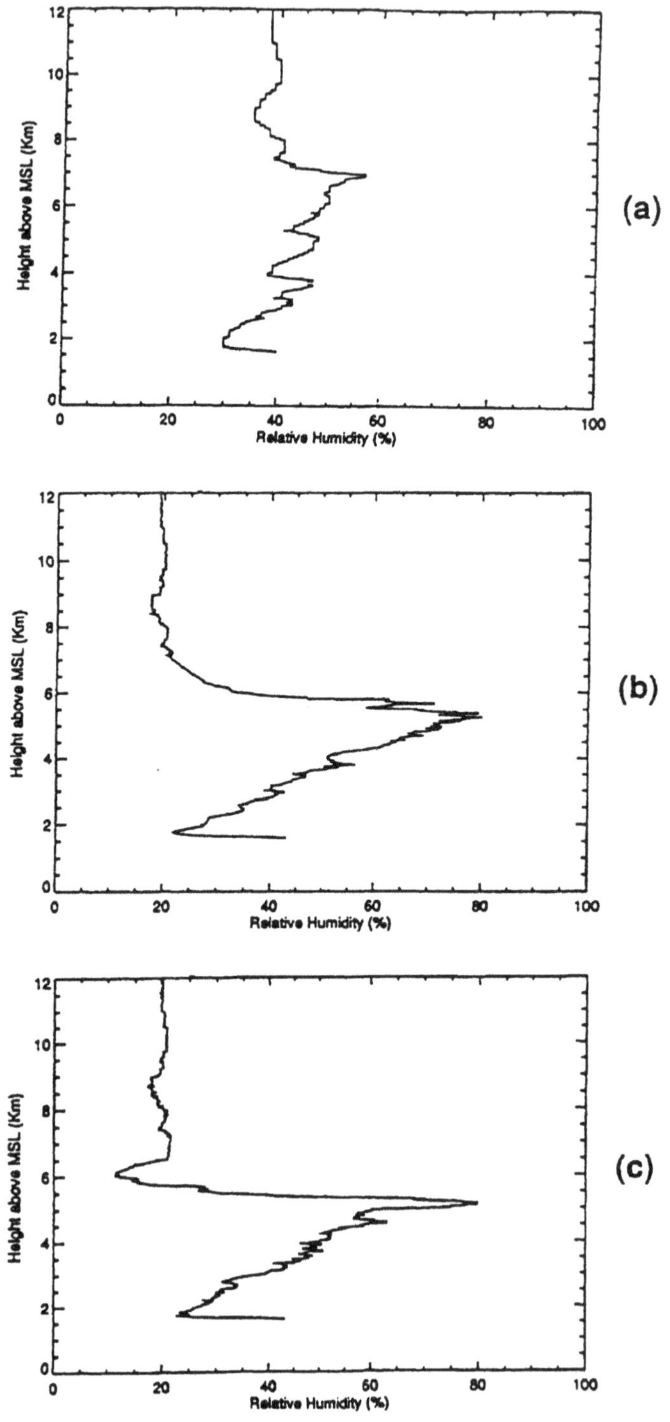

Figure 4. Humidity profile measured by National Weather Service radiosonde, 24 January 1992, 11:00 UTC (a), by NOAA King Air aircraft at 12:06 UTC (b), and by NOAA King Air aircraft at 12:29 UTC (c).

during the ascent and descent. This result clearly illustrates the importance of accounting for atmospheric inhomogeneity in correction of atmospheric measurements by satellite.

The antenna system of the WPL airborne microwave radiometer has been modified and is being tested to make downward-viewing observations. In addition to the microwave radiometer, the NOAA King Air and WP-3D aircraft are being equipped with upward- and downward-viewing narrowband infrared (IR) radiometers operating at approximately 10.7 μm.

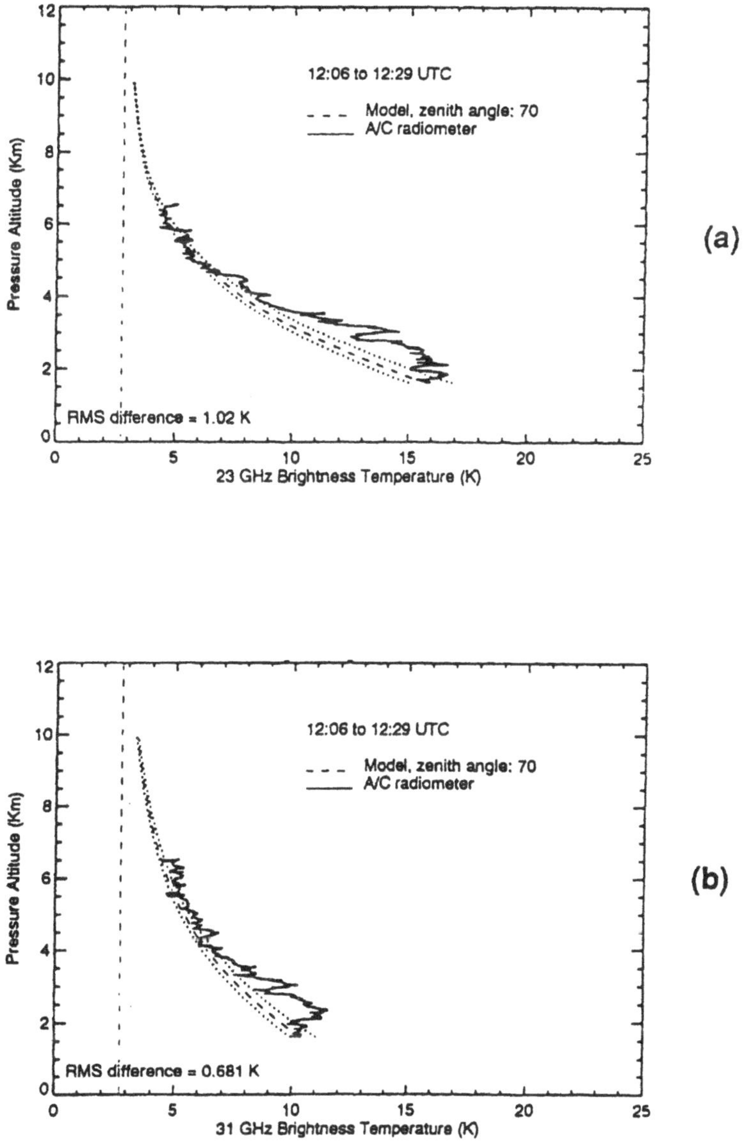

Figure 5. Brightness temperature profiles measured by airborne microwave radiometer at (a) 23.87 GHz, and at (b) 31.65 GHz, 24 January 1992, 12:06 to 12:29 UTC. Dotted and dashed lines (left to right) are brightness temperatures calculated from radiosonde data at 90°, 70°, and 50° zenith angles.

5. PROPOSED EXPERIMENT FOR MEASUREMENT OF EMISSIVITY OF LAND SURFACE

Microwave emissivity of the earth's surface varies with soil moisture, surface roughness, and other characteristics such as vegetation cover. Because a wide variation in these parameters can occur over areas of the order of a satellite footprint, over-land retrievals of water vapor and liquid water from satellites such as SSM/I have been largely unsuccessful [8]. To obtain accurate overland retrievals, it is important for the spatial distribution of emissivity over such areas to be known as a function of time. We now propose an experimental method to obtain such data over an area of approximately 900 km^2 from a combination of ground-based and airborne radiometers.

Microwave radiative transfer for a downward-viewing radiometer is illustrated graphically in Fig. 6. The measured brightness temperature, T_b, is given by

$$T_b = \varepsilon_s T_s e^{-\tau(0,R)} + (1 - \varepsilon_s) e^{-\tau(0,R)} T_b^{\downarrow} + T_b^{\uparrow}, \tag{4}$$

where ε_s equals the emissivity of the surface, T_s equals the surface temperature, and $T_b^{\uparrow}$, $T_b^{\downarrow}$ equals the upward, downward atmospheric emission. The quantity

$$\tau(0,R) = \int_0^R \alpha(s')\,ds' \tag{5}$$

is the absorption at frequency, ν, from the surface to height R.

The first term in (4) is the emission from the surface, while the second term is the downward atmospheric emission reflected from the surface and attenuated by the intervening atmosphere. The third term is the upward emission from the atmosphere, given by

$$T_b^{\uparrow} = \left[1 - e^{-\tau(0,R)}\right] T_{mr}, \tag{6}$$

where T_{mr} is the mean radiating temperature of the atmosphere at frequency ν_n. Near the surface, i.e. as $\tau \to 0$, the brightness temperature becomes

$$T_b = \varepsilon_s T_s + (1 - \varepsilon_s) T_b^{\downarrow}. \tag{7}$$

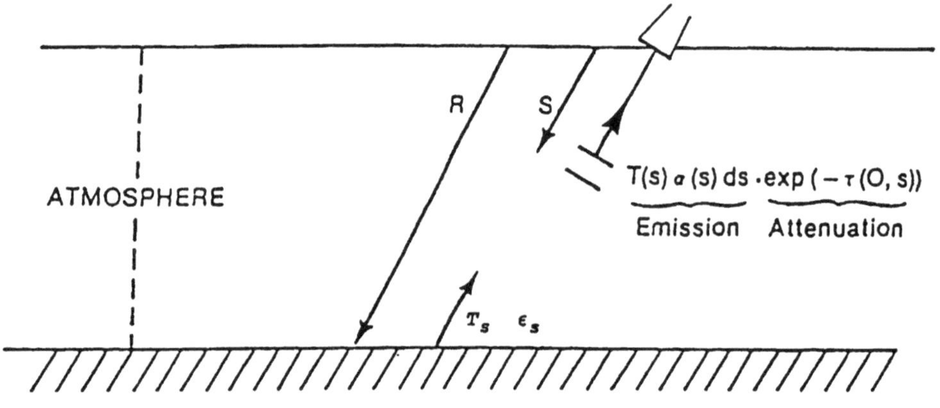

Figure 6. Graphical illustration of microwave radiative transfer for a downward viewing radiometer.

Solving (7) for the emissivity yields

$$\varepsilon_s = \frac{T_b - T_b^{\downarrow}}{T_s - T_b^{\downarrow}}. \tag{8}$$

In the proposed experiment, the surface temperature[1] in (7) is measured by a downward-looking IR radiometer on the aircraft. The upwelling microwave emission, T_b, is measured by the downward-looking dual-channel microwave radiometer while the downwelling emission, $T_b^{\downarrow}$, is measured by the airborne microwave radiometer with its antenna directed upward. The upward and downward airborne measurements would be obtained alternately as the aircraft traversed the test area. The flight pattern would be chosen to permit measurements to be performed at each antenna polarization and at each frequency.

At altitudes where $t > 0$, (7) is no longer valid and the surface emissivity must be determined from (4). Solving (4) for the emissivity and utilizing (6) gives

$$\varepsilon_s = \frac{T_b - T_b^{\downarrow} e^{-\tau(0,R)} - \left(1 + e^{-\tau(0,R)}\right) T_{mr}}{(T_s - T_b^{\downarrow}) e^{-\tau(0,R)}}. \tag{9}$$

The combined measurements from aircraft and surface instruments provide the information needed to evaluate all terms in (4). For example, $\tau(0, R)$ can be calculated from data provided by aircraft and balloon soundings of temperature and humidity, $T_b^{\downarrow}$ is measured by the array of ground-based, upward-looking radiometers, and T_{mr} can be estimated from *in situ* aircraft measurements. Other terms are measured by the aircraft radiometer as described in the preceeding paragraph.

By flying traverses at different altitudes, the footprint observed by the aircraft would change. For example, the nominal antenna beamwidth of the airborne microwave radiometer is 5°; this angular resolution produces a footprint diameter that varies from 4.4 m at an altitude of 50 m to 873 m at 10 000 m altitude. Through comparison of the measured emissivity over different footprint sizes, the effect of spatial averaging could be determined. Similarly, if data were collected during times of satellite overflights, it would be possible to determine the effects of spatial averaging of emissivity over the entire footprint observed by the satellite.

Simultaneously with the aircraft overflights, the downwelling radiation would be measured by a number of ground-based dual-channel radiometers located in the test area. Data from these instruments will be valuable in detecting the presence of inhomogeneity of the water vapor and cloud field above the test area. The WPL transportable radiometer with steerable antenna would be employed in the center of the network. Its position would be selected so that it could observe the atmosphere above each of the dual-channel radiometers. The steerable system would operate in both zenith and azimuthal scan modes to monitor the homogeneity of the moisture distribution over the test area. The combined measurements from all instruments will provide the information needed to correct the emissivity data for the atmospheric absorption by water vapor and cloud liquid.

[1]Actually an effective surface temperature, since the IR emission will be dependent on the soil temperature, soil moisture, and vegetation cover.

6. LOCATION OF EXPERIMENT

Although several site locations for the experiment are possible, e.g. the Boulder–Denver area, other considerations indicate the use of a site being established by the Atmospheric Radiation Measurements (ARM) program of the US Department of Energy (DOE). DOE/ARM has begun operation of a Cloud And Radiation Testbed (CART) site in north central Oklahoma. The site is planned to be operated for 8–10 years. The Oklahoma CART will be equipped with many remote sensors, both passive and active, including Doppler radar wind profilers, a Radio Acoustic Sounding System (RASS) temperature sounder, a lidar ceilometer, microwave, and infrared radiometers. The radiometers are deployed at a central location and at three outlying sites, and thus satisfy the experimental detail outlined above. In addition, CLASS radiosondes are launched twice daily. Finally, the state of Oklahoma also operates a network of rain gauges, and a NEXRAD radar covers the CART site. The wealth of instrumentation, either in place or planned for the relatively near future, is a compelling argument for use of the Oklahoma site for the study of surface emissivity. Another argument is the seasonal variability in weather prevailing in the area. For example, the area is mainly agricultural and benefits from an annual rainfall of 70 cm. Winter snowfalls are also common, ensuring that emissivity from a wide variety of surface conditions can be investigated.

REFERENCES

1. F. O. Guiraud, J. Howard and D. C. Hogg. A dual-channel microwave radiometer for measurement of precipitable water vapor and liquid. *IEEE Trans. Geosci. Electron.*, **GE-17**, 129–136 (1979).
2. D. C. Hogg, F. O. Guiraud, J. B. Snider, M. T. Decker and E. R. Westwater. Steerable dual-channel microwave radiometer for measurement of water vapor and liquid in the troposphere. *J. Appl. Meteor.*, **22**, 789–806 (1983).
3. H. J. Liebe and D. H. Layton. *Millimeter-Wave Properties of the Atmosphere: Laboratory Studies and Propagation Modeling*. Nat. Telecomm. and Inform. Admin., Boulder, CO, NTIA Report 87-24 (1987).
4. E. R. Westwater, J. B. Snider and M. J. Falls. Ground-based radiometric observations of atmospheric emission and attenuation at 20.6, 31.65, and 90.0 GHz: a comparison of measurements and theory. *IEEE Trans. Ant. Prop.*, **AP-38**, 1569–1580 (1990).
5. B. A. Albrecht, C. W. Fairall, D. W. Thomson, A. B. White, J. B. Snider and W. H. Schubert. Surface-based remote sensing of the observed and the adiabatic liquid water content of stratocumulus clouds. *Geophys. Res. Lett.*, **17**, 89–92 (1990).
6. G. Schiavon, D. Solimini and E. R. Westwater. Performance analysis of a multi-frequency radiometer for predicting atmospheric propagation parameters. *Radio Science*, **28**, 63–76 (1993).
7. D. Birkenheuer. An algorithm for operational water vapor analyses integrating GOES and dual-channel microwave radiometer data on the local scale. *J. Appl. Meteor.*, **30**, 834–843 (1991).
8. J. C. Alishouse, J. B. Snider, E. R. Westwater, C. T. Swift, C. S. Ruf, S. A. Snyder, J. Vongsathorn and R. R. Ferraro. Determination of cloud liquid water content using the SSM/I. *IEEE Trans. Geosci. Rem. Sens.*, **GE-28**, 817–822 (1990).

Instrumentation

ESA/NASA International Workshop, pp. 637–646
B. J. Choudhury, Y. H. Kerr, E. G. Njoku and P. Pampaloni (Eds)

The design of the ESA Multifrequency Imaging Microwave Radiometer (MIMR)

Y. MENARD

ESA/ESTEC, PO Box 299, 2200 AG Noordwijk, The Netherlands

Abstract—The Multifrequency Imaging Microwave Radiometer (MIMR) is a 6 frequencies, 20 receivers, dual polarisation passive microwave radiometer instrument, designed by Alenia for the European Space Agency (ESA) to be flown on a Polar Platform due to be launched in the year 2000.

As an introduction this paper describes the operational and functional parameters of MIMR and its mission objectives. Then the performances, the functional concept, the design and the calibration aspects are presented. MIMR is a leap forward in microwave radiometry with respect to already flown radiometers and will provide high performance microwave measurements for determination of global weather and critical oceanographic, atmospheric, cryospheric and land parameters, for operational forecasts and the research community for the years beyond 2000.

1. INTRODUCTION

MIMR will fly on a spacecraft (Fig. 1 as example) in a circular sun synchronous near polar orbit at an altitude of approximately 705 km with an inclination of 98.2° and an orbit period of 99 min. The orbit produces 14.5 full orbit revolutions per day, its nodal ascending crossing time at the equator is in the range 1:15 to 1:45 p.m. The conical scan geometry of MIMR (Fig. 2) views the earth scene with an incidence angle of 50°.

The active portion of the scan covers a swath of 1230 km which results in a 24 h global coverage of 82% of the earth surface (Fig. 3). Only the black areas are missed. Table 1 provides the Earth surface coverage for one, two and three days. The small polar caps at North and South Poles are never covered due to the orbit inclination and represent 0.2% of the Earth surface. But the diamond shaped regions will be fully covered after 72 hours.

MIMR rotates continuously about an axis parallel to the local spacecraft vertical. The scan direction is from the right to the left when looking in the fore direction of the spacecraft with the active scan measurement lying ±60° about the fore direction,

Table 1.
MIMR earth coverage capabilities

Coverage frequency (day)	Minimal latitude of full coverage	% of the earth surface covered
1	±55°	82%
2	±17°	98.8%
3	0°	99.8%

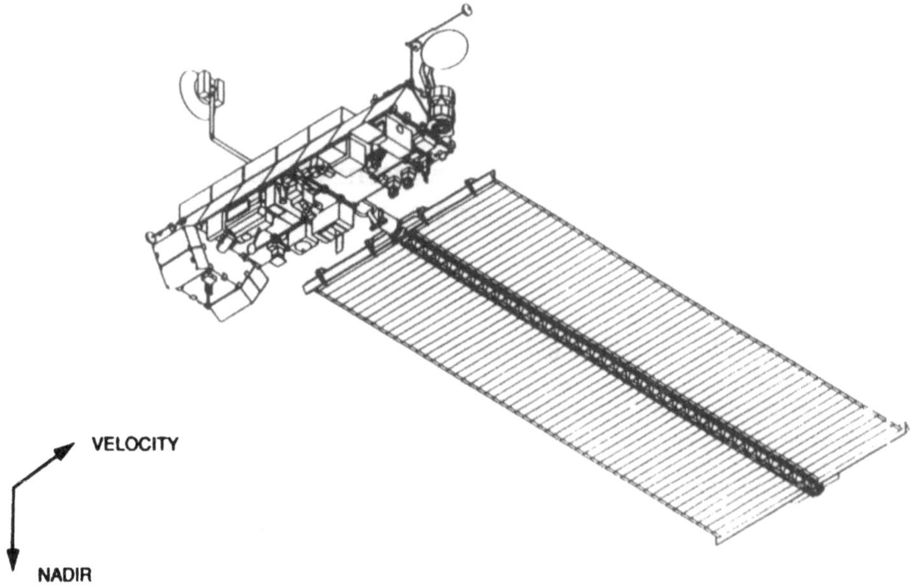

Figure 1. EOS-A observatory.

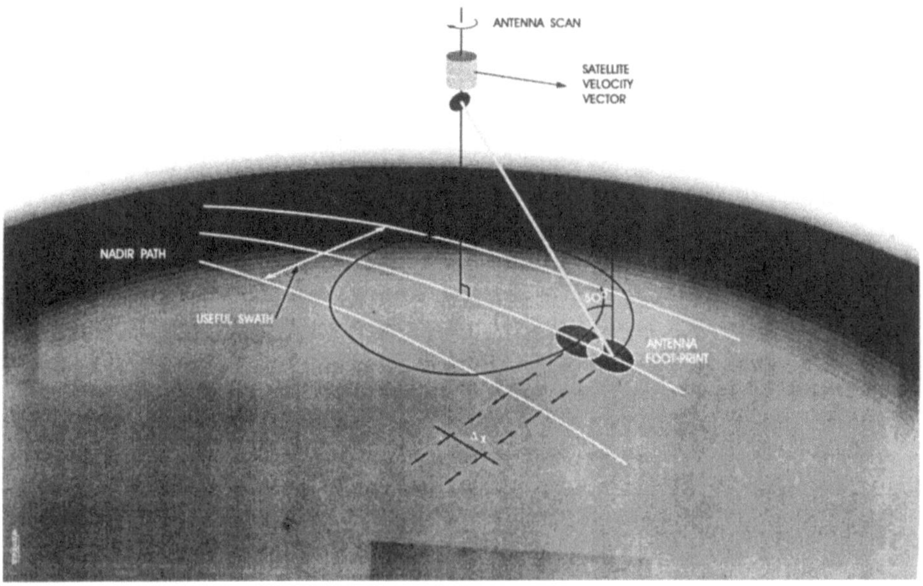

Figure 2. Conical scanning.

resulting in a swath width of 1230 km. The spin rate of 26 rpm provides a period of 2.3 seconds during which the spacecraft sub-satellite point moving at 6.75 km/s travels 15.5 km, which corresponds to an overlap of the antenna footprint along the track of

Figure 3. Daily coverage of the Earth surface.

nearly 30% for the 18.7 GHz channel. The 36.5 and 89 GHz channels use a pushbroom technique with two beams for the former one and four beams for the latter, in order to cover the 18.7 GHz footprint along the track. The intertrack distances between footprints for these channels are respectively 7.8 km and 3.9 km. The sampling interval is set to 8, 4, 2, 1 ms and equals the time for the beam to travel 15.5, 7.8, 3.9 and 1.9 km (depending upon the beam) in the cross track direction. This sampling strategy allows resampling of the data on standard gridded projections (Mercator, polar) to be used as an International Standard by the scientific community.

2. SCIENTIFIC OBJECTIVES

The MIMR mission objectives are the measurement and full-time monitoring of geophysical parameters related to the atmosphere, the ocean, the cryosphere and the land.

The most important geophysical parameters for each application area to be measured by MIMR are:

Atmosphere:
- total water vapour content in the atmosphere
- total liquid water content
- rain rate
- cloud extent
- typhoon monitoring
- ice content in clouds

Ocean:
- wind speed at the sea surface
- sea surface temperature

Cryosphere: – sea ice concentration
– ice type
– ice mapping

Land: – snow cover
– snow water content
– permafrost
– soil moisture, vegetation characteristics (semi-mid areas)

Ocean/atmosphere interaction: – carbon dioxide exchange.

Table 2.
MIMR environmental products performance

Parameter	Sampling distance (km)	Range of values	Quantisation levels	Absolute accuracy
Ocean temperature	15.5	0 to 30	0.2	±1 °C
Ocean surface wind speed	15.5	3 to 25	1	±2 m/s
Ice	15.5			
Area covered	15.5	0 to 100	5	±12%
Age		1st year multiyear	1 yr, > 2 yr	None
Edge location		N/A	N/A	±2 km
Cloud water	15.5	0 to 1	0.05	0.1 kg/m
Liquid water vapour	15.5	0 to 60	0.10	±0.2 kg/m
Precipitation over ocean	15.5	0 to 25	0, 5, 10, 15, 20, ⩽ 25	±5 mm/hr
Soil moisture	15.5	0–60%	1	None
Snow water content	15.5	0–50 cm	1	±3 cm
Surface type	15.5	12 types	N/A	N/A
Cloud amount	15.5	0–100%	1	±20%

3. INSTRUMENT CONCEPT

The MIMR instrument (Fig. 4) consists of an offset parabolic reflector of dimension 1.6 m × 1.4 m, illuminated by a set of 10 feed horns arranged as close as possible from the focal point of the antenna. The reflector and feed horns antennae are mounted on a drum enclosing the receivers, the data handling equipment, the command and control function, and the power equipment. The entire drum assembly is rotated about the axis of the instrument by a coaxially mounted bearing and a brushless DC motor. Data commands, timing, telemetry signals, and power lines pass through the rotating part to the stator via a roll ring assembly.

The MIMR is a 20 receivers, 6 frequencies passive microwave radiometric system. It

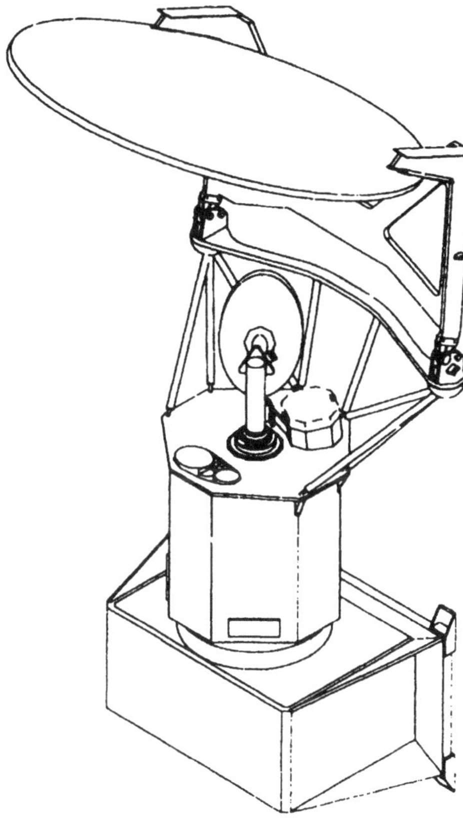

Figure 4. MIMR instrument concept.

receives both vertically and horizontally linearly polarised radiation at 6.8, 10.65, 18.7, 23.8, 36.5 and 89 GHz.

The receiver concept is based upon a total power radiometer configuration which provides a factor of two greater sensitivity over a conventional 'Dicke' switched system.

The calibration concept uses the external calibration technique where a small mirror and a hot reference absorber are mounted on the top of the drum off axis in order to occult the horn assembly once per rotation, thus providing a warm and cold (2.7 K) reference point for each channel every 2.3 seconds.

These two last concepts have been experienced with success on the SSM/I radiometer flown on board the Defense Meteorological Satellite (DMSP) and have demonstrated a considerable improvement of the performances compared to previous spaceborne radiometric systems.

4. INSTRUMENT PERFORMANCES

The scanning parameters and the altitude of the satellite govern the sampling distance along track indicated in Table 3. The data stream at the output of the instrument provide an oversampling of a factor two for the channels two to six in order to deliver data to

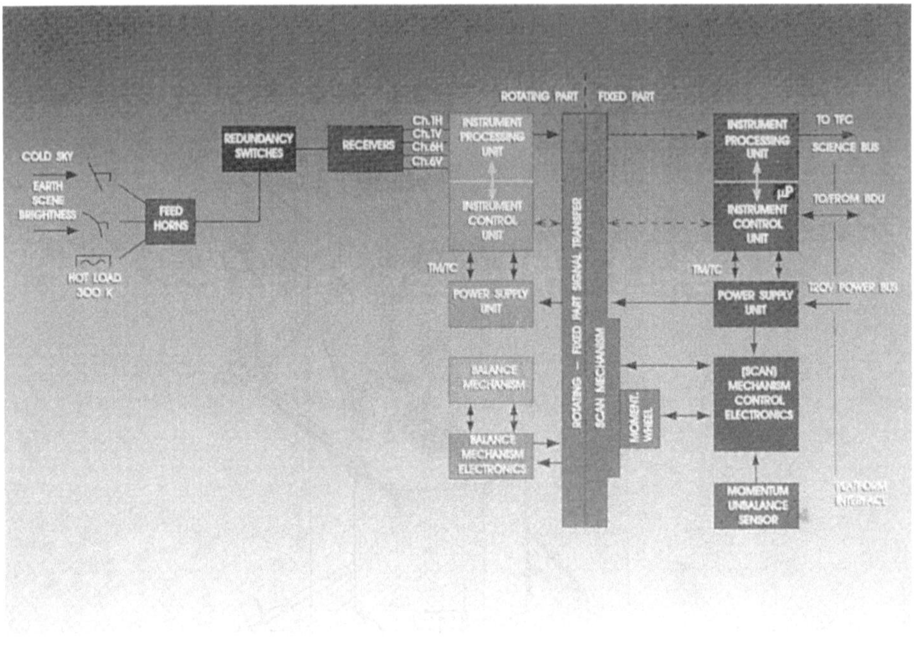

Figure 5. MIMR block diagram.

Table 3.
MIMR instrument performances

Channels		1	2	3	4	5	6
Centre frequency (GHz)		6.8	10.65	18.7	23.8	36.5	89
Pixel size	along track (km)	60	38	22	20	11.6	4.9
	across track maximum value (km)	60	38	22	20	11.6	4.9
Sampling interval	along track (km)	15.5	15.5	15.5	15.5	7.75	3.87
	across track (km)	15.5	7.75	7.75	7.75	3.87	1.94
Radiometric sensitivity (K)		0.2	0.4	0.5	0.5	0.5	0.7
Radiometric stability (K)		0.2	0.4	0.5	0.5	0.5	0.7
Radiometric accuracy (K)		1	1	1.5	1.5	1.5	1.5

the ground processing station with the minimum of aliasing allowing a resampling on a normalised grid (5, 10, 20, 40 km) with an optimum image quality.

The sensitivity provided in Table 3 indicates the sensitivity of the MIMR instrument when the signal is integrated over a distance equal to the along track footprint size. The sensitivity for each sample is slightly worse, but users are able to use either the best sampling size or the nominal sensitivity by integrating samples. The radiometric stability indicates the variations of the measurements along an orbit, and over a year due to temperature variations inherent to day and night variations and to solar flux variations.

The radiometric accuracy indicates a maximum limit of the bias due to the instrument which needs to be corrected by proper characterisations and validations.

The ground segment will include a pointing error correction algorithm in order to correct for incidence angle variations over the orbit and geolocalisation errors. A robust radiometric error correction software package will be included using instrument characterisation data collected during the pre-launch testing campaign, instrument thermal sensors telemetries, and hot and cold calibration data as provided by the calibration subsystem. The data will be resampled on a standard grid projection (e.g. Boulder University standard grids) and the co-registration of the six channels of MIMR will be made on these grids.

5. INSTRUMENT DESIGN

5.1. Antenna

The antenna sub-system retains the concept of a single 1.6 m offset antenna with a main reflector whose core is made of aluminium honeycomb with carbon fibre reinforced plastic (CFRP) face sheets layers on each side. The reflector face looking at the Earth is covered by a thin layer of aluminium vapour deposited, for good emissivity performances.

The measured surface roughness is less than 34 μm rms, which ensures very good performance at 90 GHz. The reflector illuminates a set of 9 feed horns, one for each lower frequency channel, two for the 35 GHz and four for the 90 GHz channel.

The multihorn concept used for the two 36 and 89 GHz channels has been selected in order to provide an optimum footprint size, as provided by the large antenna size, with a good sampling interval along track and a good coverage, while maintaining the scanning speed compatible with the state of the art of existing scanning mechanisms.

The antenna parameters (Table 4) have been optimized so as to provide a beam efficiency for each channel better than 93% and a cross polarisation rejection performance better than 23 dB.

5.2. Receivers

The 20 receivers operate in total power mode and three configuration schemes have been retained. Direct detection is expected for the 6.8 and 10.65 GHz channels, single side band (SSB) heterodyne receivers are preferred for the 18.7, 23.8 and 36.5 GHz channels, and a double side band (DSB) receiver with beam lead diode front end mixer is envisaged for the 90 GHz channels. HEMT amplifiers are foreseen as low noise amplifiers for channels one to five. They provide very good performance, and noise figures have been measured as low as 1.7 dB at 10.65 GHz and 3.8 dB at 36 GHz.

Table 4.
Antenna characteristics per channel

Channels		1	2	3	4	5	6
Antenna beamwidth (deg)		2.158	1.382	0.797	0.732	0.416	0.174
Footprint sizes goal (km)	along track	60.25	38.6	22.3	20.5	11.7	4.9
	across track	38.7	24.8	14.3	13.1	7.5	3.2

The local oscillators of channels 3 to 6 will use Dielectric Resonator Oscillators (DRO) as fundamental source, and will be coupled to multipliers for the highest frequencies. This technology will improve the power consumption, the stability, and the reliability of the oscillator with respect to a classical design using gun diodes. The topology of the receivers provides the capability of reconfiguration of the receiver in the event of failure of one chain. In this case, the vertical or the horizontal signal can be switched towards the operating receiver.

5.3. Scanning mechanism

The scan mechanism is based on a well-improved-by-testing design made for AMSU-B instrument.

The scan mechanism function is used primarily to command and control the rotation of the rotating part of the instrument at a speed of 26 revolutions per minute. In addition, it will provide information about the angular position of the antenna, transfer power, command, telemetry and scientific data from the payload to the rotating drum and vice versa. It will also provide compensation for the induced torque, and mechanically support the mass of the antenna and its associated horns and receivers. The scan mechanism itself is built around a 12 poles brushless DC motor mounted directly on the main shaft. The scanning part, or rotor, is supported within the mechanism by two lead-lubricated angular contact bearings assembled in a back-to-back configuration. The upper main bearings are housed in a spherical seat housing, dry lubricated with lead, providing accurate self alignment of the bearings. The mechanism rotor position coding is provided by a 50 pitch inductosyn sensor.

The reaction wheel assembly which counterbalances the torque, induced by the rotation of the antenna system (35 cm diameter), is an upgraded version of traditional design. It exhibits friction torque as low as 0.006 to 0.01 Nm. The wheel is hermetically sealed and wet lubricated with KG80 hydrocarbon oil.

The power and signal transfer across the mechanism rotating interface is provided by a roll-ring assembly, built and life tested by Honeywell, which provides low friction noise.

The balance and control system is required to provide on orbit static and dynamic balancing of the instrument to minimise disturbances to the spacecraft platform.

The balance control system comprises four piezoelectric force sensors and two balance mechanisms. Each balance mechanism employs a stepper motor which acts as an integral sliding mass.

The offload device ensures that the load of the instrument during launch bypasses the scanning mechanism. The device is locked by marmon clamp blocks which will be released by means of pyrotechnics. The scan control electronics drive the scanning mechanism motor and control the speed of the momentum wheel. It drives the balance mechanism on request such as to minimise the static and dynamic unbalance, and interfaces with the instrument Intelligent Control Unit (ICU) in order to command and monitor properly the operation of the scanning mechanism.

5.4. Command and control

The ICU is almost standard equipment which interfaces the instrument with the On Board Data Handling Bus. This unit allows command of the instrument modes, acknowledgement of the command and monitoring of the proper functioning of the instrument. It provides housekeeping data and transfers the spacecraft clock for the datation of the scientific measurements. In addition, patch and dump functions are integrated in the ICU.

5.5. Science data interface

The science data processor interfaces the output of each receiver with the payload science data handling system. The functions of this unit are mainly to sample, hold and quantify incoming receiver signals. The resulting 16-bit digital signals are then buffered and multiplexed. The science datastream is merged with telemetry, auxiliary and calibration data into standard frame packets and sent to the payload data handling system.

5.6. Power interface

The power supply unit (PSU) will interface with the spacecraft power bus and will provide to the various subsystems and equipment the necessary regulated power with a high efficiency.

5.7. Calibration

The onboard calibration concept is relying on two calibration devices, a small reflector looking at the 2.7 K cosmic background radiation and a warm reference absorber, mounted on the top of the instrument (Fig. 4), occulting each scan period the main reflector, feeding the horns with an accurately known radiation power.

This calibration scheme provides an overall end-to-end absolute calibration which includes the horns and their associated receivers.

In addition highly sensitive thermal sensors (0.05 K) will be used to monitor the antenna main reflector, the hot calibration target and the calibration reflector temperature. This information will be used during the processing of the data on ground for radiometric corrections. This calibration concept based on SSM/I experience will ensure an adequate calibration of the instrument gain variations due to the temperature cycles over one orbit and during the extreme variations of solar heating which occur between summer and winter solstices.

5.8. Accommodation budgets

In its present design, the mass of MIMR, including the support structure to adapt the instrument on the spacecraft, does not exceed 200 kg and its power consumption is nearly 190 W. The data rate including auxiliary and calibration data and headers is less than 67 kbit/s with some oversampling included. The uncompensated momentum as measured is better than 0.3 Nm and with in-orbit balance the dynamic and static imbalance are respectively better than 5800 $kg\,mm^2$ and 3.54 kg mm. The scanning position accuracy is better than 0.01 degree.

6. CONCLUSIONS

The MIMR instrument design presented in this paper is tailored to the scientific mission performances and accommodation objectives outlined in the first sections of this paper.

In its present concept the instrument is limited by the accommodation specifications (residual dynamic and static torques) and the state of the art of the in-orbit balance and control mechanism. This constrains the scan rate and the rotating mass of the rotor, and as a consequence provides boundary conditions on the antenna diameter, the footprint size and the pixel interval as well as the number of channels to be accommodated in the rotating drum. Considering image quality, further studies are necessary for the definition of the ground segment pre-processing and in the field of scientific parameters algorithms.

A good background exists in Europe [3, 5] in the field of inversion methods and radiometer instrument validation based on the work performed for the ATSR instrument on board ERS-1 launched in 1991; nevertheless further work is necessary in this area when considering the upgraded performances of MIMR with respect to ATSR.

REFERENCES

1. *The Multifrequency Imaging Microwave Radiometer*, ESA SP 1138 and F.31.
2. Selenia MIMR Phase B Study Report – ESA Contract 9676/91/NL/PR.
3. J. Y. Lojou. Algorithmie et methodes de validation des instruments en radiometrie hyperfrequence. Thesis submitted to University Paris 7 (1990).
4. E. A. Francis. Calibration of the Nimbus-7 SMMR, Thesis submitted to Oregon State University (1988).
5. Study of Microwave Interaction with the Earth's Surface – ESTEC Contract 8447/89/NL/PB(SC) Dec. 90 by Helsinki University of Technology (1990).

Working group reports

ESA/NASA International Workshop, pp. 649–685
B. J. Choudhury, Y. H. Kerr, E. G. Njoku and P. Pampaloni (Eds)

ESA/NASA St.-Lary Workshop 11–15 January 1993 Working Group Reports

EXECUTIVE SUMMARY

A major component of the workshop was the discussion which ensued on selected topics of direct importance to the land community. In the first session, there were four concurrent working groups discussing specific parameters — snow, vegetation/temperature, soil moisture, and atmospheric correction/precipitation. In the second session, the participants re-grouped to discuss passive microwave satellite data and products in relation to modelling requirements and potential synergism with other sensors.

The first session included discussion on problems, solutions, future research, and modelling in relation to discrepancies between algorithms, discrepancies between experimental results and discrepancies with *in situ* data. Data issues, including gridding and archiving were addressed. Geophysical products which are, or can be produced, were identified. Issues of scaling, calibration/validation and the potential need for large-scale experiments for algorithm development and validation were summarized by each of the four working groups. All groups discussed the question of data/product requirements for long term environmental modelling/monitoring, especially with respect to global change.

The second session involved discussion between participants from different disciplines and allowed the remote sensing community to interact with the modelling community. Those with modelling interests focused on identifying model requirements, including initialization and validation, and discussing scaling issues (GCM/mesoscale/hydrological basin scale). Others discussed potential synergism between passive microwave and other remote sensing measurements. Particular focus was given to SAR/MIMR/SSM/I issues.

Keeping with the purpose of the workshop, discussions focused very much on what has to be done to make effective use of MIMR data, especially for land applications, and on how to exploit best SMMR and SSM/I data and information in models. A detailed summary of the discussions, conclusions and recommendations for each working group is provided. It is expected that these summaries will provide a focus for further discussion and stimulate research necessary to further the use of passive microwave satellite data for land applications. There were several issues which were identified independently by each group. Some of these are summarized below. They are not in any particular order of priority, but all are deemed important to the remote sensing and modelling communities for the continuing advancement of passive microwave remote sensing into the next century.

* The frequencies and polarizations, and the associated resolution and sampling interval, proposed for MIMR will allow significant advancement in the use of passive microwave data for land applications.
* Calibration differences between satellite sensors must be determined and corrected and known drifts in sensors must be corrected to prepare the best time series possible for long-term monitoring and modelling.
* Good quality existing satellite data (SMMR 1978–1987, SSM/I 1987–...) must be made available.
* The proposed EASE-GRID product to provide global gridded SSM/I brightness temperatures to the scientific community on CD-ROM was widely supported. The need to produce the SMMR data in the same format was identified.
* More research is required on the mixed pixel problem. This is a problem for coarse resolution data, such as SSM/I, since individual pixels will have a signal from a mix of land cover/vegetation types. Variability of the radiance within a pixel must be quantified to define what phenomena we are really seeing from satellites. Airborne data are useful for this work.
* Ground based radiometry is essential. It offers a synergism of different frequencies and polarizations. It provides information on temporal variability which is essential to understand satellite data acquired at different times during the day. It will allow the development of a better physical understanding of the relationships between the microwave signal and the element being studied.
* There is a need for continuous ground based, airborne, and satellite experiments at different sites up to and through the MIMR period. Experiments should include linkages with NWP and GCM modellers to ensure interaction with forecast model outputs. This also includes a compilation of previous, on-going and planned large-scale field experiments in order to identify gaps in data sets and to have a large 'library' of data.
* Interaction between the passive microwave remote sensing science community and the GCM and NWP modelling community is necessary. The modellers need to state what parameters they want and at which scales. The retrieval groups need to determine which parameters they can produce at those scales, including an assessment of their accuracy and precision.
* There is a need to determine the optimum methods for assimilating the satellite observations (along with conventional data) into the GCMs in order to derive 'best estimates' of the parameters that describe the soil/snow/vegetation/atmosphere system.
* There is a continuing requirement to take advantage of synergism between passive microwave sensors and other satellite sensors (visible/near infra-red, thermal infra-red, active microwave) to provide the best possible product.

Working group A1: Snow

Chairman: H. Rott, R. Armstrong
Rapporteur: B. Goodison

Participants: J. Foster, N. Grody, Jin Ya-Qiu, C. Mätzler, P. R. Rowntree. J. B. Stewart, L. Tsang

INTRODUCTION

The snow working group had a balance of participant experience in theoretical modelling of passive microwave and snow cover, ground based experimentation of snow cover properties, passive microwave signatures for snow properties, algorithm development for snow extent, water equivalent and wetness, production of historical and real-time satellite products, and use of passive microwave derived snow cover products for GCM validation and global change analysis. The group focused on defining the need for snow data which could be derived from passive microwave satellite data, on summarizing the current status of snow products, including data availability, delivery and associated problems and limitations, the status of snow cover algorithms, and future work required to provide reliable and accurate snow cover products. Discussion considered SMMR, SSM/I and MIMR opportunities. Research to date has shown passive microwave data can be used to produce snow cover products on a regional and global scale.

NEED FOR SNOW DATA: APPLICATIONS

The following uses for snow data were identified by the group; in all cases passive microwave derived products would be useful:

- Large-scale climate models (model input and verification).
- Numerical weather forecasting.
- Synoptic scale and mesoscale modelling.
- Hydrology: monitoring and modelling (e.g. snowmelt runoff) design (dams) and engineering.
- Agriculture: frost damage soil moisture recharge.
- Climate monitoring and change detection.
- Ice sheets (melt/dry): indicator of melt areas and change.
- Geophysics: study of Earth's wobble.
- Lake ice freeze-up/break-up as a climate indicator.

STATUS OF SNOW COVER PRODUCTS FROM MICROWAVE RADIOMETRY

Areal extent

Several algorithms have been developed, some based on a single frequency (i.e. 37 GHz), others on spectral difference (i.e. 19–37; 37–85 GHz) with possible inclusion of polarization difference. Differences in the derived areal extent product are due to the method of data ingest, especially at high latitudes where spatial sampling and averaging of multiple orbit data may produce results different from individual swath data, and to the inversion/classification procedure used and the treatment of the effects of shallow snow, melting snow, frozen ground, precipitation, forest cover and extreme elevational differences.

Snow water equivalent

Several regional snow water equivalent algorithms have been developed and tested for dry snow conditions. Snow water equivalent cannot be determined quantitatively when the snowpack is wet. No global algorithm yet exists, and use of regional algorithms with a single set of global coefficients does not produce an accuracy which is acceptable. Problems which must be overcome and their effect included in a global algorithm include:

- snow metamorphism (cold, shallow snow; melt/refreeze);
- vegetation cover, especially varying forest cover;
- snow on warm ground (above 0 °C), usually in early winter;
- signal saturation (about 150–200 mm);
- mixed pixels (varying land and snow cover over a single 25×25 km pixel).

Surface liquid water content

The liquid water content, including the extent of wet snow, is an important hydrological characteristic. No algorithm exists for quantitative estimates of liquid water, but a regional algorithm (polarization difference for 37 GHz) to indicate areas of wet snow has been developed and tested using SMMR and SSM/I data over the Canadian Prairie region. No single global algorithm to map wet snow areas yet exists. Ground based studies suggest 6 and 10 GHz data would be useful for defining wet snow areas.

> Although the pixel size is large, it is suggested that SMMR data could be used to test the applicability of 6 and 10 GHz data. Results would be applicable to MIMR data where the pixel resolution will be significantly improved over SMMR.

Snow properties

Knowledge of snow properties is important in order to improve the estimates of areal extent and snow water equivalent. The melting and refreezing of wet snow involves the transfer of heat and is important for GCM synoptic scale modelling. There is also a relation here to optical albedo which is also very important in modelling. The question was raised as to whether there was a relation between optical albedo and microwave signature for these different snow conditions. Ripening and refreezing of the surface results in a high polarization difference in early spring, but as the pack becomes ripe, the polarization difference decreases, with the biggest change in the 10–21 GHz range. Ripeness is an important snowpack property which would be useful to be able to determine.

FUTURE RESEARCH AND DEVELOPMENT ACTIVITIES

Signature research

Signature research is done using ground based radiometry and involves development of the relationships between brightness temperature and individual snow types and conditions, such as depth hoar, surface hoar. This research involves characterizing the snow in terms of its structure and impurities (dielectric losses can be significant). Currently there are no measurements in the 37 to 90 GHz range to do this signature research.

> For signature research, measurements are required in the 37–90 GHz range; for applications work to verify SSM/I and MIMR algorithms, measurements at 19 and 37 GHz are required.
>
> Ground based radiometry is essential. It offers a synergism of different frequencies and polarizations. The ground sites require the best possible equipment. Although the systems look at only one point and do not provide spatial information that airborne systems can, they provide information on temporal variability which is essential to understand satellite data acquired at different times during the day.
>
> There is a need for ground based radiometric measurements above forest canopies to study the snowpack signature from forests with and without a snow cover and to study the effect of snow in forest canopies on the signal. The effect of riming on trees and the development of hoar frost needs definition.
>
> Portable hand held radiometers should be designed for use in characterizing grain size and snow structure.
>
> There is a need to establish the signature of snow on lake ice and of snowpacks at high latitudes (e.g. Siberia, Canada).
>
> Passive and active systems looking at the same target are highly desirable. Scatterometer data are available from ERS-1 and ground based systems allow investigation of the potential synergism between the different sensors.

Microwave interaction models

There is a need for continued theoretical modelling to characterize snowpack properties. For model input parameters, how does one relate the phase function, absorption and extinction properties to physical snowpack properties? Currently, dense media radiative transfer theory is used. The size distribution of spherical particles is assumed for the current model.

Future modelling will involve numerical simulation (some analytical modelling) where the phase function, absorption and extinction properties will be determined for complex snowpack morphologies (structures). It will include consideration of grain clusters, layers and non-spherical grains.

There is a need to develop parameterized models for inversion algorithms. Emission models for snow cover in forests are required, where the problem of phase function of the trees must be considered. Monte Carlo simulations may be used.

Inversion/classification of satellite microwave data

Calibration differences between satellite sensors must be determined and corrected. SMMR data are to be recalibrated to correct for drift over time. The best possible data set is required to develop time series for climate research. There is some benefit in analyzing 6 and 10 GHz data from SMMR since MIMR will have these frequencies and have a similar time of overpass.

There is a need for intercomparison of existing algorithms for areal extent and snow water equivalent in various climatic regions and for several years using SMMR and SSM/I data.

It is the goal of the snow community to develop within the next two to three years and agree on a single 'community' algorithm to map global snow cover extent accurately enough for climate research applications. To achieve this, land cover information (vegetation type and density information) to be used interactively in the algorithm is required. There is a need to define and then derive a suitable vegetation type/density data base useful for snow retrievals.

New algorithms will continue to be developed, drawing on experience from ground based studies. For example polarization differences could have a better relation with snow water equivalent since layers are picked up by H polarization. Snow type is important as well. Powder snow does not scatter and in fact absorbs and re-emits, thus increasing the brightness temperature rather than having it decrease with increasing snow water equivalent. This can occur in regions of thick snow cover, but must be addressed in the development of new algorithms.

Key target areas where there are different snow conditions and good ground data must be identified. Initial targets include: Finland, Canadian Prairies and forests (BOREAS), Russia (US/NOAA Data Recovery Project), Arctic (tundra), and Tibet (?).

Wet snow areas — testing of the only available wet snow discriminator (wet snow/bare ground) developed and tested for the open Canadian Prairies is necessary. Modification of the current algorithm or development of a new one may be required for more universal application. Synergism of passive radiometry with scatterometer data must be investigated for wet snow conditions.

Mixed pixels — snow distribution and associated patchiness is related to wind and topography, and the mix of forest cover, open land, lakes (frozen, unfrozen) and wetlands. A combination of the above factors results in many pixels having a mixed snow signal depending on their topographic/land cover mix. Research is required to define the snow signal from mixed pixels. It should be noted that this problem applies to the derivation of other geophysical variables as well. The problem will be reduced somewhat with the higher resolution MIMR data.

Scaling — variability of radiance within a pixel must be quantified to define what phenomena we are really seeing from satellites. Airborne data are best suited for this study.

Multi-temporal procedures for data analysis must be developed to reduce noise in the signal caused by the instrument or atmospheric effects and to exclude short-term periods (2–5 days) when there is melting. There is a need to monitor over longer periods to assess changes in snowpack morphology which is important in snow water equivalent retrieval. Corrections for atmospheric transmission must be assessed.

There is a need for improved information on albedo and its changes, particularly in the spring. This also relates to the definition of old vs. new snow and the aging of a snowpack.

Currently, global gridded brightness temperatures are being prepared, not snow cover products, since accepted, global algorithms do not exist. The future geophysical products should have full space/time resolution; they do not have to be degraded to 3–5 day averages or monthly means.

There must be a continuing effort to demonstrate the usefulness of remote sensing information to managers and scientists. This is best done by producing actual regional products which can be evaluated for their accuracy and utility by others (e.g. severe snow storms in Scotland and northern England in January 1993).

Large-scale experiments

It was agreed that a single experiment in one location would not be of great value. There is a need for continuous ground based, airborne and satellite experiments at different sites up to and through the MIMR period. These do not have to cost a lot. To ensure compatibility, a set of ground standards and instruments should be defined. The NASA SPWT could serve as a vehicle for this initiative.

Experiments should include linkages with NWP and GCM modellers to ensure interaction with forecast model outputs. Links with major forecast offices should be developed so that model snow output products are kept for comparison with satellite estimates. In recent years ECMWF, for example, has been keeping the snow products from the NWP.

DATA ISSUES: ARCHIVING, GRIDDING, DATA ASSIMILATION

NSIDC (National Snow and Ice Data Centre, CIRES, WDC-A, University of Colorado) has extensive experience in archiving and gridding passive microwave satellite data and cryosphere products (SSM/I). They have produced SSM/I gridded sea-ice products on CD-ROM for widespread community use. Gridded data are best suited for temporal and spatial analyses and combination with other information. The recommended grid is EASE-grid, which is in prototype testing by the passive microwave satellite community. The initial grid is at 40, 20 and 10 km spacing, although the revised grid may revert to 25 and 12.5 km. The gridded product contains brightness temperatures which contain antenna pattern correction. Each grid contains the ascending and descending brightness temperature. When there are multiple passes at high latitudes, data are not averaged as in past gridded sea-ice products, but two passes only are kept.

> Top priority is to have gridded brightness temperature data for application and use by the community at large. Day/night data must be kept separate and not averaged. Swath data must be archived and available for research purposes.
>
> For global change analyses it is imperative that the SMMR data be corrected and gridded in the same manner as SSM/I. The data should not be corrected to match the SSM/I data (there are absolute differences in the brightness temperature between the sensors), but the consensus differences between the sensors must be well documented. The single channel ESMR data for 1973–78 may be useful to extend the satellite record back further.
>
> Conventional snow data are a critical complementary data set for algorithm development, validation and assimilation in models. Data assimilation procedures must be developed for mixing remotely sensed information and conventional data in models. Assimilation procedures should be reviewed for their application to remotely sensed information.

The NASA SSM/I Products Working Group (SPWT) advises NSIDC on archiving, gridding and processing of SSM/I data into products. This group is willing to coordinate many of the activities proposed above with necessary linkages to ESA and the MIMR Science Team.

SUPPORT FOR REQUIRED ACTIVITIES

To carry out the proposed Intercomparisons and to compile the satellite and conventional data sets some support will be required. It is recommended that common recommendations be incorporated in the reports of the NASA SSM/I Products Working Team (SPWT) and that to the ESA/MIMR Science Team.

Working group A2: Vegetation and temperature

Chairman: P. Pampaloni
Rapporteur: E. Njoku

Participants: F. Becker, J.-C. Calvet, A. Chanzy, B. Choudhury, P. Ferrazzoli, N. de Noblet, S. Paloscia, K. Rao, P. Rowntree, A. Shutko, J. Stewart, H. Suess, U. Wegmüller, J.-P. Wigneron

INTRODUCTION

Microwave radiometry provides a useful capability for sensing land surface temperature and vegetation characteristics from space. Vegetation scatters, absorbs, and emits microwave radiation in a manner that depends on the dielectric constant, geometry, spatial distribution, and physical temperature of the vegetation components. The dielectric and structural characteristics of the vegetation are, in turn, related to such quantities as leaf area index (LAI), plant water content, fractional coverage, etc., which are of interest for modelling and observational studies in fields as diverse as hydrology, agriculture, ecology, and climate.

Many studies have focused on the forward problem, i.e. that of expressing the vegetation opacity, scattering phase function, and/or brightness temperature as functions of the vegetation structural, dielectric (water content), and temperature parameters. These studies include theoretical modelling and ground-based and aircraft experiments, which have been carried out mostly over agricultural crops and mostly at frequencies below 10 GHz. Additional considerations such as soil surface roughness, soil texture, soil moisture, and vegetation fractional cover have also been studied. Absorption and emission by clouds and atmospheric water vapour have been found to be important in accounting for atmospheric effects on passive microwave spaceborne observations, especially in the tropics and at frequencies above 10 GHz. Developing accurate and tractable theoretical models of vegetation is a complex problem, requiring trade-offs between model accuracy, physical insight, and computational efficiency. On the other hand, designing and carrying out observational experiments to determine the sensitivity of brightness temperature to variations in the canopy and soil characteristics requires significant resources in equipment and manpower. Nevertheless, much progress has been made in understanding the microwave signatures of the soil–vegetation–atmosphere medium, and in relating these to observations from satellite radiometers.

> Since most measurements have been made of agricultural crops, additional studies are required to characterize the brightness temperature signatures of various types of natural vegetation (e.g. forests, scrub, savannah), and to acquire measurements at the full complement of frequencies and polarizations that will be relevant to MIMR (i.e. 6.8 to 90 GHz).

The relationships between the characteristics of vegetation affecting the microwave signal and the quantities of interest to ecologists and climate modellers such as plant water content, leaf area index, biomass, etc., are generally not well understood. Precise formulations for the dependence of brightness temperature on these quantities are not available, and much work remains to be done in this area.

> A systematic program of field observations is desirable to develop better physical understanding of these relationships. The problem of transferring the models and relationships from the spatial scales typical of ground-based or aircraft measurements (1 to 100 m) to the satellite scales (10 to 100 km) has also not been adequately addressed.

Spaceborne radiometer footprints generally include mixtures of surface types. The number of model parameters used to characterize the scene components may be large. Thus, the inverse problem of determining vegetation parameters from a limited number of brightness temperature channels is a challenging one. Additional work is required on algorithms for retrieving vegetation parameters from spaceborne data. However this is an important step in using satellite data on a quantitative basis.

At a previous MIMR Workshop held in San Miniato, Italy (October 1990), it was concluded that there were no currently available algorithms for deriving quantitative vegetation or surface temperature information from satellite data (e.g. SMMR or SSM/I data). However, in the EOS time-frame, it was felt that research could progress sufficiently to be able to retrieve from EOS MIMR data up to three levels of vegetation biomass, surface temperature with an accuracy of 2–3 °C, and a number of qualitative indices (e.g. vegetation type, extent), on spatial scales of 10–50 km. The Saint-Lary Working Group discussion outlined research issues, presented in the following sections, that should be addressed in order to meet and exceed these goals.

APPLICATIONS AND MEASURABLE PARAMETERS

Five major application/problem areas were identified that would benefit from remotely sensed vegetation and temperature information. For each of these problems, parameters were identified which could potentially be derived from satellite data. These problems and parameters are shown as a matrix in Table 1. The ticks indicate the parameters of primary interest to each problem.

The extent to which microwave radiometry can retrieve the parameters in Table 1 needs additional study. In the table, the term albedo refers to the shortwave (solar radiation) albedo and not the microwave albedo. Temperature, as used by the flux and energy balance models, generally refers to the infrared radiating temperature of the surface skin layer and not the microwave radiating temperature (which originates from deeper in the medium, as determined by the dielectric properties, and may be referred to as an effective surface temperature). Biomass and leaf area index are only indirectly related to brightness temperature via the plant water content and plant type or structure. Classification schemes may be able to identify major vegetation types using microwave data, from which the shortwave albedo may be derived, but albedo cannot be measured directly. Soil type is difficult to estimate directly due to its small signal. Cover fraction and intercepted water are important in that they affect the microwave signal but, as with the other parameters, may be difficult to estimate from microwave data without *a priori* information either from ground truth or synergistic remote sensors.

Table 1.
Vegetation and temperature parameters potentially derived from remotely sensed satellite data, and the scientific problem areas to which they contribute

Parameters problem	Soil/plant type	Biomass	Leaf area index	Cover fraction	Intercepted water	Tempera-ture	Albedo
Climate evolution		✓	✓	✓		✓	✓
Atmospheric models	✓		✓	✓	✓	✓	✓
Desertification	✓	✓	✓	✓		✓	✓
Productivity/ plant growth		✓	✓	✓		✓	
Ecosystem diversity	✓			✓			

STATUS OF EXPERIMENTAL DATA AND RADIATIVE TRANSFER MODELS

Experimental data have been gathered over agricultural crops (corn, oats, alfalfa, wheat, sunflower, sugarbeet, soybean, canola), grass, and bare soil, over the frequency range 1.4 to 94 GHz. These data have been gathered in measurement programs including those in the US, France, Italy, Switzerland, Germany, and India. The measurements have been used to investigate the relationships between opacity and plant water content, biomass, and LAI. Considering the logistical difficulties in conducting ground-based and aircraft field measurements, the progress in understanding the relationships between brightness temperature and soil moisture, soil texture, soil roughness, frozen soil, surface temperature, vegetation water content, vegetation type, and vegetation growth stage, has been remarkable. However, a number of limitations still exist. For example, unique relationships for brightness temperature of different vegetation types, as functions of wavelength, are still undeveloped. Few experimental data exist over natural (i.e. non-agricultural) vegetation. Investigations have been made of the dependence of microwave single scattering albedo on vegetation type and water content as a function of frequency, but further studies are necessary to fully characterize this dependence. Most measurements have been made at L-, X-, and K-band (~1.4, 10.7, and 37 GHz), with some measurements also at 94 GHz, and in many cases the soil moisture, roughness, biomass, and structure of the plants were not independently varied in these measurement programs. Thus, these experimental data are as yet incomplete for deriving unique relationships between brightness temperature and vegetation parameters. Some discrepancies between theoretical models and experimental data have not been resolved, especially for the scattering effects of vegetation.

Radiative transfer models for vegetation range from simple absorption models to more complex random media and discrete scattering models. These models do an excellent job of modelling the complex vegetation structures that exist in nature, but they need to be further tested against observations. The scattering models involve numerous parameters to characterize the vegetation canopy, which can make inversion (i.e. retrieval of vegetation parameters from measured data) difficult.

> There is a further need to develop specialized but simplified scattering models that can gather the vegetation parameters into a few retrievable

parameters that are related to the desired geophysical quantities. Effects of spatial heterogeneity and spatial scaling need to be studied to understand coarse resolution satellite data.

STATUS OF RETRIEVAL ALGORITHMS AND VALIDATION

Promising studies have been made on retrieval of vegetation and temperature parameters from satellite data. Global 'vegetation index' maps have been generated from polarization difference at 37 GHz using SMMR and SSM/I data. However, the quantitative meaning of this index has not been fully understood. Surface type classification, and effective surface temperature have also been derived from SSM/I data. It has been difficult to validate these retrievals since there are usually no correlative *in situ* data on the same spatial and temporal scale as the satellite data with which to perform validations. A possible approach is to compare the satellite global retrievals with output from General Circulation Models (GCMs), where the same or similar parameters are involved (e.g. surface temperature, vegetation index). The GCMs provide data output on similar or coarser scales as compared to the satellite data, e.g. weekly or monthly, and ~100–250 km. Areas of agreement or disagreement between the GCM simulations and the satellite retrievals may lead to a consensus on improvements to both the GCM parameterizations and satellite methods.

Another approach to retrieval is to develop parameterizations that can describe both the GCMs and the radiative transfer models. The GCM and radiative transfer models can be run in coupled fashion using the same initial parameters, to compute microwave brightness temperatures that can be directly compared with the satellite observations. A subset of the initial parameters may then be fine-tuned to obtain agreement with the observed data. Variations of this approach may lead to development of optimum methods for assimilating the satellite observations into the GCMs in order to derive 'best estimates' of the parameters that describe the states of the soil/vegetation/atmosphere system. A similar approach can be used for assimilating microwave data into mesoscale meteorological models, and soil hydrologic models. This requires, where possible, specification of the relationships between parameters used in the radiative transfer models and parameters used in the flux models. Research in this area is just beginning, but appears very promising.

One reason for the current limited use of satellite data is its general unavailability. Multiyear global data have been collected by the Nimbus-7 SMMR and DMSP SSM/I instruments. However, these data have not been widely distributed on easily usable media and formats. The EOS Pathfinder projects should shortly make both these data sets more widely available for use.

Working group A3: Soil moisture

Chairman: T. J. Jackson
Rapporteur: Y. H. Kerr

Participants: J.-C. Calvet, A. England, D. Entekhabi, L. Laguerre, P. Van Oevelen, K. S. Rao, S. Saatchi, A. Shutko, T. Schmugge, M. Suarez, E. Wood, A. Chanzy, J. Stewart

INTRODUCTION

Of all the applications of passive microwaves over land surfaces, soil moisture is probably one of the best understood. However, it is important to note that up to now:

* there has been no opportunity to perform moisture retrieval from space,
* the link with the GCM community is very weak,
* retrieving soil moisture in the root extraction zone still poses problems.

The working group thus tried to address these different issues in order to identify the reasons why this was the situation, and, if possible, to identify solutions. For this purpose, after reviewing the state of the art and exchanging views on the different problems, an attempt was made to organize and coordinate our efforts.

The charter issues to be addressed were defined as:

* a need for new research to develop algorithms to optimize the use of future satellite data;
* to improve models for microwave emission;
* to investigate the coupling of remote sensing observables (i.e. soil moisture) and large or mesoscale models; and,
* to sort out the data issues.

CRITICAL ISSUES FOR SOIL MOISTURE

Definition of soil moisture
The first point addressed was what we mean by soil moisture. It was decided to call soil moisture 'the amount of water in the first five centimetres of the soil'. A second point, related to scale, was not specifically addressed. However, there was discussion about interactions with GCMs and validation. Moisture at depths greater than 5 cm will be called profile moisture.

> The group agreed that an instrument of the MIMR class would certainly give information on soil moisture, but interpretation will be difficult in

some cases since, at the lowest frequency (6.8 GHz), the influence of vegetation is far from negligible (see Fig. 1).

Consequently, soil moisture estimates will only be possible over land surfaces with relatively low vegetation covers unless sophisticated multifrequency algorithms or indirect methods (thermal inertia for example) are used. Various experiments have shown the possibilities of extracting soil moisture from L-band measurements.

In order to have a more global coverage of soil moisture, the group recommends that the possibility of flying an L-band instrument be considered. It will greatly improve the monitoring of soil moisture on a global basis, and the accuracy of the retrieval with greater simplicity.

Such a possibility is quite feasible since with the interferometric concept (e.g. ESTAR) reasonable spatial resolutions are possible; there are no big technological problems except possibly for the antenna (such an instrument exists and has already been used in an aircraft prototype: ESTAR).

Specific work has yet to be done to prepare algorithms for interpreting MIMR data. It was noted that the best algorithms would probably be developed from post launch analysis of MIMR data (especially for the temporal aspects, which cannot be efficiently studied with aircraft campaigns and should improve retrievals).

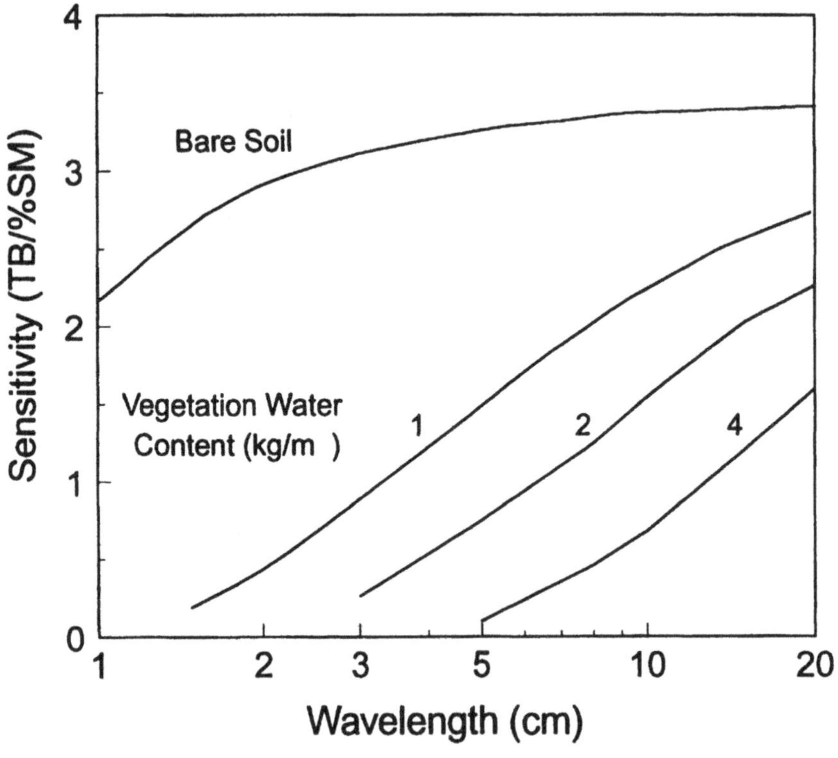

Figure 1. Sensitivity of brightness temperature to soil moisture as a function of wavelength for bare and vegetated conditions (H polarization, 10 °C look angle).

> The group recommends the development of airborne MIMR simulators and to perform experiments with such simulators and ESTAR type instruments.

As far as the modelling aspects are concerned, more work still needs to be done in conjunction with groups working on vegetation.

The problem of RFI was also clearly identified.

> The group strongly recommends that at international level, pressure should be exerted to protect the 1.4 GHz band for our future research and applications.

Retrieval from space

Even though it seems clear that L-band is the optimal wavelength for soil moisture, more research on the use of temporal series and radio brightness thermal inertia (RTI) should be conducted, since such approaches may significantly improve retrievals with MIMR type data. The need to take advantage of synergism with other sensors (visible/near infra-red, thermal infra-red, active microwave), was also discussed. Even if an L-band sensor is flown, MIMR should prove to be a significant tool for correcting perturbating factors. The retrievals would also be more accurate if other sensors are used.

It was also stated that, using a MIMR type instrument, at least two channels may be necessary to efficiently retrieve soil moisture (C-band is strongly related to vegetation amount) and that at high frequencies (X-band) the signal measured emanates from the very surface and might be 'dry' almost all the time in arid/semiarid areas.

> The group suggests that the accuracies required for soil moisture for various fields of application should be assessed.

> The group recommends that efforts be continued in algorithm development. It was also suggested that when MIMR is operational, field campaigns with the airborne ESTAR are carried out. Synergistic use of other sensors should also be addressed.

> The group also expressed that, only under optimal conditions, active instruments would give information on soil moisture. It was generally felt that, used by themselves, they would not be of much use for that purpose. However, because active instruments are more sensitive than passive instruments to the scattering characteristics of vegetation canopies, combination of the two data sets could help unravel competing canopy and moisture signatures.

Outputs

Microwaves will give information on soil moisture in the first centimetres of the soil. This is useful for soil evaporation estimation and runoff models. Models for monitoring water balance require information on the actual profile in the unsaturated zone. We could utilize the emissivity to describe the soil's surface and then use models to derive the actual profile. Microwaves do not seem to be able to give this product with the current state of the art.

More work is also required on the mixed pixel problem. The group feels that we should join forces with other groups having acquired experience in this domain ('rain' group, modellers).

Another big issue is validation. How is it possible to compare local ground measurements with satellite retrievals over a 2500 km^2 area? The only viable solution at the time being is through the comparison with model outputs.

CLIMATE MODELLING

What is soil moisture?

At the moment, models use an approach (see for example, the BATS model) which is quite different from the type of data we can retrieve. The definition of soil moisture depends on the model used. The 'bucket model' is probably the most common approach currently used. It is also necessary to make a distinction between GCM, mesoscale models, weather forecast models and hydrological models. Generally speaking, models need soil moisture in at least two layers (near surface and root zone), but this is model dependent. GCMs and mesoscale models need soil moisture and other information to derive fluxes. Soil moisture is a state variable in these models, it is predicted by the models and can be used to 'validate' satellite retrievals. Weather forecast models require soil moisture as a boundary condition. Hydrological models require soil moisture for at least two layers.

> It is very important that we begin to bring together modellers and groups collecting soil moisture. Data assimilation is very dependent upon the models, which currently do not always use the type of soil moisture we can derive.
>
> The use of model outputs for retrieval validation should be investigated.
>
> An estimate of soil moisture at a deeper depth than that obtainable directly with remote sensing is desired. Research on indirect methods (e.g. RTI) should be carried on with closer links with the modellers' community.

Spatial resolution

At the moment, GCMs run with a grid of about 200–300 km, mesoscale models with a scale of 20 km and operational hydrological models with a typical resolution of 1 km (flash floods for instance), although hydrological models may eventually be able to operate at a scale of 20 km.

It must be noted that even though GCM grids are quite large at the moment, they may eventually utilize resolutions of 50 to 100 km, and even now some models allow each grid box to be divided into smaller units, allowing the characterisation of the contents as proportions (i.e. forest, low vegetation, bare soil, free water, etc...).

> A typical spatial resolution of the order of 20 km should satisfy most of the modelling data requirements (i.e. excluding perhaps some hydrology work). However, this scale issue is not yet fully settled, especially in hydrology and water resources, and clearly requires more research.

Temporal resolution
Here again, the needs must be tailored to the different users. Moreover, the actual time of acquisition is also of equal importance and can condition the frequency of observation. Weather forecast models require the synoptic temporal resolution. It must be noted, however, that root zone soil moisture varies slowly and does not require high temporal resolution. For most geographic regions one to four times a day seems reasonable. In many cases once every other day may be sufficient.

> The group felt that, provided two MIMR sensors are flying (METOP and EOS pm), most of the data needs should be fulfilled.

How to enhance interactions with modellers
The Saint-Lary Workshop is a first step in the interaction of these communities (soil moisture, modellers, hydrologists). The group felt that this was very important for the future. We should also work together and run models to simulate existing satellite data and run models using retrievals from satellite data.

> We need to learn about each other (especially with people working in data assimilation or validation), what each community requires and is able to produce.

> We should intercompare models and satellite data (assimilation, validation).

> Good quality existing data (SMMR 1978–1987, SSM/I 1987–...) must be made available. The initiative with SSM/I is thus to be fully supported and complemented with similar work with SMMR data in a standardized fashion (formats, sampling).

> The problem of field experimentation is probably more complex since here very large areas would have to be covered during long periods of time. This would require additional study.

In summary, several problems have to be addressed before we can talk about a general assimilation procedure; however, it may be achievable in the relatively near future.

PROFILE SOIL MOISTURE ALGORITHMS

The retrieval of profile soil moisture is much less developed than surface soil moisture algorithms, even though some significant advances have been made recently. Profile retrieval is important to the ultimate application. However, at this point it is not critical in the research program. Advances in this field are dependent upon very good data sets including extensive measurements over long periods of time. An initial goal should be the estimation of water in the root extraction zone. Up to now, most work has been done in specific locations over one column (1-D).

> The group recommends that a compilation of existing data sets be performed and these data be used for intercomparison.

Field experiments should be undertaken with ground measurements 24 hrs a day during several weeks (with sampling rates adapted to temporal evolution [i.e. roughness]), followed by well designed airborne campaigns. Here again a long time series is required, especially on soil moisture profiles.

From existing data sets and future experiments, algorithms for MIMR should be developed and tested with MIMR aircraft based simulators.

Scaling up of the site specific profiles should be addressed.

Research work should be performed on soil hydrodynamic properties and water transport modelling.

NEAR TERM EXPERIMENTAL AND DATA COLLECTION ACTIVITIES

The problems addressed here are quite common in many fields of remote sensing. The group agreed that in the near future there is a need to:

Compile previous, on going and planned field experiments (characteristics of the area, instruments, data quality) in order to identify gaps in data sets, and to have a large 'library' of data. Some examples are SIR-C, EFFEDA, HAPEX SAHEL, GCP Storm, Monsoon, Washita , Mac-Europe, (EMAC), Mac-hydro.

Compile information on the existing ground and airborne instruments (characteristics, availability, ...).

Develop MIMR simulators (such as PORTOS, SLMR, PBMR, ESTAR).

Develop or refurbish L-band radiometers.

Generate and maintain a good, reliable archive of satellite passive microwave data (SSM/I, SMMR).

In order to make full use of data collected during field experiments, the group felt that is was necessary to standardize some of the key measurements using reliable techniques. The group suggested the following minimum items:

* soil moisture for the 0–5 cm, soil layer: gravimetric methods, (combined with bulk density measurements [core, gamma ray probe]); TDR measurements every two days correlated to near surface soil moisture measurements; soil moisture should be expressed in volumetric units;
* bulk density for the 0–5 cm layer using gamma ray probes (30 cm → 2.5 cm);
* soil temperature at 5 and 15 cm (possibly at 1 cm?);
* soil type (including granulometry);
* surface roughness (standard deviation of height, grid, needles, or laser profiler);
* proper quantification of the vegetation layer (type, water content, height);

and, eventually:

* skin temperature (radiothermometer);
* temperature and moisture profiles in the unsaturated zone.

There are some concerns about dielectric constant models. These should be improved and validated over a large range of frequencies and conditions.

CRITICAL ISSUES AND FUTURE DIRECTIONS

There are several critical issues which should be addressed. Many are similar to those identified by other working groups. They have been discussed above and are summarized below.

We will need C-band sensors for conducting controlled condition experiments necessary for algorithm development. SMMR is a very good means of preparing for MIMR. The 9-year data set should be made available after recalibration in EASE GRID form (on CD-ROMs) with its own 'calibration' and information on how to intercalibrate with SSM/I. Such a data set should include gridded brightness temperature and separate day and night data. A swath data archive should be maintained.

L-band is a necessity for operational large scale soil moisture and we should support the interferometric systems development and use, develop operational algorithms, have L-band acquisition during any experiment, make the sponsoring agencies aware of the risks of RFI and support fully band protection from 'trespassers'.

Modelling:

* work on synergism (vegetation),
* interactions with GCM /mesoscale modellers, hydrologists,
* progress on the mixed pixels issue,
* improve validation methods,
* continue efforts to develop methods to estimate water in the second layer,
* work on inversion algorithm to be ready in the MIMR era, scaling issues being also addressed.

Data sets:

* satellite data (see above),
* ground data — develop a data base of existing results, intercalibrate them, standardize minimum measurement requirements, develop MIMR simulators, keep up modelling efforts,
* improve dielectric constant models.

Working group A4: Atmospheric correction/precipitation

Chairman: E. A. Smith
Rapporteur: B. A. Burns

Participants: E. Barrett, L. Eymard, J. B. Snider, F. Becker, N. Grody

INTRODUCTION

The task of the working groups as defined by the workshop organizers and working group chairmen was to produce recommendations for future research efforts to be carried out in the next 2–3 years, and to address specifically the issue of what needs to be done to prepare for MIMR to make best use of its data. Topics discussed within this framework were: Modelling, Data Products, Calibration/Validation, and Future Research. High priority recommendations in each of these areas are summarized at the end of the report.

MODELLING

Two aspects of modelling were considered. The first focused on the required models, model output, and modelling studies needed to improve atmospheric retrievals, especially precipitation retrieval. The second was the interaction of the passive microwave/remote sensing scientists with the large scale atmospheric modellers and modelling centres.

Modelling to improve atmospheric retrievals

A major limitation in retrievals of atmospheric parameters over the land surface is the specification of surface boundary conditions. Therefore, improvement in retrieval accuracy will require better calibrated models, *in situ* measurements and climatologies of surface temperature and emissivity. These parameters are needed in many retrieval algorithms, particularly physically-based schemes. Emissivities responding to variability in soil moisture, soil texture, snow cover, and vegetation are required to an accuracy of 2–3%.

Monthly climatologies, including statistical measures of the temporal and spatial variability, would be very valuable for global processing problems and quality control checks on the retrievals themselves. Such information could be immediately evaluated against semi-empirical methods now used, for example, in rain/no-rain screening.

This issue was seen as a point of interaction with the land surface community, requiring further intercomparison workshops as well as coordinated field experiments (see

Calibration/Validation). These activities would offer both a means to test retrieval sensitivities to surface boundary conditions, and to check on the adequacy of emissivity and temperature models.

Improvement of retrieval techniques can also evolve from computer simulation studies.

> Particularly with the advent of MIMR, studies should be carried out to assess the impact of improved spatial resolution and wider frequency range on currently used algorithms.

The requirements for surface temperature and emissivity, for example, might be revised with the availability of the 6 and 10 GHz channels. Such studies can be made with presently available data sets.

> Studies based on data obtained from an airborne simulator of MIMR would be extremely valuable in this regard.

Interaction with atmospheric modellers

Interaction between the passive microwave (PMW) remote sensing science community and the GCM modelling community, was discussed at great lengths. The general goal of this interaction is the development and improvement of passive microwave products by way of studying their impact on validation, initialization, and parameterization of GCM, NWP, and mesoscale models. Definition of the required products in terms of spatial/temporal resolution and coverage is viewed as an iterative process.

> The modelling groups need to state what parameters they want and at which scales; the retrieval groups need to determine which parameters they can produce at those scale including an assessment of their accuracy and precision.

> Most importantly, the two groups need to agree on an acceptable first-step data set for the intercomparisons to begin.

Over land, this is most likely to start with precipitation products, because land-based algorithms which have been tested and validated now exist. Other key atmospheric and cloud properties, such as the moisture profile and cloud liquid water content, remain as difficult over-land retrieval problems. Notably, validation of cloud liquid water products over ocean have yet to take place.

Initial intercomparisons would consist of validating model predictions of rainfall against the microwave estimates, but then should attempt assimilating rainfall products in diabatic initialization schemes (e.g. latent heat nudging). This implies that passive microwave-derived products spanning a time period of several years are required now. Such intercomparisons should help point to the strengths and weaknesses in both models and retrieval algorithms. Although the results of the initial intercomparisons may not be completely successful, because of remaining problems with the retrieved data sets and limitations in the models (e.g. overly simplified cloud parameterizations), the interaction between the two groups will help stimulate the type of research needed to overcome the shortcomings.

Although in the case of GCMs it may be too early for direct assimilation of PMW-retrieved atmospheric parameters, their use as inputs to regional scale models should be feasible now. Models at all scales still need improvements in the parameterization of clouds, and here, close collaboration between the modelling and PMW communities would be very helpful. The present trend for models to incorporate ever more varied types of data would be furthered by such interaction, so that they may finally consider incorporating microwave brightness temperature measurements directly.

DATA PRODUCTS

This discussion mainly considered PMW-derived atmospheric parameters of interest to the wider land–atmosphere research community, and the aspects which need to be done to ensure the usefulness and validity of these data products. The present state of research is such that algorithms are now in place to retrieve precipitation, cloud liquid water, and water vapour on a global scale from passive microwave data. Of these three parameters, algorithms for rain rate and rain accumulation are at a stage to offer complete land–ocean coverage.

Problems that do need further research involve the reduction of errors due to incomplete spatial coverage and to limited temporal sampling. Because of the diffraction-limit effect on footprint size inherent to the antenna systems of microwave radiometers, passive microwave measurements presently, and in the foreseeable future, will be made by low orbiting satellites (either sun synchronous, such as MIMR, or non-sun synchronous, such as TRMM). This means that retrievals over much of the earth are typically based on two or fewer samples per day. It should be noted that for sun synchronous orbits, scan limitations produce recurring coverage gaps which are maximum at low latitudes. In addition, non-sun synchronous orbits do not provide any coverage at high latitudes.

For valid intercomparisons with most current GCM or NWP models, however, the data products should provide complete global coverage and represent the diurnal variability of precipitation. Many studies have already been carried out on statistical averaging methods to achieve realistic temporal and spatial representation of precipitation based on limited data samples.

> Such statistical averaging schemes, as well as new synergistic algorithms making use of the higher frequency coverage of the VIS/IR satellite sensors, need to be implemented and evaluated as part of the intercomparison of rain retrievals with GCM and NWP models.

For other parameters, such as the water vapour profile and cloud liquid water, to become global products, validation of retrievals over the ocean as well as further development of retrieval methods over land are required (see discussion on Future Research, below).

> To aid this process, coordinated experiments at various designated, widely distributed test sites are recommended.

The outcome of such validation efforts will be the production of precision surface and boundary layer measurements of interest to a wide microwave remote sensing research community.

Atmospheric parameters derived from satellite data at the regional scale were also identified as useful products, potentially as a model input and possibly for model validation. These products should be produced with the end users in mind. For example, precipitation products would be useful to the soil moisture, vegetation, and possibly snow and ice research communities, because of the direct physical links between rainfall and these surface properties. Furthermore, satellite derived water vapour products would be clearly useful in regions where good meteorological data are unavailable.

CALIBRATION AND VALIDATION

The working group determined that careful validation of precipitation algorithms has become such an important issue, that it should be placed in the hands of an international steering committee. The foremost responsibilities of this body would be to design and implement a comprehensive over-land algorithm validation plan, which would involve international participation and coordination, and would take advantage of currently designed field programs that could aid the validation process. If deemed necessary, this committee would also be charged with designing additional field experiments needed for the validation effort and take the responsibility of getting these experiments sanctioned by WMO/WCRP and ICSU.

> Since precipitation estimation from space has become such a critical component of the global climate research program, and because it has great potential for improving operational weather forecasts through diabatic initialization schemes, the working group felt that algorithm validation should be given the utmost consideration.

It will be necessary in the implementation of a validation project to make use of well-instrumented test sites. These test sites should be equipped to provide accurate information on all aspects of the surface and atmosphere relevant to precipitation estimation and its verification (e.g. surface rainfall, rain profile, water vapour profile, surface temperature and emissivity, suspended cloud liquid water and ice water contents). These test sites would also have to provide such information at scales consistent with the characteristic footprint sizes of current and future microwave radiometers (i.e. at least 10 km). There would likely have to be 3 or 4 such sites on different continents representing distinct environmental and climatological conditions. Careful thought should be given to coordinating such a program with currently operating sites and future planned sites designed for other weather and climate programs, such as the DOE-ARM or the NOAA radiation test-site programs.

FUTURE RESEARCH

The working group recognized that researchers in the atmospheric correction/precipitation retrieval community have identified a number of future goals. One of the most important is:

> the development of a community rainfall algorithm which can be used as an international standard as the various independent algorithms continue to change and improve.

This algorithm should be designed to use microwave measurements for the basic rainfall estimates, but synergistically combine infrared measurements from geosynchronous platforms to help account for diurnal variability.

It was also noted that retrieval of precipitation and other cloud properties, for their own sake, as well as providing essential atmospheric correction information for surface algorithms, has reached a point where fully instrumented test sites are needed to provide calibration information on a continuous basis. One such possibility for a test site is the recently implemented DOE-ARM site in Norman, Oklahoma. This site is already equipped with a number of precision sensors that are monitoring surface and atmospheric variables important to atmospheric and land-surface microwave remote sensing. The addition of a few additional sensors would create a first order calibration site for microwave remote sensing. The establishment of sites such as this, strategically placed at different locations around the world, would expedite the development of reliable microwave algorithms and would enable a continuous check on their performance.

The working group also recognized that since precipitation retrieval has reached a relatively advanced stage,

> it is now reasonable to begin examining the problem of retrieving snowfall rate, as opposed to estimating accumulated snow depth after it has reached the surface.

This is considered a very difficult retrieval problem, but so was rainfall fifteen years ago.

RECOMMENDATIONS

The following recommendations were identified by the group as having the highest priority. There is an urgent need to:

> Interface with GCM/NWP/mesoscale modellers to define required PMW atmospheric products;
>
> Interface with surface microwave modelling groups to specify what is needed from these models by the atmospheric algorithms, and develop global temperature and emissivity climatologies needed by both atmospheric and surface retrieval groups;
>
> Carry out computer simulation studies of the impact of MIMR space and time sampling resolution, footprint size and frequency range on atmospheric retrieval algorithms;
>
> Implement and test temporal/spatial averaging techniques in order to achieve a reliable global precipitation data product;
>
> Convene an international working group to develop a viable international research plan for calibration and validation of precipitation algorithms and other PMW atmospheric product algorithms, and to design validation/calibration experiments in concert with other planned field programs;
>
> Designate various, distributed test sites for multi-purpose studies of land-atmosphere effects on microwave retrievals at the scale of the microwave measurement footprints. This is a clear opportunity for coordinated NASA/ESA efforts;

Develop a community precipitation algorithm which makes use of micro-wave-infrared synergism and can serve as a comparison standard for future algorithm developments.

Working group B1: Modelling

Chairman: P. Rowntree, M. Suarez
Rapporteur: D. Entekhabi

Participants: B. Goodison, B. Choudhury, E. Njoku, K. S. Rao, A. Shutko, U. Wegmüller, N. de Noblet, M. Suarez, T. England, J. Noilhan, C. Mätzler, J. Foster, J.-P. Wigneron, Y. Jin, P. Rowntree, L. Eymard

INTRODUCTION

The general aim of this working group was three-fold:

* to identify passive microwave products needed by modellers;
* to assess the utility and quality of data relative to those from other sources; and,
* to determine the scale and level of accuracy of the data needs.

Passive microwave observations and derived data products have potential applications in climate, numerical weather prediction (NWP), mesoscale and hydrological models. Data assimilation procedures are therefore needed for both land surface and atmospheric processes.

There several technical issues that had to be addressed; these issues were identified based on problems encountered in the current application and use of products and information derived from passive microwave remote sensing. They include:

* catalogue practices and standardized handling and storage of data;
* retrieval algorithms and geophysical calibrations;
* gridding and format protocols;
* role of and interaction with EOS-DIS;
* maintenance of continuous multi-year records;
* error analysis and estimation of confidence limits.

To achieve a solution to many of these problems, the modelling and end-user community must interact and cooperate with the field scientists and instrument engineers.

MODELLING APPROACHES

The working group provided the opportunity for such cooperation; it resulted in the definition of a list of hydrological and meteorological variables (Table 2) that are of interest in modelling studies and hydrological investigations. The key questions relate to

Table 2.
Hydrological and meteorological variables of use in modelling studies

Surface temperature
Soil moisture
Snow extent
Snow age
Snow mass
Snow depth
Subsurface soil temperature
Permafrost
Wetland and open water
Precipitation
Liquid water profile in atmosphere
Ice water profile in atmosphere
Water vapour profile in atmosphere
Vegetation type
Surface albedo
Surface roughness
Soil type
Leaf area index (LAI)
Fractional vegetation coverage
Water content of vegetation canopy
Water retention on vegetation canopy
Vegetation biomass

how observable or retrievable these variables (Table 2) are from space-borne, airborne or truck-mounted microwave radiometer platforms. Depending on the variable and the required instrument, are the measurements valid for a point or are they averaged over certain scales of heterogeneity? These questions must be answered for each variable. Some of the variables are neither directly nor indirectly sensed in the microwave region. Some may only be estimated in a retrieval/assimilation setting that includes a model of the land–biosphere–atmosphere system.

In this respect, one suggestion is to utilize land–biosphere–atmosphere representations in mesoscale meteorological models. The dynamic states of this model are used in a radiative transfer model to estimate the emitted radio-brightness. The estimated values would then be compared with the observed values and the necessary corrections are made to the model states to bring the observed and modelled radio-brightness values to agreement. This is an alternative to the traditional statistical or empirical retrieval of states in geophysical calibration. The remaining scientific and operational issues are concerned with the choice of models and methodology for guiding them to convergent solutions.

The variables in Table 2 include the type of GCM and NWP initialization fields that may be potentially derived from passive microwave observations. Near-surface soil moisture, snow extent, liquid/ice/vapour profile in the atmosphere, depth-integrated soil temperatures, canopy water storage and other variables are necessary as initialization fields and they appear to have the potential to be derived operationally from microwave instrument readings.

Another related and important methodological issue identified by this group concerned the issue of validation. The comparison of model forecasts with observations suffers from the fact that the atmospheric system is inherently characterized by limited predictability. Part of the divergence between model forecast based on observed initial conditions and the observed evolution of the variable may be due to this implicit limitation.

> The use of remotely sensed initial and boundary conditions for models of the land–biosphere–atmosphere has severe under-sampling problems that have to be solved jointly by modellers, field scientists and instrument engineers.

Some of the variables in Table 2 have significant interannual variabilities. Since the design-life of most satellite platforms is generally less than five years, data sets available from a single spacecraft are of limited value in establishing climatologies by themselves. It is necessary to establish a longer passive microwave time series from sensors on different platforms, recognizing the need to assess data compatibility from sensor to sensor and the possible need to correct for data inhomogeneities. An example is the availability of SSM/I and SMMR data for the 1978–present period. With the longer time series it is possible to investigate more thoroughly the issue of interannual variability.

Furthermore, some of the variables in Table 2 have significant diurnal cycles and the sun-synchronous satellite orbits are not capable of providing information on this major component of variability required in models.

RECOMMENDATION

The Modelling Working Group recommended that:

> end-user modellers, remote sensing scientists who are involved in the geophysical calibration and retrieval, field scientists and instrument designers define the limits of data quality in every step of processing the microwave radio-brightness readings.

The effective use of passive microwave observations follows the precise definition of the scale of the measurement, the level of natural heterogeneity and the error-tolerance in the end-application.

Working group B2: Synergism

Chairman: F. Becker
Rapporteur: T. Schmugge

Participants: S. Paloscia, P. Pampaloni, R. Armstrong, E. Smith, B. Choudhury, J. Steward, Y. Menard, A. Shutko, B. Burns, P. Ferrazzoli, H. Rott, A. Chanzy, S. Saatchi, E. Barrett, J.-C. Calvet, J. Snider, P. Van Oevelen, H. Suess

INTRODUCTION

From the dictionary definition of synergism, we take the meaning to be the case where the combined result of two data sources is greater than the sum of the two used independently. It is another way of saying the use of data from combined sources where one source provides information needed for the interpretation of the other or for improving its interpretation. An example could be the use of TIR and PMW data to yield a surface temperature and microwave emissivity which can be related more directly to soil moisture than microwave brightness temperature alone.

POTENTIAL SYNERGISM

Table 3 shows the possible combinations of data from passive microwave (PMW) sensors with other sensors for a wide range of land surface parameters. In Table 3 the '+' sign indicates that we believe that there is a synergism, a '?' means that we are uncertain, and a blank means that we believe that there is no benefit from the combined use of the two sources of data. In addition, other possible combinations with such factors as models and diurnal observations (day/night) are considered. For models it was observed that they almost always add information content to the PMW observation and in fact they probably should not be included in the table. Some parameters such as surface albedo were left out because PMW contributes nothing to its determination.

The Visible and Near-Infrared (VNIR) data provide high spatial resolution information on land cover/land use and vegetation to aid in the interpretation of PMW data. The synergism between the two spectral bands should be studied more intensely. Active microwave or radar responds to the same surface dielectric properties as the PMW, but it is much more sensitive to geometrical or structural factors such as surface roughness and woody vegetative structures. Again the high resolution available with SAR (synthetic aperture radar) techniques offers another mechanism for interpreting the coarse resolution PMW signals.

Thermal infrared (TIR) primarily adds surface temperature information which is important for parameters such as soil moisture, vegetation and snow. However, it may

Table 3.
Synergism between passive microwave and other remotely sensed and conventional data

Synergistic parameters	Active/ passive	Ancillary info.	MW + Model	MW/ TIR	MW/ VNIR	Day/ Night
Wetland	+		+	+	+	+
Soil type	?		+	+	+	
Soil moisture						+
–surface	+	+				
–profile	+	+	+	+	+	
Soil surface roughness	+		+			
Rain						
–yes/no	+	+	?	+	?	
–rate area	?	+	+	?	?	
Intercepted water			?			
Plant cover				+	+	
Plant type	?		+	+	+	
Vegetation water content	+		+	+		
Vegetation structure	+				+	
Albedo						
E.T. incoming solar flux surface		+		+		
Temperature profile		+	+	+		
Humidity –water vapour			+	+		
Total profile	+					
Land ice	+			+		
Snow						
–extent	+				?	
–water equivalent	+				?	
Cloudiness			+	+	+	

also be of interest for the analysis of the soil characteristics and properties through the spectral analysis of the emissivity. Another example of synergism between TIR and PMW comes from their difference of penetrability of the media. Since the atmosphere is more transparent to PMW, a combination of PMW and TIR may be used to interpolate TIR data between cloud covered periods. A possible approach for this is described below.

An example of synergism for precipitation
Precipitation retrieval from space observations is a good example where synergistic retrieval techniques could be exploited. Current satellite based precipitation retrieval algorithms can be divided into two distinct categories: (1) thermal infrared techniques

that utilize cloud cover and cloud cover change as the basic variables related to rainfall; and (2) passive microwave techniques in which the measurements are sensitive to the presence of precipitation size liquid drops and ice particles and the vertical distribution of these hydrometeors.

TIR algorithms are necessarily indirect, since the upwelling radiances measured by the satellite contain no direct information on cloud microphysics. Instead, successful TIR algorithms have always depended on the underlying correlation between the area enclosed by cold cloud boundaries (and possibly the time-rate-of-change of these boundaries) and rainfall accumulation. These techniques are generally restricted to convective rainfall, and have usually not been successful with stratiform rain or warm rain systems.

On the other hand, PMW retrieval algorithms are now generally successful in directly detecting and estimating rainfall because of the strong physical controls precipitation size hydrometeors have on influencing the behaviour of microwave brightness temperatures. Generally, rain drops appear hot against radiometrically cool surfaces such as water, but cool against radiometrically hot surfaces such as found in low latitude continental areas. Although there are non-uniqueness problems associated with the PMW methods (that is, different rainrate situations can give rise to the same brightness temperature at a given frequency), multiple channel techniques tend to overcome most of these problems.

Notably, IR algorithms have one key advantage over PMW algorithms in that the geosynchronous weather satellite platforms that generate the high resolution IR measurements (GOES, METEOSAT, GMS, INSAT) provide continuous diurnal sampling, whereas the DMSP (with the SSM/I sensor) low earth orbiters in sun-synchronous orbit configurations, only sample a given earth coordinate twice per day at most (except in polar regions) and due to orbit configuration, swath coverage will miss a given spot periodically at lower latitudes.

Thus, a PMW-TIR synergistic algorithm would exploit the diurnal sampling capability of the geosynchronous-type IR measurements, coupled with the strong physical underpinnings contained within the PMW measurements. A synergistic algorithm would generally seek scaling coefficients (such as a scheme proposed by Adler *et al.*, 1993), that would adjust IR rainfall estimates taken simultaneously with MW estimates, so as to match the PMW estimates. Then at all other times over a 12 hour period surrounding a PMW measurement, these scaling coefficients could be used to adjust the IR estimates so that there would be a statistical consistency between the two sets of estimates. Such a synergistic approach would eliminate much of the rainfall bias due sampling obtained from a set of PMW rain estimates alone.

This is one case where cross-spectral synergistic methods could take advantage of both the high resolution/high sampling rates associated with geostationary platforms and the basic physical measurement clarity of microwave measurements at the lower space-time resolutions of the sun synchronous polar orbiting platforms.

Synergism for other parameters:

Tables 4–7 explore the possible combinations/synergism for particular parameters in more detail. The parameters considered are: snow, vegetation, soil moisture and precipitation. In these tables we tried to identify the areas where problems remain in the use of data combinations and where more work is needed.

Table 4.
Precipitation synergism/combinations and areas for further research

	Active/passive	MW/TIR	MW/VNIR
Topics – field improved		going to the diurnal cycle with GOES	
Physical problems to solve – meaning of measurements/models	– microphysical assumptions – boundary conditions – algorithms		
Experiment to perform	comparison with other measurements – sodars – various scales – surface characterizations		
Modelling improvement	necessary algorithms and cloud models		
Technical problems to solve	– spatial resolution – temporal sampling – calibration of radar		
Mixed pixel issues	beam filling problem		
Data base – registration – GIS	registration		
Instruments necessary	MIMR on EOS TRMM	geostationary MW	
Inversion improvement	much to be done for retrieval from MIMR		
Validation – averaging	yes	yes	
– scale embedding	yes	yes	

Table 5.
Snow synergism/combinations and areas for further research

	Active/passive	MW/TIR	MW/VNIR
Topics – field improved	– wet snow – morphology for snow water equivalent – use of longer λ for wet snow		improves spatial resolution
Physical problems to solve – meaning of measurements/models	– characterization of natural snow pack		
Experiment to perform	– large scale verification – further signature research		
Modelling improvement			
Technical problems to solve	– layered snow – depth hoar		
Mixed pixel issues	– improvement of spatial resolution		
Data base – registration – GIS	– vegetation – mountains/topography		improves resolution
Instruments necessary	scatterometer; wide swath SAR K_u- or X-Band		
Inversion improvement	MIMR will improve SSM/I – resolution + frequency		
Validation – averaging – scale embedding	yes yes		improves

Table 6.
Soil moisture synergism/combinations and areas for further research

	Active/passive	MW/TIR	MW/VNIR
Topics – field improved	ε, σ^0 to soil moisture and dielectric constant		
Physical problems to solve – meaning of measurements/models	$\varepsilon \rightarrow \rho_{b1}$, σ_0	– sampling depth surface & profile – emissivity estimation	
Experiment to perform	simultaneous aircraft and *in situ*		
Modelling improvement			
Technical problems to solve	θ (rad) θ (sar) for absolute calibration of SAR + radiometer		
Mixed pixel issues	spatial integration		
Data base – registration – GIS	swath/pixel size		
Instruments necessary	long wavelength imager in space suitable instrument for *in situ* comparison to MIMR		
Inversion improvement		emissivity $T_{surface}$	
Validation – averaging – scale embedding	spatial variability		variability problems

Table 7.
Vegetation synergism/combinations and areas of further research

	Active/passive	MW/TIR	MW/VNIR
Topics – field improved	structure morphology (height)	– emissivity determination – detection of stress	– classification – temporal change – morphology
Physical problems to solve – meaning of measurements/models	modelling improvement EM models – scattering – radiative transfer morphology → roughness length, z_0	surface to depth	radiative transfer LAI → biomass
Experiment to perform	simultaneous field experiments	same	same
Technical problems to solve	spatial resolution/FOV/calibration		pixel difference and coregistration
Mixed pixel issues	integration process		
Data base – registration – GIS	yes high to low resolution		
Instruments necessary	MIMR simulation, calibrated scatterometer		
Inversion improvement	yes		
Validation – averaging – scale embedding	problem	yes	

Bei Fragen zur Produktsicherheit wenden Sie sich bitte an:
If you have any questions regarding product safety,
please contact:

Walter de Gruyter GmbH
Genthiner Straße 13
10785 Berlin
productsafety@degruyterbrill.com